Lecture Notes in Computer Science 2161

Edited by G. Goos, J. Hartmanis, and J. van Leeuwen

T0226161

Springer
Berlin
Heidelberg
New York
Barcelona
Hong Kong
London
Milan
Paris
Singapore
Tokyo

Friedhelm Meyer auf der Heide (Ed.)

Algorithms – ESA 2001

9th Annual European Symposium
Århus, Denmark, August 28-31, 2001
Proceedings

 Springer

Series Editors

Gerhard Goos, Karlsruhe University, Germany
Juris Hartmanis, Cornell University, NY, USA
Jan van Leeuwen, Utrecht University, The Netherlands

Volume Editor

Friedhelm Meyer auf der Heide
Paderborn University
Department of Mathematics and Computer Science
and Heinz Nixdorf Institute
33095 Paderborn, Germany
E-mail: fmadh@upb.de

Cataloging-in-Publication Data applied for

Die Deutsche Bibliothek - CIP-Einheitsaufnahme

Algorithms : 9th annual European symposium ; proceedings / ESA 2001,
Aarhus, Denmark, August 28 - 31, 2001. Friedhelm Meyer auf der Heide (ed.).
- Berlin ; Heidelberg ; New York ; Barcelona ; Hong Kong ; London ; Milan ;
Paris ; Singapore ; Tokyo : Springer, 2001
    (Lecture notes in computer science ; Vol. 2161)
    ISBN 3-540-42493-8

CR Subject Classification (1998): F.2, G.1-2, E.1, F.1.3, I.3.5, C.2.4

ISSN 0302-9743
ISBN 3-540-42493-8 Springer-Verlag Berlin Heidelberg New York

Springer-Verlag Berlin Heidelberg New York
a member of BertelsmannSpringer Science+Business Media GmbH

http://www.springer.de

© Springer-Verlag Berlin Heidelberg 2001
Printed in Germany

Typesetting: Camera-ready by author, data conversion by PTP-Berlin, Stefan Sossna
Printed on acid-free paper   SPIN: 10840305   06/3142   5 4 3 2 1 0

Preface

This volume contains the 41 contributed papers and three invited papers presented at the 9th Annual European Symposium on Algorithms, held in Århus, Denmark, 28–31 August 2001. ESA 2001 continued the sequence:

- 1993 Bad Honnef (Germany)
- 1994 Utrecht (The Netherlands)
- 1995 Corfu (Greece)
- 1996 Barcelona (Spain)

- 1997 Graz (Austria)
- 1998 Venice (Italy)
- 1999 Prague (Czech Republic) and
- 2000 Saarbrücken (Germany).

The proceedings of these previous meetings were published as Springer LNCS volumes 726, 855, 979, 1136, 1284, 1461, 1643, and 1879.

Papers were solicited in all areas of algorithmic research, including approximations algorithms, combinatorial optimization, computational biology, computational geometry, databases and information retrieval, external-memory algorithms, graph and network algorithms, machine learning, online algorithms, parallel and distributed computing, pattern matching and data compression, randomized algorithms, and symbolic computation. Algorithms could be sequential, distributed, or parallel, and should be analyzed either mathematically or by rigorous computational experiments. Experimental and applied research were especially encouraged.

Each extended abstract submitted was read by at least three referees, and evaluated on its quality, originality, and relevance to the symposium. The entire Program Committee met at Paderborn University on 12–13 May 2001 and selected 41 papers for presentation from the 102 submissions. These, together with three invited papers by Susanne Albers, Lars Arge, and Uri Zwick, are included in this volume.

The Program Committee consisted of:

Friedhelm Meyer auf der Heide
(Paderborn; Chair)
Micah Adler (Amherst)
Pankaj Kumar Agarwal (Duke)
Mark de Berg (Utrecht)
Gerth Stølting Brodal (Århus)
Tom Cormen (Dartmouth)
Martin Dyer (Leeds)
Stefano Leonardi (Rome)

Peter Bro Miltersen (Århus)
Ian Munro (Waterloo)
Petra Mutzel (Wien)
Stefan Näher (Trier)
Yuval Rabani (Technion)
Jörg Rüdiger Sack (Carleton)
Alistair Sinclair (Berkeley)
Dorothea Wagner (Konstanz)

ESA 2001 was held as a combined conference (ALGO 2001) together with the Workshop on Algorithmic Engineering (WAE 2001) and the Workshop on Algorithms in BioInformatics (WABI 2001), and it was preceded by the Workshop

on Approximation and Randomized Algorithms in Communication Networks
(ARACNE 2001).

The seven distinguished invited speakers of ALGO 2001 were:

- Susanne Albers (Universität Freiburg),
- Lars Arge (Duke University),
- Andrei Broder (AltaVista),
- Herbert Edelsbrunner (Duke University),
- Jotun Hein (University of Århus),
- Gene Myers (Celera Genomics), and
- Uri Zwick (Tel Aviv University).

The Organizing Committee of both ALGO 2001 and ESA 2001, consisted
of Gerth Stølting Brodal, Rolf Fagerberg, Karen Kjær Møller, Erik Meineche
Schmidt, and Christian Nørgaard Storm Pedersen, all from BRICS, University
of Århus. I am grateful to Gerth Stølting Brodal (BRICS), Tanja Bürger, and
Rolf Wanka (Paderborn University) for their support of the program committee work and the preparation of the proceedings. ESA 2001 was sponsored by
the European Association for Theoretical Computer Science (EATCS), and by
BRICS. We thank ACM SIGACT for providing us with the software used for
handling the the electronic submissions.

Paderborn, June 2001   Friedhelm Meyer auf der Heide

Referees

We thank the referees for their timely and invaluable contribution.

Oswin Aichholzer	Geeta Chaudhry	Rudolf Fleischer
Susanne Albers	Otfried Cheong	Peter Forsyth
Lyudmil Aleksandrov	Andrea Clementi	Zihui Ge
Arne Andersson	Sabine Cornelsen	Joakim Gudmundsson
Lars Arge	George Cybenko	Anupam Gupta
Estie Arkin	Artur Czumaj	Dan Gusfield
Matthias Bäsken	Sanjiv Das	Torben Hagerup
Luca Becchetti	Suprakash Datta	Sariel Har-Peled
Petra Berenbrink	Frank Dehne	Clint Hepner
Daniel Bernstein	Erik Demaine	Bill Hesse
Therese Biedl	Martin Dietzfelbinger	Xianglong Huang
Hans Bodlaender	Jeff Edmonds	Alon Itai
Andrej Brodnik	Thomas Erlebach	Pino Italiano
Alberto Caprara	Esteban Feuerstein	Riko Jacob

Table of Contents

Session 5: Sequences

Session 6: Scheduling

Session 7: Shortest Paths

Session 8: Geometry I

Session 9: Data Structures II

Session 10: Geometry II

Session 11: Distributed Algorithms

Session 12: Graph Algorithms

Session 13: Pricing

Session 14: Broadcasting and Multicasting

Session 15: Graph Labeling and Graph Drawing

Session 16: Graphs

External Memory Data Structures
(Invited Paper)

Lars Arge*

Department of Computer Science, Duke University, Durham, NC 27708, USA

Abstract. Many modern applications store and process datasets much larger than the main memory of even state-of-the-art high-end machines. Thus massive and dynamically changing datasets often need to be stored in data structures on external storage devices, and in such cases the Input/Output (or I/O) communication between internal and external memory can become a major performance bottleneck. In this paper we survey recent advances in the development of worst-case I/O-efficient external memory data structures.

1 Introduction

Many modern applications store and process datasets much larger than the main memory of even state-of-the-art high-end machines. Thus massive and dynamically changing datasets often need to be stored in space efficient data structures on external storage devices such as disks, and in such cases the Input/Output (or I/O) communication between internal and external memory can become a major performance bottleneck. Many massive dataset applications involve geometric data (for example, points, lines, and polygons) or data which can be interpreted geometrically. Such applications often perform queries which correspond to searching in massive multidimensional geometric databases for objects that satisfy certain spatial constraints. Typical queries include reporting the objects intersecting a query region, reporting the objects containing a query point, and reporting objects near a query point.

While development of practically efficient (and ideally also multi-purpose) external memory data structures (or *indexes*) has always been a main concern in the database community, most data structure research in the algorithms community has focused on worst-case efficient internal memory data structures. Recently, however, there has been some cross-fertilization between the two areas. In this paper we survey recent advances in the development of worst-case efficient external memory data structures. We will concentrate on data structures for geometric problems—especially the important one- and two-dimensional range searching problems—but mention other structures when appropriate. A more comprehensive discussion can be found in a recent survey by the author [16].

---

* Supported in part by the National Science Foundation through ESS grant EIA–9870734, RI grant EIA–9972879, and CAREER grant EIA–9984099.

F. Meyer auf der Heide (Ed.): ESA 2001, LNCS 2161, pp. 1–29, 2001.

Model of computation. Accurately modeling memory and disk systems is a complex task [131]. The primary feature of disks we want to model is their extremely long access time relative to that of internal memory. In order to amortize the access time over a large amount of data, typical disks read or write large blocks of contiguous data at once and therefore the standard two-level disk model has the following parameters [13,153,104]:

N = number of objects in the problem instance;

T = number of objects in the problem solution;

M = number of objects that can fit into internal memory;

B = number of objects per disk block;

where $B < M < N$. An *I/O operation* (or simply *I/O*) is the operation of reading (or writing) a block from (or into) disk. Refer to Figure 1. Computation can only be performed on objects in internal memory. The measures of performance in this model are the number of I/Os used to solve a problem, as well as the amount of space (disk blocks) used and the internal memory computation time.

Several authors have considered more accurate and complex multi-level memory models than the two-level model. An increasingly popular approach to increase the performance of I/O systems is to use several disks in parallel so work has especially been done in multi disk models. See e.g. the recent survey by Vitter [151]. We will concentrate on the two-level one-disk model, since the data structures and data structure design techniques developed in this model often work well in more complex models. For brevity we will also ignore internal computation time.

Outline of paper. The rest of this paper is organized as follows. In Section 2 we discuss the B-tree, the most fundamental (one-dimensional) external data structure, as well as recent variants and extensions of the structure, and in Section 3 we discuss the so-called buffer trees. In Section 4 we illustrate some of the important techniques and ideas used in the development of provably I/O-efficient data structures for higher-dimensional problems through a discussion of the external priority search tree for 3-sided planar range searching. We also discuss a general method for obtaining a dynamic data structure from a static

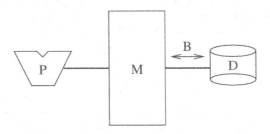

Fig. 1. Disk model; An I/O moves B contiguous elements between disk and main memory (of size M).

one. In Section 5 we discuss data structures for general (4-sided) planar range searching, and in Section 6 we survey results on external data structures for interval management, point location, range counting, higher-dimensional range searching, halfspace range searching, range searching among moving points, and proximity queries. Several of the worst-case efficient structures we consider are simple enough to be of practical interest. Still, there are many good reasons for developing simpler (heuristic) and general purpose structures without worst-case performance guarantees, and a large number of such structures have been developed in the database community. Even though the focus of this paper is on provably worst-case efficient data structures, in Section 7 we give a short survey of some of the major classes of such heuristic-based structures. The reader is referred to recent surveys for a more complete discussion [12,85,121]. Finally, in Section 8 we discuss some of the efforts which have been made to implement the developed worst-case efficient structures.

2   B-Trees

The B-tree is the most fundamental external memory data structure, corresponding to an internal memory balanced search tree [35,63,104,95]. It uses linear space—$O(N/B)$ disk blocks—and supports insertions and deletions in $O(\log_B N)$ I/Os. One-dimensional range queries, asking for all elements in the tree in a query interval $[q_1, q_2]$, can be answered in $O(\log_B N + T/B)$ I/Os.

The space, update, and query bounds obtained by the B-tree are the bounds we would like to obtain in general for more complicated problems. The bounds are significantly better than the bounds we would obtain if we just used an internal memory data structure and virtual memory. The $O(N/B)$ space bound is obviously optimal and the $O(\log_B N + T/B)$ query bound is optimal in a comparison model of computation. Note that the query bound consists of an $O(\log_B N)$ search-term corresponding to the familiar $O(\log N)$ internal memory search-term, and an $O(T/B)$ reporting-term accounting for the $O(T/B)$ I/Os needed to report T elements. Recently, the above bounds have been obtained for a number of problems (e.g [30,26,149,5,47,87]) but higher lower bounds have also been established for some problems [141,26,93,101,106,135,102]. We discuss these results in later sections.

B-trees come in several variants, like B$^+$ and B* trees (see e.g. [35,63,95,30, 104,3] and their references). A basic B-tree is a $\Theta(B)$-ary tree (with the root possibly having smaller degree) built on top of $\Theta(N/B)$ leaves. The degree of internal nodes, as well as the number of elements in a leaf, is typically kept in the range $[B/2 \ldots B]$ such that a node or leaf can be stored in one disk block. All leaves are on the same level and the tree has height $O(\log_B N)$—refer to Figure 2. In the most popular B-tree variants, the N data elements are stored in the leaves (in sorted order) and each internal node holds $\Theta(B)$ "routing" (or "splitting") elements used to guide searches.

To answer a range query $[q_1, q_2]$ on a B-tree we first search down the tree for q_1 and q_2 using $O(\log_B N)$ I/Os, and then we report the elements in the

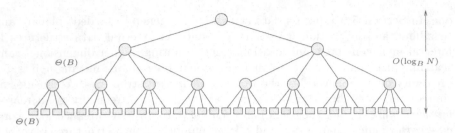

$\Theta(B)$

$O(\log_B N)$

$\Theta(B)$

Fig. 2. B-tree; All internal nodes (except possibly the root) have fan-out $\Theta(B)$ and there are $\Theta(N/B)$ leaves. The tree has height $O(\log_B N)$.

$O(T/B)$ leaves between the leaves containing q_1 and q_2. We perform an insertion in $O(\log_B N)$ I/Os by first searching down the tree for the relevant leaf l. If there is room for the new element in l we simply store it there. If not, we *split* l into two leaves l' and l'' of approximately the same size and insert the new element in the relevant leaf. The split of l results in the insertion of a new routing element in the parent of l, and thus the need for a split may propagate up the tree. Propagation of splits can often be avoided by *sharing* some of the (routing) elements of the full node with a non-full sibling. A new (degree 2) root is produced when the root splits and the height of the tree grows by one. Similarly, we can perform a deletion in $O(\log_B N)$ I/Os by first searching for the relevant leaf l and then removing the deleted element. If this results in l containing too few elements we either *fuse* it with one of its siblings (corresponding to deleting l and inserting its elements in the sibling), or we perform a *share* operation by moving elements from a sibling to l. As splits, fuse operations may propagate up the tree and eventually result in the height of the tree decreasing by one.

B-tree variants and extensions. Recently, several important variants and extensions of B-trees have been considered.

Arge and Vitter [30] developed the *weight-balanced B-trees*, which can be viewed as an external version of BB[α] trees [120]. Weight-balanced B-trees are very similar to B-trees but with a weight constraint imposed on each node in the tree instead of a degree constraint. The weight of a node v is defined as the number of elements in the leaves of the subtree rooted in v. Like B-trees, weight-balanced B-trees are rebalanced using split and fuse operations, and a key property of weight-balanced B-trees is that after performing a rebalance operation on a weight $\Theta(w)$ node v, $\Omega(w)$ updates have to be performed below v before another rebalance operation needs to be performed on v. This means that an update can still be performed in $O(\log_B N)$ I/Os (amortized) even if v has a large associated secondary structure that needs to be updated when a rebalance operation is performed on v, provided that the secondary structure can be updated in $O(w)$ I/Os. Weight-balanced B-trees have been used in numerous efficient data structure (see e.g. [30,26,89,90,38,3,28]).

In some applications we need to be able to traverse a path in a B-tree from a leaf to the root. To do so we need a *parent-pointer* from each node to its parent.

Such pointers can easily be maintained efficiently in normal B-trees or weight-balanced B-trees, but cannot be maintained efficiently if we also want to support *divide* and *merge* operations. A divide operation at element x constructs two trees containing all elements less than and greater than x, respectively. A merge operation performs the inverse operation. Without parent pointers, B-trees and weight-balanced B-trees supports the two operations in $O(\log_B N)$ I/Os. Agarwal et al. [3] developed the *level-balance B-trees* in which divide and merge operations can be supported I/O-efficiently while maintaining parent pointers. In level-balanced B-trees a global balance condition is used instead of the local degree or weight conditions used in B-trees or weight-balanced B-trees; a constraint is imposed on the number of nodes on each level of the tree. When the constraint is violated the whole subtree at that level and above is rebuilt. Level-balanced B-trees e.g. have applications in dynamic maintenance of planar *st*-graphs [3].

Partial persistent B-trees, that is, B-trees where each update results in a new *version* of the structure, and where both the current and older versions can be queried (in the database community often called *multiversion B-trees*), are useful not only in database applications where previous versions of the database needs to be stored and queried, but (as we will discuss in Section 4) also in the solution of many geometric problems. Using standard persistent techniques [137, 70], a persistent B-tree can be designed such that updates can be performed in $O(\log_B N)$ I/Os and such that any version of the tree can be queried in $O(\log_B N + T/B)$ I/Os. Here N is the number of updates and the tree uses $O(N/B)$ disk blocks [36,147].

In string applications a data element (string of characters) can often be arbitrarily long or different elements can be of different length. Such elements cannot be manipulated efficiently in standard B-trees, which assumes that elements (and thus routing elements) are of unit size. Ferragina and Grossi developed the elegant *string B-tree* where a query string q is routed through a node using a so-called *blind trie* data structure [78]. A blind trie is a variant of the compacted trie [104,117], which fits in one disk block. In this way a query can be answered in $O(\log_B N + |q|/B)$ I/O. See [64,80,77,20] for other results on string B-trees and external string processing.

3   Buffer Trees

In internal memory, an N element search tree can be constructed in optimal $O(N \log N)$ time simply by inserting the elements one by one. This construction algorithm can also be used as on optimal sorting algorithm. In external memory, we would use $O(N \log_B N)$ I/Os to build a B-tree using the same method. Interestingly, this is not optimal since Aggarwal and Vitter showed that sorting N elements in external memory takes $\Theta(\frac{N}{B} \log_{M/B} \frac{N}{B})$ I/Os [13]. We can of course build a B-tree in the same bound by first sorting the elements and then building the tree level-by-level bottom-up.

In order to obtain an optimal sorting algorithm based on a search tree struc-
ture, we would need a structure that supports updates in $O(\frac{1}{B} \log_{M/B} \frac{N}{B})$ I/Os.
The inefficiency of the B-tree sorting algorithm is a consequence of the B-tree
being designed to be used in an "on-line" setting where queries should be answered immediately—updates and queries are handled on an individual basis.
This way we are not able to take full advantage of the large internal memory.
It turns out that in an "off-line" environment where we are only interested in
the overall I/O use of a series of operations and where we are willing to relax
the demands on the query operations, we can develop data structures on which
a series of N operations can be performed in $O(\frac{N}{B} \log_{M/B} \frac{N}{B})$ I/Os in total. To
do so we use the *buffer tree* technique developed by Arge [14].

Basically the buffer tree is just a fan-out $\Theta(M/B)$ B-tree where each internal
node has a buffer of size $\Theta(M)$. The tree has height $O(\log_{M/B} \frac{N}{B})$; refer to
Figure 3. Operations are performed in a "lazy" manner: In order to perform an
insertion we do not (like in a normal B-tree) search all the way down the tree for
the relevant leaf. Instead, we wait until we have collected a block of insertions
and then we insert this block in the buffer of the root (which is stored on disk).
When a buffer "runs full" its elements are "pushed" one level down to buffers
on the next level. We can do so in $O(M/B)$ I/Os since the elements in the
buffer fit in main memory and the fan-out of the tree is $O(M/B)$. If the buffer
of any of the nodes on the next level becomes full by this process, the buffer-
emptying process is applied recursively. Since we push $\Theta(M)$ elements one level
down the tree using $O(M/B)$ I/Os (that is, we use $O(1)$ I/Os to push one block
one level down), we can argue that every block of elements is touched a constant
number of times on each of the levels of the tree. Thus, not counting rebalancing,
inserting N elements requires $O(\frac{N}{B} \log_{M/B} \frac{N}{B})$ I/Os in total, or $O(\frac{1}{B} \log_{M/B} \frac{N}{B})$
amortized. Arge showed that rebalancing can be handled in the same bound [14].

The basic buffer tree supporting insertions only can be used to construct a B-
trees in $O(\frac{N}{B} \log_{M/B} \frac{N}{B})$ I/Os (without explicitly sorting). This of course means
that buffer trees can be used to design an optimal sorting algorithm. Note that
unlike other sorting algorithm, the N elements to be sorted do not all need to
be given at the start of this algorithm. Deletions and (one-dimensional) range

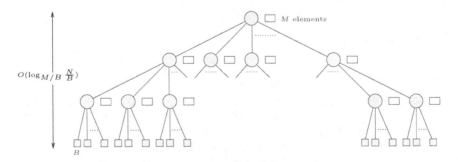

Fig. 3. Buffer tree; Fan-out M/B tree where each node has a buffer of size M. Operations are performed in a lazy way using the buffers.

queries can also be supported I/O-efficiently using buffers [14]. The range queries are *batched* in the sense that we do not obtain the result of a query immediately. Instead parts of the result will be reported at different times as the query is pushed down the tree. This means that the data structure can only be used in algorithms where future updates and queries do not depend on the result of the queries. Luckily this is the case in many plane-sweep algorithms [73,14]. In general, problems where the entire sequence of updates and queries is known in advance, and the only requirement on the queries is that they must all eventually be answered, are known as *batched dynamic problems* [73].

As mentioned, persistent B-trees are often used in geometric data structures. Often, a data structure is constructed by performing N insertion and deletions on an initially empty persistent B-tree, and then the resulting (static) structure is used to answer queries. Using the standard update algorithms the construction would take $O(N \log_B N)$ I/Os. A straightforward application of the buffer tree technique improves this to the optimal $O(\frac{N}{B} \log_{M/B} \frac{N}{B})$ I/Os [145,21] (another optimal, but not linear space, algorithm can be designed using the *distribution-sweeping* technique [86]). Several other data structures can be constructed efficiently using buffers, and the buffer tree technique has been used to develop several other data structures which in turn have been used to develop algorithms in many different areas [25,29,20,21,107,15,76,46,144,145,96,44,136].

Priority queues. External buffered *priority queues* have been extensively researched because of their applications in graph algorithms. Arge showed how to perform deletemin operations on a basic buffer tree in amortized $O(\frac{1}{B} \log_{M/B} \frac{N}{B})$ I/Os [14]. Note that in this case the deletemin occurs right away, that is, it is not batched. This is accomplished by periodically computing the $O(M)$ smallest elements in the structure and storing them in internal memory. Kumar and Schwabe [107] and Fadel et al. [76] developed similar buffered heaps. Using a partial rebuilding idea, Brodal and Katajainen [45] developed a worst-case efficient external priority queue. Using the buffer tree technique on a tournament tree, Kumar and Schwabe [107] developed a priority queue supporting update operations in $O(\frac{1}{B} \log \frac{N}{B})$ I/Os. They also showed how to use their structure in several efficient external graph algorithms (see e.g [2,7,18,22,27,46,59,81,97,107, 110,111,116,118,122,142,156] for other results on external graph algorithms and data structures). Note that if the priority of an element is known, an update operation can be performed in $O(\frac{1}{B} \log_{M/B} \frac{N}{B})$ I/Os on a buffer tree using a delete and an insert operation.

4   3-Sided Planar Range Searching

In internal memory many elegant data structures have been developed for higher-dimensional problems like range searching—see e.g. the recent survey by Agarwal and Erickson [12]. Unfortunately, most of these structures are not efficient when mapped directly to external memory—mainly because they are normally based on binary trees. The main challenge when developing efficient external structures is to use B-trees as base structures, that is, to use multiway trees instead of binary trees. Recently, some progress has been made in the development of provably I/O-efficient data structures based on multi-way trees. In this section we consider a special case of two-dimensional range searching, namely the *3-sided planar range searching* problem: Given a set of points in the plane, the solution to a 3-sided query (q_1, q_2, q_3) consists of all points (x, y) with $q_1 \leq x \leq q_2$ and $y \geq q_3$. The solution to this problem is not only an important component of the solution to the general planar range searching problem we discuss in Section 5, but it also illustrate many of the techniques and ideas utilized in the development of other external data structures.

The static version of the 3-sided problem where the points are fixed can easily be solved I/O-efficiently using a sweeping idea and a persistent B-tree; consider sweeping the plane with a horizontal line from $y = \infty$ to $y = -\infty$ and inserting the x-coordinate of points in a persistent B-tree as they are met. To answer a query (q_1, q_2, q_3) we simply perform a one-dimensional range query $[q_1, q_2]$ on the version of the persistent B-tree we had when the sweep-line was at $y = q_3$. Following the discussion in Section 2, the structure obtained this way uses linear space and queries can be answered in $O(\log_B N + T/B)$ I/Os. The structure can be constructed in $O(\frac{N}{B} \log_{M/B} \frac{N}{B})$ I/Os using the buffer tree technique.

From the static solution we can obtain a linear space dynamic structure which answers a query in $O(\log_B^2 N + T/B)$ I/Os and can be updated in $O(\log_B^2 N)$ I/Os using an external version of the *logarithmic method* for transforming a static structure into a dynamic structure [39,125]. This technique was developed by Arge and Vahrenhold as part of the design of a dynamic external planar point location structure (See Section 6). Due to its general interest, we describe the technique in Section 4.1 below. An optimal $O(\log_B N)$ query structure can be obtained in a completely different way, and we discuss this result in Section 4.2.

4.1   The Logarithmic Method

In internal memory, the main idea in the logarithmic method is to partition the set of N elements into $\log N$ subsets of exponentially increasing size 2^i, $i = 0, 1, 2, \ldots$, and build a static structure $\mathcal{D}_i$ for each of these subsets. Queries are then performed by querying each $\mathcal{D}_i$ and combining the answers, while insertions are performed by finding the first empty $\mathcal{D}_i$, discarding all structures $\mathcal{D}_j$, $j < i$, and building $\mathcal{D}_i$ from the new element and the $\sum_{l=0}^{i-1} 2^l = 2^i - 1$ elements in the discarded structures.

To make the logarithmic method I/O-efficient we need to decrease the number of subsets to $\log_B N$, which in turn means increasing the size of $\mathcal{D}_i$ to B^i. When doing so $\mathcal{D}_j$, $j < i$, does not contain enough objects to build $\mathcal{D}_i$ (since $1 + \sum_{l=0}^{i-1} B^l < B^i$). However, it turns out that if we can build a static structure I/O-efficiently enough, this problem can be resolved and we can make a modified version of the method work in external memory.

Consider a static structure $\mathcal{D}$ that can be constructed in $O(\frac{N}{B} \log_B N)$ I/Os and that answers queries in $O(\log_B N)$ I/Os (note that $O(\frac{N}{B} \log_{M/B} \frac{N}{B}) =$

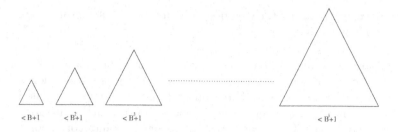

Fig. 4. Logarithmic method; $\log_B N$ structures—$\mathcal{D}_i$ contains less than $B^i + 1$ elements. $\mathcal{D}_1, \mathcal{D}_2, \ldots, \mathcal{D}_j$ do not contain enough elements to build $\mathcal{D}_{j+1}$ of size B^{j+1}.

$O(\frac{N}{B} \log_B N)$ if $M > B^2$). We partition the N elements into $\log_B N$ sets such that the ith set has size *less than* $B^i + 1$ and construct an external memory static data structure $\mathcal{D}_i$ for each set—refer to Figure 4. To answer a query, we simply query each $\mathcal{D}_i$ and combine the results using $O(\sum_{j=1}^{\log_B N} \log_B |\mathcal{D}_j|) = O(\log_B^2 N)$ I/Os. We perform an insertion by finding the first structure $\mathcal{D}_i$ such that $\sum_{j=1}^{i} |\mathcal{D}_j| \leq B^i$, discarding all structures $\mathcal{D}_j$, $j \leq i$, and building a new $\mathcal{D}_i$ from the elements in these structures using $O((B^i/B) \log_B B^i) = O(B^{i-1} \log_B N)$ I/Os. Now because of the way $\mathcal{D}_i$ was chosen, we know that $\sum_{j=1}^{i-1} |\mathcal{D}_j| > B^{i-1}$. This means that at least B^{i-1} objects are moved from lower indexed structures to $\mathcal{D}_i$. If we divide the $\mathcal{D}_i$ construction cost between these objects, each object is charged $O(\log_B N)$ I/Os. Since an object never moves to a lower indexed structure we can at most charge it $O(\log_B N)$ times during N insertions. Thus the amortized cost of an insertion is $O(\log_B^2 N)$ I/Os. Note that the key to making the method work is that the factor of B we lose when charging the construction of a structure of size B^i to only B^{i-1} objects is offset by the $1/B$ factor in the construction bound. Deletions can also be handled I/O-efficiently using a global rebuilding idea.

4.2   Optimal Dynamic Structure

Following several earlier attempts [101,127,141,43,98], Arge et al. [26] developed an optimal dynamic structure for the 3-sided planar range searching problem. The structure is an external version of the internal memory *priority search tree* structure [113]. The external priority search tree consists of a *base B-tree* on the x-coordinates of the N points. A range X_v (containing all points below v) can be associated with each node v in a natural way. This range is subdivided into $\Theta(B)$ subranges associated with the children of v. For illustrative purposes we call the subranges *slabs*. In each node v we store $O(B)$ points for each of v's $\Theta(B)$ children v_i, namely the B points with the highest y-coordinates in the x-range of v_i (if existing) that have not been stored in ancestors of v. We store the $O(B^2)$ points in the linear space static structure discussed above (the "$O(B^2)$–structure") such that a 3-sided query on them can be answered in $O(\log_B B^2 + T/B) = O(1 + T/B)$

I/Os. Since every point is stored in precisely one $O(B^2)$–structure, the structure uses $O(N/B)$ space in total.

To answer a 3-sided query (q_1, q_2, q_3) we start at the root of the external priority search tree and proceed recursively to the appropriate subtrees; when visiting a node v we query the $O(B^2)$–structure and report the relevant points, and then we advance the search to some of the children of v. The search is advanced to child v_i if v_i is either along the leftmost search path for q_1 or the rightmost search path for q_2, or if the entire set of points corresponding to v_i in the $O(B^2)$–structure were reported—refer to Figure 5. The query procedure reports all points in the query range since if we do not visit child v_i corresponding to a slab completely spanned by the interval $[q_1, q_2]$, it means that at least one of the points in the $O(B^2)$–structure corresponding to v_i does not satisfy the query. This in turn means that none of the points in the subtree rooted at v_i can satisfy the query. That we use $O(\log_B N + T/B)$ I/Os to answer a query can be seen as follows. In every internal node v visited by the query procedure we spend $O(1 + T_v/B)$ I/Os, where T_v is the number of points reported. There are $O(\log_B N)$ nodes visited on the search paths in the tree to the leaf containing q_1 and the leaf containing q_2 and thus the number of I/Os used in these nodes adds up to $O(\log_B N + T/B)$. Each remaining visited internal node v is not on the search path but it is visited because $\Theta(B)$ points corresponding to it were reported when we visited its parent. Thus the cost of visiting these nodes adds up to $O(T/B)$, even if we spend a constant number of I/Os in some nodes without finding $\Theta(B)$ points to report.

To insert a point $p = (x, y)$ in the external priority search tree we search down the tree for the leaf containing x, until we reach the node v where p needs to be inserted in the $O(B^2)$–structure. The $O(B^2)$–structure is static but since it has size $O(B^2)$ we can use a global rebuilding idea to make it dynamic [125]; we simply store the update in a special "update block" and once B updates have been collected we rebuild the structure using $O(\frac{B^2}{B} \log_{M/B} \frac{B^2}{B})$ I/Os. Assuming $M > B^2$, that is, that the internal memory is capable of holding B blocks, this is $O(B)$ and we obtain an $O(1)$ amortized update bound. Arge et al. [26] showed how to make this worst-case, even without the assumption on the main

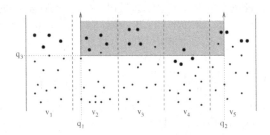

Fig. 5. Internal node v with children $v_1, v_2, \ldots, v_5$. The points in bold are stored in the $O(B^2)$–structure. To answer a 3-sided query we report the relevant of the $O(B^2)$ points and answer the query recursively in v_2, v_3, and v_5. The query is not extended to v_4 because not all of the points from v_4 in the $O(B^2)$–structure satisfy the query.

memory size. Insertion of p in v (may) result in the $O(B^2)$–structure containing one too many points from the slab corresponding to the child v_j containing x. Therefore, apart from inserting p in the $O(B^2)$–structure, we also remove the point p' with the lowest y-coordinate among the points corresponding to v_j. We insert p' recursively in the tree rooted in v_j. Since we use $O(1)$ I/Os in each of the nodes on the search path, the insertion takes $O(\log_B N)$ I/Os. We also need to insert x in the base B-tree. This may result in split and/or share operations and each such operation may require rebuilding an $O(B^2)$–structure (as well as movement of some points between $O(B^2)$–structures). Using weight-balanced B-tress, Arge et al. [26] showed how the rebalancing after an insertion can be performed in $O(\log_B N)$ I/Os worst case. Deletions can be handled in $O(\log_B N)$ I/Os in a similar way [26].

The above solution to the 3-sided planar range searching problem illustrates some of the problems encountered when developing I/O-efficient dynamic data structures, as well as the techniques commonly used to overcome these problems. As already discussed, the main problem is that in order to be efficient, external tree data structures need to have large fan-out. In the above example this resulted in the need for what we called the $O(B^2)$–structure. This structure solved a static version of the problem on $O(B^2)$ points. The structure was necessary since to "pay" for a visit to a child node v_i, we needed to find $\Theta(B)$ points in the slab corresponding to v_i satisfying the query. The idea of charging some of the query cost to the output size is often called *filtering* [51], and the idea of using a static structure on $O(B^2)$ elements in each node has been called the *bootstrapping* paradigm [151,152]. Finally, the ideas of weight-balancing and global rebuilding were used to obtain worst-case efficient update bounds. All these ideas have been used in the development of other efficient external data structures.

5   General Planar Range Searching

After discussing 3-sided planar range searching we are now ready to consider general planar range searching; given a set of points in the plane we want to be able to find all points contained in a query rectangle. While linear space and $O(\log_B N + T/B)$ query structures exist for special cases of this problem— like the 3-sided problem described in Section 4—Subramanian and Ramaswamy showed that one cannot obtain an $O(\log_B N + T/B)$ query bound using less than $\Theta(\frac{N \log(N/B)}{B \log \log_B N})$ disk blocks [141].[1] This lower bound holds in a natural external memory version of the *pointer machine model* [53]. A similar bound in a slightly different model where the search component of the query is ignored was proved by Arge et al. [26]. This *indexability model* was defined by Hellerstein et al. [93] and considered by several authors [101,106,135].

Based on a sub-optimal but linear space structure for answering 3-sided queries, Subramanian and Ramaswamy developed the *P-range tree* that uses optimal $O(\frac{N \log(N/B)}{B \log \log_B N})$ space but uses more than the optimal $O(\log_B N + T/B)$

---

[1] In fact, this bound even holds for a query bound of $O(\log_B^c N + T/B)$ for any constant c.

I/Os to answer a query [141]. Using their optimal structure for 3-sided queries, Arge et al. obtained an optimal structure [26]. We discuss the structure in Section 5.1 below. In practical applications involving massive datasets it is often crucial that external data structures use linear space. We discuss this further in Section 7. Grossi and Italiano developed the elegant linear space *cross-tree* data structure which answers planar range queries in $O(\sqrt{N/B}+T/B)$ I/Os [89, 90]. This is optimal for linear space data structures—as e.g. proven by Kanth and Singh [102]. The *O-tree* of Kanth and Singh [102] obtains the same bounds using ideas similar to the ones used by van Kreveld and Overmars in *divided k-d trees* [146]. In Section 5.2 below we discuss the cross-tree further.

5.1   Logarithmic Query Structure

The $O(\log_B N + T/B)$ query data structure is based on ideas from the corresponding internal memory data structure due to Chazelle [51]. The structure consists of a fan-out $\log_B N$ base tree over the x-coordinates of the N points. As previously an x-range is associated with each node v and it is subdivided into $\log_B N$ slabs by v's children $v_1, v_2, \ldots, v_{\log_B N}$. We store *all* the points in the x-range of v in four secondary data structures associated with v. The first structure store the points in a linear list sorted by y-coordinate. The three other structures are external priority search trees. Two of these structures are used for answering 3-sided queries with the opening to the left and to the right, respectively. For the third priority search tree, we consider for each child v_i the points in the x-range of v_i in y-order, and for each pair of consecutive points (x_1, y_1) and (x_2, y_2) we store the point (y_1, y_2) in the tree. With each constructed point we also store pointers to the corresponding two original point in the sorted list of points in a child node. Since we use linear space on each of the $O(\log_{\log_B N}(N/B)) = O(\log(N/B)/\log\log_B N)$ levels of the tree, the structure uses $O(\frac{N\log(N/B)}{B\log\log_B N})$ disk blocks in total.

To answer a 4-sided query $q = (q_1, q_2, q_3, q_4)$ we first find the topmost node v in the base tree where the x-range $[q_1, q_2]$ of the query contains a boundary between two slabs. Consider the case where q_1 lies in the x-range of v_i and q_2 lies in the x-range of v_j—refer to Figure 6. The query q is naturally decomposed into three parts, consisting of a part in v_i, a part in v_j, and a part completely spanning nodes v_k, for $i < k < j$. The points contained in the first two parts can be found in $O(\log_B N + T/B)$ I/Os using the right opening priority search tree in v_i and the left opening priority search tree in v_j. To find the points in the third part we query the third priority search tree associated with v with $(-\infty, q_2, q_2)$, that is, we find all points (y_1, y_2) in the structure with $y_1 \leq q_2$ and $y_2 \geq q_2$. Since a point (y_1, y_2) corresponds to a consecutive pair (x_1, y_1) and (x_2, y_2) of the original points in a slab, we in this way obtain the $O(\log_B N)$ bottommost point contained in the query for each of the nodes $v_{i+1}, v_{i+2}, \ldots, v_{j-1}$. Using the pointers to the same points in these children nodes, we then traverse the $j - i - 1 = O(\log_B N)$ relevant sorted lists and output the remaining points using $O(\log_B N + T/B)$ I/Os.

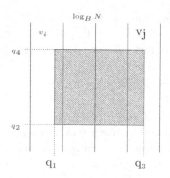

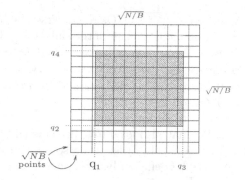

Fig. 6. The slabs corresponding to a node v in the base tree. To answer a query (q_1, q_2, q_3, q_4) we need to answer 3-sided queries on the points in slab v_i and slab v_j, and a range query on the points in the $O(\log_B N)$ slabs between v_i and v_j.

Fig. 7. Basic squares. To answer a query (q_1, q_2, q_3, q_4) we check points in two vertical and two horizontal slabs, and report points in basic squares completely covered by the query.

To insert or delete a point, we need to perform $O(1)$ updates on each of the $O(\log(N/B)/\log\log_B N)$ levels of the base tree. Each of these updates takes $O(\log_B N)$ I/Os. We also need to update the base tree. Using a weight-balanced B-tree, Arge et al. showed how this can be done in $O((\log_B N)(\log \frac{N}{B})/\log\log_B N)$ I/Os [26].

5.2   Linear Space Structure

The linear space cross-tree structure of Grossi and Italiano consists of two levels [89,90]. The lower level partitions the plane into $\Theta(\sqrt{N/B})$ vertical slabs and $\Theta(\sqrt{N/B})$ horizontal slabs containing $\Theta(\sqrt{NB})$ points each, forming an irregular grid of $\Theta(N/B)$ *basic squares*—refer to Figure 7. Each basic square can contain between 0 and $\sqrt{N/B}$ points. The points are grouped and stored according to the vertical slabs—points in vertically adjacent basic squares containing less than B points are grouped together to form groups of $\Theta(B)$ points and stored in blocks together. The points in a basic square containing more than B points are stored in a B-tree. Thus the lower level uses $O(N/B)$ space. The upper level consists of a linear space search structure which can be used to determine the basic square containing a given point—for now we can think of the structure as consisting of a fan-out $\sqrt{B}$ B-tree $\mathcal{T}_V$ on the $\sqrt{N/B}$ vertical slabs and a separate fan-out $\sqrt{B}$ B-tree $\mathcal{T}_H$ on the $\sqrt{N/B}$ horizontal slabs.

In order to answer a query (q_1, q_2, q_3, q_4) we use the upper level search tree to find the vertical slabs containing q_1 and q_3 and the horizontal slabs containing q_2 and q_4 using $O(\log_{\sqrt{B}} N) = O(\log_B N)$ I/Os. We then explicitly check all points in these slabs and report all the relevant points. In doing so we use $O(\sqrt{NB}/B) = O(\sqrt{N/B})$ I/Os to traverse the vertical slabs and $O(\sqrt{NB}/B +$

$\sqrt{N/B}) = O(\sqrt{N/B})$ I/Os to traverse the horizontal slabs (the $\sqrt{N/B}$-term in the latter bound is a result of the slabs being blocked vertically—a horizontal slab contains $\sqrt{N/B}$ basic squares). Finally, we report all points corresponding to basic squares fully covered by the query. To do so we use $O(\sqrt{N/B}+T/B)$ I/Os since the slabs are blocked vertically. In total we answer a query in $O(\sqrt{N/B}+T/B)$ I/Os.

In order to perform an update we need to find and update the relevant basic square. We may also need to split slabs (insertion) or merge slabs with neighbor slabs (deletions). In order to do so efficiently while still being able to answer a range query I/O-efficiently, the upper level is actually implemented using a *cross-tree* $\mathcal{T}_{HV}$. $\mathcal{T}_{HV}$ can be viewed as a cross product of $\mathcal{T}_V$ and $\mathcal{T}_H$: For each pair of nodes $u \in \mathcal{T}_H$ and $v \in \mathcal{T}_V$ on the same level we have a node (u, v) in $\mathcal{T}_{HV}$, and for each pair of edges $(u, u') \in \mathcal{T}_H$ and $(v, v') \in \mathcal{T}_V$ we have an edge $((u, v), (u', v'))$ in $\mathcal{T}_{HV}$. Thus the tree has fan-out $O(B)$ and uses $O((\sqrt{N/B})^2) = O(N/B)$ space. Grossi and Italiano showed how we can use the cross-tree to search for a basic square in $O(\log_B N)$ I/Os and how the full structure can be used to answer a range query in $O(\sqrt{N/B} + T/B)$ I/Os [89,90]. They also showed that if $\mathcal{T}_H$ and $\mathcal{T}_V$ are implemented using weight-balanced B-trees, the structure can be maintained in $O(\log_B N)$ I/Os during an update.

6   Survey of Other External Data Structures

After having discussed planar range searching is some detail, in this section we survey other results on worst-case efficient external data structures.

Interval management. The interval management (or stabbing query) problem is the problem of maintaining a dynamically changing set of (one-dimensional) intervals such that given a query point q all intervals containing q can be reported efficiently. By mapping each interval $[x, y]$ to the point (x, y) in the plane, a query corresponds to finding all points such that $x \leq q$ and $y \geq q$, which means that the external priority search tree describe in Section 4.2 can be used to solve this problem optimally. However, the problem was first solved optimally by Arge and Vitter [30], who developed an external version of the *interval tree* of Edelsbrunner [71,72].

The external interval tree was the first to use several of the ideas also utilized in the external priority search tree; filtering, bootstrapping, and weight-balanced B-trees. The structure also utilized the notion of *multislabs*, which is useful when storing objects (like intervals) with a spatial extent. Recall that a slab is a subranges of the range associated with a node v of a base tree defined by the range of one of v's children. A multislabs is simply a contiguous sets of slabs. A key idea in the external interval tree is to decrease the fanout of the base tree to $\sqrt{B}$, maintaining the $O(\log_B N)$ tree height, such that each node has $O(\sqrt{(B)}^2) = O(B)$ associated multislabs. This way a constant amount of information can be stored about each multislab in $O(1)$ blocks. Similar ideas have been utilized in several other external data structures [14,25,3,28]. Variants of the external interval tree structure, as well as applications of it in isosurface extraction, have been considered by Chiang and Silva [29,60,62,61] (see also [7]).

Planar point location. The planar point location problem is defined as follows: Given a planar subdivision with N vertices (i.e., a decomposition of the plane into polygonal regions induced by a straight-line planar graph), construct a data structure so that the face containing a query point p can be reported efficiently. In internal memory, a lot of work has been done on this problem—see e.g. the survey by Snoeyink [140]. Goodrich et al. [86] described the first query optimal $O(\log_B N)$ I/O static solution to the problem, and several structures which can answer a batch of queries I/O-efficiently have also been developed [86,29,25,65, 143].

Recently, progress has been made in the development of I/O-efficient dynamic point location structures. In the dynamic version of the problem one can change the subdivision dynamically (insert and delete edges and vertices). Based on the external interval tree structure and ideas also utilized in several internal memory structures [56,34], Agarwal et al. [3] developed a dynamic structure for *monotone* subdivisions. Utilizing the logarithmic method and a technique similar to dynamic fractional cascading [54,114], Arge and Vahrenhold improved the structure to work for general subdivisions. Their structure uses linear space and supports updates and queries in $O(\log_B^2 N)$ I/Os.

Range counting. Given a set of N points in the plane, a range counting query asks for the number of points within a query rectangle. Based on ideas utilized in an internal memory counting structure due to Chazelle [52], Agarwal et al. [6] designed an external data structure for the range counting problem. Their structure use linear space and answers a query in $O(\log_B N)$ I/Os. Based on a reduction due to Edelsbrunner and Overmars [74], they also designed a linear space and $O(\log_B N)$ query structure for the rectangle counting problem. In this problem, given a set of N rectangles in the plane, a query asks for the number of rectangles intersecting a query rectangle. Finally, they extended their structures to the d-dimensional versions of the two problems. See also [157] and references therein.

Higher-dimensional range searching. Vengroff and Vitter [149] presented a data structure for 3-dimensional range searching with a logarithmic query bound. With recent modifications their structure answers queries in $O(\log_B N + T/B)$ I/Os and uses $O(\frac{N}{B} \log^3 \frac{N}{B} / \log \log_B^3 N)$ space [151]. More generally, they presented structures for answering $(3+k)$-sided queries (k of the dimensions, $0 \le k \le 3$, have finite ranges) in $O(\log_B N + T/B)$ I/Os using $O(\frac{N}{B} \log^k \frac{N}{B} / \log \log_B^k N)$ space.

As mentioned, space use is often as crucial as query time when manipulating massive datasets. The linear space cross-tree of Grossi and Italiano [89,90], as well as the O-tree of Kanth and Singh [102], can be extended to support d-dimensional range queries in $O((N/B)^{1-1/d} + T/B)$ I/Os. Updates can be performed in $O(\log_B N)$ I/Os. The cross-tree can also be used in the design of dynamic data structures for several other problems [89,90].

Halfspace range searching. Given a set of points in d-dimensional space, a halfspace range query asks for all points on one side of a query hyperplane. Halfspace range searching is the simplest form of non-isothetic (non-orthogonal) range searching. The problem was first considered in external memory by Franciosa and Talamo [83,82]. Based on an internal memory structure due to Chazelle et al. [55], Agarwal et al. [5] described an optimal $O(\log_B N + T/B)$ query and linear space structure for the 2-dimensional case. Using ideas from an internal memory result of Chan [50], they described a structure for the 3-dimensional case, answering queries in $O(\log_B N + T/B)$ expected I/Os but requiring $O((N/B)\log(N/B))$ space. Based on the internal memory *partition trees* of Matoušek [112], they also gave a linear space data structure for answering d-dimensional halfspace range queries in $O((N/B)^{1-1/d+\epsilon} + T/B)$ I/Os for any constant $\epsilon > 0$. The structure supports updates in $O((\log(N/B))\log_B N)$ expected I/Os amortized. Using an improved construction algorithm, Agarwal et al. [4] obtained an $O(\log_B^2 N)$ amortized and expected update I/O-bound for the planar case. Agarwal et al. [5] also showed how the query bound of the structure can be improved at the expense of extra space. Finally, their linear space structure can also be used to answer very general queries—more precisely, all points within a query polyhedron with m faces can be found in $O(m(N/B)^{1-1/d+\epsilon} + T/B)$ I/Os.

Range searching on moving points. Recently there has been an increasing interest in external memory data structures for storing continuously moving objects. A key goal is to develop structures that only need to be changed when the velocity or direction of an object changes (as opposed to continuously).

Kollios at al. [105] presented initial work on storing moving points in the plane such that all points inside a query range at query time t can be reported in a provably efficient number of I/Os. Their results were improved and extended by Agarwal et al. [4] who developed a linear space structure that answers a query in $O((N/B)^{1/2+\epsilon} + T/B)$ I/Os for any constant $\epsilon > 0$. A point can be updated using $O(\log_B^2 N)$ I/Os. The structure is based on partition trees and can also be used to answer queries where two time values t_1 and t_2 are given and we want to find all points that lie in the query range at any time between t_1 and t_2. Using the notion of *kinetic data structures* introduced by Basch et al. [33], as well as a persistent version of the planar range searching structure [26] discussed in Section 5, Agarwal et al. [4] also developed a number of other structures with improved query performance. One of these structures has the property that queries in the near future are answered faster than queries further away in time. Further structures with this property were developed by Agarwal et al. [10].

Proximity queries. Proximity queries such as nearest neighbor and closest pair queries have become increasingly important in recent years, for example because of their applications in similarity search and data mining. Callahan et al. [47] developed the first worst-case efficient external proximity query data structures. Their structures are based on an external version of the *topology trees* of Frederickson [84] called *topology B-trees*, which can be used to dynamically maintain arbitrary binary trees I/O-efficiently.

Using topology B-trees and ideas from an internal structure of Bespamyatnikh [42], Callahan et al. [47] designed a linear space data structure for dynamically maintaining the closest pair of a set of points in d-dimensional space. The structure supports updates in $O(\log_B N)$ I/Os. The same result was obtained by Govindarajan et al. [87] using the *well-separated pair decomposition* of Callahan and Kosaraju [48,49]. Govindarajan et al. [87] also showed how to dynamically maintain a well-separated pair decomposition of a set of d-dimensional points using $O(\log_B N)$ I/Os per update.

Using topology B-trees and ideas from an internal structure due to Arya et al. [31], Callahan et al. [47] developed a linear space data structure for the dynamic approximate nearest neighbor problem. Given a set of points in d-dimensional space, a query point p, and a parameter ϵ, the approximate nearest neighbor problem consists of finding a point q with distance at most $(1+\epsilon)$ times the distance of the actual nearest neighbor of p. The structure answers queries and supports updates in $O(\log_B N)$ I/Os. Agarwal et al. [4] designed I/O-efficient data structures for answering approximate nearest neighbor queries on a set of moving points.

In some applications we are interested in finding not only the nearest but all the k nearest neighbors of a query point. Based on their 3-dimensional halfspace range searching structure, Agarwal et al [5] described a structure that uses $O((N/B)\log(N/B))$ space to store N points in the plane such that a k nearest neighbors query can be answered in $(\log_B N + k/B)$ I/Os.

7   Practical General-Purpose Structures

Although several of the worst-case efficient (and often optimal) data structures discussed in the previous sections are simple enough to be of practical interest, they are often not the obvious choices when deciding which data structures to use in a real-world application. There are several reasons for this, one of the most important being that in real applications involving massive datasets it is practically feasible to use data structures of size cN/B only for a very small constant c. Since fundamental lower bounds often prevent logarithmic worst-case search cost for even relatively simple problems when restricting the space use to linear, we need to develop heuristic structures which perform well in most practical cases. Space restrictions also motivate us not to use structures for single specialized queries but instead design general structures that can be used to answer several different types of queries. Finally, implementation considerations often motivate us to sacrifice worst-case efficiency for simplicity. All of these considerations have led to the development of a large number of general-purpose data structures that often work well in practice, but which do not come with worst-case performance guarantees. Below we quickly survey the major classes of such structures. The reader is referred to more complete surveys for details [12,85,121,88,124,134].

Range searching in d-dimensions is the most extensively researched problem. A large number of structures have been developed for this problem, including space filling curves (see e.g. [123,1,32]), grid-files [119,94], various quad-trees [133, 134], kd-B tress [128]—and variants like Buddy-trees [138], hB-trees [109,75] and cell-trees [91]—and various R-trees [92,88,139,37,100]. Often these structures are broadly classified into two types, namely *space driven* structures (like quad-trees and grid-files), which partition the embedded space containing the data points and *data driven* structures (like kd-B trees and R-trees), which partition the data points themselves. Agarwal et al. [9] describe a general framework for efficient construction and updating of many of the above structures.

As mentioned above, we often want to be able to answer a very diverse set of queries, like halfspace range queries, general polygon range queries, and point location queries, on a single data structure. Many of the above data structures can easily be used to answer many such different queries and that is one main reason for their practical success. Recently, there has also been a lot of work on extensions—or even new structures—which also support e.g. moving objects (see e.g [155,132,154,126] and references therein) or proximity queries (see e.g. [41, 129,103,85,12,121] and references therein). However, as discussed, most often no guarantee on the worst-case query performance is provided for these structures.

So far we have mostly discussed point data structures. In general, we are interested in storing objects such as lines and polyhedra with a spatial extent. Like in the point case, a large number of heuristic structures, many of which are variations of the ones mentioned above, have been proposed for such objects. However, almost no worst-case efficient structures are known. In practice a *filtering/refinement* method is often used when managing objects with spatial extent. Instead of directly storing the objects in the data structure we store the *minimal bounding (axis-parallel) rectangle* containing each object together with a pointer to the object itself. When answering a query we first find all the minimal bounding rectangles fulfilling the query (the *filtering* step) and then we retrieve the objects corresponding to these rectangles and check each of them to see if they fulfill the query (the *refinement* step). One way of designing data structures for rectangles (or even more general objects) is to transform them into points in higher-dimensional space and store these points in one of the point data structures discussed above (see e.g. [85,121] for a survey). However, a structure based on another idea has emerged as especially efficient for storing and querying minimal bounding rectangles. Below we further discuss this so-called R-tree and its many variants.

R-trees. The R-tree, originally proposed by Guttman [92], is a multiway tree very similar to a B-tree; all leaf nodes are on the same level of the tree and a leaf contains $\Theta(B)$ data rectangles. Each internal node v (except maybe for the root) has $\Theta(B)$ children. For each of its children v_i, v contains the minimal bounding rectangle of all the rectangles in the tree rooted in v_i. An R-tree has height $O(\log_B N)$ and uses $O(N/B)$ space. An example of an R-tree is shown in Figure 8. Note that there is no unique R-tree for a given set of data rectangles and that minimal bounding rectangles stored within an R-tree node can overlap.

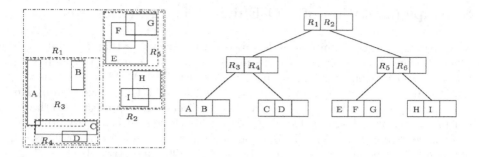

Fig. 8. R-tree constructed on rectangles A, B, C, ... , I ($B = 3$).

In order to query an R-tree to find, say, all rectangles containing a query point p, we start at the root and recursively visit all children whose minimal bounding rectangle contains p. This way we visit all internal nodes whose minimal bounding rectangle contains p. There can be many more such nodes than actual data rectangles containing p and intuitively we want the minimal bounding rectangles stored in an internal node to overlap as little as possible in order to obtain a query efficient structure.

An insertion can be performed in $O(\log_B N)$ I/Os like in a B-tree. We first traverse the path from the root to the leaf we choose to insert the new rectangle into. The insertion might result in the need for node splittings on the same root-leaf path. As insertion of a new rectangle can increase the overlap in a node, several heuristics for choosing which leaf to insert a new rectangle into, as well as for splitting nodes during rebalancing, have been proposed [88,139,37,100]. The R*-tree variant of Beckmann et al. [37] seems to result in the best performance in many cases. Deletions are also performed similarly to deletions in a B-tree but we cannot guarantee an $O(\log_B N)$ bound since finding the data rectangle to delete may require many more I/Os. Rebalancing after a deletion can be performed by fusing nodes like in a B-tree but some R-tree variants instead delete a node when it underflows and reinsert its children into the tree (often referred to as "forced reinsertion"). The idea is to try to obtain a better structure by forcing a global reorganization of the structure instead of the local reorganization a node fuse constitutes.

Constructing an R-tree using repeated insertion takes $O(N \log_B N)$ I/Os and does not necessarily result in a good tree in terms of query performance. Therefore several sorting based $O(\frac{N}{B} \log_{M/B} \frac{N}{B})$ I/O construction algorithms have been proposed [130,99,69,108,40]. Several of these algorithms produce an R-tree with practically better query performance than an R-tree built by repeated insertion. Still, no better than a linear worst-case query I/O-bound has been proven for any of them. Very recently, however, de Berg et al. [68] and Agarwal et al. [11] presented R-tree construction algorithms resulting in R-trees with provably efficient worst-case query performance measured in terms of certain parameters describing the input data. They also discussed how these structures can be efficiently maintained dynamically.

8   Implementation of I/O-Efficient Data Structures

Two ongoing projects aim at developing software packages that facilitates implementation of I/O-efficient algorithms and data structures in a high-level, portable and efficient way. These projects are the LEDA-SM project at MPI in Germany [66,67] and the TPIE (Transparent Parallel I/O Programming Environment) project at Duke [17,148]. We briefly discuss these projects and the experiments performed within them below. Outside these projects, a few other authors have reported on stand-alone implementations of geometric algorithms [57, 58], external interval trees [60,61,62], buffer trees [96], and string B-trees [79].

LEDA-SM. LEDA-SM is an extension of the LEDA library [115] of efficient algorithms and data structures. It consists of a kernel that gives an abstract view of external memory as a collection of disks, each consisting of a collection of blocks. The kernel provides a number of primitives for manipulating blocks, which facilitate efficient implementation of external memory algorithms and data structures. The LEDA-SM distribution also contains a collection of fundamental data structures, such as stacks, queues, heaps, B-trees and buffer-trees, as well as a few fundamental algorithms such as external sorting and matrix operations. It also contains algorithms and data structures for manipulating strings and massive graphs. Results on the practical performance of LEDA-SM implementations of external priority queues and I/O-efficient construction of suffix arrays can be found in [44] and [64], respectively.

TPIE. The first part of the TPIE project took a stream-based approach to computation [148,17], where the kernel feeds a continuous stream of elements to the user programs in an I/O-efficient manner. This approach is justified by theoretical research on I/O-efficient algorithms, which show that a large number of problems can be solved using a small number of streaming paradigms, all implemented in TPIE. This part of TPIE also contains fundamental data structures such as queues and stacks, algorithms for sorting and matrix operations, as well as a few more specialized geometric algorithms. It has been used in I/O-efficient implementations of several scientific computation [150], spatial join [24,25,23], and terrain flow [27,19] algorithms.

Since most external data structures cannot efficiently be implemented in a stream-based framework, the second part of the TPIE project adds kernel support for a block oriented programming style. Like in LEDA-SM, the external memory is viewed as a collection of blocks and primitives for manipulating such blocks are provided. Fundamental data structure such as B-trees and R-trees are also provided with this part of TPIE. The block oriented part of TPIE is integration with the stream oriented part, and together the two parts have been used to implement I/O-efficient algorithms for R-trees construction [21] based on the buffer technique, to implement an I/O-efficient algorithms for constriction and updating of kd-tree [8,9], as well as to implement the recently developed structures for range counting [6]. Other external data structures currently being implemented includes persistent B-trees, structures for planar point location, and the external priority search tree.

9   Conclusions

In this paper we have discussed recent advances in the development of provably efficient external memory dynamic data structures, mainly for geometric objects and especially for the one- and two-dimensional range searching problems. A more detailed survey by the author can be found in [16]. Even though a lot of progress has been made, many problems still remain open. For example, $O(\log_B N)$-query and space efficient structures still need to be found for many higher-dimensional problems.

Acknowledgments. The author thanks Tammy Bailey, Tavi Procopiuc, and Jan Vahrenhold for comments on earlier drafts of this paper.

References

1. D. J. Abel and D. M. Mark. A comparative analysis of some two-dimensional orderings. *Intl. J. Geographic Informations Systems*, 4(1):21–31, 1990.
2. J. Abello, A. L. Buchsbaum, and J. R. Westbrook. A functional approach to external graph algorithms. In *Proc. Annual European Symposium on Algorithms, LNCS 1461*, pages 332–343, 1998.
3. P. K. Agarwal, L. Arge, G. S. Brodal, and J. S. Vitter. I/O-efficient dynamic point location in monotone planar subdivisions. In *Proc. ACM-SIAM Symp. on Discrete Algorithms*, pages 1116–1127, 1999.
4. P. K. Agarwal, L. Arge, and J. Erickson. Indexing moving points. In *Proc. ACM Symp. Principles of Database Systems*, pages 175–186, 2000.
5. P. K. Agarwal, L. Arge, J. Erickson, P. Franciosa, and J. Vitter. Efficient searching with linear constraints. *Journal of Computer and System Sciences*, 61(2):194–216, 2000.
6. P. K. Agarwal, L. Arge, and S. Govindarajan. CRB-tree: An optimal indexing scheme for 2d aggregate queries. Manuscript, 2001.
7. P. K. Agarwal, L. Arge, T. M. Murali, K. Varadarajan, and J. S. Vitter. I/O-efficient algorithms for contour line extraction and planar graph blocking. In *Proc. ACM-SIAM Symp. on Discrete Algorithms*, pages 117–126, 1998.
8. P. K. Agarwal, L. Arge, O. Procopiuc, and J. S. Vitter. Dynamic kd-trees on large data sets. Manuscript, 2001.
9. P. K. Agarwal, L. Arge, O. Procopiuc, and J. S. Vitter. A framework for index bulk loading and dynamization. In *Proc. Annual International Colloquium on Automata, Languages, and Programming*, 2001.
10. P. K. Agarwal, L. Arge, and J. Vahrenhold. A time responsive indexing scheme for moving points. In *Proc. Workshop on Algorithms and Data Structures*, 2001.
11. P. K. Agarwal, M. de Berg, J. Gudmundsson, M. Hammer, and H. J. Haverkort. Box-trees and R-trees with near-optimal query time. In *Proc. ACM Symp. on Computational Geometry*, pages 124–133, 2001.
12. P. K. Agarwal and J. Erickson. Geometric range searching and its relatives. In B. Chazelle, J. E. Goodman, and R. Pollack, editors, *Advances in Discrete and Computational Geometry*, volume 223 of *Contemporary Mathematics*, pages 1–56. American Mathematical Society, Providence, RI, 1999.
13. A. Aggarwal and J. S. Vitter. The Input/Output complexity of sorting and related problems. *Communications of the ACM*, 31(9):1116–1127, 1988.

14. L. Arge. The buffer tree: A new technique for optimal I/O-algorithms. In *Proc. Workshop on Algorithms and Data Structures, LNCS 955*, pages 334–345, 1995. A complete version appears as BRICS technical report RS-96-28, University of Aarhus.

15. L. Arge. The I/O-complexity of ordered binary-decision diagram manipulation. In *Proc. Int. Symp. on Algorithms and Computation, LNCS 1004*, pages 82–91, 1995. A complete version appears as BRICS technical report RS-96-29, University of Aarhus.

16. L. Arge. External memory data structures. In J. Abello, P. M. Pardalos, and M. G. C. Resende, editors, *Handbook of Massive Data Sets*. Kluwer Academic Publishers, 2001. (To appear).

17. L. Arge, R. Barve, O. Procopiuc, L. Toma, D. E. Vengroff, and R. Wickremesinghe. *TPIE User Manual and Reference (edition 0.9.01a)*. Duke University, 1999. The manual and software distribution are available on the web at http://www.cs.duke.edu/TPIE/.

18. L. Arge, G. S. Brodal, and L. Toma. On external memory MST, SSSP and multiway planar graph separation. In *Proc. Scandinavian Workshop on Algorithms Theory, LNCS 1851*, pages 433–447, 2000.

19. L. Arge, J. Chase, P. Halpin, L. Toma, D. Urban, J. Vitter, and R. Wickremesinghe. Flow computation on massive grids. In *16'th Annual Symposium of the International Association of Landscape Ecology (US-IALE 2001)*, 2001.

20. L. Arge, P. Ferragina, R. Grossi, and J. Vitter. On sorting strings in external memory. In *Proc. ACM Symp. on Theory of Computation*, pages 540–548, 1997.

21. L. Arge, K. H. Hinrichs, J. Vahrenhold, and J. S. Vitter. Efficient bulk operations on dynamic R-trees. In *Proc. Workshop on Algorithm Engineering, LNCS 1619*, pages 328–347, 1999.

22. L. Arge, U. Meyer, L. Toma, and N. Zeh. On external-memory planar depth first search. In *Proc. Workshop on Algorithms and Data Structures*, 2001.

23. L. Arge, O. Procopiuc, S. Ramaswamy, T. Suel, J. Vahrenhold, and J. S. Vitter. A unified approach for indexed and non-indexed spatial joins. In *Proc. Conference on Extending Database Technology*, 1999.

24. L. Arge, O. Procopiuc, S. Ramaswamy, T. Suel, and J. S. Vitter. Scalable sweeping-based spatial join. In *Proc. International Conf. on Very Large Databases*, 1998.

25. L. Arge, O. Procopiuc, S. Ramaswamy, T. Suel, and J. S. Vitter. Theory and practice of I/O-efficient algorithms for multidimensional batched searching problems. In *Proc. ACM-SIAM Symp. on Discrete Algorithms*, pages 685–694, 1998.

26. L. Arge, V. Samoladas, and J. S. Vitter. On two-dimensional indexability and optimal range search indexing. In *Proc. ACM Symp. Principles of Database Systems*, pages 346–357, 1999.

27. L. Arge, L. Toma, and J. S. Vitter. I/O-efficient algorithms for problems on grid-based terrains. In *Proc. Workshop on Algorithm Engineering and Experimentation*, 2000.

28. L. Arge and J. Vahrenhold. I/O-efficient dynamic planar point location. In *Proc. ACM Symp. on Computational Geometry*, pages 191–200, 2000.

29. L. Arge, D. E. Vengroff, and J. S. Vitter. External-memory algorithms for processing line segments in geographic information systems. In *Proc. Annual European Symposium on Algorithms, LNCS 979*, pages 295–310, 1995. To appear in special issues of *Algorithmica* on Geographical Information Systems.

30. L. Arge and J. S. Vitter. Optimal dynamic interval management in external memory. In *Proc. IEEE Symp. on Foundations of Comp. Sci.*, pages 560–569, 1996.

31. S. Arya, D. M. Mount, N. S. Netanyahu, R. Silverman, and A. Wu. An optimal algorithm for approximate nearest neighbor searching. In *Proc. 5th ACM-SIAM Sympos. Discrete Algorithms*, pages 573–582, 1994.

32. T. Asano, D. Ranjan, T. Roos, E. Welzl, and P. Widmayer. Space-filling curves and their use in the design of geometric data structures. *Theoret. Comput. Sci.*, 181(1):3–15, July 1997.

33. J. Basch, L. J. Guibas, and J. Hershberger. Data structures for mobile data. *Journal of Algorithms*, 31(1):1–28, 1999.

34. H. Baumgarten, H. Jung, and K. Mehlhorn. Dynamic point location in general subdivisions. *Journal of Algorithms*, 17:342–380, 1994.

35. R. Bayer and E. McCreight. Organization and maintenance of large ordered indexes. *Acta Informatica*, 1:173–189, 1972.

36. B. Becker, S. Gschwind, T. Ohler, B. Seeger, and P. Widmayer. An asymptotically optimal multiversion B-tree. *VLDB Journal*, 5(4):264–275, 1996.

37. N. Beckmann, H.-P. Kriegel, R. Schneider, and B. Seeger. The R*-tree: An efficient and robust access method for points and rectangles. In *Proc. SIGMOD Intl. Conf. on Management of Data*, pages 322–331, 1990.

38. M. A. Bender, E. D. Demaine, and M. Farach-Colton. Cache-oblivious B-trees. In *Proc. IEEE Symp. on Foundations of Comp. Sci.*, pages 339–409, 2000.

39. J. L. Bentley. Decomposable searching problems. *Information Processing Letters*, 8(5):244–251, 1979.

40. S. Berchtold, C. Böhm, and H.-P. Kriegel. Improving the query performance of high-dimensional index structures by bulk load operations. In *Proc. Conference on Extending Database Technology, LNCS 1377*, pages 216–230, 1998.

41. S. Berchtold, B. Ertl, D. A. Keim, H.-P. Kriegel, and T. Seidl. Fast nearest neighbor search in high-dimensional spaces. In *Proc. IEEE International Conference on Data Engineering*, pages 209–218, 1998.

42. S. N. Bespamyatnikh. An optimal algorithm for closets pair maintenance. *Discrete and Computational Geometry*, 19:175–195, 1998.

43. G. Blankenagel and R. H. Güting. XP-trees—External priority search trees. Technical report, FernUniversität Hagen, Informatik-Bericht Nr. 92, 1990.

44. K. Brengel, A. Crauser, P. Ferragina, and U. Meyer. An experimental study of priority queues in external memory. In *Proc. Workshop on Algorithm Engineering, LNCS 1668*, pages 345–358, 1999.

45. G. S. Brodal and J. Katajainen. Worst-case efficient external-memory priority queues. In *Proc. Scandinavian Workshop on Algorithms Theory, LNCS 1432*, pages 107–118, 1998.

46. A. L. Buchsbaum, M. Goldwasser, S. Venkatasubramanian, and J. R. Westbrook. On external memory graph traversal. In *Proc. ACM-SIAM Symp. on Discrete Algorithms*, pages 859–860, 2000.

47. P. Callahan, M. T. Goodrich, and K. Ramaiyer. Topology B-trees and their applications. In *Proc. Workshop on Algorithms and Data Structures, LNCS 955*, pages 381–392, 1995.

48. P. B. Callahan and S. R. Kosaraju. Algorithms for dynamic closest-pair and *n*-body potential fields. In *Proc. 6th ACM-SIAM Sympos. Discrete Algorithms*, pages 263–272, 1995.

49. P. B. Callahan and S. R. Kosaraju. A decomposition of multidimensional point sets with applications to k-nearest-neighbors and n-body potential fields. *Journal of the ACM*, 42(1):67–90, 1995.

50. T. M. Chan. Random sampling, halfspace range reporting, and construction of ($\leq k$)-levels in three dimensions. *SIAM Journal of Computing*, 30(2):561–575, 2000.

51. B. Chazelle. Filtering search: a new approach to query-answering. *SIAM J. Comput.*, 15(3):703–724, 1986.

52. B. Chazelle. A functional approach to data structures and its use in multidimensional searching. *SIAM J. Comput.*, 17(3):427–462, June 1988.

53. B. Chazelle. Lower bounds for orthogonal range searching: I. the reporting case. *Journal of the ACM*, 37(2):200–212, Apr. 1990.

54. B. Chazelle and L. J. Guibas. Fractional cascading: I. A data structuring technique. *Algorithmica*, 1:133–162, 1986.

55. B. Chazelle, L. J. Guibas, and D. T. Lee. The power of geometric duality. *BIT*, 25(1):76–90, 1985.

56. S. W. Cheng and R. Janardan. New results on dynamic planar point location. *SIAM J. Comput.*, 21(5):972–999, 1992.

57. Y.-J. Chiang. *Dynamic and I/O-Efficient Algorithms for Computational Geometry and Graph Problems: Theoretical and Experimental Results.* PhD thesis, Brown University, August 1995.

58. Y.-J. Chiang. Experiments on the practical I/O efficiency of geometric algorithms: Distribution sweep vs. plane sweep. In *Proc. Workshop on Algorithms and Data Structures, LNCS 955*, pages 346–357, 1995.

59. Y.-J. Chiang, M. T. Goodrich, E. F. Grove, R. Tamassia, D. E. Vengroff, and J. S. Vitter. External-memory graph algorithms. In *Proc. ACM-SIAM Symp. on Discrete Algorithms*, pages 139–149, 1995.

60. Y.-J. Chiang and C. T. Silva. I/O optimal isosurface extraction. In *Proc. IEEE Visualization*, pages 293–300, 1997.

61. Y.-J. Chiang and C. T. Silva. External memory techniques for isosurface extraction in scientific visualization. In J. Abello and J. S. Vitter, editors, *External memory algorithms and visualization*, pages 247–277. American Mathematical Society, DIMACS series in Discrete Mathematics and Theoretical Computer Science, 1999.

62. Y.-J. Chiang, C. T. Silva, and W. J. Schroeder. Interactive out-of-core isosurface extraction. In *Proc. IEEE Visualization*, pages 167–174, 1998.

63. D. Comer. The ubiquitous B-tree. *ACM Computing Surveys*, 11(2):121–137, 1979.

64. A. Crauser and P. Ferragina. On constructing suffix arrays in external memory. In *Proc. Annual European Symposium on Algorithms, LNCS, 1643*, pages 224–235, 1999.

65. A. Crauser, P. Ferragina, K. Mehlhorn, U. Meyer, and E. Ramos. Randomized external-memory algorithms for some geometric problems. In *Proc. ACM Symp. on Computational Geometry*, pages 259–268, 1998.

66. A. Crauser and K. Mehlhorn. *LEDA-SM: A Platform for Secondary Memory Computation.* Max-Planck-Institut für Informatik, 1999. The manual and software distribution are available on the web at
http://www.mpi-sb.mpg.de/~crauser/leda-sm.html.

67. A. Crauser and K. Mehlhorn. LEDA-SM: Extending LEDA to secondary memory. In *Proc. Workshop on Algorithm Engineering*, 1999.

68. M. de Berg, J. Gudmundsson, M. Hammar, and M. Overmars. On R-trees with low stabbing number. In *Proc. Annual European Symposium on Algorithms*, pages 167–178, 2000.

69. D. J. DeWitt, N. Kabra, J. Luo, J. M. Patel, and J.-B. Yu. Client-server paradise. In *Proc. International Conf. on Very Large Databases*, pages 558–569, 1994.

70. J. R. Driscoll, N. Sarnak, D. D. Sleator, and R. Tarjan. Making data structures persistent. *Journal of Computer and System Sciences*, 38:86–124, 1989.

71. H. Edelsbrunner. A new approach to rectangle intersections, part I. *Int. J. Computer Mathematics*, 13:209–219, 1983.

72. H. Edelsbrunner. A new approach to rectangle intersections, part II. *Int. J. Computer Mathematics*, 13:221–229, 1983.

73. H. Edelsbrunner and M. Overmars. Batched dynamic solutions to decomposable searching problems. *Journal of Algorithms*, 6:515–542, 1985.

74. H. Edelsbrunner and M. H. Overmars. On the equivalence of some rectangle problems. *Inform. Process. Lett.*, 14:124–127, 1982.

75. G. Evangelidis, D. Lomet, and B. Salzberg. The hb^π-tree: A multi-attribute index supporting concurrency, recovery and node consolidation. *The VLDB Journal*, 6(1):1–25, 1997.

76. R. Fadel, K. V. Jakobsen, J. Katajainen, and J. Teuhola. Heaps and heapsort on secondary storage. *Theoretical Computer Science*, 220(2):345–362, 1999.

77. M. Farach, P. Ferragina, and S. Muthukrishnan. Overcoming the memory bottleneck in suffix tree construction. In *Proc. IEEE Symp. on Foundations of Comp. Sci.*, pages 174–183, 1998.

78. P. Ferragina and R. Grossi. A fully-dynamic data structure for external substring search. In *Proc. ACM Symp. on Theory of Computation*, pages 693–702, 1995.

79. P. Ferragina and R. Grossi. Fast string searching in secondary storage: Theoretical developments and experimental results. In *Proc. ACM-SIAM Symp. on Discrete Algorithms*, pages 373–382, 1996.

80. P. Ferragina and F. Luccio. Dynamic dictionary matching in external memory. *Information and Computation*, 146(2):85–99, 1998.

81. E. Feuerstein and A. Marchetti-Spaccamela. Memory paging for connectivity and path problems in graphs. In *Proc. Int. Symp. on Algorithms and Computation*, *LNCS 762*, pages 416–425, 1993.

82. P. Franciosa and M. Talamo. Time optimal halfplane search on external memory. Unpublished manuscript, 1997.

83. P. G. Franciosa and M. Talamo. Orders, k-sets and fast halfplane search on paged memory. In *Proc. Workshop on Orders, Algorithms and Applications (ORDAL'94)*, *LNCS 831*, pages 117–127, 1994.

84. G. N. Frederickson. A structure for dynamically maintaining rooted trees. In *Proc. ACM-SIAM Symp. on Discrete Algorithms*, pages 175–184, 1993.

85. V. Gaede and O. Günther. Multidimensional access methods. *ACM Computing Surveys*, 30(2):170–231, 1998.

86. M. T. Goodrich, J.-J. Tsay, D. E. Vengroff, and J. S. Vitter. External-memory computational geometry. In *Proc. IEEE Symp. on Foundations of Comp. Sci.*, pages 714–723, 1993.

87. S. Govindarajan, T. Lukovszki, A. Maheshwari, and N. Zeh. I/O-efficient well-separated pair decomposition and its applications. In *Proc. Annual European Symposium on Algorithms*, pages 220–231, 2000.

88. D. Greene. An implementation and performance analysis of spatial data access methods. In *Proc. IEEE International Conference on Data Engineering*, pages 606–615, 1989.

89. R. Grossi and G. F. Italiano. Efficient cross-tree for external memory. In J. Abello and J. S. Vitter, editors, *External Memory Algorithms and Visualization*, pages 87–106. American Mathematical Society, DIMACS series in Discrete Mathematics and Theoretical Computer Science, 1999. Revised version available at ftp://ftp.di.unipi.it/pub/techreports/TR-00-16.ps.Z.

90. R. Grossi and G. F. Italiano. Efficient splitting and merging algorithms for order decomposable problems. *Information and Computation*, 154(1):1–33, 1999.

91. O. Günther. The design of the cell tree: An object-oriented index structure for geometric databases. In *Proc. IEEE International Conference on Data Engineering*, pages 598–605, 1989.

92. A. Guttman. R-trees: A dynamic index structure for spatial searching. In *Proc. SIGMOD Intl. Conf. on Management of Data*, pages 47–57, 1984.

93. J. M. Hellerstein, E. Koutsoupias, and C. H. Papadimitriou. On the analysis of indexing schemes. In *Proc. ACM Symp. Principles of Database Systems*, pages 249–256, 1997.

94. K. H. Hinrichs. *The grid file system: Implementation and case studies of applications*. PhD thesis, Dept. Information Science, ETH, Zürich, 1985.

95. S. Huddleston and K. Mehlhorn. A new data structure for representing sorted lists. *Acta Informatica*, 17:157–184, 1982.

96. D. Hutchinson, A. Maheshwari, J.-R. Sack, and R. Velicescu. Early experiences in implementing the buffer tree. In *Proc. Workshop on Algorithm Engineering*, pages 92–103, 1997.

97. D. Hutchinson, A. Maheshwari, and N. Zeh. An external-memory data structure for shortest path queries. In *Proc. Annual Combinatorics and Computing Conference, LNCS 1627*, pages 51–60, 1999.

98. C. Icking, R. Klein, and T. Ottmann. Priority search trees in secondary memory. In *Proc. Graph-Theoretic Concepts in Computer Science, LNCS 314*, pages 84–93, 1987.

99. I. Kamel and C. Faloutsos. On packing R-trees. In *Proc. International Conference on Information and Knowledge Management*, pages 490–499, 1993.

100. I. Kamel and C. Faloutsos. Hilbert R-tree: An improved R-tree using fractals. In *Proc. International Conf. on Very Large Databases*, pages 500–509, 1994.

101. P. C. Kanellakis, S. Ramaswamy, D. E. Vengroff, and J. S. Vitter. Indexing for data models with constraints and classes. *Journal of Computer and System Sciences*, 52(3):589–612, 1996.

102. K. V. R. Kanth and A. K. Singh. Optimal dynamic range searching in non-replicating index structures. In *Proc. International Conference on Database Theory, LNCS 1540*, pages 257–276, 1999.

103. N. Katayama and S. Satoh. The SR-tree: An index structure for high-dimensional nearest-neighbor queries. In *Proc. SIGMOD Intl. Conf. on Management of Data*, pages 369–380, 1997.

104. D. E. Knuth. *Sorting and Searching*, volume 3 of *The Art of Computer Programming*. Addison-Wesley, Reading MA, second edition, 1998.

105. G. Kollios, D. Gunopulos, and V. J. Tsotras. On indexing mobile objects. In *Proc. ACM Symp. Principles of Database Systems*, pages 261–272, 1999.

106. E. Koutsoupias and D. S. Taylor. Tight bounds for 2-dimensional indexing schemes. In *Proc. ACM Symp. Principles of Database Systems*, pages 52–58, 1998.

107. V. Kumar and E. Schwabe. Improved algorithms and data structures for solving graph problems in external memory. In *Proc. IEEE Symp. on Parallel and Distributed Processing*, pages 169–177, 1996.

108. S. T. Leutenegger, M. A. López, and J. Edgington. STR: A simple and efficient algorithm for R-tree packing. In *Proc. IEEE International Conference on Data Engineering*, pages 497–506, 1996.

109. D. Lomet and B. Salzberg. The hB-tree: A multiattribute indexing method with good guaranteed performance. *ACM Transactions on Database Systems*, 15(4):625–658, 1990.

110. A. Maheshwari and N. Zeh. External memory algorithms for outerplanar graphs. In *Proc. Int. Symp. on Algorithms and Computation, LNCS 1741*, pages 307–316, 1999.

111. A. Maheshwari and N. Zeh. I/O-efficient algorithms for graphs of bounded treewidth. In *Proc. ACM-SIAM Symp. on Discrete Algorithms*, pages 89–90, 2001.

112. J. Matoušek. Efficient partition trees. *Discrete Comput. Geom.*, 8:315–334, 1992.

113. E. McCreight. Priority search trees. *SIAM Journal of Computing*, 14(2):257–276, 1985.

114. K. Mehlhorn and S. Näher. Dynamic fractional cascading. *Algorithmica*, 5:215–241, 1990.

115. K. Mehlhorn and S. Näher. *LEDA: A Platform for Combinatorial and Geometric Computing*. Cambridge University Press, Cambridge, UK, 2000.

116. U. Meyer. External memory bfs on undirected graphs with bounded degree. In *Proc. ACM-SIAM Symp. on Discrete Algorithms*, pages 87–88, 2001.

117. D. R. Morrison. PATRICIA: Practical algorithm to retrieve information coded in alphanumeric. *Journal of the ACM*, 15:514–534, 1968.

118. K. Munagala and A. Ranade. I/O-complexity of graph algorithm. In *Proc. ACM-SIAM Symp. on Discrete Algorithms*, pages 687–694, 1999.

119. J. Nievergelt, H. Hinterberger, and K. Sevcik. The grid file: An adaptable, symmetric multikey file structure. *ACM Transactions on Database Systems*, 9(1):38–71, 1984.

120. J. Nievergelt and E. M. Reingold. Binary search tree of bounded balance. *SIAM Journal of Computing*, 2(1):33–43, 1973.

121. J. Nievergelt and P. Widmayer. Spatial data structures: Concepts and design choices. In M. van Kreveld, J. Nievergelt, T. Roos, and P. Widmayer, editors, *Algorithmic Foundations of GIS*, pages 153–197. Springer-Verlag, LNCS 1340, 1997.

122. M. H. Nodine, M. T. Goodrich, and J. S. Vitter. Blocking for external graph searching. *Algorithmica*, 16(2):181–214, 1996.

123. J. Orenstein. Spatial query processing in an object-oriented database system. In *Proc. ACM SIGMOD Conf. on Management of Data*, pages 326–336, 1986.

124. J. Orenstein. A comparison of spatial query processing techniques for native and parameter spaces. In *Proc. SIGMOD Intl. Conf. on Management of Data*, pages 343–352, 1990.

125. M. H. Overmars. *The Design of Dynamic Data Structures*. Springer-Verlag, LNCS 156, 1983.

126. D. Pfoser, C. S. Jensen, and Y. Theodoridis. Novel approaches to the indexing of moving objects trajectories. In *Proc. International Conf. on Very Large Databases*, pages 395–406, 2000.

127. S. Ramaswamy and S. Subramanian. Path caching: A technique for optimal external searching. In *Proc. ACM Symp. Principles of Database Systems*, pages 25–35, 1994.

128. J. Robinson. The K-D-B tree: A search structure for large multidimensional dynamic indexes. In *Proc. SIGMOD Intl. Conf. on Management of Data*, pages 10–18, 1981.

129. N. Roussopoulos, S. Kelley, and F. Vincent. Nearest neighbor queries. In *Proc. SIGMOD Intl. Conf. on Management of Data*, pages 71–79, 1995.

130. N. Roussopoulos and D. Leifker. Direct spatial search on pictorial databases using packed R-trees. In *Proc. SIGMOD Intl. Conf. on Management of Data*, pages 17–31, 1985.

131. C. Ruemmler and J. Wilkes. An introduction to disk drive modeling. *IEEE Computer*, 27(3):17–28, 1994.

132. B. Salzberg and V. J. Tsotras. A comparison of access methods for time evolving data. *ACM Computing Surveys*, 31(2):158–221, 1999.

133. H. Samet. *Applications of Spatial Data Structures: Computer Graphics, Image Processing, and GIS*. Addison Wesley, MA, 1990.

134. H. Samet. *The Design and Analyses of Spatial Data Structures*. Addison Wesley, MA, 1990.

135. V. Samoladas and D. Miranker. A lower bound theorem for indexing schemes and its application to multidimensional range queries. In *Proc. ACM Symp. Principles of Database Systems*, pages 44–51, 1998.

136. P. Sanders. Fast priority queues for cached memory. In *Proc. Workshop on Algorithm Engineering and Experimentation, LNCS 1619*, pages 312–327, 1999.

137. N. Sarnak and R. E. Tarjan. Planar point location using persistent search trees. *Communications of the ACM*, 29:669–679, 1986.

138. B. Seeger and H.-P. Kriegel. The buddy-tree: An efficient and robust access method for spatial data base systems. In *Proc. International Conf. on Very Large Databases*, pages 590–601, 1990.

139. T. Sellis, N. Roussopoulos, and C. Faloutsos. The R$^+$-tree: A dynamic index for multi-dimensional objects. In *Proc. International Conf. on Very Large Databases*, pages 507–518, 1987.

140. J. Snoeyink. Point location. In J. E. Goodman and J. O'Rourke, editors, *Handbook of Discrete and Computational Geometry*, chapter 30, pages 559–574. CRC Press LLC, Boca Raton, FL, 1997.

141. S. Subramanian and S. Ramaswamy. The P-range tree: A new data structure for range searching in secondary memory. In *Proc. ACM-SIAM Symp. on Discrete Algorithms*, pages 378–387, 1995.

142. J. D. Ullman and M. Yannakakis. The input/output complexity of transitive closure. *Annals of Mathematics and Artificial Intellegence*, 3:331–360, 1991.

143. J. Vahrenhold and K. H. Hinrichs. Planar point-location for large data sets: To seek or not to seek. In *Proc. Workshop on Algorithm Engineering*, 2000.

144. J. van den Bercken, B. Seeger, and P. Widmayer. A generic approach to bulk loading multidimensional index structures. In *Proc. International Conf. on Very Large Databases*, pages 406–415, 1997.

145. J. van den Bercken, B. Seeger, and P. Widmayer. A generic approach to processing non-equijoins. Technical Report 14, Philipps-Universität Marburg, Fachbereich Matematik und Informatik, 1998.

146. M. J. van Kreveld and M. H. Overmars. Divided *k*-d trees. *Algorithmica*, 6:840–858, 1991.

147. P. J. Varman and R. M. Verma. An efficient multiversion access structure. *IEEE Transactions on Knowledge and Data Engineering*, 9(3):391–409, 1997.

148. D. E. Vengroff. A transparent parallel I/O environment. In *Proc. DAGS Symposium on Parallel Computation*, 1994.

149. D. E. Vengroff and J. S. Vitter. Efficient 3-D range searching in external memory. In *Proc. ACM Symp. on Theory of Computation*, pages 192–201, 1996.
150. D. E. Vengroff and J. S. Vitter. I/O-efficient scientific computation using TPIE. In *Proceedings of the Goddard Conference on Mass Storage Systems and Technologies*, NASA Conference Publication 3340, Volume II, pages 553–570, 1996.
151. J. S. Vitter. External memory algorithms and data structures. In J. Abello and J. S. Vitter, editors, *External Memory Algorithms and Visualization*, pages 1–38. American Mathematical Society, DIMACS series in Discrete Mathematics and Theoretical Computer Science, 1999.
152. J. S. Vitter. Online data structures in external memory. In *Proc. Annual International Colloquium on Automata, Languages, and Programming, LNCS 1644*, pages 119–133, 1999.
153. J. S. Vitter and E. A. M. Shriver. Algorithms for parallel memory, I: Two-level memories. *Algorithmica*, 12(2–3):110–147, 1994.
154. S. Šaltenis, C. S. Jensen, S. T. Leutenegger, and M. A. López. Indexing the positions of continuously moving objects. In *Proc. SIGMOD Intl. Conf. on Management of Data*, pages 331–342, 2000.
155. O. Wolfson, A. P. Sistla, S. Chamberlain, and Y. Yesha. Updating and querying databases that track mobile units. *Distributed and Parallel Databases*, 7(3):257–287, 1999.
156. N. Zeh. I/O-efficient planar separators and applications. Manuscript, 2001.
157. D. Zhang, A. Markowetz, V. Tsotras, D. Gunopulos, and B. Seeger. Efficient computation of temporal aggregates with range predicates. In *Proc. ACM Symp. Principles of Database Systems*, pages 237–245, 2001.

Some Algorithmic Problems in Large Networks

Susanne Albers

Dept. of Computer Science, Freiburg University, 79110 Freiburg, Germany
albers@informatik.uni-freiburg.de

Abstract. We will investigate a number of algorithmic problems that arise in large networks such as the world-wide web. We will mostly concentrate on the following problems.

General Caching Problems: Caching is a very well-studied problem. Consider a two-level memory system consisting of a small fast memory, that can store up to k bits, and a large slow memory, that can store potentially infinitely many bits. The goal is to serve a sequence of memory accesses with low total cost. In standard caching it is assumed that all memory pages have a uniform size and a uniform fault cost. In a general caching problem, however, pages or *documents* have varying sizes and varying costs. This problem arises, among other places, in the cache design for networked file systems or the world-wide web. In the web, for instance, the transmission time of a web page depends on the location of the corresponding server and also on transient conditions such as the server load or network congestion.

The offline variant of general caching problems is NP-hard. Irani [7] recently presented polynomial time algorithms that achieve an approximation ratio of $O(\log(k/s))$, where s is the size of the smallest document ever requested. We present polynomial time constant factor approximation algorithms that use a small amount of additional memory. The values of the approximation ratios depend on the exact cost model assumed. In the most general setting we achieve a $(4 + \varepsilon)$-approximation, for any $\varepsilon > 0$. The main idea of our solutions is to formulate caching problems as integer programs and solve linear relaxations. A fractional solution is transformed into a feasible solution using a new rounding technique. Our results were published in [2].

Management of persistent TCP connections: Communication between clients and servers in the web is performed using HTTP (Hyper Text Transfer Protocol), which in turn uses TCP (Transmission Control Protocol) to transfer data. If data has to be transmitted between two network nodes, then there has to exist an open TCP connection between these two nodes. While the earlier HTTP/1.0 opened and closed a separate TCP connection for each transmission request, the new HTTP/1.1. permits *persistent connections*, i.e. a TCP connection may be kept open and idle until a new transmission request arrive or the connection is explicitly closed by the server or the client. The problem is to maintain a limited number of open TCP connections at each network node, not knowing which connections will be required in the future.

Cohen et al. [4,5] recently initiated the study of connection caching in the web and presented optimal competitive online algorithms assuming

F. Meyer auf der Heide (Ed.): ESA 2001, LNCS 2161, pp. 30–32, 2001.

that the establishment cost is uniform for all the connections. They also analyzed various models of communication. We investigate the setting where connections may incur different establishment costs and present online algorithms that achieve an optimal competitiveness. Our algorithms use extra communication between network nodes while managing open connections. We can develop refined algorithms that allow trade-offs between extra communication and competitive performance. We also consider problem extensions where connections may have time-out values or asymmetric establishment costs. The results appeared in [1]

Dynamic TCP acknowledgment: Consider a sequence of data packets that arrives at a network node over an open TCP connection. The node has to acknowledge the receipt of the packets by sending acknowledgments to the sending site. Most implementations of TCP use some acknowledgment delay mechanism, i.e. multiple incoming data packets are acknowledged with a single acknowledgment. A reduction of the number of acknowledgments sent leads to a smaller network congestion and to a smaller overhead incurred in sending and receiving acknowledgments. On the other hand, by sending fewer acknowledgments, we increase the latency of the TCP connection. The goal is to acknowledge dynamically a sequence of data packets, that arrives over time, so that the number of acknowledgments and the acknowledgment delays for the individual packets is as small as possible.

The study of dynamic TCP acknowledgment was initiated by Dooly et al. [6]. They considered the objective function of minimizing the sum of the number of acknowledgments and the total acknowledgment delays for all the packets. They presented deterministic online algorithms that achieve an optimal competitive ratio of 2. Recently Karlin et al. [8] gave randomized algorithms that achieve a competitiveness of $e/(e-1) \approx 1.58$. We consider a different objective function that penalizes long acknowledgment delays. In practice if a data packet is not acknowledged within a certain amount of time, the packet is resent by the sending site, which increases again the network congestion. We investigate an objective function that minimizes the sum of the number of acknowledgments and the maximum delay that ever occurs for any of the data packets. We present optimal online algorithms [3].

References

1. S. Albers. Generalized connection caching. In *Proc. 12th Annual ACM Symposium on Parallel Algorithms and Architectures*, pages 70-78, 2000.
2. S. Albers, S. Arora and S. Khanna. Page replacement for general caching problems. In *Proc. 10th Annual ACM-SIAM Symposium Discrete Algorithms*, pages 31-40, 1999.
3. S. Albers and H. Bals. Dynamic TCP acknowledgment. Manuscript, July 2001.
4. E. Cohen, H. Kaplan and U. Zwick. Connection caching. In *Proc. of the 31st Annual ACM Symposium on Theory of Computing*, pages 612–621, 1999.
5. E. Cohen, H. Kaplan and U. Zwick. Connection caching under various models of communication. In *Proc. 12th Annual ACM Symposium on Parallel Algorithms and Architectures*, pages 54–63, 2000.

6. D.R. Dooly, S.A. Goldman and S.D. Scott. TCP dynamic acknowledgment delay: theory and practice. In *Proc. of the 30th Annual ACM Symposium on Theory of Computing*, pages 389–398, 1998.
7. S. Irani. Page replacement with multi-size pages and applications to web caching. In *Proc. 29th Annual ACM Symposium on Theory of Computing*, pages 701–710, 1997.
8. A.R. Karlin, C. Kenyon and D. Randall. Dynamic TCP acknowledgement and other stories about $e/(e-1)$. In *Proc. Thirty-Third Annual ACM Symposiumon Theory of Computing*, 2001.

Exact and Approximate Distances in Graphs – A Survey

Uri Zwick[*]

School of Computer Science
Tel Aviv University, Tel Aviv 69978, Israel
zwick@cs.tau.ac.il
http://www.cs.tau.ac.il/~zwick/

Abstract. We survey recent and not so recent results related to the computation of exact and approximate distances, and corresponding shortest, or almost shortest, paths in graphs. We consider many different settings and models and try to identify some remaining open problems.

1 Introduction

The problem of finding distances and shortest paths in graphs is one of the most basic, and most studied, problems in algorithmic graph theory. A great variety of intricate and elegant algorithms were developed for various versions of this problem. Nevertheless, some basic problems in this area of research are still open. In this short survey, I will try to outline the main results obtained, and mention some of the remaining open problems.

The input to all versions of the problem is a graph $G = (V, E)$. The graph G may be *directed* or *undirected*, and it may be *weighted* or *unweighted*. If the graph is weighted, then each edge $e \in E$ has a weight, or length, $w(e)$ attached to it. The edge weights are either arbitrary *real* numbers, or they may be *integers*. In either case, the weights may be *nonnegative*, or may allowed to be *negative*.

We may be interested in the distances and shortest paths from a single source vertex s to all other vertices of the graph, this is known as the *Single-Source Shortest Paths (SSSP)* problem, or we may be interested in the distances and shortest paths between all pairs of vertices in the graph, this is known as the *All-Pairs Shortest Paths (APSP)* problem. We may insist on getting *exact* distances and genuine shortest paths, or we may be willing to settle for *approximate* distances and almost shortest paths. The errors in the approximate distances that we are willing to accept may be of an *additive* or *multiplicative* nature.

If we insist on explicitly obtaining the distances between any pair of vertices in the graph, then the size of the output is $\Omega(n^2)$, where n is the number of vertices of the graph. Perhaps this is not what we had in mind. We may be

---

[*] Work supported in part by **The Israel Science Foundation** founded by The Israel Academy of Sciences and Humanities.

F. Meyer auf der Heide (Ed.): ESA 2001, LNCS 2161, pp. 33–48, 2001.

interested, for example, only in a concise *implicit* approximation to all the distances. This may be achieved, for example, by finding a sparse subgraph that approximates all the distances in G. Such a subgraph is called a *spanner*.

Finally, perhaps the implicit approximation of all the distances offered by spanners is not enough. We may want a concise representation of approximate distances, together with quick means of extracting these approximations when we need them. This leads us to the study of *approximate distance oracles*.

This summarizes the different problems considered in this survey. We still need to specify, however, the computational models used. We use two different variants of the unit cost *Random Access Machine* (RAM) model (see [1]). When the edge weights are real numbers, the only operations we allow on the edge weights, and the numbers derived from them, are *addition* and *comparison*. These operations are assumed to take $O(1)$ time. No other operations on real numbers are allowed. We call this the *addition-comparison model*. It is reminiscent of the algebraic computation tree model (see, e.g., [10]), though we are counting all operations, not only those that manipulate weights, and want a single concise program that works for any input size. When the edge weights are integral, we adopt the *word RAM* model that opens the way for more varied algorithmic techniques. In this model, each word of memory is assumed to be w-bit wide, capable of holding an integer in the range $\{-2^{w-1}, \ldots, 2^{w-1} - 1\}$. We assume that every *distance* in the graph fits into one machine word. We also assume that $w \geq \log n$, so that, for example, the name of vertex can be stored in a single machine word. Other than that, no assumptions are made regarding the relation between n and w. We are allowed to perform additions, subtractions, comparisons, *shifts*, and various logical *bit operations*, on machine words. Each such operation takes only $O(1)$ time. (Surprisingly, shifts, ANDs, XORs, and the other such operations, that seem to have little to do with the problem of computing shortest paths, do speed up algorithms for the problem, though only by sub-logarithmic factors.) In some cases, we allow operations like *multiplication*, but generally, we try to avoid such non-AC^0 operations. See [39] for a further discussion of this model.

Most of the algorithmic techniques used by the algorithms considered are *combinatorial*. There are, however, some relations between distance problems and *matrix multiplication*. Thus, some of the algorithms considered, mostly for the APSP problem, rely on fast matrix multiplication algorithms (see, e.g., [18]). Some of the algorithms considered are *deterministic* while others are *randomized*.

There are many more interesting variants of problems related to distances and shortest paths that are *not* considered in this short survey. These include: Algorithms for restricted families of graphs, such as planar graphs (see, e.g., [43,71]); Parallel algorithms (see, e.g., [49,13,15]); Algorithms for dynamic versions of the problems (see, e.g., [48,21]); Routing problems (see, e.g., [20,57,73]); Geometrical problems involving distances (see, e.g., [28,54,55]); and many more.

Also, this short survey adopts a *theoretical* point of view. Problems involving distances and shortest paths in graphs are not only great mathematical problems, but are also very practical problems encountered in everyday life. Thus, there is great interest in developing algorithms for these problems that work well in practice. This, however, is a topic for a different survey that I hope someone else would write. For some discussion of practical issues see, e.g., [11,12].

2   Basic Definitions

Let $G = (V, E)$ be a graph. We let $|V| = n$ and $|E| = m$. We always assume that $m \geq n$. The *distance* $\delta(u, v)$ from u to v in the graph is the smallest length of a path from u to v in the graph, where the length of a path is the sum of the weights of the edges along it. If the graph is unweighted then the weight of each edge is taken to be 1. If the graph is directed, then the paths considered should be directed. If there is no (directed) path from u to v in the graph, we define $\delta(u, v) = +\infty$. If all the edge weights are nonnegative, then all distances are well defined. If the graph contains (directed) cycles of negative weight, and there is a path from u to v that passes through such a negative cycle, we let $\delta(u, v) = -\infty$.

Shortest paths from a source vertex s to all other vertices of the graph can be compactly represented using a *tree of shortest paths*. This is a tree, rooted at s, that spans all the vertices reachable from s in the graph, such that for every vertex v reachable from s in the graph, the unique path in the tree from s to v is a shortest path from s to v in the graph. Almost all the algorithms we discuss return such a tree (or a tree of almost shortest paths), or some similar representation. In most cases, producing such a tree is straightforward. In other cases, doing so without substantially increasing the running time of the algorithm is a non-trivial task. Due to lack of space, we concentrate here on the computation of exact or approximate distances, and only briefly mention issues related to the generation of a representation of the corresponding paths.

3   Single-Source Shortest Paths

We begin with the single-source shortest paths problem. The input is a graph $G = (V, E)$ and a *source* $s \in V$. The goal is to compute all the distances $\delta(s, v)$, for $v \in V$, and construct a corresponding shortest paths tree. The following subsections consider various versions of this problem.

3.1   Nonnegative Real Edge Weights

If the input graph $G = (V, E)$ is unweighted, then the problem is easily solved in $O(m)$ time using *Breadth First Search (BFS)* (see, e.g., [19]). We suppose, therefore, that each edge $e \in E$ has a nonnegative real edge weight $w(e) \geq 0$ associated with it. The problem can then be solved using the classical *Dijkstra's algorithm* [22]. For each vertex of G, we hold a *tentative distance* $d(v)$. Initially $d(s) = 0$, and $d(v) = +\infty$, for every $v \in V - \{s\}$. We also keep a set T of *unsettled vertices*. Initially $T = V$. At each stage we choose an unsettled vertex u with minimum tentative distance, make it settled, and explore the edges emanating from it. If $(u, v) \in E$, and v is still unsettled, we update the tentative distance of v as follows: $d(v) \leftarrow \min\{d(v), d(u) + w(u, v)\}$. This goes on until all vertices are settled. It is not difficult to prove that when a vertex u becomes settled we have $d(u) = \delta(s, u)$, and that $d(u)$ would not change again.

An efficient implementation of Dijkstra's algorithm uses a *priority queue* to hold the unsettled vertices. The *key* associated with each unsettled vertex is its tentative distance. Vertices are inserted into the priority queue using *insert* operations. An unsettled vertex with minimum tentative distance is obtained using an *extract-min* operation. Tentative distances are updated using *decrease-key* operations.

A simple heap-based priority queue can perform insertions, extract-min and decrease-key operations in $O(\log n)$ worst case time per operation. This gives immediately an $O(m \log n)$ time SSSP algorithm. *Fibonacci heaps* of Fredman and Tarjan [30] require only $O(1)$ *amortized* time per insert and decrease-key operation, and $O(\log n)$ *amortized* time per extract-min operation. This gives an $O(m + n \log n)$ time SSSP algorithm. (*Relaxed heaps* of Driscoll *et al.* [24] may also be used to obtain this result. They require $O(1)$ worst case (not amortized) time for per decrease-key operation, and $O(\log n)$ worst case time per extract-min operation.) The only operations performed by these algorithms on edge weights are additions and comparisons. Furthermore, every sum of weights computed by these algorithms is the length of a path in the graph. The $O(m + n \log n)$ time algorithm is the fastest known algorithm in this model.

Dijkstra's algorithm produces the distances $\delta(s, v)$, for $v \in V$, in *sorted order*. It is clear that this requires, in the worst case, $\Omega(n \log n)$ comparisons. (To sort n elements, form a star with n leaves and attach the elements to be sorted as weights to the edges.) However, the definition of the SSSP problem does not require the distances to be returned in sorted order. This leads us to our first open problem: Is there an algorithm for the SSSP problem in the addition-comparison model that beats the information theoretic $\Omega(n \log n)$ lower bound for sorting? Is there such an algorithm in the algebraic computation tree model?

3.2 Nonnegative Integer Edge Weights – Directed Graphs

We consider again the single-source shortest paths problem with nonnegative edge weights. This time we assume, however, that the edge weights are integral and that we can do more than just add and compare weights. This leads to improved running times. (The improvements obtained are sub-logarithmic, as the running time of Dijkstra's algorithm in the addition-comparison model is already almost linear.)

The fact that the edge weights are now integral opens up new possibilities. Some of the techniques that can be applied are *scaling*, *bucketing*, *hashing*, *bit-level parallelism*, and more. It is also possible to tabulate solutions of small subproblems. The description of these techniques is beyond the scope of this survey. We merely try to state the currently best available results.

Most of the improved results for the SSSP problem in the word RAM model are obtained by constructing improved priority queues. Some of the pioneering results here were obtained by van Emde Boas *et al.* [74,75] and Fredman and Willard [31,32]. It is enough, in fact, to construct *monotone* priority queues, i.e., priority queues that are only required to support sequences of operations in which the value of the minimum key never decreases.

Thorup [70] describes a priority queue with $O(\log \log n)$ expected time per operation. This gives immediately an $O(m \log \log n)$ expected time algorithm for the SSSP problem. (Randomization is needed here, and in most other algorithms, to reduce the work space needed to linear.) He also shows that the SSSP problem is not harder than the problem of sorting the m edge weights. Han [42] describes a deterministic sorting algorithm that runs in $O(n \log \log n \log \log \log n)$ time. This gives a deterministic, $O(m \log \log n \log \log \log n)$ time, linear space algorithm for the SSSP problem. Han's algorithm uses multiplication. Thorup [67] describes a deterministic, $O(n(\log \log n)^2)$ time, linear space sorting algorithm that does not use multiplication, yielding a corresponding SSSP algorithm.

Improved results for graphs that are not too sparse may be obtained by constructing (monotone) priority queues with constant (amortized) time per decrease-key operation. Thorup [70] uses this approach to obtain an $O(m + (n \log n)/w^{1/2-\epsilon})$ expected time algorithm, for any $\epsilon > 0$. (Recall that w is the width of the machine word.) Raman [60], building on results of Ahuja *et al.* [2] and Cherkassky *et al.* [12], obtains an $O(m + nw^{1/4+\epsilon})$ expected time algorithm, for any $\epsilon > 0$. Note that the first algorithm is fast when w is large, while the second algorithm is fast when w is small. By combining these algorithms, Raman [60] obtains an $O(m + n(\log n)^{1/3+\epsilon})$ expected time algorithm, for any $\epsilon > 0$. Building on his results from [59], he also obtains deterministic algorithms with running times of $O(m + n(w \log w)^{1/3})$ and $O(m + n(\log n \log \log n)^{1/2})$.

It is interesting to note that w may be replaced in the running times above by $\log C$, where C is the largest edge weight in the graph.

Finally, Hagerup [40], extending a technique of Thorup [68] for undirected graphs, obtains a deterministic $O(m \log w)$ time algorithm.

Is there a linear time algorithm for the directed SSSP problem in the word RAM model? Note that a linear time sorting algorithm would give an affirmative answer to this question. But, the SSSP problem may be easier than sorting.

3.3   Nonnegative Integer Edge Weights – Undirected Graphs

All the improved algorithms mentioned above (with one exception) are 'just' intricate implementations of Dijkstra's algorithm. They produce, therefore, a sorted list of the distances and do not avoid, therefore, the sorting bottleneck.

In a sharp contract, Thorup [68,69] developed recently an elegant algorithm that avoids the rigid settling order of Dijkstra's algorithm. Thorup's algorithm bypasses the sorting bottleneck and runs in optimal $O(m)$ time! His algorithm works, however, only on *undirected* graphs. It remains an open problem whether extensions of his ideas could could be used to obtain a similar result for directed graphs. (Some results along these lines were obtained by Hagerup [40].)

3.4 Positive and Negative Real Edge Weights

We now allow, for the first time, negative edge weights. The gap, here, between the best upper bound and the obvious lower bound is much wider.

Dijkstra's algorithm breaks down in the presence of negative edge weights. The best algorithm known for the problem in the addition-comparison model is the simple $O(mn)$ time Bellman-Ford algorithm (see, e.g., [19]): Start again with $d(s) = 0$ and $d(v) = +\infty$, for $v \in V - \{s\}$. Then, perform the following n times: For every edge $(u, v) \in E$, let $d(v) \leftarrow \min\{d(v), d(u) + w(u, v)\}$. If any of the tentative distances change during the last iteration, then the graph contains a negative cycle. Otherwise, $d(v) = \delta(s, v)$, for every $v \in V$.

The problem of deciding whether a graph contains a negative cycles is a special case of the problem for finding a *minimum mean weight* cycle in a graph. Karp [47] gives an $O(mn)$ time algorithm for this problem.

Is there a $o(mn)$ time algorithm for the single-source shortest paths problem with positive and negative weights in the addition-comparison model?

3.5 Positive and Negative Integer Edge Weights

Goldberg [38], improving results of Gabow [33] and of Gabow and Tarjan [34], uses scaling to obtain an $O(mn^{1/2} \log N)$ time algorithm for this version of the problem, where N is the absolute value of the smallest edge weight. (It is assumed that $N \geq 2$.) Goldberg's algorithm, like most algorithms dealing with negative edge weights, uses *potentials* (see Section 4.2). No progress on the problem was made in recent years. Is there a better algorithm?

4   All-Pairs Shortest Paths – Exact Results

We now move to consider the all-pairs shortest paths problem. The input is again a graph $G = (V, E)$. The required output is a matrix holding all the distances $\delta(u, v)$, for $u, v \in V$, and some concise representation of all shortest paths, possibly a shortest path tree rooted at each vertex.

4.1 Nonnegative Real Edge Weights

If all the edge weights are nonnegative, we can simply run Dijkstra's algorithm independently from each vertex. The running time would be $O(mn + n^2 \log n)$. Karger, Koller and Phillips [46] and McGeoch [52] note that by orchestrating the operation of these n Dijkstra's processes, some unnecessary operations may saved. This leads to an $O(m^* n + n^2 \log n)$ time algorithm for the problem, where m^* is the number of *essential* edges in G, i.e., the number of edges that actually participate in shortest paths. In the worst case, however, $m^* = m$.

Karger *et al.* [46] also introduce the notion of *path-forming* algorithms. These are algorithms that work in the addition-comparison model, with the additional requirement that any sum of weights computed by the algorithm is the length of some path in the graph. (All the addition-comparison algorithms mentioned so far satisfy this requirement.) They show that any path-forming algorithm for the APSP problem must perform, in the worst case, $\Omega(mn)$ operations.

4.2   Positive and Negative Real Edge Weights

When some of the edge weights are negative, Dijkstra's algorithm cannot be used directly. However, Johnson [45] observed that if there are no negative weight cycles, then new nonnegative edge weights that preserve shortest paths can be computed in $O(mn)$ time. The idea is very simple. Assign each vertex $v \in V$ a *potential* $p(v)$. Define new edge weights as follows $w_p(u, v) = w(u, v) + p(u) - p(v)$, for every $(u, v) \in E$. It is easy to verify that the new distances satisfy $\delta_p(u, v) = \delta(u, v) + p(u) - p(v)$, for every $u, v \in V$. (The potentials along any path from u to v cancel out, except those of u and v.) Thus, the shortest paths with respect to the new edge weights are also shortest paths with respect to the original edge weights, and the original distances can be easily extracted. Now, add to G a new vertex s, and add zero weight edges from it to all other vertices of the graph. Let $p(v) = \delta(s, v)$. If there are no negative weight cycles in the graph then these distances are well defined and they could be found in $O(mn)$ time using the Bellman-Ford algorithm. The *triangle inequality* immediately implies that $w_p(u, v) \geq 0$, for every $u, v \in V$. Now we can run Dijkstra's algorithm from each vertex with the new nonnegative weights. The total running time is again $O(mn + n^2 \log n)$.

As m may be as high as $\Omega(n^2)$, the running time of Johnson's algorithm may be as high as $\Omega(n^3)$. A running time of $O(n^3)$ can also be achieved using the simple Floyd-Warshall algorithm (see [19]). We next consider the possibility of obtaining faster algorithms for dense graphs.

The all-pairs shortest paths problem is closely related to the $\{\min, +\}$-product of matrices. If $A = (a_{ij})$ and $B = (b_{ij})$ are $n \times n$ matrices, we let $A \star B$ be the $n \times n$ matrix whose (i, j)-th element is $(A \star B)_{ij} = \min_k \{a_{ik} + b_{kj}\}$. We refer to $A \star B$ as the *distance product* of A and B.

Let $G = (V, E)$ be a graph. We may assume that $V = \{1, 2, \ldots, n\}$. Let $W = (w_{ij})$ be an $n \times n$ matrix with $w_{ij} = w(i, j)$, if $(i, j) \in E$, and $w_{ij} = +\infty$, otherwise. It is easy to see that W^n, where the exponentiation is done with respect to distance product, gives the distance between any pair of vertices in the graph. Furthermore, the graph contains a negative cycle if and only if there are negative elements on the diagonal of W^n. Thus, the APSP problem can be easily solved using $O(\log n)$ distance products. In fact, under some reasonable assumptions, this logarithmic factor can be saved, and it can be shown that the APSP problem is not harder than the problem of computing a single distance product of two $n \times n$ matrices (see [1, Theorem 5.7 on p. 204]).

Distance products could be computed naively in $O(n^3)$ time, but this is of no help to us. Algebraic, i.e., $\{+, \times\}$-products of matrices could be computed much faster. Strassen [64] was the first to show that it could be done using $o(n^3)$ operations. Many improvements followed. We let ω be the *exponent* of matrix multiplication, i.e., the smallest constant for which matrix multiplication can be performed using only $O(n^{\omega + o(1)})$ algebraic operations, i.e., additions, *subtractions*, and multiplications. (For brevity, we 'forget' the annoying $o(1)$ term, and use ω as a substitute for $\omega + o(1)$.) Coppersmith and Winograd [18] showed that $\omega < 2.376$. The only known lower bound on ω is the trivial lower bound $\omega \geq 2$.

Could similar techniques be used, directly, to obtain $o(n^3)$ algorithms for computing distance products? Unfortunately not. The fast matrix multiplication algorithms rely in an essential way on the fact that addition operations could be reversed, via subtractions. This opens the way for clever cancellations that speed up the computation. It is known in fact, that matrix multiplication requires $\Omega(n^3)$ operations, if only additions and multiplications are allowed. This follows from lower bounds on monotone circuits for *Boolean matrix multiplication* obtained by Mehlhorn and Galil [53] and by Paterson [56].

Yuval [76] describes a simple transformation from distance products to standard algebraic products. He assumes, however, that exact exponentiations and logarithms of infinite precision real numbers could be computed in constant time. His model, therefore, is very unrealistic. His ideas could be exploited, however, in a more restricted form, as would be mentioned in Section 4.3. (Several erroneous follow-ups of Yuval's result appeared in the 80's. They are not cited here.)

Fredman [29] describes an elegant way of computing distance products of two $n \times n$ matrices using $O(n^{5/2})$ additions and comparisons in the algebraic computation tree model. There does not seem to be any an efficient way of implementing his algorithm in the RAM model, as it require programs of exponential size. However, by running his algorithm on small matrices, for which short enough programs that implement his algorithm could be precomputed, he obtains an $O(n^3(\log\log n/\log n)^{1/3})$ time algorithm for computing distance products. Takaoka [65] slightly improves his bound to $O(n^3(\log\log n/\log n)^{1/2})$.

Is there a genuinely sub-cubic algorithm for the APSP problem in the addition-comparison model, i.e., an algorithm that runs in $O(n^{3-\epsilon})$ time, for some $\epsilon > 0$?

4.3   Integer Edge Weights – Directed Graphs

We next consider the all-pairs shortest paths problem in directed graphs with integer edge weights. Even the unweighted case is interesting here. The first to obtain a genuinely sub-cubic algorithm for the unweighted problem were Alon, Galil and Margalit [4]. Their algorithm runs in $\tilde{O}(n^{(3+\omega)/2})$ time.[1] Their result also extends to the case in which the edge weights are in the range $\{0, \ldots, M\}$. The running time of their algorithm is then $\tilde{O}(M^{(\omega-1)/2}n^{(3+\omega)/2})$, if $M \leq n^{(3-\omega)/(\omega+1)}$, and $\tilde{O}(Mn^{(5\omega-3)/(\omega+1)})$, if $M \geq n^{(3-\omega)/(\omega+1)}$ (see Galil and Margalit [36,37]). Takaoka [66] obtained an algorithm whose running time is $\tilde{O}(M^{1/3}n^{(6+\omega)/3})$. The bound of Takaoka is better than the bound of Alon *et al.* [4] for larger values of M. The running time of Takaoka's algorithm is sub-cubic for $M < n^{3-\omega}$.

The algorithms of Galil and Margalit [36,37] and of Takaoka [66] were improved by Zwick [77,78]. Furthermore, his algorithm works with edge weights to be in the range $\{-M, \ldots, M\}$. The improvement is based on two ingredients. The first is an $\tilde{O}(Mn^\omega)$ algorithm, mentioned in [4], for computing distance

---

[1] We use $\tilde{O}(f)$ as a shorthand for $f \cdot (\log n)^{O(1)}$. In the SSSP problem we are fighting to shave off sub-logarithmic factors. In the APSP problem the real battle is still over the right exponent of n, so we use the $\tilde{O}(\cdot)$ notation to hide not so interesting polylogarithmic factors.

products of $n \times n$ matrices whose finite elements are in the range $\{-M, \ldots, M\}$. This algorithm is based on the idea of Yuval [76]. It is implemented this time, however, in a realistic model. The algorithm uses both the fast matrix multiplication algorithm of [18], and the *integer* multiplication algorithm of [61]. (Note that an $\tilde{O}(Mn^\omega)$ time algorithm for distance products does not give immediately an $\tilde{O}(Mn^\omega)$ time algorithm for the APSP problem, as the range of the elements is increased by each distance product.) The second ingredient is a sampling technique that enables the replacement of a distance product of two $n \times n$ matrices by a smaller *rectangular* product. The algorithm uses, therefore, the fast rectangular matrix multiplication algorithm of [17] (see also [44]).

To state the running time of Zwick's algorithm, we need to introduce exponents for rectangular matrix multiplication. Let $\omega(r)$ be the smallest constant such that the product of an $n \times n^r$ matrix by an $n^r \times n$ matrix could be computed using $O(n^{\omega(r)+o(1)})$ algebraic operations. Suppose that $M = n^t$. Then, the running time of his algorithm his $\tilde{O}(n^{2+\mu(t)})$, where $\mu = \mu(t)$ satisfies $\omega(\mu) = 1 + 2\mu - t$. The best available bounds on $\omega(r)$ imply, for example, that $\mu(0) < 0.575$, so that the APSP problem for directed graphs with edge weights taken from $\{-1, 0, 1\}$ can be solved in $O(n^{2.575})$ time. The algorithm runs in sub-cubic time when $M < n^{3-\omega}$, as was the case with Takaoka's algorithm.

The algorithms mentioned above differ from almost all the other algorithms mentioned in this survey in that augmenting them to produce a compact representation of shortest paths, and not only distances, is a non-trivial task. This requires the computation of *witnesses* for Boolean matrix multiplications and distance products. A simple randomized algorithm for computing witnesses for Boolean matrix multiplication is given by Seidel [62]. His algorithm was *derandomized* by Alon and Naor [6] (see also [5]). An alternative, somewhat slower deterministic algorithm was given by Galil and Margalit [35].

Obtaining improved algorithms, and in particular sub-cubic algorithms for larger values of M for this version of the problem is a challenging open problem.

Finally, Hagerup [40] obtained an $O(mn + n \log \log n)$ time algorithm for the problem in the word RAM model. Could this be reduced to $O(mn)$?

4.4   Integer Edge Weights – Undirected Graphs

Galil and Margalit [36,37] and Seidel [62], obtained $\tilde{O}(n^\omega)$ time algorithms for solving the APSP problem for unweighted *undirected* graphs. Seidel's algorithm is much simpler. Both algorithms show, in fact, that this version of the problem is harder then the *Boolean* matrix multiplication problem by at most a logarithmic factor. (Seidel's algorithm, as it appears in [62], uses integer matrix products, but it is not difficult to obtain a version of it that uses only Boolean products.) Again, witnesses for Boolean matrix products are needed, if paths, and not only distances, are to be found.

Seidel's algorithm is extremely simple and elegant. There seems to be no simple way, however, of using his ideas to obtain a similar algorithm for weighted graphs. The algorithm of Galil and Margalit can be extended, in a fairly straightforward way, to handle small integer edge weights. The running time of their algorithm, when the edge weights are taken from $\{0, 1, \ldots, M\}$, is $\tilde{O}(M^{(\omega+1)/2}n^\omega)$.

An improved time bound of $\tilde{O}(Mn^\omega)$ for the problem was recently obtained by Shoshan and Zwick [63]. They show, in fact, that the APSP problem for undirected graphs with edge weights taken from $\{0, 1, \ldots, M\}$ is harder than the problem of computing the distance product of two $n \times n$ matrices with elements taken from the *same* range by at most a logarithmic factor. (As mentioned in Section 4.3, this is not known for directed graphs.)

Obtaining improved algorithms, and in particular sub-cubic algorithms for larger values of M for this version of the problem is again a challenging open problem. For undirected graphs this is equivalent, as mentioned, to obtaining faster algorithms for distance products of matrices with elements in the range $\{0, 1, \ldots, M\}$.

5   All-Pairs Shortest Paths – Approximate Results

The cost of exactly computing all distances in a graph may be prohibitively large. In this section we explore the savings that may be obtained by settling for approximate distances and almost shortest paths. Throughout this section, we assume that the edge weights are nonnegative.

We say that an estimated distance $\hat{\delta}(u, v)$ is of *stretch t* if and only if $\delta(u, v) \leq \hat{\delta}(u, v) \leq t \cdot \delta(u, v)$. We say that an estimated distance $\hat{\delta}(u, v)$ is of *surplus t* if and only if $\delta(u, v) \leq \hat{\delta}(u, v) \leq \delta(u, v) + t$. All our estimates correspond to actual paths in the graph, and are thus upper bounds on the actual distances.

5.1   Directed Graphs

It is not difficult to see [23] that for any finite t, obtaining stretch t estimates of all distances in a graph is at least as hard as Boolean matrix multiplication. On the other hand, Zwick [77] shows that for any $\epsilon > 0$, approximate distances of stretch $1 + \epsilon$ of all distances in a directed graph may be computed in time $\tilde{O}((n^\omega/\epsilon) \log(W/\epsilon))$, where W is the largest edge weight in the graph, after the edge weights are scaled so that the smallest nonzero edge weight is 1.

5.2   Unweighted Undirected Graphs

Surprisingly, perhaps, when the graph is undirected and unweighted, estimated distance with small *additive* error may be computed rather quickly, *without* using fast matrix multiplication algorithms. This was first shown by Aingworth et al. [3]. They showed that surplus 2 estimates of the distances between k *specified* pairs of vertices may be computed in $O(n^{3/2}(k \log n)^{1/2})$ time. In particular, surplus 2 estimates of all the distances in the graph, and corresponding paths, may be computed in $O(n^{5/2}(\log n)^{1/2})$ time. Aingworth et al. [3] also give a 2/3-approximation algorithm for the *diameter* of a weighted directed graph that runs in $O(m(n \log n)^{1/2} + n^2 \log n)$ time.

Elkin [25] describes an algorithm for computing estimated distances from a set S of sources to all other vertices of the graph. He shows that for any $\epsilon > 0$, there is a constant $b = b(\epsilon)$, such that an estimated distance $\hat{\delta}(u, v)$ satisfying

$\delta(u,v) \le \hat{\delta}(u,v) \le (1+\epsilon)\delta(u,v)+b$, for every $u \in S$ and $v \in V$, may be computed in $O(mn^\epsilon + |S|n^{1+\epsilon})$ time. Furthermore, the corresponding shortest paths, use only $O(n^{1+\epsilon})$ edges of the graph. (See also Section 6.2.) Note, however, that although the term multiplying $\delta(u,v)$ above can be made arbitrarily close to 1, the errors in the estimates obtained are not purely additive.

Dor *et al.* [23] obtained improved algorithms for obtaining finite surplus estimates of *all* distances in the graph. They show that surplus 2 estimates may be computed in $\tilde{O}(n^{3/2}m^{1/2})$ time, and also in $\tilde{O}(n^{7/3})$ time. Furthermore, they exhibit a surplus-time tradeoff showing that surplus $2(k-1)$ estimates of all distances may be computed in $\tilde{O}(kn^{2-1/k}m^{1/k})$ time. In particular, surplus $O(\log n)$ estimates of all distances may be obtained in almost optimal $\tilde{O}(n^2)$ time.

5.3   Weighted Undirected Graphs

Cohen and Zwick [16] adapted the techniques of Dor *et al.* [23] for weighted graphs. They obtain stretch 2 estimates of all distances in $\tilde{O}(n^{3/2}m^{1/2})$ time, stretch 7/3 estimates in $\tilde{O}(n^{7/3})$ time, and stretch 3 estimates, and corresponding stretch 3 paths, in almost optimal $\tilde{O}(n^2)$ time. Algorithms with additive errors are also presented. They show, for example, that if p is a any path between u and v, then the estimate $\hat{\delta}(u,v)$ produced by the $\tilde{O}(n^{3/2}m^{1/2})$ time algorithm satisfies $\hat{\delta}(u,v) \le w(p) + 2w_{\max}(p)$, where $w(p)$ is the length of p, and $w_{\max}(p)$ is the weight of the heaviest edge on p.

6   Spanners

6.1   Weighted Undirected Graphs

Let $G = (V, E)$ be an undirected graph. In many applications, many of them related to distributed computing (see [57]), it is desired to obtain a sparse subgraph $H = (V, F)$ of G that approximates, at least to a certain extent, the distances in G. Such a subgraph H is said to be a *t-spanner* of G if and only if for every $u, v \in V$ we have $\delta_H(u,v) \le t \cdot \delta_G(u,v)$. (This definition, implicit in [8], appears explicitly in [58].)

Althöfer *et al.* [7] describe the following simple algorithm for constructing a *t*-spanner of an undirected graph $G = (V, E)$ with nonnegative edge weights. The algorithm is similar to Kruskal's algorithm [51] (see also [19]) for computing minimum spanning trees: Let $F \leftarrow \phi$. Consider the edges of G in nondecreasing order of weight. If $(u,v) \in E$ is the currently considered edge and $w(u,v) < t \cdot \delta_F(u,v)$, then add (u,v) to F. It is easy to see that at the end of this process $H = (V, F)$ is indeed a *t*-spanner of G. It is also easy to see that the *girth* of H is greater than $t + 1$. (The girth of a graph G is the smallest number of edges on a cycle in G.) It is known that any graph with at least $n^{1+1/k}$ edges contains a cycle with at most $2k$ edges. It follows that any weighted graph on n vertices has a $(2k - 1)$-spanner with $O(n^{1+1/k})$ edges. This result is believed to be tight for any $k \ge 1$. It is proved, however, only for $k = 1, 2, 3$ and 5. (See, e.g., [72].)

The fastest known implementation of the algorithm of Althöfer *et al.* [7] runs in $O(mn^{1+1/k})$ time. If the graph is unweighted, then a $(2k-1)$-spanner of size $O(n^{1+1/k})$ can be easily found in $O(m)$ time [41]. Thorup and Zwick [72], improving a result of Cohen [14], give a randomized algorithm for computing a $(2k-1)$-spanner of size $O(n^{1+1/k})$ in $O(kmn^{1/k})$ expected time.

Approximation algorithms and hardness results related to spanners were obtained by Kortsarz and Peleg [50] and Elkin and Peleg [27].

6.2   Unweighted Undirected Graphs

Following Elkin and Peleg [26], we say that a subgraph H of an unweighted graph G is an (a,b)-spanner of G if and only if $\delta_H(u,v) \le a \cdot \delta(u,v) + b$, for every $u, v \in V$. (For a related notion of k-emulators, see [23].) Elkin and Peleg [26] and Elkin [25], improving and extending some preliminary results of Dor *et al.* [23], show that any graph on n vertices has a $(1,2)$-spanner with $O(n^{3/2})$ edges, and that for any $\epsilon > 0$ and $\delta > 0$ there exists $b = b(\epsilon, \delta)$, such that every graph on n vertices has a $(1+\epsilon, b)$-spanner with $O(n^{1+\delta})$ edges.

The intriguing open problem here is whether the $(1+\epsilon, b)$-spanners of [26,25] could be turned into $(1, b)$-spanners, i.e., purely additive spanners. In particular, it is still open whether there exists a $b > 0$ such that any graph on n vertices has a $(1, b)$-spanner with $o(n^{3/2})$ edges. (In [23] it is shown that any graph on n vertices has a *Steiner* $(1, 4)$-*spanner* with $\tilde{O}(n^{4/3})$ edges. A Steiner spanner, unlike standard spanners, is not necessarily a subgraph of the approximated graph. Furthermore, the edges of the Steiner spanner may be weighted, even if the original graph is unweighted.)

7   Distance Oracles

In this section we consider the following problem: We are given a graph $G = (V, E)$. We would like to *preprocess* it so that subsequent *distance queries* or *shortest path queries* could be answered very quickly. A naive solution is to solve the APSP problem, using the best available algorithm, and store the $n \times n$ matrix of distances. Each distance query can then be answered in constant time. The obvious drawbacks of this solution are the large preprocessing time and large space requirements. Much better solutions exist when the graph is undirected, and when we are willing to settle for approximate results.

The term *approximate distance oracles* is coined in Thorup and Zwick [72], though the problem was considered previously by Awerbuch *et al.* [9], Cohen [14] and Dor *et al.* [23]. Improving the results of these authors, Thorup and Zwick [72] show that for any $k \ge 1$, a graph $G = (V, E)$ on n vertices can be preprocessed in $O(kmn^{1/k})$ expected time, constructing a data structure of size $O(kn^{1+1/k})$, such that a stretch $2k-1$ answer to any distance query can be produced in $O(k)$ time. The space requirements of these approximate distance oracles are optimal for $k = 1, 2, 3, 5$, and are conjectured to be optimal for any value of k. (This is related to the conjecture regarding the size of $(2k-1)$-spanners made in Section 6.1. See discussion in [72].)

Many open problems still remain regarding the possible tradeoffs between the preprocessing time, space requirement, query answering time, and the obtained stretch of approximate distance oracles. In particular, is it possible to combine the techniques of [16] and [72] to obtain a stretch 3 distance oracle with $\tilde{O}(n^2)$ preprocessing time, $\tilde{O}(n^{3/2})$ space, and constant query time? Which tradeoffs are possible when no randomization is allowed? Finally, all the distance oracles currently available have multiplicative errors. Are there non-trivial distance oracles with additive errors?

References

1. A.V. Aho, J.E. Hopcroft, and J.D. Ullman. *The design and analysis of computer algorithms*. Addison-Wesley, 1974.
2. R.K. Ahuja, K. Mehlhorn, J.B. Orlin, and R.E. Tarjan. Faster algorithms for the shortest path problem. *Journal of the ACM*, 37:213–223, 1990.
3. D. Aingworth, C. Chekuri, P. Indyk, and R. Motwani. Fast estimation of diameter and shortest paths (without matrix multiplication). *SIAM Journal on Computing*, 28:1167–1181, 1999.
4. N. Alon, Z. Galil, and O. Margalit. On the exponent of the all pairs shortest path problem. *Journal of Computer and System Sciences*, 54:255–262, 1997.
5. N. Alon, Z. Galil, O. Margalit, and M. Naor. Witnesses for boolean matrix multiplication and for shortest paths. In *Proceedings of the 33rd Annual IEEE Symposium on Foundations of Computer Science, Pittsburgh, Pennsylvania*, pages 417–426, 1992.
6. N. Alon and M. Naor. Derandomization, witnesses for Boolean matrix multiplication and construction of perfect hash functions. *Algorithmica*, 16:434–449, 1996.
7. I. Althöfer, G. Das, D. Dobkin, D. Joseph, and J. Soares. On sparse spanners of weighted graphs. *Discrete & Computational Geometry*, 9:81–100, 1993.
8. B. Awerbuch. Complexity of network synchronization. *Journal of the ACM*, 32:804–823, 1985.
9. B. Awerbuch, B. Berger, L. Cowen, and D. Peleg. Near-linear time construction of sparse neighborhood covers. *SIAM Journal on Computing*, 28:263–277, 1999.
10. M. Ben-Or. Lower bounds for algebraic computation trees. In *Proceedings of the 15th Annual ACM Symposium on Theory of Computing, Boston, Massachusetts*, pages 80–86, 1983.
11. B.V. Cherkassky, A.V. Goldberg, and T. Radzik. Shortest paths algorithms: theory and experimental evaluation. *Mathematical Programming (Series A)*, 73(2):129–174, 1996.
12. B.V. Cherkassky, A.V. Goldberg, and C. Silverstein. Buckets, heaps, lists, and monotone priority queues. *SIAM Journal on Computing*, 28(4):1326–1346, 1999.
13. E. Cohen. Using selective path-doubling for parallel shortest-path computations. *Journal of Algorithms*, 22:30–56, 1997.
14. E. Cohen. Fast algorithms for constructing t-spanners and paths with stretch t. *SIAM Journal on Computing*, 28:210–236, 1999.
15. E. Cohen. Polylog-time and near-linear work approximation scheme for undirected shortest paths. *Journal of the ACM*, 47(1):132–166, 2000.
16. E. Cohen and U. Zwick. All-pairs small-stretch paths. *Journal of Algorithms*, 38:335–353, 2001.

17. D. Coppersmith. Rectangular matrix multiplication revisited. *Journal of Complexity*, 13:42–49, 1997.
18. D. Coppersmith and S. Winograd. Matrix multiplication via arithmetic progressions. *Journal of Symbolic Computation*, 9:251–280, 1990.
19. T.H. Cormen, C.E. Leiserson, and R.L. Rivest. *Introduction to algorithms*. The MIT Press, 1990.
20. L.J. Cowen. Compact routing with minimum stretch. *Journal of Algorithms*, 38:170–183, 2001.
21. C. Demetrescu and G. F. Italiano. Fully dynamic transitive closure: breaking through the $O(n^2)$ barrier. In *Proceedings of the 41th Annual IEEE Symposium on Foundations of Computer Science, Redondo Beach, California*, pages 381–389, 2000.
22. E.W. Dijkstra. A note on two problems in connexion with graphs. *Numerische Mathematik*, 1:269–271, 1959.
23. D. Dor, S. Halperin, and U. Zwick. All pairs almost shortest paths. *SIAM Journal on Computing*, 29:1740–1759, 2000.
24. J.R. Driscoll, H.N. Gabow, R. Shrairman, and R.E. Tarjan. Relaxed heaps: an alternative to Fibonacci heaps with applications to parallel computation. *Communications of the ACM*, 31(11):1343–1354, 1988.
25. M.L. Elkin. Computing almost shortest paths. Technical Report MCS01-03, Faculty of Mathematics and Computer Science, The Weizmann Institute of Science, Rehovot, Israel, 2001.
26. M.L. Elkin and D. Peleg. $(1 + \epsilon, \beta)$-Spanner constructions for general graphs. In *Proceedings of the 33th Annual ACM Symposium on Theory of Computing, Crete, Greece*, 2001. To appear.
27. M.L. Elkin and D. Peleg. Approximating k-spanner problems for $k > 2$. In *Proceedings of the 8th Conference on Integer Programming and Combinatorial Optimization, Utrecht, The Netherlands*, 2001. To appear.
28. D. Eppstein. Spanning trees and spanners. In Jörg-Rudiger Sack and Jorge Urrutia, editors, *Handbook of Computational Geometry*, chapter 9, pages 425–461. Elsevier, 2000.
29. M.L. Fredman. New bounds on the complexity of the shortest path problem. *SIAM Journal on Computing*, 5:49–60, 1976.
30. M.L. Fredman and R.E. Tarjan. Fibonacci heaps and their uses in improved network optimization algorithms. *Journal of the ACM*, 34:596–615, 1987.
31. M.L. Fredman and D.E. Willard. Surpassing the information-theoretic bound with fusion trees. *Journal of Computer and System Sciences*, 47(3):424–436, 1993.
32. M.L. Fredman and D.E. Willard. Trans-dichotomous algorithms for minimum spanning trees and shortest paths. *Journal of Computer and System Sciences*, 48:533–551, 1994.
33. H.N. Gabow. Scaling algorithms for network problems. *Journal of Computer and System Sciences*, 31(2):148–168, 1985.
34. H.N. Gabow and R.E. Tarjan. Faster scaling algorithms for network problems. *SIAM Journal on Computing*, 18:1013–1036, 1989.
35. Z. Galil and O. Margalit. Witnesses for boolean matrix multiplication. *Journal of Complexity*, 9:201–221, 1993.
36. Z. Galil and O. Margalit. All pairs shortest distances for graphs with small integer length edges. *Information and Computation*, 134:103–139, 1997.
37. Z. Galil and O. Margalit. All pairs shortest paths for graphs with small integer length edges. *Journal of Computer and System Sciences*, 54:243–254, 1997.

38. A.V. Goldberg. Scaling algorithms for the shortest paths problem. *SIAM Journal on Computing*, 24:494–504, 1995.
39. T. Hagerup. Sorting and searching on the word RAM. In *Proceedings of the 15th Annual Symposium on Theoretical Aspects of Computer Science, Paris, France*, pages 366–398, 1998.
40. T. Hagerup. Improved shortest paths on the word RAM. In *Proceedings of the 27st International Colloquium on Automata, Languages and Programming, Geneva, Switzerland*, pages 61–72, 2000.
41. S. Halperin and U. Zwick. Unpublished result, 1996.
42. Y. Han. Improved fast integer sorting in linear space. In *Proceedings of the 12th Annual ACM-SIAM Symposium on Discrete Algorithms, Washington, D.C.*, pages 793–796, 2001.
43. M.R. Henzinger, P. Klein, S. Rao, and S. Subramanian. Faster shortest-path algorithms for planar graphs. *Journal of Computer and System Sciences*, 55:3–23, 1997.
44. X. Huang and V.Y. Pan. Fast rectangular matrix multiplications and applications. *Journal of Complexity*, 14:257–299, 1998.
45. D.B. Johnson. Efficient algorithms for shortest paths in sparse graphs. *Journal of the ACM*, 24:1–13, 1977.
46. D.R. Karger, D. Koller, and S.J. Phillips. Finding the hidden path: time bounds for all-pairs shortest paths. *SIAM Journal on Computing*, 22:1199–1217, 1993.
47. R.M. Karp. A characterization of the minimum cycle mean in a digraph. *Discrete Mathematics*, 23:309–311, 1978.
48. V. King. Fully dynamic algorithms for maintaining all-pairs shortest paths and transitive closure in digraphs. In *Proceedings of the 40th Annual IEEE Symposium on Foundations of Computer Science, New York, New York*, pages 81–89, 1999.
49. P.N. Klein and S. Subramanian. A randomized parallel algorithm for single-source shortest paths. *Journal of Algorithms*, 25(2):205–220, 1997.
50. G. Kortsarz and D. Peleg. Generating sparse 2-spanners. *Journal of Algorithms*, 17(2):222–236, 1994.
51. J.B. Kruskal. On the shortest spanning subtree of a graph and the traveling salesman problem. *Proceedings of the American Mathematical Society*, 7:48–50, 1956.
52. C.C. McGeoch. All-pairs shortest paths and the essential subgraph. *Algorithmica*, 13:426–461, 1995.
53. K. Mehlhorn and Z. Galil. Monotone switching circuits and Boolean matrix product. *Computing*, 16(1-2):99–111, 1976.
54. J.S.B. Mitchell. Shortest paths and networks. In Jacob E. Goodman and Joseph O'Rourke, editors, *Handbook of Discrete and Computational Geometry*, chapter 24, pages 445–466. CRC Press LLC, Boca Raton, FL, 1997.
55. J.S.B. Mitchell. Geometric shortest paths and network optimization. In Jörg-Rüdiger Sack and Jorge Urrutia, editors, *Handbook of Computational Geometry*, pages 633–701. Elsevier Science Publishers B.V. North-Holland, Amsterdam, 2000.
56. M.S. Paterson. Complexity of monotone networks for Boolean matrix product. *Theoretical Computer Science*, 1(1):13–20, 1975.
57. D. Peleg. *Distributed computing – A locality-sensitive approach*. Society for Industrial and Applied Mathematics (SIAM), Philadelphia, PA, 2000.
58. D. Peleg and A.A. Schäffer. Graph spanners. *J. of Graph Theory*, 13:99–116, 1989.
59. R. Raman. Priority queues: small, monotone and trans-dichotomous. In *Proceedings of the 4nd European Symposium on Algorithms, Barcelona, Spain*, pages 121–137, 1996.

60. R. Raman. Recent results on the single-source shortest paths problem. *SIGACT News*, 28:81–87, 1997.
61. A. Schönhage and V. Strassen. Schnelle multiplikation grosser zahlen. *Computing*, 7:281–292, 1971.
62. R. Seidel. On the all-pairs-shortest-path problem in unweighted undirected graphs. *Journal of Computer and System Sciences*, 51:400–403, 1995.
63. A. Shoshan and U. Zwick. All pairs shortest paths in undirected graphs with integer weights. In *Proceedings of the 40th Annual IEEE Symposium on Foundations of Computer Science, New York, New York*, pages 605–614, 1999.
64. V. Strassen. Gaussian elimination is not optimal. *Numerische Mathematik*, 13:354–356, 1969.
65. T. Takaoka. A new upper bound on the complexity of the all pairs shortest path problem. *Information Processing Letters*, 43:195–199, 1992.
66. T. Takaoka. Subcubic cost algorithms for the all pairs shortest path problem. *Algorithmica*, 20:309–318, 1998.
67. M. Thorup. Faster deterministic sorting and priority queues in linear space. In *Proceedings of the 8th Annual ACM-SIAM Symposium on Discrete Algorithms, San Francisco, California*, pages 550–555, 1998.
68. M. Thorup. Undirected single-source shortest paths with positive integer weights in linear time. *Journal of the ACM*, 46:362–394, 1999.
69. M. Thorup. Floats, integers, and single source shortest paths. *Journal of Algorithms*, 35:189–201, 2000.
70. M. Thorup. On RAM priority queues. *SIAM Journal on Computing*, 30(1):86–109, 2000.
71. M. Thorup. Compact oracles for reachability and approximate distances in planar digraphs. manuscript, 2001.
72. M. Thorup and U. Zwick. Approximate distance oracles. In *Proceedings of the 33th Annual ACM Symposium on Theory of Computing, Crete, Greece*, 2001.
73. M. Thorup and U. Zwick. Compact routing schemes. In *Proc. of the 13th Annual ACM Symposium on Parallel Algorithms and Architectures, Crete, Greece*, 2001.
74. P. van Emde Boas. Preserving order in a forest in less than logarithmic time and linear space. *Information Processing Letters*, 6(3):80–82, 1977.
75. P. van Emde Boas, R. Kaas, and E. Zijlstra. Design and implementation of an efficient priority queue. *Mathematical Systems Theory*, 10:99–127, 1977.
76. G. Yuval. An algorithm for finding all shortest paths using $N^{2.81}$ infinite-precision multiplications. *Information Processing Letters*, 4:155–156, 1976.
77. U. Zwick. All pairs shortest paths in weighted directed graphs – exact and almost exact algorithms. In *Proceedings of the 39th Annual IEEE Symposium on Foundations of Computer Science, Palo Alto, California*, pages 310–319, 1998. Journal version submitted for publicaiton under the title *All-pairs shortest paths using bridging sets and rectangular matrix multiplication*.
78. U. Zwick. All pairs lightest shortest paths. In *Proceedings of the 31th Annual ACM Symposium on Theory of Computing, Atlanta, Georgia*, pages 61–69, 1999.

Strongly Competitive Algorithms for Caching with Pipelined Prefetching

Alexander Gaysinsky, Alon Itai, and Hadas Shachnai

Dept. of Computer Science, Technion, Haifa, Israel
{csalex@tx, itai@cs, hadas@cs}.technion.ac.il

Abstract. Prefetching and caching are widely used for improving the performance of file systems. Recent studies have shown that it is important to *integrate* the two. In this model we consider the following problem. Suppose that a program makes a sequence of m accesses to data blocks. The cache can hold k blocks, where $k < m$. An access to a block in the cache incurs one time unit, and fetching a missing block incurs d time units. A fetch of a new block can be initiated while a previous fetch is in progress. Thus, d block fetches can be in progress simultaneously. The locality of references to the cache is captured by the *access graph* model of [2]. The goal is to find a policy for prefetching and caching, which minimizes the overall execution time of a given reference sequence. This problem is called *caching with locality and pipelined prefetching (CLPP)*. Our study is motivated from the pipelined operation of modern memory controllers, and program execution on fast processors. For the offline case we show that an algorithm introduced in [4] is *optimal*. In the online case we give an algorithm which is within factor of 2 from the optimal in the set of online deterministic algorithms, for *any* access graph, and $k, d \geq 1$. Improved ratios are obtained for several important classes of access graphs, including *complete graphs* and *directed acyclic graphs (DAG)*. Finally, we study the CLPP problem assuming a Markovian access model, on branch trees, which often arise in applications. We give algorithms whose expected performance ratios are within factor 2 from the optimal.

1 Introduction

1.1 Problem Statement

Caching and prefetching have been studied extensively in the past decades; however, the interaction between the two was not well understood until the important work of Cao et. al [4], who proposed to integrate caching with prefetching. They introduced the following execution model. Suppose that a program makes a sequence of m accesses to data blocks and the cache can hold $k < m$ blocks. An access to a block in the cache incurs one time unit, and fetching a missing block incurs d time units. While accessing a block in the cache, the system can fetch a block from secondary storage, either in response to a cache miss (*caching by demand*), or before it is referenced, in anticipation of a miss (*prefetching*); at

F. Meyer auf der Heide (Ed.): ESA 2001, LNCS 2161, pp. 49–61, 2001.

most one fetch can be in progress at any given time. The *Caching with Prefetching (CP)* problem is to determine the sequence of block evictions/prefetches, so as to minimize the overall time required for accessing all blocks.

Motivated from the operation of modern memory controllers, and from program execution on fast processors,[1] we consider the problem of caching integrated with *pipelined* prefetching. Here, a fetch of a new block can be initiated while a previous fetch is in progress. Thus, d block fetches can be in progress simultaneously. We adopt the *access graph* model developed by Borodin et al. [2], which captures locality of reference in memory access patterns of real programs; thus, we assume, that any sequence of block references is a walk on an *access graph G*. As before, our measure is the overall execution time of a given reference sequence.

Formally, suppose that a set of n data blocks $b_1, b_2, \ldots, b_n$ is held in secondary storage. The access graph for the program that reads/writes into $b_1, b_2, \ldots, b_n$, is given by a directed graph $G = (V, E)$, where each node corresponds to a block in this set. Any sequence of block references has to obey the locality constraints imposed by the edges of G: following a request to a block (node) u, the next request has to be either to block u or to a block v, such that $(u, v) \in E$. The usage of pipelined prefetching implies that if a prefetch is initiated at time t, then the next prefetch can be initiated at time $(t + 1)$.

Given a reference sequence $\sigma = \{r_1, \ldots, r_m\}$, r_i can be satisfied immediately at time t, incurring one time unit, if r_i is in the cache; otherwise, if a prefetch of r_i was initiated at time $t_i \leq t$, there is a *stall* for $d(r_i) = d - (t - t_i)$ time units. The total execution time of σ is the time to access the m blocks plus the stall time, i.e., $C(\sigma) = m + \sum_{i=1}^{m} d(r_i)$. The problem of *Caching with Locality and Pipelined Prefetching (CLPP)* can be stated as follows. Given a cache of size $k \geq 1$, a delivery time $d \geq 1$, and a reference sequence $\sigma = \{r_1, \ldots, r_m\}$, find a policy for pipelined prefetching and caching that minimizes $C(\sigma)$.

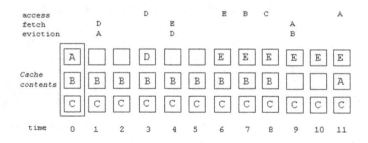

Fig. 1. An example of caching ($k = 3$ and $d = 2$)

Example 1. Consider a program whose block accesses are given by the sequence "DEBCA" (the entire sequence is known at time 0). The cache can hold three

---

[1] A detailed survey of these applications is given in [9].

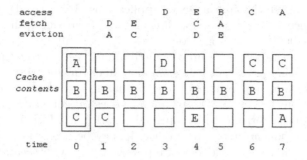

Fig. 2. An example of caching with pipelined prefetching, using Algorithm AGG ($k = 3$ and $d = 2$)

blocks; fetching a block takes two time units. Initially A,B and C are in the cache. Figure 1 shows the execution of the optimal caching-by-demand algorithm [3] that incurs 11 time units[2]. Figure 2 shows an execution of the **Aggressive (AGG)** algorithm (see Section 3), which combines caching with pipelined prefetching; thus, the reference sequence is completed within 7 time units, which is optimal.

The above example suggests that pipelined prefetching can be helpful. When future accesses are fully known, the challenge of a good algorithm is to achieve maximum overlap between accesses to blocks in the cache and fetches. Without this knowledge, achieving maximum overlap may be harmful, due to evictions of blocks that will be requested in the near future. Thus, more careful decisions need to be made, based on past accesses, and the structure of the underlying access graph.

1.2 Our Results

We study the CLPP problem both in the *offline* case, where the sequence of cache accesses is known in advance, and in the *online* case, where r_{i+1} is revealed to the algorithm only when r_i is accessed. In the offline case (Section 3), we show that algorithm **Aggressive (AGG)** introduced in [4] is optimal, for *any* access graph and any $d, k \geq 1$.

In the online case (Section 4), we give an algorithm which is within factor of 2 from the optimal in the set of online deterministic algorithms, for *any* access graph, and $k, d \geq 1$. Improved ratios are obtained (in Section 5) for several important classes of access graphs, including *complete graphs* and *directed acyclic graphs (DAG)*. In particular, for complete graphs we obtain a ratio of $1 + 2/k$, for DAGs $\min(2, 1 + k/d)$, and for branch trees $1 + o(1)$. Finally, we study (in Section 6) the CLPP problem assuming a Markovian access model,

---

[2] Note that if a block b_i is replaced in order to bring the block b_j, then b_i becomes unavailable for access when the fetch is initiated; b_j can be accessed when the fetch terminates, i.e., after d time units.

on branch trees, which often arise in applications. We give algorithms whose expected performance ratios are within factor 2 from the optimal for general trees, and $(1 + o(1))$ for homogeneous trees.

Our results contain two technical contributions. We present (in Section 4) a general proof technique, for deriving upper bounds on the competitive ratios of online algorithms for our problem. (The technique is used also for deriving the results in Sections 5 and 6). The technique relies on comparing a *lazy* version of a given online algorithm to an optimal algorithm, which is allowed to use *parallel* (rather than *pipelined*) prefetches; it may be useful for other problems that involve online *pipelined* service of input sequences. Our second contribution is an *asymptotic* analysis of the q-distance Fibonacci numbers, for any $q \geq 2$. We use asymptotic estimates of these numbers for solving the recursion formula for algorithm DEE, in the Markovian model (see Section 6.2).

1.3   Related Work

The concept of cooperative prefetching and caching was first investigated by Cao et al. [4]: this paper studies offline prefetching and caching algorithms, where fetches are *serialized*, i.e., at most one fetch can be in progress at any given time. An algorithm, called Aggressive (AGG), was shown to yield a $min(1 + d/k, 2)$-approximation to the optimal. Karlin and Kimbrel [12] extended this study to storage systems which consist of r units (e.g., disks); fetches are serialized on each storage unit, thus, up to r block fetches can be processed in parallel. The paper gives performance bounds for several offline algorithms in this setting. Algorithm AGG is shown to achieve a ratio of $(1 + rd/k)$ to the optimal. Later papers [18, 21] present experimental results for cooperative prefetching and caching, in the presence of optional program-provided hints of future accesses.

Note that the classic paging problem, where the cost of an access is zero and the cost of a fault is 1, is a special case of our problem, in which $d \gg 1$.[3] There is a wide literature on the caching (paging) problem. (Comprehensive surveys appear in [3,10,16,5].) Borodin et al. [2] introduced the *access graph* model. The paper presents an online algorithm that is strongly competitive on any access graph. Later works (e.g., [11,6,8]) consider extensions of the access graph model, or give experimental results for some heuristics for paging in this model [7].

Karlin et al. [13] introduced the Markov paging problem, in which the access graph model is combined with the generation of reference sequences by a *Markov chain*. Specifically, the transition from a reference to page u to the reference to page v (both represented as nodes in the access graph of the program) is done with some fixed probability. The paper presents an algorithm whose *fault rate* is at most a constant factor from the optimal, for any Markov chain.

There has been some earlier work on the Markovian CLPP, on branch trees in which $k > 1$ and $d = k - 1$. We examine here two of the algorithms proposed in theses works, namely, the algorithms Eager Execution (EE) [1,20] and

---

[3] Thus, when normalizing (by factor d) we get that the delivery time equals to one, while the access time, $1/d$, asymptotically tends to 0.

Disjoint Eager Execution (DEE) [20]. EE was shown to perform well in practice; however, no theoretical performance bounds were derived. Raghavan et al. [17] showed, that for the special case of a homogeneous branch tree, where the transition parameter p is close to 1, DEE is optimal to within a constant factor. We improve this result, and show (in Section 6.2) that DEE is nearly optimal on branch trees, for *any* $p \in [1/2, 1]$.

Due to space limitations, we omit most of the proofs. Detailed proofs can be found in [9].

2   Preliminaries

Let G denote an access graph. Any reference sequence σ is a path in G. (When G is a directed graph, σ forms a directed path in G.) We assume that $r_i \neq r_{i+1}$, for all $1 \leq i < |\sigma|$. Clearly, this makes the problem no easier. Denote by $Paths(G)$ the set of paths in G. Let OPT be an optimal offline algorithm for the CLPP. We refer to an optimal offline algorithm and the source of requests together as the *adversary*, who generates the sequence and serves the access requests offline.

We use competitive analysis (see e.g. [3]) to establish performance bounds for online algorithms for our problem. The competitive ratio of an online algorithm $\mathcal{A}$ on a graph G, for fixed k and d, is given by $c_{\mathcal{A},k,d}(G) = \sup_{\sigma \in Paths(G)} \mathcal{A}(\sigma)/OPT(\sigma)$, where $\mathcal{A}(\sigma)$, $OPT(\sigma)$ are the costs incurred by $\mathcal{A}$ and OPT, respectively, for the execution of σ.

In the *Markovian* model, each reference sequence, σ, is associated with a probability, such that the probability of $\sigma = r_1, r_2, \ldots, r_m$ is given by $Pr(\sigma) = \prod_{i=1}^{m-1} p_{r_i, r_{i+1}}$, where $p_{r_i, r_{i+1}}$ is the *transition probability* from r_i to r_{i+1}. The expected performance ratio of an algorithm, $\mathcal{A}$, in the *Markovian* model on a graph G, for fixed k and d, is $c_{\mathcal{A},k,d}(G) = \sum_{\sigma \in Paths(G)} Pr(\sigma) \cdot \mathcal{A}(\sigma)/OPT(\sigma)$.

We abbreviate the formulation of our results, using the following notation: $c_{k,d}(G)$ ($c_{k,d}^{\det}(G)$) is the competitive ratio of an optimal (deterministic) online algorithm for the CLPP, on an access graph G, for fixed k and d; $c_k(G)$ ($c_k^{\det}(G)$) is the competitive ratio of an optimal (deterministic) online algorithm on G, for a cache of size k and *arbitrary* d. We say that (deterministic) algorithm $\mathcal{A}$ is *strongly competitive*, if $c_{\mathcal{A},k,d}(G) = O(c_k(G))$ ($c_{\mathcal{A},k,d}(G) = O(c_{k,d}^{\det}(G))$). In the Markovian model we replace in the above notation *competitive ratio* with *expected performance ratio*.

Finally, an access graph G is called a *branch tree*, if G is an ordered binary out-tree in which every internal node has a *left* child and a *right* child. Let T be a branch tree rooted at r. In the Markovian CLPP, the transition probability $p_{u,v}$ from any node u to its child v is called the *local probability* of v. Denote by $\{v_0 = r, v_1, \ldots, v_n = v\}$ the path from r to some node v, then the *accumulated probability* of v is given by $p_a(v) = \prod_{i=1}^{n} p_{v_{i-1}, v_i}$.

3   The Offline CLPP Problem

In the offline case we are given the reference sequence, and our goal is to achieve maximal overlap between prefetching and references to blocks in the cache, so as to minimize the overall execution time of the sequence. The next lemma shows that a set of rules formulated in [4], to characterize the behavior of optimal algorithms for the CP problem, applies also for the offline CLPP problem.

Lemma 1. *[No harm rules] There exists an optimal algorithm $\mathcal{A}$, which satisfies the following rules : (i) $\mathcal{A}$ fetches the next block in the reference sequence that is missing in the cache; (ii) $\mathcal{A}$ evicts the block whose next reference is furthest in the future. (iii) $\mathcal{A}$ never replaces a block B by a block C, if B will be referenced before C.*

In the remainder of this section we consider only optimal algorithms that follow the "no harm" rules. Clearly, once an algorithm $\mathcal{A}$ decides to fetch a block, these rules uniquely define the block that should be fetched, and the block that will be evicted. Thus, the only decision to be made by any algorithm is *when* to start the next fetch.

Algorithm AGG, proposed by Cao et al. [4], follows the "no harm" rules; in addition, it fetches each block at the *earliest* opportunity, i.e., whenever there is a block in the cache, whose next reference is after the first reference to the block that will be fetched. (An example of the execution of AGG is given in Figure 2.) As stated in our next result, this algorithm is the best possible for the CLPP.

Theorem 1. AGG *is an optimal offline algorithm for the CLPP problem.*

It can be shown by induction on i, that for $i \geq 1$, any optimal offline algorithm $\mathcal{A}$ which satisfies the "no harm" rules, can be modified to act like AGG in the first i steps, without harming $\mathcal{A}$'s optimality.   ∎

The greediness of AGG plays an important role when $d < k$. In this case, one can show that for any $i \geq 1$, when AGG accesses r_i in the cache, each of the blocks $r_{i+1}, \ldots, r_{i+d-1}$ is either in the cache or is being fetched; thus, in any reference sequence AGG incurs a single miss: in the first reference.

4   The Online CLPP Problem

Let $G = (V, E)$ be an access graph. Suppose that $\sigma = \{r_1, \ldots, r_l\}$, $1 \leq l \leq k$, is a reference sequence to blocks, where only r_1 is known.

Recall that σ is a path in G. For a given subgraph $G' \subseteq G$, let $PREF_\sigma(G')$ denote the maximal set of vertices in G' that form a *prefix* of σ. Let $\mathcal{A}$ be an online deterministic algorithm. Initially, $\mathcal{A}$ knows only that $r_1 \in \sigma$. In the *Single Phase CLPP (S_CLPP)*, $\mathcal{A}$ needs to select a subgraph $G_{\mathcal{A}} \subseteq G$ of l vertices, such that $PREF_\sigma(G_{\mathcal{A}})$ is maximal. Formally, denote by $B(G_{\mathcal{A}}) = \min_{\sigma \in Paths(G), |\sigma|=l} |PREF_\sigma(G_{\mathcal{A}})|$ the *benefit* of $\mathcal{A}$ from the selection of $G_{\mathcal{A}}$. We seek an algorithm $\mathcal{A}$, whose benefit is maximal.

Consider algorithm AGG_o which mimics the operation of AGG in an online fashion. Denote by $dist(v, u)$ the length of the shortest path from v to u in G.[4] Let IN_t denote the blocks that are in the cache or are being fetched at time t, and $OUT_t = V \setminus IN_t$. The following is a pseudocode description for AGG_o.

Algorithm AGG_o
for $t = 1, \ldots, l$ do
  Let $u = \arg \min\{dist(r_t, v) : v \in OUT_t\}$.
  Let $w = \arg \max\{dist(r_t, v)v : v \in IN_t\}$.
  If $dist(r_t, u) < dist(r_t, w)$
    Evict w from the cache and initiate a fetch for u.

Lemma 2. *Algorithm AGG_o is optimal in the set of deterministic online algorithms for the S_CLPP problem, for any graph G, and $k, d \geq 1$.*

Theorem 2. *For any graph G, and $k, d \geq 1$ there exists an algorithm, $\mathcal{A}$, such that $c_{\mathcal{A}, k, d}(G) \leq 2 \cdot c_{k,d}^{det}(G)$.*

Proof: Consider first Algorithm Lazy-AGG_o which operates in phases. Phase i starts at some time t_i, with a stall of d time units, for fetching a missing block— r_i. Each phase is partitioned into sub-phases. The first sub-phase of phase i starts at time $t_{i,1} = t_i$. At sub-phase j, Lazy-AGG_o uses the rules of AGG_o for selecting a subgraph, $G_{i,j}$, of $k' = \min(d, k)$ blocks. Some of these blocks are already in the cache: Lazy-AGG_o initiates pipelined fetching of the remaining blocks. Let r_{i_j} be the block that is accessed first in sub-phase j of phase i. Then, $\sigma_d = \{r_{i_j}, \ldots, r_{i_j + d - 1}\}$ is the set of first d block accesses in sub-phase j. Let $t_{i,j}$ be the start time of sub-phase j. Let $Good(i, j) = PREF_{\sigma_d}(IN_{t_{i,j} + d})$. We handle separately two cases:

- If $|Good(i, j)| = d$, then Lazy-AGG_o waits until d blocks were accessed in the cache; at time $t_{i,j} + 2d$ the j-th sub-phase terminates, and Lazy-AGG_o starts sub-phase $j + 1$ of phase i.
- If $g_j = |Good(i, j)| < d$ then at time $t_{i,j} + d + g_j$ phase i terminates and the first missing block in the cache becomes r_{i+1}.

Consider now Algorithm AGG'_o that operates like Lazy-AGG_o; however, AGG'_o has the advantage that in each sub-phase, j, all the prefetches are initiated in parallel and d time units after this sub-phase starts AGG'_o knows the value of g_j and the first missing block in the cache. If $g_j = d$ then AGG'_o proceeds to the next sub-phase of phase i; if $g_j < d$ phase i terminates. Note that combining Lemma 2 with the parallel fetching property we get that AGG'_o outperforms any deterministic online algorithm.

To compute $c_{\text{Lazy-}AGG_o, k, d}(G)$ it suffices to compare the length of phase i of Lazy-AGG_o and AGG'_o, for any $i \geq 1$. Suppose that there are $sp(i)$ sub-phases in phase i. For Lazy-AGG_o each of the first $sp(i) - 1$ sub-phases incurs $2d$ time

---

[4] When u is unreachable from v $dist(v, u) = \infty$.

units, while the last sub-phase incurs $d + g$ time units, for some $1 \leq g \leq d$. For AGG'$_o$ each sub-phase (including the last one) incurs d time units; thus, we get the ratio $(d + g + (sp(i) - 1)2d)/(d \cdot sp(i)) \leq 2$. ∎

5   Online CLPP on DAGs and Complete Graphs

5.1   Directed Acyclic Access Graphs

When G is a DAG, AGG$_o$ acts exactly like algorithm EE, defined as follows. At any time $t \geq 1$, fetch a missing block that is closest to the currently accessed block; discard a block that is unreachable from the current position. From Lemma 2 we get that EE is optimal in the set of online deterministic algorithms for the S_CLPP on DAGs. Now, consider the operation of EE on a DAG. Our next result improves the bound in Theorem 2 in the case where $k < d$.

Theorem 3. *If G is a DAG then for any cache size $k \geq 1$ and delivery time $d \geq 1$, $c_{EE,k,d}(G) \leq \min(1 + k/d, 2) \cdot c_{k,d}^{det}(G)$.*

Proof: As in the proof of Theorem 2, we use Algorithms Lazy-AGG$_o$ and AGG'$_o$. When G is a DAG, we call these algorithms Lazy-EE and EE', respectively. Note that from Theorem 2 we immediately get that $c_{EE,k,d}(G) \leq 2 \cdot c_{k,d}^{det}(G)$. This follows from the fact that EE performs at least as well as Lazy-EE, since EE initiates a block fetch whenever possible, while Lazy-EE waits for a sequence of fetches to be completed before it starts fetching the next sequence.

When $d > k$ we can improve the ratio of 2: in this case each phase consists of a single sub-phase. Indeed, each block is accessed exactly once; thus, for any $i \geq 1$, during the stall for r_i both EE' and Lazy-EE can access at most $k < d$ blocks in the cache (i.e., $g_1 < d$), and phase i terminates. It follows that the ratio between the length of phase i for Lazy-EE and EE' is at most $(d + k)/d$. This completes the proof. ∎

Branch Trees. The case where G is a branch tree and $d = k - 1$ is of particular interest, in the application of the CLPP to pipeline execution of programs on fast processors (see in [9]). For this case, we can tighten the bound in Theorem 3.

Theorem 4. *If G is a branch tree then $c_{EE,k,k-1}(G) \leq (1 + o(1))c_{k,k-1}(G)$, where the $o(1)$ term refers to a function of k.*

5.2   Complete Graphs

Suppose that G is a complete graph. On these graphs the lower bound obtained for deterministic algorithms for the classic paging problem, remains valid for the CLPP; that is, if G is a complete graph, then for any cache size $k \geq 1$ $c_k^{det}(G) \geq k - 1$ (see in [9]).

Consider the set of *marking* algorithms proposed for the classical caching (paging) problem (see, e.g., [2]). A marking algorithm proceeds in phases. At the beginning of a phase all the blocks in the cache are unmarked. Whenever a block is requested, it is marked. On a cache fault, the marking algorithm evicts an unmarked block from the cache and fetches the requested one. A phase ends on the first 'miss' in which all the blocks in the cache are marked. At this point all the blocks become unmarked, and a new phase begins.

Lemma 3. *For any access graph G, cache size k and delivery time $d \geq 1$, if $\mathcal{A}$ is a marking algorithm, then $c_{\mathcal{A},k,d}(G) \leq k + 1$.*

The proof is based on showing that if the j-th phase of $\mathcal{A}$ consists of n_j references, then the cost of $\mathcal{A}$ on the phase is at most $kd + n_j$, while the cost of OPT is at least $\max(n_j, d)$. ∎

We summarize the above discussion in the next result.

Theorem 5. *On a complete graph, any marking algorithm has competitive ratio $1 + 2/k$ with respect to the set of deterministic on-line algorithms for the CLPP.*

6   The Markovian CLPP on Branch Trees

The following algorithm, known as DEE, is a natural greedy algorithm for the CLPP in the Markovian model: at any time $t \geq 1$, fetch a missing block that is *most likely* to appear in σ from the current position; discard a block that is unreachable from the current position.

In this section we analyze the performance of DEE on branch trees, and show that its expected performance ratio is ≤ 2. This ratio is reduced to $(1 + o(1))$ for a special class of Markov chains that we call *homogeneous* (see Section 6.2).

6.1   Performance of DEE on Branch Trees

Let T be a branch tree, and $k \geq 1$ an integer. Suppose that σ is a reference sequence of length k. In the *Markovian S_CLPP* we need to choose a subtree $T_{\mathcal{A}}$ of T of size k, such that the *expected* size of $PREF_\sigma(T_{\mathcal{A}})$ is maximal. Formally, let $B(T_{\mathcal{A}}) = \sum_{\sigma \in Paths(T)} Pr(\sigma)|PREF_\sigma(T_{\mathcal{A}})|$ be the *expected benefit* of an on-line algorithm $\mathcal{A}$ from $T_{\mathcal{A}}$. We seek an algorithm, $\mathcal{A}$, whose expected benefit is maximal. We proceed to obtain a performance bound for DEE, when applied for the CLPP.

Theorem 6. DEE *is optimal to within factor 2 in the set of online algorithms on branch trees.*

For showing the theorem, we use the optimality of DEE for the Markovian S_CLPP, as shown in [17]. The proof is based on defining two variants of DEE: DEE' and Lazy-DEE, which operate in phases, similar to the proof of Theorem 3. We omit the details.

6.2 Homogeneous Branch Trees

We call a branch tree T *homogeneous*, if all the left children in T have the *same* local probability, $p \in [1/2, 1)$. In other words, the transition probabilities from any vertex u to its its left child is p and to its right child is $1 - p$.

In the following we derive an explicit asymptotic expression for the expected benefit of DEE, when solving the Markovian S_CLPP on a homogeneous branch tree, with any parameter $1/2 \leq p < 1$. Our computations are based on a Fibonacci-type analysis, which well suits the homogeneous case. For any integer $q \geq 2$, the n-th number of the *q-distance Fibonacci sequence* is given by

$$g(n) = \begin{cases} 0 & \text{if } n < q - 1 \\ 1 & \text{if } n = q - 1 \\ g(n-1) + g(n-q) & \text{otherwise.} \end{cases}$$

Note that in the special case where $q = 2$, we get the well known Fibonacci numbers [14].[5]

Lemma 4. *For any $n \geq 1$ and a given $q \geq 2$, $g(n) = O((r^{n-q})/q)$, where $1 < r = r(q) \leq 2$.*

Proof. To find an expression for $g(n)$, $n \geq q$, we need to solve the equation generated from the recursion formula, i.e.,

$$x^n = x^{n-1} + x^{n-q}. \tag{1}$$

In other words, we need to find the roots of the polynomial $p(x) = x^q - x^{q-1} - 1$. Note that $p(x)$ has no multiple roots. Hence, the general form of our sequence is $g(n) = \sum_{i=1}^{q} b_i r_i^n$, where $r_1, r_2, ..., r_q$ are the roots of the polynomial $p(x)$. Also, for $1 \leq x \leq 2$ $p(x)$ is monotone and non-decreasing. As $p(1) = -1$, and $p(2) = 2^q - 2^{q-1} - 1 = 2^{q-1} - 1 \geq 0$, we get that $p(x)$ has a single root $r_q \in R^+$ in the interval $[1, 2]$.

Denote by $|x|$ the module of a complex number x. We claim that $|r_i| < r_q$ for all $i < q$. The claim trivially holds for any r_i satisfying $|r_i| \leq 1$, thus we may assume that $|r_i| > 1$. If r_i is a root of $p(x)$, then $0 = p(r_i) = |r_i^q - r_i^{q-1} - 1| \geq |r_i|^q - |r_i|^{q-1} - 1 = p(|r_i|)$, and since $p(x)$ is non-decreasing for $x \geq 1$, we conclude that $|r_i| \leq r_q$. In fact, it can be shown that the last inequality is strong, i.e., $|r_i| < r_q$, for any $1 \leq i \leq q - 1$. We can write

$$g(n) = \sum_{i=1}^{q} b_i r_i^n = b_q r_q^n \left(1 + \sum_{i=1}^{q-1} \frac{b_i}{b_q} \left(\frac{r_i}{r_q}\right)^n\right) \tag{2}$$

and since $|r_i/r_q| < 1$, the sum in the rhs of (2) exponentially tends to zero, i.e., $\lim_{n \to \infty} \sum_{i=1}^{q-1} \frac{b_i}{b_q} \left(\frac{r_i}{r_q}\right)^n = 0$ Hence, we get that

$$g(n) = b_q r_q^n (1 + o(1)). \tag{3}$$

---

[5] This sequence can be viewed as a special case of the q-order Fibonacci sequence, in which $g(n) = \sum_{j=1}^{q} g(n - j)$ (see, e.g., [14,19]).

Now, b_q can be calculated by solving a linear system for the first q elements of the sequence $g(n)$.

$$
\begin{pmatrix}
1 & 1 & \ldots & 1 \\
r_1 & r_2 & \ldots & r_q \\
\vdots & & & \vdots \\
r_1^{q-1} & r_2^{q-1} & \ldots & r_q^{q-1}
\end{pmatrix}
\begin{pmatrix}
b_1 \\
b_2 \\
\vdots \\
b_q
\end{pmatrix}
=
\begin{pmatrix}
0 \\
0 \\
\vdots \\
1
\end{pmatrix}
$$

The determinant of the above matrix is known as Vandermondé determinant [15]. The general solution of such a system is $b_i = \prod_{\alpha=1,\alpha\neq i}^{q}(r_i - r_\alpha)^{-1}$. Our polynomial is $p(x) = \prod_{i=1}^{q}(x - r_i)$, and $p'(r_i) = \prod_{\alpha=1,\alpha\neq i}^{q}(r_i - r_\alpha)$, therefore we get that $b_i = \frac{1}{p'(r_i)}$. We can calculate all the coefficients. In particular, $b_q = 1/(qr_q^{q-1} - (q-1)r_q^{q-2})$. Substituting into (3) we get the statement of the lemma. ∎

Lemma 5. *Let $\frac{1}{2} \leq p < 1$ satisfy that for some natural q*

$$p^q = 1 - p \text{ and } q \in N, \tag{4}$$

and let $\alpha = (1-p)q + p$, $\delta \in \{0,1\}$. Then the height of the subtree chosen by DEE *for the Markovian S_CLPP is given by $height(T_{DEE}) = \log_{1/p}((1-p)\alpha k + p) + \delta + o(1)$. The $o(1)$ term refers to a function of k.*

Proof sketch: Let q be defined as in (4), and denote by $f(n)$ the number of vertices in T with accumulated probability p^n. Then, $f(n)$ can be computed recursively as follows. For $1 \leq n < q$, since $p^n > 1-p$, there is a single node with accumulated probability p^n; for $n \geq q$, we get a node with accumulated probability p^n, either by taking the left child of a node with accumulated probability p^{n-1}, or by taking the right child of a node, whose accumulated probability is p^{n-q}. Hence, $f(n) = g(n+q-1)$, where $g(n)$ is the nth q-distance Fibonacci number. From (3), $f(n) = b_q r_q^{n+q-1}(1 + o(1))$, where r_q is the single root of equation (1) in the interval $[1,2]$. Using (4), it is easy to verify that $r_q = 1/p$ satisfies (1). Hence, we get that $f(n) = (1 + o(1))\alpha p^{-n}$. Let h be the maximal integer satisfying $\sum_{n=0}^{h} f(n) \leq k$. Then,

$$k \geq \frac{1 + o(1)}{\alpha(1-p)}\left(\frac{1}{p^h} - p\right). \tag{5}$$

Also, there exists a vertex with accumulated probability p^{h+1} which is not in T_{DEE}. Thus,

$$k \leq \sum_{n=0}^{h+1} f(n) = \frac{1 + o(1)}{\alpha(1-p)}\left(\frac{1}{p^{h+1}} - p\right). \tag{6}$$

Combining (5) and (6) we get the statement of the lemma. ∎

Lemma 5 yields an asymptotic expression for the expected benefit of DEE.

Corollary 1. *For any* $\frac{1}{2} \leq p < 1$, *the expected benefit of* DEE *in solving the Markovian S_CLPP problem is*

$$B(T_{DEE}) = \frac{1 + \lg_{\frac{1}{p}}((1 - p)\alpha k + p)}{\alpha}(1 + o(1)), \tag{7}$$

where the $o(1)$ *term refers to a function of* k.

Corollary 1 is essential for proving the next theorem (see [9]).

Theorem 7. DEE *is within a factor* $1 + o(1)$ *from the optimal in the set of algorithms for the Markovian CLPP on homogeneous branch trees, for any* $p \in [1/2, 1]$.

Acknowledgments. The authors thank Anna Karlin, Rajeev Motwani and Prabhakar Raghavan, for stimulating discussions on this paper.

References

1. D. Bhandarkar, J. Ding, "Performance Characterization of the Pentium Pro Processor", 3rd International Symposium on High Performance Computer Architecture, San Antonio, 1997.
2. A. Borodin, S. Irani, P. Raghavan and B. Schieber, "Competitive Paging with Locality of Reference", *Journal of Computer and System Science*, 1995, pp. 244-258.
3. A. Borodin and R. El-Yaniv, "Competitive Analysis and Online Computation", Cambridge University Press, 1998.
4. P. Cao, E. Felton, A. Karlin, K. Li, "A Study of Integrated Prefetching and Caching Strategies", *SIGMETRICS/PERFORMANCE*, 1995.
5. A. Fiat. and G. J. Woeginger,, "Online Algorithms, The State of the Art", Springer, 1998 (LNCS #1442).
6. A. Fiat. and A. R. Karlin, "Randomized and Multipointer Paging with Locality of Reference", *STOC*, 1995.
7. A. Fiat. and Z. Rosen, "Experimental Studies of access Graph Based Heuristics: Beating the LRU Standard?", *SODA*, 1997.
8. A. Fiat. and M. Mendel, "Truly Online Paging with Locality of Reference", *FOCS*, 1997.
9. A. Gaysinsky, A. Itai and H. Shachnai. "Strongly Competitive Algorithms for Caching with Pipelined Prefetching", full version.
   http://www.cs.technion.ac.il/ ~hadas/PUB/clpp.ps.gz.
10. D.S. Hochbaum, *Approximation Algorithms for NP-Hard Problems*, PUS Publishing Company, 1995.
11. S. Irani, A. R. Karlin and S. Phillips, "Strongly competitive algorithms for paging with locality of reference", *SIAM Journal Comput.*, June 1996, pp. 477-497.
12. A. R. Karlin, T. Kimbrel, "Near-optimal parallel prefetching and caching", CS TR, Washington Univ., 1996.
13. A. R. Karlin, S. Phillips and P. Raghavan, "Markov paging", *FOCS* 1992.
14. D. E. Knuth, *The Art of Computer Programming*, Vol. 3, Addison Wesley, 1973.
15. S. Lang, *Algebra, Columbia University, New York*, 153-155, 1965.

16. R. Motwani, P. Raghavan, *Randomized Algorithms*, Cambridge University Press, 1995.
17. P. Raghavan, H. Shachnai, M. Yaniv, "Dynamic Schemes for Speculative Execution of Code", *MASCOTS*, 1998.
18. A. Tomkins, R. H. Patterson and G. Gibson. "Informed Multi-Process Prefetching and Caching", *SIGMETRICS*, 1997.
19. R. Sedgewick and P. Flajolet, *An Introduction to the Analysis of Algorithms*, Addison-Wesley Publishing, 1996.
20. A.K. Uht and V. Sindagi, "Disjoint Eager Execution: An optimal form of speculative execution", *MICRO-28*, 1995.
21. G. M. Voelker, E. J. Anderson, T. Kimbrel, M. J. Feeley, J. S. Chase, A. R. Karlin and H. M. Levy, "Implementing Cooperative Prefetching and Caching in Globally-Managed Memory System". *SIGMETRICS*, 1998.
22. S. S. H. Wang and A. K. Uht, "Ideograph/Ideogram: Framework/Architecture for Eager Evaluation", *MICRO-23*, 1990.
23. M. Yaniv, "Dynamic Schemes for Speculative Execution of Code", M.Sc. Thesis, Dept. of Computer Science, The Technion, 1998.

Duality between Prefetching and Queued Writing with Parallel Disks

David A. Hutchinson[1]*, Peter Sanders[2]**, Jeffrey Scott Vitter[1]***

[1] Department of Computer Science, Duke University, Durham, NC 27708–0129
hutchins,jsv@cs.duke.edu
[2] Max-Planck-Institute for Computer Science, Stuhlsatzenhausweg 85, 66123
Saarbrücken, Germany, sanders@mpi-sb.mpg.de

Abstract. Parallel disks promise to be a cost effective means for achieving high bandwidth in applications involving massive data sets, but algorithms for parallel disks can be difficult to devise. To combat this problem, we define a useful and natural duality between writing to parallel disks and the seemingly more difficult problem of prefetching. We first explore this duality for applications involving read-once accesses using parallel disks. We get a simple linear time algorithm for computing optimal prefetch schedules and analyze the efficiency of the resulting schedules for randomly placed data and for arbitrary interleaved accesses to striped sequences. Duality also provides an optimal schedule for the integrated caching and prefetching problem, in which blocks can be accessed multiple times. Another application of this duality gives us the first parallel disk sorting algorithms that are provably optimal up to lower order terms. One of these algorithms is a simple and practical variant of multiway merge sort, addressing a question that has been open for some time.

1 Introduction

External memory (EM) algorithms are designed to be efficient when the problem data do not fit into the high-speed random access memory (RAM) of a computer and therefore must reside on external devices such as disk drives [17]. In order to cope with the high latency of accessing data on such devices, efficient EM algorithms exploit locality in their design. They access a large *block* of B contiguous data elements at a time and perform the necessary algorithmic operations on the elements in the block while in the high-speed memory. The speedup can be significant. However, even with blocked access, a single disk provides much less bandwidth than the internal memory. This problem can be mitigated by using multiple disks in parallel. For each input/output operation, one block is

* Supported in part by the NSF through research grant CCR–0082986.
** Partially supported by the IST Programme of the EU under contract number IST-1999-14186 (ALCOM-FT)
*** Supported in part by the NSF through research grants CCR–9877133 and EIA–9870724 and by the ARO through MURI grant DAAH04–96–1–0013

F. Meyer auf der Heide (Ed.): ESA 2001, LNCS 2161, pp. 62–73, 2001.

transferred between memory and each of the D disks. The algorithm therefore transfers D blocks at the cost of a single-disk access delay.

A simple approach to algorithm design for parallel disks is to employ large logical blocks, or *superblocks* of size $B \cdot D$ in the algorithm. A superblock is split into D physical blocks—one on each disk. We refer to this as *superblock striping*. Unfortunately, this approach is suboptimal for EM algorithms like sorting that deal with many blocks at the same time. An optimal algorithm for sorting and many related EM problems requires *independent access* to the D disks, in which each of the D blocks in a parallel I/O operation can reside at a different position on its disk [19,17]. Designing algorithms for independent parallel disks has been surprisingly difficult [19,14,15,3,8,9,17,16,18].

In this paper we consider parallel disk output and input separately, in particular as the *output scheduling problem problem* and the *prefetch scheduling problem* respectively. The (online) *output scheduling (or queued writing) problem* takes as input a fixed size pool of m (initially empty) memory buffers for storing blocks, and the sequence $\langle w_0, w_1, \dots, w_{L-1} \rangle$ of block *write requests* as they are issued. Each write request is labeled with the disk it will use. The resulting schedule specifies when the blocks are output. The buffer pool can be used to reorder the outputs with respect to the logical writing order given by Σ so that the total number of output steps is minimized.

The (offline) *prefetch scheduling problem* takes as input a fixed size pool of m (empty) memory buffers for storing blocks, and the sequence $\langle r_0, r_1, \dots, r_{L-1} \rangle$ of distinct block *read requests* that will be issued. Each read request is labeled with the disk it will use. The resulting *prefetch schedule* specifies when the blocks should be fetched so that they can be consumed by the application in the right order.

The central theme in this paper is the newly discovered duality between these two problems. Roughly speaking, an output schedule corresponds to a prefetch schedule with reversed time axis and vice versa. We illustrate how computations in one domain can be analyzed via duality with computations in the other domain.

Sect. 2 introduces the duality principle formally for the case of distinct blocks to be written or read (*write-once* and *read-once* scheduling). Then Sect. 3 derives an optimal write-once output scheduling algorithm and applies the duality principle to obtain an optimal read-once prefetch scheduling algorithm.

Even an optimal schedule might use parallel disks very inefficiently because for difficult inputs most disks might be idle most of the time. In Sect. 4 we therefore give performance guarantees for randomly placed data and for arbitrarily interleaved accesses to a number of data streams. In particular, we discuss the following allocation strategies:

Fully Randomized (FR): Each block is allocated to a random disk.
Striping (S): Consecutive blocks of a stream are allocated to consecutive disks in a simple, round-robin manner.
Simple Randomized (SR): Striping where the disk selected for the first block of each stream is chosen randomly.

Randomized Cycling (RC): Each stream i chooses a random permutation π_i of disk numbers and allocates the j-th block of stream i on disk $\pi_i(j \bmod D)$.

In Sect. 5 we relax the restriction that blocks are accessed only once and allow repetitions (*write-many* and *read-many* scheduling). Again we derive a simple optimal algorithm for the writing case and obtain an optimal algorithm for the reading case using the duality principle. A similar result has recently been obtained by Kallahalla and Varman [11] using more complicated arguments.

Finally, in Sect. 6 we apply the results from Sects. 3 and 4 to parallel disk sorting. Results on online writing translate into improved sorting algorithms using the distribution paradigm. Results on offline reading translate into improved sorting algorithms based on multi-way merging. By appending a 'D' for distribution sort or an 'M' for mergesort to an allocation strategy (FR, S, SR, RC) we obtain a descriptor for a sorting algorithm (FRD, FRM, SD, SM, SRD, SRM, RCD, RCM). This notation is an extension of the notation used in [18]. RCD and RCM turn out to be particularly efficient. Let

$$\text{Sort}(N) = \frac{N}{DB}\left(1 + \log_{\frac{M}{B}} \frac{N}{M}\right)$$

and note that $2 \cdot \text{Sort}(N)$ appears to be the lower bound for sorting N elements on D disks [1]. Our versions of RCD and RCM are the first algorithms that provably match this bound up to a lower order term $\mathcal{O}(BD/M)\text{Sort}(N)$. The good performance of RCM is particularly interesting. The question of whether there is a simple variant of mergesort that is asymptotically optimal has been open for some time.

Related Work

Prefetching and caching has been intensively studied and can be a quite difficult problem. Belady [5] solves the caching problem for a single disk using our machine model. Cao et al. [7] propose a model that additionally allows overlapping of I/O and computation. Albers et al. [2] were the first to find an optimal polynomial time offline algorithm for the single-disk case in this model but it does not generalize well to multiple disks. Kimbrel and Karlin [12] devised a simple algorithm called *reverse aggressive* that obtains good approximations in the parallel disk case if the buffer pool is large and the failure penalty F is small. However, in our model, which corresponds to $F \to \infty$, the approximation ratio that they show goes to infinity. Reverse aggressive is very similar to our algorithm so that it is quite astonishing that the algorithm is optimal in our model. Kallahalla and Varman [10] studied online prefetching of read-once sequences for our model. They showed that very large lookahead $L \gg mD$ is needed to obtain good competitiveness against an optimal offline algorithm. They proposed an $\mathcal{O}(L^2 D)$ time algorithm with this property, and yielding optimal schedules for the offline case. A practical disadvantage of this algorithm is that some blocks may be fetched and discarded several times before they can be delivered to the application.

There is less work on performance guarantees. A (slightly) suboptimal writing algorithm is analyzed in [16] for FR allocation and extended to RC-allocation in [18]. These results are the basis for our results in Sect. 4. For reading there is an algorithm for SR allocation that is close to optimal if $m \gg D \log D$ [3].

There are asymptotically optimal deterministic algorithms for external sorting [15], but the constant factors involved make them unattractive in practice. Barve et al. [3] introduced a simple and efficient randomized sorting algorithm called *Simple Randomized Mergesort (SRM)*. For each run, SRM allocates blocks to disks using the SR allocation discipline. SRM comes within $\gamma \cdot \mathrm{Sort}(N)$ of the apparent lower bound if $M/B = \Omega\left(D \log(D)/\gamma^2\right)$ but for $M = o(D \log D)$ the bound proven is not asymptotically optimal. It was an open problem whether SRM or another variant of striped mergesort could be asymptotically optimal for small internal memory. Knuth [13, Exercise 5.4.9-31] gives the question of a tight analysis of SR a difficulty of 48 on a scale between 0 and 50.

To overcome the apparent difficulty of analyzing SR, Vitter and Hutchinson [18] analyzed RC allocation, which provides more randomness but retains the advantages of striping. RCD is an asymptotically optimal distribution sort algorithm for multiple disks that allocates successive blocks of a bucket to the disks according to the RC discipline and adapts the approach and analysis of Sanders, Egner, and Korst [16] for write scheduling of blocks. However, the question remained whether such a result can be obtained for mergesort and how close one can come to the lower bound for small internal memory.

2   The Duality Principle

Duality is a quite simple concept once the model is properly defined. Therefore, we start with a more formal description of the model:

Our machine model is the parallel disk model of Vitter and Shriver [19] with a single[1] processor, D disks and an internal memory of size M. All blocks have the same size B. In one *I/O step*, one block on each disk can be accessed in a synchronized fashion. We consider either a queued writing or a buffered prefetching arrangement, where a pool of m block buffers is available to the algorithm (see Fig. 1).

A *write-once output scheduling problem* is defined by a sequence $\Sigma = \langle b_0, \ldots, b_{L-1} \rangle$ of distinct blocks. Let $\mathrm{disk}(b_i)$ denote the disk on which block b_i is located. An application process *writes* these blocks in the order specified by Σ. We use the term *write* for the logical process of moving a block from the responsibility of the application to the responsibility of the scheduling algorithm. The scheduling algorithm orchestrates the physical *output* of these blocks to disks. An output schedule is specified by giving a function $\mathrm{oStep} : \{b_0, \ldots, b_{L-1}\} \to \mathbb{N}$ that specifies for each disk block $b_i \in \Sigma$ the time step when it will be output. An output schedule is *correct* if the following conditions hold: (i) No disk is referenced more than once in a single time step, i.e., if $i \neq j$ and $\mathrm{disk}(b_i) = \mathrm{disk}(b_j)$

---

[1] A generalization our results to multiple processors is relatively easy as long as data exchange between processors is much faster than disk access.

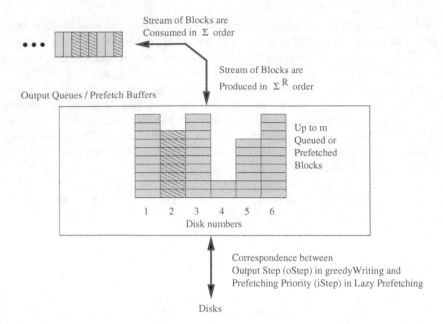

Fig. 1. Duality between the prefetching priority and the output step. The hashed blocks illustrate how the blocks of disk 2 might be distributed.

then $\text{oStep}(b_i) \neq \text{oStep}(b_j)$. (ii) The buffer pool is large enough to hold all the blocks b_j that are written before a block b_i but not output before b_i, i.e.,

$$\forall 0 \leq i < L : \text{o}\mathcal{B}\text{acklog}(b_i) := |\{j < i : \text{oStep}(b_j) \geq \text{oStep}(b_i)\}| < m \ .$$

The number of steps needed by an output schedule is $T = \max_{0 \leq i < L} \text{oStep}(b_i)$. A schedule is *optimal* if it minimizes T among all correct schedules.

It will turn out that our write-once output scheduling algorithms even work if they are given the blocks *online*, i.e., one at a time without specifying Σ explicitly.

A *read-once prefetch scheduling problem* is defined analogously. Now *reading* means the logical process of moving a block from the responsibility of the scheduling algorithm to the application and *fetching* means the physical disk access. A prefetch schedule is defined using a function $\text{iStep} : \{b_0, \dots, b_{L-1}\} \to \mathbb{N}$. The limited buffer pool size requires the correctness condition

$$\forall 0 \leq i < L : \text{i}\mathcal{B}\text{acklog}(b_i) := |\{j > i : \text{iStep}(b_j) \leq \text{iStep}(b_i)\}| < m$$

(all blocks b_j that are fetched no later than a block b_i but consumed after b_i must be buffered).

It will turn out that our prefetch scheduling algorithms work *offline*, i.e., they need to know Σ in advance.

The following theorem shows that the reading and writing not only have similar models but are equivalent to each other in a quite interesting sense:

Theorem 1. (Duality Principle) *Consider any sequence $\Sigma = \langle b_0, \ldots, b_{L-1} \rangle$ of distinct write requests. Let* oStep *denote a correct output schedule for Σ that uses T output steps. Then we get a correct prefetch schedule* iStep *for $\Sigma^R = \langle b_{L-1}, \ldots, b_0 \rangle$ that uses T fetch steps by setting* $\text{iStep}(b_i) = T - \text{oStep}(b_i) + 1$.*

Vice versa, every correct prefetch schedule iStep *for Σ^R that uses T fetch steps yields a correct output schedule* $\text{oStep}(b_i) = T - \text{iStep}(b_i) + 1$ *for Σ, using T output steps.*

Proof. For the first part, consider $\text{iStep}(b_i) = T - \text{oStep}(b_i) + 1$. The resulting fetch steps are between 1 and T and all blocks on the same disk get different fetch steps. It remains to show that $\text{iBacklog}(b_i) < m$ for $0 \le i < L$. With respect to Σ^R, we have

$$
\begin{aligned}
\text{iBacklog}(b_i) &= |\{j < i : \text{iStep}(b_j) \le \text{iStep}(b_i)\}| \\
&= |\{j < i : T - \text{oStep}(b_j) + 1 \le T - \text{oStep}(b_i) + 1\}| \\
&= |\{j < i : \text{oStep}(b_j) \ge \text{oStep}(b_i)\}| \quad .
\end{aligned}
$$

the latter value is $\text{oBacklog}(b_i)$ with respect to Σ and hence smaller than m.

The proof for the converse case is completely analogous. $\qquad\square$

3   Optimal Write-Once and Read-Once Scheduling

We give an optimal algorithm for writing a write-once sequence, prove its optimality and then apply the duality principle to transform it into a read-once prefetching algorithm.

Consider the following algorithm *greedyWriting* for writing a sequence $\Sigma = \langle b_0, \ldots, b_{L-1} \rangle$ of distinct blocks. *Let Q denote the set of blocks in the buffer pool, so initially $Q = \emptyset$. Let $Q_d = \{b \in Q : \text{disk}(b) = d\}$. Write the blocks b_i in sequence as follows: (1) If $|Q| < m$ then simply insert b_i into Q. (2) Otherwise, each disk with $Q_d \neq \emptyset$ outputs the block in Q_d that appears first in Σ. The blocks output are then removed from Q and b_i is inserted into Q. (3) Once all blocks are written the queues are flushed, i.e., additional output steps are performed until Q is empty,*

Any schedule where blocks are output in arrival order on each disk, is called a *FIFO schedule*. The following lemma tells us that it is sufficient to consider FIFO schedules when we look for optimal schedules. The proof is based on transforming a non-FIFO schedule into a FIFO schedule by exchanging blocks in the schedule of a disk that are output out of order.

Lemma 1. *For any sequence of blocks Σ and every correct output schedule* oStep' *there is a FIFO output schedule* oStep *consisting of at most the same number of output steps.*

Algorithm greedyWriting is one way to compute a FIFO schedule. The following lemma shows that greedyWriting outputs every block as early as possible.

Lemma 2. *For any sequence of blocks Σ and any FIFO output schedule* oStep$'$, *let* oStep *denote the schedule produced by algorithm greedyWriting. Then for all $b_i \in \Sigma$, we have* oStep$(b_i) \leq$ oStep$'(b_i)$.

Proof. (Outline) The proof is by induction over the number of blocks. There are two nontrivial cases. One case corresponds to the situation where the output step of a block immediately follows an output of a previous block on the same disk. The other case corresponds to the situation where no earlier step is possible because otherwise its o$\mathcal{B}$acklog would be too large.

Combining Lemmas 1 and 2 we see that greedyWriting gives us optimal schedules for write-once sequences:

Theorem 2. *Algorithm greedyWriting gives a correct, minimum length output schedule for any write-once reference sequence Σ.*

Combining the duality principle and the optimality of greedyWriting, we get an optimal algorithm for read-once prefetching that we call *lazy prefetching*:

Corollary 1. *An optimal prefetch schedule* iStep *for a sequence Σ can be obtained by using greedyWriting to get an output schedule* oStep *for Σ^R and setting* iStep$(b_i) = T -$ oStep$(b_i) + 1$.

Note that the schedule can be computed in time $\mathcal{O}(L + D)$ using very simple data structures.

4   How Good Is Optimal?

When we are processing several streams concurrently, the knowledge that we have an optimal prefetching algorithm is often of little help. We also want to know "how good is optimal?" In the worst case, all requests may go to the same disk and no prefetching algorithm can cure the dreadful performance caused by this bottleneck. However, the situation is different if blocks are allocated to disks using striping, randomization[2] or both.

Theorem 3. *Consider a sequence of L block requests. and a buffer pool of size $m \geq D$ blocks. The number of I/O steps needed by greedyWriting or lazy prefetching is given by the following bounds, depending on the allocation discipline. For striping and randomized cycling, an arbitrary interleaving of sequential accesses to S sequences is allowed.*

$$\text{Striping: } \tfrac{L}{D} + S, \quad \text{if } m > S(D-1);$$

$$\text{Fully Random (FR): } \left(1 + \mathcal{O}\left(\tfrac{D}{m}\right)\right)\tfrac{L}{D} + \mathcal{O}\left(\tfrac{m}{D}\log m\right) \quad \text{(expected);}$$

$$\text{Randomized Cycling (RC): } \left(1 + \mathcal{O}\left(\tfrac{D}{m}\right)\right)\tfrac{L}{D} + \min\left\{S + \tfrac{L}{D}, \tfrac{m}{D}\log m\right\} \text{(expected)}$$

---

[2] In practice, this will be done using simple hash functions. However, for the analysis we assume that we have a perfect source of randomness.

Proof. (Outline) The bound for striped writing is based on the observation that $L/D + S$ is the maximum number of blocks to be handled by any disk and that the oBacklog of any block can never exceed m if $m > S(D - 1)$.

For fully random placement the key idea is that greedyWriting dominates the "throttled" algorithm of [16], which admits only $(1 - \Theta(D/M))D$ blocks per output step into the queues.

The bound for RC is a combination of the two previous bounds. The bound for FR applies to RC writing using the observation of [18] that the throttled algorithm of [16] performs at least as well for RC as for FR.

The results for writing transfer to offline prefetching via duality. For the RC case we also need the observation that the reverse of a sequence using RC is indistinguishable from a nonreversed sequence.   □

For writing, the trailing additive term for each case enumerated in Theorem 3 can be dropped if the final contents of the buffer pool is not flushed.

5 Integrated Caching and Prefetching

We now relax the condition that the read requests in Σ are for distinct blocks, permitting the possibility of saving disk accesses by keeping previously accessed blocks in memory. For this *read-many* problem, we get a tradeoff for the use of the buffer pool because it has to serve the double purposes of keeping blocks that are accessed multiple times, and decoupling physical and logical accesses to equalize transient load imbalance of the disks. We define the *write-many* problem in such a way that the duality principle from Theorem 1 transfers: *The latest instance of each block must be kept either on its assigned disk, or in the buffer pool. The final instance of each block must be written to its assigned disk.*[3]

We prove that the following offline algorithm *manyWriting* minimizes the number of output operations for the write-many problem: *Let Q and Q_d be defined as for greedyWriting. To write block b_i, if $b_i \in Q$, the old version is overwritten in its existing buffer. Otherwise, if $|Q| < m$, b_i is inserted into Q. If this also fails, an output step is performed before b_i is inserted into Q. The output analogue of Belady's* MIN *rule [5] is used on each disk, i.e., each disk with $Q_d \neq \emptyset$ outputs the block in Q_d that is accessed again farthest in the future.*

Applying duality, we also get an optimal algorithm for integrated prefetching and caching of a sequence Σ: using the same construction as in Cor. 1 we get an optimal prefetching and caching schedule. It remains to prove the following theorem:

Theorem 4. *Algorithm manyWriting solves the* write-many *problem with the fewest number of output steps.*

---

[3] The requirement that the latest versions have to be kept might seem odd in an offline setting. However, this makes sense if there is a possibility that there are reads at unknown times that need an up-to-date version of a block.

Proof. (Outline) We generalize the optimality proof of Belady's algorithm by Borodin and El-Yaniv [6] to the case of writing and multiple disks. Let $\Sigma = \langle b_0, \ldots, b_{L-1} \rangle$ be any sequence of blocks to be written. The proof is based on the following claim.

Claim: Let ALG be any algorithm for cached writing. Let d denote a fixed disk. For any $0 \leq i < L$ it is possible to construct an offline algorithm ALG_i that satisfies the following properties: (i) ALG_i processes the first $i - 1$ block write requests exactly as ALG does. (ii) If block b_i is written immediately after output step s, then immediately before step s we had $b_i \notin Q$ and $|Q| = m$. (iii) If b_i is written immediately after output step s, then ALG_i performs this output according to the MIN rule on disk d. (iv) ALG_i takes no more steps than ALG.

Once this claim is established, we can transform an optimal scheduling algorithm OPT into an algorithm S that uses the MIN rule by iteratively applying the claim for each disk and each $0 \leq i < L$ without increasing the number of output steps used.

To prove the claim we modify any algorithm ALG to get an algorithm ALG_i that fulfills all the properties. If property (ii) is violated, it suffices to write b_i earlier. If property (iii) is violated, the output step s preceding the write of b_i is modified on disk d to follow the MIN rule. Suppose, ALG_i outputs block b in step s. It remains to explain how ALG_i can mimic ALG in the subsequent steps despite this difference in step s. A problem can only arise if ALG later overwrites the current version of b. ALG_i exploits the fact that ALG either outputs nothing or something that is accessed again before the next access to b. Either way, ALG_i can arrange to have an unused buffer block available when ALG overwrites the current version of b. □

6   Application to Sorting

Optimal algorithms for read-once prefetching or write-once output scheduling can be used to analyze or improve a number of interesting parallel disk sorting algorithms. We start by discussing multiway mergesort using randomized cycling allocation (RCM) in some detail and then survey a number of additional results.

Multiway mergesort is a frequently used external sorting algorithm. We describe a variant that is similar to the SRM algorithm in [3]. Originally the N input elements are stored as a single data stream using any kind of striping. During *run formation* the input is read in chunks of size M, that are sorted internally and then written out in runs allocated using RC allocation. Neglecting trivial rounding issues, run formation is easy to do using $2N/(DB)$ I/O steps. By investing another $\mathcal{O}\big(N/(DB^2)\big)$ I/O steps we can keep *triggers*, the largest keys of each block, in a separate file. Then we set aside a buffer pool of size $m = cD$ for some constant c and perform $\lceil \log_{M/B - \mathcal{O}(D)} \frac{N}{M} \rceil$ *merge phases*. In a merge phase, groups of $k = \frac{M}{B} - \mathcal{O}(D)$ runs are merged into new sorted runs, i.e., after the last merge phase, only one sorted run is left. Merging k runs of size sB can be performed using s block reads by keeping one block of each run in the internal memory of the sorting application. The order of these reads for an entire phase can be exactly predicted using the trigger information and $\mathcal{O}(N/B^2)$ I/Os for merging trigger files [3]. Hence, if we use optimal prefetching, Theorem 3 gives an upper bound of $(1 + \mathcal{O}(1/c))\frac{N}{BD} + \cdots$ for the number of fetch steps of a phase. The number of output steps for a phase is $N/(BD)$ if we have an additional output buffer of D blocks. The final result is written using any striped allocation strategy, i.e., the application calling the sorting routing need not be able to handle RC allocation. We can write the resulting total number of I/O steps as $\mathrm{Sort}_{m+D}^{2+\mathcal{O}(1/c),\min(\frac{N}{BD},\frac{\log D}{\mathcal{O}(\gamma)})}(N)$ where

$$\mathrm{Sort}_m^{a,f}(N) = \frac{2N}{DB} + a \cdot \frac{N}{DB} \cdot \left\lceil \log_{\frac{M}{B}-m} \frac{N}{M} \right\rceil + f + o\left(\frac{N}{DB}\right).$$

Table 1 compares a selection of sorting algorithms using this generalized form of the I/O bound for parallel disk sorting. (In the full paper we present additional results for example for FR allocation.) The term $\frac{2N}{DB}$ represents the reading and writing of the input and the final output respectively. The factor a is a constant that dominates the I/O complexity for large inputs. Note that for $a = 2$ and $f = m = 0$ this expression is the apparent lower bound for sorting. The additive offset f may dominate for small inputs. The reduction of the memory by m blocks in the base of the logarithm is due to memory that is used for output or prefetching buffer pools outside the merging or distribution routines, and hence reduces the number of data streams that can be handled concurrently. One way to interpret m is to view it as the amount of *additional* memory needed to match the performance of the algorithm on the multihead I/O model [1] (where load balancing disk accesses is not an issue).[4]

Even without any randomization, Theorem 3 shows that mergesort with deterministic striping and optimal prefetching (SM) is at least as efficient as the common practice of using superblock striping. However, both algorithms achieve good performance only if a lot of internal memory is available.

Using previous work on distribution sort and the duality between prefetching and writing, all results obtained for mergesort can be extended to distribution sort (e.g., SD, SRD, FRD, RCD+). There are several sorting algorithms based on the distribution principle, e.g. radix sort. The bounds given here are based on a generalization of quicksort where $k - 1$ splitter elements are chosen to split an unsorted input stream into k approximately equal sized output streams with disjoint ranges of keys. After $\lceil \log_{M/B-\mathcal{O}(D)} \frac{N}{M} \rceil$ splitting phases, the remaining streams can be sorted using internal sorting.

A simple variant of distribution sort with randomized cycling (RCD) was already analyzed in [18]. The new variant, RCD+, has some practical improvements (fewer tuning parameters, simpler application interface) and, it turns out

---

[4] If we assume a fixed memory size we cannot discriminate between some of the algorithms using the abstract I/O model. One algorithm may have a smaller factor a yet need an extra distribution or merging phase for some input sizes N. In practice, one could use a smaller block size for these input sizes. The abstract I/O model does not tell us how this affects the total I/O time needed.

Table 1. Summary of Main Results for I/O Complexity of Parallel Disk Sorting Algorithms. Algorithms with **boldface** names are asymptotically optimal: M/D = Merge /Distribution sort. SM/SD = merge / distribution sort with any striping (S) allocation. SRM and SRD use Simple Randomized striping (SR). RCD, RCD+ and RCM use Randomized Cycling (RC) allocation.

	$\text{Sort}_m^{a,f}(N)$	I/Os	Algorithm	Source
a	f	$\Theta(m)$		
Deterministic algorithms				
$2+\gamma$	0	$(2D)^{1+\frac{2}{\gamma}}$	M, superblock striping	
$2+\gamma$	0	$(2D)^{1+\frac{2}{\gamma}}$	SM	here
$2+\gamma$	0	$(2D)^{1+\frac{2}{\gamma}}$	SD	here
Randomized algorithms				
$2+\gamma$	0	$D\log(D)/\gamma^2$	SRM	[3]
$2+\gamma$	0	$D\log(D)/\gamma^2$	SRD	here
$3+\gamma$	0	D/γ	**RCD**	[18]
$2+\gamma$	$\min(\frac{N}{BD}, \log(D)/\mathcal{O}(\gamma))$	D/γ	**RCM**	here
$2+\gamma$	0	D/γ	**RCD+**	here

that the additive term f can also be eliminated. Using a careful formulation of the algorithmic details it is never necessary to flush the write buffers. All in all, RCD+ is currently the parallel disk sorting algorithm with the best I/O performance bounds known.

Acknowledgments. We would like to thank Jeffrey Chase, Andreas Crauser, S. Mitra, Nitin Rajput, Erhard Rahm, and Berthold Vöcking for valuable discussions.

References

1. A. Aggarwal and J. S. Vitter. The Input/Output complexity of sorting and related problems. *Communications of the ACM*, 31(9):1116–1127, 1988.
2. S. Albers, N. Garg, and S. Leonardi. Minimizing stall time in single and parallel disk systems. In *Proceedings of the 30th Annual ACM Symposium on Theory of Computing (STOC-98)*, pages 454–462, New York, May 23–26 1998. ACM Press.
3. R. D. Barve and J. S. Vitter. A simple and efficient parallel disk mergesort. In *Proceedings of the 11th Annual ACM Symposium on Parallel Algorithms and Architectures*, pages 232–241, St. Malo, France, June 1999.
4. Rakesh D. Barve, Edward F. Grove, and Jeffrey Scott Vitter. Simple randomized mergesort on parallel disks. *Parallel Computing*, 23(4):601–631, 1997.
5. A. L. Belady. A study of replacement algorithms for virtual storage computers. *IBM Systems Journal*, 5:78–101, 1966.
6. Allan Borodin and Ran El-Yaniv. *Online Computation and Competitive Analysis*. Cambridge University Press, Cambridge, 1998.
7. Pei Cao, Edward W. Felten, Anna R. Karlin, and Kai Li. Implementation and performance of integrated application-controlled file caching, prefetching and disk scheduling. *ACM Transactions on Computer Systems*, 14(4):311–343, Nov. 1996.

8. F. Dehne, W. Dittrich, and D. Hutchinson. Efficient external memory algorithms by simulating coarse-grained parallel algorithms. In *Proceedings of the 9th ACM Symposium on Parallel Algorithms and Architectures*, pages 106–115, June 1997.

9. F. Dehne, D. Hutchinson, and A. Maheshwari. Reducing I/O complexity by simulating coarse grained parallel algorithms. In *Proc. of the Intl. Parallel Processing Symmposium*, pages 14–20, April 1999.

10. M. Kallahalla and P. J.Varman. Optimal read-once parallel disk scheduling. In *IOPADS*, pages 68–77, 1999.

11. M. Kallahalla and P.J. Varman. Optimal prefetching and caching for parallel I/O systems. In *Proc. of the ACM Symposium on Parallel Algorithms and Architectures*, 2001. To appear.

12. Tracy Kimbrel and Anna R. Karlin. Near-optimal parallel prefetching and caching. *SIAM Journal on Computing*, 29(4):1051–1082, 2000.

13. D. E. Knuth. *The Art of Computer Programming — Sorting and Searching*, volume 3. Addison Wesley, 2nd edition, 1998.

14. M. H. Nodine and J. S. Vitter. Deterministic distribution sort in shared and distributed memory multiprocessors. In *Proceedings of the 5th Annual ACM Symposium on Parallel Algorithms and Architectures*, pages 120–129, Velen, Germany, June–July 1993.

15. M. H. Nodine and J. S. Vitter. Greed Sort: An optimal sorting algorithm for multiple disks. *Journal of the ACM*, 42(4):919–933, July 1995.

16. P. Sanders, S. Egner, and J. Korst. Fast concurrent access to parallel disks. In *11th ACM-SIAM Symposium on Discrete Algorithms*, pages 849–858, 2000.

17. J. S. Vitter. External memory algorithms and data structures: Dealing with massive data. *ACM Computing Surveys*, in press. An earlier version entitled "External Memory Algorithms and Data Structures" appeared in *External Memory Algorithms and Visualization*, DIMACS Series in Discrete Mathematics and Theoretical Computer Science, American Mathematical Society, 1999, 1–38.

18. J. S. Vitter and D. A. Hutchinson. Distribution sort with randomized cycling. In *Proceedings of the 12th ACM-SIAM Symposium on Discrete Algorithms*, Washington, January 2001.

19. J. S. Vitter and E. A. M. Shriver. Algorithms for parallel memory I: Two-level memories. *Algorithmica*, 12(2–3):110–147, 1994.

Online Bin Coloring

Sven O. Krumke[*1], Willem E. de Paepe[2], Jörg Rambau[1], and Leen Stougie[**3]

[1] Konrad-Zuse-Zentrum für Informationstechnik Berlin
Department Optimization, Takustr. 7, 14195 Berlin-Dahlem, Germany.
{krumke,rambau}@zib.de
[2] Department of Technology Management, Technical University of Eindhoven,
P. O. Box 513, 5600MB Eindhoven, The Netherlands.
w.e.d.paepe@tm.tue.nl
[3] Department of Mathematics, Technical University of Eindhoven, P. O. Box 513,
5600MB Eindhoven, The Netherlands and Centre for Mathematics and Computer
Science (CWI), P. O. Box 94079, NL-1090 GB Amsterdam, The Netherlands.
leen@win.tue.nl

Abstract. We introduce a new problem that was motivated by a (more complicated) problem arising in a robotized assembly environment. The *bin coloring problem* is to pack unit size colored items into bins, such that the maximum number of different colors per bin is minimized. Each bin has size $B \in \mathbb{N}$. The packing process is subject to the constraint that at any moment in time at most $q \in \mathbb{N}$ bins are partially filled. Moreover, bins may only be closed if they are filled completely. An online algorithm must pack each item without knowledge of any future items.

We investigate the existence of competitive online algorithms for the bin coloring problem. We prove an upper bound of $3q - 1$ and a lower bound of $2q$ for the competitive ratio of a natural greedy-type algorithm, and show that surprisingly a trivial algorithm which uses only one open bin has a strictly better competitive ratio of $2q - 1$. Moreover, we show that any deterministic algorithm has a competitive ratio $\Omega(q)$ and that randomization does not improve this lower bound even when the adversary is oblivious.

1   Introduction

One of the commissioning departments in the distribution center of Herlitz PBS AG, Falkensee, one of the main distributors of office supply in Europe, is devoted to greeting cards. The cards are stored in parallel shelving systems. Order pickers on automated guided vehicles collect the orders from the storage systems, following a circular course through the shelves. At the loading zone, which can hold q vehicles, each vehicle is logically "loaded" with B orders which arrive online. The goal is to avoid congestion among the vehicles (see [1] for details). Since the vehicles are unable to pass each other and the "speed" of a vehicle

---

[*] Research supported by the German Science Foundation (DFG, grant GR 883/9-10)
[**] Supported by the TMR Network DONET of the European Community ERB TMRX-CT98-0202

F. Meyer auf der Heide (Ed.): ESA 2001, LNCS 2161, pp. 74–85, 2001.

is correlated to the number of different stops it must make, this motivates to assign the orders in such a way that the vehicles stop as few times as possible.

The above situation motivated the following *bin coloring problem*: One receives a sequence of unit size items $r_1, \ldots, r_m$ where each item has a *color* $r_i \in \mathbb{N}$, and is asked to pack them into bins with size B. The goal is to pack the items into the bins "most uniformly", that is, to minimize the maximum number of different colors assigned to a bin. The packing process is subject to the constraint that at any moment in time at most $q \in \mathbb{N}$ bins may be partially filled. Bins may only be closed if they are filled completely. (Notice that without these strict bounded space constraints the problem is trivial since in this case each item can be packed into a separate bin). In the *online version* of the problem, denoted by OLBCP, each item must be packed without knowledge of any future items. An online algorithm is *c-competitive*, if for all possible request sequences the maximum colors in the bins packed by the algorithm and the optimal offline solution is bounded by c. The OLBCP can be viewed as a variant of the bounded space binpacking problem in (see [3,4] for recent surveys on binpacking problems).

Our investigations of the OLBCP reveal a curiosity of competitive analysis: a truly stupid algorithm achieves essentially a (non-trivial) best possible competitive ratio whereas a seemingly reasonable algorithm performs provably worse in terms of competitive analysis.

We first analyze a natural greedy-type strategy, and show that this strategy has a competitive ratio no greater than $3q$ but no smaller than $2q$, where q is the maximum number of open bins. We show that a trivial strategy that only uses one open bin, has a strictly better competitive ratio of $2q - 1$. Then we show that surprisingly no deterministic algorithm can be substantially better than the trivial strategy. More specifically, we prove that no deterministic algorithm can, in general, have a competitive ratio less than q. Even more surprising, the lower bound of q for the competitive ratio continues to hold for randomized algorithms against an oblivious adversary. Finally, not even "resource augmentation", which means that the online algorithm is allowed to use a fixed number $q' \geq q$ of open bins can help to overcome the lower bound of $\Omega(q)$ on the competitive ratio.

2   Problem Definition

Definition 2.1 (Online Bin Coloring Problem). *In the* Online Bin Coloring Problem (OLBCP$_{B,q}$) *with parameters* $B, q \in \mathbb{N}$ *($B, q \geq 2$), one is given a sequence* $\sigma = r_1, \ldots, r_m$ *of unit size items (requests), each with a color* $r_i \in \mathbb{N}$, *and is asked to pack them into bins with size* B, *that is, each bin can accommodate exactly* B *items. The packing is subject to the following constraints:*

1. *The items must be packed according to the order of their appearance, that is, item i must be packed before item k for all* $i < k$.
2. *At most q partially filled bins may be open to further items at any point in the packing process.*
3. *A bin may only be closed if it is filled completely, i.e., if it has been assigned exactly B items.*

The objective is to minimize the maximum number of different colors assigned to a bin. An online algorithm for $\text{OLBCP}_{B,q}$ must pack each item r_i (irrevocably) without knowledge of requests r_k with $k > i$.

In the sequel it will be occasionally helpful to use the following view on the bins used by an arbitrary algorithm ALG to process an input sequence σ. Each open bin has an *index* x, where $1 \leq x \leq q$. Each time a bin with index x is closed and a new bin is opened the new bin will also have index x. If no confusion can occur, we will refer to a bin with index x as *bin x*.

We denote by $\text{ALG}(\sigma)$ the objective function value of the solution produced by algorithm ALG on input σ. OPT denotes an optimal offline algorithm which has complete knowledge about the input sequence σ in advance. However, the packing must still obey the constraints 1 to 3 specified in Definition 2.1.

Definition 2.2. *A deterministic online algorithm* ALG *for* $\text{OLBCP}_{B,q}$ *is called c-competitive, if there exists a constant c such that* $\text{ALG}(\sigma) \leq c \cdot \text{OPT}(\sigma)$ *holds for any request sequence σ.*

The competitive ratio of an algorithm ALG is the smallest number c such that ALG is c-competitive. The size of the bins B is a trivial upper bound on the competitive ratio of *any* algorithm for $\text{OLBCP}_{B,q}$.

A randomized online algorithm is a probability distribution over a set of deterministic online algorithms. The objective value produced by a randomized algorithm is therefore a random variable. In this paper we analyze the performance of randomized online algorithms only against an *oblivious adversary*. An oblivious adversary does not see the realizations of the random choices made by the online algorithm and therefore has to generate a request sequence in advance. We refer to [2] for details on the various adversary models.

Definition 2.3. *A randomized online algorithm* RALG *is c-competitive against an oblivious adversary if* $\mathbb{E}\left[\text{RALG}(\sigma)\right] \leq c \cdot \text{OPT}(\sigma)$ *for any request sequence σ.*

3   The Algorithm greedyfit

In this section we introduce a natural greedy-type strategy, which we call GREEDYFIT, and show that the competitive ratio of this strategy is at most $3q$ but no smaller than $2q$ (provided the capacity B is sufficiently large).

GREEDYFIT: If upon the arrival of request r_i the color r_i is already contained in one of the currently open bins, say bin b, then put r_i into bin b. Otherwise put item r_i into a bin that contains the least number of different colors (which means opening a new bin if currently less than q bins are non-empty).

The analysis of the competitive ratio of GREEDYFIT is essentially via a pigeon-hole principle argument. We first show a lower bound on the number of bins that *any* algorithm can use to distribute a the items in a contiguous subsequence and then relate this number to the number of colors in the input sequence.

Lemma 3.1. *Let* $\sigma = r_1, \ldots, r_m$ *be any request sequence and let* $\sigma' = r_i, \ldots, r_{i+\ell}$ *be any contiguous subsequence of* σ. *Then any algorithm packs the items of* σ' *into at most* $2q + \lfloor (\ell - 2q)/B \rfloor$ *different bins.*

Proof. Let ALG be any algorithm and let $b_1, \ldots, b_t$ be the set of open bins for ALG just prior to the arrival of the first item of σ'. Denote by $f(b_j) \in \{1, \ldots, B-1\}$ the empty space in bin b_j at that moment in time. To close an open bin b_j, ALG needs $f(b_j)$ items. Opening and closing an additional new bin needs B items. To achieve the maximum number of bins ($\geq 2q$), ALG must first close each open bin and put at least one item into each newly opened bin. From this moment in time, opening a new bin requires B new items. It follows that the maximum number of bins ALG can use is bounded from above as claimed in the lemma. $\square$

Theorem 3.2. GREEDYFIT *is c-competitive for* $\text{OLBCP}_{B,q}$ *with* $c = \min\{2q + \lfloor (qB - 3q + 1)/B \rfloor, B\} \leq \min\{3q - 1, B\}$.

Proof. Let σ be any request sequence and suppose $\text{GREEDYFIT}(\sigma) = w$. It suffices to consider the case $w \geq 2$. Let s be minimum with the property that $\text{GREEDYFIT}(r_1, \ldots, r_{s-1}) = w - 1$ and $\text{GREEDYFIT}(r_1, \ldots, r_s) = w$. By the construction of GREEDYFIT, after processing $r_1, \ldots, r_{s-1}$ each of the currently open bins must contain exactly $w - 1$ different colors. Moreover, since $w \geq 2$, after processing additionally request r_s, GREEDYFIT has exactly q open bins (where as an exception we count here the bin where r_s is packed as open even if by this assignment it is just closed). Denote those bins by $b_1, \ldots, b_q$.

Let bin b_j be the bin among $b_1, \ldots, b_q$ that has been opened last by GREEDYFIT. Let r'_s be the first item that was assigned to b_j. Then, the subsequence $\sigma' = r_{s'}, \ldots, r_s$ consists of at most $qB - (q-1)$ items, since between $r_{s'}$ and r_s no bin is closed and at the moment $r_{s'}$ was processed, $q - 1$ bins already contained at least one item. Moreover, σ' contains items with at least w different colors. By Lemma 3.1 OPT distributes the items of σ' into at most $2q + \lfloor (qB - 3q + 1)/B \rfloor$ bins. Consequently, $\text{OPT}(\sigma) \geq \frac{w}{2q + \lfloor (qB - 3q + 1)/B \rfloor}$. $\square$

We continue to prove a lower bound on the competitive ratio of GREEDYFIT.

Theorem 3.3. GREEDYFIT *has a competitive ratio greater or equal to* $2q$ *for the* $\text{OLBCP}_{B,q}$ *if* $B \geq 2q^3 - q^2 - q + 1$.

Proof. We construct a request sequence σ that consists of a finite number M of phases in each of which qB requests are given. The sequence is constructed in such a way that after each phase the adversary has q empty bins.

Each phase consists of two steps. In the first step q^2 items are presented, each with a new color which has not been used before. In the second step $qB - q^2$ items are presented, all with a color that has occurred before. We will show that we can choose the items given in Step 2 of every phase such that the following properties hold for the bins of GREEDYFIT:

Property 1 The bins with indices $1, \ldots, q-1$ are never closed.

Property 2 The bins with indices $1, \ldots, q-1$ contain only items of different colors.

Property 3 There is an $M \in \mathbb{N}$ such that during Phase M GREEDYFIT assigns for the first time an item with a new color to a bin that already contains items with $2q^2 - 1$ different colors.

Property 4 There is an assignment of the items of σ such that no bin contains items with more than q different colors.

We analyze the behavior of GREEDYFIT by distinguishing between the items assigned to the bin (with index) q and the items assigned to bins (with indices) 1 through $q-1$. Let L_k be the set of colors of the items assigned to bins $1, \ldots, q-1$ and let R_k be the set of colors assigned to bin q during Step 1 of Phase k.

We now describe a general construction of the request sequence given in Step 2 of a phase. During Step 1 of Phase k there are items with $|R_k|$ different colors assigned to bin q. For the moment, suppose that $|R_k| \geq q$ (see Lemma 3.6 (iv)). We now partition the at most q^2 colors in $|R_k|$ into q disjoint non-empty sets $S_1, \ldots, S_q$. We give $qB - q^2 \geq 2q^2$ items with colors from $|R_k|$ such that the number of items with colors from S_j is $B - q$ for every j, and the last $|R_k|$ items all have a different color. GREEDYFIT will pack all items given in Step 2 into bin q (Lemma 3.6 (iii)). Hence bins $1, \ldots, q-1$ only get assigned items during Step 1, which implies the properties 1 and 2.

The adversary assigns the items of Step 1 such that every bin receives q items, and the items with colors in the color set S_j go to bin j. Clearly, the items in every bin have no more than q different colors. The items given in Step 2 can by construction of the sequence be assigned to the bins of the adversary such that all bins are completely filled, and the number of different colors per bin does not increase (this ensures that property 4 is satisfied). Due to lack of space we omit the proofs of the following lemmas.

Lemma 3.4. *At the end of Phase $k < M$, bin q of* GREEDYFIT *contains exactly $B - \sum_{j \leq k} |L_j|$ items, and this number is at least q^2.*   □

Corollary 3.5. *For any Phase $k < M$, bin q is never closed by* GREEDYFIT *before the end of Step 1 of Phase k.*   □

Lemma 3.6. *For $k \geq 1$ the following statements are true:*

(i)   *At the beginning of Phase k bin q of* GREEDYFIT *contains exactly the colors from R_{k-1} (where $R_0 := \emptyset$).*

(ii) *After Step 1 of Phase k, each of the bins $1, \ldots, q-1$ of* GREEDYFIT *contains at least $|R_k| + |R_{k-1}| - 1$ different colors.*

(iii) *In Step 2 of Phase k* GREEDYFIT *packs all items into bin q.*

(iv) $|R_k| \geq q$.   □

To this point we have shown that we can actually construct the sequence as suggested, and that the optimal offline cost on this sequence is no more than q.

Now we need to prove that there is a number $M \in \mathbb{N}$ such that after M phases there is a bin from GREEDYFIT that contains items with $2q^2$ different colors. We will do this by establishing the following lemma:

Lemma 3.7. *In every two subsequent Phases k and $k+1$, either $|L_k \cup L_{k+1}| > 0$ or bin q contains items with $2q^2$ different colors during one of the phases.* □

We can conclude from Lemma 3.7 that at least once every two phases the number of items in the bins 1 through $q - 1$ grows. Since these bins are never closed (property 1), and all items have a unique color (property 2), after a finite number M of phases, one of the bins of GREEDYFIT must contain items with $2q^2$ different colors. This completes the proof of the Theorem. □

4   The Trivial Algorithm onebin

This section is devoted to arguably the simplest (and most trivial) algorithm for the OLBCP, which surprisingly has a better competitive ratio than GREEDYFIT. Moreover, as we will see later that this algorithm achieves essentially the best competitive ratio for the problem.

ONEBIN: The next item r_i is packed into the (at most one) open bin. A new bin is opened only if the previous item has closed the previous bin by filling it up completely.

The proof of the upper bound on the competitive ratio of ONEBIN is along the same lines as that of GREEDYFIT.

Lemma 4.1. *Let $\sigma = r_1, \ldots, r_m$ be any request sequence. Then for $i \geq 0$ any algorithm packs the items $r_{iB+1}, \ldots, r_{(i+1)B}$ into at most $\min\{2q - 1, B\}$ bins.*

Proof. Omitted in this abstract.

Theorem 4.2. *Algorithm ONEBIN is c-competitive for the OLBCP$_{B,q}$ with $c = \min\{2q - 1, B\}$.*

Proof. Omitted in this abstract.

The competitive ratio proved in the previous theorem is tight as the following example shows. Let $B \geq 2q - 1$. First we give $(q - 1)B$ items, after which by definition ONEBIN has only empty bins. The items have q different colors, every color but one occurs $B-1$ times, one color occurs only $q-1$ times. The adversary assigns all items of the same color to the same bin, using one color per bin. After this, q items with all the different colors used before are requested. The adversary can now close $q - 1$ bins, still using only one color per bin. ONEBIN ends up with q different colors in its bin. Then $q-1$ items with new (previously unused) colors are given. The adversary can assign every item to an empty bin, thus still having only one different color per bin, while ONEBIN puts these items in the bin where already q different colors where present.

5   A General Lower Bound for Deterministic Algorithms

In this section we prove a general lower bound on the competitive ratio of any deterministic online algorithm for the OLBCP. We establish a lemma which immediately leads to the desired lower bound but which is even more powerful. In particular, this lemma will allow us to derive essentially the same lower bound for randomized algorithms in Section 6.

In the sequel we will have to refer to the "state" of (the bins managed by) an algorithm ALG after processing a prefix of a request sequence σ. To this end we introduce the notion of a $\mathcal{C}$-*configuration*.

Definition 5.1 ($\mathcal{C}$-configuration). *Let $\mathcal{C}$ a set of colors. A $\mathcal{C}$-configuration is a packing of items with colors from $\mathcal{C}$ into at most q bins. More formally, a $\mathcal{C}$-configuration can be defined as a mapping $K \colon \{1, \dots, q\} \to \mathcal{S}_{\leq B}$, where*

$$\mathcal{S}_{\leq B} := \{\, S : S \text{ is a multiset over } \mathcal{C} \text{ containing at most } B \text{ elements from } \mathcal{S} \,\}$$

with the interpretation that $K(j)$ is the multiset of colors contained in bin j. We omit the reference to the set $\mathcal{C}$ if it is clear from the context.

Lemma 5.2. *Let $B, q, s \in \mathbb{N}$ such that $s \geq 1$ and the inequality $B/q \geq s - 1$ holds. There exists a finite set $\mathcal{C}$ of colors and a constant $L \in \mathbb{N}$ with the following property. For any deterministic algorithm ALG and any $\mathcal{C}$-configuration K there exists an input sequence $\sigma_{\mathrm{ALG},K}$ of $\mathrm{OLBCP}_{B,q}$ such that*

(i)   *The sequence $\sigma_{\mathrm{ALG},K}$ uses only colors from $\mathcal{C}$ and $|\sigma_{\mathrm{ALG},K}| \leq L$, that is, $\sigma_{\mathrm{ALG},K}$ consists of at most L requests.*
(ii) *If ALG starts with initial $\mathcal{C}$-configuration K then $\mathrm{ALG}(\sigma_{\mathrm{ALG},K}) \geq (s-1)q$.*
(iii) *If OPT starts with the empty configuration (i.e., all bins are empty), then $\mathrm{OPT}(\sigma_{\mathrm{ALG},K}) \leq s$. Additionally, OPT can process the sequence in such a way that at the end again the empty configuration is attained.*

Moreover, all of the above statements remain true even in the case that the online algorithm is allowed to use $q' \geq q$ bins instead of q (while the offline adversary still only uses q bins). In this case, the constants $|\mathcal{C}|$ and K depend only on q' but not on the particular algorithm ALG.

Proof. Let $\mathcal{C} = \{c_1, \dots, c_{(s-1)^2 q^2 q'}\}$ be a set of $(s-1)^2 q^2 q'$ colors and ALG be any deterministic online algorithm which starts with some initial $\mathcal{C}$-configuration K.

The construction of the request sequence $\sigma_{\mathrm{ALG},K}$ works in *phases*, where at the beginning of each phase the offline adversary has all bins empty. During the run of the request sequence, a subset of the currently open bins of ALG will be *marked*. We will denote by P_k the subset of marked bins at the beginning of Phase k. $P_1 = \emptyset$ and during some Phase M, one bin in P_M will contain at least $(s-1)q$ colors. In order to assure that this goal can in principle be achieved, we keep the invariant that each bin $b \in P_k$ has the property that the number of different colors in b plus the free space in b is at least $(s - 1)q$. In other words, each bin $b \in P_k$ could potentially still be forced to contain at least $(s - 1)q$ different colors. For technical reasons, P_k is only a subset of the bins with this property.

For bin j of ALG we denote by $n(j)$ the number of different colors currently in bin j and by $f(j)$ the space left in bin j. Then every bin $j \in P_k$ satisfies $n(j) + f(j) \geq (s - 1)$. By $\min P_k := \min_{j \in P_k} n(j)$ we denote the minimum number of colors in a bin from P_k.

We now describe Phase k with $1 \leq k \leq q(s - 1)q'$. The adversary selects a set of $(s - 1)q$ new colors $C_k = \{c_1, \ldots, c_{(s-1)q}\}$ from $\mathcal{C}$ not used in any phase before and starts to present one item of each color in the order

$$c_1, c_2, \ldots, c_{(s-1)q}, c_1, c_2, \ldots, c_{(s-1)q}, c_1, c_2, \ldots \tag{1}$$

until one of the following cases appears:

Case 1 ALG puts an item into a bin $p \in P_k$. In this case we let $Q := P_k \setminus \{j \in P_k : n(j) < n(p)\}$, that is, we remove all bins from P_k which have less than $n(p)$ colors. Notice that $\min_{j \in Q} n(j) > \min P_k$, since the number of different colors in bin p increases.

Case 2 ALG puts an item into some bin $j \notin P_k$ which satisfies

$$n(j) + f(j) \geq (s - 1)q. \tag{2}$$

In this case we set $Q := P_k \cup \{j\}$ (we tentatively add bin j to the set P_k).

Notice that after a finite number of requests one of these two cases must occur: Let $b_1, \ldots, b_t$ be the set of currently open bins of ALG. If ALG never puts an item into a bin from P_k then at some point all bins of $\{b_1, \ldots, b_t\} \setminus P_k$ are filled and a new bin, say bin j, must be opened by ALG by putting the new item into bin j. But at this moment bin j satisfies satisfies $n(j) = 1$, $f(j) = B - 1$ and hence $n(j) + f(j) = B \geq (s - 1)q$ which gives (2).

Since the adversary started the phase with all bins empty and during the current phase we have given no more than $(s - 1)q$ colors, the adversary can assign the items to bins such that no bin contains more than $s - 1$ different colors (we will describe below how this is done precisely). Notice that due to our stopping criterions from above (case 1 and case 2) it might be the case that in fact so far we have presented less than $(s - 1)q$ colors.

In the sequel we imagine that each currently open bin of the adversary has an index x, where $1 \leq x \leq q$. Let $\beta : C_k \to \{1, \ldots, q\}$ be any mapping of the colors from C_k to the offline bin index such that $|\beta^{-1}(\{x\})| \leq s - 1$ for $j = 1, \ldots, q$. We imagine color c_r to "belong" to the bin with index $\beta(c_r)$ even if no item of this color has been presented (yet). For those items presented already in Phase k, each item with color c_r goes into the currently open bin with index $\beta(c_r)$. If there is no open bin with index $\beta(c_r)$ when the item arrives a new bin with index $\beta(c_r)$ is opened by the adversary to accommodate the item.

Our goal now is to clear all open offline bins so that we can start a new phase. During our clearing loop the offline bin with index x might be closed and replaced by an empty bin multiple times. Each time a bin with index x is replaced by an empty bin, the new bin will also have index x. The bin with index x receives a color not in $\beta^{-1}(\{x\})$ at most once, ensuring that the optimum offline cost still remains bounded from above by s. The clearing loop works as follows:

1. (Start of clearing loop iteration) Choose a color $c^* \in C_k$ which is not contained in any bin from Q. If there is no such color, goto the "good end" of the clearing loop (Step 4).
2. Let $F \leq qB$ denote the current total empty space in the open offline bins. Present items of color c^* until one of the following things happens:
   Case (a): At some point in time ALG puts the ℓth item with color c^* into a bin $j \in Q$ where $1 \leq \ell < F$. Notice that the number of different colors in j increases. Let $Q' := Q \setminus \{b \in Q : n(b) < n(j)\}$, in other words, we remove all bins b from Q which currently have less than $n(j)$ colors. This guarantees that $\min_{b \in Q'} n(b) > \min_{b \in Q} n(b) \geq \min P_k$. The adversary puts all t items of color c^* into bins with index $\beta(c^*)$. Notice that during this process the open bin with index $\beta(c^*)$ might be filled up and replaced by a new empty bin with the same index.
   Set $Q := Q'$ and go to the start of the next clearing loop iteration (Step 1). Notice that the number of colors from C_k which are contained in Q decreases by one, but $\min_{b \in Q} n(b)$ increases.
   Case (b): F items of color c^* have been presented, but ALG has not put any of these items into a bin from Q.
   In this case, the offline adversary processes these items differently from case (a): The F items of color c^* are used to fill up the exactly F empty places in all currently open offline bins. Since up to this point, each offline bin with index x had received colors only from the $s - 1$ element set $\beta^{-1}(\{x\})$, it follows that no offline bin has contained more than s different colors. We close the clearing loop by proceeding as specified in the next step.
3. (Standard end of clearing loop iteration)
   In case we have reached this step, we are in the situation that all offline bins have been cleared (we can originate only from case (b) above). We set $P_{k+1} := Q$ and end the clearing loop and the current Phase k.
4. (Good end of clearing loop iteration)
   We have reached the point that all colors from C_k are contained in a bin from Q. Before the first iteration, exactly one color from C_k was contained in Q. The number of colors from C_k which are contained in bins from Q can only increase by one (which is in case (a) above) if $\min_{b \in Q} n(b)$ increases. Hence, if all colors from C_k are contained in bins from Q, $\min_{b \in Q} n(v)$ must have increased $(s - 1)q - 1$ times, which implies $\min_{b \in Q} n(b) = (s - 1)q$. In other words, one of ALG's bins in Q contains at least $(s - 1)q$ different colors. The only thing left to do is append a suitable suffix to our sequence constructed so far such that all open offline bins are closed. Clearly this can be done without increasing the offline-cost.

In case the clearing loop finished with a "good end" we have achieved our goal of constructing a sufficiently bad sequence for ALG. What happens if the clearing loop finishes with a "standard end"?

Claim. If Phase k completes with a "standard end", then $\min P_{k+1} > \min P_k$ or $|P_{k+1}| > |P_k|$.

Before we prove the above claim, let us show how this claim implies the result of the lemma. Since the case $|P_{k+1}| > |P_k|$ can happen at most q' times, it follows that after at most q' phases $\min P_k$ must increase. On the other hand, since $\min P_k$ never decreases by our construction and the offline costs remain bounded from above by s, after at most $q(s-1)q'$ phases we must be in the situation that $\min P_k \geq (s-1)q$, which implies a "good end". Since in each phase at most $(s-1)q$ new colors are used, it follows that our initial set $\mathcal{C}$ of $(s-1)^2 q^2 q'$ colors suffices to construct the sequence $\sigma_{\mathrm{ALG},K}$. Clearly, the length of $\sigma_{\mathrm{ALG},K}$ can be bounded by a constant L independent of ALG and K.

Proof (of Claim). Suppose that the sequence (1) at the beginning of the phase was ended because case 1 occurred, i.e., ALG put one of the new items into a bin from P_k. In this case $\min_{b \in Q} n(b) > \min P_k$. Since during the clearing loop $\min_{b \in Q} n(b)$ can never decrease and P_{k+1} is initialized with the result of Q at the "standard end" of the clearing loop, the claim follows.

The remaining case is that the sequence (1) was ended because of a case 2-situation. Then $|Q| = |P_k \cup \{j\}|$ for some $j \notin P_k$ and hence $|Q| > |P_k|$. During the clearing loop Q can only decrease in size if $\min_{i \in Q} n(i)$ increases. It follows that either $|P_{k+1}| = |P_k| + 1$ or $\min P_{k+1} > \min P_k$ which is what we claimed. $\square$

This completes the proof of the lemma. $\square$

As an immediate consequence of Lemma 5.2 we obtain the following lower bound result for the competitive ratio of any deterministic algorithm:

Theorem 5.3. *Let $B, q, s \in \mathbb{N}$ such that $s \geq 1$ and the inequality $B/q \geq s-1$ holds. No deterministic algorithm for $\mathrm{OLBCP}_{B,q}$ can achieve a competitive ratio less than $(s-1)/s \cdot q$. Hence, the competitive ratio of any deterministic algorithm for fixed B and q is at least $\left(1 - \frac{q}{B+q}\right) q$. In particular, for the general case with no restrictions on the relation of the capacity B to the number of bins q, there can be no deterministic algorithm for $\mathrm{OLBCP}_{B,q}$ that achieves a competitive ratio less than q. All of the above claims remain valid, even if the online algorithm is allowed to use an arbitrary number $q' \geq q$ of open bins.* $\square$

6   A General Lower Bound for Randomized Algorithms

In this section we show lower bounds for the competitive ratio of any randomized algorithm against an oblivious adversary for $\mathrm{OLBCP}_{B,q}$. The basic method for deriving such a lower bound is Yao's principle (see also [2,7]). Let X be a probability distribution over input sequences $\Sigma = \{\sigma_x : x \in \mathcal{X}\}$. We denote the *expected cost* of the deterministic algorithm ALG according to the distribution X on Σ by $\mathbb{E}_X [\mathrm{ALG}(\sigma_x)]$. Yao's principle can now be stated as follows.

Theorem 6.1 (Yao's principle). *Let* $\{\,\mathrm{ALG}_y : y \in \mathcal{Y}\,\}$ *denote the set of deterministic online algorithms for an online minimization problem. If* X *is a probability distribution over input sequences* $\{\,\sigma_x : x \in \mathcal{X}\,\}$ *such that*

$$\inf_{y \in \mathcal{Y}} \mathbb{E}_X\left[\mathrm{ALG}_y(\sigma_x)\right] \geq \bar{c}\,\mathbb{E}_X\left[\mathrm{OPT}(\sigma_x)\right]. \tag{3}$$

for some real number $\bar{c} \geq 1$, *then* $\bar{c}$ *is a lower bound on the competitive ratio of any randomized algorithm against an oblivious adversary.* □

Theorem 6.2. *Let* $B, q, s \in \mathbb{N}$ *such that* $s \geq 1$ *and the inequality* $B/q \geq s - 1$ *holds. Then no randomized algorithm for* $\mathrm{OLBCP}_{B,q}$ *can achieve a competitive ratio less than* $(s-1)/s \cdot q$ *against an oblivious adversary. In particular for fixed* B *and* q, *the competitive ratio against an oblivious adversary is at least* $\left(1 - \frac{q}{B+q}\right) q$. *All of the above claims remain valid, even if the online algorithm is allowed to use an arbitrary number* $q' \geq q$ *of open bins.*

Proof. Let $\mathcal{A} := \{\,\mathrm{ALG}_y : y \in \mathcal{Y}\,\}$ the set of deterministic algorithms for $\mathrm{OL\text{-}BCP}_{B,q}$. We will show that there is a probability distribution X over a certain set of request sequences $\{\,\sigma_x : x \in \mathcal{X}\,\}$ such that for any $\mathrm{ALG}_y \in \mathcal{A}$ we have $\mathbb{E}_X\left[\mathrm{ALG}_y(\sigma_x)\right] \geq (s-1)q$, and, moreover, $\mathbb{E}_X\left[\mathrm{OPT}(\sigma_x)\right] \leq s$. The claim of the theorem then follows by Yao's principle.

Let us recall the essence of Lemma 5.2. The lemma establishes the existence of a finite color set $\mathcal{C}$ and a constant L such that for a fixed configuration K any deterministic algorithm can be "fooled" by one of at most $|\mathcal{C}|^L$ sequences. Since there are no more than $|\mathcal{C}|^{qB}$ configurations, a *fixed finite* set of at most $N := |\mathcal{C}|^{L+qB}$ sequences $\Sigma = \{\sigma_1, \ldots, \sigma_N\}$ suffices to "fool" *any* deterministic algorithm provided the initial configuration is known.

Let X be a probability distribution over the set of finite request sequences $\{\,\sigma_{i_1}, \sigma_{i_2}, \ldots, \sigma_{i_k} : k \in \mathbb{N}, 1 \leq i_j \leq N\,\}$ such that σ_{i_j} is chosen from Σ uniformly and independently of all previous subsequences $\sigma_{i_1}, \ldots, \sigma_{i_{j-1}}$. We call subsequence σ_{i_k} the *kth phase*. Let $\mathrm{ALG}_y \in \mathcal{A}$ be arbitrary. Define ϵ_k by

$$\epsilon_k := \Pr\nolimits_X\left[\mathrm{ALG}_y \text{ has one bin with at least } (s-1)q \text{ colors during Phase } k\right]. \tag{4}$$

The probability that ALG_y has one bin with at least $(s-1)q$ colors on any given phase is at least $1/N$, whence $\epsilon_k \geq 1/N$ for all k. Let

$$p_k := \Pr\nolimits_X\left[\mathrm{ALG}_y(\sigma_{i_1} \ldots \sigma_{i_{k-1}}\sigma_{i_k}) \geq (s-1)q\right]. \tag{5}$$

Then the probabilities p_k satisfy the recursion: $p_0 = 0$ and $p_k = p_{k-1} + (1 - p_{k-1})\epsilon_k$. The first term in the latter equation corresponds to the probability that ALG_y has already cost at least $(s-1)q$ after Phase $k-1$, the second term accounts for the probability that this is not the case but cost at least $(s-1)q$ is achieved in Phase k. By construction of X, these events are independent. Since $\epsilon_k \geq 1/N$ we get that $p_k \geq p_{k-1} + (1 - p_{k-1})/N$. It is easy to see that any sequence of real numbers $p_k \in [0,1]$ with this property must converge to 1. Hence, also the expected cost $\mathbb{E}_X\left[\mathrm{ALG}_y(\sigma_x)\right]$ converges to $(s-1)q$. On the other hand, the offline costs remain bounded by s by the choice of the σ_{i_j} according to Lemma 5.2. □

7   Conclusions

We have studied the online bin coloring problem OLBCP, which was motivated by applications in a robotized assembly environment. The investigation of the problem from a competitive analysis point of view revealed a number of odds. A natural greedy-type strategy (GREEDYFIT) achieves a competitive ratio strictly worse than arguably the most stupid algorithm (ONEBIN). Moreover, no algorithm can be substantially better than the trivial strategy (ONEBIN). Even more surprising, neither randomization nor "resource augmentation" helps to overcome the $\Omega(q)$ lower bound on the competitive ratio (see [9,8] for successful applications to scheduling problems) can help to overcome the $\Omega(q)$ lower bound on the competitive ratio. Intuitively, the strategy GREEDYFIT should perform well "on average" (which we could sort of confirm by preliminary experiments with random data).

An open problem remains the existence of a deterministic (or randomized) algorithm which achieves a competitive ratio of q (matching the lower bound of Theorems 5.3 and 6.2). However, the most challenging issue raised by our work seems to be an investigation of OLBCP from an average-case analysis point of view.

Acknowledgements. The authors would like to thank Errol Lloyd (University of Delaware), Remco Peters (University of Amsterdam) and S. S. Ravi (SUNY at Albany) for fruitful discussions in the early stages of this work.

References

1. N. Ascheuer, M. Grötschel, S. O. Krumke, and J. Rambau, *Combinatorial online optimization*, Proceedings of the International Conference of Operations Research (OR'98), Springer, 1998, pp. 21–37.
2. A. Borodin and R. El-Yaniv, *Online computation and competitive analysis*, Cambridge University Press, 1998.
3. E. G. Coffman, M. R. Garey, and D. S. Johnson, *Approximation algorithms for bin packing: a survey*, In Hochbaum [6].
4. J. Csirik and G. J. Woeginger, *On-line packing and covering problems*, In Fiat and Woeginger [5].
5. A. Fiat and G. J. Woeginger (eds.), *Online algorithms: The state of the art*, Lecture Notes in Computer Science, vol. 1442, Springer, 1998.
6. D. S. Hochbaum (ed.), *Approximation algorithms for NP -hard problems*, PWS Publishing Company, 20 Park Plaza, Boston, MA 02116-4324, 1997.
7. R. Motwani and P. Raghavan, *Randomized algorithms*, Cambridge University Press, 1995.
8. C. Phillips, C. Stein, E. Torng, and J. Wein, *Optimal time-critical scheduling via resource augmentation*, Proceedings of the 29th Annual ACM Symposium on the Theory of Computing, 1997, pp. 140–149.
9. K. Pruhs and B. Kalyanasundaram, *Speed is as powerful as clairvoyance*, Proceedings of the 36th Annual IEEE Symposium on the Foundations of Computer Science, 1995, pp. 214–221.

A General Decomposition Theorem for the k-Server Problem

Steven S. Seiden*

Department of Computer Science
298 Coates Hall
Louisiana State University
Baton Rouge, LA 70803, USA
sseiden@acm.org

Abstract. The first general decomposition theorem for the k-server problem is presented. Whereas all previous theorems are for the case of a finite metric with $k + 1$ points, the theorem given here allows an arbitrary number of points in the underlying metric space. This theorem implies $O(\text{polylog}(k))$-competitive randomized algorithms for certain metric spaces consisting of a polylogarithmic number of widely separated sub-spaces, and takes a first step towards a general $O(\text{polylog}(k))$-competitive algorithm. The only other cases for which polylogarithmic competitive randomized algorithms are known are the uniform metric space, and the weighted cache metric space with two weights.

1 Introduction

The *k-server problem* is one of the most intriguing problems in the area of online algorithms [10,18]. Furthermore, it has as special cases several important and practical problems. The most prominent of these is weighted caching, which has applications in the management of web browser caches. We investigate the randomized variant of this problem, for which very few results are known. Our main result is a theorem which allows us to construct $O(\text{polylog}(k))$-competitive randomized algorithms for a broad class of metric spaces and provides a first step towards a general solution.

Central to the k-server problem is the *k-server conjecture*, which proposes that there is a deterministic k-server algorithm which is k-competitive for all metrics. Although it has received less attention, the *randomized k-server conjecture* is just as intriguing. This conjecture puts forward that there is a $O(\log k)$-competitive randomized k-server algorithm for all metrics.

There has been much work on the k-server conjecture since it was proposed by Manasse, McGeoch and Sleator [24]. It is easy to show a lower bound of k [24]. The best upper bound result for an arbitrary metric space is $2k - 1$, due to Koutsoupias and Papadimitriou [22]. In addition, k-competitive algorithms

---

* This research was partially supported by an LSU Council on Research summer stipend and by the Research Competitiveness Subprogram of the Louisiana Board of Regents.

F. Meyer auf der Heide (Ed.): ESA 2001, LNCS 2161, pp. 86–97, 2001.

have been exhibited for a number of special cases [27,12,13,6,23]. We refer the reader to Chapter 4 of [18] for a more comprehensive treatment of the status of the deterministic conjecture.

The situation for randomized algorithms is less satisfactory. Only a small number of algorithms are known for specific metrics. These are as follows: For the uniform metric space, matching upper and lower bounds of $H_k = 1 + \frac{1}{2} + \cdots + \frac{1}{k}$ are known. The lower bound is due to Fiat $et\ al.$ [16], while the upper bound is presented by McGeoch and Sleator [25]. A $O(\log k)$-competitive algorithm for the weighted cache problem with 2 weights has recently been exhibited by Irani [19]. For the case of 2 servers on the isosceles triangle, Karlin, Manasse, McGeoch and Owicki [20] show matching upper and lower bounds of $\frac{e}{e-1}$. For the case of 2 servers on the real line, a $\frac{155}{78} < 1.98717$-competitive algorithm has been developed by Bartal, Chrobak and Larmore [5]. Finally, the case where we have a finite metric with $k + 1$ points is closely related to the metrical task system (MTS) problem [11,3,17]. The results on that problem imply that there is a $O(\text{polylog}(k))$-competitive algorithm for every finite space with $k+1$ points. In summary, the only two metrics for which a polylogarithmic competitive algorithm exists for general k are the uniform metric [16] and the 2-weighted cache metric [19]. In particular, we are lacking a good randomized algorithm for the general weighted cache problem.

The status of randomized lower bounds is also displeasing. Kaloff, Rabani and Ravid [21] showed the first lower bound, namely $\Omega(\min\{\log k, \log \log n\})$, where n is the number of points in the metric space. Karloff, Rabani and Saks [9] improved this by showing a lower bound of $\Omega(\sqrt{\log k / \log \log k})$ for all spaces. This has recently been further improved to $\Omega(\log k / \log^3 \log k)$ by Bartal, Bollobás and Mendel [4]. For $k = 2$, a lower bound of $1 + 1/\sqrt{e} > 1.60653$ is presented by Chrobak, Larmore, Lund and Reingold [15]. As mentioned before, a lower bound of H_k holds for the uniform space [16]. As pointed out by Seiden [26], the work of Blum $et\ al.$ implies a lower bound of $\log_2(k + 1)$ for a certain family of metric spaces. Note that whereas in the deterministic setting we conjecture that k is the correct bound for all spaces, the preceding discussion implies that in the randomized setting the competitive ratio depends on the underlying space.

Our main theorem gives an $O(\text{polylog}(k))$-competitive algorithm for metric spaces which can be decomposed into a small number of widely separated subspaces. The theorem may be applied recursively. As we shall argue in the next section, we feel that this is the first step towards a resolution of the randomized conjecture. Another important contribution of this work is in illustrating the usefulness of $unfair\ metrical\ task\ systems$ (UMTS) [26,3,17] as a general algorithmic design tool. Unfair metrical task systems allow us to design "divide and conquer" online algorithms. As far as we know, this work is the first application of the UMTS technique outside of the metrical task system problem.

2   A Line of Attack on the k-Server Conjecture

As we have already mentioned, the k-server problem and metrical task system problem are in certain special cases equivalent in terms of competitive ratio.

An important contribution of this paper is in revealing a new and important connection between these problems. While it has long been known that the finite k-server problem can in general be modeled as an MTS, the unsatisfying result is a polylog($\binom{n}{k}$)-competitive algorithm, where n is the number of points in the metric space. Our main contribution is in recognizing that if we instead model only the most important components of the k-server problem *using an unfair MTS*, we get a much better result.

Blum, Burch and Kalai [8] argue that the line of attack which leads to randomized $O(\text{polylog}(n))$-competitive algorithms for the MTS problem is a logical one to use for the k-server problem. We also feel that this strategy is likely to be fruitful, and the result provided here brings us one step closer to realizing it.

There are two components to the MTS line of attack:

- The metric space approximation technique developed by Bartal [1,2]. This technique allows one to approximate any metric space using a specific type of space called an h-hierarchical well-separated tree (h-HST). An h-HST is a metric space with diameter Δ, which recursively consists of k-HST sub-spaces of diameter at most Δ/h. The distance between any two points in separate sub-spaces is Δ. Specifically, Bartal gives a method of finding a probability distribution over h-HST's such that the expected distance in the HST is with a factor of $O(h \log n \log \log n)$ of the distance in the original space.

- Randomized algorithms for HST metric spaces [3,17]. The key subroutines in these algorithms are algorithms for spaces which consist of two or more widely separated sub-spaces. *Decomposition theorems*, providing upper and lower bounds for the MTS problem on such spaces are presented Blum *et al.* [9] and Seiden [26].

The first component carries over directly to the k-server problem. However, because distances are distorted by a factor polylogarithmic in the number of points in the space, it can (currently) only yield polylogarithmic competitive algorithms for spaces with a polynomial number of points.

Except in the case of $k + 1$ points, MTS algorithms for HST spaces cannot be adapted to the k-server problem—it would seem that a new approach is needed. Along these lines, Blum, Burch and Kalai [8] have given a first step towards a decomposition theorem for the k-server problem. However, their result is incomplete, in that no way of modeling the distribution of servers in the sub-spaces is proposed. In this paper, we correct this problem, and give the first working decomposition result for the k-server problem where the number of points in the underlying metric space is unrestricted.

Our approach is to carefully model the k-server problem using an UMTS, and show that the costs in the model differ insignificantly from the actual costs. Unfortunately, while the basic idea is not too hard, there are many messy technical details to be resolved.

3   Preliminaries

For a general introduction to *competitive analysis* we refer the reader to [10]. We define $\text{cost}_{\mathcal{A}}(\sigma, C)$ be the cost incurred by $\mathcal{A}$ on request sequence σ starting at initial configuration C. Let $\text{cost}(\sigma, C)$ be the cost incurred by the optimal offline algorithm on request sequence σ starting at initial configuration C. We need the following terminology, which is akin to the definition of *constrainedness* introduced by Fiat and Mendel [17]. Randomized online algorithm $\mathcal{A}$ is said to be *c-competitive and f-inhibited* if

$$E[\text{cost}_{\mathcal{A}}(\sigma, C)] \leq c\,\text{cost}(\sigma, C) + f,$$

for all σ and C. Since this is a worst case measure, for the purposes of analysis, we assume that the input sequence is generated by a malicious *adversary*, who forces the algorithm to perform as badly as possible. where the expectation is taken over the random choices of the algorithm. There are several types of adversaries in the randomized scenario; we use exclusively the *oblivious adversary* [7].

Let (P, d_P) be a metric space. We define the *diameter* of a non-empty set of points $X \subset P$ to be $\Delta(X) = \sup_{x,y \in X} d_P(x, y)$. For each positive integer i and $X, Y \in P^i$ we define the distance between X and Y to be

$$d_P(X, Y) = \min_{Z \in \mu(X,Y)} \sum_{(x,y) \in Z} d_P(x, y),$$

where $\mu(X, Y)$ is the set of all maximal matchings on the complete bipartite graph induced on X and Y.

Let $\mathcal{U} = \{U_1, \ldots, U_t\}$ be a partition of P. Define $\nabla = \max_{U \in \mathcal{U}} \Delta(U)$ and

$$\theta(U, V) = \inf_{u \in U} \inf_{v \in V} \frac{d_P(u, v)}{\nabla}.$$

We are interested in metrics where

$$\theta = \min_{\substack{U, V \in \mathcal{U} \\ U \neq V}} \theta(U, V)$$

is large. I.e. the subspaces are widely separated and have small diameter relative to the distances between them. We call such a space θ-*decomposable*. We give an example of a space decomposable into three sub-spaces in Figure 1.

Note that any h-HST is recursively h-decomposable; a k-HST is just a special type of decomposable space.

In the k-server problem, we have k mobile *servers*, each of which is located at some point in a metric space (P, d_P). We are given $C_0 \in P^k$, the *initial configuration* of our servers. We are then confronted with $\sigma = p_1, p_2, \ldots, p_n$, a sequence of *request points* in P. Each request point must be *served*, by moving some server to it (if one is not already there). Formally, a *configuration* is a member of P^k. A service is a sequence of configurations $\pi = C_1, \ldots, C_n$ where $p_i \in C_i$ for $1 \leq i \leq n$. The cost of the service π is

$$\text{cost}(\pi, C_0) = \sum_{i=1}^{n} d_P(C_{i-1}, C_i).$$

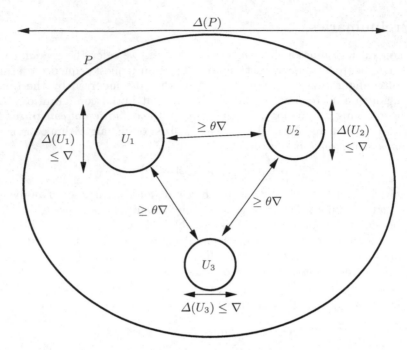

Fig. 1. A θ-decomposable space P with sub-spaces $\mathcal{U} = \{U_1, U_2, U_3\}$.

Our goal is to produce a service with low cost. An algorithm is *online* if, in the service it produces, C_i is a function of only $p_1, \ldots, p_i$. Since it is not in general possible to produce the optimal service online, we consider approximation algorithms.

Let (P, d_P) be a metric space with P finite. Let $n = |P|$. In the *metrical task system (MTS) problem* on (P, d_P), there is a single mobile server. We refer to the points of an MTS as *states*. We are given an *initial state* s_0, and a sequence of *tasks*, $\sigma = T_1, \ldots, T_N$. Each task is a function $T_i : P \mapsto \mathbb{R}_{\geq 0} \cup \{\infty\}$. For $s \in P$, $T_i(s)$ is the *local cost* for serving task i in state s. The goal is to minimize the sum of the local costs and distances moved by the server.

Formally, an (MTS) *service* is a sequence $\varpi = s_0, s_1, \ldots, s_N$ of states. The cost of the service ϖ is

$$\text{cost}(\varpi, s_0) = \sum_{i=1}^{n} d_P(s_{i-1}, s_i) + T_i(s_i). \tag{1}$$

An algorithm is *online* if s_i is a function of only $T_1, \ldots, T_i$.

With respect to competitive analysis, it is also possible to consider the *unfair metrical task system problem* on P. In this case, we are given real numbers $\alpha \geq 1$ and $\beta \geq 1$. The adversary pays (1) for its service, whereas the online algorithm pays

$$\text{cost}^*(\varpi, s_0) = \sum_{i=1}^{n} \beta \, d_P(s_{i-1}, s_i) + \alpha \, T_i(s_i).$$

A more general definition is possible, for instance, α might be a function of s_i. However, this simple definition shall suffice for our purposes.

The following result proves to be extremely useful:

Theorem 1. *There exists an $\alpha + O(\beta \log^2 n \log^2 \log n)$-competitive randomized algorithm for the UMTS problem on any metric space of n points.*

This follows almost directly from [17]. For certain specific metrics, the upper bound in the preceding theorem can be improved.

4   A k-Server Algorithm for Decomposable Spaces

We now develop an algorithm for any metric space that is θ-decomposable. To begin, we study the structure of the optimal offline service within an arbitrary sub-space. We then study the "big picture," showing how to put together optimal offline services for sub-spaces to get a complete service. This big picture problem is formulated as an UMTS. Finally, we show how to use an algorithm for UMTS to get an algorithm for the original k-server problem.

We use the following notation: We denote the empty sequence by ϵ. For $p \in P$, define $\sigma \wedge p$ to be the concatenation of σ by p. For $1 \le i \le j \le n$, we denote $\sigma_i^j = p_i, p_{i+1}, \ldots, p_j$. For $i > j$ let $\sigma_i^j = \epsilon$.

Let X be some arbitrary non-empty subset of P. For $p \in X$, define $X^i(p)$ to be the subset of X^i whose members all contain p. We define $\sigma \cap X$ to be the maximal subsequence of σ containing points only from X.

For $1 \le i \le k$, the *i-server work function on X* is defined as follows:

$$w_I^C(\epsilon, i, X) = d_P(C, I \cap X);$$

$$w_I^C(\sigma \wedge p, i, X) = \begin{cases} w_I^C(\sigma, i, X), & \text{if } p \notin X; \\ \inf_{D \in X^i(p)}\{w_I^D(\sigma, i, X) + d_P(C, D)\}, & \text{if } p \in X; \end{cases}$$

$$w_I(\sigma, i, X) = \inf_{C \in X^i} w_I^C(\sigma, i, X).$$

This is a generalization of the work function introduced by Chrobak and Larmore [14]. Intuitively, $w_I^C(\sigma, i, X)$ is the optimal offline cost to serve requests in $\sigma \cap X$ using i servers that stay inside X, starting at configuration I and ending at configuration C. We make use of the following facts, which follow directly from the definition of w:

Fact 41 *For any non-empty $X \subset P$; $i \ge 1$; $I, F, C, D \in X^i$ and σ we have*

$$|w_I^C(\sigma, i, X) - w_I^D(\sigma, i, X)| \le d_P(C, D),$$
$$|w_C^F(\sigma, i, X) - w_D^F(\sigma, i, X)| \le d_P(C, D).$$

These are known as the *slope conditions* [14]. Intuitively, the slope conditions follow from the fact that to serve a sequence and end up in a certain configuration, one could serve the sequence ending in a different configuration and then switch configurations.

Fact 42 *For any non-empty $X \subset P$; $i \geq 1$; $I, F \in X^i$; σ and $0 \leq j \leq n$ there exists a $C \in X^i$ such that $w_I^F(\sigma, i, X) = w_C^F(\sigma_{j+1}^n, i, X) + w_I^C(\sigma_1^j, i, X)$.*

In essence, this just says that there is some configuration at each step of the optimal offline service.

For $0 \leq i \leq k$, the *i-server cost on X* of the request p is defined to be

$$\tau(\sigma \wedge p, i, X, I) = \begin{cases} 0, & \text{if } p \notin X; \\ \infty, & \text{if } i = 0 \text{ and } p \in X; \\ w_I(\sigma \wedge p, i, X) - w_I(\sigma, i, X), & \text{otherwise.} \end{cases}$$

The total optimal offline cost for serving σ is $w_{C_0}(\sigma, k, P)$. However, rather than using $w_{C_0}(\sigma, k, P)$ directly, we compute a lower bound on it. Using this approach, our evaluation of the optimal offline cost is not perfect, however, the "big picture" is much more clear. This turns out to be the key to designing an algorithm with good competitive ratio.

Given these definitions, we now explain our approach to modeling the k-server problem using an UMTS. To begin, using (P, d_P) and the partition $\mathcal{U}$, we define a new metric space $(S, d_S) = \Phi(P, d_P, \mathcal{U})$. Consider the set of all configurations of k-servers on P. We consider two configurations C and D to be equivalent if $|C \cap U| = |D \cap U|$ for all $U \in \mathcal{U}$. The set of *states* is the set of equivalence classes of configurations. The number of such classes is $s = \binom{k+t-1}{t-1}$. We shall also consider each state to be a function $\phi : \mathcal{U} \mapsto \mathbb{Z}_{\geq 0}$. $\phi(U)$ is the number of servers located in U. We index these functions using the non-negative integers and denote the set of all states as S. ϕ_0 is the state containing C_0.

In order for (S, d_S) to be well defined as a metric space, we require $\theta > 2k$. If ϕ and φ are states then

$$d_S(\phi, \varphi) = \left(1 - \frac{2k}{\theta}\right) \inf_{C \in \phi, D \in \varphi} d_P(C, D).$$

I.e. this is $1 - 2k/\theta$ times the minimum cost for moving between a configuration in ϕ and one in φ.

We can compute a lower bound on the optimal offline cost to serve the k-server request sequence σ as follows: If ϕ is a state, we fix $\zeta(\phi)$ to be some arbitrary k-server configuration in the equivalence class of ϕ. When calculating the optimal offline cost for state ϕ, we use $\zeta(\phi)$ as our starting configuration. This seems a bit crude at first, but does not affect the overall result. Define the *task induced by σ* to be

$$\boldsymbol{T}(\sigma) = \langle T(\phi_0, \sigma), T(\phi_1, \sigma), \ldots, T(\phi_{s-1}, \sigma) \rangle,$$

where

$$T(\phi, \sigma) = \sum_{U \in \mathcal{U}} \tau(\sigma, \phi(U), U, \zeta(\phi)).$$

The *task sequence* is $\varrho = \boldsymbol{T}(\sigma_1^1), \boldsymbol{T}(\sigma_1^2), \ldots, \boldsymbol{T}(\sigma_1^n)$.

Define

$$W_\varphi^\phi(\epsilon) = d_S(\phi, \varphi),$$

$$W_\varphi^\phi(\sigma \wedge p) = \min_{\psi \in S} \left\{ W_\varphi^\psi(\sigma) + T(\psi, \sigma \wedge p) + d_S(\phi, \psi) \right\},$$

$$W_\varphi(\sigma) = \min_{\phi \in S} W_\varphi^\phi(\sigma).$$

Note that this is a s state metrical task system, with task sequence ϱ and initial state ϕ_0. The optimal offline cost for serving the tasks induced by σ is $W_{\phi_0}(\sigma)$, which is the *MTS work function*.

Lemma 1. *If $\theta > 2k$ then for all input sequences σ, $w_{C_0}(\sigma, k, P) \geq W_{\phi_0}(\sigma) - O(1)$.*

Proof. Omitted due to space considerations.   □

We have shown how to use the MTS (S, d_S) to lower bound the optimal offline cost for the original k-server problem. Now we show how to design a randomized k-server algorithm for (P, d_P), given that we have competitive algorithms for its sub-spaces. Specifically, we assume that we have α-competitive, $\psi j \nabla$-inhibited algorithms for the j-server problems for all $U \in \mathcal{U}$ and all $j \leq k$. In order to combine these algorithms, we shall need to make (S, d_S) into an UMTS as follows: While the optimal offline algorithm pays $T(\phi, \sigma)$ for serving $\boldsymbol{T}(\sigma)$ in state ϕ and $d_S(\phi, \varphi)$ for moving from state ϕ to state φ, we charge the online algorithm $\alpha T(\phi, \sigma)$ and $\beta d_S(\phi, \varphi)$, respectively.

Our algorithm, which we call DECO, operates as follows: We simulate some algorithm UNFAIR for the UMTS problem on (S, d_S). At each request we compute $w_{\zeta(\phi)}(\sigma, \phi, U)$ for all ϕ and all U. From these, we compute the task vector $\boldsymbol{T}(\sigma)$. We feed this vector to UNFAIR. Whenever UNFAIR changes state, we move servers between sub-spaces, maintaining the invariant that if UNFAIR is in state ϕ, then the servers are in some configuration in ϕ.

To further describe the algorithm, we define some notation. We break the request sequence into *phases*. The first phase starts at the beginning of the sequence, and ends when the state of UNFAIR changes. In general, the ith phase begins immediately after the end of phase $i - 1$, and is completed when UNFAIR changes state for the ith time. The last phase is terminated by the end of the sequence. We define λ_i to be the state during the ith phase. We can assume without loss of generality that when the state changes, exactly one server moves between sub-spaces; any multiple server configuration change can be seen as a sequence of single server moves.

At the beginning of the ith phase, we have some configuration I_i of servers. If $i = 1$ this is just the initial configuration, while if $i > 1$ this is dictated by the behavior of the algorithm at the end of phase $i - 1$.

During the phase, we use the algorithm $\mathcal{A}(U, \lambda_i(U))$ to control the movements of servers within U, for each $U \in \mathcal{U}$. At the beginning of the phase, these algorithms are initialized, and given the starting configurations $I_i \cap U, U \in \mathcal{U}$. During the phase, we give each request to a point in U to $\mathcal{A}(U, \lambda_i(U))$. Within each $U \in \mathcal{U}$, the servers of DECO mimic those of $\mathcal{A}(U, \lambda_i(U))$ exactly.

At the end of the phase, a server moves between sub-spaces. We move it to an arbitrary starting point in the destination sub-space. We define F_i to be the final configuration of servers, before this occurs.

Lemma 2. *If*

$$\beta \geq \frac{\theta + 2}{\theta - 2k} + \frac{k\psi + 2k}{\theta},$$

then for all σ, $\mathrm{E}[\mathrm{cost}_{\mathrm{DECO}}(\sigma, C_0)] \leq \mathrm{E}[\mathrm{cost}_{\mathrm{UNFAIR}}(\varrho, s_0)] + O(1)$.

Proof. Fix the set of random choices made by UNFAIR. Once this is done, the behavior of UNFAIR is deterministic. For all sets of random choices, we show that the cost incurred by UNFAIR is within an additive constant of the cost incurred by DECO. Taking expectations over all possible random choices gives the desired result.

Define ℓ_i be the index of the last request of the ith phase, and m be the number of phases. Further define $f_i = \ell_{i-1}+1$, $\ell_0 = 0$, $\ell_{m+1} = n$ and $\lambda_{m+1} = \lambda_m$.

Let Z_i be the cost incurred by DECO during the ith phase. For the sake of readability, we shall drop subscripts of i in the following arguments. By the definition of DECO we have

$$Z \leq d_P(F, I_{i+1}) + \sum_{U \in \mathcal{U}} \mathrm{cost}_{\mathcal{A}(U, \lambda(U))}(\sigma_f^\ell \cap U, I \cap U)$$

$$\leq \frac{\theta + 2}{\theta - 2k} d_S(\lambda, \lambda_{i+1}) + k\psi\nabla + \alpha \sum_{U \in \mathcal{U}} w_I(\sigma_f^\ell, \lambda(U), U).$$

By Fact 42, for some configuration D, the cost incurred by UNFAIR during this phase is

$$\alpha \sum_{j=f}^{l} T(\lambda, \sigma_1^j) + \beta \, d_S(\lambda, \lambda_{i+1})$$

$$= \alpha \sum_{U \in \mathcal{U}} \left(w_{\zeta(\lambda)}(\sigma_1^\ell, \lambda(U), U) - w_{\zeta(\lambda)}(\sigma_1^{f-1}, \lambda(U), U) \right) + \beta \, d_S(\lambda, \lambda_{i+1})$$

$$= \alpha \sum_{U \in \mathcal{U}} \left(w_D(\sigma_f^\ell, \lambda(U), U) + w_{\zeta(\lambda)}^D(\sigma_1^{f-1}, \lambda(U), U) - w_{\zeta(\lambda)}(\sigma_1^{f-1}, \lambda(U), U) \right)$$

$$\quad + \beta \, d_S(\lambda, \lambda_{i+1})$$

$$\geq \alpha \sum_{U \in \mathcal{U}} \left(w_D(\sigma_f^\ell, \lambda(U), U) - \lambda(U)\nabla \right) + \beta \, d_S(\lambda, \lambda_{i+1})$$

$$\geq \alpha \sum_{U \in \mathcal{U}} \left(w_I(\sigma_f^\ell, \lambda(U), U) - 2\lambda(U)\nabla \right) + \beta \, d_S(\lambda, \lambda_{i+1})$$

$$= \alpha \sum_{U \in \mathcal{U}} w_I(\sigma_f^\ell, \lambda(U), U) - 2k\nabla + \beta \, d_S(\lambda, \lambda_{i+1}).$$

The two inequality steps are by the slope conditions. If $\lambda_i \neq \lambda_{i+1}$ then this is at least Z. $\lambda_i = \lambda_{i+1}$ is only true in the last phase. $\qquad\square$

Putting this all together, we get our main theorem:

Theorem 2. *Let* (P, d_P) *be a metric space and define*

$$\beta = 1 + \frac{2k+2}{\theta - 2k} + \frac{k\psi + 2k}{\theta}, \qquad s = \binom{k+t-1}{t-1}.$$

Suppose that

1. (P, d_P) *is* θ-*decomposable into* $U_1, \ldots, U_t$ *with* $\theta > 2k$;
2. *we have an* $\alpha + \beta g(s)$ *competitive algorithm for the UMTS problem on the metric* $(S, d_S) = \Phi(P, d_P, \mathcal{U})$;
3. *we have* α-*competitive* $\psi k \nabla$ *inhibited algorithms for* $U_1, \ldots, U_t$;

then DECO *is* $\alpha + \beta g(s)$-*competitive for the k-server problem on* (P, d_P).

Using Theorem 1, we get a $\alpha + O(\beta t \log^2(k+t) \log^2(t \log(k+t)))$-competitive algorithm. The term $t \log^2(k+t) \log^2(t \log(k+t))$ is polylogarithmic in k when t is polylogarithmic in k. For the case of $t - 2$, the metric (S, d_S) is isomorphic to $k+1$ evenly spaced points on the line. By exploiting the special structure of this metric, we get an $\alpha + O(\beta \log^2 k)$-competitive k-server algorithm.

5   Application to Specific Metrics

In addition to having potential application towards the eventual solution of the randomized k-server problem, our main theorem can be applied to get randomized algorithms for a number of specific metric spaces. We give a few examples here:

Suppose we have a finite metric space which is $\Omega(k \log k)$-decomposable into $O(\log k)$ uniform sub-spaces, each with diameter 1. We use the MARK algorithm within each sub-space. MARK is $2H_k$-competitive and $O(k \log k)$-inhibited. $\beta = O(1)$ and therefore DECO is $2H_k + O(\log^3 k \log^2 \log^2 k) = O(\text{polylog}(k))$-competitive.

As a second example, the balanced metric space $B(2^i, \theta)$ is defined as follows: 1) $B(1, \theta)$ consists of a single point. 2) $B(2^i, \theta)$ consists of two copies of $B(2^{i-1}, \theta)$, call them T and U, such that $d(t, u) = \theta^i$, for all $t \in T$ and $u \in U$. DECO is $O(\log^3 k)$-competitive for balanced spaces for sufficiently large θ.

6   Conclusions

We have presented the first general decomposition theorem for the k-server problem. We feel that such theorems will inevitably be a part of the final resolution of the randomized k-server conjecture. It is our hope that the result presented here provides the impetus for further work on this fascinating problem.

Some steps in the right direction are as follows: 1) The construction of Bartal [1,2], allows us to $O(h \log n \log \log n)$ approximate any metric with h-HST's. While our decomposition theorem can be applied to give algorithms for h-HST's, we require that $h = \theta > 2k$. To get an algorithm for general spaces, we need a decomposition theorem which can be applied to h-HST's with h polylogarithmic in k. 2) Another direction of progress would be to improve the dependence of the competitive ratio on the number of sub-spaces in the decomposition. Bartal's construction gives no upper limit on the number of such sub-spaces. I.e. we might potentially have an h-HST which decomposes into $O(n)$ sub-spaces. Our decomposition theorem has a competitive ratio which is super-linear in the number of sub-spaces. Thus we need either to modify Bartal's technique to guarantee a restricted number of sub-spaces at each h-HST level, or improve the competitive ratio guarantee of the decomposition theorem to be polylogarithmic in the number of sub-spaces. If we could overcome these two short-comings, we would have a $O(\text{polylog}(k))$-competitive k-server algorithm for any metric space on $n = O(\text{poly}(k))$ points.

We also feel that the result presented here is important in that it illustrates how unfair metrical task systems can be used to design randomized online algorithms. The only previous application was in the design of MTS algorithms. We believe that the UMTS technique should have much wider applicability.

References

1. BARTAL, Y. Probabilistic approximation of metric spaces and its algorithmic applications. In *Proceedings of the 37th IEEE Symposium on Foundations of Computer Science* (1996), pp. 183–193.
2. BARTAL, Y. On approximating arbitrary metrics by tree metrics. In *Proceedings of the 30th ACM Symposium on Theory of Computing* (1998), pp. 161–168.
3. BARTAL, Y., BLUM, A., BURCH, C., AND TOMKINS, A. A polylog(n)-competitive algorithm for metrical task systems. In *Proceedings of the 29th ACM Symposium on Theory of Computing* (1997), pp. 711–719.
4. BARTAL, Y., BOLLOBÁS, B., AND MENDEL, M. Ramsey-type theorems for metric spaces and their application for metrical task systems. Manuscript, 2001.
5. BARTAL, Y., CHROBAK, M., AND LARMORE, L. L. A randomized algorithm for two servers on the line. *Information and Computation 158*, 1 (Apr 2000), 53–69.
6. BEIN, W., CHROBAK, M., AND LARMORE, L. The 3-server problem in the plane. In *Proceedings of the 7th Annual European Symposium on Algorithms* (Jul 1999), pp. 301–312.
7. BEN-DAVID, S., BORODIN, A., KARP, R., TARDOS, G., AND WIGDERSON, A. On the power of randomization in on-line algorithms. *Algorithmica 11*, 1 (Jan 1994), 2–14.
8. BLUM, A., BURCH, C., AND KALAI, A. Finely-competitive paging. In *Proceedings of the 40th IEEE Symposium on Foundations of Computer Science* (1999), pp. 450–457.
9. BLUM, A., KARLOFF, H., RABANI, Y., AND SAKS, M. A decomposition theorem for task systems and bounds for randomized server problems. *SIAM Journal on Computing 30*, 5 (Dec 2000), 1624–1661.
10. BORODIN, A., AND EL-YANIV, R. *Online Computation and Competitive Analysis.* Cambridge University Press, 1998.
11. BORODIN, A., LINIAL, N., AND SAKS, M. An optimal online algorithm for metrical task systems. *Journal of the ACM 39*, 4 (Oct 1992), 745–763.

12. CHROBAK, M., KARLOFF, H., PAYNE, T., AND VISHWANATHAN, S. New results on server problems. *SIAM Journal on Discrete Mathematics 4*, 2 (May 1991), 172–181.

13. CHROBAK, M., AND LARMORE, L. An optimal on-line algorithm for k-servers on trees. *SIAM Journal on Computing 20*, 1 (Feb 1991), 144–148.

14. CHROBAK, M., AND LARMORE, L. L. Server problems and on-line games. In *Proceedings of the DIMACS Workshop on On-line Algorithms* (Feb 1991), pp. 11–64.

15. CHROBAK, M., LARMORE, L. L., LUND, C., AND REINGOLD, N. A better lower bound on the competitive ratio of the randomized 2-server problem. *Information Processing Letters 63*, 2 (1997), 79–83.

16. FIAT, A., KARP, R., LUBY, M., McGEOCH, L., SLEATOR, D., AND YOUNG, N. Competitive paging algorithms. *Journal of Algorithms 12*, 4 (Dec 1991), 685–699.

17. FIAT, A., AND MENDEL, M. Better algorithms for unfair metrical task systems and applications. In *Proceedings of the 32nd Annual ACM Symposium on Theory of Computing* (May 2000), pp. 725–734.

18. FIAT, A., AND WOEGINGER, G., Eds. *On-Line Algorithms—The State of the Art.* Lecture Notes in Computer Science. Springer-Verlag, 1998.

19. IRANI, S. Randomized weighted caching with two page weights. Manuscript, 1999.

20. KARLIN, A., MANASSE, M., McGEOCH, L., AND OWICKI, S. Competitive randomized algorithms for nonuniform problems. *Algorithmica 11*, 6 (Jun 1994), 542–571.

21. KARLOFF, H., RABANI, Y., AND RAVID, Y. Lower bounds for randomized k-server and motion-planning algorithms. *SIAM Journal on Computing 23*, 2 (Apr 1994), 293–312.

22. KOUTSOUPIAS, E., AND PAPADIMITRIOU, C. On the k-server conjecture. *Journal of the ACM 42* (1995), 971–983.

23. KOUTSOUPIAS, E., AND PAPADIMITRIOU, C. The 2-evader problem. *Information Processing Letters 57*, 5 (Mar 1996), 249–252.

24. MANASSE, M., McGEOCH, L., AND SLEATOR, D. Competitive algorithms for server problems. *Journal of Algorithms 11*, 2 (Jun 1990), 208–230.

25. McGEOCH, L., AND SLEATOR, D. A strongly competitive randomized paging algorithm. *Algorithmica 6*, 6 (1991), 816–825.

26. SEIDEN, S. S. Unfair problems and randomized algorithms for metrical task systems. *Information and Computation 148*, 2 (Feb 1999), 219–240.

27. SLEATOR, D., AND TARJAN, R. Amortized efficiency of list update and paging rules. *Communications of the ACM 28*, 2 (Feb 1985), 202–208.

Buying a Constant Competitive Ratio for Paging

János Csirik[1], Csanád Imreh[1*], John Noga[2**], Steve S. Seiden[3***], and
Gerhard J. Woeginger[2**]

[1] University of Szeged, Department of Computer Science, Árpád tér 2, H-6720
Szeged, Hungary. {csirik,cimreh}@inf.u-szeged.hu
[2] Institut für Mathematik, Technische Universität Graz, Steyrergasse 30, A-8010
Graz, Austria. {noga,gwoegi}@opt.math.tu-graz.ac.at
[3] Department of Computer Science, 298 Coates Hall, Louisiana State University,
Baton Rouge, LA 70803, USA. sseiden@acm.org

Abstract. We consider a variant of the online paging problem where
the online algorithm may buy additional cache slots at a certain cost.
The overall cost incurred equals the total cost for the cache plus the
number of page faults. This problem and our results are a generalization
of both, the classical paging problem and the ski rental problem.

We derive the following three tight results: (1) For the case where the
cache cost depends linearly on the cache size, we give a λ-competitive
online algorithm where $\lambda \approx 3.14619$ is a solution of $\lambda = 2 + \ln \lambda$. This
competitive ratio λ is best possible. (2) For the case where the cache cost
grows like a polynomial of degree d in the cache size, we give an online
algorithm whose competitive ratio behaves like $d/\ln d + o(d/\ln d)$. No
online algorithm can reach a competitive ratio better than $d/\ln d$. (3)
We exactly characterize the class of cache cost functions for which there
exist online algorithms with finite competitive ratios.

1 Introduction

The classical problem. The paging problem considers a two level memory system
where the first level (the *cache*) can hold k pages, and where the second level
(the *slow memory*) can store n pages. The n pages ($n \gg k$) in slow memory represent virtual memory pages. A *paging algorithm* is confronted with a sequence
of requests to virtual memory pages. If the page requested is in the cache (a page
hit), no cost is incurred; but if the page is not in the cache (a page *fault*), then
the algorithm must bring it into the cache at unit cost. Moreover, the algorithm
must decide which of the k pages currently in cache to evict in order to make
room for the newly requested page.

The paging problem has inspired several decades of theoretical and applied
research and has now become a classical problem in computer science. This is

---

* Supported by the Hungarian National Foundation for Scientific Research, Grant
T030074
** Supported by the START program Y43-MAT of the Austrian Ministry of Science.
*** Supported by the Research Competitiveness Subprogram of the Louisiana Board of
Regents.

F. Meyer auf der Heide (Ed.): ESA 2001, LNCS 2161, pp. 98–108, 2001.

due to the fact that managing a two level store of memory has long been, and continues to be, a fundamentally important problem in computing systems. The paging problem has also been one of the cornerstones in the development of the area of online algorithms. Starting with the seminal work of Sleator & Tarjan [7] which initiated the recent interest in the competitive analysis of online algorithms, the paging problem has motivated the development of many important innovations in this area.

In the *offline* version of the paging problem the request sequence is a priori known to the algorithm. Belady [1] gives a polynomial time optimal offline algorithm for paging. In the *online* version of the paging problem each request must be served without any knowledge of future requests. An online algorithm is *R-competitive* if on all possible request sequences the ratio of the algorithm's cost to the optimal offline cost is bounded by the constant R. The *competitive ratio* of an online paging algorithm is the smallest such constant R. Sleator & Tarjan [7] show that for cache size k, the online algorithm LRU (which always evicts the *least recently used* page) has a competitive ratio of k and that no better ratio is possible. In fact, they prove the following more general result that we will use many times in this paper.

Proposition 1. *(Sleator & Tarjan [7])*
If LRU with cache size k is compared against an optimal offline algorithm with cache size ℓ, then LRU has a competitive ratio of $k/(k - \ell + 1)$.

No better competitive ratio is possible: For every online algorithm A with cache size k, there exist arbitrarily long request sequences σ such that A faults on every page of σ whereas the optimal offline algorithm with cache size ℓ only faults on a fraction $(k - \ell + 1)/k$ of σ.   □

For more information on online paging we refer the reader to the survey chapter of Irani [5] and to the book of Borodin & El Yaniv [2].

The problem considered in this paper. In all previous work on online paging a basic assumption was that the cache size k is fixed a priori and cannot be changed. In this paper, we consider the situation where at any moment in time the algorithm may increase its cache size by purchasing additional cache slots. If its final cache size is x, then it is charged a cost $c(x)$ for it. Here $c : \mathbb{N} \rightarrow \mathbb{R}^+$ is a non-decreasing, non-negative cost function that is a priori known to the algorithm. An equivalent way of stating this problem is the following: The algorithm starts with no cache. At each request, the algorithm may increase its cache from its current size x_1 to a larger size x_2 at a cost of $c(x_2) - c(x_1)$.

We have three basic motivations for looking at this problem. First, and most importantly, the classical online analysis of paging is criticized for being overly pessimistic. For instance, LRU (and all other deterministic algorithms) cannot have a competitive ratio smaller than the size of the cache, which can be arbitrarily large. However, in practice LRU typically performs within a small constant ratio of optimal for all cache sizes. For any particular instance of our problem any algorithm will either have a constant competitive ratio or not be competitive. Second, we see this version as a first approximation to the problem of deciding when and how to upgrade memory systems. Though our version ignores many practical concerns, it incorporates enough to illustrate ideas like amortizing cost over time and balancing cost against performance. Third, we can model a system where a portion of memory is reserved for the use of one high priority process (or set of processes) and balance the faults incurred by this process against the decreasing system performance for all other processes.

The offline version of this problem is quite easy to solve: For a given cost function $c(k)$ and a given request sequence σ, there are only a polynomial number of possible optimal cache sizes (from 0 to the length of the request sequence). All of them can be checked in polynomial time by computing the cost of Belady's algorithm [1] on the sequence σ with this cache size, and by adding the cache cost to the result. The best such solution yields the global optimum. Note that for many special cost functions $c(k)$ (e.g., convex functions) there are even faster offline methods for finding the optimal cache size.

Now let us turn to the online version of this problem. If the function $c(x)$ has the form $c(x) = 0$ for $x \leq k$ and $c(x) = \infty$ for $x > k$, then we are back at classical paging with cache size k. Moreover, our problem captures some features of classical renting-versus-buying problems (the most simple instance of which is the well known ski rental problem (see Karlin, Manasse, McGeoch & Owicki [6])). Imreh & Noga [4] analyze machine scheduling problems where the online algorithm may adjust its resources at an additional cost.

Notation. For a fixed algorithm A and a given request sequence σ, we denote by $\mathrm{fault}_A(\sigma)$ the number of page faults that A incurs on σ, and we denote by $\mathrm{cache}_A(\sigma)$ the cost $c(x)$ where x is the final cache size of A on σ. Moreover, we define $\mathrm{cost}_A(\sigma) = \mathrm{fault}_A(\sigma) + \mathrm{cache}_A(\sigma)$. An online algorithm A is called *R-competitive* if there is a fixed constant b such that for all request sequences σ we have

$$\mathrm{cost}_A(\sigma) \leq R \cdot \mathrm{cost}_{OPT}(\sigma) + b. \tag{1}$$

The smallest R for which an online algorithm is R-competitive is its competitive ratio. The optimal offline algorithm will be denoted by *OPT*.

Organization of the paper. In Section 2 we summarize and explain our three main results. The proof of the result on linear cost functions can be found in Sections 3 (proof of the positive result) and 4 (proof of the negative result), respectively. The results on polynomial cost functions are proved in Section 5. The characterization of cost functions that allow finite competitive online algorithms is proved in Section 6. Finally, we end with a short conclusion in Section 7.

2   Our Results

We start by discussing several 'natural' cost functions $c(x)$. In the simplest case the cost of the cache is proportional to its size and each cache slot costs much more than a single page fault.

Theorem 1. *Assume that the cache cost is $c(x) = \alpha x$ where α is some positive real. Let $\lambda \approx 3.14619$ be the largest real solution of*

$$\lambda \;=\; 2 + \ln \lambda. \tag{2}$$

Then there exists an online algorithm for the cache purchase problem with competitive ratio λ. For any $r < \lambda$ there exists an α such that no online algorithm can be r-competitive.

The equation (2) has two real solutions, where the smaller one is approximately 0.15859 and where the larger one is $\lambda \approx 3.14619$. For users of MAPLE we remark that $\lambda = -W(-e^{-2})$ where $W(\cdot)$ is the well known Lambert W function [3].

We also note that in the (easier, but somewhat unreasonable) case where cache slots are not much more expensive than a single fault (i.e., where α is close to 1), an online algorithm can be $(1 + \alpha)$-competitive by simply purchasing a cache location for each new item requested and never evicting any item. Besides linear cost functions, another fairly natural special case considers polynomial cost functions.

Theorem 2. *Assume that the cache cost is $c(x) = x^d$ with $d \geq 2$. Then there exists an online algorithm for the cache purchase problem whose competitive ratio grows like $d/\ln d + o(d/\ln d)$. Moreover, no online algorithm can reach a competitive ratio that is better than $d/\ln d$.*

Finally, we will exactly characterize the class of cost functions c for which there exist online algorithms with finite competitive ratios.

Theorem 3. *There exists an online algorithm with finite competitive ratio for online paging with cache purchasing cost $c(x)$ if and only if*

$$\exists q > 1 \; \exists p > 0 \; \exists s \geq 0 \; \exists X > 0 \; \forall x \geq X : \quad c(qx) \;\leq\; p \cdot c(x) + s. \tag{3}$$

One way to interpret the condition described above is that the cost function $c(x)$ must be polynomially bounded.

3 The Positive Result for Linear Cost Functions

In this section we prove the positive result claimed in Theorem 1 for linear cost functions of the form $c(x) = \alpha x$ with $\alpha > 0$. We consider the online algorithm *BETA* that uses *LRU* as its paging strategy, and that increases its cache size to ℓ as soon as fault$_{BETA} \geq \alpha\beta(\ell - 1)$ holds. Here the critical parameter β equals $1/\ln \lambda$ where λ was defined in equation (2). Then $\beta \approx 0.872455$, and it can be checked that β is the unique positive root of $2 + 1/\beta = e^{1/\beta}$.

Lemma 1. *If λ is defined as in (2) and $\beta = 1/\ln \lambda$, then any real $y \geq 0$ satisfies the inequality*

$$(\beta + 1)y \;\leq\; \lambda(1 + \beta y - \beta - \beta \ln y).$$

Proof. First note that by the definition of β we have $\ln(\frac{\beta\lambda}{\beta\lambda-\beta-1}) = 1/\beta$. Next we will apply the well known inequality $z + 1 \le e^z$ for real numbers z. Setting $z \doteq y \cdot \frac{\beta\lambda-\beta-1}{\beta\lambda} - 1$ in this inequality, taking logarithms, and applying some algebra yields

$$\ln y \le \ln\left(\frac{\beta\lambda}{\beta\lambda-\beta-1}\right) + y \cdot \frac{\beta\lambda-\beta-1}{\beta\lambda} - 1 = \frac{1}{\beta} + y \cdot \frac{\beta\lambda-\beta-1}{\beta\lambda} - 1.$$

It is easily verified that this inequality leads to the claimed inequality. $\square$

Now consider an arbitrary request sequence σ. Let k denote the final cache size of *BETA* when fed with σ, and let ℓ denote the optimal offline cache size for σ. We denote by f_i the total number of page faults that *BETA* incurs up to the moment when it purchases the ith cache slot. Then between buying the $(i+1)$th and the ith slot, *BETA* has $f_{i+1} - f_i = \alpha\beta$ faults.

Lemma 2. *For a sequence σ on which BETA purchases a cache of size k and OPT uses a cache of size ℓ*

$$\text{fault}_{OPT}(\sigma) \ge \sum_{i=\ell}^{k-1}(f_{i+1} - f_i)(i - \ell + 1)\frac{1}{i}.$$

Proof. If $\ell \ge k$ then the lemma is trivial.

For $\ell < k$, we prove the lemma by modifying the standard technique of dividing the request sequence into phases. The first phase will begin with the first request. If at some point i is the current size of *BETA*'s cache, i distinct items have been requested during the current phase, and an item distinct from the i items requested during this phase is the next request then end the current phase and begin the next phase with the next request.

Consider a given phase which ends with *BETA*'s cache size equal to i. Since *BETA* uses *LRU* as its paging strategy and because of the way a phase is ended, during this phase *BETA* will fault on any item at most once. On the other hand between the second request in this phase and the first request of the next phase the optimal algorithm must fault at least $i - \ell + 1$ times. If the size of *BETA*'s cache was also i at the beginning of the phase then this phase contributes exactly what is needed to the sum. If however *BETA*'s cache size was smaller then this phase contributes (slightly) more than is necessary to the sum. $\square$

We now prove the positive result in Theorem 1.

Proof. First assume that $k \le \ell$. Since offline purchases ℓ cache slots the offline cost is at least $\alpha\ell$. Since *BETA* did not purchase the $(k+1)$-th cache slot, at the end $\text{fault}_{BETA}(\sigma) < \alpha\beta k$ and $\text{cache}_{BETA}(\sigma) = \alpha k$. Since $\beta + 1 \le \lambda$, the cost of *BETA* is at most a factor of λ above the optimal offline cost.

Now assume that $k > \ell$. Using the previous lemma:

$$\mathrm{cost}_{OPT}(\sigma) = \mathrm{cache}_{OPT}(\sigma) + \mathrm{fault}_{OPT}(\sigma)$$

$$\geq \alpha\ell + \sum_{i=\ell}^{k-1}(f_{i+1} - f_i)(i - \ell + 1)\frac{1}{i}$$

$$\geq \alpha\ell + \alpha\beta\sum_{i=\ell}^{k-1}(i - \ell)\frac{1}{i}$$

$$\geq \alpha\ell + \alpha\beta\left(k - \ell - \ell\ln\frac{k}{\ell}\right).$$

Since the online cache cost is αk and the online fault cost is at most $\alpha\beta k$, by substituting $y \doteq k/\ell > 1$ we have

$$\mathrm{cost}_{BETA}(\sigma)/\mathrm{cost}_{OPT}(\sigma) \leq \alpha(\beta + 1)k \;/\; \left(\alpha\ell + \alpha\beta\left(k - \ell - \ell\ln\frac{k}{\ell}\right)\right)$$

$$= (\beta + 1)y \;/\; (1 + \beta y - \beta - \beta\ln y) \quad \leq \quad \lambda.$$

Here we used Lemma 1 to get the final inequality. Therefore, algorithm $BETA$ indeed is a λ-competitive online algorithm for paging with cache purchasing cost $c(x) = \alpha x$. This completes the proof of the positive statement in Theorem 1. $\square$

4   The Negative Result for Linear Cost Functions

In this section we prove the negative result claimed in Theorem 1 for linear cost functions of the form $c(x) = \alpha x$ with $\alpha \gg 1$. The proof is done by an adversary argument.

Assume that there is an online algorithm $\mathcal{A}$ which is r-competitive for some $1 < r < \lambda$ and that α is very large. Further, assume that the pages are numbered $1,2,3,\ldots$. The adversary always requests the smallest numbered page which is not in the online cache. Thus, the online algorithm faults on every request. Let σ be the infinite sequence of requests generated by this procedure.

In order to be finitely competitive, the online algorithm cannot have any fixed upper bound on the size of its cache; hence, the number of purchased slots is unbounded. Let f_i be the number of requests (respectively, the number of page faults) which precede the purchase of the ith cache slot. Note that $f_1 = 0$, and that $f_1, f_2, \ldots$ is a monotone non-decreasing integer sequence. The cost to $\mathcal{A}$ for the requests $1, \ldots, f_i$ equals $i\alpha + f_i$.

We will now consider the adversary's cost for requests $\sigma_i = 1, \ldots, f_i$. Guided by the results of Sleator & Tarjan [7] (see Proposition 1) we upper bound the adversary's cost in the following lemma:

Lemma 3. *If $\mathcal{A}$ purchases the jth cache slot after f_j faults and OPT uses a cache of size ℓ then*

$$\mathrm{fault}_{OPT}(\sigma_i) \;\leq\; i + \sum_{j=\ell}^{i-1}(f_{j+1} - f_j)\frac{j - \ell + 1}{j}.$$

Proof. Any algorithm will fault the first time a particular page is requested. These i faults correspond to the first term in the right hand side of the above inequality. Below we will explicitly exclude these faults from consideration.

As in Lemma 2 we divide the request sequence into phases. However, this time we do this in a slightly different way. The first phase begins with the first request. If j is the size of $\mathcal{A}$'s cache and j requests have been made during the current phase then the current phase ends and the next request begins a new phase. Note that this differs slightly from Lemma 2, since the first request in a phase need not be distinct from all requests in the previous phase.

Consider a phase which begins with the size of $\mathcal{A}$'s cache equal to j and ends with the size of $\mathcal{A}$'s cache equal to j'. During this phase there will be j' requests from items $1, 2, \ldots, j' + 1$. Note that none of the items labeled $j + 2, \ldots, j' + 1$ will have been requested in any previous phase, since the size of $\mathcal{A}$'s cache prior to this phase was j. Recall that the first fault on each of these $j' - j$ items has already been counted.

From this point we proceed in a fairly standard manner. We claim that an optimal offline algorithm will incur at most $j' - \ell + 1$ faults. Note that this is equivalent to stating that the optimal offline algorithm will *not* fault on $\ell - 1$ requests.

If possible, whenever the offline algorithm faults it evicts an item which will not be requested in the remainder of this phase. If at some point it is not possible for the offline algorithm to evict such an item then all ℓ items in its cache will be requested later in this phase. In this case, it is easy to see that there will be at least $\ell - 1$ requests on which the offline algorithm will not fault. On the other hand, if the offline algorithm is able to evict $j' - \ell + 1$ items which will not be requested later in this phase then its cache contains all of the (at most ℓ distinct) items which will be requested during the remainder of the phase.

Of the $j' - \ell + 1$ faults which the offline algorithm will incur during this phase, $j' - j$ faults correspond to the first time an item is requested (these $j' - j$ faults have already been counted). So this phase will contribute $j - \ell + 1$ faults to the sum. If $j' = j$ then this phase contributes precisely what is claimed. If instead $j' > j$ this phase contributes (slightly) less.   □

We now prove the lower bound of λ on any online algorithm.

Proof. Fix an online algorithm $\mathcal{A}$. For a given α, if f_i/i is not bounded above then $\mathcal{A}$ cannot have a constant competitive ratio. Clearly, f_i/i is bounded below by 0 (for $i \geq 1$). So $L = \liminf_i \frac{f_i}{i}$ exists. Suppose that the adversary initially purchases i/λ cache locations and serves σ_i with only these locations. From the definition of L, we know that for any ϵ there are arbitrarily large M such that $f_i/i \geq L - \epsilon$ for all $i \geq M/\lambda$. Further for sufficiently large M, $\left| \sum_{j=M/\lambda}^{M-1} 1/j - \ln(\lambda) \right| \leq \epsilon$. Using the previous lemma we get

$$\frac{\text{cost}_{\mathcal{A}} - b}{\text{cost}_{OPT}} \geq \frac{\alpha M + f_M - b}{\alpha M/\lambda + M + \sum_{j=M/\lambda}^{M-1} (f_{j+1} - f_j)\frac{j - M/\lambda + 1}{j}}$$

$$= \frac{\alpha M + f_M - b}{\alpha M/\lambda + M + f_M \frac{M-M/\lambda}{M-1} + \sum_{j=M/\lambda+1}^{M-1} f_j \frac{1-M/\lambda}{j(j-1)} - f_{M/\lambda} \frac{\lambda}{M}}$$

$$\geq \frac{\alpha + L - b/M - \epsilon}{\alpha/\lambda + 1 + (L + 2\epsilon)(1 - 1/\lambda - 1/\lambda(\ln(\lambda) - \epsilon)) + (\ln(\lambda) + \epsilon)/M)}$$

$$\geq \frac{\alpha + L - b/M - \epsilon}{\alpha/\lambda + 1 + (L + 2\epsilon)/\lambda + (L + 2\epsilon)(\ln(\lambda) + \epsilon)/M}.$$

In the final inequality we have used that $1/\lambda = 1 - 1/\lambda - 1/\lambda \ln(\lambda)$. For α and M sufficiently large, the final value can be made arbitrarily close to λ.   □

5   The Results for Polynomial Cost Functions

In this section we prove the two results (one positive and one negative) that are claimed in Theorem 2 for cost functions of the form $c(x) = x^d$ with $d \geq 2$.

We start with the proof of the positive result.

Proof. Let $\varepsilon > 0$ be a small real number. Similar to Section 3 we consider an online algorithm *BETA2* that uses *LRU* as its paging strategy. This time the online algorithm increases its cache size to k as soon as it has incurred at least $d^\varepsilon (k-1)^d$ page faults. We will show that this algorithm has a competitive ratio of at most $d(1 + d^{-\varepsilon})/[(1 - \varepsilon) \ln d]$.

Consider an arbitrary request sequence σ. Let k denote the final cache size of *BETA2*, and let ℓ denote the optimal offline cache size for σ. If $\ell \geq k+1$, the offline cost is at least $(k+1)^d$ and the online cost is at most $k^d + d^\varepsilon (k+1)^d$. Then the online cost is at most a factor of $1 + d^\varepsilon$ above the offline cost. From now on we assume that $\ell \leq k$. For $i = \ell, \ldots, k$ we denote by f_i the total number of page faults that *BETA2* incurs until the moment when it purchases the ith cache slot. Using an argument similar to that of Lemma 2, we get that the optimal offline algorithm incurs at least $\sum_{i=\ell}^{k-1} (f_{i+1} - f_i)(i - \ell + 1)\frac{1}{i}$ page faults during this time. Therefore,

$$\text{cost}_{OPT}(\sigma) \geq \ell^d + \sum_{i=\ell}^{k-1} (f_{i+1} - f_i) \frac{i - \ell + 1}{i}$$

$$\geq \ell^d + d^\varepsilon \sum_{i=\ell}^{k-1} ((i+1)^d - i^d) \frac{i - \ell}{i}$$

$$= \ell^d + d^\varepsilon (k^d - \ell^d) - d^\varepsilon \ell \sum_{i=\ell}^{k-1} ((i+1)^d - i^d) \frac{1}{i}$$

$$\geq \ell^d + d^\varepsilon (k^d - \ell^d) - d^\varepsilon \ell \int_\ell^k \frac{d \cdot x^{d-1}}{x} dx$$

$$= \ell^d + d^\varepsilon(k^d - \ell^d) - \ell\frac{d^{1+\varepsilon}}{d-1}(k^{d-1} - \ell^{d-1})$$

$$\approx \ell^d + d^\varepsilon(k - \ell)k^{d-1}.$$

The online cache cost is approximately $\text{cost}_{BETA2}(\sigma) \approx (1 + d^\varepsilon)k^d$. Substituting $y := \ell/k \leq 1$ and putting things together we get

$$\text{cost}_{BETA2}(\sigma)/\text{cost}_{OPT}(\sigma) \leq \frac{(1 + d^\varepsilon)k^d}{\ell^d + d^\varepsilon(k - \ell)k^{d-1}} = \frac{1 + d^\varepsilon}{y^d + d^\varepsilon(1 - y)}. \quad (4)$$

The denominator in the right hand side of (4) is minimized for $y = d^{(\varepsilon-1)/(d-1)}$ and hence is at least

$$d^{(\varepsilon-1)d/(d-1)} + d^\varepsilon\left(1 - d^{(\varepsilon-1)/(d-1)}\right) \geq d^\varepsilon\left(1 - d^{(\varepsilon-1)/(d-1)}\right).$$

By applying some (tedious) calculus we get that as d tends to infinity, the function $1 - d^{(\varepsilon-1)/(d-1)}$ grows like $(1 - \varepsilon)\frac{\ln d}{d}$. By combining these observations with the inequality in (4), we conclude that the competitive ratio R of $BETA2$ is bounded by

$$R \leq \frac{(1 + d^{-\varepsilon})d}{(1 - \varepsilon)\ln d}.$$

This completes the proof of the positive statement in Theorem 2.   □

We turn to the proof of the negative statement in Theorem 2 which is done by an adversary argument.

Proof. Consider an r-competitive online algorithm for cost functions of the form $c(x) = x^d$. The pages are numbered 1,2,3,..., and the adversary always requests the smallest numbered page which is not in the online cache. Thus, the online algorithm faults on every request. In order to have a finite competitive ratio, the online algorithm cannot run forever with the same number of slots. Hence, the number of purchased slots must eventually exceed gr where g is a huge integer. Suppose that after i_g requests the online cache is extended for the first time to a size $k \geq gr$. Then the corresponding online cost at this moment is at least $i_g + (gr)^d$.

Now consider the optimal offline algorithm for cache size $gr - g + 1$. Since the online algorithm was serving the first i_g requests with a cache size of at most $gr - 1$, the results of Sleator & Tarjan [7] (see Proposition 1) yield that the number of offline faults is at most

$$i_g \cdot \frac{(gr - 1) - (gr - g + 1) + 1}{gr - 1} = \frac{i_g(g - 1)}{gr - 1} \leq \frac{i_g}{r}.$$

The offline cache cost is $(gr - g + 1)^d$. Since the online algorithm is r-competitive, there exists a constant b such that the following inequality is fulfilled for all integers g; cf. equation (1):

$$i_g + (gr)^d \leq r \cdot \left(\frac{i_g}{r} + (gr - g + 1)^d\right) + b.$$

Since g can be arbitrarily large, this implies $r^d \leq r(r-1)^d$ which is equivalent to

$$\frac{1}{r} \leq \left(1 - \frac{1}{r}\right)^d. \tag{5}$$

Now suppose that $r < d/\ln d$. Then the left hand side in (5) is at least $\ln d/d$, whereas the right hand side is at most $1/d$. This contradiction completes the proof of the negative statement in Theorem 2.   □

6   The Results for the General Case

In this section we prove the two results (one positive and one negative) that are claimed in Theorem 3.

We start with the proof of the positive result.

Proof. So we assume that the cost function $c(x)$ satisfies condition (3). Fix a request sequence σ. We may assume that the optimal offline algorithm for σ uses a cache of size $x_{OPT} \geq X$. The case when $x_{OPT} < X$ can be disregarded by making b in the definition of competitiveness sufficiently large; in fact, any b greater than $c(X)$ will do.

Consider the algorithm BAL which uses LRU as its paging strategy and which tries to balance its cache cost and its fault cost. In other words, BAL increases its cache size to x as soon as $\text{fault}_{BAL} \geq c(x)$. Until the time where BAL purchases a cache of size $q \cdot x_{OPT}$, the cost ratio of online to offline is at most $2p$: At this time $\text{cache}_{BAL} = c(qx_{OPT}) \leq p \cdot c(x_{OPT})$ and $\text{fault}_{BAL} \approx \text{cache}_{BAL}$, whereas the offline cost is at least $c(x_{OPT})$. From the time where BAL purchases a cache of size $q \cdot x_{OPT}$ onwards, the ratio is at most $2q/(q-1)$: By using the result of Sleator & Tarjan as stated in Proposition 1 with $\ell = x_{OPT}$ and $k = q \cdot x_{OPT}$, we get that $\text{fault}_{BAL}/\text{fault}_{OPT} \leq qx_{OPT}/(qx_{OPT} - x_{OPT} + 1)$. Therefore,

$$\text{cost}_{BAL} \approx 2\,\text{fault}_{BAL} \leq 2\frac{q \cdot x_{OPT} \cdot \text{fault}_{OPT}}{qx_{OPT} - x_{OPT} + 1} < \frac{2q}{q-1}\,\text{cost}_{OPT}.$$

To summarize, we have shown that BAL is $\max\{2p, 2q/(q-1)\}$-competitive. This completes the argument for the positive result.   □

Now let us turn to the negative result.

Proof. So we assume that the cost function $c(x)$ does not satisfy the condition (3), and that therefore

$$\forall q > 1 \; \forall p > 0 \; \forall s \geq 0 \; \forall X > 0 \; \exists x \geq X : \quad c(qx) > p \cdot c(x) + s. \tag{6}$$

The idea is quite simple. If OPT uses a cache of size x then an online algorithm which wants to be R competitive must eventually purchase a cache of size px for some $p \approx R/(R-1)$. The result of Sleator and Tarjan as stated in Proposition 1 requires that $R \cdot \text{fault}_{OPT} - \text{fault}_A$ cannot be too large. On the other hand, $Rc(x) - c(px)$ can be made arbitrarily large, since c is not polynomially bounded.

Now for the details. We will proceed by contradiction. Assume that there is an algorithm A which is R-competitive for some $R > 1$ and fix b as in the definition of competitiveness. By using (6) we can choose x to satisfy

$$c\left(x\frac{2R-1}{2R-2}\right) > R \cdot c(x) + x(2R-1)/2 + R + b.$$

If we use the lower bound sequence from Proposition 1 for $k = x(2R-1)/(2R-2) - 1$ and $\ell = x$ until A purchases a cache of size $x(2R-1)/(2R-2)$, then $R \cdot \text{fault}_{BEL}(\sigma) - \text{fault}_A(\sigma) \le x(2R-1)/2$. Note that A must eventually purchase a cache of this size, since otherwise cost_A will tend to ∞ while $\text{cost}_{OPT} \le c((2R-1)x/(2R-2)) + (2R-1)x/(2R-2)$. Therefore,

$$\begin{aligned} \text{cost}_A(\sigma) &= \text{cache}_A(\sigma) + \text{fault}_A(\sigma) \\ &> R \cdot \text{cache}_{BEL}(\sigma) + R \cdot \text{fault}_{BEL}(\sigma) + b \\ &= R \cdot \text{cost}_{OPT}(\sigma) + b. \end{aligned}$$

This contradiction completes the proof of the negative result in Theorem 3.   □

7   Conclusion

We have considered a simple model of caching which integrates the ability to add additional cache locations. A number of results for linear, polynomial, and arbitrary cost functions have been found. One possible direction for further study of this problem is to consider the degree to which randomization can further reduce the competitive ratios found in this paper. The primary difficulty when attacking the randomized version of the problem is finding relationships between the costs of the online and offline algorithms when the cache sizes are unequal.

References

1. L.A. BELADY. A study of replacement algorithms for virtual storage computers. *IBM Systems Journal 5*, 78–101, 1966.
2. A. BORODIN AND R. EL YANIV. *Online Computation and Competitive Analysis.* Cambridge University Press, 1998.
3. R.M. CORLESS, G.H. GONNET, D.E.G. HARE, D.J. JEFFREY, AND D.E. KNUTH. On The Lambert W Function. Maple Share Library.
4. CS. IMREH AND J. NOGA. Scheduling with Machine Cost. Proceedings 2nd International Workshop on Approximation Algorithms, Springer LNCS 1671, 168–176, 1999.
5. S. IRANI. Competitive analysis of paging. In A. Fiat and G.J. Woeginger (eds.) *Online Algorithms – The State of the Art.* Springer LNCS 1442, 52–73, 1998.
6. A. KARLIN, M. MANASSE, L. MCGEOCH, AND S. OWICKI. Competitive randomized algorithms for nonuniform problems. *Algorithmica 11*, 542–571, 1994.
7. D. SLEATOR AND R.E. TARJAN. Amortized efficiency of list update and paging rules. *Communications of the ACM 28*, 202–208, 1985.

Simple Minimal Perfect Hashing in Less Space

Martin Dietzfelbinger[1] and Torben Hagerup[2]

[1] Fakultät für Informatik und Automatisierung, Technische Universität Ilmenau,
Postfach 100565, D–98684 Ilmenau.
martin.dietzfelbinger@theoinf.tu-ilmenau.de

[2] Institut für Informatik, Johann Wolfgang Goethe-Universität Frankfurt,
D–60054 Frankfurt am Main. hagerup@ka.informatik.uni-frankfurt.de

Abstract. A *minimal perfect hash function* for a set S is an injective mapping from S to $\{0, \ldots, |S| - 1\}$. Taking as our model of computation a unit-cost RAM with a word length of w bits, we consider the problem of constructing minimal perfect hash functions with constant evaluation time for arbitrary subsets of $U = \{0, \ldots, 2^w - 1\}$. Pagh recently described a simple randomized algorithm that, given a set $S \subseteq U$ of size n, works in $O(n)$ expected time and computes a minimal perfect hash function for S whose representation, besides a constant number of words, is a table of at most $(2 + \epsilon)n$ integers in the range $\{0, \ldots, n - 1\}$, for arbitrary fixed $\epsilon > 0$. Extending his method, we show how to replace the factor of $2 + \epsilon$ by $1 + \epsilon$.

Keywords: Data structures, randomized algorithms, dictionaries, hashing, hash tables, minimal perfect hash functions, space requirements.

1 Introduction

A *minimal perfect hash function* for a set S is an injective mapping from S to $\{0, \ldots, |S| - 1\}$. Given a minimal perfect hash function h for the set of keys of a collection $\mathcal{R}$ of n records, we can store $\mathcal{R}$ efficiently to enable the retrieval of records by their keys. If a record with key x is stored in $B[h(x)]$ for some array B of size n, the only additional space needed is that required for the representation of h, and the time needed to access a record in $\mathcal{R}$ is essentially the evaluation time of h.

Taking as our model of computation the unit-cost RAM with a word length of w bits, for some fixed positive integer w, we study the construction of minimal perfect hash functions for arbitrary given subsets S of the universe $U = \{0, \ldots, 2^w - 1\}$ of keys representable in single computer words. We will only be interested in hash functions that can be evaluated in constant time. The remaining parameters of interest are the time needed to find a minimal perfect hash function for a given set S and the space needed to store it.

Tarjan and Yao [12] investigated the compression of a sparsely used two-dimensional table A and considered a class of *displace-and-project* functions that shift each row of A horizontally by an amount associated with the row and called

F. Meyer auf der Heide (Ed.): ESA 2001, LNCS 2161, pp. 109–120, 2001.

its *displacement* and then project the shifted rows vertically to a one-dimensional table B. We call displacements for a subset of the rows of A *compatible* if no two used entries in the shifted rows under consideration reside in the same column. In order for the compression from A to B to work as intended, the displacements of all the rows must be compatible.

Define the *weight* of a row in A as the number of used entries in that row. Tarjan and Yao showed that if A contains n used entries and satisfies a certain *harmonic-decay* property, then compatible row displacements in the range $\{0, \ldots, n-1\}$ can be found by a simple *first-fit-decreasing* (*FFD*) algorithm that processes the rows one by one in an order of nonincreasing weight and, for each row, chooses the smallest nonnegative displacement compatible with all previously chosen row displacements. They also demonstrated that the harmonic-decay property can be enforced by a preprocessing phase that shifts each column of A vertically by a suitable *column displacement*.

Pagh [10] observed that for a table A with n columns, the row shifts of Tarjan and Yao can be replaced by cyclic shifts (he considered a more general operation of no concern here). More significantly, he introduced a new approach to enforcing the harmonic-decay property. In his setting, the table entries in A are not the original universe. Instead, the column and row indices of a used entry in A are the values obtained by applying suitable functions f and g to a key in a set S with $|S| = n$ for which a minimal perfect hash function is to be constructed. Pagh identified a set of conditions concerning f and g, one of which is that (f, g) is injective on S, and showed that if these conditions are satisfied, then a minimal perfect hash function for S can be computed in $O(n)$ expected time by a *random-fit-decreasing* (*RFD*) algorithm that operates like the FFD algorithm, except that the displacements tried for each row of weight at least 2 are chosen randomly, rather than in the order $0, 1, \ldots$, until a displacement compatible with the previously chosen displacements is encountered. Complementing this, Pagh showed that if f and g are chosen at random from suitable classes of hash functions and A has $m \geq (2 + \epsilon)n$ rows, for some fixed $\epsilon > 0$, then the conditions required by the analysis of the RFD algorithm hold with a probability bounded from below by a positive constant. Altogether, this yields a construction in $O(n)$ expected time of a simple minimal perfect hash function for S that can be evaluated in constant time and whose representation consists of a constant number of words (to specify f and g) together with the m row displacements. If the chosen displacements are $d_0, \ldots, d_{m-1}$, the hash value of a key x is simply $(f(x) + d_{g(x)}) \bmod n$.

Taking Pagh's construction as our starting point, we reduce the minimal number of row displacements from $(2 + \epsilon)n$ to $(1 + \epsilon)n$, for arbitrary fixed $\epsilon > 0$, thus essentially halving the space requirements of the minimal perfect hash function. In order to achieve this, we introduce a new algorithm for computing row displacements, the *Undo-One* algorithm. Except for the initial sorting of the rows by nonincreasing weight, the FFD and RFD algorithms are both *online* in the sense that they choose each displacement taking into account only rows whose displacements were already chosen. Informally, a displacement, once fixed, is never again changed. The Undo-One algorithm deviates slightly from this principle by allowing the computation of each displacement to change a single displacement that was chosen earlier. Informally, if it is difficult to place the row at hand, it is permissible to make room for it by relocating a single previously placed row. This added flexibility allows the Undo-One algorithm to cope with a higher level of crowding in the array A, which translates into a smaller value of m.

Concretely, we formulate new conditions concerning f and g, more permissive than those of Pagh, and show that these conditions enable the Undo-One algorithm to succeed in $O(n)$ expected time. On the other hand, we show that functions f and g that satisfy the new conditions can be found in $O(n)$ expected time as long as $m \geq (1 + \epsilon)n$, for arbitrary fixed $\epsilon > 0$. Our proof of this makes heavier demands on the class of hash functions from which g is drawn than does Pagh's analysis—the values under g should behave approximately as if they were uniformly distributed and independent random variables. We show, extending results of [3] slightly, that a standard class of hash functions—remainders of polynomials of sufficiently high degree over a finite field—behaves sufficiently randomly.

Both parts of our argument, analyzing the Undo-One algorithm and establishing the required properties of f and g, are more involved than the corresponding parts of Pagh's analysis. The form of the resulting minimal perfect hash function, however, $x \mapsto (f(x) + d_{g(x)}) \bmod n$, is exactly the same as in Pagh's construction if n is odd, except that g must now be chosen from a class of hash functions that meets additional requirements.

A comprehensive discussion of related work can be found in a survey by Czech et al. [2]. In a seminal paper, Fredman et al. [6] described a randomized construction of a static dictionary with constant access time for a given set of n keys. The construction works in $O(n)$ expected time and is easily modified to yield a minimal perfect hash function. The space requirements are $O(1)$ words of w bits plus $O(n)$ words of $O(\log n)$ bits, for a total of $O(w + n \log n)$ bits. Fredman and Komlos [5] showed that the minimal number of bits needed to represent minimal perfect hash functions is $n \log_2 e + \log_2 w - O(\log n)$. Schmidt and Siegel [11] proved that a constant access time can be achieved together with space requirements of $O(n + \log w)$, but they did not describe an efficient construction algorithm. Hagerup and Tholey [8] closed this gap by exhibiting a randomized construction that works in $O(n + \log w)$ expected time, and they also reduced the space requirements further to $n \log_2 e + \log_2 w + o(n + \log w)$, i.e., to within lower-order terms of the optimum.

The work of Pagh cited above as well as other efforts in the area of minimal perfect hashing do not emphasize the space minimization quite so heavily, but rather try to keep the number of probes into the data structure as small as possible. Pagh's algorithm, as well as the new algorithm described here, needs one—input-independent—access to the description of f and g and one access to the table of displacements. In a different line of research, building upon a long series of earlier work, Czech et al. [1] and Havas et al. [9] studied the construction of minimal perfect hash functions of the form $h(x) = (\sum_{i=1}^{t} d_{g_i(x)}) \bmod n$, where t is a constant, g_i maps U to $\{0, \ldots, m-1\}$, for $i = 1, \ldots, t$, and $d_0, \ldots, d_{m-1}$ are suitably chosen values in $\{0, \ldots, n-1\}$. Storing the description of such a function essentially amounts to storing $d_0, \ldots, d_{m-1}$. Assuming an idealized situation in which $g_1, \ldots, g_t$ behave fully randomly on S, the authors utilized results on threshold values for random (hyper)graphs to be acyclic to show that $m = (2+\epsilon)n$ is a suitable choice for $t = 2$, and they demonstrated experimentally that $m \approx 1.23n$ is one for $t = 3$. In the latter case, evaluating h requires three accesses to the table containing $d_0, \ldots, d_{m-1}$. Similarly, functions of the form $h(x) = (f(x) + d_{g_1(x)} + d_{g_2(x)}) \bmod n$, where $f : U \to \{0, \ldots, n-1\}$, $g_1 : U \to \{0, \ldots, m/2-1\}$, and $g_2 : U \to \{m/2, \ldots, m-1\}$, were proposed by Fox et al. [4] and stated to work well (meaning successful construction of minimal perfect hash functions in $O(n)$ expected time) for m as small as $0.6n$. However, only experimental evidence and no formal arguments were offered in support of this claim.

2   The Undo-One Algorithm

Let n and m be positive integers and suppose that S is a set of size n and that f and g are functions mapping S to $\{0, \ldots, n-1\}$ and to $\{0, \ldots, m-1\}$, respectively, so that (f, g) is injective on S. In terms of the informal description of the previous section, $f(x)$ and $g(x)$ are the column and row indices, respectively, of the cell in A to which $x \in S$ is mapped by (f, g). For $0 \leq j < m$, let $R_j = \{f(x) : x \in S \text{ and } g(x) = j\}$. Informally, R_j is the set of used positions in row j of A. Correspondingly, we call R_j a *row set*. Given a row set R and an integer d, we define $R \oplus d$ to be the set $\{(r+d) \bmod n : r \in R\}$ obtained by "rotating R by d positions to the right". The goal of the Undo-One algorithm (and of the RFD algorithm) is to compute integers $d_0, \ldots, d_{m-1}$ in the range $\{0, \ldots, n-1\}$, called *(row) displacements*, for which the sets $R_0 \oplus d_0, \ldots, R_{m-1} \oplus d_{m-1}$ are disjoint.

For $0 \leq j < m$ and $0 \leq d < n$, by "placing R_j at (displacement) d" we will mean fixing the value of d_j to be d. At a time when exactly the row sets R_j with indices j in some set J have been placed, we call $V = \bigcup_{j \in J}(R_j \oplus d_j)$ the set of *occupied positions* and $W = \{0, \ldots, n-1\} \setminus V$ the set of *free positions*. Two row sets R and R', placed at d and d', *collide* if $(R \oplus d) \cap (R' \oplus d') \neq \emptyset$. A row set R will be called *large* if $|R| \geq 3$, and it is a *pair* if $|R| = 2$ and a *singleton* if $|R| = 1$.

The Undo-One algorithm begins by computing a permutation σ of $\{0, \ldots, m-1\}$ with $|R_{\sigma(0)}| \geq |R_{\sigma(1)}| \geq \cdots \geq |R_{\sigma(m-1)}|$ and then places $R_{\sigma(l)}$ for $l = 0, \ldots, m-1$. Observe than once all larger row sets have been placed, it is trivial to place the singletons in linear time. This is because, at that time, the number of free positions exactly equals the number of singletons. For this reason, we need not describe how to place singletons (nor how to place empty row sets).

Large row sets are simply placed according to the RFD algorithm. When attempting to place a pair P, the Undo-One algorithm works harder than for large row sets. It first tries to place P using a random displacement, just as the RFD algorithm. If P collides with exactly one previously placed row set R, however, the algorithm does not give up right away, but instead tries to make room for P at the displacement chosen for it by relocating R to a random displacement. Only if this also fails, a new attempt to place P is started. It turns out to be essential for the analysis that the alternative displacement used for the tentative relocation of R is not chosen according to the uniform distribution over $\{0, \ldots, n-1\}$. If the old displacement of R is d and $q = \max P - \min P$, the tentative new displacement is found as follows: With probability $1/2$, a random value is chosen from the uniform distribution over $\{(d-q) \bmod n, (d+q) \bmod n\}$, and with probability $1/2$, a random value is chosen from the uniform distribution over $\{0, \ldots, n-1\}$.

3   Analysis of the Undo-One Algorithm

We will analyze the Undo-One algorithm assuming n to be odd. The analysis centers on the distribution of row-set sizes. For $k = 0, 1, \ldots$, let $T_k = |\{0 \leq j < m : |R_j| = k\}|$ be the number of row sets of size k. In this section, the following conditions will be assumed to hold for some $\nu > 0$:

$$\text{The function } (f, g) \text{ is injective on } S; \tag{A}$$

$$(1 + \nu) \sum_{k \geq 3} k^2 T_k \leq n; \tag{B}$$

$$T_1 + \sum_{k \geq 1} T_k \geq (1 + \nu)n. \tag{C}$$

Define a *trial* to be a single attempt by the Undo-One algorithm, as described in the previous section, to place a row set. We consider the placing of large row sets and the placing of pairs separately (as observed in the previous section, we need not concern ourselves with the placing of smaller row sets). The analysis of the placing of large row sets is exactly as Pagh's analysis, except that our Condition (B) pertains only to large row sets and therefore enables the placing of large row sets only.

Lemma 1. *For constant ν, the Undo-One algorithm places all large row sets in $O(n)$ expected time.*

Proof. Consider the placing of a fixed large row set R and let $k_0 = |R| \geq 3$. Each trial chooses $d \in \{0, \ldots, n-1\}$ uniformly at random and succeeds unless $(R \oplus d) \cap V \neq \emptyset$, where V is the set of occupied positions. The latter happens with probability at most $k_0 |V|/n$. Since $|R'| \geq k_0$ for all previously placed rows sets R', we have $|V| \leq \sum_{k \geq k_0} k T_k$ and $k_0 |V|/n \leq \sum_{k \geq k_0} k^2 T_k/n$. By Condition (B) and because $k_0 \geq 3$, the latter quantity is bounded by $1/(1 + \nu) < 1$. For constant ν, the expected number of trials to place R is therefore constant. Each trial can be carried out in $O(k_0)$ time, so that R is placed in $O(k_0)$ expected time. Over all large row sets, this sums to $O(n)$ expected time.

From now on we consider the placing of a fixed pair P and define V and W as the set of free and occupied positions, respectively, before the placing of P.

Lemma 2. *A trial can be executed in constant expected time.*

Proof. Associate a random variable K with a trial in the following way: If P collides with a single row set R, then let $K = |R|$; otherwise take $K = 1$. If we maintain for each position in $\{0, \ldots, n-1\}$ an indication of the row set by which it is occupied, if any, it is easy to execute a trial in $O(K)$ time. For $k \geq 2$, $\Pr(K = k) \leq |P|kT_k/n$. Therefore, by Condition (B), $E(K) = O\big(1 + (1/n)\sum_{k \geq 3} k^2 T_k\big) = O(1)$.

Let $q = \max P - \min P$. We will say that two elements v and v' of $\{0, \ldots, n-1\}$ are q-*spaced* if $v' = (v + q) \bmod n$ or $v' = (v - q) \bmod n$, i.e., if v and v' are at a "cyclic distance" of q in the additive group $\mathbb{Z}_n$. Let $G = (V \cup W, E)$ be the undirected bipartite graph on the vertex sets V and W that contains an edge $\{v, v'\}$ between $v \in V$ and $v' \in W$ precisely if v and v' are q-spaced. The maximum degree of G is bounded by 2. We will call the vertices in V and W *black* and *white*, respectively.

Lemma 3. *G is acyclic.*

Proof. Assume that G contains a simple cycle of length l. Then l is the order of the element q in the additive group $\mathbb{Z}_n$. By Lagrange's theorem, l divides $|\mathbb{Z}_n| = n$, which is odd by assumption. On the other hand, l is even because G is bipartite, a contradiction.

Lemma 4. *For constant ν, the expected number of trials is constant.*

Proof. It suffices to bound the success probability of a trial from below by a positive constant. We will assume without loss of generality that $\nu \leq 1/2$. Observe first that since $\sum_{k \geq 1} T_k \leq n$, Condition (C) implies that $|W| \geq T_1 \geq \nu n$. We consider two cases, depending on the number of edges in E.

Case 1: $|E| \leq 2(1 - \nu)|W|$. Let $E' = \{\{v, v'\} : v, v' \in W$ and v and v' are q-spaced$\}$ and suppose that we inserted the edges in E' in G. This would raise the degree of each white vertex to exactly 2; i.e., $|E| + 2|E'| = 2|W|$ and $|E'| = |W| - |E|/2 \geq \nu|W| \geq \nu^2 n$. It is easy to see that there is a 1-1 correspondence between E' and the set of displacements at which P can be placed without collisions and without relocating another row set. Therefore a trial succeeds with probability at least ν^2.

Case 2: $|E| > 2(1 - \nu)|W|$. For each row set R placed at a displacement d, we call the subgraph of G spanned by the edges incident on vertices in $R \oplus d$ the *row graph* of R. A row graph with k black vertices can have at most $2k$ edges. It is called *good* if it has $2k - 1$ or $2k$ edges. In order to establish a lower bound on the number of good row graphs, let us imagine that we remove

$\min\{2k - 2, r\}$ edges from each row graph with exactly k black vertices and r edges. The number of edges removed is at most $\sum_{k \geq 2}(2k - 2)T_k$, and therefore the number of remaining edges is larger than

$$2(1 - \nu)|W| - \sum_{k \geq 2}(2k - 2)T_k \geq 2\Big((1 - \nu)T_1 - \sum_{k \geq 1}(k - 1)T_k\Big)$$

$$= 2\Big(T_1 + \sum_{k \geq 1}T_k - \sum_{k \geq 1}kT_k - \nu T_1\Big) \overset{(C)}{\geq} 2((1 + \nu)n - n - \nu T_1)$$

$$= 2\nu(n - T_1) \geq 2\nu|V| \geq \nu|E| > 2\nu(1 - \nu)|W| \geq \nu^2 n.$$

Each remaining edge belongs to a good row graph, and no row graph contains more than two of the remaining edges. Therefore the number of good row graphs is at least $(\nu^2/2)n$.

Since G is acyclic and has maximum degree at most 2, every connected component of a row graph is a simple path. We call such a path *perfect* if its first and last vertices are white. It is easy to see that a good row graph can have at most one connected component that is not a perfect path. We call a row set *perfect* if its row graph is good and has at least one perfect path. Let us consider two subcases.

Subcase 2.1: There are at least $(\nu^2/4)n$ perfect row sets. It is easy to see that a perfect row set R_j can be rotated either left or right by q positions without colliding with any other row set. Moreover, this necessarily leaves two free q-spaced positions at the end of a former perfect path, in which P can be placed (see Fig. 1). A trial that attempts to place P in these two positions will succeed with probability at least $1/4$, namely if the displacement used for the tentative relocation of R_j is chosen to be one particular value in $\{(d_j - q) \bmod n, (d_j + q) \bmod n\}$. Thus each trial succeeds with probability at least $\nu^2/16$.

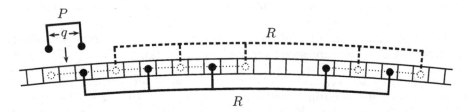

Fig. 1. A perfect row set R is rotated right by q positions to make room for P.

Subcase 2.2: There are fewer than $(\nu^2/4)n$ perfect row sets. Assume first that $n \geq 2(4/\nu^2)$. We call a row set R *regular* if its row graph is a simple path with one black and one white end vertex. If the row graph of a row set R is good yet R is not perfect, then R is regular. Since the number of good row graphs is at least $(\nu^2/2)n$, at least $(\nu^2/4)n$ row sets are regular. The key observation is that every regular row set can be relocated to "mesh with" any other regular row set that is at least as large, and that this leaves at least two free q-spaced positions, in which P can be placed (see Fig. 2). This shows that a trial succeeds with probability at least

$$\frac{1}{n^2}\binom{(\nu^2/4)n}{2} \geq \tfrac{1}{4}(\nu^2/4)^2.$$

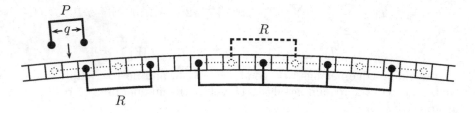

Fig. 2. Two regular row sets are "meshed" to make room for P.

Assume finally that $n < 2(4/\nu^2)$. Since the number of good row graphs is positive, at least one white vertex is not of degree 2. But then a trial succeeds with probability at least $1/n \geq \tfrac{1}{2}(\nu^2/4)$.

4   Analysis of the Hash Functions

In this section we show how to find functions f and g that satisfy Conditions (A)–(C). Both f and g will be picked from a family of the form

$$\mathcal{H}_{p,s}^t = \left\{ x \mapsto \left(\left(\sum_{i=0}^{t-1} a_i x^i \right) \bmod p \right) \bmod s : 0 \leq a_0, \ldots, a_{t-1} < p \right\},$$

where t is a positive integer, $p > \max S$ is a prime, and $s = n$ in the case of f, while $s = m$ in the case of g.

It is well-known that the unique-interpolation property of polynomials over arbitrary fields can be expressed as follows in our setting.

Proposition 5. *Let $x_1, \ldots, x_t \in S$ be distinct and choose $a_0, \ldots, a_{t-1}$ independently from the uniform distribution over $\{0, \ldots, p-1\}$. Then the random variables $\left(\sum_{i=0}^{t-1} a_i x_j^i\right) \bmod p$, for $j = 1, \ldots, t$, are independent and uniformly distributed over $\{0, \ldots, p-1\}$.*

We consider the random experiment of obtaining a function $h \in \mathcal{H}_{p,s}^t$ by choosing the coefficients $a_0, \ldots, a_{t-1}$ independently from the uniform distribution over $\{0, \ldots, p-1\}$. Using Proposition 5, it is easy to see that for arbitrary $x \in S$ and $j \in \{0, \ldots, s-1\}$, $\Pr(h(x) = j) = \pi_{s,j}$, where

$$\pi_{s,j} = \begin{cases} 1/p \cdot \lceil p/s \rceil, & \text{if } j < p \bmod s; \\ 1/p \cdot \lfloor p/s \rfloor, & \text{if } j \geq p \bmod s. \end{cases}$$

For every integer i, denote by $\binom{S}{i}$ the set of all subsets of S of size i. Given a subset X of S, let $\delta(X)$ be the random variable that takes the value 1 if h is constant on X, and 0 otherwise. By another application of Proposition 5, for arbitrary $i \in \{1, \ldots, t\}$ and $X \in \binom{S}{i}$,

$$E(\delta(X)) = \mu_{s,i},$$

where $\mu_{s,i} = \sum_{j=0}^{s-1} \pi_{s,j}^i$. Since $\sum_{j=0}^{s-1} \pi_{s,j} = 1$, Jensen's inequality shows that $\mu_{s,i} \geq s^{1-i}$, and since $\pi_{s,j} \leq (1/p)(1+p/s) = (1/s)(1+s/p)$ for $j = 0, \ldots, s-1$, we also have $\mu_{s,i} \leq s^{1-i}(1+s/p)^i$. For every fixed t, we can choose p sufficiently large to make $c = (1+m/p)^t$ arbitrarily close to 1.

We next prove that if g has been chosen to satisfy Conditions (B) and (C) (which depend only on g), then in $O(n)$ additional expected time, we can find a function f so that f and g together satisfy Condition (A). The procedure is to choose f repeatedly from $\mathcal{H}_{p,n}^2$ until Condition (A) holds. For the analysis, note first that by virtue of Condition (C),

$$2T_1 + T_2 + \sum_{k \geq 3} T_k \geq n = T_1 + 2T_2 + \sum_{k \geq 3} kT_k,$$

from which we can conclude that $T_1 \geq T_2$ and, subsequently, that $T_2 \leq n/3$. With $b_j = |\{x \in S : g(x) = j\}|$, for $j = 0, \ldots, m-1$, the expected number of collisions under (f, g) (i.e., of pairs $\{x, y\} \subseteq S$ with $f(x) = f(y)$ and $g(x) = g(y)$) is

$$\sum_{0 \leq j < m} \binom{b_j}{2} \mu_{n,2} \leq \frac{c}{n}\left(T_2 + \sum_{\substack{0 \leq j < m \\ b_j \geq 3}} \frac{b_j^2}{2}\right)$$

$$= \frac{c}{n}\left(T_2 + \frac{1}{2}\sum_{k \geq 3} k^2 T_k\right) \leq \frac{c}{n}\left(\frac{n}{3} + \frac{n}{2}\right) = \frac{5c}{6},$$

where Condition (B) was used in the second-to-last step. Provided that p is chosen sufficiently large to make $c < 6/5$, the expected number of collisions is therefore less than one. By Markov's inequality, this means that the expected number of trials needed to satisfy Condition (A) is bounded by a constant.

We now turn to Conditions (B) and (C). Take $b_j = |\{x \in S : g(x) = j\}|$ as above, for $j = 0, \ldots, m-1$, and define $C_i = \sum_{j=0}^{m-1} \binom{b_j}{i} = \sum_{X \in \binom{S}{i}} \delta(X)$, for $i = 0, 1, \ldots$ The following relation between the quantities T_k and C_i is obtained from the "classical" Bonferroni inequalities [7, Inequality I2] by summation over all $j \in \{0, \ldots, m-1\}$.

Proposition 6. *For all integers $k, l \geq 0$,*

$$(-1)^l \cdot \left(T_k - \sum_{i=k}^{k+l} (-1)^{i-k} \binom{i}{k} C_i\right) \leq 0.$$

We now find

$$\sum_{k\geq 3} k^2 T_k = \sum_{k\geq 1} k^2 T_k - T_1 - 4T_2$$

$$= \sum_{j=0}^{m-1} \left(2\binom{b_j}{2} + \binom{b_j}{1}\right) - T_1 - 4T_2 = 2C_2 + C_1 - T_1 - 4T_2.$$

Continuing using Proposition 6, we obtain

$$\sum_{k\geq 3} k^2 T_k \leq 2C_2 + C_1 - (C_1 - 2C_2 + 3C_3 - 4C_4 + 5C_5 - 6C_6)$$

$$- 4(C_2 - 3C_3 + 6C_4 - 10C_5) = 9C_3 - 20C_4 + 35C_5 + 6C_6.$$

Consider repeated trials that pick g randomly from $\mathcal{H}_{p,m}^t$. For $i = 1, \dots, t$,

$$E(C_i) = \binom{n}{i}\mu_{m,i} \leq \frac{n^i}{i!} \cdot cm^{1-i} = c \cdot \frac{\alpha^{i-1}}{i!} n,$$

where $\alpha = n/m$ is the *load factor*. Moreover, for every fixed t,

$$E(C_i) \geq \left(1 - O\left(\tfrac{1}{n}\right)\right) \cdot \frac{\alpha^{i-1}}{i!} n.$$

The upper and lower bounds on $E(C_i)$ show that for fixed $t \geq 6$,

$$E\left(\sum_{k\geq 3} k^2 T_k\right) \leq \left(1 + O\left(\tfrac{1}{n}\right)\right) \cdot cn\phi(\alpha),$$

where $\phi(\alpha) = \frac{9}{3!}\alpha^2 - \frac{20}{4!}\alpha^3 + \frac{35}{5!}\alpha^4 + \frac{6}{6!}\alpha^5$. We are interested only in the range $0 < \alpha \leq 1$, corresponding to $m \geq n$. It is easy to see that ϕ is increasing in this range, so that $\phi(\alpha) \leq \phi(1) = \frac{29}{30}$. Thus, provided that ν is sufficiently small and p is sufficiently large to make $\frac{29}{30}(1+\nu)c < 1$, the expected number of trials needed to satisfy Condition (B) is bounded by a constant for sufficiently large values of n.

The analysis in the case of Condition (C) is similar. We first observe that $T_1 + \sum_{k\geq 1} T_k = T_1 + (m - T_0)$. By Proposition 6, for all even $l \geq 2$ we have

$$T_1 + \sum_{k\geq 1} T_k \geq \sum_{i=1}^{l} (-1)^{i-1}(i+1)C_i$$

and, if $t \geq l$,

$$E\left(T_1 + \sum_{k\geq 1} T_k\right) \geq (1 - O\left(\tfrac{1}{n}\right)) n \cdot \sum_{i=1}^{l} (-1)^{i-1}\frac{i+1}{i!}\alpha^{i-1}.$$

Consider the infinite series

$$\psi(\alpha) = \sum_{i=1}^{\infty} (-1)^{i-1} \frac{i+1}{i!} \alpha^{i-1} = e^{-\alpha} + \frac{1}{\alpha}(1 - e^{-\alpha}).$$

We have $\psi(1) = 1$, and differentiation shows that $\psi'(\alpha) < 0$ for $0 < \alpha \leq 1$, so that $\psi(\alpha) > 1$ for all α with $0 < \alpha < 1$. It follows that if ν is sufficiently small and n and t are sufficiently large, the expected number of trials needed to satisfy Condition (C) is bounded by a constant.

We have shown how to satisfy Conditions (B) and (C) separately. However, we need to satisfy the two conditions simultaneously. We argue that this is possible by showing that for $i \leq t/2$, the random quantity C_i is sharply concentrated around its mean, so that picking g at random from $\mathcal{H}_{p,m}^t$ for suitable p and t satisfies (B) and (C) not only with a probability that is bounded away from zero, but in fact with a probability that tends to 1 as n tends to infinity. By Chebyshev's inequality, it suffices to prove that $\mathrm{Var}(C_i) = O(n)$. This fact, which we establish below, completes the overall argument.

Theorem 7. *Let n, s and t be positive integers, p a prime, $i \leq t/2$ a nonnegative integer and S a subset of $\{0, \ldots, p-1\}$ of size n such that $\alpha = n/s$, i and ts/p are all bounded by constants. Suppose that h is drawn uniformly at random from $\mathcal{H}_{p,s}^t$ and define $b_j = |\{x \in S : h(x) = j\}|$, for $j = 0, \ldots, s-1$, and $C_i = \sum_{j=0}^{s-1} \binom{b_j}{i}$. Then $\mathrm{Var}(C_i) = O(n)$.*

Proof. Assume that $i \geq 1$, since the claim is obvious for $i = 0$. As argued near the beginning of this section, the value of $E(\delta(X))$, where X is a nonempty subset of S of size $k \leq t$, depends only on k; we denote this quantity by μ_k. Now

$$\mathrm{Var}(C_i) = E\big((C_i - E(C_i))^2\big) = E\bigg(\bigg(\sum_{X \in \binom{S}{i}} (\delta(X) - \mu_i)\bigg)^2\bigg)$$

$$= \sum_{X,Y \in \binom{S}{i}} E\big((\delta(X) - \mu_i)(\delta(Y) - \mu_i)\big) = \sum_{X,Y \in \binom{S}{i}} \big(E(\delta(X)\delta(Y)) - \mu_i^2\big)$$

$$\leq \sum_{X,Y \in \binom{S}{i}} E(\delta(X \cup Y)) = \sum_{k=i}^{2i} \sum_{\substack{X,Y \in \binom{S}{i} \\ |X \cup Y| = k}} \mu_k.$$

The term-by-term validity of the last inequality above is immediate in the case of overlapping sets X and Y, for which $\delta(X)\delta(Y) = \delta(X \cup Y)$. When X and Y are disjoint, on the other hand, $\delta(X)$ and $\delta(Y)$ are independent, since $2i \leq t$, so that the left-hand term is zero. Let $i \leq k \leq 2i$. Since i is bounded by a constant, the number of pairs of subsets $X, Y \in \binom{S}{i}$ with $|X \cup Y| = k$ is $O(n^k)$. Moreover, $\mu_k \leq cs^{1-k}$, where $c = (1 + s/p)^t \leq e^{ts/p}$ is bounded by a constant. Hence

$$\mathrm{Var}(C_i) = O\bigg(\sum_{k=i}^{2i} n^k \cdot s^{1-k}\bigg) = O\bigg(n \cdot \sum_{k=i}^{2i} \alpha^{k-1}\bigg) = O(n).$$

If n is not odd, as required by our analysis in Section 3, and $S \neq U$, we can choose an element $x \in U \setminus S$ and compute a minimal perfect hash function h for the key set $S \cup \{x\}$. Subsequently, by adding the same constant to all displacements, modulo $n + 1$, we "rotate" h to obtain a minimal perfect hash function h' for $S \cup \{x\}$ with $h'(x) = n$. The function h' is also a minimal perfect hash function for S. We summarize our findings as follows.

Theorem 8. *For every fixed $\epsilon > 0$, there is an integer constant $t \geq 1$ such that for every integer $w \geq 1$, every subset S of $\{0, \ldots, 2^w - 1\}$ of size $n \geq 1$ and every given prime $p > \max(S \cup \{tn\})$ of $O(w)$ bits, we can, in $O(n)$ expected time on a unit-cost RAM with a word length of w bits, compute a positive integer $m \leq (1 + \epsilon)n$, the coefficients of functions $f \in \mathcal{H}^2_{p,n'}$ and $g \in \mathcal{H}^t_{p,m}$ and integers $d_0, \ldots, d_{m-1}$ such that the function $x \mapsto (f(x) + d_{g(x)}) \bmod n'$ maps S injectively to $\{0, \ldots, n - 1\}$. Here $n' = n$ if n is odd, and $n' = n + 1$ if n is even.*

References

1. Z. J. Czech, G. Havas, and B. S. Majewski, An optimal algorithm for generating minimal perfect hash functions, *Inform. Process. Lett.* **43** (1992), pp. 257–264.
2. Z. J. Czech, G. Havas, and B. S. Majewski, Perfect hashing, *Theoret. Comput. Sci.* **182** (1997), pp. 1–143.
3. M. Dietzfelbinger, J. Gil, Y. Matias, and N. Pippenger, Polynomial hash functions are reliable, *in* Proc. 19th International Colloquium on Automata, Languages and Programming (ICALP 1992), Lecture Notes in Computer Science, Springer, Berlin, Vol. 623, pp. 235–246.
4. E. A. Fox, L. S. Heath, Q. F. Chen, and A. M. Daoud, Practical minimal perfect hash functions for large databases, *Comm. Assoc. Comput. Mach.* **35** (1992), pp. 105–121.
5. M. L. Fredman and J. Komlós, On the size of separating systems and families of perfect hash functions, *SIAM J. Alg. Disc. Meth.* **5** (1984), pp. 61–68.
6. M. L. Fredman, J. Komlós, and E. Szemerédi, Storing a sparse table with $O(1)$ worst case access time, *J. Assoc. Comput. Mach.* **31** (1984), pp. 538–544.
7. J. Galambos and I. Simonelli, *Bonferroni-type Inequalities with Applications*, Springer, New York, 1996.
8. T. Hagerup and T. Tholey, Efficient minimal perfect hashing in nearly minimal space, *in* Proc. 18th Annual Symposium on Theoretical Aspects of Computer Science (STACS 2001), Lecture Notes in Computer Science, Springer, Berlin, Vol. 2010, pp. 317–326.
9. G. Havas, B. S. Majewski, N. C. Wormald, and Z. J. Czech, Graphs, hypergraphs and hashing, *in* Proc. 19th International Workshop on Graph-Theoretic Concepts in Computer Science (WG 1993), Lecture Notes in Computer Science, Springer, Berlin, Vol. 790, pp. 153–165.
10. R. Pagh, Hash and displace: Efficient evaluation of minimal perfect hash functions, *in* Proc. 6th International Workshop on Algorithms and Data Structures (WADS 1999), Lecture Notes in Computer Science, Springer, Berlin, Vol. 1663, pp. 49–54.
11. J. P. Schmidt and A. Siegel, The spatial complexity of oblivious k-probe hash functions, *SIAM J. Comput.* **19** (1990), pp. 775–786.
12. R. E. Tarjan and A. C.-C. Yao, Storing a sparse table, *Comm. Assoc. Comput. Mach.* **22** (1979), pp. 606–611.

Cuckoo Hashing

Rasmus Pagh* and Flemming Friche Rodler

BRICS**
Department of Computer Science
University of Aarhus, Denmark
{pagh,ffr}@brics.dk

Abstract. We present a simple and efficient dictionary with worst case constant lookup time, equaling the theoretical performance of the classic dynamic perfect hashing scheme of Dietzfelbinger et al. The space usage is similar to that of binary search trees, i.e., three words per key on average. The practicality of the scheme is backed by extensive experiments and comparisons with known methods, showing it to be quite competitive also in the average case.

1 Introduction

The *dictionary* data structure is ubiquitous in computer science. A dictionary is used to maintain a set S under insertion and deletion of elements (referred to as *keys*) from a universe U. Membership queries ("$x \in S$?") provide access to the data. In case of a positive answer the dictionary also provides a piece of *satellite data* that was associated with x when it was inserted.

A large literature, briefly surveyed in Sect. 1.1, is devoted to practical and theoretical aspects of dictionaries. It is common to study the case where keys are bit strings in $U = \{0,1\}^w$ and w is the word length of the computer (for theoretical purposes modeled as a RAM). Section 2 briefly discusses this restriction. It is usually, though not always, clear how to return associated information once membership has been determined. E.g., in all methods discussed in this paper, the associated information of $x \in S$ can be stored together with x in a hash table. Therefore we disregard the time and space used to handle associated information and concentrate on the problem of maintaining S. In the following we let n denote $|S|$.

The most efficient dictionaries, in theory and in practice, are based on hashing techniques. The main performance parameters are of course lookup time, update time, and space. In theory there is no trade-off between these. One can simultaneously achieve constant lookup time, expected amortized constant update time, and space within a constant factor of the information theoretical minimum of $B = \log\binom{|U|}{n}$ bits [3]. In practice, however, the various constant factors are crucial for many applications. In particular, lookup time is a critical parameter. It is well known that the expected time for all operations can

---

* Partially supported by the IST Programme of the EU under contract number IST-1999-14186 (ALCOM-FT). Work initiated while visiting Stanford University.
** Basic Research in Computer Science (www.brics.dk), funded by the Danish National Research Foundation.

F. Meyer auf der Heide (Ed.): ESA 2001, LNCS 2161, pp. 121–133, 2001.
© Springer-Verlag Berlin Heidelberg 2001

be made a factor $(1 + \epsilon)$ from optimal (one universal hash function evaluation, one memory lookup) if space $O(n/\epsilon)$ is allowed. Therefore the challenge is to combine speed with a reasonable space usage. In particular, we only consider schemes using $O(n)$ words of space.

The contribution of this paper is a new, simple hashing scheme called *cuckoo hashing*. A description and analysis of the scheme is given in Sect. 3, showing that it possesses the same theoretical properties as the dynamic dictionary of Dietzfelbinger et al. [7]. That is, it has worst case constant lookup time and amortized expected constant time for updates. A special feature of the lookup procedure is that (disregarding accesses to a small hash function description) there are just two memory accesses, which are *independent* and can be done in parallel if this is supported by the hardware. Our scheme works for space similar to that of binary search trees, i.e., three words per key in S on average.

Using weaker hash functions than those required for our analysis, cuckoo hashing is very simple to implement. Section 4 describes such an implementation, and reports on extensive experiments and comparisons with the most commonly used methods, having no worst case guarantee on lookup time. Our experiments show the scheme to be quite competitive, especially when the dictionary is small enough to fit in cache. We thus believe it to be attractive in practice, when a worst case guarantee on lookups is desired.

1.1   Previous Work

Hashing, first described by Dumey [9], emerged in the 1950s as a space efficient heuristic for fast retrieval of keys in sparse tables. Knuth surveys the most important classical hashing methods in [14, Sect. 6.4]. These methods also seem to prevail in practice. The most prominent, and the basis for our experiments in Sect. 4, are CHAINED HASHING (with separate chaining), LINEAR PROBING and DOUBLE HASHING. We refer to [14, Sect. 6.4] for a general description of these schemes, and detail our implementation in Sect. 4.

Theoretical Work. Early theoretical analysis of hashing schemes was typically done under the assumption that hash function values were uniformly random and independent. Precise analyses of the average and expected worst case behaviors of the abovementioned schemes have been made, see e.g. [11]. We mention just that for LINEAR PROBING and DOUBLE HASHING the expected *longest* probe sequence is of length $\Omega(\log n)$. In DOUBLE HASHING there is even no bound on the length of unsuccessful searches. For CHAINED HASHING the expected maximum chain length is $\Theta(\log n / \log \log n)$.

Though the results seem to agree with practice, the randomness assumptions used for the above analyses are questionable in applications. Carter and Wegman [4] succeeded in removing such assumptions from the analysis of chained hashing, introducing the concept of *universal* hash function families. When implemented with a random function from Carter and Wegman's universal family, chained hashing has constant expected time per dictionary operation (plus an amortized expected constant cost for resizing the table).

A dictionary with worst case constant lookup time was first obtained by Fredman, Komlós and Szemerédi [10], though it was *static*, i.e., did not support updates. It was later augmented with insertions and deletions in amortized expected constant time by Dietzfelbinger et al. [7]. Dietzfelbinger and Meyer auf der Heide [8] improved the update performance by exhibiting a dictionary in which operations are done in constant time with high probability, i.e., probability at least $1 - n^{-c}$, where c is any constant of our choice. A simpler dictionary with the same properties was later developed [5]. When $n = |U|^{1-o(1)}$ a space usage of $O(n)$ words is not within a constant factor of the information theoretical minimum. The dictionary of Brodnik and Munro [3] offers the same performance as [7], using $O(B)$ bits in all cases.

Experimental Work. Although the above results leave little to improve from a theoretical point of view, large constant factors and complicated implementation hinder direct practical use. For example, the "dynamic perfect hashing" scheme of [7] uses more than $35n$ words of memory. The authors of [7] refer to a more practical variant due to Wenzel that uses space comparable to that of binary search trees. According to [13] the implementation of this variant in the LEDA library [17], described in [21], has average insertion time larger than that of AVL trees for $n \leq 2^{17}$, and more than four times slower than insertions in chained hashing[1]. The experimental results listed in [17, Table 5.2] show a gap of more than a factor of 6 between the update performance of chained hashing and dynamic perfect hashing, and a factor of more than 2 for lookups[2].

Silverstein [20] explores ways of improving space as well as time of the dynamic perfect hashing scheme of [7], improving both the observed time and space by a factor of roughly three. Still, the improved scheme needs 2 to 3 times more space than linear probing to achieve similar time per operation. It should be noted that emphasis in [20] is very much on space efficiency. For example, the hash tables of both methods are stored in a packed representation, presumably slowing down linear probing considerably.

A survey of experimental work on dictionaries that do not have worst case constant lookup time is beyond the scope of this paper. However, we do remark that Knuth's selection of algorithms seems to be in agreement with current practice for implementation of general purpose dictionaries. In particular, the excellent cache usage of LINEAR PROBING makes it a prime choice on modern architectures.

2   Preliminaries

Our algorithm uses hash functions from a *universal* family.

Definition 1. *A family $\{h_i\}_{i \in I}$, $h_i : U \to R$, is (c,k)-universal if, for any k distinct elements $x_1, \ldots, x_k \in U$, any $y_1, \ldots, y_k \in R$, and uniformly random $i \in I$, $\Pr[h_i(x_1) = y_1, \ldots, h_i(x_k) = y_k] \leq c/|R|^k$.*

---

[1] On a Linux PC with an Intel Pentium 120 MHz processor.

[2] On a 300 MHz SUN ULTRA SPARC.

A standard construction of a $(2, k)$-universal family for $U = \{0, \ldots, p-1\}$ and range $R = \{0, \ldots, r-1\}$, where p is prime, contains, for every choice of $0 \le a_0, a_1, \ldots, a_{k-1} < p$, the function $h(x) = ((\sum_{l=0}^{k-1} a_l x^l) \bmod p) \bmod r$.

We assume that keys from U fit in a single machine word, i.e., $U = \{0, 1\}^w$. This is not a serious restriction, as long keys can be mapped to short keys by choosing a random function from a $(O(1), 2)$-universal family for each word of the key, mapping the key to the bitwise exclusive or of the individual function values [4]. A function chosen in this way can be used to map S injectively to $\{0, 1\}^{2 \log n + O(1)}$, thus effectively reducing the universe size to $O(n^2)$. In fact, with constant probability the function is injective on a given *sequence* of n consecutive sets in a dictionary (see [7]). A result of Siegel [19] says that for any constant $\epsilon > 0$, if the universe is of size $n^{O(1)}$ there is an $(O(1), O(\log n))$-universal family that can be evaluated in *constant* time, using space and initialization time $O(n^\epsilon)$. However, the constant factor of the evaluation time is rather high.

We reserve a special value $\perp \in U$ to signal an empty cell in hash tables. For DOUBLE HASHING an additional special value is used to indicate a deleted key.

3   Cuckoo Hashing

Cuckoo hashing is a dynamization of a static dictionary described in [18]. The dictionary uses two hash tables, T_1 and T_2, of length r and two hash functions $h_1, h_2 : U \to \{0, \ldots, r-1\}$. Every key $x \in S$ is stored in cell $h_1(x)$ of T_1 or $h_2(x)$ of T_2, but never in both. Our lookup function is

> **function** lookup(x)
>   **return** $T_1[h_1(x)] = x \;\vee\; T_2[h_2(x)] = x$.
> **end**;

We remark that the idea of storing keys in one out of two places given by hash functions previously appeared in [12] in the context of PRAM simulation, and in [1] for a variant of chained hashing. It is shown in [18] that if $r \ge (1+\epsilon)\,n$ for some constant $\epsilon > 0$ (i.e., the tables are to be a bit less than half full), and h_1, h_2 are picked uniformly at random from an $(O(1), O(\log n))$-universal family, the probability that there is no way of arranging the keys of S according to h_1 and h_2 is $O(1/n)$. A slightly weaker conclusion, not sufficient for our purposes, was derived in [12]. A suitable arrangement was shown in [18] to be computable in linear time by a reduction to 2-SAT.

We now consider a simple dynamization of the above. Deletion is of course simple to perform in constant time, not counting the possible cost of shrinking the tables if they are becoming too sparse. As for insertion, it turns out that the "cuckoo approach", kicking other keys away until every key has its own "nest", works very well. Specifically, if x is to be inserted we first see if cell $h_1(x)$ of T_1 is occupied. If not, we are done. Otherwise we set $T_1[h_1(x)] \leftarrow x$ anyway, thus making the previous occupant "nestless". This key is then inserted in T_2 in the same way, and so forth. As it may happen that this process loops, the number of iterations is bounded by a value "MaxLoop" to be specified below.

If this number of iterations is reached, everything is rehashed with new hash functions, and we try once again to accommodate the nestless key. Using the notation $x \leftrightarrow y$ to express that the values of variables x and y are swapped, the following code summarizes the insertion procedure.

```
procedure insert(x)
    if lookup(x) then return;
    loop MaxLoop times
        if T₁[h₁(x)] = ⊥ then { T₁[h₁(x)] ← x; return; }
        x ↔ T₁[h₁(x)];
        if T₂[h₂(x)] = ⊥ then { T₂[h₂(x)] ← x; return; }
        x ↔ T₂[h₂(x)];
    end loop
    rehash(); insert(x);
end;
```

The above procedure assumes that the tables remain larger than $(1 + \epsilon)n$ cells. When no such bound is known, a test must be done to find out when a rehash to larger tables is needed. Note that the insertion procedure is biased towards inserting keys in T_1. As seen in Section 4 this leads to faster successful lookups.

3.1   Analysis

We first show that if the insertion procedure loops for MaxLoop $= \infty$, it is not possible to accommodate all the keys of the new set using the present hash functions. Consider the sequence $a_1, a_2, \ldots$ of nestless keys in the infinite loop. For $i, j \geq 1$ we define $A_{i,j} = \{a_i, \ldots, a_j\}$. Let j be the smallest index such that $a_j \in A_{1,j-1}$. At the time when a_j becomes nestless for the second time, the change in the tables relative to the configuration before the insertion is that a_k is now in the previous location of a_{k+1}, for $1 \leq k < j$. Let $i < j$ be the index such that $a_i = a_j$. We now consider what happens when a_j is nestless for the second time. If $i > 1$ then a_j reclaims its previous location, occupied by a_{i-1}. If $i > 2$ then a_{i-1} subsequently reclaims its previous position, which is occupied by a_{i-2}, and so forth. Thus we have $a_{j+z} = a_{i-z}$ for $z = 0, 1, \ldots, i - 1$, and end up with a_1 occurring again as a_{i+j-1}. Define $s_k = |h_1[A_{1,k}]| + |h_2[A_{1,k}]|$, i.e., the number of table cells available to $A_{1,k}$. Obviously $s_k \leq s_{k-1} + 1$, as every key a_i, $i > 1$, has either $h_1(a_i) = h_1(a_{i-1})$ or $h_2(a_i) = h_2(a_{i-1})$. In fact, $s_{j-1} = s_{j-2} \leq j - 1$, because the key a_j found in $T_1[h_1(a_{j-1})]$ or $T_2[h_2(a_{j-1})]$ occurred earlier in the sequence. As all of the keys $a_j, \ldots, a_{j+i-1}$ appeared earlier in the sequence, we have $s_{j+i-2} = s_{j-2}$. Let j' be the minimum index such that $j' > j$ and $a_{j'} \in A_{1,j'-1}$. Similar to before we have $s_{j'-1} = s_{j'-2}$. In conclusion, $|A_{1,j'-1}| = j' - i$ and $s_{j'-1} = s_{j'-2} \leq s_{j+i-2} + (j' - 2) - (j + i - 2) = s_{j-2} + j' - j - i < j' - i$. Thus, there are not sufficiently many cells to accommodate $A_{i,j'-1}$ for the current choice of hash functions.

In conjunction with the result from [18], the above shows that the insertion procedure loops without limit with probability $O(1/n)$. We now turn to the analysis for the case where there is no such loop, showing that the insertion procedure terminates in $O(1)$ iterations, in the expected sense. Consider a prefix $a_1, a_2, \ldots, a_l$ of the sequence of nestless keys. The crucial fact is that there must be a subsequence of at least $l/3$ keys without repetitions, starting with an occurrence of the key a_1, i.e., the inserted key. As earlier, we pick i and j, $i < j$, such that $a_i = a_j$ and j is minimal, and once again we have $a_{j+z} = a_{i-z}$ for $z = 0, 1, \ldots, i-1$. There can be no index $j' > j+i-1$ such that $a_{j'} \in A_{1,j'-1}$, in that our earlier argument showed that the set cannot be accommodated when such indices i, j and j' can be chosen. This means that both of the sequences $a_1, \ldots, a_{j-1}$ and $a_{j+i-1}, \ldots, a_l$ have no repetitions. As $a_1 = a_{j+i-1}$ and $i < j$, one of the sequences must be the desired one of length at least $l/3$.

Suppose that the insertion loop runs for at least t iterations. By the above there is a sequence of distinct keys $b_1, \ldots, b_m$, $m \geq (2t-1)/3$, such that b_1 is the key to be inserted, and such that for some $\beta \in \{0, 1\}$

$$h_{2-\beta}(b_1) = h_{2-\beta}(b_2), \; h_{1+\beta}(b_2) = h_{1+\beta}(b_3), \; h_{2-\beta}(b_3) = h_{2-\beta}(b_4), \ldots \quad (1)$$

Given b_1 there are at most n^{m-1} sequences of m distinct keys. For any such sequence and any $\beta \in \{0, 1\}$, if the hash functions were chosen from a (c, m)-universal family, the probability that (1) holds is bounded by $c\, r^{-(m-1)}$. Thus, the probability that there is *any* sequence of length m satisfying (1) is bounded by $2c\,(n/r)^{m-1} \leq 2c\,(1+\epsilon)^{-(2t-1)/3+1}$. Suppose we use a $(c, 6\log_{1+\epsilon} n)$-universal family, for some constant c (e.g., Siegel's family with constant time evaluation [19]). Then the probability of more than $3\log_{1+\epsilon} n$ iterations is $O(1/n^2)$. Thus, we can set MaxLoop $= 3\log_{1+\epsilon} n$ with a negligible increase in the probability of a rehash. When there is no rehash the expected number of iterations is at most

$$1 + \sum_{t=2}^{\infty} 2c\,(1+\epsilon)^{-(2t-1)/3+1} = O(1 + 1/\epsilon) \; .$$

A rehash has no failed insertions with probability $1 - O(1/n)$. In this case, the expected time per insertion is constant, so the expected time is $O(n)$. As the probability of having to start over with new hash functions is bounded away from 1, the total expected time for a rehash is $O(n)$. This implies that the expected time for insertion is constant if $r \geq (1+\epsilon)(n+1)$. Resizing of tables can be done in amortized expected constant time per update by the usual doubling/halving technique.

4   Experiments

To examine the practicality of CUCKOO HASHING we experimentally compare it to three well known hashing methods, CHAINED HASHING (with separate chaining), LINEAR PROBING and DOUBLE HASHING, as described in [14, Sect. 6.4]. We also consider TWO-WAY CHAINING [1], implemented in a cache-friendly way, as recently suggested in [2].

4.1   Data Structure Design and Implementation

We consider positive 32 bit signed integer keys and use 0 as $\bot$. The data structures are *robust* in that they correctly handle attempts to insert an element already in the set, and attempts to delete an element not in the set. A slightly faster implementation can be obtained if this is known not to occur.

Our focus is on achieving high performance dictionary operations with a reasonable space usage. By the *load factor* of a dictionary we will understand the size of the set relative to the memory used[3]. As seen in [14, Fig. 44] there is not much to be gained in terms of average number of probes for the classic schemes by going for load factor below, say, 1/2 or 1/3. As CUCKOO HASHING only works when the size of each table is larger than the size of the set, we can only perform a comparison for load factors less than 1/2. To allow for doubling and halving of the table size, we allow the load factor to vary between 1/5 and 1/2, focusing especially on the "typical" load factor of 1/3. For CUCKOO HASHING and TWO-WAY CHAINING there is a chance that an insertion may fail, causing a "forced rehash". If the load factor is larger than a certain threshold, somewhat arbitrarily set to 5/12, we use the opportunity to double the table size. By our experiments this only slightly decreases the average load factor.

Apart from CHAINED HASHING, the schemes considered have in common the fact that they have only been analyzed under randomness assumptions that are currently, or inherently, unpractical to implement ($O(\log n)$-wise independence or n-wise independence). However, experience shows that rather simple and efficient hash function families yield performance close to that predicted under stronger randomness assumptions. We use a function family from [6] with range $\{0, 1\}^q$ for positive integer q. For every odd a, $0 < a < 2^w$, the family contains the function $h_a(x) = (ax \bmod 2^w) \operatorname{div} 2^{w-q}$. Note that evaluation can be done by a 32 bit multiplication and a shift. This choice of hash function restricts us to consider hash tables whose sizes are powers of two. A random function from the family (chosen using C's `rand` function) appears to work fine with all schemes except CUCKOO HASHING. For CUCKOO HASHING we found that using a $(1, 3)$-universal family resulted in fewer forced rehashes than when using a $(1, 2)$-universal family. However, it turned out that the exclusive or of three independently chosen functions from the family of [6] was faster and worked equally well. We have no good explanation for this phenomenon. For all schemes, various other families were tried, with a decrease in performance.

All methods have been implemented in C. We have striven to obtain the fastest possible implementation of each scheme. Details differing from the references and specific choices made are:

CHAINED HASHING. We store the first element of each linked list directly in the hash table. This often saves one cache miss, and slightly decreases memory usage, in the expected sense, as every non-empty chained list is one element shorter. C's `malloc` and `free` functions were found to be a performance

---

[3] For CHAINED HASHING, the notion of load factor traditionally disregards the space used for chained lists, but we desire equal load factors to imply equal memory usage.

bottleneck, so a simple "free list" memory allocation scheme is used. Half of the allocated memory is used for the hash table, and half for list elements. If the data structure runs out of free list elements, its size is doubled.

DOUBLE HASHING. Deletions are handled by putting a "deleted" marker in the cell of the deleted key. Queries skip over deleted cells, while insertions over- write them. To prevent the tables from clogging up with deleted cells, re- sulting in poor performance for unsuccessful lookups, all keys are rehashed when 2/3 of the hash table is occupied by keys and "deleted" markers.

TWO-WAY CHAINING. We allow four keys in each bucket. This is enough to keep the probability of a forced rehash low for hundreds of thousands of keys, by the results in [2]. For larger collections of keys one should allow more keys in each bucket, resulting in general performance degradation.

CUCKOO HASHING. The architecture on which we experimented could not par- allelize the two memory accesses in lookups. Therefore we only evaluate the second hash function after the first memory lookup has shown unsuccessful.

Some experiments were done with variants of CUCKOO HASHING. In partic- ular, we considered ASYMMETRIC CUCKOO, in which the first table is twice the size of the second one. This results in more keys residing in the first table, thus giving a slightly better average performance for successful lookups. For example, after a long sequence of alternate insertions and deletions at load factor 1/3, we found that about 76% of the elements resided in the first table of ASYMMET- RIC CUCKOO, as opposed to 63% for CUCKOO HASHING. There is no significant slowdown for other operations. We will describe the results for ASYMMETRIC CUCKOO when they differ significantly from those of CUCKOO HASHING.

4.2   Setup and Results

Our experiments were performed on a PC running Linux (kernel version 2.2) with an 800 MHz Intel Pentium III processor, and 256 MB of memory (PC100 RAM). The processor has a 16 KB level 1 data cache and a 256 KB level 2 "advanced transfer" cache. Our results can be explained in terms of processor, cache and memory speed in our machine, and are thus believed to have signif- icance for other configurations. An advantage of using the Pentium processor for timing experiments is its `rdtsc` instruction which can be used to measure time in clock cycles. This gives access to very precise data on the behavior of functions. Programs were compiled using the `gcc` compiler version 2.95.2, using optimization flags `-O9 -DCPU=586 -march=i586 -fomit-frame-pointer -finline-functions -fforce-mem -funroll-loops -fno-rtti`. As mentioned earlier, we use a global clock cycle counter to time operations. If the number of clock cycles spent exceeds 5000, and there was no rehash, we conclude that the call was interrupted, and disregard the result (it was empirically observed that no operation ever took between 2000 and 5000 clock cycles). If a rehash is made, we have no way of filtering away time spent in interrupts. However, all tests were made on a machine with no irrelevant user processes, so disturbances should be minimal.

Our first test was designed to model the situation in which the size of the dictionary is not changing too much. It considers a sequence of mixed operations generated at random. We constructed the test operation sequences from a collection of high quality random bits publicly available on the Internet [15]. The sequences start by insertion of n distinct random keys, followed by $3n$ times four operations: A random unsuccessful lookup, a random successful lookup, a random deletion, and a random insertion. We timed the operations in the "equilibrium", where the number of elements is stable. For load factor $1/3$ our results appear in Fig. 1, which shows an average over 10 runs. As LINEAR PROBING was consistently faster than DOUBLE HASHING, we chose it as the sole open addressing scheme in the plots. Time for forced rehashes was added to the insertion time. Results had a large variance for sets of size 2^{12} to 2^{16} – outside this range the extreme values deviated from the average by less than about 7%.

As can be seen, the time for lookups is almost identical for all schemes as long as the entire data structure resides in level 2 cache. After this the average number of random memory accesses (with the probability of a cache miss approaching 1) shows up. Filling a cache line seems to take around 160 clock cycles, with the memory location looked up arriving at the processor after about 80 clock cycles on average. This makes linear probing an average case winner, with CUCKOO HASHING and TWO-WAY CHAINING following about half a cache miss behind. For insertion the number of random memory accesses again dominates the picture for large sets, while the higher number of in-cache accesses and more computation makes CUCKOO HASHING, and in particular TWO-WAY chaining, relatively slow for small sets. The cost of forced rehashes sets in for TWO-WAY CHAINING for sets of more than a million elements, at which point better results may have been obtained by a larger bucket size. For deletion CHAINED HASHING lags behind for large sets due to random memory accesses when freeing list elements, while the simplicity of CUCKOO HASHING makes it the fastest scheme. We believe that the slight rise in time for the largest sets in the test is due to saturation of the bus, as the machine runs out of memory and begins swapping. It is interesting to note that all schemes would run much faster if the random memory accesses could bypass the cache (using perhaps 20 clock cycles per random memory access on our machine).

The second test concerns the cost of insertions in growing dictionaries and deletions in shrinking dictionaries. Together with Fig. 1 this should give a fairly complete picture of the performance of the data structures under general sequences of operations. The first operation sequence inserts n distinct random keys, while the second one deletes them. The plot is shown in Fig. 2. For small sets the time per operation seems unstable, and dominated by memory allocation overhead (if minimum table size 2^{10} is used, the curves become monotone). For sets of more than 2^{12} elements the largest deviation from the averages over 10 runs was about 6%. Disregarding the constant minimum amount of memory used by any dictionary, the average load factor during insertions was within 2% of $1/3$ for all schemes except CHAINED HASHING whose average load factor was

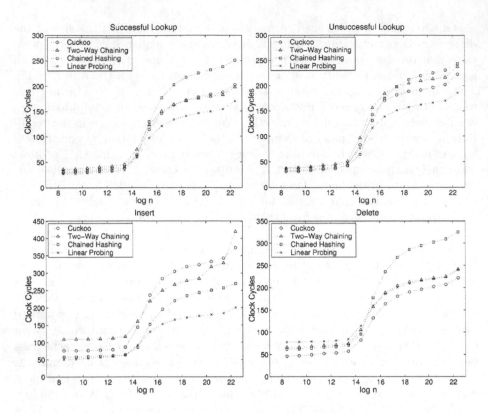

Fig. 1. The average time per operation in equilibrium for load factor 1/3.

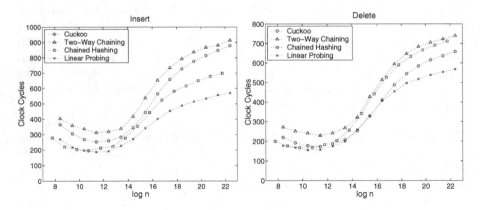

Fig. 2. The average time per insertion/deletion in a growing/shrinking dictionary for average load factor ≈ 1/3.

about 0.31. During deletions all schemes had average load factor 0.28. Again the winner is LINEAR PROBING. We believe this is largely due to very fast rehashes.

Access to data in a dictionary is rarely random in practice. In particular, the cache is more helpful than in the above random tests, for example due to repeated lookups of the same key, and quick deletions. As a rule of thumb, the time for such operations will be similar to the time when all of the data structure is in cache. To perform actual tests of the dictionaries on more realistic data, we chose a representative subset of the dictionary tests of the 5th DIMACS implementation challenge [16]. The tests involving string keys were preprocessed by hashing strings to 32 bit integers, preserving the access pattern to keys. Each test was run six times – minimum and maximum average time per operation can be found in Table 1, which also lists the average load factor. Linear probing is again the fastest, but mostly only 20-30% faster than the CUCKOO schemes.

Table 1. Average clock cycles per operation and load factors for the DIMACS tests.

	Joyce	Eddington	3.11-Q-1	Smalltalk-2	3.2-Y-1
LINEAR	42 - 45 (.35)	26 - 27 (.40)	99 - 103 (.30)	68 - 72 (.29)	85 - 88 (.32)
DOUBLE	48 - 53 (.35)	32 - 35 (.40)	116 - 142 (.30)	77 - 79 (.29)	98 - 102 (.32)
CHAINED	49 - 52 (.31)	36 - 38 (.28)	113 - 121 (.30)	78 - 82 (.29)	90 - 93 (.31)
A.CUCKOO	47 - 50 (.33)	37 - 39 (.32)	166 - 168 (.29)	87 - 95 (.29)	95 - 96 (.32)
CUCKOO	57 - 63 (.35)	41 - 45 (.40)	139 - 143 (.30)	90 - 96 (.29)	104 - 108 (.32)
TWO-WAY	82 - 84 (.34)	51 - 53 (.40)	159 - 199 (.30)	111 - 113 (.29)	133 - 138 (.32)

We have seen that the number of random memory accesses (i.e., cache misses) is critical to the performance of hashing schemes. Whereas there is a very precise understanding of the probe behavior of the classic schemes (under suitable randomness assumptions), the analysis of the expected time for insertions in Sect. 3.1 is rather crude, establishing just a constant upper bound. Figure 3 shows experimentally determined values for the average number of probes during insertion for various schemes and load

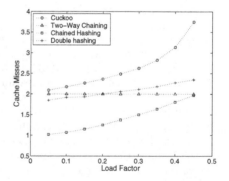

Fig. 3. The average number of random memory accesses for insertion.

factors below 1/2. We disregard reads and writes to locations known to be in cache, and the cost of rehashes. Measurements were made in "equilibrium" after 10^5 insertions and deletions, using tables of size 2^{15} and truly random hash function values. It is believed that this curve is independent of the table size (up to vanishing terms). The curve for LINEAR PROBING does not appear, as the number of non-cached memory accesses depends on cache architecture (length of the cache line), but it is typically very close to 1. It should be remarked that the highest load factor for TWO-WAY CHAINING is $O(1/\log\log n)$.

5   Conclusion

We have presented a new dictionary with worst case constant lookup time. It is very simple to implement, and has average case performance comparable to the best previous dictionaries. Earlier schemes with worst case constant lookup time were more complicated to implement and had considerably worse average case performance. Several challenges remain. First of all an explicit practical hash function family that is provably good for the scheme has yet to be found. Secondly, we lack a precise understanding of why the scheme exhibits low constant factors. In particular, the curve of Fig. 3 and the fact that forced rehashes are rare for load factors quite close to $1/2$ need to be explained.

References

[1] Yossi Azar, Andrei Z. Broder, Anna R. Karlin, and Eli Upfal. Balanced allocations. *SIAM J. Comput.*, 29(1):180–200 (electronic), 1999.

[2] Andrei Broder and Michael Mitzenmacher. Using multiple hash functions to improve IP lookups. To appear in INFOCOM 2001.

[3] Andrej Brodnik and J. Ian Munro. Membership in constant time and almost-minimum space. *SIAM J. Comput.*, 28(5):1627–1640 (electronic), 1999.

[4] J. Lawrence Carter and Mark N. Wegman. Universal classes of hash functions. *J. Comput. System Sci.*, 18(2):143–154, 1979.

[5] Martin Dietzfelbinger, Joseph Gil, Yossi Matias, and Nicholas Pippenger. Polynomial hash functions are reliable (extended abstract). In *Proceedings of the 19th International Colloquium on Automata, Languages and Programming (ICALP '92)*, volume 623 of *Lecture Notes in Computer Science*, pages 235–246. Springer-Verlag, Berlin, 1992.

[6] Martin Dietzfelbinger, Torben Hagerup, Jyrki Katajainen, and Martti Penttonen. A reliable randomized algorithm for the closest-pair problem. *Journal of Algorithms*, 25(1):19–51, 1997. doi:10.1006/jagm.1997.0873.

[7] Martin Dietzfelbinger, Anna Karlin, Kurt Mehlhorn, Friedhelm Meyer auf der Heide, Hans Rohnert, and Robert E. Tarjan. Dynamic perfect hashing: Upper and lower bounds. *SIAM J. Comput.*, 23(4):738–761, 1994.

[8] Martin Dietzfelbinger and Friedhelm Meyer auf der Heide. A new universal class of hash functions and dynamic hashing in real time. In *Proceedings of the 17th International Colloquium on Automata, Languages and Programming (ICALP '90)*, volume 443 of *Lecture Notes in Computer Science*, pages 6–19. Springer-Verlag, Berlin, 1990.

[9] Arnold I. Dumey. Indexing for rapid random access memory systems. *Computers and Automation*, 5(12):6–9, 1956.

[10] Michael L. Fredman, János Komlós, and Endre Szemerédi. Storing a sparse table with $O(1)$ worst case access time. *J. Assoc. Comput. Mach.*, 31(3):538–544, 1984.

[11] Gaston Gonnet. *Handbook of Algorithms and Data Structures*. Addison-Wesley Publishing Co., London, 1984.

[12] Richard M. Karp, Michael Luby, and Friedhelm Meyer auf der Heide. Efficient PRAM simulation on a distributed memory machine. *Algorithmica*, 16(4-5):517–542, 1996.

[13] Jyrki Katajainen and Michael Lykke. Experiments with universal hashing. Technical Report DIKU Report 96/8, University of Copenhagen, 1996.

[14] Donald E. Knuth. *Sorting and Searching*, volume 3 of *The Art of Computer Programming*. Addison-Wesley Publishing Co., Reading, Mass., second edition, 1998.

[15] George Marsaglia. The Marsaglia random number CDROM including the diehard battery of tests of randomness. http://stat.fsu.edu/pub/diehard/.

[16] Catherine C. McGeoch. The fifth DIMACS challenge dictionaries. http://cs.amherst.edu/~ccm/challenge5/dicto/.

[17] Kurt Mehlhorn and Stefan Näher. *LEDA. A platform for combinatorial and geometric computing*. Cambridge University Press, Cambridge, 1999.

[18] Rasmus Pagh. On the Cell Probe Complexity of Membership and Perfect Hashing. In *Proceedings of the 33rd Annual ACM Symposium on Theory of Computing (STOC '01)*. ACM Press, New York, 2001. To appear.

[19] Alan Siegel. On universal classes of fast high performance hash functions, their time-space tradeoff, and their applications. In *Proceedings of the 30th Annual Symposium on Foundations of Computer Science (FOCS '89)*, pages 20–25. IEEE Comput. Soc. Press, Los Alamitos, CA, 1989.

[20] Craig Silverstein. A practical perfect hashing algorithm. Manuscript, 1998.

[21] M. Wenzel. Worterbucher für ein beschränktes universum. Diplomarbeit, Fachbereich Informatik, Universität des Saarlandes, 1992.

Coupling Variable Fixing Algorithms
for the Automatic Recording Problem*

Meinolf Sellmann and Torsten Fahle

University of Paderborn
Department of Mathematics and Computer Science
Fürstenallee 11, D-33102 Paderborn
{sello,tef}@uni-paderborn.de

Abstract. Variable fixing is an important technique when solving combinatorial optimization problems. Unique profitable variable values are detected with respect to the objective function and to the constraint structure of the problem. Relying on that specific structure, effective variable fixing algorithms (VFAs) are only suited for the problems they have been designed for. Frequently, new combinatorial optimization problems evolve as a combination of simpler structured problems. For such combinations, we show how VFAs for linear optimization problems can be coupled via Lagrangian relaxation. The method is applied on a multimedia problem incorporating a knapsack and a maximum weighted stable set problem.

1 Introduction

Reduction algorithms are of great importance when combinatorial optimization problems have to be solved exactly. The tightening of problem formulations within a branch-and-bound approach improves on the quality of the bounds computed as well as on the approach's robustness. Given a maximization problem $P(x)$ where $x \in \{0,1\}^n$, $n \in \mathbb{N}$, the idea of variable fixing is to use upper bound information to detect unique profitable assignments for a variable: If an upper bound on $P(x_{|x_i=k})$, $k \in \{0,1\}$, drops below the best known solution value, then we can set $x_i \leftarrow 1 - k$.

Frequently, constraints of optimization problems can be grouped such that the overall problem can be viewed as a combination of two or more simpler structured problems. Assuming that efficient variable fixing algorithms (VFAs) for these subproblems exist, their independent application usually does not yield an effective algorithm to perform variable fixing for the combined problem. The reason for this is that tight bounds on the objective cannot be obtained by taking only a subset of the constraints into account.

For a multimedia application incorporating a knapsack problem (KP) and a maximum weighted stable set problem (MWSSP) on an interval graph, we show exemplary how two VFAs for linear optimization problems can be coupled via Lagrangian relaxation to achieve an effective reduction algorithm for the combined problem.

---

* This work was partly supported by the German Science Foundation (DFG) project SFB-376, by the UP-TV project, partially funded by the IST program of the Commission of the European Union as project number 1999-20 751, and by the IST Programme of the EU under contract number IST-1999-14186 (ALCOM-FT).

F. Meyer auf der Heide (Ed.): ESA 2001, LNCS 2161, pp. 134–145, 2001.
© Springer-Verlag Berlin Heidelberg 2001

The paper is structured as follows: In Section 2, we introduce the Automatic Recording Problem, that can be viewed as a combination of a knapsack problem and a MWSSP on interval graphs. In Section 3, we introduce an efficient VFA for the latter problem. Then, in Section 4, we show how it can be coupled with a previously developed VFA for KP via Lagrangian relaxation. Finally, in Section 5 we give numerical results.

2   The Automatic Recording Problem

The *Automatic Recording Problem (ARP)* is an example of a problem that is constituted by two simpler substructures. We focus on algorithms that solve the problem exactly and give a tightened formulation of the ARP as an integer program (IP).

The technology of digital television offers new possibilities for individualized services that cannot be provided by nowadays analog broadcasts. Additional information like classification of content, or starting and ending times can be submitted within the digital broadcast stream. With those informations at hand, new services can be provided that make use of individual profiles and maximize customer satisfaction.

One service – which is available already today – is an "intelligent" digital video recorder that is aware of its users' preferences and records automatically (see [2]). The recorder tries to match a given user profile with the information submitted by the different TV channels. E.g., a user may be interested in thrillers, the more recent the better. The digital video recorder is supposed to record movies such that the users' satisfaction is maximized. As the number of channels may be enormous (more than 100 digital channels are possible), a service that automatically provides an individual selection is highly appreciated and subject of current research activities (for example within projects like *UP-TV* funded by the European Union or the *TV-Anytime Forum*).

In this context, two restrictions have to be met. First, the storage capacity is limited (10h of MPEG-2 video need about 18 GB). And second, only one video can be recorded at a time. More formally, we define the problem as follows:

Definition 1. *Let $n \in \mathbb{N}$, $V = \{0, \ldots, n-1\}$ the set of movies, $start(i) < end(i) \ \forall i \in V$ the corresponding starting and ending times. $w = (w_i)_{0 \leq i < n} \in \mathbb{R}_+^n$ the storage requirements, $K \in \mathbb{R}_+$ the storage capacity, and $p = (p_i)_{0 \leq i < n} \in \mathbb{N}^n$ the profit vector.*

We say that the interval $I_i := [start(i), end(i)]$ corresponds to movie $i \in V$, and call two movies $i, j \in V$ overlapping whose corresponding intervals overlap, i.e. $I_i \cap I_j \neq \emptyset$. For $X \subseteq V$ we call $p_X := \sum_{i \in X} p_i$ the user satisfaction (with respect to X).

*The **Automatic Recording Problem (ARP)** then is to find a subset $X \subseteq V$ such that*

(a) X can be stored within the given disc size, i.e. $\sum_{i \in X} w_i \leq K$.
(b) at most one movie must be recorded at a time, i.e. $I_i \cap I_j = \emptyset \ \forall \ i \neq j \in X$
(c) X maximizes the user satisfaction, i.e. $p_X \geq p_Y \ \forall \ Y \subseteq V, Y$ respecting (a),(b).

Obviously, even if all movies are pairwise non-overlapping (i.e., if restriction (b) is obsolete), it remains to solve a knapsack problem. Thus, the ARP is NP-hard. However, there is an FPTAS for the ARP. For details, we refer to [12].

2.1 A Mathematical Programming Formulation

Because the problem of finding and proving optimal solutions is of interest in its own right, and also because the FPTAS we developed is far too memory consuming to be of practical relevance, we focus on exact approaches to solve the ARP. Using mathematical programming, the problem can be stated as an integer program (IP):

$$\text{Maximize} \quad \sum_{0 \le i < n} p_i x_i \qquad\qquad (IP\ 1)$$
$$\text{subject to} \quad x_i + x_j \le 1 \qquad \forall\, 0 \le i < j < n,\ I_i \cap I_j \ne \emptyset$$
$$\sum_{0 \le i < n} w_i x_i \le K$$
$$x \in \{0,1\}^n$$

The objective function maximizes the user satisfaction. Constraints of the form $x_i + x_j \le 1$ ensure that for overlapping intervals I_i, I_j at most one movie can be selected. Storage restrictions are enforced by the last row. The formulation can be tightened by replacing the overlapping constraints with maximal clique constraints.

Definition 2. *A set $C \subseteq V$ is called a* conflict clique, *iff $I_i \cap I_j \ne \emptyset\ \forall\, i, j \in C$. A conflict clique C is called* maximal, *iff $\forall\, D \subseteq V$, D conflict clique: $C \subseteq D \Rightarrow C = D$. Let $M := \{C_0, \ldots, C_{m-1}\} \subseteq 2^V$ the set of maximal conflict cliques.*

On interval graphs, the computation of maximal cliques can be performed in time $\Theta(n \log n)$. Then, restrictions of the form $\sum_{i \in C_p} x_i \le 1\ \forall\, 0 \le p < m$ imply that $x_i + x_j \le 1$ for all nodes $i, j \in V$ whose corresponding intervals overlap. On the other hand, if $x_i + x_j \le 1$ for all overlapping intervals, it is also true that $\sum_{i \in C_p} x_i \le 1\ \forall\, 0 \le p < m$. Thus, IP (1) is equivalent to

$$\text{Maximize} \quad \sum_{0 \le i < n} p_i x_i \qquad\qquad (IP\ 2)$$
$$\text{subject to} \quad \sum_{i \in C_p} x_i \le 1 \qquad \forall\, 0 \le p < m$$
$$\sum_{0 \le i < n} w_i x_i \le K$$
$$x \in \{0,1\}^n$$

To solve a (mixed) integer program, branch-and-bound approaches have proofed to be efficient, widely applicable and thus are most commonly used. In every search node, a bound based on some (often continuous) relaxation is being computed. If that bound is worse than the objective value B of the incumbent solution, backtracking occurs. A successful application of the branch-and-bound paradigm relies heavily on tight bounds that can be computed quickly. Fixing variables can help to improve on the performance of a branch-and-bound search if the VFA is both, effective and efficient. Effective means, that the algorithm must have an impact, i.e., it has to be able to fix many variables, whereas the efficiency measures how quickly the routine works.

The effectivity of a VFA mainly depends on the quality of bounds it uses to estimate the impact of fixing a variable to one of its values. For the ARP, our experiments show that the continuous relaxation bound yields a good estimate on the solution quality that can be reached. Thus, it can be used for pruning purposes in a branch-and-bound approach. But it is not straight forward to see, how this bound could be used for fixing variables effectively, that is, other than by probing via full reoptimization, which is inefficient. On the other hand, the fixing of variables with respect to reduced cost information can be done quickly, but is not very effective.

The ARP can be viewed as a combination of two simpler optimization problems: a knapsack problem, and a MWSSP on an interval graph. For the knapsack problem, an efficient VFAs exists [4]. It runs in time $\Theta(n \log n)$, and in amortized linear time for $\Omega(\log n)$ search nodes. In the following, we develop a VFA for the MWSSP substructure of the ARP.

3   Maximum Weighted Stable Set on Interval Graphs

On interval graphs, the problem of finding a maximum weighted stable set [6] can be solved easily in time $\Theta(n \log n)$ [7]. However, the existing algorithms based on sweep line or dynamic programming approaches neither provide dual values for the maximal clique constraints (which is important for the coupling of VFAs as we shall see later), nor do they suggest how variable fixing could be performed efficiently.

3.1   A Mathematical Programming Approach

We present an algorithm based on mathematical programming for the MWSSP on interval graphs that provides us with dual information as a by product, and that can be extended to an efficient VFA for the problem. Due to space restrictions, we omit the proofs here. For details, please refer to [12].

Remark 1. $\bigcup_{0 \leq p < m} C_p = V$, because $\{i\}$ is a conflict clique $\forall\ i \in V$. Thus, there exists a maximal conflict clique C_p, $0 \leq p < m$, such that $i \in C_p$.

Definition 3. *We set $max\_start : M \to \mathbb{N}$, $max\_start(C) := \max_{i \in C}\{start(i)\}$.*

Lemma 1. *The function max_start is injective.*

Without loss of generality, we may thus assume that the conflict cliques are ordered with respect to max_start, i.e. $max\_start(C_p) < max\_start(C_q) \ \forall\ 0 \leq p < q < m$.

Lemma 2. *Let $0 \leq p < r < m$ and $i \in C_p \cap C_r$. Then, $i \in C_q\ \forall\ p < q < r$.*

Corollary 1. $m \leq n$.

Definition 4. *We set $R_p := C_p \setminus C_{p+1}\ \forall\ 0 \leq p < m-1$ and $R_{m-1} := C_{m-1}$, and call every such R_p a (max_start) rest clique.*

Remark 2. The rest cliques form a partition of V.

Let $C_0, \dots, C_{m-1}$ denote the maximal conflict cliques ordered according to max_start, and consider IP (3) that evolves from IP (2) by dropping the capacity restriction.

The maximal conflict clique restrictions imply that $x_i + x_j \leq 1$ for all nodes $i, j \in V$ whose corresponding intervals overlap. On the other hand, if $x_i + x_j \leq 1$ for all overlapping intervals I_i and I_j, it is also true that $\sum_{i \in C_p} x_i \leq 1\ \forall\ 0 \leq p < m$. Thus, the IP (3) solves the MWSSP on interval graphs.

In the following, with $A \in \{0, 1\}^{m \times n}$ we denote the corresponding matrix to IP (3), i.e. $A = (a_{pi})_{0 \leq p < m, 0 \leq i < n}$ with $a_{pi} = 1$ iff $i \in C_p$.

Theorem 1. *The corresponding matrix A of IP (3) is an interval matrix.*

Corollary 2. *IP (3) is totally unimodular.*

Let $c_i := -p_i$ for all $0 \leq i < n$. With Corollary 2 it is now possible to solve the MWSSP on interval graphs as a Linear Program. With Remark 1, the maximal conflict clique restrictions imply that $x \leq 1$.

A Pivot Selection Strategy. We use the Simplex method for solving LP (4). Let $R_0, \ldots, R_{m-1}$ denote the (max_start) rest cliques. In iteration $0 \leq t < m$, we choose $q := t$ as pivot row and $j \in R_q$ with the highest reduced costs as pivot column. If the reduced costs of j are lower than 0, we perform a pivot step. Otherwise we proceed with the next iteration immediately.

Theorem 2. *After m such iterations, the Simplex tableau is primal and dual feasible.*

Before we can proof the above Theorem 2, we need to state two more Lemmata first. In the following, we refer to the matrix $A^t \in \{-1, 0, 1\}^{m \times n}$ after $0 \leq t \leq m$ Simplex iterations with $(a_{pi}^t)_{0 \leq p < m, 0 \leq i < n}$, to the right hand side $b^t \in \mathbb{R}^m$ with $(b_p^t)_{0 \leq p < m}$, and to the reduced costs $\hat{c}^t$ with $(\hat{c}_i^t)_{0 \leq i < n}$.

We first show that our pivot selection preserves primal feasibility. We observe that $x = 0$ is primal feasible as $b_p^0 = 1 \geq 0$, $\forall 0 \leq p < m$. To assure the maintenance of primal feasibility, we must show that $b^t \geq 0$, $\forall 0 \leq t \leq m$. To do so, we can proof the following

Lemma 3. *Let $0 \leq p < m$, $0 \leq t \leq m$, $0 \leq i < n$. Then,*

(a) $p \geq t - 1$ implies that $a_{pi}^t = a_{pi}^0$ and $b_p^t = b_p^0 = 1$
(b) $b_p^t = 0$, $i \in \bigcup_{p \geq t} R_p$ implies $a_{pi}^t \in \{-1, 0\}$
(c) $b_p^t = 1$, $i \in \bigcup_{p \geq t} R_p$ implies $a_{pi}^t \in \{0, 1\}$
(d) $b_p^t \in \{0, 1\}$

The following Lemma implies that after at most m iterations we achieve dual feasibility.

Lemma 4. *Let $1 \leq t \leq m$. Then,*

(a) $\hat{c}_i^t \geq 0$ for all $i \in R_{t-1}$.
(b) $i \in \bigcup_{0 \leq p < t < m} R_p$ implies $\hat{c}_i^{t+1} = \hat{c}_i^t$.

Proof. **of Theorem 2** In Lemma 3 and Lemma 4, we have shown that after $m \leq n$ iterations the Simplex tableau is primal and dual feasible.   □

We have shown how the MWSSP on interval graphs can be stated as a totally unimodular LP. Further, we have proven a feasible pivot selection strategy that yields an optimal tableau after at most n Simplex iterations.

An Efficient Simplex Realization. In the following, we develop an efficient $\Theta(n \log n)$-time algorithm to calculate a set $Q \subseteq \{0, \ldots, n-1\}$ with $I_i \cap I_j = \emptyset \; \forall \, i, j \in Q$, such that $c_Q = \sum_{i \in Q} c_i$ is minimal. Most importantly, the algorithm also provides us with dual information as a byproduct. To establish that algorithm, we show how the Simplex calculations according to the pivot strategy developed can be performed efficiently.

Again, due to limited space we have to omit all proofs. They can be found in [12].

Theorem 3. *Let $0 \leq l < m$, and let $(j_0, \ldots, j_l) \in \{0, \ldots, n-1\}^l$ denote the sequence of pivot columns according to Section 3.1 for which a pivot step has been performed. Further, for all $0 \leq k \leq l$ let $0 \leq p_k < m$ with $j_k \in R_{p_k}$. Set $Q := \{j_k \mid 0 \leq k \leq l \text{ and } I_{j_k} \cap I_{j_r} = \emptyset \; \forall \, k < r \leq l\}$. Then c_Q is minimal.*

The above Theorem 3 allows us to construct an optimal solution if we know the sequence of pivot elements. But according to Section 3.1, calculating the sequence of pivot elements is an easy task, if only we can determine the reduced costs quickly:

Lemma 5. *Let* $0 \leq t < m$, $(j_0, \ldots, j_{m-1}) \in V^m$ *be the sequence of pivot columns according to our pivot selection strategy, and let* $d : \{0, \ldots, m - 1\} \to \{0, 1\}$ *with* $d(t) := 1 \Leftrightarrow \hat{c}_{j_t}^t < 0$. *Further, we set* $q_t := t$, $0 \leq q_t < m$, *and let* $j_t \in R_{q_t}$.

Let $z^t \in \mathbb{R}$ *denote the objective function value and* $a_{q_t j_t}^t$ *be the pivot element in iteration* t. *Further, let* $f_i := \min\{p \mid 0 \leq p < m, i \in C_p\} \forall 0 \leq i < n$ *denote the index of the first maximal conflict clique that node* i *belongs to, i.e.* $a_{f_i i}^0 = 1$, *and* $a_{pi}^0 = 0 \forall 0 \leq p < f_i$. *Finally, let* $g_i^t := z^{f_i} - z^t$ *if* $f_i \leq t$, *and otherwise* $g_i^t := 0 \forall 0 \leq i < n$. *Then,*

(a) $z^{t+1} = z^t + \hat{c}_{j_t}^t \cdot d(t) \leq z^t$.

(b) $z^t = \sum_{0 \leq r < t} \hat{c}_{j_r}^r \cdot d(r)$.

(c) $\hat{c}_i^t = c_i + g_i^t \qquad \forall i \in \bigcup_{t \leq p < m} R_p$

An Algorithm providing Dual Information. With Theorem 3 and Lemma 5, we can formulate an efficient algorithm solving the MWSSP on interval graphs that provides us with dual values as a byproduct. In phase 1, we determine the (max_start) rest cliques R_p, with $0 \leq p < m$, and the according values $f_i \forall 0 \leq i < n$. This can be done in time $\Theta(n \log n)$.

Phase 2 consists of m iterations: First, we set $z^0 := 0$. In each iteration $0 \leq t < m$ we calculate $\hat{c}_i^t = c_i + z^{f_i} - z^t \forall i \in R_t$, and $j_t \in R_t$ with $\hat{c}_{j_t}^t = \min_{i \in R_t}\{\hat{c}_i^t\}$. If $\hat{c}_{j_t}^t \geq 0$, we set $z^{t+1} := z^t$, otherwise $z^{t+1} := z^t + \hat{c}_{j_t}^t$. Finally, we set $t := t + 1$ and proceed with the next iteration.

With Remark 2, we know that the sets R_p form a partition of V. Thus, all nodes $0 \leq i < n$ are being looked at exactly once to calculate the reduced costs. Also, in all computations of the pivot columns, each node is incorporated only once. Therefore, phase 2 takes time $\Theta(n)$.

After m iterations, we know the shortest path value z^m, as well as the sequences $(j_0, \ldots, j_l) \in V^l$ and $(p_0, \ldots, p_l) \in \{0, \ldots, m - 1\}^l$, $0 \leq l < m$, of pivot columns and rows for which a pivot step has been performed. By applying Theorem 3, we can construct a shortest path out of this information in linear time. Moreover, as the rest cliques of the underlying interval graph are independent of the objective function, we achieve an incremental linear time algorithm for $\Omega(\log n)$ calls with different objectives.

Most importantly, we get dual values as a byproduct. By looking at the optimal tableau, we find that the optimal dual variable for each maximal clique constraint $0 \leq t < m$ for which a pivot step has been carried out has value $-\hat{c}_{j_t}^m$. All other dual values are 0.

3.2   Variable Fixing

Now, we want to use the previously developed algorithm to perform variable fixing. To develop a VFA, we reinterpret the problem as finding a shortest path in a node-weighted co-interval graph. We introduce an artificial source σ and sink τ before and after all other nodes, and define $Path_{\sigma,\tau}$ as the set of paths from σ to τ. The value c_P of a path $P \in Path_{\sigma,\tau}$ is defined as $c_P := \sum_{i \in P \setminus \{\sigma,\tau\}} c_i$.

Definition 5. *Given an upper bound* $Z \in \mathbb{R}$, *we define* $Rem(Z)$ *and* $Req(Z)$ *by* $Rem(Z) := \{0 \leq i < n \mid \forall P \in Path_{\sigma,\tau}, i \in P : c_P \geq Z\}$ $Req(Z) := \{0 \leq i < n \mid \forall P \in Path_{\sigma,\tau}, i \notin P : c_P \geq Z\}$. *The variable fixing problem based on shortest path information (VFSPP) then is to determine* $Rem(Z)$ *and* $Req(Z)$.

Removing Nodes. To compute $Rem(Z)$, we need to find out the value of the shortest path from the source σ via node i to the sink τ for all $i \in \{0, \dots, n-1\}$. This can be done by calculating the shortest path distances from the source and to the sink. The algorithm in Section 3.1 can easily be adapted to determine the shortest path distances from the source to each node $j \in \{0, \dots, n-1\}$. For $0 \leq i < n$, let $f_i := \min\{p \mid 0 \leq p < m, i \in C_p\}$ denote the index of the first maximal conflict clique that node i belongs to. Then, we have to solve the following Linear Program

$$
\begin{array}{ll}
\text{Minimize} & \sum_{0 \leq i < j} c_i x_i \qquad\qquad\qquad (LP\,5) \\
\text{subject to} & \sum_{i \in C_p} x_i \leq 1 \quad \forall\, 0 \leq p < f_j \\
& x \geq 0
\end{array}
$$

According to the previously developed theory, the minimal objective for the above LP (5) is exactly z^{f_j}. Thus, the shortest path distance of node j is exactly $d_j = z^{f_j} + c_j$.

A similar theory shows that the shortest path distances to the sink can be determined by applying the algorithm of Section 3.1 using the last clique belongings $l_i := \max\{p \mid 0 \leq p < m, i \in C_p\} \, \forall\, 0 \leq i < n$, and the min_end rest cliques, where $min\_end : M \to \mathbb{N}$, $min\_end(C) := \min_{i \in C}\{end(i)\}$. Solving LP (4) in this inverse manner yields objective function values $^\tau z^t$ for all iterations $0 \leq t \leq m$.

Then, the shortest path distance to the sink is $^\tau d_j = {}^\tau z^{m-l_j-1} + c_j$. With those values at hand, we can determine the shortest path value e_j through node $0 \leq j < n$ by $e_j = d_j + {}^\tau d_j - c_j$. Then, $Rem(Z) = \{0 \leq i < n \mid e_i \geq Z\}$.

We conclude that $Rem(Z)$ can be computed in time $\Theta(n \log n)$ and in amortized linear time for $\Omega(\log n)$ calls of the VFA.

Now we know how to determine nodes that should be removed from the graph. Of course, other constraints may as well remove nodes. In both cases, we must be able to take these changes into consideration efficiently for the next call to our routine. Without going into implementation details, we note that the data structures storing the rest cliques as well as the first and last clique belongings can be compressed in linear time to delete any number of nodes from the graph.

Requiring Nodes. To compute $Req(Z)$, we need to identify all nodes that must be an element of any path having a value lower than Z. Obviously, only nodes on the shortest path S potentially have this property. For every node $j \in S$, we thus need to find out what the value of a shortest path P with $j \notin P$ is.

Remark 3. Let $0 \leq f_j \leq l_j < m$ denote the first and the last clique belongings of j. Further, let P be the shortest path with $j \notin P$. Obviously, it either holds that $I_j \cap I_i = \emptyset \, \forall\, i \in P$, or $I_j \cap I_i \neq \emptyset \, \exists\, i \in P$. In the first case, we know that the value of the shortest path not using the time interval I_j has the value $c_P = z^{f_j} + {}^\tau z^{l_j} - 2c_j$. In the second case, we do not know how to determine the value of P efficiently. However, after we have determined and deleted $Rem(Z)$ from the graph, we only have to check if there exists any node $i \in \{0, \dots, n-1\}$, $i \neq j$, with $I_j \cap I_i \neq \emptyset$. Because if such a node i

exists, there also exists a path $\bar{P}$ with $i \in \bar{P}$ and $c_{\bar{P}} < Z$ (otherwise i would have been deleted before). As i and j overlap, we further know that $j \notin \bar{P}$. Thus, in $\bar{P}$ we have found a path not covering j with a value lower than Z. Therefore, $j \notin Req(Z)$. On the other hand, if no such node i exists, the second case is obsolete, and we only need to consider the first case.

By making that observation, we can determine $Req(Z)$ in linear time: First, for all $j \in P$ we check whether there exists a node in the shrunk graph that overlaps with j. Without specifying the implementation details here, we just note that this can be done in constant time for each node j. If no overlapping node exists, we compute $z^{f_j} + {}^{\tau}z^{l_j} - 2c_j$ and check whether this value is lower than Z. If not, we add j to $Req(Z)$, otherwise we do not.

Now, we have an efficient algorithm at hand to compute $Req(Z)$. Obviously, other constraints and branching decisions must be taken into account when our procedure is being called next. Thus, we have to be able to transform our graph in such a way that from now on every path covers the new required nodes. At a first glance this sounds problematic, as an easy approach would delete all arcs going around the required nodes. This procedure would cause the resulting graph not to have the co-interval property anymore.

We can force the shortest path to visit the required nodes by making them extremely cheap: Let $Req \subseteq \{0, \dots, n-1\}$ the set of (currently) required nodes. Further, let $M \gg 0$ sufficiently large[1]. Then, we set $\bar{c}_j := c_j - M \; \forall \; j \in Req$, and $\bar{c}_j := c_j \; \forall \; j \notin Req$. We use $\bar{c}$ instead of c as our objective and check whether the shortest path value is lower than $Z - |Req| \cdot M$. If not, either two required nodes overlap, or the shortest path value in the original graph exceeds Z. Moreover, by determining $Rem(Z - |Req| \cdot M)$, we find all nodes that overlap with some required node plus all nodes that would cause the shortest path in the original graph to exceed the threshold Z. We conclude:

Theorem 4. *The VFSPP can be solved in time $\Theta(n \log n)$ or in amortized linear time for $\Omega(\log n)$ incremental calls of the variable fixing algorithm.*

4   Coupling Variable Fixing Algorithms via Lagrangian Relaxation

An obvious approach to solve the ARP exactly is to apply a branch-and-bound algorithm using linear relaxation bounds for pruning and the existing VFAs for knapsack and MWSSP on interval graphs for the fixing of variables.

Although the existing VFAs are efficient and effective for the substructures they have been designed for, their application for the ARP is not. This is because neither assuming an unlimited storage capacity nor ignoring the fact that only one movie can be recorded at a time yields a tight bound on the objective for the combined problem. But the accuracy of the upper bound is essential for the effectiveness of a VFA. An accurate bound can only be computed by looking at the entire problem, i.e., it cannot be achieved by looking at either one constraint family only.

The linear relaxation bound can easily be obtained by applying a standard LP solver or by using specialized methods tailored for this specific application. That bound yields

---

[1] Assuming that $\min_{i \in V}\{c_i\} < 0$, a valid setting for M is for example $M := n \cdot (1 + \max_{i \in V}\{c_i\} - \min_{i \in V}\{c_i\})$

a good estimate on the performance that can (still) be reached. However, it is not straight forward to see how it could be exploited for the fixing of variables. Applying conventional reduced cost variable fixing techniques only indirectly exploits the structure of the problem and is therefore not effective enough, whereas to perform probing via full reoptimization is very costly and inefficient.

Lagrangian relaxation (see e.g. [1] for an introduction) allows us to bring together the advantages of a tight continuous global bound and the existing VFAs that exploit the special structure of their respective constraint families. As the stable set VFA allows us to incorporate changing objectives at a low computational cost, we decide to relax the capacity constraint. We introduce a non-negative Lagrange multiplier $\lambda \geq 0$ and define the Lagrangian subproblem

$$
\begin{array}{lll}
\text{Maximize} & z(\lambda) := z & \qquad\qquad L(\lambda) \\
\text{subject to} & z = \sum_{0 \leq i < n}(p_i - \lambda w_i)x_i + \lambda K & \\
& \sum_{i \in C_p} x_i \leq 1 & \forall\, 0 \leq p < m \\
& x \in \{0,1\}^n &
\end{array}
$$

The Lagrange multiplier problem then is to solve Minimize $z(\lambda)$, such that $\lambda \geq 0$. For every $\lambda \geq 0$, $z(\lambda)$ is a valid upper bound on the objective. Therefore, we can apply the VFA for MWSSP on interval graphs each time we solve the Lagrangian subproblem. After we have found an optimal Lagrange multiplier λ^*, i.e. $z(\lambda^*) \leq z(\lambda) \ \forall\ \lambda \geq 0$, we can use the (optimal) dual information $\pi \in \mathbb{R}^m$ from the corresponding stable set subproblem to perform variable fixing with respect to the knapsack substructure now. By Lagrange relaxing the maximal clique constraints with multipliers $\pi \geq 0$, we obtain a knapsack problem:

Let $\mu_i := \sum_{j\, :\, i \in C_j} \pi_j \ \forall\ 0 \leq i < n$ and $\overline{\pi} := \sum_{0 \leq j < m} \pi_j$. The problem then is to

$$
\begin{array}{ll}
\text{Maximize} & \sum_{0 \leq i < n}(p_i + \mu_i)x_i - \overline{\pi} \\
\text{subject to} & \sum_{0 \leq i < n} w_i x_i \leq K \\
& x \in \{0,1\}^n
\end{array}
$$

Relaxations of this problem again yield an upper bound on the objective, and we can apply the knapsack VFA from [4].

In general, two linear optimization constraint families for which efficient VFAs are known can be combined effectively by computing Lagrangian multipliers for the first, using the second for fixing variables in each Lagrangian subproblem $L(\lambda)$, and then handing back dual information of the optimal $L(\lambda^*)$ to fix variables with respect to the first constraint family with the corresponding (optimal) reduced cost objective. This procedure even strengthens the bound on the objective, as variable fixing is also done during the bound computation. However, if variables are being fixed during the process of finding optimal Lagrange multipliers, the algorithm that solves the Lagrangian dual must be aware of this. It is subject to further research, how e.g. subgradient methods must be adapted to be able to cope with that situation.

5   Numerical Results

We used four different approaches for our experiments: the first is a pure branch-and-bound algorithm without any variable fixing (referred to as *no fixing (F-0)*). The second uses the VFAs for knapsack and MSSP on the original objective (*fixing 1 (F-1)*). The third and the fourth approach (*fixing 2 (F-2)* and *fixing 3 (F-3)*) realize the idea of coupling VFAs for linear optimization problems via Lagrangian relaxation. *F-2* calls for the fixing of variables just once after the Lagrangian dual has been solved, whereas *F-3* also performs maximum weighted stable set variable fixing during the search for optimal Lagrange multipliers. For details on the computation of the Lagrange multipliers and the choice of the branching variable, we refer to [12].

5.1   Experiments

All experiments were performed on a PC with an AMD-Athlon 600 processor and 256 MB ram running Linux 2.2. The implementation was done in C++ and compiled by gcc 2.95. The algorithms were built on top of ILOG SOLVER 5.0 [8].

The ARP test instances were generated by specifying the time horizon and the number of channels. The generator sequentially fills the channels by starting each new movie one minute after the last. First, a class for the next movie is being chosen randomly. That class then determines the intervals from which the length and the profit are chosen randomly (for now, we have only been using 3 different classes of movies). The disc space necessary to store each movie equals its length, and the storage capacity is randomly chosen as 40%–60% of the entire time horizon.

The experiment consists of 50 random instances per test set. For each instance, the approaches *F-0* – *F-3* were run to find and prove an optimal solution. The minutes for the time horizon equal 6 hours, 12 hours, one day, 3 days, and 5 days of digital television, respectively.

It should be noted that all approaches find a first solution rather early in the search. Therefore, the main work lies in the proof of optimality rather than in the construction of the solution. We conclude that the branching variable selection we used efficiently supports finding near-optimal solutions in a non-exhaustive search.

Obviously, the linear continuous bound we use for pruning is rather tight. Except for the 4320-50 and 7200-20 instances, even *F-0* is able to cope with all problem classes fairly well. Thus, it is justified to base pruning and also variable fixing on that bound.

The positive effect of applying a VFA can be seen when comparing *F-0* and *F-1*. Although the latter is not very effective, it is already able to reduce the number of choice points by a factor of 1.5 – 2. This results in a speed-up of up to 3 for the 7200-20 instances.

F-2 and *F-3* yield a dramatic further reduction of choice points. For the 7200-20 instances, *F-3* only needs 10% of the choice points of *F-1*. For the 4320-50 instances, it even reduces the number of choice points to 7% compared to *F-1*, and to 3% with respect to the number of choice points visited by *F-0*.

The gain in time is not that drastic, as the greater effectivity must be paid for by a higher computational effort per search node. Thus, the maximum speed-up we get between *F-0* and *F-2* is "only" 10 – 11. And although *F-3* is more effective and outperforms *F-2* regarding the number of choice points, it is slower on all instances. The reason for both is that it performs maximum weighted stable set variable fixing for each Lagrangian subproblem. Of course, if the absolute time spent per choice point was higher (due to other variable fixing algorithms or more costly bound computations for example), the

Table 1. Numerical results for the ARP. The first three columns characterize each group of 50 instances by giving the planning horizon in minutes, the number of channels, and the average number of movies. Columns 4 to 7 present running times for the different approaches. In brackets, we give the number of search nodes, in the following referred to as *choice points*.

Test set			runtime in sec. (choice points)			
min.	#ch	mov.	no xing F-0	xing F-1	xing F-2	xing F-3
360	5	29.0	0.1 (27.2)	0.1 (15.6)	0.0 (14.6)	0.0 (10.9)
360	20	115.6	0.4 (49.9)	0.2 (32.3)	0.1 (24.2)	0.2 (17.3)
360	50	285.2	1.1 (52.5)	0.8 (37.3)	0.3 (25.7)	0.5 (22.1)
360	100	569.0	2.9 (96.7)	2.1 (67.9)	0.8 (36.7)	1.2 (30.0)
720	5	52.4	0.1 (69.3)	0.1 (33.3)	0.1 (26.0)	0.1 (17.6)
720	20	208.3	0.6 (91.7)	0.4 (69.0)	0.3 (36.5)	0.4 (29.4)
720	50	523.4	4.2 (305.5)	2.4 (166.6)	1.3 (58.7)	2.0 (52.3)
720	100	1048.3	18.8 (680.8)	9.1 (326.3)	5.1 (108.6)	7.6 (95.5)
1440	5	101.4	0.8 (130.4)	0.7 (98.2)	0.2 (48.0)	0.4 (38.5)
1440	20	402.9	7.9 (479.6)	5.3 (306.1)	1.6 (103.0)	2.5 (81.9)
1440	50	1013.4	37.8 (1461.1)	19.5 (734.4)	11.1 (187.9)	13.8 (178.5)
1440	100	2019.8	291.0 (5351.2)	108.2 (2028.1)	55.6 (530.2)	71.0 (439.9)
4320	5	291.7	4.7 (508.2)	6.0 (327.3)	2.0 (138.9)	3.1 (121.6)
4320	20	1175.0	177.4 (3003.1)	111.7 (1897.2)	36.6 (468.4)	42.9 (401.2)
4320	50	2949.7	12227.8 (149877.2)	5647.5 (67864.5)	1191.5 (4744.5)	1197.3 (4703.5)
7200	5	487.8	64.0 (3272.2)	43.5 (2705.4)	17.5 (878.0)	17.9 (434.5)
7200	20	1956.7	5210.3 (60839.2)	1734.4 (30676.4)	455.9 (3433.9)	490.1 (2945.1)

increased effectivity should also result in a faster overall computation. By using a parameter ζ that determines the frequency of maximum weighted stable set variable fixing, we can trade time for effectivity. Like that, we can tune our coupled VFA towards our specific application. To achieve a most effective filtering algorithm, the VFAs for both substructures could be applied for each Lagrangian subproblem.

A topic of further investigation is the comparison of test sets with an equal number of movies: E.g., instances in 720-100, 1440-50, and 4320-20 all contain about 1000 movies, and the solution time decreases with the number of channels increasing. In a test set with a high number of channels, many nodes are overlapping, hence the maximum weighted stable set VFA can easily eliminate many variables from consideration. With a lower number of channels and a wider time horizon, the knapsack constraint gains more importance. We assume that a vice versa approach (i.e., Lagrange relaxing the non-overlapping constraints and solving a knapsack as subproblem) would be a suitable choice for such a situation.

6   Conclusions and Future Work

For the automatic recording problem, we exemplary introduced the idea of coupling VFAs via Lagrangian relaxation. It allows to combine existing variable fixing routines for linear optimization problems to obtain effective and efficient filtering algorithms based on tight global bounds. We believe, that this idea is generic and independent of the specific application we presented to base an empirical evaluation on. The numerical results show a significant improvement due to the coupling method with respect to the computation time and other algorithmic measures such as search nodes. The method is suited for linear optimization problems for which bounds based on continuous or Lagrangian relaxations can be used effectively.

In order to be able to couple VFAs for knapsack and MWSSPs, we first had to develop a VFA for the latter. Although for the problem itself algorithms running in time $\Theta(n \log n)$ have been developed before, they did not provide dual information that is needed for the coupling method we introduced. Moreover, to our knowledge no VFA for the MWSSP on interval graphs running in amortized linear time existed before. As the problem occurs as a substructure in many optimization problems, especially in scheduling contexts, the VFA we developed in our view is a contribution that is of general interest and goes beyond the application we presented.

For the multimedia application we introduced, we developed a refined IP formulation. The continuous relaxation of that IP yields a tight upper bound as our experiments showed. Several extensions are possible for that application. A digital video recorder could have more than one recording unit which allows the recording of a limited number of channels simultaneously. In an IP context, this modification can be introduced easily. For the new exact approach presented, a fast and efficient VFA for this type of relaxed non-overlapping constraint is subject to further research.

Acknowledgement. We would like to thank Burkhard Monien for helpful comments.

References

1. R.K. Ahuja, T.L. Magnati, J.B. Orlin. *Network Flows.* Prentice Hall, 1993.
2. *mediaTV*, technical description, Axcent AG. http://www.axcent.de.
3. E. Balas and E. Zemel. An algorithm for large-scale zero-one knapsack problems. *Operations Research*, 28:119–148, 1980.
4. T. Fahle and M. Sellmann. Constraint Programming Based Column Generation with Knapsack Subproblems. *Proc. of the CP-AI-OR'00*, Paderborn Center for Parallel Computing, Technical Report tr-001-2000:33–44, 2000.
5. F. Focacci, A. Lodi, M. Milano. Cutting Planes in Constraint Programming: An Hybrid Approach. *Proc. of the CP-AI-OR'00*, Paderborn Center for Parallel Computing, Technical Report tr-001-2000:45–51, 2000.
6. M.C. Golumbic. Algorithmic Graph Theory and Perfect Graphs. *Academic Press, New York*, 1991.
7. J.Y. Hsiao, C.Y. Tang, R.S. Chang. An efficient algorithm for finding a maximum weight 2-independent set on interval graphs. *Information Processing Letters*, 43(5):229–235, 1992.
8. ILOG. ILOG SOLVER. Reference manual and user manual. V5.0, ILOG, 2000.
9. M. Lehradt. Basisalgorithmen für ein TV Anytime System, *Diploma Thesis*, University of Paderborn, 2000.
10. S. Martello and P. Toth. An upper bound for the zero-one knapsack problem and a branch and bound algorithm. *European Journal of Operational Research*, 1:169–175, 1977.
11. G.L. Nemhauser, L.A. Wolsey. *Integer and Combinatorial Optimization*. Wiley Interscience, 1988.
12. M. Sellmann and T. Fahle. Coupling Variable Fixing Algorithms for the Automatic Recording Problem. Technical Report, University of Paderborn, 2001.

Approximation Algorithms for Scheduling Malleable Tasks under Precedence Constraints

Renaud Lepère[1], Denis Trystram[1], and Gerhard J. Woeginger[2]

[1] Laboratoire Informatique et Distribution - IMAG, France
[2] Institut für Mathematik, TU Graz, Austria

Abstract. This work presents approximation algorithms for scheduling the tasks of a parallel application that are subject to precedence constraints. The considered tasks are malleable which means that they may be executed on a varying number of processors in parallel. The considered objective criterion is the makespan, i.e., the largest task completion time.

We demonstrate a close relationship between this scheduling problem and one of its subproblems, the allotment problem. By exploiting this relationship, we design a polynomial time approximation algorithm with performance guarantee arbitrarily close to $(3 + \sqrt{5})/2 \approx 2.61803$ for the special case of series parallel precedence constraints and for the special case of precedence constraints of bounded width. These special cases cover the important situation of tree structured precedence constraints. For the general case with arbitrary precedence constraints, we give a polynomial time approximation algorithm with performance guarantee $3 + \sqrt{5} \approx 5.23606$.

1 Introduction

Scheduling and load-balancing are central issues in the parallelization of large scale applications. One of the main problems in this area concerns efficient scheduling of the tasks of a parallel program. This problem asks to determine at what time and on which processor all the tasks should be executed. Among the various possible approaches, the most commonly used is to consider the tasks of the program at the finest level of granularity, and to apply some adequate clustering heuristics for reducing the relative communication overhead; see Gerasoulis & Yang [8]. Several models have been developed for modeling the communication and the parallelization overhead in these problems. In models with a finer communication representation (like in the LogP model [3]), the impact of the parallelization overhead is usually ignored.

Recently, a new computational model called *Malleable tasks* (MT) has been proposed by Turek, Wolf & Yu [14] as an alternative to the usual delay model. Malleable tasks are computational units which may be themselves executed in parallel. The influence of communications is taken into account implicitly by a penalty factor which may be determined more or less precisely for each application; cf. Blayo, Debreu, Mounié & Trystram [1]. As the granularity in the MT

F. Meyer auf der Heide (Ed.): ESA 2001, LNCS 2161, pp. 146–157, 2001.

model is large, the communications between the malleable tasks are usually neglected. We refer the reader to Lepère, Mounié, Robič & Trystram [11] for more details and for motivations of the MT model. MT are closely related to two other models, namely to the model of *multiprocessor tasks* (see e.g. Drozdowski [6]) and to the model of *divisible tasks* (Prasanna & Musicus [12]). The difference between these models lies in the freedom allowed to the task allotment, that is, the number of processors which execute each task: A multiprocessor task requires to be executed by a fixed integer number of processors, whereas divisible tasks share the processors as a continuously divisible resource.

1.1   The Malleable Tasks Model

Throughout this paper we assume that the parallel program is represented by a set of generic malleable tasks, that is, computational units that may be parallelized and that are linked by precedence constraints. The precedence constraints are determined a priori by the analysis of the data flow between the tasks. More formally, let $G = (V, E)$ be a directed graph where $V = \{1, 2, \ldots, n\}$ represents the set of malleable tasks, and where $E \subseteq V \times V$ represents the set of precedence constraints among the tasks. If there is an arc from task i to task j in E, then task i must be processed completely before task j can begin its execution. This situation will be denoted by $i \to j$; i is called a *predecessor* of j, and j is called a *successor* of i. All tasks are available at time 0 for execution, and they are to be scheduled on an overall number of m processors. Every task j is specified by m positive integers $p_{j,q}$ $(1 \leq q \leq m)$ where $p_{j,q}$ denotes the execution time of task j when it is executed in parallel on q processors.

Motivated by the usual behavior of parallel programs (cf. Cosnard & Trystram [2]), we make the following assumptions on the task execution times. Blayo, Debreu, Mounié & Trystram [1] have shown these assumptions to be realistic while implementing actual parallel applications.

Assumption 1 *(Monotonous penalty assumptions)*

(a) The execution time $p_{j,q}$ of a malleable task j is a non-increasing function of the number q of processors executing the task.
(b) The work $w_{j,q} \doteq q \cdot p_{j,q}$ of a malleable task j is a non-decreasing function of the number q of processors executing the task.

Assumption (a) means that adding some processors for executing a malleable task cannot increase its execution time. In practice, the execution time even goes down in this situation, at least until a threshold from which onwards there is no more parallelism. Assumption (b) reflects that the total overhead for managing and administrating the parallelism usually increases with the number of processors.

A *schedule* σ is specified by two functions $start_\sigma : V \to \mathbb{N}$ and $allot_\sigma : V \to [1, m]$ where the function $start_\sigma$ associates to each task a date of execution (or starting time), and where the function $allot_\sigma$ specifies the number of processors to execute a task. In schedule σ, the task j completes at time $C_\sigma(j) = start_\sigma(j) + p_{j,allot_\sigma(j)}$. We say that task j is *active* during the time interval from $start_\sigma(j)$ to $C_\sigma(j)$, and we denote by $active(t)$ the set of all tasks that are active at time t. A schedule σ is a *feasible* schedule, if at any moment t in time at most m processors are engaged in the computation

$$\sum_{j \in active(t)} allot_\sigma(j) \leq m \qquad \text{for all } t \geq 0,$$

and if all the precedence constraints are respected:

$$start_\sigma(i) + p_{i,allot_\sigma(i)} \leq start_\sigma(j) \qquad \text{for all } i \to j.$$

The *makespan* $C_{\max}$ of a schedule σ is the maximum of all task completion times $C_\sigma(j)$. We now introduce the central problem of this paper.

Problem 1. MAKESPAN FOR MALLEABLE TASKS (MT-MAKESPAN)

INSTANCE: A directed graph $G = (V, E)$ that represents a set of n precedence constrained malleable tasks; the number m of processors; positive integers $p_{j,q}$ with $1 \leq j \leq n$ and $1 \leq q \leq m$ that specify the task execution times.

GOAL: Find a feasible schedule that minimizes the makespan $C_{\max}$.

Consider an instance of MT-MAKESPAN, and assume that some processor allotment α has been prespecified for all tasks, and that task j is to be executed on exactly α_j processors. Then the execution time of task j is p_j, and its work is $\alpha_j p_j$. With every directed path through the precedence graph $G = (V, E)$, we associate the total execution time p_j of the vertices on this path. The longest path under this definition of length is called the *critical* path of the allotment α, and its length is denoted by L^α. Moreover, we denote by $W^\alpha = \sum_{j=1}^n \alpha_j p_j$ the overall work in allotment α. Clearly,

$$c(\alpha) \doteq \max\{L^\alpha, \frac{1}{m} W^\alpha\} \leq C_{\max} \tag{1}$$

holds for the makespan $C_{\max}$ of any feasible schedule σ under allotment α: Since the schedule must obey the precedence constraints, the tasks along the critical path form a chain that forces the makespan to at least L^α. Since the total work W^α can only be distributed across m processors, some processor will run for at least W^α/m time units. The value $c(\alpha)$ in equation (1) will be called the *cost* of allotment α. With this discussion, it is fairly natural to consider the following auxiliary problem.

Problem 2. ALLOTMENT FOR MALLEABLE TASKS (MT-ALLOTMENT)

INSTANCE: A directed graph $G = (V, E)$ that represents a set of n precedence constrained malleable tasks; the number m of processors; positive integers $p_{j,q}$ with $1 \leq j \leq n$ and $1 \leq q \leq m$ that specify the task execution times.

GOAL: Find an allotment $\alpha : V \to [1, m]$ that minimizes the cost $c(\alpha)$.

1.2   Known Results

The complexity of the makespan problem for malleable tasks has been studied in the paper of Du and Leung [7]: The problem with arbitrary precedence constraints is strongly NP-hard for $m = 2$ processors, and the problem of scheduling independent malleable tasks is strongly NP-hard for $m = 5$ processors.

Only a few positive results are available for scheduling malleable tasks. Prasanna & Musicus [12] proposed an algorithm for some specially structured precedence task graphs for the so-called continuous version of the problem; in the continuous version, a non-integer number of processors may be alloted to any task. Moreover, they assume the same speed-up function for all tasks. The results of Lenstra & Rinnooy Kan [10] for makespan minimization of precedence constrained sequential tasks imply that unless $P=NP$, makespan minimization of precedence constrained malleable tasks cannot have a polynomial time approximation algorithm with worst case performance guarantee better than 4/3.

The ALLOTMENT PROBLEM FOR MALLEABLE TASKS is closely related to the *discrete time-cost tradeoff* problem, a well-known problem from the project management literature; see e.g. De, Dunne, Ghosh & Wells [4]. The discrete time-cost tradeoff problem is a bicriteria problem for projects, where a project essentially is a system of precedence constrained tasks. Every task may be executed according to several different alternatives, where each alternative takes a certain amount of time and costs a certain amount of money. By selecting one alternative for every task, one fixes the cost (= total cost of all tasks) and the duration (= length of the longest chain) of the project. In the budget variant of the discrete time-cost tradeoff problem, the instance consists of such a project together with a cost bound C. The goal is to select alternatives for all tasks such that the project duration is minimized subject to the condition that the project cost is at most C; the corresponding optimal duration is denoted by $D^*(C)$. By rounding the solutions of a linear programming relaxation, Skutella [13] derives a polynomial time algorithm for this budget variant that finds a solution with project cost at most $2C$ and project duration at most $2D^*(C)$.

Now let us discuss the connection between the allotment problem MT-ALLOTMENT and the discrete time-cost tradeoff problem. In the allotment problem MT-ALLOTMENT, every task j can be executed in m alternative ways by assigning α_j machines to it, where $1 \leq \alpha_j \leq m$. In the language of the discrete time-cost tradeoff problem, the resulting duration of task j is p_{j,α_j} and the resulting cost of task j is $\alpha_j p_{j,\alpha_j}/m$, i.e., its contribution to the value $\frac{1}{m}W^\alpha$. Then the corresponding project cost equals $\frac{1}{m}W^\alpha$, the corresponding project duration equals L^α, and the maximum of these two values equals the cost $c(\alpha)$ of allotment α. By combining the above mentioned result of Skutella [13] with a binary search procedure, we now get the following proposition.

Proposition 1. *The ALLOTMENT PROBLEM FOR MALLEABLE TASKS possesses a polynomial time 2-approximation algorithm.* ∎

We furthermore note that the arguments of De, Dunne, Ghosh & Wells [5] imply that the ALLOTMENT PROBLEM FOR MALLEABLE TASKS is NP-complete in the strong sense.

2   Results and Outline of the Paper

We stress that all results in this paper are based on the monotonous penalty Assumption 1. In this paper, we derive polynomial time approximation algorithms for various cases of the MAKESPAN PROBLEM FOR MALLEABLE TASKS and of the ALLOTMENT PROBLEM FOR MALLEABLE TASKS. Let us first define for $m \geq 3$ the real numbers $r(m)$ by

$$r(m) = \min_{1 \leq \mu \leq (m+1)/2} \max \left\{ \frac{m}{\mu}, \frac{2m - \mu}{m - \mu + 1} \right\}. \tag{2}$$

Moreover, let $\mu(m)$ be the integer μ with $1 \leq \mu \leq (m+1)/2$ for which this minimum is attained. The following lemma provides the reader with some intuition on the (somewhat erratic) behaviour of the values $r(m)$ and $\mu(m)$. For small m, the values of $\mu(m)$ and $r(m)$ are listed in Figure 1.

Lemma 1. *The real numbers $r(m)$ and the integers $\mu(m)$ satisfy the following properties.*

(i)   *For all $m \geq 2$, we have $r(m) < (3 + \sqrt{5})/2 \approx 2.61803$.*
(ii) *As m tends to infinity, $r(m)$ tends to $(3 + \sqrt{5})/2$.*
(iii) *For every $m \geq 2$, the value $\mu(m)$ either equals the integer above or the integer below $\frac{1}{2}(3m - \sqrt{5m^2 + 4m})$.*
(iv) *For every $m \geq 2$ with $m \neq 3$ and $m \neq 5$, we have $\mu(m) \leq m/2$.*
(v)   *As m tends to infinity, $\mu(m)/m$ tends to $(3 - \sqrt{5})/2 \approx 0.38196$.* ∎

m	$\mu(m)$	$r(m)$	m	$\mu(m)$	$r(m)$	m	$\mu(m)$	$r(m)$	m	$\mu(m)$	$r(m)$
2	1	2.0000	10	4	2.5000	18	8	2.5454	26	10	2.5625
3	2	2.0000	11	5	2.4285	19	8	2.5000	27	11	2.5294
4	2	2.0000	12	5	2.4000	20	8	2.5000	28	11	2.5454
5	3	2.3333	13	6	2.5000	21	9	2.5384	29	12	2.5555
6	3	2.2500	14	6	2.4444	22	9	2.5000	30	12	2.5263
7	3	2.3333	15	6	2.5000	23	9	2.5555	31	13	2.5789
8	4	2.4000	16	7	2.5000	24	10	2.5333	32	13	2.5500
9	4	2.3333	17	7	2.4545	25	10	2.5000	33	13	2.5384

Fig. 1. A listing of the values $\mu(m)$ and $r(m)$ for $2 \leq m \leq 33$.

The straightforward proof of Lemma 1 is omitted. The following theorem summarizes our structural main result on the problems MT-MAKESPAN and MT-ALLOTMENT; its proofs can be found in Section 3. The theorem demonstrates that these two problems are strongly interlocked and interrelated. Moreover, up to some small constant factor it is sufficient to deal with the approximability of the – seemingly easier – problem MT-ALLOTMENT.

Theorem 2. *If there exists a polynomial time ϱ-approximation algorithm A for problem* MT-ALLOTMENT *on m processors, then there exists a polynomial time $\varrho \cdot r(m)$-approximation algorithm B for problem* MT-MAKESPAN *on m processors.*

An immediate consequence of Proposition 1, Theorem 2, and Lemma 1(i) is the following corollary.

Corollary 1. *The MAKESPAN PROBLEM FOR MALLEABLE TASKS possesses a polynomial time approximation algorithm with performance guarantee* $3 + \sqrt{5} \approx 5.23606$. ∎

The following Theorem 3 will be a strong and helpful tool for handling specially structured precedence constraints. Its proof can be found in Section 4.

Theorem 3. *Consider the decision version of problem* MT-ALLOTMENT *where for a given instance I of* MT-ALLOTMENT *and for a positive integer bound X, one must decide whether there exists an allocation of cost at most X. If there exists a pseudo-polynomial time exact algorithm for this decision version with running time polynomially bounded in the size of I and in the value of X, then there does exist a fully polynomial time approximation scheme for problem* MT-ALLOTMENT.

A directed precedence graph $G = (V, E)$ is *series parallel* if (i) it is a single vertex, (ii) it is the series composition of two series parallel graphs, or (iii) it is the parallel composition of two series parallel graphs. Only graphs that can be constructed via rules (i)–(iii) are series parallel. Here the *series composition* of two directed graphs $G_1 = (V_1, E_1)$ and $G_2 = (V_2, E_2)$ with $V_1 \cap V_2 = \emptyset$ is the graph that results from G_1 and G_2 by making all vertices in V_1 predecessors of all vertices in V_2, whereas the *parallel composition* of G_1 and G_2 simply is their disjoint union. Series parallel precedence constraints are a proper generalization of tree precedence constraints. We have the following result for series parallel precedence constraints.

Theorem 4. *There exists a pseudo-polynomial time exact algorithm for the decision version of the restriction of problem* MT-ALLOTMENT *to series parallel precedence graphs.*

Two tasks i and j are called *independent* if neither i is a predecessor of j nor j is a predecessor of i. A set of tasks is *independent*, if the tasks in it are pairwise independent. The *width* of the precedence graph G is the cardinality of its largest independent set. We have the following result for precedence graphs of bounded width.

Theorem 5. *There exists a pseudo-polynomial time exact algorithm for the decision version of the restriction of problem* MT-ALLOTMENT *to precedence graphs whose width is bounded by a constant d.*

The proofs of Theorems 4 and 5 are sketched in Section 5. Finally, by combining the statements in Theorems 2, 3, 4, and 5, we derive the following corollary.

Corollary 2. *For the restriction of the MAKESPAN PROBLEM FOR MALLEABLE TASKS to (a) series parallel precedence graphs and to (b) precedence graphs of bounded width, there exist polynomial time approximation algorithms whose performance guarantee can be made arbitrarily close to* $(3 + \sqrt{5})/2$. ■

3   From Allotments to Makespans

In this section we will prove Theorem 2. Consider an instance I of the malleable tasks problem as defined in Problems 1 and 2. Consider an optimal allotment α^+ and a ϱ-approximate allotment α^A for instance I with respect to problem MT-ALLOTMENT. Denote by W^+ and W^A the total work in these two allotments, and by L^+ and L^A the lengths of their critical paths, respectively. Since α^A is a ϱ-approximate allotment, we have

$$\max\{L^A, \frac{1}{m}W^A\} \leq \varrho \cdot \max\{L^+, \frac{1}{m}W^+\}. \tag{3}$$

Moreover, consider an optimal feasible schedule for instance I with respect to problem MT-MAKESPAN, and let $C^*_{\max}$ denote the optimal makespan. By applying equation (1) to $C^*_{\max}$ and to the allotment induced by the optimal schedule, and by using the fact that α^+ minimizes the allotment cost, we get that

$$\max\{L^+, \frac{1}{m}W^+\} \leq C^*_{\max}. \tag{4}$$

We will now define and analyze an approximation algorithm B for problem MT-MAKESPAN. This approximation algorithm is based on the value $\mu(m)$ with $1 \leq \mu(m) \leq (m+1)/2$ as we defined in the paragraph after equation (2). To simplify the presentation, we will from now on briefly write μ for $\mu(m)$, and omit the dependence on m. Algorithm B is a generalization of Graham's [9] well-known list scheduling algorithm for sequential tasks. The algorithm is described in Figure 2. The resulting schedule is denoted σ^B, the corresponding makespan is $C^B_{\max}$, the underlying allotment is α^B, the total work in α^B is W^B, and the length of the critical path in α^B is L^B. The only difference between allotments α^A and α^B is that the tasks using more than μ processors in α^A are compressed to μ processors in α^B. By the monotonous penalty Assumption 1(b), reducing the number of alloted processors cannot increase the work of a task. Together with inequalities (3) and (4) this yields

$$W^B \leq W^A \leq m\varrho C^*_{\max}. \tag{5}$$

The time interval from 0 to $C_{\max}^B$ is partitioned into three types of time slots: During the first type of time slot, at most $\mu - 1$ processors are busy. During the second type, at least μ and at most $m - \mu$ processors are busy, and during the third type at least $m - \mu + 1$ processors are busy. The corresponding sets of time slots are denoted by T_1, T_2, and T_3, respectively. The overall length of the time slots in set T_i, $1 \leq i \leq 3$, is denoted by $|T_i|$. If $\mu \leq m/2$, then every time slot from 0 to $C_{\max}^B$ belongs to exactly one of the three types, and all three types may actually occur. In the boundary case where $\mu = (m + 1)/2$ every time slot from 0 to $C_{\max}^B$ either belongs to the first or to the third type. In this boundary case there are no time slots of second type, since this would require that at least $(m+1)/2$ and at most $(m-1)/2$ processors are busy, which clearly is impossible. Since in either case these three types of time slots cover the whole interval from 0 to $C_{\max}^B$, we get that

$$C_{\max}^B \;=\; |T_1| + |T_2| + |T_3|. \tag{6}$$

Since during time slots of the first (respectively second and third) type at least one (respectively μ and $m - \mu + 1$) processors are busy, we get that

$$W^B \;\geq\; |T_1| + \mu |T_2| + (m - \mu + 1)|T_3|. \tag{7}$$

---

1. Initialization.
– Allot to task j ($j = 1, \ldots, n$) exactly $\alpha_j^B = \min\{\alpha_j^A, \mu\}$ processors.
– This fixes the execution time p_j^B and the work $w_j^B = \alpha_j^B \cdot p_j^B$ of every task j.

2. Repeat the following step until all tasks have been scheduled.
– Let READY denote the set of tasks whose predecessors all have already been scheduled.
– Compute for each task $j \in$ READY the earliest possible start time under the allotment α^B.
– Schedule the task in READY with the smallest computed earliest start time (ties are broken in favor of tasks with smaller indices).

Fig. 2. Approximation algorithm B for problem MT-MAKESPAN.

Lemma 2. *The sets T_1 and T_2 of time slots satisfy the following inequality with respect to the length L^A of the critical path in allotment α^A.*

$$|T_1| + \frac{\mu}{m}|T_2| \;\leq\; L^A. \tag{8}$$

Proof. The idea is to construct a 'heavy' directed path $\mathcal{P}$ in the transitive closure of the graph $G = (V, E)$. The last task in the path $\mathcal{P}$ is any multiprocessor task j_1 that completes at time $C_{\max}^B$ in the schedule σ^B. After we have defined the last $i \geq 1$ tasks $j_i \to j_{i-1} \to \cdots \to j_2 \to j_1$ on the path $\mathcal{P}$, we find the next task j_{i+1} as follows: Consider the latest time slot t in $T_1 \cup T_2$ that lies before the starting time of task j_i in σ^B. Consider the set V' of tasks that consists of task j_i and of all its predecessor tasks that start after time t in σ^B. Since during time slot t at most $m - \mu$ processors are busy, and since σ^B allots at most μ processors to any task in V', all the tasks in V' cannot be ready for execution during the time slot t. Hence, for every task in V' some predecessor is being executed during the time slot t. As the next task j_{i+1} on path $\mathcal{P}$, we select any predecessor of task j_i that is running during slot t. This procedure terminates when $\mathcal{P}$ contains a task that starts before all time slots in $T_1 \cup T_2$.

Now consider a task j on the resulting path $\mathcal{P}$. If α^B allots less than μ processors to task j, then α^A and α^B both allot the same number of processors to j. In this case the execution times of j in α^A and α^B are identical. In schedule σ_B such a task j may be executed during any time slot in $T_1 \cup T_2$. If α^B allots exactly μ processors to task j, then α^A may allot any number k of processors to j, where $\mu \le k \le m$. By the monotonous penalty Assumption 1(b), the work $\mu \cdot p_j^B$ in α^B is less or equal to the work $k \cdot p_j^A$ in α^A. Therefore, the execution time p_j^A of task j in allotment α^A is at least $\mu/k \ge \mu/m$ times the execution time p_j^B of j in allotment α^B. In schedule σ_B such a task j may be executed during any time slot in T_2, but not during a time slot in T_1.

By our construction, the tasks on the directed path $\mathcal{P}$ cover all time slots in $T_1 \cup T_2$ in schedule σ_B. Let us estimate the length $L^A(\mathcal{P})$ of the path $\mathcal{P}$ under the allotment α^A. The tasks that are executed during time slots in T_1 contribute a total length of at least $|T_1|$ to $L^A(\mathcal{P})$. The tasks that are executed during time slots in T_2 contribute a total length of at least $|T_2|\mu/m$ to $L^A(\mathcal{P})$. Since the length L^A of the critical path in α^A is an upper bound on $L^A(\mathcal{P})$, our proof is complete. ∎

Now let us complete the proof of Theorem 2. Multiplying (6) by $m - \mu + 1$ and subtracting (7) from it yields

$$(m - \mu + 1)C_{\max}^B \le W^B + (m - \mu)|T_1| + (m - 2\mu + 1)|T_2|. \tag{9}$$

We distinguish two cases. In the first case we assume that $m/\mu \le (2m-\mu)/(m-\mu+1)$. Then (2) yields $r(m) = (2m - \mu)/(m - \mu + 1)$. Moreover, the assumed inequality is equivalent to $(m - 2\mu + 1) \le \mu(m - \mu)/m$. Plugging this into (9), using (8) to bound $|T_1| + \mu|T_2|/m$, using (5) to bound W^B, and using (3) and (4) to bound L^A by $\varrho C_{\max}^*$ alltogether yields that

$$\begin{aligned}
(m - \mu + 1)C_{\max}^B &\le W^B + (m - \mu)|T_1| + \mu(m - \mu)|T_2|/m \\
&\le W^B + (m - \mu)L^A \\
&\le m\varrho C_{\max}^* + (m - \mu)\varrho C_{\max}^* = (2m - \mu)\varrho C_{\max}^*.
\end{aligned}$$

Hence, in this case schedule σ^B indeed yields a $\varrho \cdot r(m)$-approximation for $C_{\max}^*$. In the second case we assume that the inequality $m/\mu \ge (2m - \mu)/(m - \mu + 1)$ holds. Then (2) yields $r(m) = m/\mu$. Moreover, the assumed inequality is equivalent to $(m - \mu) \le (m - 2\mu + 1)m/\mu$. By plugging this into (9) and by using similar arguments as in the first case, we conclude that

$$(m - \mu + 1)C_{\max}^B \leq W^B + (m - 2\mu + 1)m|T_1|/\mu + (m - 2\mu + 1)|T_2|$$
$$\leq W^B + (m - 2\mu + 1)mL^A/\mu$$
$$\leq m\varrho\, C_{\max}^* + (m - 2\mu + 1)m\varrho\, C_{\max}^*/\mu$$
$$= (m - \mu + 1)m\varrho\, C_{\max}^*/\mu.$$

Hence, also in the second case schedule σ^B yields a $\varrho \cdot r(m)$-approximation for $C_{\max}^*$. The proof of Theorem 2 is complete.

4   From a Pseudo-Polynomial Time Algorithm to an FPTAS

In this section we will prove Theorem 3. Our first goal is to get a fast algorithm for the following auxiliary allotment problem MT-ALLOTMENT on series parallel precedence graphs: We assume that we are given an instance I of MT-ALLOTMENT, a positive real ε, and an a priori bound X such that there exists an allotment for I with cost at most X. Our goal is to find within polynomial time an allotment α that satisfies $c(\alpha) \leq (1 + \varepsilon)X$.

Define $Z = \varepsilon X/n$. Furthermore, define a scaled instance I' by setting $p'_{j,q} = \lfloor p_{j,q}/Z \rfloor$ for all tasks j and all $1 \leq q \leq m$ while keeping the same precedence constraints as in instance I. Note that $p_{j,q} \leq Z(p'_{j,q} + 1)$. Moreover, note that instance I' must have an allotment of cost at most X/Z, since the original instance I had some allotment of cost at most X. Take the pseudo-polynomial time algorithm that exists according to the assumption of Theorem 3, and apply it to the scaled instance I' with bound $\lfloor X/Z \rfloor$. Denote the resulting allotment by α with $c(\alpha) \leq \lfloor X/Z \rfloor$, and interpret allotment α for I' as an allotment β for the original instance I. Consider an arbitrary path $\mathcal{P}$ with $|\mathcal{P}|$ tasks in allotment β. Then

$$\sum_{j \in \mathcal{P}} p_{j,\beta(j)} \leq \sum_{j \in \mathcal{P}} Z(p'_{j,\beta(j)} + 1) = Z|\mathcal{P}| + Z \sum_{j \in \mathcal{P}} p'_{j,\alpha(j)} \leq Z\,n + Z\,L^\alpha. \quad (10)$$

This implies $L^\beta \leq Z\,n + Z\,L^\alpha$. Moreover,

$$\sum_{j \in V} \beta(j) \cdot p_{j,\beta(j)} \leq \sum_{j \in V} Z \cdot \alpha(j) \cdot (p'_{j,\alpha(j)} + 1) \leq Z\,mn + Z\,W^\alpha. \quad (11)$$

This implies $W^\beta \leq Z\,mn + Z\,W^\alpha$. Putting things together we conclude that

$$c(\beta) = \max\{L^\beta, \frac{1}{m}W^\beta\} \leq \max\{Z\,n + Z\,L^\alpha, Z\,n + Z\,\frac{1}{m}W^\beta\}$$
$$= Z\,n + Z\,c(\alpha) \leq \varepsilon X + Z\,(X/Z) = (1 + \varepsilon)X.$$

Hence, the cost of allotment β for I is at most $(1 + \varepsilon)X$ as desired. By the assumption of Theorem 3, the time to find β is polynomially bounded in the size of I and in $X/Z = n\varepsilon$. To summarize, we can solve our auxiliary problem and find the desired allotment within a running time that is polynomially bounded in the size of I and in $1/\varepsilon$.

It remains to get rid of the assumption that we do have an a priori knowledge of the bound X. Let $P = \sum_{j=1}^{n} p_{j,1}$ denote the total execution time of all tasks in I when they are executed on a single processor. By the monotonous penalty Assumption 1, every critical path in every allotment for I has length at most P, and also the average work of every allotment is at most P. Therefore, the cost of the optimal allotment is at most P, and we can find an $(1+\varepsilon)$-approximation by performing a binary search over the interval from 1 to P. This completes the proof of Theorem 3.

5   Allotments for Special Classes of Precedence Constraints

In this section we will prove Theorem 4. Hence, we are given an instance I of MT-ALLOTMENT where the precedence graph $G = (V, E)$ is series parallel, together with a positive integer bound X. Our goal is to decide within pseudo-polnomial time, whether there exists an allotment α with cost $c(\alpha) \leq X$.

---

1. Initialization of leaf vertices.
– For every leaves v of the decomposition tree and for every ℓ with $0 \leq \ell \leq X$, set
$F[v, \ell] := \min_{1 \leq q \leq m} \{q \cdot p_{v,q} \mid p_{v,q} \leq \ell\}$.

2. Handling interior vertices of the decomposition tree.
– For every interior vertex v with left child v_1 and right child v_2 and for every ℓ with $0 \leq \ell \leq X$ do the following:
– If v is a p vertex, then $F[v, \ell] := F[v_1, \ell] + F[v_2, \ell]$
– If v is an s vertex, then $F[v, \ell] := \min_{1 \leq k \leq \ell-1} F[v_1, k] + F[v_2, \ell - k]$

3. Termination.
– Answer YES if there exists some $1 \leq \ell \leq X$ with $F[root, \ell]/m \leq X$. Otherwise, answer NO.

Fig. 3. A dynamic programming algorithm for computing $F[v, \ell]$.

It is well known that a series parallel graph can be decomposed in polynomial time into its atomic parts according to the series and parallel compositions. Essentially, such a decomposition corresponds to a rooted, ordered, binary tree where all interior vertices are labeled by s or p (series or parallel composition) and where all leaves correspond to single vertices of the precedence graph G. We associate with every interior vertex v of the decomposition tree the series parallel graph $G(v)$ induced by the leaves of the subtree below v. Note that for the root vertex we have $G(root) = G$. For a vertex v in the decomposition tree, and for an integer ℓ with $1 \leq \ell \leq X$, we denote by $F[v, \ell]$ the smallest possible value w with the following property: There exists an allotment α for the tasks in $G(v)$ with $L^\alpha \leq \ell$ and $W^\alpha \leq w$. It is easy to compute all such values $F[v, \ell]$ by a dynamic programming approach that starts in the leaves of the decomposition tree, and then moves upwards towards the root. This algorithm is sketched in Figure 3. The time complexity of this dynamic programming algorithm is $O(nmX^2)$.

This completes the proof of Theorem 4. Theorem 5 can be proved by a similar dynamic programming approach. The exact arguments are omitted from this extended abstract.

References

1. E. BLAYO, L. DEBREU, G. MOUNIÉ, AND D. TRYSTRAM [1999]. Dynamic load balancing for ocean circulation with adaptive meshing. *Proceedings of the 5th European Conference on Parallel Computing*, Springer LNCS 1685, 303–312.
2. M. COSNARD AND D. TRYSTRAM [1995]. *Parallel Algorithms and Architectures.* International Thomson Publishing.
3. D. CULLER, R.KARP, D.PATTERSON, A.SAHAY, E. SANTOS, K. SCHAUSER, R. SUBRAMANIAN, AND T. VON EICKEN [1996]. LogP: A practical model of parallel computation. *Communications of the ACM 39*, 78–85.
4. P. DE, E.J. DUNNE, J.B. GOSH, AND C.E. WELLS [1995]. The discrete time-cost tradeoff problem revisited. *European Journal of Operational Research 81*, 225–238.
5. P. DE, E.J. DUNNE, J.B. GOSH, AND C.E. WELLS [1997]. Complexity of the discrete time-cost tradeoff problem for project networks. *Operations Research 45*, 302–306.
6. M. DROZDOWSKI [1996]. Scheduling multiprocessor tasks – An overview. *European Journal of Operational Research 94*, 215–230.
7. J. DU AND J.Y.-T. LEUNG [1989]. Complexity of scheduling parallel task systems. *SIAM Journal on Discrete Mathematics 2*, 473–487.
8. A. GERASOULIS AND T. YANG [1992]. PYRROS: Static scheduling and code generation for message passing multiprocessors. *Proceedings of the 6th ACM International Conference on Supercomputing*, 428–437.
9. R.L. GRAHAM [1966]. Bounds for certain multiprocessing anomalies. *Bell System Technical Journal 45*, 1563–1581.
10. J.K. LENSTRA AND A.H.G. RINNOOY KAN [1978]. Complexity of scheduling under precedence constraints. *Operations Research 26*, 22–35.
11. R. LEPÈRE, G. MOUNIÉ, B. ROBIČ, AND D. TRYSTRAM [1999]. Malleable tasks: An electromagnetic efficient model for solving actual parallel applications. *Proceedings of the International Conference on Parallel Computing 99 (Parco'99)*, Imperial College Press, 598–605.
12. G.N.S. PRASANNA AND B.R. MUSICUS [1991]. Generalized multiprocessor scheduling using optimal control. *Proceedings of the 3rd Annual Symposium on Parallel Algorithms and Architectures (SPAA'91)*, 216–228.
13. M. SKUTELLA [1998]. Approximation algorithms for the discrete time-cost tradeoff problem. *Mathematics of Operations Research 23*, 909–929.
14. J. TUREK, J. WOLF, AND P. YU [1992]. Approximate algorithms for scheduling parallelizable tasks. *Proceedings of the 4th Annual Symposium on Parallel Algorithms and Architectures (SPAA'92)*, 323–332.

On the Approximability of the Minimum Test Collection Problem

(Extended Abstract)

Bjarni V. Halldórsson[*1], Magnús M. Halldórsson[2,3], and R. Ravi[**4]

[1] Dept. of Math. Sciences, Carnegie Mellon University. `bjarni@cmu.edu`
[2] Dept. of Computer Science, University of Iceland. `mmh@hi.is`
[3] Iceland Genomics Corp. `mmh@uvs.is`
[4] GSIA, Carnegie Mellon University. `ravi@cmu.edu`

Abstract. The minimum test collection problem is defined as follows. Given a ground set S and a collection C of tests (subsets of S), find the minimum subcollection C' of C such that for every pair of elements (x, y) in S there exists a test in C' that contains exactly one of x and y. It is well known that the greedy algorithm gives a $1 + 2\ln n$ approximation for the test collection problem where $n = |S|$, the size of the ground set. In this paper, we show that this algorithm is close to the best possible, namely that there is no $o(\log n)$-approximation algorithm for the test collection problem unless $P = NP$.

We give approximation algorithms for this problem in the case when all the tests have a small cardinality, significantly improving the performance guarantee achievable by the greedy algorithm. In particular, for instances with test sizes at most k we derive an $O(\log k)$ approximation. We show APX-hardness of the version with test sizes at most two, and present an approximation algorithm with ratio $\frac{7}{6} + \epsilon$ for any fixed $\epsilon > 0$.

1 Introduction and Motivation

The test collection problem arises naturally in the following general setting of identification problems: Given a set of individuals (database entries) and a set of binary attributes that may or may not occur in each individual, the goal is to find the minimal subset of attributes (a test collection) such that each individual can be uniquely identified from the information on which of this subset of attributes it contains. In this way, the incidence vector of any individual with the test collection is a unique binary signature for it distinguishing it from other individuals, and thus uniquely identifying it from the list of individuals. This problem is also commonly known in the literature as the minimum test set problem [16] and the minimum test cover problem [3] and arises commonly in fault analysis, medical diagnostics and pattern recognition (see, e.g., [16]).

[*] Supported by a Merck Computational Biology and Chemistry Program Graduate Fellowship from the Merck Company Foundation.

[**] Supported in part by subcontract No. 16082-RFP-00-2C in the area of "Combinatorial Optimization in Biology (XAXE)," Los Alamos National Laboratories.

F. Meyer auf der Heide (Ed.): ESA 2001, LNCS 2161, pp. 158–169, 2001.

The test collection problem first came to our attention in the setting of protein identification in computational biology [7]. A major thrust in post-genome biology is proteomics, where one analyzes all cellular proteins in a high throughput manner. We are concerned with the protein identification aspect of proteomics.

Halldórsson, Minden and Ravi [7] proposed a new approach of using an array of antibodies that recognize and bind specifically to short peptide sequences (called epitopes) - such an epitope can distinguish proteins that contain this epitope from those that do not. Each antibody binds its cognate epitope by virtue of complementary three dimensional structure. To observe the occurrence of epitope binding, the epitopes will be fluorescently tagged. The binding of antibodies in the chip to the given unidentified protein can then be measured using a fluorescence detector. Thus the final output is a binary indicator vector of dimension equal to the number of antibodies in the chip, each bit indicating whether or not the protein is bound to the corresponding antibody.

The proposal in [7] is to generate a set of antibodies that recognize a set of epitopes that are shared by many proteins in such a way that the entire set of epitopes covers all possible proteins in the organism's proteome. Moreover, the set of antibodies will be selected such that each protein will be recognized by a unique subset of antibodies, i.e., each protein will have a unique signature of binding antibodies. This problem can hence be viewed as a test collection problem where the elements of the ground set are the proteins and there is a test for each antibody, determined by the binding of the antibodies to the proteins.

The novelty of our approach is the utilization of antibodies that bind to a large subset of proteins. However, most currently known antibodies are very specific in their binding characteristics. Hence a special case of particular interest is one where each antibody binds only to a few of the proteins, i.e. when the test sets are small. In Section 5 we give two results for this case.

Our Results

The test collection problem is NP-hard, as was shown by Garey and Johnson [6] via a reduction from three-dimensional matching. By a natural reduction of the problem to the set covering problem (see e.g., [16]) and an application of the results of Johnson [11] or Lovász [14], the natural greedy algorithm gives a $1 + 2\ln n$ approximation for the test collection problem, where n is the number of elements in the ground set. We show that the greedy algorithm has optimal approximation ratio up to a constant multiple unless $P = NP$. Motivated by moving towards a practical solution, we consider the case when all test sizes are at most k, and give a new algorithm with $O(\log k)$ approximation guarantee, significantly improving the straightforward $O(\log(n - k)k)$ approximation guarantee of the greedy algorithm. We further consider the case where $k = 2$ and give a APX-hardness proof and an algorithm with approximation ratio $\frac{7}{6} + \epsilon$, for any $\epsilon > 0$.

Previous Work

Subsequent to the acceptance of this extended abstract, we discovered that many of the results we present here have been obtained independently earlier. Moret and Shapiro [16] already described the crux of our reduction of a set covering problem to a test collection problem outlined in the proof of Theorem 1. Their result, combined with the hardness of approximation results on set covering [2, 4,15] already imply a logarithmic lower bound on the approximation ratio of minimum test collection (Theorems 2 and 3).

For the cases with small test sizes, we recently discovered that previous work by C.A.J. Hurkens, J. K. Lenstra and L. Stougie (unpublished, but documented in [1,3] - personal communication with L. Stougie, June 2001) obtained many of our approximation results independently including the following: an $O(\log k)$-approximation algorithm for problem instances with test sizes at most k (Theorem 4) and for instances with test sizes at most two, (i) a $\frac{11}{8}$-appoximation guarantee for the greedy algorithm, (ii) a sequence of improved heuristics employing increasingly extensive local search: a case analysis for determining the approximation ratio of these heuristics using linear programming [3] shows that the fourth member in the family achieves an approximation ratio of $\frac{7}{6}$ while the fifth one has an even better ratio. The approximation ratio of this sequence of heuristics is conjectured to converge to one. Our APX-hardness result of the problem with test sizes two shows that even if the conjecture held true, the trade-off between performance ratio and running time (the convergence "rate") is not represented by a PTAS, unless $P = NP$ [9]. These authors also provide an NP-hardness proof for the case with test sizes at most two by a direct reduction from a two-path packing problem (We use a similar reduction in Algorithm 3)

In Section 2, we give different formulations of the test collection problem and prove their equivalence. In Section 3 we show how the test collection problem relates to the set covering problem. In Section 4 we show the hardness of approximation of test collection. Finally, in Section 5 we prove our results for the case of small test sets.

2   Formulations

We give three equivalent formulations of the test collection problem.

Definition 1. *Given a collection $\mathcal{C}$ of subsets of a finite set S find a minimum subcollection $\mathcal{C}'$ such that for each $x, y \in S$ there exists $C \in \mathcal{C}'$ that contains exactly one of x and y.*

Definition 2. *Given a complete graph on the node set S, and a set of cuts $\mathcal{C} = \{C_1, C_2, \ldots, C_m\}$, find a minimum subcollection $\mathcal{C}' \subseteq \mathcal{C}$ the union of whose edges is the entire edge set of the complete graph.*

Definition 3. *Given a collection $\mathcal{C}$ of subsets of a finite set S find a minimum subcollection $\{C_1, C_2, \ldots, C_m\} = \mathcal{C}'$ such that the vectors*

$$I_x = (\delta(C_1, x), \delta(C_2, x), \ldots, \delta(C_m, x))$$

where $\delta(C_i, x)$ is one if x is in C_i and zero otherwise, are unique for all $x \in S$.

The first definition allows for a natural reduction of the problem to the set covering problem. The second definition allows for a graphical representation of the tests as cuts in a graph. We will henceforth refer to the test C as the *cut* C as a shorthand for the cut $C : (S - C)$, i.e. the set of edges with exactly one endpoint in C.

The last definition relates best to our motivating example; Here we can think of S to be our set of proteins, C_i to be the set of proteins that contain the ith epitope (i.e. , those proteins that the ith antibody recognizes).

The following lemmas are immediate.

Lemma 1. *All three Definitions 1, 2 and 3 of the test collection problem are equivalent.*

Lemma 2. *The size of any solution to the test collection problem is at least* $\lceil \log_2 |S| \rceil$.

Proof. We note by Definition 3 that all the elements of S must have a unique binary incidence pattern with the solution. Therefore if m is the size of a solution to a test collection problem $2^m \geq |S|$ or $m \geq \lceil \log_2 |S| \rceil$.   □

A special case of the test collection problem as defined in Definition 2 is one where *any* cut can be chosen in the given complete graph [13]. It is easy to show that this problem has a simple optimal solution of size $\lceil \log_2 n \rceil$, where n is the cardinality of the node set S (An interpretation of the bits in a binary labeling of the nodes in S as defining the shores of cuts gives this result).

3   Relation to Set Covering

We first define the set covering problem.

Definition 4. *Given a ground set S and a collection of subsets $\{C_1, C_2, \ldots, C_m\}$ $= C$ find a minimum cardinality subcollection C' of C such that for all $e \in S$ there exists a $C \in C'$ such that $e \in C$.*

The following lemma is independently due to Johnson [11] and Lovász [14].

Lemma 3. *The greedy algorithm gives a $1 + \ln k$ approximation for the set covering problem, where k is the size of the largest set.*

Considerable work has been done on the approximation of set covering. Lund and Yannakakis [15] showed that set covering cannot be approximated within $c \ln n$ for any $c \leq \frac{1}{4}$ unless $NP \subset DTIME(n^{poly \ln n})$. Feige [4] improved on their result and showed that set covering cannot be approximated within $(1 - \epsilon) \ln n$ unless $NP \subset DTIME(n^{\log \log n})$. Finally, Arora and Sudan [2] showed that set cover cannot be approximated to within $o(\log n)$ unless $P = NP$.

We note a natural reduction of test collection to set covering (see also [16]). This reduction is depicted in Figure 1.

Lemma 4. *The test collection problem is reducible to a set covering problem.*

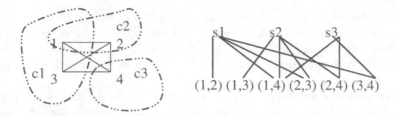

Fig. 1. Left figure shows a test collection instance and the right its reduction to a set covering problem. Each pair of elements in the test collection problem gets mapped to an element in the ground set of the set covering problem. A test set C gets mapped to a set that covers all the pairs in the cut $C : (S - C)$.

Proof. Let C_{tc}, S_{tc} be an instance of the test collection problem. We will now construct an instance C_{sc}, S_{sc} of the set covering problem. For all unordered pairs of elements $(x, y) \in S_{tc}$, we define an element $e_{xy} \in S_{sc}$. For each test $C_{tc} \in C_{tc}$ we construct a corresponding a set $C_{sc} \in C_{sc}$ by letting $e_{xy} \in C_{sc}$ when exactly one of x or y is in C_{tc}. It is not hard to verify that any solution to the resulting set cover problem is also a feasible solution to the test collection problem using the equivalence of Definitions 1 and 3. □

Note the quadratic increase in the size of the ground set $|S_{sc}|$ of the resulting set cover problem as compared with $|S_{tc}|$. If there are n elements in the original test collection problem there will be $\binom{n}{2}$ elements its set covering formulation. Notice also that given a partial solution to the test collection problem, we can also map the residual problem to a set covering problem using the above method.

As noted by Moret and Shapiro [16], combined with Lemma 3, Lemma 4 implies a $1 + 2 \ln n$ approximation guarantee for the greedy algorithm.

Recall (Definition 2) that the test collection problem can also be viewed as a form of cut-covering problem in a complete graph. Given a partial solution to the test collection problem, as in Definition 2, we can view the remaining problem on a graph whose node set is the elements of the test collection problem and edge set only the pairs of elements that have not been distinguished by the partial solution. The use of a partial solution leads us to generalize the test collection problem.

Definition 5. *Given a graph $G = (V, E)$, and a set of cuts $C = \{C_1, C_2, \ldots, C_m\}$, find a minimum subcollection $C' \subseteq C$ the union of whose edges is E.*

We note that this is a generalization as the original test collection problem in Definition 2 occurs as the special case when the edge set is the complete graph. As before, in the special case when the tests can be *any* subset of S (i.e., when all cuts in the graph are allowed in the cover), this problem is known simply as the cut cover problem [13]. Motwani and Naor [17] have shown that the cut cover problem is NP-hard by showing a relation to the graph coloring problem. Using an algorithm of Halldórsson [8] they give an algorithm which has solution which is no larger than the size of the optimal solution plus $\ln n - 3\ln\ln n$. Using a hardness of approximation result of Lund and Yannakakis [15] for graph coloring they show that unless $P = NP$ cut cover is hard to approximate within an additive $\epsilon \ln n$ for some $\epsilon > 0$. Combining their result with the more resent work of Feige and Killian [5] it can be shown that unless $ZPP = NP$ no poly-time algorithm can approximate cut cover to any better than within an additive $(1 - \epsilon)\ln n$ for any $\epsilon > 0$.

The NP-hardness of the test collection problem is shown in [6] via a reduction from 3-dimensional matching. We now give a reduction of the test collection problem from the set covering problem, that will also be useful in showing a near-optimal approximation hardness results for the test collection problem.

Theorem 1. *The set covering problem is reducible in polynomial time to the test collection problem.*

Proof. Let $\mathcal{C}_{sc}, \mathcal{S}_{sc}$ be an instance of a set covering problem, where $\mathcal{S}_{sc}$ is the ground set and $\mathcal{C}_{sc}$ is a collection of subsets of $\mathcal{S}_{sc}$ and let $n = |\mathcal{S}_{sc}|$.

Let us now construct $\mathcal{C}_{tc}, \mathcal{S}_{tc}$. Let there be two elements $e_l, e_r \in \mathcal{S}_{tc}$ for every element $e \in \mathcal{S}_{sc}$. Now arbitrarily assign distinct numbers from 1 through n to the elements of $\mathcal{S}_{sc}$ and construct the sets $G_{tc}^i \in \mathcal{C}_{tc}, i \in \{1, 2, \ldots \lceil \log_2 n \rceil\}$ where for every element $e \in \mathcal{S}_{sc}$ both $e_l, e_r \in G_{tc}^i$ if the binary representation of e has 1 in the i-th bit. Furthermore for all sets $C_{sc}^j \in \mathcal{C}_{sc}$ construct the corresponding set $C_{tc}^j \in \mathcal{C}_{tc}$ such that $e_r \in C_{tc}^j$ if and only if $e \in C_{sc}^j$. This reduction is depicted in Figure 2.

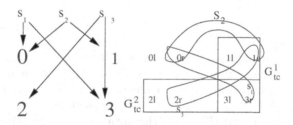

Fig. 2. The reduction of set cover to test collection. Each node in the set covering instance is mapped to two nodes in the test collection instance and $\lceil \log_2 n \rceil$ new tests are added.

In the previous construction we can think of the sets G_{tc}^i as forming a hypercube-like pattern, splitting the ground set into two approximately equal parts. If the cardinality of the original set covering ground set is n then there are $\lceil \log_2 n \rceil$ sets G_{tc}^i; The necessity of including them all then follows from having to distinguish the n e_l elements from each other with these sets being the only ones in the collection that can accomplish this task. Clearly the sets G_{tc}^i are sufficient to distinguish between all the e_l elements.

Let $\mathcal{C}'_{SC}$ be any set cover for SC and let $\mathcal{C}'_{TC}$ be the collection of tests C_{TC} where the test $C_{TC} \in \mathcal{C}'_{TC}$ if and only if the set $C_{SC} \in \mathcal{C}'_{SC}$. Then $\mathcal{C}'_{TC} \cup$

$\cup_{i=1}^{\lceil \log_2 n \rceil} G_{tc}^i$ is a solution to the reduced test collection problem: The sets G_{tc}^i will distinguish between all pairs except for the (e_l, e_r) pairs; Since the sets in $\mathcal{C}'_{SC}$ cover all the elements in the ground set of SC, each e_r element will be contained in least one of the tests in $\mathcal{C}'_{TC}$ hence distinguishing it from its e_l neighbor.

Now let $\mathcal{C}'_{TC}$ be any test collection for a reduced set cover problem SC. By the necessity of including the sets G_{tc}^i, note that $\cup_{i=1}^{\lceil \log_2 n \rceil} G_{tc}^i \subset \mathcal{C}'_{TC}$. We can construct a solution $\mathcal{C}'_{SC}$ to SC by letting $C_{SC}^j \in \mathcal{C}'_{SC}$ if and only if $C_{TC}^j \in \mathcal{C}_{TC} \setminus \cup_{i=1}^{\lceil \log_2 n \rceil} G_{tc}^i$. Each element e_r is distinguished from its neighbor e_l in the test collection solution and hence is contained in one of the sets $C_{tc}^j \in \mathcal{C}'_{TC} \setminus \cup_{i=1}^{\lceil \log_2 n \rceil} G_{tc}^i$. The sets $\mathcal{C}'_{SC}$ therefore cover the ground set of SC.

By the arguments in the two previous paragraphs we have the following lemma, relating the size of the reduction to the size of the original set cover problem.

Lemma 5. *An instance of a set covering problem has a solution of size k if and only if its reduction to a test collection problem described above has a solution of size $k + \lceil \log_2 n \rceil$.*

It is easy to verify that all the steps in the reduction described above are polynomial-time implementable.

$\square$

4 Hardness of Approximating Test Collection

In this section we will prove that test collection is as hard to approximate as the set covering problem.

We will show an approximation hardness for test collection from the corresponding hardness of approximating set cover. Let ϕ be a polynomial time algorithm that can approximate test collection. Let SC be an instance of the set covering problem and n be the number of elements in the ground set of SC. From Lemma 5, we see that the size of the optimal solution to the test collection reduction of the set covering problem is $\lceil \log_2 n \rceil$ larger than the optimal solution to SC. Hence this reduction does not guarantee hardness of approximation directly. To translate the hardness, we make multiple copies of the original set covering problem, and apply the reduction to the copied instance to drive down the logarithmic additive error.

Let k be a positive integer. The following is our approximation algorithm for the set covering problem using a call to an approximation algorithm for test collection.

Algorithm 1 $TCtoSC(SC, \phi, k)$

Multiply *the original set covering problem k times to construct a set covering instance SC_k.*

Reduce *SC_k to a test collection problem using the reduction in Lemma 1; Denote this reduced problem TC_k.*

Solve TC_k *using the approximation algorithm ϕ for the test collection problem.*
Construct *a solution to SC_k from the solution to TC_k.*
Divide *the solution of SC_k into solutions, one for each of the individual copies of SC, and output the best (minimum size) solution among them.*

$\square$

Observation 1 *Algorithm 1 runs in time polynomial in n and k.*

It follows that if k is a polynomial function of n, Algorithm 1 runs in time polynomial in n.

We now prove a technical lemma to be used in our approximation hardness result.

Lemma 6. *Let ϕ by an algorithm with performance guarantee $A_\phi(n)$, where n is the number of elements in the ground set of the test collection problem. The size of the solution returned by the algorithm $TCtoSC(SC, \phi, k)$ is at most $A_\phi(2kn)(\frac{1}{k}\lceil \log_2 kn \rceil + Opt(SC))$, where $Opt(SC)$ denotes the size of the optimal solution to the set covering instance SC.*

The proof follows by observing that the test collection instance TC_k has $2kn$ entries, applying Lemma 5, and bounding the size of the output solution to SC by the average size of the solutions induced on the k copies.

Theorem 2. *Test collection is hard to approximate within $o(\log n)$ unless $P = NP$, where n is the number of elements in the ground set.*

Proof. Suppose there exists an algorithm ϕ that can approximate test collection within $o(\log n)$. Let SC be any instance of the set covering problem. Solve SC using $TCtoSC(SC, \phi, \lceil \ln n \rceil)$. By Observation 1, $TCtoSC$ runs in time polynomial in n. Using Lemma 6 and the performance guarantee definition, the size of the solution returned by $TCtoSC$ is $o(\log(n\lceil \log n \rceil)) \cdot (\frac{\lceil \log_2 \lceil \ln n \rceil n \rceil}{\lceil \ln n \rceil} + Opt(SC)) = o(\log n \cdot Opt(SC))$ since $Opt(SC)$ is a positive integer. This contradicts the results of Arora and Sudan [2] giving the claim. $\square$

A slight refinement of our main theorem follows.

Theorem 3. *If there some $\epsilon > 0$ such that a polynomial time algorithm can approximate test collection within $(1 - \epsilon) \ln n$, then $NP \subset DTIME(n^{\log \log n})$. Here n is the number of elements in the ground set.*

5   Test Collection with Small Tests

In this section, we consider the special case when all the tests are small. We first give an algorithm for the case when the size of the largest set is bounded by k. The performance guarantee is $O(\ln k)$. We then look at the special case when the size of the set is bounded by 2, first verifying the NP-hardness of the problem restricted to this special case and then giving a $\frac{7}{6} + \epsilon$-approximation algorithm, for any $\epsilon > 0$.

Test Collection with Tests of Size at Most k

We consider the general case where the size of each test is bounded above by a constant k. We notice here that using the approximation guarantee of Lemma 3 and the reduction of the the test collection problem to the set covering problem in Lemma 4 gives a $O(\log((n-k)k))$ approximation guarantee for the test collection problem when the size of the maximum set is bounded by k. This is due to the fact that a test of size k will be reduced to a set of size $(n-k)k$ in the set covering problem. In other words, following Definition 2, such a test set will be converted to a cut in the complete graph in the cut cover version where the size of a shore is bounded by k, thus having all the induced edges in such a cut, which is at most $k(n-k)$. We give a new algorithm that improves this guarantee to $O(\log k)$.

We first prove a relation between the test collection problem and the set covering problem on the same *identical* instance where we think of the test sets as covering sets in the set cover instance. This is not a problem reduction as we did earlier in Lemma 4.

Lemma 7. *If $SC(\mathcal{S},\mathcal{C})$ is an instance of the set covering obtained from $TC(\mathcal{S},\mathcal{C})$, an instance of a test collection problem on the same ground set by defining the sets in SC as tests in TC, then $Opt(SC(\mathcal{S},\mathcal{C})) \leq Opt(TC(\mathcal{S},\mathcal{C}))+1$, where $Opt(x)$ denotes the size of an optimal solution to problem x.*

Proof. By Definition 3 all the elements of the ground set $\mathcal{S}$ must have a different incidence pattern in any solution to TC. As a consequence, in a valid solution to TC all but at most one elements of the ground set $\mathcal{S}$ must have incidence with at least one member of any valid solution to TC. This implies that at most one of the elements of $\mathcal{S}$ is not covered by a valid solution to TC. Thus any solution to TC is a set cover for all but at most one member of $\mathcal{S}$.   □

We give a new algorithm for the test collection problem that performs particularity well when the size of the largest set is small.

Algorithm 2 *SmallTC$(\mathcal{S},\mathcal{C})$*

Phase I *(Cover) Identity map $TC(\mathcal{S},\mathcal{C})$ to a set covering problem $SC_I(\mathcal{S},\mathcal{C})$, i.e. both the test collection and the set cover instance have the same ground set and the tests in $TC(\mathcal{S},\mathcal{C})$ become sets in $SC(\mathcal{S},\mathcal{C})$ (Lemma 7).*
  Solve SC_I using the greedy algorithm for set covering (Lemma 3) and let $\mathcal{C}_I$ be the solution obtained. Note that $\mathcal{C}_I$ is not necessarily a test collection and hence is a partial solution to the TC instance.

Phase II *(Reduce and Cover) Reduce the remaining test collection problem along with the partial solution $\mathcal{C}_I$ to a set covering problem, SC_R, using the reduction outlined in Lemma 4.*
  Solve SC_R using the greedy algorithm and map the solution to a subset of $\mathcal{C}_R \in \mathcal{C}$.

Return $\mathcal{C}_I \cup \mathcal{C}_R$.

Theorem 4. *The approximation guarantee of algorithm SmallTC$(\mathcal{S},\mathcal{C})$ is $O(\ln k)$, where k is the size of the largest test in $\mathcal{C}$.*

Proof. In the first phase of the algorithm we identity map the test collection problem to a set cover problem where each *test* in the test collection problem becomes a *set* in the set cover problem. The maximal set size in SC_I will hence be k and by Lemma 3 we have that $|C_I| \leq (1 + \ln k)\, Opt(SC_I)$ and by Lemma 7 we have $Opt(SC_I) \leq (Opt(TC) + 1)$. Thus, $|C_I| = O(\ln k)Opt(TC)$ observing that $Opt(TC) \geq \lceil \log_2 |\mathcal{S}| \rceil \geq \log_2 k$.

We now apply the reduction given in Lemma 4 for reducing an instance of the test collection problem to an instance of the set covering problem. We recall from our previous discussion that the reduction of the test collection problem to the set cover problem corresponds to looking at all the elements of the test collection problem as nodes and the task is to cover all the edges in the induced complete graph using edges of the available cuts. The partial solution has covered a large part of the edges and the remaining graph consists of connected components, each having size at most k. The size of each set in the set covering reduction in Lemma 4 is now at most $k(k-1)$, since each set has at most k nodes and each node has degree at most $k-1$. By Lemma 3 we have $|C_R| \leq (1 + \ln k(k-1))\, Opt(SC_R) \leq O(\ln k)\, Opt(TC)$.

The result follows by combining the contributions from the two phases. ☐

Test Collection with Tests of Size at Most Two

Using a reduction from 3-dimensional matching similar to the one in Garey and Johnson's book [6] it is easy to show that the test collection problem remains NP-hard even if the sets are limited to have at most three elements. Garey and Johnson also state ([6], p. 222) that when the size of the sets is limited to two the problem is solvable in polynomial-time. However, in a last-page update added to later reprints, this is corrected and the problem is claimed to be NP-hard.

Theorem 5. *The minimum test collection problem where the size of the largest set is limited to two is APX-complete.*

We omit a (routine) proof of the above theorem that uses a reduction from maximum bounded 3-dimensonal matching [12] due to lack of space. The above theorem implies that there is no polynomial-time approximation scheme for this even unless $P = NP$ [9].

We will use the notation 2-TC for test collection problems where the size of the sets is bounded by two. It is easy to show that we can assume without loss of generality that 2-TC contains only tests of size two, we may therefore refer to the tests of 2-TC as edges. Next we give an approximation algorithm for 2-TC. Before that, we make an observation which changes our perspective of 2-TC and motivates the algorithm.

Lemma 8. *2-TC is equivalent to the following problem: Given a graph, find a edge-subset that defines a subgraph with at most one isolated node, and in which every other connected component has at least three nodes, and the number of components is maximized.*

Proof. Let $TC2$ be an instance of 2-TC. In a solution to $TC2$, at most one node has no adjacent edges; Every other node must be adjacent to at least one edge in the solution. Furthermore, the solution cannot have single edge components since the endpoints of such an edge have the same incidence pattern (Definition 3); Therefore each component must have size at least 3. Minimizing the number of edges in the graph will be the same as maximizing the number of components in the graph induced by the solution since each component will be a tree in the optimal (minimal) solution.   □

By the previous lemma we see that we can use an algorithm for set packing with 3-sets as a subroutine in solving 2-TC. In particular, we can use the algorithm of Hürkens and Schrijver [10] which has a performance guarantee of $\frac{3}{2} + \epsilon$ for any $\epsilon > 0$.

Algorithm 3 *Approx(TC2)*

Reduce *Given an instance $TC2$ of 2-TC construct an instance SP of set packing with 3-sets; Let the ground sets be the same; For every triple x, y, z in the ground set of $TC2$, if at least two of the edges $(x, y), (x, z), (y, z)$ occur in $TC2$, include $\{x, y, z\}$ in the set packing instance SP.*

Solve *Fix an $\epsilon > 0$ and use the algorithm for set packing with 3-sets of [10] on SP with performance guarantee $\frac{3}{2} + \epsilon$.*

Construct *a solution s_{sp}. For each 3-set in the solution to the set packing problem add a corresponding pair of edges to s_{sp}.*

Complete *the solution by adding edges to s_{sp} to connect all but an isolated node to other larger connected components in the graph to form s_{tc}.*

Return s_{tc}.

Lemma 9. *If $TC2$ is feasible then Algorithm 3 returns a valid solution to 2-TC and runs in polynomial time.*

Proof. The **Reduce** step can be done in time $O(m^2)$, where m is the number of edges in $TC2$. The algorithm of [10] runs in polynomial time for any fixed ϵ. The construct step can be done in time $O(m)$. This algorithm guarantees returning a maximal solution. If $TC2$ is feasible then the **Complete** step can be done in time $O(n^2)$ by iteratively looking at all nodes in the graph that are not to incident to edges in the current solution. As the number of components in the set partitioning solution is maximal all but one isolated node will be connectable to the components of the current solution. We can therefore iteratively add one node at a time to connect it to the current solution.   □

Lemma 10. *Algorithm 3 gives a $\frac{7}{6} + \epsilon$ approximation to 2-TC, for any $\epsilon > 0$.*

Proof. We notice that the size of the solution to the test collection problem is same as the number of edges in the solution. This is the same as the number of nodes in the graph less the number of components in the solution.

Let SP be the reduced set packing problem. The algorithm for finding a set packing with 3-sets [10] has a $\frac{3}{2} + \epsilon$-approximation guarantee, for any fixed $\epsilon > 0$. Exactly one edge will be added for each node not in a 3 set. The size of the solution returned by the algorithm will therefore be at most $2Apx(SP) +$

$(n - 3Apx(SP) = n - Apx(S) \leq n - \frac{1}{\frac{3}{2}+\epsilon}Opt(SP)$. Since we can recover a set packing solution from a test cover solution in a natural way, $Opt(TC) \geq Opt(SP) + (n - 2Opt(SP)) = (n - Opt(SP))$. The approximation guarantee of τ is then $Apx_\tau \leq \frac{n - \frac{1}{\frac{3}{2}+\epsilon}Opt(SP)}{n - Opt(SP)}$. Notice that $Opt(SP) \leq \frac{n}{3}$ and we see that the performance ratio of $Approx(TC2)$ is at most $\frac{7}{6} + \epsilon'$, for any fixed $\epsilon' > 0$. $\square$

References

1. M. Abbink. *The test cover problem*. Master's thesis, University of Amsterdam, 1995.
2. S. Arora and M. Sudan. Improved low degree testing and its applications. In *Proceedings of the Twenty-Ninth Annual ACM Symposium on Theory of Computing*, pages 485–495, El Paso, Texas, 4–6 May 1997.
3. K.M.J. De Bontridder. Methods for solving the test cover problem. Master's thesis, Eindhoven University of Technology, 1997.
4. U. Feige. A threshold of $\ln n$ for approximating set cover. *Journal of the ACM*, 45:634–652, 1998.
5. U. Feige and J. Killian. Zero knowledge and the chromatic number. *J. Comput. System Sci.*, 57:187–199, 1998.
6. M.R. Garey and D.S. Johnson. *Computers and Intractability: A Guide to the Theory of NP-completeness*. W.H. Freeman and Company, 1979.
7. B.V. Halldórsson, J.S. Minden, and R. Ravi. PIER: Protein identification by epitope recognition. In N. El-Mabrouk, T. Lengauer, and D. Sankoff, editors, *Currents in Computational Molecular Biology 2001*, pages 109–110, 2001.
8. M.M. Halldórsson. A still better performance guarentee for approximate graph coloring. *Information Processing Letters*, 45:19–23, 1993.
9. D. S. Hochbaum *Approximation Algorithms for NP-hard problems*. PWS, 1997.
10. C.A.J. Hurkens and A. Schrijver. On the size of systems of sets every t of which have an SDR, with an application to the worst-case ratio of heuristics for packing problems. *SIAM J. Discrete Mathematics*, 2:68–72, 1989.
11. D.S. Johnson. Approximation algorithms for combinatorial problems. *J. Comput. System Sci*, 9:256–278, 1972.
12. V. Kann Maximum bounded 3-dimensional matching is MAX SNP-complete. *Information Processing Letters*, 37:27-35, 1991.
13. R. Loulou. Minimal cut cover of a graph with an application to the testing of electronic boards. *Operations Research Letters*, 12:301–305, 1992.
14. L. Lovász. On the ratio of optimal integral and fractional covers. *Discrete Mathematics*, 13:383–390, 1975.
15. C. Lund and M. Yannakakis. On the hardness of approximating minimization problems. *Journal of the ACM*, 41(5):960–981, 1994.
16. B.M.E. Moret and H.D. Shapiro. On minimizing a set of tests. *SIAM Journal on Scientific and Statistical Computing*, 6:983–1003, 1985.
17. R. Motwani and J. Naor. On exact and approximate cut covers of graphs. Technical report, Stanford University, 1994.

Finding Approximate Repetitions under Hamming Distance

Roman Kolpakov* and Gregory Kucherov

LORIA/INRIA-Lorraine, 615, rue du Jardin Botanique, B.P. 101, 54602
Villers-lès-Nancy France, {roman,kucherov}@loria.fr

Abstract. The problem of computing tandem repetitions with K possible mismatches is studied. Two main definitions are considered, and for both of them an $O(nK \log K + S)$ algorithm is proposed (S the size of the output). This improves, in particular, the bound obtained in [17].

1 Introduction

Repetitions (periodicities) play a central role in word combinatorics [16]. On the other hand, repetitions are important from the application perspective. As an example, their properties allow to speed up pattern matching algorithms [9,5,4].

The problem of efficiently identifying repetitions in a given word is one of the classical pattern matching problems [6,24]. A *tandem repeat* or a *square* is a pair of consecutive occurrences of a subword in a word. For example, *baba* is a tandem repeat in word *cbacbabacba*. Since the beginning of 80s [7], it is known that checking whether a word contains no tandem repeat (or is *square-free*) can be done in time $O(n)$ (n length of the word). If one wants to find *all* tandem repeats, their number comes into consideration. Word a^n contains $O(n^2)$ tandem repeats. If we restrict ourselves to *primitive squares* (i.e. subwords uu where u is not itself a repetition v^k for $k \geq 2$), then a word may contain $O(n \log n)$ of them and this bound is tight. All primitive squares can be found in time $O(n + S)$ where S is their number [15,23,12], hence in the worst-case time $O(n \log n)$.

In [13,12], we studied *maximal repetitions* (see also [20,19]). Those can be viewed as maximal *runs* of squares [11,23], i.e. series of squares of equal length shifted by one letter one with respect to another. For example, *bcbacacacaab* contains a maximal repetition *acacaca* which is a succession of four squares : *acac, caca, acac, caca*. Thus, the set of maximal repetitions can be regarded as an encoding of all tandem repeats in the string. We showed [13] that this encoding is more compact in the worst case, as there are only $O(n)$ maximal repetitions in words of length n. Moreover, all of them can be found in time $O(n)$ [12].

More recently, searching for repetitions in a string received a new motivation, due to biosequence analysis [10]. Successive occurrences of a fragment often bear

---

* on leave from the French-Russian Institute for Informatics and Applied Mathematics at Moscow University

F. Meyer auf der Heide (Ed.): ESA 2001, LNCS 2161, pp. 170–181, 2001.

important information in DNA sequences and their presence is characteristic for many genomic structures (such as telomer regions for example). From practical viewpoint, satellites and alu-repeats are involved in chromosome analysis and genotyping, and thus are of great interest to genomic researchers. Tools for finding successive repeats are nowadays an obligatory part of integrated systems for analyzing and annotating whole genomes [2].

The major difficulty in finding biologically relevant repetitions in genomic sequences is a certain variation that must be admitted between the copies of the repeated subword. In other words, biologists are interested in *approximate repetitions* and not necessarily in exact repetitions only. The first natural definition of approximate repetition is an *approximate tandem repeat* which is a subword uv where u and v are within a given distance k and the notion of distance could be one of those usually used in biological applications, such as Hamming distance or edit distance. The problem of finding approximate tandem repeats for both these distances has been studied by G. Landau and J. Schmidt [17]. They showed that in case of the Hamming distance (respectively edit distance), all approximate tandem repeats can be found in time $O(nK \log(n/K) + S)$ (respectively $O(nK \log K \log n + S)$), where S is the number of repeats found. Several other approaches to finding approximate tandem repeats in DNA sequences have been proposed in the bioinformatics community – some of them use statistical framework [1,2], some require to specify the size of repeated motif [3,21], some use very general framework and have to make use of some heuristic filtering steps to avoid exponential blow-up [25].

This paper deals with finding approximate repetitions using exact combinatorial methods of string matching. We focus on the Hamming distance case when the variability between repeated copies can be only letter replacements. An important motivation is to define structures encoding families of approximate tandem repeats, analogous to maximal repetitions in the exact case. In Section 2, we define two fundamental structures : *globally-defined approximate repetitions* and *runs of approximate tandem repeats*. In Section 3, we show that all globally-defined approximate repetitions can be found in time $O(nK \log K + S)$, where S is their number. In Section 4 we show that the same bound holds for runs of approximate tandem repeats: all of them can be found in time $O(nK \log K + R)$, where R is their number. This result implies, in particular, that all approximate tandem repeats can be found in time $O(nK \log K + T)$ (T their number), improving the $O(nK \log(n/K) + T)$ bound of G. Landau and J. Schmidt for the most interesting case of small K. Finally, in the last section we give some concluding remarks and mention possible extensions of presented results.

Due to space limitations, we present only a high-level description of the algorithms, and we refer to the extended version [14] for algorithm pseudo-codes.

2 K-Mismatch Globally-Defined Repetitions and Runs of K-Mismatch Tandem Repeats

Quoting [1], *one difficulty in dealing with* (approximate) *tandem repeats is accurately defining them.* Even if we concentrate only on mismatches, as it is the case in this paper, different definitions of approximate repetitions can be thought of. Here we introduce two basic notions of approximate repetitions.

We start by recalling briefly some facts about exact repetitions. The period of a word $w[1 : n]$ is the minimal natural number p such that $w[i] = w[i + p]$ for all $1 \leq i, i + p \leq n$. The ratio n/p is called the *exponent* of w. A *repetition* is any word with the exponent greater or equal to 2 [13]. A *tandem repeat*, or a *square*, is a word which is a catenation of another word with itself. Equivalently, a tandem repeat is a repetition the exponent of which is an even natural number. In the case when the exponent is equal to 2, the tandem repeat (square) is called *primitive*. The following proposition is well-known (see [16]).

Proposition 1. *A word $r[1 : n]$ is a repetition with period $p \leq n/2$ if and only if one of the following conditions holds:*

(i) $r[1..n - p] = r[p + 1..n]$, and p is the minimal number with this property,
(ii) any subword of r of length $2p$ is a tandem repeat, and p is the minimal number with this property.

When considering repetitions as subwords of a bigger word, the notion of maximality turns out to be very useful: a repetition is *maximal* iff it cannot be extended (by one letter) to the right or left while keeping the same period. Formally, given a word $w[1 : n]$ and a subword $w[i..j]$ which is a repetition of period p, this repetition is called *maximal* if the period of both $w[i..j + 1]$ (provided that $j < n$) and $w[i - 1..j]$ (provided that $i > 1$) is strictly larger than p. For example, $acaabaababc$ contains repetition (tandem repeat) $aabaab$ which is not maximal, as the a which follows it respects the periodicity. On the other hand, $aabaaba$ occurs as a maximal repetition. Maximal repetitions were studied in [19,13,12,23].

We now turn to defining approximate repetitions. Similar to the exact case, the basic notion here is the approximate tandem repeat. Assume $h(\cdot, \cdot)$ is the Hamming distance between two words of equal length, that is $h(w_1, w_2)$ is the number of mismatches (letter differences at corresponding positions) between w_1 and w_2. For example, $h(baaacb, bcabcb) = 2$.

Definition 1. *A word $\alpha = \alpha'\alpha''$, such that $|\alpha'| = |\alpha''|$, is called a K-mismatch tandem repeat iff $h(\alpha', \alpha'') \leq K$. Reusing the terminology of the exact case [12], we call number $p = |\alpha'| = |\alpha''|$ the period of α, and words α', α'' left and right root of α respectively.*

We now want to define a more global structure which would be able to capture "long approximate periodicities", generalizing repetitions with arbitrary exponent in the exact case. As opposed to the exact case, Conditions (i)-(ii) of Proposition 1 generalize to different notions of approximate repetition. Condition (i) gives rise to the strongest of them:

Definition 2. *A word* $r[1:n]$ *is called a* K-mismatch globally-defined repetition *with period* p, $p \leq n/2$, *iff* $h(r[1..n-p], r[p+1..n]) \leq K$.

Equivalently, $r[1:n]$ is a K-mismatch globally-defined repetition with period p, if the number of i such that $r[i] \neq r[i+p]$ is at most K. For example, *abaa abba cbba cb* is a 2-mismatch globally-defined repetition with period 4. *abc abc abc abb abc abc abc abc* is a 1-mismatch globally-defined repetition with period 3 but *abc abc abc abb abc abc abc abb* is not.

Another viewpoint, expressed by Condition (ii) of Proposition 1, considers a repetition as an encoding of squares it contains [11,23]. Projecting this to the approximate case, we come up with the notion of *run of approximate tandem repeats*:

Definition 3. *A word* $r[1:n]$ *is called a* run *of* K-mismatch tandem repeats of *period* p, $p \leq n/2$, *iff for every* $i \in [1..n-2p+1]$, *subword* $\alpha = r[i..i+2p-1] = r[i..i+p-1]r[i+p..i+2p-1]$ *is a* K-mismatch tandem repeat of period p.

Similarly to the exact case, when we are looking for approximate repetitions occurring in a word, it is natural to consider *maximal* approximate repetitions. These are repetitions extended to the right and left as far as possible provided that the corresponding definition is still verified. Note that the notion of maximality applies to both definitions of approximate repetition considered above : in both cases we can extend a repetition to the right/left as long as the obtained subword remains a repetition according to the considered definition. Throughout this paper we will be *always* interested in maximal repetitions, without mentioning it explicitly. Note that for both notions of approximate repetitions defined above, the maximality requirement implies that if $w[i:j]$ is a repetition of period p in $w[1:n]$, then $w[j+1] \neq w[j+1-p]$ (provided $j < n$) and $w[i-1] \neq w[i-1+p]$ (provided $i > 1$). Furthermore, if $w[i:j]$ is a maximal globally-defined repetition, it contains *exactly* K mismatches $w[l] \neq w[l+p]$, $i \leq l, l+p \leq j$, unless the whole word w contains less than K mismatches (to simplify the presentation, we always exclude this latter case from consideration).

Example 1. The following Fibonacci word contains three runs of 3-mismatch tandem repeats of period 6. They are shown in regular font, in positions aligned with their occurrences. Two of them are identical, and contain each four 3-mismatch globally-defined repetitions, shown in italic for the first run only. The third run is a 3-mismatch globally-defined repetition in itself.

010010 100100 101001 010010 010100 1001

10010 100100 101001
10010 100100 10
0010 100100 101
10 100100 10100
0 100100 101001

1001 010010 010100 1
10 010100 1001

In general, each K-mismatch globally-defined repetition is a subword of a run of K-mismatch tandem repeats. On the other hand, a run of tandem repeats in a word is the union of all globally-defined repetitions it contains. However, a run of tandem repeats may contain as many as a linear number of globally-defined repetitions. For example, the word $(000\,100)^n$ of length $6n$ is a run of 1-mismatch tandem repeats of period 3, which contains $(2n-1)$ 1-mismatch globally-defined repetitions. In general, the following observation holds.

Lemma 1. *Let $w[1:n]$ be a run of K-mismatch tandem repeats of period p and let s be the number of mismatches $w[i] \neq w[i+p]$, $1 \leq i, i+p \leq n$ (equivalently, $s = h(w[1..n-p], w[p+1..n])$). Then w contains $s - K + 1$ globally-defined repetitions.*

Note that both definitions can be criticized as for their relevance to practical situations. An obvious property of runs is that the repeated pattern can change completely along the run regardless the value of K. For example, $aaa\,aba\,abb\,abb\,bbb$ is a run of 1-mismatch tandem repeats of period 3, although 3-letter patterns aaa and bbb have nothing in common. On the other hand, globally-defined repetitions put a global limit on the number of mismatches and therefore may not capture some repetitions that one would possibly like to consider as such, in particular repetitions with big exponents where the total number of mismatches can exceed K while the relative number of mismatches remains low. However, these two structures are of primary importance as they provide respectively the weakest and strongest notions of repetitions with K mismatches, and therefore "embrace" all practically relevant repetitions. In what follows we propose efficient algorithms to find both those types of repetitions.

3   Finding K-Mismatch Globally-Defined Repetitions

In this section we describe how to find, in a given word w, all maximal K-mismatch globally-defined repetitions occurring in w (K is a given constant). Our algorithm extends, on the one hand, the one for exact maximal repetitions [19,12] and on the other hand, generalizes the one of [17] (see also [10]) by using a special factorization of the word to speed-up the algorithm.

To proceed, we need more definitions. Consider a globally-defined repetition $r = w[i..j]$ of period p in a word $w[1:n]$. $w[i..i+p-1]$ is called the *left root* of r and $w[j-p+1..j]$ its *right root*. r is said to *contain* the character $w[l]$ iff $i \leq l \leq j$, and is said to *touch* $w[l]$ iff r contains $w[l]$, or contains one of characters $w[l-1]$, $w[l+1]$.

Our first basic technique is described by the following auxiliary problem: Given a word $w[1:n]$ and a distinguished character $w[l]$, $l \in [2..n-1]$, we wish to find all K-mismatch globally-defined repetitions in w which touch $w[l]$. We distinguish two disjoint classes of repetitions according to whether their right root starts to the left or to the right to $w[l]$. We concentrate on repetitions of the first class, those of the second class are found similarly.

For each $p \in [1..l-1]$, and for all $k \in [0..K]$, we compute the following functions :

$$LP_k(p) = \max\{j \mid h(w[l-p..l-p+j-1], w[l..l+j-1]) \leq k\}, \tag{1}$$
$$LS_k(p) = \max\{j \mid h(w[l-p-j..l-p-1], w[l-j..l-1]) \leq k\}. \tag{2}$$

Informally, $LP_k(p)$ is the length of the longest subword in w starting at position $l - p$ and equal, within k mismatches, to a subword starting at l. Similarly, $LS_k(p)$ is the length of the longest subword ending at position $l - p - 1$ equal, within k mismatches, to a subword ending at position $l - 1$. These functions are variants of *longest common extension functions* [17,10] and can be computed in time $O(nK)$ using suffix trees combined with the lowest common ancestor computation in a tree. We refer to [10] for a detailed description of the method.

Consider now a K-mismatch globally-defined repetition r with a period p which has its right root starting to the left of $w[l]$. Note that character $w[l - p]$ is contained in r, and that r is uniquely defined by the number of mismatches $w[i - p] \neq w[i]$, $i \geq l$, contained in r. Let k be the number of those mismatches. Then

$$LP_k(p) + LS_{K-k}(p) \geq p. \tag{3}$$

Conversely, (3) can be used to detect a repetition. The following theorem holds (see [17,10]), which is a generalization of the corresponding result of [20,19].

Theorem 1. *Let $w[1 : n]$ be a word and $w[l]$, $1 < l < n$, a distinguished character. There exists a K-mismatch globally-defined repetition with period p which contains $w[l]$, and has its right period starting to the left to $w[l]$, iff for some $k \in [0..K]$,*

$$LP_k(p) \leq p, \tag{4}$$

and inequation (3) holds. In this case, this repetition starts at position $l - p - LS_{K-k}(p)$ and ends at position $l + LP_k(p) - 1$.

Inequation 4 ensures that the right root starts to the left of $w[l]$.

Theorem 1 provides an $O(nK)$ algorithm for finding all considered globally-defined repetitions: compute longest extension function (1) (2) (this takes time $O(nK)$) and then check inequations (3), (4) for all $p \in [1..l-1]$ and all $k \in [0..K]$ (this takes time $O(nK)$ too). Each time the inequations are verified, a new repetition is identified. Finding repetitions with the right root starting to the right of $w[l]$ is a symmetric problem, which is solved within the same time bound.

The algorithm solving the auxiliary problem described above will be referred to as Algorithm 1.

The second important tool is Lempel-Ziv factorization used in the well-known compression method. Let w be a word and assume that the last symbol of w does not occur elsewhere. In this paper, we need two variants of the Lempel-Ziv factorization, that we call *with copy overlap* and *without copy overlap*[1].

Definition 4. *The Lempel-Ziv factorization of w with copy overlap (respectively without copy overlap) is the factorization $w = f_1 f_2 \ldots f_m$, where f_i's are defined inductively as follows:*

---

[1] The s-factorization used in [19,12] is a minor modification of the Lempel-Ziv factorization with copy overlap. The difference is that the s-factorization considers the longest factor occurring earlier, while the Lempel-Ziv factorization considers the shortest factor which does not occur earlier (see [10] for a related discussion). In this paper, we use the Lempel-Ziv factorization which suits better to our purposes.

(i) $f_1 = w[1]$,

(ii) for $i \geq 2$, f_i is the shortest word occurring in w immediately after $f_1 f_2 \ldots f_{i-1}$ which does not occur in $f_1 f_2 \ldots f_i$ other than in prefix (respectively, does not occur in $f_1 f_2 \ldots f_{i-1}$).

As an example, the Lempel-Ziv factorization with copy overlap of the word $aabbabababbbc$ is $a|ab|ba|babab|bc$; the factorization without copy overlap is $a|ab|ba|bab|abbb|c$. Both variants of Lempel-Ziv factorization can be computed in linear time [22,10]. If $w = f_1 f_2 \ldots f_m$ is the Lempel-Ziv factorization, we call f_i's Lempel-Ziv factors or simply factors's of w. The last character of f_i will be called the head of f_i.

We are now ready to describe the algorithm for finding all K-mismatch globally-defined repetitions. Consider the Lempel-Ziv factorization of w with copy overlap. The algorithm consists of three stages. The key to the **first stage** is the following lemma.

Lemma 2. The right root of a K-mismatch globally-defined repetition cannot contain as subword $K + 1$ consecutive Lempel-Ziv factors.

Proof. Each factor contained in the right root contains a character mismatching the one located one period to the left. Indeed, if it does not contain a mismatch, it has an exact copy occurring earlier, which contradicts the definition of factorization. As the right root contains at most K mismatches, it cannot contain $K + 1$ or more factors.

We divide w into consecutive *blocks* of $K + 2$ Lempel-Ziv factors. Let $w = B_1 \ldots B_{m'}$ be the partition of w into such blocks. The last character of B_i will be called the *head character* of this block. At the first stage, we find, for each block B_i, those repetitions which touch the head character of B_i but do not touch that of B_{i+1}. First, concentrate on those of such repetitions with the right root starting before the head character of B_i.

Lemma 3. Assume a K-mismatch globally-defined repetition r touches the head character of B_i but not that of B_{i+1}. Then $|r| < 2|B_i B_{i+1}|$.

Proof. Lemma 2 implies that the right root of r cannot start before the first character of B_i. Therefore, the period of r is bounded by $|B_i B_{i+1}|$. On the other hand, by the argument of the proof of Lemma 2, r cannot extend by more than a period to the left of B_i. Therefore, the total length of r is bounded by $2|B_i B_{i+1}|$.

Lemma 3 allows us to apply Algorithm 1: Consider the word $w_i = v B_i B_{i+1}$, where v is the suffix of $B_1 \ldots B_{i-1}$ of length $|B_i B_{i+1}|$. Then find, using Algorithm 1, all repetitions in w_i touching the head character of B_i and discard those which touch the head character of B_{i+1}. The resulting complexity is $O(K(|B_i| + |B_{i+1}|))$.

After processing all blocks, we find all repetitions touching block head characters. Observe that repetitions resulting from processing different blocks are distinct. Summing up over all blocks, the resulting complexity of the first stage is $O(nK)$. The repetitions which remain to be found are those which lie entirely within a block – this is done at the next two stages.

At the **second stage** we find all repetitions inside each block B_i which touch factor heads other than the block head (=last character of the block). For each B_i, we proceed by simple binary division approach:

(i)   divide current block of factors $B = f_i f_{i+1} \ldots f_{i+m}$ into two sub-blocks
      $B' = f_i \ldots f_{\lfloor m/2 \rfloor}$ and $B'' = f_{\lfloor m/2 \rfloor + 1} \ldots f_{i+m}$,
(ii) using Algorithm 1, find the repetitions in B which touch the head character
      of $f_{\lfloor m/2 \rfloor}$, but discard those which touch the head character of f_{i+m} or
      contain the first character of f_i,
(iii) process recursively B' and B''.

The above algorithm has $\lceil \log K \rceil$ levels of recursion, and since at each step the
word is split into disjoint sub-blocks, the whole complexity of the second stage
is $O(nK \log K)$.

   Finally, at the **third stage**, it remains to find the repetitions which occur
entirely inside each Lempel-Ziv factor, namely which don't contain its first character and don't touch its head character. By definition of factorization with copy
overlap (Definition 4), each factor without its head character has another (possibly overlapping) occurrence to the left. Therefore, each of these repetitions has
another occurrence to the left too. Using this observation, these repetitions can
be found using the same technique as the one of [12]: When constructing the
Lempel-Ziv factorization we keep for each factor wa a pointer to a copy of w
to the left. Then processing factors from left to right, recover repetitions inside
the factor from its pointed copy. We refer to [12] for algorithmic details. The
complexity of this stage is $O(n + S)$, where S is the number of repetitions found.
The following theorem summarizes this section.

Theorem 2. *All K-mismatch globally-defined repetitions can be found in time
$O(nK \log K + S)$ where n is the word length and S is the number of repetitions
found.*

4   Finding Runs of K-Mismatch Tandem Repeats

In this section we describe an algorithm for finding all runs of K-mismatch
tandem repeats in a word.

   The general structure of the algorithm is the same as for globally-defined
repetitions – it has the three stages playing similar roles. At the first and second
stages, the key difference is the type of objects we are looking for: instead of
computing globally-defined repetitions we now compute *subruns* of K-mismatch
tandem repeats. Formally, a subrun is a run of K-mismatch tandem repeats,
which is not necessarily maximal. At each point of the first and second stage when
we search for repetitions touching some head character $w[l]$, we now compute
subruns of those K-mismatch tandem repeats which touch $w[l]$. This can be
seen as outputting by Algorithm 1 only the part of the globally-defined repetition
falling to the interval $l - 2p..l + 2p$.

   The major additional difficulty of computing runs is *assembling* subruns into
runs. To perform the assembling, we need to store subruns in an additional data
structure and to carefully manage merging of subruns into bigger runs. We have
to ensure that the number of subruns we come up with and the work spent
on processing them do not increase the resulting complexity bound. Below we
describe the three stages of the algorithm in more details.

We identify a subrun with the interval of *end positions* of the tandem repeats it contains.

For the input word w, we compute the Lempel-Ziv factorization *without copy overlap* and divide it into blocks $B_1 \ldots B_{m'}$, each containing $K + 2$ consecutive Lempel-Ziv factors. Note that Lemma 2 still holds for the factorization without copy overlap. **At the first stage**, we find subruns of all those tandem repeats which touch block head characters. For each block B_i, we find the tandem repeats which touch the head character of B_i but not that of B_{i+1}. Let l_i be the position of the head character of B_i. Then the subruns of period p, found at this step, belong to the interval $[l_i - 1 .. \min\{l_i + 2p, l_{i+1} - 2\}]$. We call this interval the *explored interval* for $w[l_i]$ and p. The subruns found at this step can be seen as subintervals of this explored interval. These subruns are stored into a double-linked list in increasing order of positions. position $l_i - 2$, it is merged with the explored interval for $w[l_i]$, thus forming a bigger explored interval. Accordingly, the lists of subruns associated with these intervals are merged into a single list. All additional operations take constant time, and the resulting complexity of the first stage is $O(nK)$.

The second stage is modified in a similar way. Recall that at each call of modified Algorithm 1 we are searching for repetitions occurring between some factor head, say $w[l']$, another factor head $w[l'']$, and touching some factor head $w[l]$ ($l' < l < l''$). Assuming that recursive calls are executed in preorder (see the description of the second stage in the previous section), no factor head between $w[l']$ and $w[l'']$ has been processed yet. In this case, the explored interval is $[\max\{l' + 2p + 1, l - 1\} .. \min\{l + 2p, l'' - 2\}]$, and we may have to merge it either with the previous explored interval, or with the next one, or both. The complexity of the second stage stays $O(nK \log K)$.

After accomplishing the first and second stages, we have, *for each period p*, a set of non-intersecting explored intervals. Each interval is associated with a sequence of successive head characters $w[l_i], w[l_{i+1}], \ldots, w[l_m]$ such that $l_{j+1} - l_j \leq 2p+2$ for $j \in [i..m-1]$, and the interval itself is $[l_i-1..l_m+2p]$. In particular, the interval is associated to $w[l_i]$ and $w[l_m]$ - the first and the last head characters of this sequence. Those subruns of tandem repeats which have been actually found within this interval, are stored in a double-linked list associated to the interval.

At **the third stage**, we have to find subruns of those tandem repeats which lie entirely inside Lempel-Ziv factors. For each period, potential occurrences of these subruns correspond precisely to the gaps between explored intervals. Thus, the third stage can be also seen as closing up the gaps between explored intervals for this period.

As in the previous section, the key observation here is the fact that Lempel-Ziv factors without their head character have a copy to the left (here required to be non-overlapping), and the idea is again to process w from left to right and to retrieve the subruns inside each factor from its copy. However, the situation here is different in comparison to globally-defined repetitions: we may have to "cut out" a chain of subruns belonging to the factor copy from a longer list and then to "fit" it into the gap between two explored intervals. The "cutting out" may entail splitting subruns which span over the borders of the factor copy, and "fitting into" may entail merging those subruns with subruns from the neighboring explored intervals. Below we sketch the algorithm for the third stage, which copes with these difficulties.

During the computation of the Lempel-Ziv factorization, for each Lempel-Ziv factor $f_i = va$ we choose a copy of v occurring earlier and point from the end position of this copy to the head character a of f_i. It may happen that one position has to have several pointers, in which case we organize them in a list. We traverse w from left to right and maintain for the current position the last runs (of all possible periods) which start before this character. To this purpose, we also maintain the following invariant: at the moment we arrive at a position, we know the list of all subruns which start at this position. This information is collected according to the following general rule: for each subrun starting at the current position, we assign the starting position of the next subrun in the list. Of course, there may be no next subrun if the current subrun is the last one in the explored interval. In this case, the starting position of the subrun following the current subrun will be set at the moment we fill the gap after this explored interval.

When we arrive at the end position of a copy of a Lempel-Ziv factor, we need to copy into the factor all the subruns which this copy contains. Therefore, we scan *backwards* the subruns contained in the copy and copy them to the factor. Copying the subruns closes up two explored intervals into one interval, and links together two lists of subruns, possibly inserting a new list of runs in between. Copying subruns in the backward direction is important for the correction of the algorithm – this guarantees that no subruns are missed. It is also for this reason that we need the copy to be non-overlapping with the factor.

After the whole word has been traversed, no more gaps between explored intervals exist anymore. This means that for each period, we have a list of subruns with this period occurring in the word, which are actually the searched runs. The complexity of the third stage is $O(n + S)$, where S is the number of resulting runs. Putting together the three stages, we obtain the main result of this section.

Theorem 3. *All runs of K-mismatch tandem repeats can be found in time $O(nK \log K + S)$ where n is the word length and S is the number of runs found.*

Once all runs have been found, we can easily output all tandem repeats. We then have the following result improving the result of [17].

Corollary 1. *All K-mismatch tandem repeats can be found in time $O(nK \log K + S)$ where n is the word length and S is the number of tandem repeats found.*

5 Concluding Remarks

We proposed $O(nK \log K + S)$ algorithms for finding K-mismatch globally-defined repetitions and runs of K-mismatch tandem repeats (S the output size). Note that if K is considered constant, we have $O(n + S)$ algorithms for finding each of these structures. This is an interesting result, which had been long time unknown even for the exact case [15,23,12].

Globally-defined repetitions and runs of tandem repeats provide respectively the strongest and the weekest notion of approximate repetitions. For practical applications, such as genome analysis, it might be interesting to consider intermediate definitions with respect to the two "extreme" cases. In the extended version [14], we introduced two such types of approximate repetitions, so-called K-mismatch uniform repetitions and K-mismatch consensus repetitions, and analyzed their relationship to the notions of repetition considered in this paper. However, designing an efficient algorithm for finding repetitions of those types remains an open problem.

In the final stage of preparation of this paper, we got known of paper [18]. In this paper, yet another definition of approximate repetitions is considered, which is weaker than globally-defined repetitions, but stronger than both uniform repetitions and consensus repetitions from [14]. The algorithm presented in [18] runs in time $O(nKE\log(n/K))$, where E is the maximal exponent of reported repetitions.

The algorithms presented in this paper are now being implemented within the mreps software[2]. Currently, mreps implements the algorithm of finding exact maximal repetitions [12]. Some interesting experiments have been done by applying mreps to genomic sequences [8].

Acknowledgments. We thank Mathieu Giraud, with whom we had first discussions on the subject, and anonymous referees for their comments.

References

1. G. Benson. An algorithm for finding tandem repeats of unspecified pattern size. In S. Istrail, P. Pevzner, and M. Waterman, editors, *Proceedings of the 2nd Annual International Conference on Computational Molecular Biology (RECOMB 98)*, pages 20–29. ACM Press, 1998.
2. G. Benson. Tandem repeats finder: a program to analyze DNA sequences. *Nucleic Acids Research*, 27(2):573–580, 1999.
3. G. Benson and M. Waterman. A method for fast database search for all k-nucleotide repeats. *Nucleic Acids Research*, 22:4828–4836, 1994.
4. R. Cole and R. Hariharan. Approximate string matching: A simpler faster algorithm. In *Proceedings of the Ninth Annual ACM-SIAM Symposium on Discrete Algorithms*, pages 463–472, San Francisco, California, 25–27 January 1998.
5. M. Crochemore and W. Rytter. Squares, cubes, and time-space efficient string searching. *Algorithmica*, 13:405–425, 1995.
6. M. Crochemore. An optimal algorithm for computing the repetitions in a word. *Information Processing Letters*, 12:244–250, 1981.
7. M. Crochemore. Recherche linéaire d'un carré dans un mot. *Comptes Rendus Acad. Sci. Paris Sér. I Math.*, 296:781–784, 1983.
8. M. Giraud and G. Kucherov. Maximal repetitions and application to DNA sequences. In *Proceedings of the* Journées Ouvertes : Biologie, Informatique et Mathématiques, pages 165–172, Montpelier, 3-5 mai 2000.

---

[2] http://www.loria.fr/~kucherov/SOFTWARE/MREPS/index.html

9. Z. Galil and J. Seiferas. Time-space optimal string matching. *Journal of Computer and System Sciences*, 26(3):280–294, 1983.
10. D. Gusfield. *Algorithms on Strings, Trees, and Sequences. Computer Science and Computational Biology.* Cambridge University Press, 1997.
11. C.S. Iliopoulos, D. Moore, and W.F. Smyth. A characterization of the squares in a Fibonacci string. *Theoretical Computer Science*, 172:281–291, 1997.
12. R. Kolpakov and G. Kucherov. Finding maximal repetitions in a word in linear time. In *Proceedings of the 1999 Symposium on Foundations of Computer Science, New York (USA)*. IEEE Computer Society, October 17-19 1999.
13. R. Kolpakov and G. Kucherov. On maximal repetitions in words. In *Proceedings of the 12-th International Symposium on Fundamentals of Computation Theory, 1999, Iasi (Romania)*, Lecture Notes in Computer Science, August 30 - September 3 1999.
14. R. Kolpakov and G. Kucherov. Finding approximate repetitions under hamming distance. Technical Report RR-4163, INRIA, avril 2001. available at http://www.inria.fr/rrrt/rr-4163.html.
15. S. R. Kosaraju. Computation of squares in string. In M. Crochemore and D. Gusfield, editors, *Proceedings of the 5th Annual Symposium on Combinatorial Pattern Matching*, number 807 in Lecture Notes in Computer Science, pages 146–150. Springer Verlag, 1994.
16. M. Lothaire. *Combinatorics on Words*, volume 17 of *Encyclopedia of Mathematics and Its Applications*. Addison Wesley, 1983.
17. G. Landau and J. Schmidt. An algorithm for approximate tandem repeats. In A. Apostolico, M. Crochemore, Z. Galil, and U. Manber, editors, *Proceedings of the 4th Annual Symposium on Combinatorial Pattern Matching*, number 684 in Lecture Notes in Computer Science, pages 120–133, Padova, Italy, 1993. Springer-Verlag, Berlin.
18. G. Landau, J. Schmidt, and D. Sokol. An algorithm for approximate tandem repeats. *Journal of Computational Biology*, 8(1):1–18, 2001.
19. M. G. Main. Detecting leftmost maximal periodicities. *Discrete Applied Mathematics*, 25:145–153, 1989.
20. M.G. Main and R.J. Lorentz. An $O(n \log n)$ algorithm for finding all repetitions in a string. *Journal of Algorithms*, 5(3):422–432, 1984.
21. E. Rivals, O. Delgrange, J-P. Delahaye, and M. Dauchet. A first step towards chromosome analysis by compression algorithms. In N.G. Bourbakis, editor, *Proceedings of the 1st IEEE Symposium on Intelligence in Neural and Biological Systems (INBS), Herndon, VA*. IEEE Computer Society, May 29-31 1995.
22. M. Rodeh, V.R. Pratt, and S. Even. Linear algorithm for data compression via string matching. *Journal of the ACM*, 28(1):16–24, Jan 1981.
23. J. Stoye and D. Gusfield. Linear time algorithms for finding and representing all the tandem repeats in a string. Technical Report CSE-98-4, Computer Science Department, University of California, Davis, 1998.
24. A.O. Slisenko. Detection of periodicities and string matching in real time. *Journal of Soviet Mathematics*, 22:1316–1386, 1983.
25. M.-F. Sagot and E.W. Myers. Identifying satellites in nucleic acid sequences. In S. Istrail, P. Pevzner, and M. Waterman, editors, *Proceedings of the 2nd Annual International Conference on Computational Molecular Biology (RECOMB 98)*, pages 234–242. ACM Press, 1998.

SNPs Problems, Complexity, and Algorithms

Giuseppe Lancia[1,2], Vineet Bafna[1], Sorin Istrail[1],
Ross Lippert[1], and Russell Schwartz[1]

[1] Celera Genomics, Rockville MD, USA,
{Giuseppe.Lancia,Vineet.Bafna,Sorin.Istrail,Ross.Lippert,
Russell.Schwartz}@celera.com
[2] D.E.I., Università di Padova, Padova 35100, Italy

Abstract. Single nucleotide polymorphisms (SNPs) are the most frequent form of human genetic variation. They are of fundamental importance for a variety of applications including medical diagnostic and drug
design. They also provide the highest–resolution genomic fingerprint for
tracking disease genes. This paper is devoted to algorithmic problems
related to computational SNPs validation based on genome assembly of
diploid organisms. In diploid genomes, there are two copies of each chromosome. A description of the SNPs sequence information from one of the
two chromosomes is called SNPs haplotype. The basic problem addressed
here is the Haplotyping, i.e., given a set of SNPs prospects inferred from
the assembly alignment of a genomic region of a chromosome, find the
maximally consistent pair of SNPs haplotypes by removing data "errors"
related to DNA sequencing errors, repeats, and paralogous recruitment.
In this paper, we introduce several versions of the problem from a computational point of view. We show that the general SNPs Haplotyping
Problem is NP–hard for mate–pairs assembly data, and design polynomial time algorithms for fragment assembly data. We give a network–flow
based polynomial algorithm for the Minimum Fragment Removal Problem, and we show that the Minimum SNPs Removal problem amounts
to finding the largest independent set in a weakly triangulated graph.

1 Introduction

1.1 Motivation

The large–scale laboratory discovery and typing of genomic sequence variation
presents considerable challenges and it is not certain that the present technologies are sufficiently sensitive and scalable for the task. Computational methods
that are intertwined with the experimental technologies are emerging, leading
the way for this discovery process.

Single nucleotide polymorphisms (SNPs) are the most frequent form of human genetic variation and provide the highest–resolution genomic fingerprint for
tracking disease genes. The SNPs discovery process has several stages: sample
collection, DNA purification, amplification of loci, sequence analysis, and data
management. The "large–scale" dimension of the analysis refers to either the

F. Meyer auf der Heide (Ed.): ESA 2001, LNCS 2161, pp. 182–193, 2001.

large number of loci or of individuals. Effective integration of these stages is important for the strategies employed in the pipeline. The choice of method, for any stage, could be influenced by the processes used in other stages.

The SNPs discovery involves the analysis of sequence differences and haplotype separation from several samples of a genomic locus from a given population. The leading two methods are based on Shotgun Genome Assembly (SGA) and on PCR amplification (PCR). The methods have complementary strengths and recognized computational bottlenecks associated with them. The SGA, in principle, generates haploid genotyping, and does not require sequence information for the loci, however, it needs good library coverage, and is computationally very challenging to distinguish paralogous repeats from polymorphism. The PCR method requires the knowledge of the genomic region of the locus, and could be done very effectively; however, it is expensive for large-scale projects.

There is need for powerful computational approaches that are sensitive enough and scalable so that they can remove noisy data and provide effective algorithmic strategies for these technologies. This paper is a first step towards such "computational SNPology". It is devoted to algorithmic problems related to computational SNPs discovery and validation based on genome assembly. The basic problem is to start from a set of SNPs prospects inferred from the assembly alignment and to find out the maximal consistent subset of SNPs by removing "errors" related to sequencing errors, repeats, and paralogous recruitment.

1.2   Preliminaries

Recent whole–genome sequencing efforts have confirmed that the genetical makeup of humans is remarkably well–conserved, and small regions of differences are responsible for our diversities. The smallest possible region consists of a single nucleotide, and is called Single Nucleotide Polymorphism, or SNP ("snip"). This is a position in our genome at which we can have one of two possible values (alleles), while in the neighborhood of this position we all have identical DNA content. Since our DNA is organized in pairs of chromosomes, for each SNP we can either be homozygous (same allele on both chromosomes) or heterozygous (different alleles). Independently of what the actual different alleles at a SNP are, in the sequel we will denote the two values that each SNP can take by the letters A and B. A chromosome content projected on a set of SNPs (or *haplotype*), is then simply a string over the alphabet {A, B}, while a *genotype* is a pair of such strings, one for each haplotype.

DNA sequencing techniques are restricted to small, overlapping fragments. Such fragments can contain errors (e.g., due to low quality reads), and can come from either one of the two chromosome copies. Further, e.g. in shotgun sequencing, some pairs of these fragments (*mate pairs*) are known to come from the same copy of a chromosome and to have a given distance between them. The basic problem is then the following: *"Given a set of fragments obtained by DNA sequencing from the two copies of a chromosome, reconstruct the two haplotypes from the SNPs values observed in the fragments."*

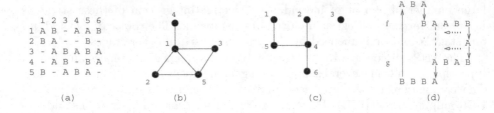

```
    1 2 3 4 5 6
1   A B - A A B
2   B A - - B -
3   - A B A B A
4   - A B - B A
5   B - A B A -

    (a)              (b)              (c)              (d)
```

Fig. 1. (a) M. (b) $G_{\mathcal{F}}(M)$. (c) $G_{\mathcal{S}}(M)$ (d) An odd cycle of fragments.

Note that it is possible, even in an error–free scenario, that the above problem cannot be solved because the information is insufficient. For instance, if a set of fragments does not share any SNPs with any of the remaining fragments, we may only be able to reconstruct partial haplotypes, but then we wouldn't know how to merge them into only two (a problem known as *phasing*). It is better then to relax the requirement from "reconstruct the two haplotypes" to "reconstruct two haplotypes that would be compatible with all the fragments observed". Stated in this form, the problem becomes trivial for the error–free case (as we will see in the sequel, it is simply the problem of determining the two shores of a bipartite graph). However, experiments in molecular biology are never error–free, and, under a general parsimony principle, we are led to reformulate the problem as *"Find the smallest number of errors in the data so that there exist two haplotypes compatible with all the (corrected) fragments observed."*

Depending on the errors considered, we will define several different combinatorial problems. "Bad" fragments can be due either to contaminants (i.e. DNA coming from a different organism than the actual target) or to read errors (i.e. a false A, a false B, or a – inside a fragment, which represents a SNP whose value was not determined). A dual point of view assigns the errors to the SNPs, i.e. a "bad" SNP is a SNP for which some fragments contain read errors. Correspondingly, we may define the following optimization problems: *"Find the minimum number of fragments to ignore"*, or *"Find the minimum number of SNPs to ignore"*, so that *"the (corrected) data is consistent with the existence of two haplotypes measured by error–free DNA sequencing. Find such haplotypes."*

1.3   Notation

The basic framework for SNPs problems is as follows. There is a set $\mathcal{S} = \{1, \ldots, n\}$ of snips and a set $\mathcal{F} = \{1, \ldots, m\}$ of fragments. Each snip is covered by some of the fragments, and can take the values A or B. Hence, a snip i is defined by a pair of disjoint subsets of fragments, A_i and B_i. There is a natural (canonical) ordering of the snips, given by their physical location on the chromosome, from left to right. Then, the data can also be thought of as an $m \times n$ matrix over the alphabet $\{A, B, -\}$, which we call the *SNP matrix*, defined in the obvious way.

The Fragment– and Snip– conflict graphs. We say that two fragments i and j are *in conflict* if there exists a snip k such that $i \in A_k$, $j \in B_k$ or $i \in B_k$, $j \in A_k$. Given a SNP matrix M, the *fragment conflict graph* is the graph $G_{\mathcal{F}}(M) = (\mathcal{F}, E_{\mathcal{F}})$ with an edge for each pair of fragments in conflict. Note that if M is error–free, $G_{\mathcal{F}}(M)$ is a bipartite graph, since each haplotype defines a shore of $G_{\mathcal{F}}(M)$, made of all the fragments coming from that haplotype. Conversely, if $G_{\mathcal{F}}(M)$ is bipartite, with shores H_1 and H_2, all the fragments in H_1 can be merged into one haplotype and similarly for H_2. We call a SNP matrix M *feasible* (and *infeasible* otherwise) if $G_{\mathcal{F}}(M)$ is bipartite, and we call the haplotypes obtained by merging the fragments on each shore *derived* from M. For K a set of rows (fragments), we denote by $M[K]$ the submatrix of M containing only the rows in K. The fundamental underlying problem in SNPs haplotyping is determining an optimal set of changes to M (e.g., row– and/or column– deletion) so that M becomes feasible. We remark that $G_{\mathcal{F}}(M)$ is the union of n complete bipartite graphs, one for each column j of M, with shores A_j and B_j.

We now turn to snip conflicts. We say that two snips i and j are in conflict if A_i, B_i, A_j, B_j are all nonempty and there exist two fragments u and v such that the submatrix defined by rows u and v and columns i and j has three symbols of one type (A or B) and one of the opposite (B or A respectively). Given a SNP matrix M, the *snip conflict graph* is the graph $G_{\mathcal{S}}(M) = (\mathcal{S}, E_{\mathcal{S}})$, with an edge for each pair of snips in conflict. In section 2.2 we will state the fundamental theorem relating the two conflict graphs.

In this paper we are going to define the following optimization problems:

- MFR (*Minimum Fragment Removal*): Given a SNP matrix, remove the minimum number of fragments (rows) so that the resulting matrix is feasible.
- MSR (*Minimum Snip Removal*): Given a SNP matrix, remove the minimum number of snips (columns) so that the resulting matrix is feasible.
- LHR (*Longest Haplotype Reconstruction*): Given a SNP matrix, remove a set of fragments (rows) so that the resulting matrix is feasible, and the sum of lengths of the derived haplotypes is maximized.

A *gapless* fragment is one covering a set of consecuitive SNPs. We say that a fragment has k gaps if it covers $k + 1$ blocks of consecutive SNPs. Such a fragment is equivalent to $k + 1$ gapless fragments with the constraint that they must all be put in the same haplotype or all discarded. Particularly important is the case $k = 1$, which is equivalent to 2 gapless fragments coming from the same chromosome. This is the case of *mate pairs*, used for shotgun sequencing [7]. In the remainder of the paper we show that the above problems are NP–hard in general. Furthermore, we show that MFR is NP–hard if even a single gap per fragment is allowed and MSR is NP–hard for fragments with two gaps. On the positive side, we study the special case of gapless fragments, and show that in this case the problems can be solved effectively. We provide polynomial algorithms for MFR, MSR and LHR. Note that the gapless case arises often in practical applications. For space limitations, some of the proofs are omitted in the sequel.

2   Getting Started: The Gapless Case

The simplest scenario for the SNPs haplotype reconstruction problem is when the fragments are consecutive, gapless genomic regions. This is not an unrealistic situation, since it arises, for example, in EST (Expressed Sequence Tags) mapping. This is in fact the case without mate pairs and with no missed SNPs inside a fragment because of thresholding or base–skipping read errors.

2.1   The Minimum Fragment Removal

In this section we show that in the gapless case, the minimum fragment removal (MFR) problem can be solved in polynomial time. For this section, we assume that there are no fragment inclusions, i.e., denoting by f_i and l_i the first and last snip of a fragment i, $f_i \leq f_j$ implies $l_i \leq l_j$. We define a directed graph $D = (\mathcal{F}, A)$ as follows. Given two fragments i and j, with $f_i \leq f_j$, there is an arc $(i, j) \in A$ if i and j can be aligned without any mismatch, i.e., they agree in all their common snips (possibly none). Note that the common snips are a suffix of i and a prefix of j.

Lemma 1. *Let M be a SNP matrix, and P_1, P_2 be node–disjoint directed paths in D such that $|V(P_1)| + |V(P_2)|$ is maximum. Let $R = \mathcal{F} - (V(P_1) \cup V(P_2))$. Then R is a minimum set of fragments to remove such that $M[\mathcal{F} - R]$ is feasible.*

Theorem 1. *There is a polynomial time algorithm for finding P_1 and P_2 in D such that $|V(P_1)| + |V(P_2)|$ is maximum.*

Proof. We will use a reduction to a maximum cost flow problem. We turn D into a network as follows. First, we introduce a dummy source s, a dummy sink t, and an arc (t, s) of capacity 2 and cost 0. s is connected to each node i with an arc (s, i) of cost 0, and each node i gets connected to t, at cost 0 and capacity 1. Then, we replace each node $i \in D$, with two nodes i' and i'' connected by an arc (i', i'') of cost 1 and capacity 1. All original arcs (u, v) of D are then replaced by arcs of type (u'', v'). A maximum cost circulation can be computed in polynomial time, by, e.g., Linear Programming. Since D is acyclic, the solution is one cycle, which uses the arc (t, s) and then splits into two s–t dipaths, saturating as many arcs of type (i', i'') as possible, i.e. going through as many nodes as possible of D. Since the capacity of arcs (i', i'') is 1, the paths are node–disjoint.

With a similar reduction, we can show that the problem LHP can also be solved in polynomial time. The problem consists in finding two haplotypes of maximum total length (where the length of an haplotype is the number of SNPs it covers). We use a similar reduction as before, with the same capacities, but different costs for the arcs. Now the arcs of type (i', i'') have cost 0, while an arc (i'', j') has cost equal to the number of SNPs in j that are not also in i (e.g., an arc $(-\text{ABB}, - - \text{BBABA})$ has cost 3). Arcs (s, i') have cost equal to the number of SNPs in i. An s–t unit flow in this network describes a path that goes through some fragments, such that the total length of the SNPs spanned (i.e. of the haplotype) is equal to the cost of the path. Hence, the max cost circulation individues two haplotypes of max total length. This proves the following

Theorem 2. *The LHP for gapless fragments is polynomial.*

2.2   The Minimum Snip Removal

Theorem 3. *Let M be a gapless SNP matrix. Then $G_{\mathcal{F}}(M)$ is a bipartite graph if and only if $G_{\mathcal{S}}(M)$ is an independent set.*

Proof. (If) Consider a cycle of fragments in a SNP matrix (see Fig. 1d). Wlog, assume the cycle involves fragments $0, 1, \ldots, k$. For each pair of consecutive fragments i, $i + 1$ (mod $(k + 1)$) there is a position u_i at which one has an **A** and the other a **B**. We associate to a fragment cycle a directed cycle between entries in the matrix, made of horizontal arcs from u_{i-1} to u_i in fragment i, and vertical arcs from u_i in fragment i to u_i in fragment $i + 1$. We call a *vertical line* a maximal run of vertical arcs in such a cycle. In a vertical line, the letters **A** and **B** alternate. By definition, an infeasible SNP matrix contains an odd cycle of fragments. Let us call *weight* of an infeasible SNP matrix the minimum number of vertical lines of any odd cycles of fragments it contains.

Assume there exists an infeasible gapless SNP matrix M such that $G_{\mathcal{S}}(M)$ is an independent set, and pick M to have minimum weight among all such M. Consider an odd cycle in M achieving this weight. Since an infeasible matrix cannot have weight 1 there are at least two vertical lines, and hence a "rightmost" vertical line, say between fragments f and g. Since the line is rightmost, $u_{f-1}, u_g \leq u_f$. Assume $u_g \geq u_{f-1}$ (same argument if $u_{f-1} \geq u_g$). Since M is gapless, there exists a symbol at row f, column u_g. The symbols M_{f,u_f} and M_{g,u_f} are the same if and only if M_{f,u_g} and M_{g,u_g} are the same (otherwise, the rows f and g individue a snip conflict of columns u_g, u_f). Now, consider the SNP matrix M' obtained by M by first deleting (i.e., replacing with gaps) all the symbols between rows f and g (excluded), and then inserting an alternating chain of **A**s and **B**s, starting with M'_{g,u_g}, between rows f and g in column u_g. M' is an infeasible gapless SNP matrix of weight at least one smaller than M. Further, there are no snip conflicts in M' that were not already in M, so $G_{\mathcal{S}}(M')$ is an independent set. Hence, M was not minimal.

(Only if) We omit the simple proof.

Note that, in the presence of gaps, only the "only if" part of the theorem holds. We now show that the Minimum SNP removal problem can be solved in polynomial time on gapless SNP matrices. In particular, we prove that $G_{\mathcal{S}}(M)$ is a perfect graph. The basic tool to do this is the following: if I are the nodes of a hole or a antihole in $G_{\mathcal{S}}(M)$, and $\{i, j\}$ is a conflict, with $i, j \in I$, then for any $k \in I$ such that column k is between columns i and j in M, some relations of k with either i or j are forced. This will allow us to forbid long holes and antiholes.

Lemma 2. *Let M be a gapless SNP matrix and c_1, c_2, c_3 be snips (columns of M) with $c_1 \leq c_2 \leq c_3$. If $\{c_1, c_3\}$ is a snip conflict, then at least one of $\{c_1, c_2\}$ and $\{c_2, c_3\}$ is also a snip conflict.*

Proof. There are two rows r_1 and r_2 such that the 2×2 submatrix induced by rows r_1, r_2 and columns c_1, c_3 has three symbols of one type and one of the opposite. In this submatrix call a 2×1 column of identical symbols type I and one of different symbols type D. Since M is gapless, c_2 must be either I or D. But one of c_1 and c_3 is I and the other D.

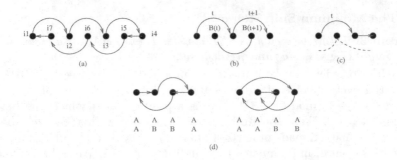

Fig. 2. (a) A cycle without long jumps. (b) two jumps in a row. (c) a jump and a shift. (d) The only possible holes.

Lemma 3. *Let M be a gapless SNP matrix and c_1, c_2, c_3, c_4 be snips with $c_1 \le c_2 \le c_3 \le c_4$. Assume $\{c_1, c_4\}$ is a snip conflict. Then, if $\{c_2, c_3\}$ is not a conflict, one of c_1 and c_4 conflicts with both c_2 and c_3. If $\{c_2, c_3\}$ is a conflict, then the conflicts in $\{c_1, c_2, c_3, c_4\}$ contain a length–4 cycle.*

Lemma 4. *If M is a gapless SNP matrix, $G_S(M)$ does not have a chordless cycle of length > 4.*

Proof. Assume $C = (i_1, \ldots, i_k, i_{k+1} = i_1)$ is a chordless cycle, $k \ge 4$, in $G_S(M)$ (i.e. the columns of M, listed in the order defined by the cycle). For $t = 1, \ldots, k$, let $B(t)$ be the set of nodes in C which lie between i_t and i_{t+1} as columns of M. We say that t *hits* i_t and i_{t+1}. We call t a *long jump* if $|B(t)| > 1$, a *jump* if $|B(t)| = 1$ and a *shift* if $|B(t)| = 0$. For a jump, we denote by b_t the node such that $B(t) = \{b_t\}$. The following facts are true: (i) If C has a long jump, it must be $k = 4$. (ii) If C has no long jump, each jump t must be followed by a shift, pointing to the node b_t.

To prove (i), consider a long jump, and let $B(t) = \{a_1, a_2, \ldots\}$. By lemma 3, if $\{a_1, a_2\}$ is not a conflict, that either i_t or i_{t+1} would have degree ≥ 3 in C, impossible. Hence it must be a conflict. So, by lemma 3, $\{i_1, a_1, a_2, i_k\}$ must contain a cycle, and hence $k = 4$, since C cannot contain other cycles.

To prove (ii), note that if t is a shift for $t = 1, \ldots, k - 1$, then k must be a long jump. Hence, if there are no long jumps, there must be jumps (see Fig. 2a for an example of a generic such cycle). Now, assume that a jump t is followed by another jump. Then (see Fig. 2b), since neither b_t nor b_{t+1} can be hit by a long jump, there must be a jump hitting b_t and b_{t+1}. But, then lemma 2 applies to b_t, i_{t+1}, b_{t+1}, so that i_{t+1} would have degree 3 in C. Hence, the jump t is followed by a shift. If the shift points away from b_t, then b_t should be hit by a long jump (see Fig. 2c), impossible. But then, the only possible hole is C_4.

Note that C_4 is actually achieved by some matrices (see Fig. 2d).

The following lemma generalizes lemma 3.

Lemma 5. *Let M be a gapless SNP matrix and c_1, c_2 be snips, with $c_1 < c_2$ and $\{c_1, c_2\}$ a snip conflict. Let $(a_1, a_2, \ldots, a_t)$ be a path in $G_S^c(M)$, such that*

$c_1 < a_i < c_2$ for all $i = 1, \ldots, t$. If $\{c_1, a_1\}$ is not a snip conflict, then $\{a_t, c_2\}$ is a snip conflict. If $\{c_2, a_1\}$ is not a snip conflict, then $\{a_t, c_1\}$ is a snip conflict.

Proof. Since M is gapless, in the $2 \times n$ submatrix of M for which (wlog) c_1 is type I and c_2 is type D, $a_1, \ldots, a_t$ must be I or D. We prove the first part of the lemma, the other being the same argument. Let $a_0 := c_1$. Then a_i is type I for all $i = 1, \ldots, t$ to avoid a conflict of a_{i-1} and a_i. This forces the conflict $\{a_t, c_2\}$.

Lemma 6. *If M is a gapless SNP matrix, $G_S^c(M)$ does not have a chordless cycle of length > 4.*

Proof. Let us call a path in $G_S^c(M)$ an *antipath* and a cycle in $G_S^c(M)$ an *anticycle*. Assume $(i_1, i_2, \ldots, i_k, i_{k+1} = 1)$ is a chordless anticycle of length $k \geq 5$. Let $i_x = \min\{i_1, \ldots, i_k\}$ and $i_y = \max\{i_1, \ldots, i_k\}$. $\{i_x, i_y\}$ cannot be a snip conflict, or otherwise the part of the anticycle from i_x to either i_{y-1} or i_{y+1}, can be used in lemma 5 to derive a contradiction. So, after possibly changing origin and orientation of the antihole, we can assume that $x = 1$ and $y = 2$. We will argue that the only possible ordering of these columns in M is $i_1 < i_3 < i_5 < \ldots < i_6 < i_4 < i_2$. In fact, we can apply the same argument from left to right and from right to left. Assume i_3 is not successive to i_1, but i_p, $p > 3$, is. Then, the antipath $(i_2, i_3, \ldots, i_{p-1})$ would be contained within i_p and i_2, and, by lemma 5, $\{i_{p-1}, i_p\}$ would be a conflict. Similarly, now assume that i_4 is not the second-to-last, but some i_p, $p > 4$ is. Then the antipath $(i_3, \ldots, i_{p-1})$ is contained within i_3 and i_p and, by lemma 5, $\{i_{p-1}, i_p\}$ would be a conflict. We can continue this way to prove the only ordering possible, but the important part is that the ordering looks like $i_1 < \ldots < i_4 < i_2$. Then, the antipath $(i_1, i_k, i_{k-1}, \ldots, i_5)$ is all contained within i_1 and i_4. By lemma 5, $\{i_5, i_4\}$ must be a conflict, contradiction.

Theorem 4. *If M is a gapless SNP matrix, then $G_S(M)$ is a perfect graph.*

Proof. Because of lemma 4 and lemma 6, $G_S(M)$ is weakly triangulated, i.e. neither $G_S(M)$ or its complement have a chordless cycle of length > 4. Since weakly triangulated graphs are perfect (Hayward, [5]), the result follows.

The next corollary follows from Theorem 3, Theorem 4, and the fact that the max independent set can be found in polynomial time on perfect graphs ([3,4]).

Corollary 1. *The Minimum SNP Removal problem can be solved in polynomial time on gapless SNP matrices.*

3   Dealing with Gaps

If gaps in the fragments are allowed, SNP problems become considerably more complex. Typically, a gap corresponds to a SNP whose value at a fragment which in fact covers it was not determined (e.g., because of thresholding of low quality reads, or for sequencing errors which missed some bases). Also, an important case of gaps occurs when fragments are paired up in the so called *mate pairs*. These, used in shotgun sequencing, are fragments taken from the same chromosome, with some fixed distance between them. A mate pair can be thought of as a single fragment, with a large gap in the middle and SNPs reads at both ends.

A class of inputs in which the SNP matrix has gaps but that can still be solved in polynomial time is the following. A 0-1 matrix is said to have the *consecutive ones property* (C1P) if the columns can be rearranged so that in each row the 1s appear consecutively. Analogously, we say that a SNP matrix is C1P if there exists a permutation π of the SNPs such that each fragment covers a consecutive (in π) subset of SNPs. Since finding such π is polynomial ([1]), it follows from Theorem 1, 2 and Corollary 1 the

Corollary 2. *The problems MFR, LSH and MSR can be solved in polynomial time on SNP matrices that are C1P.*

For those matrices that are not C1P, it is easy to show NP–hardness for many problems, by using the following lemma, that shows how to code a graph into a SNP matrix.

Lemma 7. *Let $G = (V, E)$ be a graph. Then there exists a SNP matrix M such that $G_{\mathcal{F}}(M) = G$.*

Proof. G can be made into a $|V| \times |E|$ SNP matrix by having a fragment for each node in V and a SNP for each edge $e = \{i, j\}$, with value A in i and B in j.

We can now give a simple proof that MFR is NP–hard.

Theorem 5. *MFR is NP-hard.*

Proof. We use a reduction from the following NP-hard problem [6,8]: Given a graph $G = (V, E)$, remove the fewest number of nodes to make it bipartite. This is exactly MFR when G is encoded into a SNP matrix as described in lemma 7.

In the following theorem, we show that it is the very presence of gaps in fragments, and not their quantity, that makes the problem difficult.

Theorem 6. *MFR is NP-hard for SNP matrices in which each fragment has at most 1 gap.*

The proof is through a polynomial time reduction from Max2SAT [2]. Consider an instance $\Phi(k, n)$ of Max2SAT with k clauses over n boolean variables. Denote the clauses of Φ as $C_1, C_2, \ldots, C_k$, and the variables as $x_1, x_2, \ldots x_n$. By definition, each clause contains at most 2 literals. Without loss of generality, we assume that a variable appears at most once in a clause.

We transform the instance $\Phi(n, k)$ into a SNP matrix M_{Φ} with a set $\mathcal{F}$ of $n(nk+3k+1)+3k$ fragments (rows), and $\mathcal{S}$ of $2n+5k$ SNPs (columns). See Fig. 3. Each variable x contributes $nk + 3k + 1$ fragments, which can be partitioned into 3 sets with $k, k, (nk + k + 1)$ fragments respectively. The fragments with labels $f_{x,1}, \ldots, f_{x,k}$ form the set $T(x)$(true). Similarly, the set $F(x)$(False) is the set containing fragments $f_{\bar{x},1}, \ldots, f_{\bar{x},k}$, and the set $S(x)$ (support) contains the remaining $nk + k + 1$ fragments. No two fragments from different sets can be in the same haplotype, and any number of fragments from the same set can be in the same haplotype. Denote a literal of the variable x as $x_l \in \{x, \bar{x}\}$. If $x_l = x$,

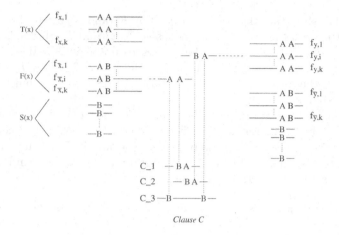

Clause C

Fig. 3. Gadget for the reduction.

then the fragment $f_{x_l,i}$ corresponds to $f_{x,i}$, and $f_{\bar{x}_l,i} = f_{\bar{x},i}$. Similarly, if $x_l = \bar{x}$, $f_{x_l,i} = f_{\bar{x},i}$, and $f_{\bar{x}_l,i} = f_{x,i}$.

Next, consider a clause $C_i = (x + \bar{y})$. There are three other fragments $C_{i,1}, C_{i,2}, C_{i,3}$ for each clause. The clause fragments are located in between the fragments for x and y, $C_{i,1}$ and $C_{i,2}$ conflict with each other. Extend fragment $f_{\bar{x},i}$ with a mate-pair, so that it shares a SNP (and a conflict) with the clause fragment $C_{i,1}$. Likewise, extend $f_{y,i}$ to conflict with $C_{i,2}$. Finally the fragment $C_{i,3}$ is a mate-pair which conflict with both $f_{\bar{x},i}$ and $f_{y,i}$. Denote the fragment conflict graph on M_Φ as $G(M_\Phi)$ (for simplicity, we drop the subscript $\mathcal{F}$).

Lemma 8. *Given a clause $C_i = (x_l + y_l)$ of a 2SAT instance Φ, the fragments $(C_{i,1}, C_{i,2}, f_{y_l,i}, C_{i,3}, f_{\bar{x}_l,i})$ form a chordless cycle in $G(M_\Phi)$.*

Lemma 9. *Each fragment in M_Φ has at most one gap.*

Lemma 10. *Let K be a set of fragments (rows) in M_Φ. The following are sufficient conditions for $M_\Phi[K]$ to be feasible.*

1. *For every variable x, $K \cap F(x) = \{\}$, or $K \cap T(x) = \{\}$.*
2. *For every clause $C_i = (x_l + y_l)$, K does not contain all the four fragments $f_{\bar{x}_l,i}, f_{\bar{y}_l,i}, C_{i,1}$, and $C_{i,2}$.*

Proof. The proof is constructive. If $G(M_\Phi[K])$ is bipartite, its nodes can be partitioned into two independent sets (shores) K_1, and K_2. We employ the following construction.

(1) For all x, add $S(x)$ to K_1, and $T(x)$ (or, $F(x)$) to K_2. (2) For all clauses $C_i = (x_l + y_l)$, add $C_{i,3}$ to K_1. (3) For all clauses $C_i = (x_l + y_l)$: (a) if $(f_{\bar{x}_l,i} \notin F)$, add $C_{i,1}$ to K_2, and $C_{i,2}$ to K_1. (b) else if $(f_{\bar{y}_l,i} \notin F)$, add $C_{i,2}$ to K_2, and $C_{i,1}$ to K_1. (c) else add $C_{i,2}$ (or, $C_{i,1}$) to K_1.

We need to show that the graphs induced by K_1 and K_2 are both independent sets. Note that $S(x)$ in K_1 only has edges to $T(x)$ and $F(x)$ which are in K_2. Likewise for all i, $C_{i,3}$ in K_1 only has edges to nodes in K_2. If both $C_{i,1}$ and $C_{i,2}$ are present, then the condition ensures that both $f_{\bar{x}_l,i}$ and $f_{\bar{y}_l,i}$ are not in K. Therefore, the construction in 3a, and 3b ensures that $C_{i,1}$ and $C_{i,2}$ are put on different shores.

Next, consider the fragments in $T(x)$, and $F(x)$ for all x. Condition 1 ensures that they can be all placed in K_2 without conflicting edges between different literals of x. Next, from 3a, $C_{i,1}$ is placed in K_2 only if $f_{\bar{x}_l,i}$ is not in K. From 3b, $C_{i,2}$ is placed in K_2 only if $f_{\bar{y}_l,i}$ is not in K. Thus, K_1 and K_2 induce independent sets, and $M_\Phi[K]$ is feasible.

Lemma 11. *An optimal solution to the MFR problem on M_Φ has at most $nk+k$ fragments.*

Proof. Consider a set of fragments R with $F(x)$ for all x, and $C_{i,1}$, for all clauses C_i. Removing R satisfies the conditions of lemma 10, implying that R is a solution to the MFR problem on M_Φ with $nk + k$ fragments.

Lemma 12. *Let R be an optimal solution to the MFR problem on M_Φ. Then, $R \cap S(x) = \phi$ for all x.*

Proof. Consider an optimal solution R that contains a fragment f from $S(x)$, for an arbitrary variable x. Let $K = \mathcal{F} - R$. As R is optimal, adding f to $G(M_\Phi[K])$ must create an odd-cycle C. Consider any other fragment $f' \in S(x)$. By construction, $C - \{f\} + \{f'\}$ is also an odd-cycle. This implies that all fragments in $S(x)$ are in R. Therefore, $|R| \geq |S(x)| = nk + k + 1 > nk + k$, a contradiction to lemma 11!

Lemma 13. *Let R be an optimal solution to the MFR problem for M_Φ. Then, for all x, either $T(x) \subseteq R$, or $F(x) \subseteq R$.*

Proof. Consider an optimal solution R with a variable x, and fragments $f_1 \in T(x) - R$, and $f_2 \in F(x) - R$. By lemma 12, there is a fragment $f \in S(x) - R$. By construction, f, f_1, and f_2 form an odd cycle, a contradiction!

Theorem 7. *Consider a Max2SAT instance Φ with n variables and k clauses, and the associated SNP matrix M_Φ. $k' \leq k$ clauses of Φ are satisfiable if and only if there exists a solution to the MFR problem for M_Φ with $nk+k-k'$ fragments.*

Proof. Consider an assignment of variables satisfying k' clauses. For each variable x set to TRUE, add all the fragments in $F(x)$ to R, and for every fragment set to FALSE, add all the fragments in $T(x)$ to R. Next, consider all the clauses that are satisfied. If $C_i = (x_l + y_l)$ is satisfied, at least one of $f_{\bar{x}_l,i}$, and $f_{\bar{y}_l,i}$ is in R, breaking the odd cycle, and we do nothing. If C_i is not satisfied, we add $C_{i,1}$ to R. The number of fragments in R due to variables is nk, and the number of fragments in R due to clauses is $k - k'$. By lemma 10, $M_\Phi[\mathcal{F} - R]$ is feasible.

Next, consider an optimal solution R of the MFR problem on M_Φ with $nk + k - k'$ fragments. For every x, by lemma 13, either $F(x) \subseteq R$ or $T(x) \subseteq R$.

If $F(x) \subseteq R$ set x to TRUE. Otherwise set x to FALSE. We need to show that exactly k' clauses are satisfied. Note that a set D of nk nodes must be in any optimal solution R to the MFR problem (lemma 13). Further, each clause is associated with 5 fragments that induce an odd cycle in the conflict graph (lemma 8). At least one of these fragments must be in R. If a clause $C_i = (x_l + y_l)$ is satisfied, then this fragment can be attributed to the set D. If however, C_i is not satisfied, the number of fragments in R increases by at least one. Thus if the total number of clauses satified is k'', then $|R| \geq nk + k - k''$. If $k'' < k'$, then $|R| > nk + k - k'$, a contradiction. On the other hand, if $k'' > k$, then by earlier argument, there is a solution to the MFR problem with $nk + k - k'' < nk + k - k'$ fragments, which is a contradiction to optimality.

We close this section with a complexity result for the snip removal problem, which follows using lemma 7 and the fact that MAXCUT is NP–hard for 3–regular graphs.

Theorem 8. *The MSR problem is NP-hard for SNP matrices with at most 2 gaps per fragment.*

Acknowledgments. We would like to thank Jinghui Zhang for many valuable and inspiring discussions about the challenges of computational SNPs discovery. We also want to thank Andy Clark for exciting discussions about SNPology.

References

1. K. S. Booth and S. G. Lueker. Testing for consecutive ones property, interval graphs and planarity using PQ–tree algorithms, *J. Comput. Syst. Sci.* 13, 335–379, 1976.
2. M. R. Garey and D. S. Johnson. Computers and Intractability: A Guide to the Theory of NP-completeness. W. Freeman and Co, SF, 1979.
3. M. C. Golumbic. *Algorithmic Graph Theory and Perfect Graphs*. Academic press, NY, 1980.
4. M. Groetschel, L. Lovasz and A. Schrijver. A polynomial algorithm for perfect graphs, *Annals of Discr. Math.* 21, 325–356, 1984.
5. R. B. Hayward. Weakly triangulated graphs, *J. Comb. Th. (B)* 39, 200–209, 1985.
6. J. M. Lewis, On the complexity of the maximum subgraph problem, *Xth ACM Symposium on Theory of Computing*, 265–274, 1978
7. J. C. Venter, M. D. Adams, E. W. Myers *et al.*, The Sequence of the Human Genome, *Science*, 291, 1304–1351, 2001.
8. M. Yannakakis, Node– and Edge– deletion NP–complete Problems, *Xth ACM Symposium on Theory of Computing*, 253–264, 1978.

A FPTAS for Approximating the Unrelated Parallel Machines Scheduling Problem with Costs*

Eric Angel[1], Evripidis Bampis[1], and Alexander Kononov[2]

[1] LaMI, CNRS-UMR 8042, Université d'Évry Val-d'Essonne, Boulevard François Mitterrand, 91025 Evry, Cedex, France,
{angel, bampis}@lami.univ-evry.fr
[2] Sobolev Institute of Mathematics, pr Koptyuga 4, Novosibirsk, Russia,
{alvenko@math.nsc.ru}

Abstract. We consider the classical problem of scheduling a set of independent jobs on a set of unrelated machines with costs. We are given a set of n monoprocessor jobs and m machines where each job is to be processed without preemptions. Executing job j on machine i requires time $p_{ij} \geq 0$ and incurs cost c_{ij}. Our objective is to find a schedule obtaining a tradeoff between the makespan and the total cost. We focus on the case where the number of machines is a fixed constant, and we propose a simple FPTAS that computes for any $\epsilon > 0$ a schedule with makespan at most $(1+\epsilon)T$ and cost at most $C_{opt}(T)$, in time $\mathcal{O}(n(n/\epsilon)^m)$, given that there exists a schedule of makespan T, where $C_{opt}(T)$ is the cost of the minimum cost schedule which achieves a makespan of T. We show that the optimal makespan-cost trade-off (Pareto) curve can be approximated by an efficient polynomial time algorithm within any desired accuracy. Our results can also be applied to the scheduling problem where the rejection of jobs is allowed. Each job has a penalty associated to it, and one is allowed to schedule any subset of jobs. In this case the goal is the minimization of the makespan of the scheduled jobs and the total penalty of the rejected jobs.

1   Introduction

We consider the problem of scheduling n independent jobs on m unrelated parallel machines. When job j is processed on machine i it requires $p_{ij} \geq 0$ time units and incurs a cost c_{ij}. Our objective is to find a schedule obtaining a trade-off between the *makespan* and the *total cost*. This kind of problems has many applications in vehicle routing [6,11], distribution systems [4], facility location [16],

* This work has been supported by the ASP "Approximabilité et Recherche Locale" of the French Ministry of Education, Technology and Research, and the Thematic Network of the European Union APPOL (IST-1999-14084).

and in optical networks that are based on the Wavelength Division Multiplexing (WDM) technology [1,2]. Notice that this problem covers as a special case the problem of scheduling jobs when rejections are allowed [3]. In this problem, a job j can either be rejected, in which case a penalty c_j is paid, or scheduled on one of the machines [3].

In the last ten years a large number of works deal with different variants of the general bi-criteria problem on unrelated parallel machines [14,17,12,18]. Lin and Vitter [14] gave a polynomial time algorithm that, given makespan T, cost C, and $\epsilon > 0$, finds a solution of makespan $(2 + \frac{1}{\epsilon})T$ and cost $(1 + \epsilon)C$, if there exists a schedule of makespan T and schedule C. This result is based on solving a linear relaxation of a particular integer programming formulation, and then rounding the fractional solution to a nearby integer solution by using the rounding theorem of Lenstra, Shmoys and Tardos [13]. Shmoys and Tardos [17] improved this result by presenting a polynomial-time algorithm that, given T and C finds a schedule of cost C and makespan $2T$, if there is a schedule of makespan T and cost C. The main difference with the previous result is the introduction of a new rounding technique of the fractional solution which does not require that the solution to be rounded be a vertex of the linear relaxation. This result cannot be substantially improved in the case where the number of machines is in the input of the problem, since Lenstra et al. [13] proved that for the single-criterion problem of minimizing the makespan, no polynomial-time $(1 + \epsilon)$-approximation algorithm with $\epsilon < 1/2$ exists, unless $\mathcal{P} = \mathcal{NP}$.

Hence, many papers deal with the natural question of how well the problem can be approximated when there is only a constant number of machines [10,12]. For the bi-criterion scheduling problem that we consider in this paper, Jansen and Porkolab [12] proposed a fully polynomial-time approximation scheme (FPTAS) that given values T and C computes for any $\epsilon > 0$, a schedule in time $O(n(m/\epsilon)^{\mathcal{O}(m)})$ with makespan at most $(1 + \epsilon)T$ and cost at most $(1 + \epsilon)C$, if there exists a schedule of makespan T and cost C. This result relies in part on linear programming and uses quite sophisticated methods such as the logarithmic potential price directive decomposition method of Grigoriadis and Khatchiyan [7].

In this paper, we propose a FPTAS that, given T, computes for any $\epsilon > 0$ a schedule with makespan T' such that $T' \leq (1 + \epsilon)T$ and cost at most $C_{opt}(T)$, in $\mathcal{O}(n(n/\epsilon)^m)$ time, with $C_{opt}(T)$ the cost of the minimum cost schedule which achieves a makespan of T.

The algorithm of Jansen and Porkolab [12], given T and C, can only guarantee that the obtained cost is at most $(1 + \epsilon)C$ while our result guarantees a cost (at the worst case) equal to C, while offering the same quality in what concerns the makespan, i.e. a schedule of makespan at most $(1 + \epsilon)T$. In addition, our methods are much more simpler since we use just a combination of well known combinatorial methods (dynamic programming and rounding) and we require only the knowledge of T. On the other hand, the complexity of our algorithm is worse than the one of [12].

In the following table, we give a comparison of our result with the most important results, for the unrelated parallel machines scheduling problem with costs.

	m general		m constant	
Reference	[14]	[17]	[12]	this paper
Quality	$(2 + \frac{1}{\epsilon})T, (1 + \epsilon)C$	$2T, C$	$(1 + \epsilon)T, (1 + \epsilon)C$	$(1 + \epsilon)T, C$
Time	$poly(m, n)$	$poly(m, n)$	$n(\frac{m}{\epsilon})^{\mathcal{O}(m)}$	$\mathcal{O}(n(\frac{n}{\epsilon})^m)$

Furthermore, we have adapted our algorithm for finding the schedule with the optimal makespan T_{opt} (within a factor $1 + \epsilon$) and the smallest total cost (with respect to T_{opt}) without knowing the value of T_{opt}, i.e. a schedule s^* such that $T_{opt} \leq T(s^*) \leq (1 + \epsilon)T_{opt}$ and $C(s^*) \leq C_{opt}(T_{opt})$, where T_{opt} is the optimal makespan and $C_{opt}(T_{opt})$ is the minimal cost of a schedule with makespan T_{opt}. The complexity of our algorithm in this case is in $\mathcal{O}((\frac{n}{\epsilon})^{m+1} \log m)$.

Pareto curves and approximation. Usually, given an *optimization problem* we are searching for a feasible solution optimizing its objective function. In *multiobjective optimization*, we are interested not in a single optimal solution, but in the set of all possible solutions whose vector of the various objective criteria is not dominated by the vector of another solution. This set known as the *Pareto curve* captures the intuitive notion of "trade-off" between the various objective criteria. For even the simplest problems (matching, minimum spanning tree, shortest path) and even for two objectives, determining whether a point belongs to the Pareto curve is an $\mathcal{NP}$-hard problem. Moreover, this set is exponential in size and so, until recently all computational approaches to multiobjective optimization are concerned with less ambitious goals, such as optimizing lexicographically the various criteria. Recently, an *approximate version* of this concept has been studied, the ϵ-*approximate Pareto curve* [15]. Informally, an ϵ-approximate Pareto curve is a set of solutions that approximately dominate all other solutions, i.e. for every other solution, the set contains a solution that is at least as good approximately (within a factor $1 + \epsilon$) in all objectives.

It has been shown in [15], that under some very general conditions there is always a polynomially (in the size of the instance and $\frac{1}{\epsilon}$) succinct ϵ-approximate Pareto curve. For discrete optimization problems with *linear objective functions*, it has been also shown that there is a fully polynomial time approximation scheme (FPTAS) for calculating the ϵ-approximate Pareto curve if the exact version of the problem is pseudo-polynomial time solvable [15]. A linear objective function means that the criterion can be expressed as a scalar product between the solution (represented by a vector) and a fixed vector of costs. This result can be applied on the problem we consider by embedding the makespan problem (which is the maximum of a fixed number of machine loads, i.e., the maximum of a fixed number of linear functions) into the multi-criteria problem where every machine load is a criterion and the total cost is a criterion.

We study in this paper a stronger version of approximate Pareto curves, the $(\epsilon, 0)$-Pareto curve and give an efficient polynomial time algorithm to construct it. Informally, we search for a set of solutions that dominates all other solutions approximately (within a factor $1 + \epsilon$) in all but one objectives, and it is optimal with respect to the last objective. For the scheduling problem that we consider, the solutions that belong to the $(\epsilon, 0)$-Pareto curve approximately dominate all other solutions with respect to the makespan criterion, but they must give the smallest possible value of the total cost. In this paper, we propose a FPTAS for constructing an $(\epsilon, 0)$-Pareto curve which runs in time $\mathcal{O}(\frac{n^2}{\epsilon}(n/\epsilon)^m)$.

Scheduling with rejections. In this kind of problems, for each job we have to decide whether to accept that job or whether to reject it. For the accepted jobs we pay the makespan of the constructed schedule and for the rejected jobs we pay the corresponding rejection penalties. The objective is to find a feasible schedule minimizing the sum of the makespan of the accepted jobs plus the total penalty. Bartal et al. [3] have considered the online and offline versions of this problem, in the case of identical machines i.e. when the processing times of the jobs are machine independent. Epstein and Sgall have considered the problem in the case of uniformly related machines [5]. More recently, Hoogeveen, Skutella and Woeginger considered the preemptive version of the unrelated machines case and presented a complete classification of the different variants of the problem [9]. It is easy to see that the results presented in this paper can be applied to the non-preemptive scheduling problem with rejections in the case of unrelated parallel machines. To take into account rejection, it is sufficient to add a *dummy* machine $m + 1$, such that the processing times of all the jobs j are zero on this machine and $c_{m+1,j} = c_j$. In fact, this machine receives all the rejected jobs. The cost of every job j on the other machines is equal to zero. Therefore, we can obtain a FPTAS for constructing an $(\epsilon, 0)$-Pareto curve in time $\mathcal{O}(\frac{n^2}{\epsilon}(n/\epsilon)^{m+1})$ for scheduling with rejection on unrelated parallel machines. This curve contains a set of solutions dominating all others solutions approximately (within a factor $1 + \epsilon$) in the makespan criteria and it is optimal with respect to the total penalty criteria.

Organization of the paper. In the next section we present an exact dynamic programming algorithm while in Section 3 we use it in order to obtain a FPTAS in the case where the value of the makespan is given. We show also how to modify this FPTAS in order to compute the optimal makespan (within a factor $1 + \epsilon$) at the minimum cost. Section 4 is devoted to the construction of a FPTAS approximating the Pareto curve. In the sequel, we shall consider the case of two machines. The generalization for $m > 2$ machines, with m fixed, is direct.

2   An Exact Dynamic Programming Algorithm

For a schedule s, possibly partial, i.e. involving only a subset of all the jobs, we denote by $C(s)$ and $T(s)$ its cost and makespan, respectively. We shall also use the notation $s = (m_1, m_2) = (J_1, J_2)$ meaning that m_1 (resp. m_2) is the total processing time on machine 1 (resp. 2), and J_1 (resp. J_2) is the set of jobs scheduled on machine 1 (resp. 2). If $s = (J_1, J_2)$ then the makespan is defined by $T(s) = \max\{m_1 = \sum_{j \in J_1} p_{1j}, m_2 = \sum_{j \in J_2} p_{2j}\}$, and the cost by $C(s) = \sum_{j \in J_1} c_{1j} + \sum_{j \in J_2} c_{2j}$.

We define $C_{opt}(T)$ as the minimum cost schedule, if any, with makespan T, i.e. $C_{opt}(T) = \min_{s, T(s)=T} C(s)$. Otherwise, $C_{opt}(T) = +\infty$.

In this section, we give a pseudo-polynomial time algorithm, which given T returns a schedule s with makespan $T(s) = T$ and cost $C(s) = C_{opt}(T)$ if such a schedule exists.

Let $J = \{1, 2, \ldots, n\}$ be the set of jobs. We shall assume, in this section only, that the processing times p_{ij} are integer ones.

The states of the dynamic programming algorithm are the tuples $E(j, m_1, m_2)$, with $1 \leq j \leq n$ and $m_1, m_2 \in \{0, 1, \ldots, T\}$.
By definition $E_c(j, m_1, m_2) = +\infty$ if a schedule involving jobs $\{1, 2, \ldots j\}$ and which achieves a completion time equal to m_1 on machine 1, and m_2 on machine 2, does not exist. Otherwise, $E_c(j, m_1, m_2)$ is the minimum cost of a schedule among all such schedules. We have

$$E_c(j, m_1, m_2) = \min\{E_c(j-1, m_1 - p_{1j}, m_2) + c_{1j}, E_c(j-1, m_1, m_2 - p_{2j}) + c_{2j}\}.$$

Initially, we set $E_c(1, 0, p_{21}) = c_{21}$, $E_c(1, p_{11}, 0) = c_{11}$, and otherwise $E_c(1, m_1, m_2) = +\infty$.

To be able to retrieve the schedule, we need an additional variable $E_s(j, m_1, m_2)$ which stores, for each state, the partial schedule constructed so far, i.e. the jobs that are scheduled on each machine. For example, if we have $E_c(j, m_1, m_2) = E_c(j-1, m_1 - p_{1j}, m_2) + c_{1j}$, it means that job j is scheduled on machine 1, and therefore we have $E_s(j, m_1, m_2) = E_s(j-1, m_1 - p_{1j}, m_2) \cup \{j \rightarrow 1\}$.

The above backward recursion can be turned into an efficient to implement forward recursion. The algorithm is the following one:

for $j = 1$ **to** $n - 1$
for $m_1 = 0$ **to** T by step of 1 (m_1 is always an integer)
for $m_2 = 0$ **to** T by step of 1 (m_2 is always an integer)
  $E_c(j+1, m_1 + p_{1j}, m_2) \leftarrow \min\{E_c(j+1, m_1 + p_{1j}, m_2), E_c(j, m_1, m_2) + c_{1j}\}$
  $E_c(j+1, m_1, m_2 + p_{2j}) \leftarrow \min\{E_c(j+1, m_1, m_2 + p_{2j}), E_c(j, m_1, m_2) + c_{2j}\}$

Initially, we set $E_c(1, 0, p_{21}) = c_{21}$, $E_c(1, p_{11}, 0) = c_{11}$, and otherwise $E_c(j, m_1, m_2) = +\infty$ for $j \geq 1$.

To obtain a solution with a makespan of T and cost $C_{opt}(T)$, we select all the states $E(n, T, m_2)$ and $E(n, m_1, T)$ if any, with $m_1, m_2 \in \{0, 1, \ldots, T\}$, and among those states we choose a state with minimum cost E_c.

The number of states is $n(T+1)^2$ and therefore this algorithm is only pseudo-polynomial. In the next section, we show how to obtain a fully polynomial time approximation scheme from it.

3 An Approximation Scheme Given the Value of the Makespan

Let us consider $0 < \epsilon \leq 1$ such that without lost of generality $1/\epsilon$ is an integer. The technique that we use, in order to obtain a fully polynomial time approximation scheme, is inspired by the classical method of Horowitz and Sahni [10].

We subdivide the interval $]0, (1+2\epsilon)T]$ into $n/\epsilon + 2n$ subintervals, $\mathcal{I}_k =](k-1)l, kl]$, $1 \leq k \leq n/\epsilon + 2n$, each one of length $l = \epsilon T/n$.

For a real $x \in]0, (1+2\epsilon)T]$, we note $[x]$ the right extremity of the subinterval in which x belongs, i.e. $[x] = kl$, with k such that $x \in \mathcal{I}_k$. We set $[0] = 0$.

The approximation algorithm consists in running the previous exact dynamic programming algorithm on the modified instance where each processing time p_{ij} is replaced by $[p_{ij}]$. The first *for loop* is replaced by "for $m_1 = 0$ to $(1+2\epsilon)T$ by step of l", idem for m_2. Later we will explain how to find the appropriate solution among all the computed states.

For a set of jobs $J' \subseteq J$ and a machine $m \in \{1,2\}$, we shall note $[J' \to m] = \sum_{j \in J'} [p_{mj}]$.

Definition 1. *We note $[T](s)$ the makespan of the schedule s assuming that the processing time of each job p_{ij} has been rounded up to $[p_{ij}]$. Under the same assumption, we note $[s] = (m_1, m_2)$ the total processing time on each machine for the solution s. In other words, if $s = (J_1, J_2)$ then $[s] = ([J_1 \to 1], [J_2 \to 2])$.*

Proposition 1. *For any schedule s, we have $0 \leq [T](s) - T(s) \leq \epsilon T$.*

Proof. Let $s = (J_1, J_2) = (m_1, m_2)$ be the initial schedule, and $[s] = (J_1, J_2) = (m'_1, m'_2)$ the schedule obtained after rounding up the processing time of each job. Let us assume that $T(s) = m_1$. Then, we can have either $[T](s) = m'_1$ or $[T](s) = m'_2$. Let us assume first we have $[T](s) = m'_1$. We have $T(s) = m_1 = \sum_{j \in J_1} p_{1j}$ and $[T](s) = m'_1 = \sum_{j \in J_1} [p_{1j}]$. Notice now, that since l is the length of each subinterval, for any $x > 0$ we have $[x] - x \leq l$. Since $|J_1| \leq n$, we have $[T](s) - T(s) \leq nl = \epsilon T$. Let us assume now we have $[T](s) = m'_2$. Since $T(s) = m_1$, we have $m_1 \geq m_2$. So $[T](s) - T(s) = m'_2 - m_1 \leq m'_2 - m_2 \leq nl = \epsilon T$ (with the same argument as above).

Definition 2. *We define by $[(m_1, m_2)]_k$ the set of all schedules s involving the jobs $\{1, 2, \ldots, k\}$, such that $[T](s) = (m_1, m_2)$.*

Proposition 2. *Let $k \in \{1,\ldots,n\}$ and let $s = E_s(k, m_1, m_2)$ the (partial, if $k < n$) schedule computed by the dynamic programming algorithm, then $\forall \tilde{s} \in [(m_1, m_2)]_k$ we have $C(s) \leq C(\tilde{s})$.*

Proof. Consider the simplified search graph of the dynamic programming depicted in Figure 1. A directed path of length l from the node r in this search graph corresponds to a set of decisions involving jobs $\{1, 2, \ldots, l\}$ in that order, which consist in choosing for each job the machine where it is scheduled.

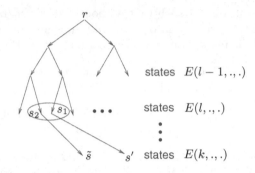

Fig. 1. The search graph of the dynamic programming with merging of states

Notice now that if we do not merge any state, then there is always a path from the node r of the search space to any solution $\tilde{s}$. If we consider an arbitrary schedule $\tilde{s}$, we can reorder the jobs on each machine such that the index of the jobs on each machine is in increasing order, without changing the makespan and the cost of the schedule. It is now clear that there is a path on the search graph from r, which leads to this new schedule which is equivalent to schedule $\tilde{s}$.

The dynamic programming algorithm at each stage computes a set of states. Namely, during the $l-1$-th stage, the states $E(l-1, m_1, m_2)$ are computed, for all the values of m_1 and m_2 which are feasible. Then, the l-th job is considered and "added" to each of the states $E(l-1, ., .)$ to obtain the states $E(l, ., .)$. During this stage, some states among the states $E(l, ., .)$ are merged; the one which has the lowest cost remains, whereas the others disappear. If the statement of Proposition 2 is not true, then there is a state that was wrongly eliminated. Let us assume that the two states $E(l-1, \alpha, \beta) = (J_1, J_2)$ and $E(l-1, \alpha', \beta') = (J_1', J_2')$ have been merged when the job l has been added in the first state, say on machine 1, and in the second state on machine 2. This means that $[J_1 \rightarrow 1] + [\{l\} \rightarrow 1] = [J_1' \rightarrow 1]$ and $[J_2 \rightarrow 2] = [J_2' \rightarrow 2] + [\{l\} \rightarrow 2]$. Let $s_1 = (J_1 \cup \{l\}, J_2)$, and $s_2 = (J_1', J_2' \cup \{l\})$. Assuming that we have $C(s_1) < C(s_2)$, the first state remained, whereas the second one was forgotten. Now let us suppose that because the partial schedule s_2 has been eliminated, there does not exist anymore a path in the search graph which leads to solution $\tilde{s}$. This is not a problem, since in that case there is a path which leads from s_1 to a better solution s' than $\tilde{s}$. Let $\tilde{s} = (J_1'', J_2'') = (m_1, m_2)$. Since there was a path from s_2 to $\tilde{s}$, the schedule

$s' = (J_1 \cup \{l\} \cup (J_1'' \setminus J_1'), J_2 \cup (J_2'' \setminus (J_2' \cup \{l\})))$ obtained from s_1 is well defined. Then clearly we have $s' = (m_1, m_2)$ and $C(s') < C(\tilde{s})$. If s' is not reachable from s_1 because of subsequent mergings, the same reasoning shows that there exist a solution reachable from s_1 which has a lower cost than s' and has the same makespan as $\tilde{s}$.

Theorem 1. *Given $T > 0$ and $\epsilon > 0$, if there exist a schedule of makespan between T and $(1 + \epsilon)T$, the schedule s_{dyna} returned by the dynamic programming algorithm verifies $(1 - \epsilon)T \leq T(s_{dyna}) \leq (1 + 2\epsilon)T$ and $C(s_{dyna}) \leq \min_{t \in [T,(1+\epsilon)T]} C_{opt}(t) \leq C_{opt}(T)$. Moreover the running time is $\mathcal{O}(n(n/\epsilon)^m)$. For the problem with rejections allowed, the running time is $\mathcal{O}(n(n/\epsilon)^{m+1})$.*

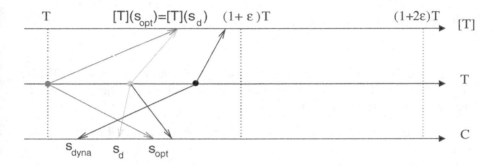

Fig. 2. Choosing s_{dyna} (in the case $T(s_{opt}) = T$)

Proof. Let $s_{opt} = (m_1, m_2)$ be a schedule with makespan between T and $(1 + \epsilon)T$ and cost $\min_{t \in [T,(1+\epsilon)T]} C_{opt}(t)$. Let $[s_{opt}] = (m_1', m_2')$ and let $s_d = E_s(n, m_1', m_2')$. We have $s_{opt} \in [(m_1', m_2')]_n$, and so using Proposition 2 we obtain $C(s_d) \leq C(s_{opt}) = C_{opt}(T)$. We have $T \leq T(s_{opt}) \leq (1 + \epsilon)T$, and using Proposition 1 we have $T \leq [T](s_{opt}) \leq (1 + 2\epsilon)T$. By definition of s_d we have $[T](s_d) = [T](s_{opt})$, and using Proposition 1 again we obtain $(1 - \epsilon)T \leq T(s_d) \leq (1 + 2\epsilon)T$.

To obtain the desired schedule s_{dyna}, we examine all the states returned by the dynamic programming algorithm with a rounded makespan between T and $(1 + 2\epsilon)T$, and we retain among them the schedule s_{dyna} which achieves the minimum cost. We unround the solution we return, i.e. each processing time is p_{ij} instead of $[p_{ij}]$. Of course we will find the schedule s_d, but perhaps a better schedule, and so we have $C(s_{dyna}) \leq C(s_d)$. The result follows, and the number of states that we need to consider is at most $n(n/\epsilon + 2n)^2$.

Remark 1. The case $T = 0$ can be easily solved, by assigning each job to a machine where its processing time is zero and on which the job incurs the minimum cost. If such a machine cannot be found for each job, then obviously the optimal makespan T_{opt} is strictly greater than 0.

Remark 2. Notice that making the rounding at the beginning of the dynamic programming algorithm is equivalent to making the rounding during the execution of the algorithm since we have the following simple proposition.

Proposition 3. *Let* $p_1, \ldots, p_n$ *be any positive real numbers, then we have*

$$[\ldots [[[p_1] + p_2] + p_3] + \ldots p_n] = [p_1] + [p_2] + \ldots [p_n].$$

Proof. We first show that $[[p_1] + p_2] = [p_1] + [p_2]$. Let us assume that $p_1 \in \mathcal{I}_{k_1}$ and $p_2 \in \mathcal{I}_{k_2}$. Then $[p_1] = k_1 l$ and $[p_2] = k_2 l$, $[p_1] + p_2 \in \mathcal{I}_{k_1 + k_2}$, and $[[p_1] + p_2] = (k_1 + k_2)l$. The proof is now by induction on the depth (number of left brackets) of the formula.

In the sequel we note $Dyna(T, \epsilon) = s_{dyna}$.

Let $T_{opt} = \min_s T(s)$ over all the schedules s.

An interesting question is to consider the case $T = T_{opt}$ and to determine whether it is necessary to know the value of T_{opt} to obtain an approximate schedule. Theorem 2 shows that this is not the case.

Let $d_j = \min_i p_{ij}$ the minimum processing time of job j, and $D = \sum_j d_j$. We assume that $D > 0$, otherwise the problem is trivial (see Remark 1). We have the following proposition.

Proposition 4. *We have* $D/m \leq T_{opt} \leq D$.

Proof. To obtain a schedule s with a makespan smaller than D, it is sufficient to put each job on the machine where its processing time is the smallest. The lower bound corresponds to the ideal case where the previous schedule s is such that each machine has the same load.

Theorem 2. *Without knowing the value* T_{opt}, *we can obtain, in* $\mathcal{O}((n/\epsilon)^{m+1} \log m)$ *time, a schedule* s^* *such that* $T_{opt} \leq T(s^*) \leq (1+\epsilon)T_{opt}$ *and* $C(s^*) \leq C_{opt}(T_{opt})$.

Proof. Using Proposition 4 we know that the makespan T_{opt} lies in the interval $[D/m, D]$. We subdivide the interval $[D/m, D[$ into $k = \lceil \log m / \log \sqrt{1 + \epsilon} \rceil = \mathcal{O}(\frac{\log m}{\epsilon})$ geometric subintervals $\mathcal{J}_1, \ldots, \mathcal{J}_k$, each one having a length $\sqrt{1 + \epsilon}$ times greater than its predecessor, with $\mathcal{J}_1 = [D/m, \sqrt{1 + \epsilon}D/m[$. Then we apply our dynamic programming algorithm with $0 < \epsilon' \leq \sqrt{1 + \epsilon} - 1$ on the left extremity x_i of each subinterval $\mathcal{J}_i$, $i = 1, \ldots, k$. Let $s_\alpha = Dyna(x_\alpha, \epsilon')$ and $s_{\alpha+1} = Dyna(x_{\alpha+1}, \epsilon')$ the first two solutions returned by our dynamic programming algorithm, i.e $Dyna(x_i, \epsilon') = \emptyset$ for $i = 1, \ldots, \alpha - 1$. Then s^* is the solution among s_α and $s_{\alpha+1}$ which has the minimum cost. To see this, consider the Figure 3. The value T_{opt} lies in some subinterval $\mathcal{J}_\alpha = [x_\alpha, \sqrt{1 + \epsilon}x_\alpha[$. Let

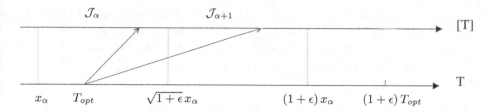

Fig. 3. An illustration of the proof of Theorem 2

s_{opt} a minimum makespan schedule, $T(s_{opt}) = T_{opt}$. Then using Proposition 1, we have $[T](s_{opt}) \leq T_{opt} + \epsilon'T \leq (1+\epsilon')T_{opt}$ (since $T = x_\alpha \leq T_{opt}) \leq \sqrt{1+\epsilon}\,T_{opt} < (1+\epsilon)x_\alpha$, since $T_{opt} < \sqrt{1+\epsilon}\,x_\alpha$. Therefore either $[T](s_{opt}) \in \mathcal{J}_\alpha$ or $[T](s_{opt}) \in \mathcal{J}_{\alpha+1}$. Now Theorem 1 shows that $C(s^*) \leq C_{opt}(T_{opt})$ and we have $T(s^*) \leq \sqrt{1+\epsilon}\,x_\alpha(1+2\epsilon') = \sqrt{1+\epsilon}\,x_\alpha(2\sqrt{1+\epsilon}-1) = 2x_\alpha(1+\epsilon) - x_\alpha\sqrt{1+\epsilon} \leq 2(1+\epsilon)T_{opt} - T_{opt} = (1+2\epsilon)T_{opt}$. Now using $\epsilon/2$ instead of ϵ yields the result.

4   A FPTAS for $(\epsilon, 0)$-Approximate Pareto Curves

In this section, we consider the notion of approximate Pareto curves [8]: Given an instance x of a minimization multi-objective problem, with objective functions f_i, $i = 1, \ldots, k$, and an $\epsilon > 0$, an ϵ-*approximate Pareto curve*, denoted $P_\epsilon(x)$, is a set of solutions s such that there is no other solution s' such that, for all $s \in P_\epsilon(x)$, $(1+\epsilon)f_i(x, s') < f_i(x, s)$ for some i.

Our goal is to get a stronger result for our bi-criteria problem: Informally, we search for a set of solutions that approximately dominate all other solutions in what concerns the makespan criterion, and absolutely dominates all other solutions in what concerns the cost criterion. In other words, for every other solution $T(s')$, $C(s')$, the set contains a solution s such that $T(s) \leq (1+\epsilon)T(s')$ and $C(s) \leq C(s')$. In order to do so, we introduce the definition of an $(\epsilon, 0)$-*approximate Pareto curve*: Given an instance x of a bi-criterion minimization problem, an $(\epsilon, 0)$-*approximate Pareto curve*, denoted $P_{\epsilon,0}(x)$, is a set of solutions s such that there is no other solution s' such that, for all $s \in P_{\epsilon,0}(x)$: $(1+\epsilon)T(s', x) < T(s, x)$ or $C(s', x) < C(s, x)$.

A solution s is said to $(\epsilon, 0)$-*approximately dominate* a solution s' if $T(s, x) \leq T(s', x)(1+\epsilon)$ and $C(s) \leq C(s')$. Equivalently s' is said to be $(\epsilon, 0)$-*approximately dominated* by s. In other words, an $(\epsilon, 0)$-approximate Pareto curve $P_{\epsilon,0}(x)$ is a set of solutions such that any solution is $(\epsilon, 0)$-approximately dominated by some solution in $P_{\epsilon,0}(x)$.

In the following we give an algorithm which constructs in $\mathcal{O}(n(\frac{n}{\epsilon})^{m+1})$ time an $(\epsilon, 0)$-approximate Pareto curve for the problem that we consider.

The algorithm. Let us denote by M_j the machine on which job j has the minimum cost. Let $C_{min} = \sum_j c_{M_j j}$ and $\tilde{D} = \sum_j p_{M_j j}$. Let s_{min} a schedule with cost C_{min}. It is clear that $T(s_{min}) \leq \tilde{D}$. We subdivide the interval $[D/m, \tilde{D}]$ into $k = \lceil \log(m\tilde{D}/D)/\log(1+\epsilon) \rceil + 1 = \mathcal{O}(n/\epsilon)$ geometric subintervals. Then we apply the algorithm $Dyna(T, \epsilon)$ on the left extremity of each subinterval and get a schedule. We also compute $Dyna(\tilde{D}, \epsilon)$. Among the schedules obtained, we keep an undominated subset Π.

Theorem 3. *The set Π is an $(\epsilon, 0)$-approximate Pareto curve, and can be obtained in $\mathcal{O}(n(\frac{n}{\epsilon})^{m+1})$ time. For the problem with rejections allowed, the running time is $\mathcal{O}(n(\frac{n}{\epsilon})^{m+2})$.*

Proof. Let s a schedule with $T(s) \leq \tilde{D}$. Let $t_k \leq T(s) < t_{k+1}$, where t_k is the left extremity of the k-th subinterval (we have $t_1 = D/m$ and $t_2 = (1+\epsilon)D/m$). By Theorem 1 there is $s^* \in \Pi$ such that $T(s^*) \leq (1+2\epsilon)t_k \leq (1+2\epsilon)T(s)$ and $C(s^*) \leq C(s)$. If $T(s) > \tilde{D}$, then $T(s) > T(s_{min})$ and $C(s) \geq C_{min}$. We know that there is $s^* \in \Pi$ such that $T(s^*) \leq (1+2\epsilon)T(s_{min}) \leq (1+2\epsilon)T(s)$ and $C(s^*) = C(s_{min}) \leq C(s)$. So by definition of $(\epsilon, 0)$-approximate Pareto curve, we get that Π is an $(2\epsilon, 0)$-approximate Pareto curve. Now using $\epsilon/2$ instead of ϵ yields the result.

Now notice that if $\tilde{D} = 0$ the Pareto curve is a single point, and if $D = 0$ we use Remark 1 to obtain the first point of the Pareto curve, and we run the algorithm using the interval $[\min_{ij \mid p_{ij} > 0} p_{ij}, \tilde{D}]$, instead of $[D/m, \tilde{D}]$, to obtain the other points.

Remark 3. Given the polynomially sized $(\epsilon, 0)$-approximate Pareto curve, it is easy to get a schedule that minimizes the sum of the makespan and the cost of rejected job, by inspecting all points on the Pareto curve and taking one which minimizes this sum.

Acknowledgment. We would like to thank Claire Kenyon for her valuable comments on this work.

References

1. I. Baldine and G. Rouskas. Dynamic load-balancing in broadcast WDM networks with tuning latencies. In *Proc. IEEE InfoCom'98*, pages 78–85, 1998.
2. E. Bampis, S. Chaoui, and S. Vial. On the performance of LPT with lookahead for the reconfiguration of WDM all optical networks. In *OptiComm'2000, Proc. SPIE Vol. 4233*, pages 141–152, 2000.
3. Y. Bartal, S. Leonardi, A. Marchetti-Spaccamela, J. Sgall, and L. Stougie. Multiprocessor scheduling with rejection. *SIAM Journal on Discrete Mathematics*, 13(1):64–78, 2000.
4. J.F. Benders and J.A.E.E. van Nunen. A property of assignment type mixed integer linear programming problems. *Operations Research Letters*, 2:47–52, 1982.
5. L. Epstein and J. Sgall. Approximation schemes for scheduling on uniformly related and identical parallel machines. In *ESA, Lecture Notes in Computer Science 1643*, pages 151–162, 1999.

6. M.L. Fisher and R. Jaikumar. A generalized assignment heuristic for vehicle routing. *Networks*, 11:109–124, 1981.
7. M.D. Grigoriadis and L.G. Khachiyan. Coordination complexity of parallel price-
   directive decomposition. *Mathematics of Operations Research*, 21:321–340, 1996.
8. P. Hansen. Bicriterion path problems. In *Proc. 3d Conf. Multiple Criteria Decision
   Making, Theory and application*, pages 109–127. Springer-Verlag, 1979. LNEMS,
   177.
9. H. Hoogeveen, M. Skutella, and G.J. Woeginger. Preemptive scheduling with rejection. In *ESA, Lecture Notes in Computer Science 1879*, pages 268–277, 2000.
10. E. Horowitz and S. Sahni. Exact and approximate algorithms for scheduling non
    identical processors. *Journal of the Association for Computing Machinery*, 23:317–
    327, 1976.
11. C.D. Jacobs. *The vehicle routing problem with backhauls*. PhD thesis, Georgia
    Institute of Technology, 1987.
12. K. Jansen and L. Porkolab. Improved approximation schemes for scheduling un-
    related parallel machines. In *STOC'99*, pages 408–417, 1999.
13. J.K. Lenstra, D.B. Shmoys, and É. Tardos. Approximation algorithms for schedul-
    ing unrelated parallel machines. *Mathematical Programming A*, 46:259–271, 1990.
14. J.-H. Lin and J.S. Vitter. ϵ-approximation algorithms with minimum packing
    constraint violation. In *STOC'92*, pages 771–782, 1992.
15. C.H. Papadimitriou and M. Yannakakis. On the approximability of trade-offs and
    optimal access of web sources. In *FOCS'2000*, pages 86–92, 2000.
16. G.T. Ross and R.M. Soland. A branch and bound algorithm for the generalized
    assignment problem. *Management Science*, 24:345–357, 1977.
17. D.B. Shmoys and É. Tardos. An approximation algorithm for the generalized
    assignment problem. *Mathematical Programming A*, 62:461–474, 1993.
18. M.A. Trick. Scheduling multiple variable-speed machines. In *1st Conference of
    Integer Programming and Combinatorial Optimization*, pages 485–494, 1990.

Grouping Techniques for Scheduling Problems: Simpler and Faster*

Aleksei V. Fishkin[1], Klaus Jansen[1], and Monaldo Mastrolilli[2]

[1] Institut für Informatik und Praktische Mathematik, Universität zu Kiel, Germany,
{avf,kj}@informatik.uni-kiel.de
[2] IDSIA-Istituto Dalle Molle di Studi sull'Intelligenza Artificiale, Manno,
Switzerland, monaldo@idsia.ch

Abstract. In this paper we describe a general grouping technique to devise faster and simpler approximation schemes for several scheduling problems. We illustrate the technique on two different scheduling problems: scheduling on unrelated parallel machines with costs and the job shop scheduling problem. The time complexity of the resulting approximation schemes is always linear in the number n of jobs, and the multiplicative constant hidden in the $O(n)$ running time is reasonably small and independent of the error ϵ.

1 Introduction

The problem of scheduling jobs on machines such that maximum completion time (makespan) is minimized has been extensively studied for various problem formulations. Recently, several polynomial time approximation schemes have been found for various shop and multiprocessor scheduling problems [1,2,7,8,10, 11]: these include scheduling on unrelated machines, multiprocessor tasks (e.g. dedicated, parallel, malleable tasks), and classical open, flow and job shops. These results were extended in [9] by providing a polynomial time approximation scheme for a general multiprocessor job shop scheduling problem (containing as special cases the above problems).

The goal of this paper is to show that a general and powerful technique can be applied to speed up and enormously simplify all the previous algorithms for the aforementioned scheduling problems. The basic idea is to reduce the number of jobs to a constant and to apply enumeration or dynamic programming afterwards. The reduced set of jobs is computed by first structuring the input and then grouping jobs that have the same structure. In order to show that our algorithm is an approximation scheme, we prove that two linear programming formulations (one for the original instance and one for the transformed instance) have a gap of at most ϵOPT, where OPT is the minimum objective value.

* Supported by Swiss National Science Foundation project 21-55778.98, "Resource Allocation and Scheduling in Flexible Manufacturing Systems", by the "Metaheuristics Network", grant HPRN-CT-1999-00106, and by the "Thematic Network APPOL", Approximation and on-line algorithms, grant IST-1999-14084.

F. Meyer auf der Heide (Ed.): ESA 2001, LNCS 2161, pp. 206–217, 2001.

Due to space limitations, we focus our attention on two scheduling problems: scheduling on unrelated machines with costs and the classical job shop scheduling problem. Although these two problems are different in nature (the first is an assignment problem while the other is an ordering problem), the reader will recognize that the underlying ideas are very similar. Furthermore, the described techniques can also be applied to many other scheduling problems including the general multiprocessor job shop scheduling problem studied in [9]. The overall running time is always $O(n) + C$, where the constant hidden in $O(n)$ is reasonably small and independent of the accuracy ϵ, whereas the additive constant C depends on the number of machines, the accuracy ε (and the number of operations for job shops). The full details of these extensions will be given in the long version of this paper. Note that the time complexity of these PTASs is best possible with respect to the number of jobs. Moreover, the existence of PTASs for these strongly NP-hard problems whose running time is also polynomial in the reciprocal value of the precision and the number of machines would imply P=NP [3].

Scheduling unrelated parallel machines with costs. We begin with the problem of scheduling a set $\mathcal{J} = \{J_1, ..., J_n\}$ of n independent jobs on a set $M = \{1, ..., m\}$ of m unrelated parallel machines. Each machine can process at most one job at a time, and each job has to be processed without interruption by exactly one machine. Processing job J_j on machine i requires $p_{ij} \geq 0$ time units and incurs a cost $c_{ij} \geq 0$, $i = 1, \ldots, m$, $j = 1, ..., n$. The *makespan* is the maximum job completion time among all jobs. We consider the problem of minimizing the objective function that is a weighted sum of the makespan and total cost.

When $c_{ij} = 0$ the problem turns into the classical makespan minimization form. Lenstra, Shmoys and Tardos [13] gave a polynomial-time 2-approximation algorithm for this problem; and this is the currently known best approximation ratio achieved in polynomial time. They also proved that for any positive $\epsilon < 1/2$, no polynomial-time $(1 + \epsilon)$-approximation algorithm exists, unless P=NP. Furthermore, Shmoys and Tardos [15] gave a polynomial-time 2-approximation algorithm for the general variant with cost. Since the problem is NP-hard even for $m = 2$, it is natural to ask how well the optimum can be approximated when there is only a constant number of machines. In contrast to the previously mentioned inapproximability result for the general case, there exists a fully polynomial-time approximation scheme for the problem when m is fixed. Horowitz and Sahni [6] proved that for any $\epsilon > 0$, an ϵ-approximate solution can be computed in $O(nm(nm/\epsilon)^{m-1})$ time, which is polynomial in both n and $1/\epsilon$ if m is constant. Lenstra, Shmoys and Tardos [13] also gave an approximation scheme for the problem with running time bounded by the product of $(n + 1)^{m/\epsilon}$ and a polynomial of the input size. Even though for fixed m their algorithm is not fully polynomial, it has a much smaller space complexity than the one in [6]. Recently, Jansen and Porkolab [7] presented a fully polynomial-time approximation scheme for the problem whose running time is $n(m/\epsilon)^{O(m)}$. Their algorithm has to solve at least $(m^3/\epsilon^2)^m$ many linear programs. In order to obtain a linear running time for the case when m is fixed, they use the price-directive decomposition method proposed by Grigoriadis and Khachiyan [4] for computing approximate solutions of block structured convex programs. The final ingredient is an intricate rounding technique based on the solution of a linear program and a partition of the job set.

In contrast to the previous approach [7], our algorithm (that works also for the general variant with cost) is extremely simple: first we preprocess the data to obtain a new instance with $\min\{n, (\log m/\varepsilon)^{O(m)}\}$ grouped jobs. The preprocessing step requires linear time. Then using dynamic programming we compute an approximate solution for the grouped jobs in $(\log m/\varepsilon)^{O(m^2)}$ time. Both steps together imply a fully polynomial-time approximation scheme that runs in $O(n) + C$ time where $C = (\log m/\varepsilon)^{O(m^2)}$. We remark that the multiplicative constant hidden in the $O(n)$ running time of our algorithm is reasonably small and does not depend on the accuracy ε.

Makespan minimization in job shops. In the job shop scheduling problem, there is a set $\mathcal{J} = \{J_1, \ldots, J_n\}$ of n jobs that must be processed on a given set $M = \{1, \ldots, m\}$ of m machines. Each job J_j consists of a sequence of μ operations $O_{1j}, O_{2j}, \ldots, O_{\mu j}$ that need to be processed in this order. Operation O_{ij} must be processed without interruption on machine $m_{ij} \in M$, during p_{ij} time units. Each machine can process at most one operation at a time, and each job may be processed by at most one machine at any time. For any given schedule, let C_{ij} be the completion time of operation O_{ij}. The objective is again to find a schedule that minimizes the *makespan* (the maximum completion time $C_{\max} = \max_{ij} C_{ij}$).

The job shop problem is strongly NP-hard even if each job has at most three operations and there are only two machines [12]. Williamson et al. [17] proved that when the number of machines, jobs, and operations per job are part of the input there does not exist a polynomial time approximation algorithm with worst case bound smaller than $\frac{5}{4}$ unless $P = NP$. When m and μ are part of the input the best known result [5] is an approximation algorithm with worst case bound $O((\log(m\mu)\log(min(m\mu, p_{max}))/\log\log(m\mu))^2)$, where p_{max} is the largest processing time among all operations. For those instances where m and μ are fixed (the restricted case we are focusing on in this paper), Shmoys et al. [16] gave approximation algorithms that compute $(2 + \varepsilon)$-approximate solutions in polynomial time for any fixed $\varepsilon > 0$. This result has recently been improved by Jansen et al. [10] who have shown that $(1 + \varepsilon)$-approximate solutions of the problem can be computed in polynomial time. The main idea is to divide the set of jobs $\mathcal{J}$ into two groups $\mathcal{L}$ and $\mathcal{S}$ formed by jobs with "large" and "small" total processing time, respectively. The total number of large jobs is bounded by a constant exponentially in m, μ and ε. Then they construct all possible schedules for the large jobs. In any schedule for the large jobs, the starting and completion times of the jobs define a set of time intervals, into which the set of small jobs have to be scheduled. Then for every possible job ordering of large jobs, a linear program is used to assign small jobs to time intervals. The rounded solution of the linear program in combination with an algorithm by Sevastianov [14], gives an approximate schedule in time polynomial in n. In order to speed up the whole algorithm, they suggest [11] a number of improvements: they use the logarithmic potential price decomposition method of Grigoriadis and Khachiyan [4] to compute an approximate solution of the linear program in linear time, and a novel rounding procedure to bring down to a constant the number of fractional assignments in any solution of the linear program. The overall running time is $O(n)$, where the multiplicative constant hidden in the $O(n)$ running time is exponentially in m, μ and ε.

In contrast to the approximation schemes introduced in [10,11], our algorithm is again extremely simple and even faster. We show that we can preprocess in linear time the input to obtain a new instance with a constant number of grouped jobs. This immediately gives a linear time approximation scheme with running time $O(n) + C$, where C is a constant that depends on m, ε and μ. Again, we remark that the multiplicative constant hidden in the $O(n)$ running time of our algorithm is reasonably small and does not depend on the accuracy ε.

Throughout this paper, we use several transformations which may potentially increase the objective function value by a factor of $1 + O(\varepsilon)$. Therefore we can perform a constant number of them while still staying within $1 + O(\varepsilon)$ of the original optimum. When we describe this type of transformation, we shall say it produces $1 + O(\varepsilon)$ *loss*.

2   Scheduling Unrelated Parallel Machines with Costs

Let $0 < \varepsilon < 1$ be an arbitrary small rational number, and let $m \geq 2$ be an integral value. Throughout this section, the values ε and m are considered to be constants and not part of the input.

The problem can be stated by using the following integer linear program ILP that represents the problem of assigning jobs to machines ($x_{ij} = 1$ means that job J_j has been assigned to machine i, and μ is any given positive weight):

$$
\begin{aligned}
\min\ & T + \mu \sum_{i=1}^{m} \sum_{j=1}^{n} x_{ij} c_{ij} \\
\text{s.t.}\ & \sum_{j=1}^{n} x_{ij} p_{ij} \leq T, && i = 1, \ldots, m; \\
& \sum_{i=1}^{m} x_{ij} = 1, && j = 1, \ldots, n; \\
& x_{ij} \in \{0, 1\}, \quad i = 1, \ldots, m,\ j = 1, \ldots, n.
\end{aligned}
$$

The first set of constraints relates the makespan T to the processing time on each of the machines, while the second set ensures that every job gets assigned.

We begin by computing some lower and upper bounds for the minimum objective value OPT. By multiplying each cost value by μ we may assume, w.l.o.g., that $\mu = 1$. Let $d_j = \min_{i=1,\ldots,m}(p_{ij} + c_{ij})$, and $D = \sum_{j=1}^{n} d_j$. Consider an optimum assignment (x_{ij}^*) of jobs to machines with makespan T^* and total cost C^*. Then, $D = \sum_{j=1}^{n} d_j \leq \sum_{i=1}^{m} \sum_{j=1}^{n} x_{ij}^* c_{ij} + \sum_{i=1}^{m} \sum_{j=1}^{n} x_{ij}^* p_{ij} \leq C^* + m \cdot T^* \leq m \cdot OPT$, where $OPT = T^* + C^*$. On the other hand, we can generate a feasible schedule according to the d_j values. Indeed, let $m_j \in \{1, \ldots, m\}$ denote any machine such that $d_j = p_{m_j,j} + c_{m_j,j}$. Assign every job J_j to machine m_j. The objective value of this schedule can be bounded by $\sum_{j \in J} c_{m_j,j} + \sum_{j \in J} p_{m_j,j} = D$.

Therefore, $OPT \in [D/m, D]$, and by dividing all processing times and cost values by D/m, we get directly the following bounds for the optimum value:

$$1 \leq OPT \leq m.$$

Fast, slow, cheap and expensive machines. For each job J_j, we define four sets of machines: the set $\mathcal{F}_j = \{i : p_{ij} \leq \frac{\varepsilon}{m}d_j\}$ of *fast*, the set $\mathcal{C}_j = \{i : c_{ij} \leq \frac{\varepsilon}{m}d_j\}$ of *cheap*, the set $\mathcal{S}_j = \{i : p_{ij} \geq \frac{m}{\varepsilon}d_j\}$ of *slow* and the set $\mathcal{E}_j = \{i : c_{ij} \geq d_j/\varepsilon\}$ of *expensive* machines, respectively. Then, we can prove the following lemma.

Lemma 1. *With $1 + 3\varepsilon$ loss, we can assume that for each job J_j the following holds:*

- $p_{ij} = 0$, *for $i \in \mathcal{F}_j$, and $c_{ij} = 0$, for $i \in \mathcal{C}_j$;*
- $p_{ij} = +\infty$, *for $i \in \mathcal{S}_j$, and $c_{ij} = +\infty$, for $i \in \mathcal{E}_j$;*
- *for any other machine i, $p_{ij} = \frac{\varepsilon}{m}d_j(1 + \varepsilon)^{\pi_i}$ and $c_{ij} = \frac{\varepsilon}{m}d_j(1 + \varepsilon)^{\gamma_i}$, where $\pi_i, \gamma_i \in \mathbb{N}$.*

Proof. Set $p_{ij} = 0$, for every $i \in \mathcal{F}_j$ and $j = 1, ..., n$; consider an optimal assignment $A : \mathcal{J} \to M$ of jobs to machines. Clearly, the optimal value corresponding to A cannot be larger than OPT. Let F denote the set of jobs which are processed on fast machines according to A. Now, if we replace the processing times of the transformed instance by the original processing times, we may potentially increase the makespan of A by at most $\sum_{J_j \in F} \frac{\varepsilon}{m}d_j \leq \sum_{j=1}^{n} \frac{\varepsilon}{m}d_j = \varepsilon \frac{D}{m} \leq \varepsilon$. The other statement for the cost values follows in a similar way.

Now we show that there exists an approximate schedule where jobs are scheduled neither on slow nor on expensive machines. This allows us to set $p_{ij} = +\infty$, for $i \in \mathcal{S}_j$, and $c_{ij} = +\infty$, for $i \in \mathcal{E}_j$. Consider an optimal assignment $A : \mathcal{J} \to M$ of jobs to machines with T^* and C^* denoting the resulting makespan and cost, respectively. Let S and E represent, respectively, the set of jobs which are processed on slow and on expensive machines according to A. Then, assign every job $J_j \in S \cup E$ to machine m_j (recall that $m_j \in \{1, ..., m\}$ denote any machine such that $d_j = p_{m_j,j} + c_{m_j,j}$). Moving jobs $J_j \in S \cup E$ onto machines m_j may potentially increase the objective value by at most $\sum_{J_j \in S \cup E} d_j \leq \frac{\varepsilon}{m} \sum_{J_j \in S} p_{A(j),j} + \varepsilon \sum_{J_j \in E} c_{A(i),j} \leq \varepsilon T^* + \varepsilon C^*$, since $p_{A(j),j} \geq \frac{m}{\varepsilon}d_j$ for $J_j \in S$, and $c_{A(j),j} \geq d_j/\varepsilon$ for $J_j \in E$.

By the above arguments, all the positive costs c_{ij} and positive processing times p_{ij} are greater than $\frac{\varepsilon}{m}d_j$. Round every positive processing time p_{ij} and every positive cost c_{ij} down to the nearest lower value of $\frac{\varepsilon}{m}d_j(1+\varepsilon)^h$, for $h \in \mathbb{N}$, $i = 1, ..., m$ and $j = 1, ..., n$. Consider the optimal value of the rounded instance. Clearly, this value cannot be greater than OPT. It follows that by replacing the rounded values with the original ones we may increase each value by a factor $1 + \varepsilon$, and consequently, the solution value potentially increases by the same factor $1 + \varepsilon$. $\square$

We define the *execution profile* of a job J_j to be an m-tuple $< \pi_1, ..., \pi_m >$ such that $p_{ij} = \frac{\varepsilon}{m}d_j(1 + \varepsilon)^{\pi_i}$. We adopt the convention that $\pi_i = +\infty$ if $p_{ij} =$

$+\infty$, and $\pi_i = -\infty$ if $p_{ij} = 0$. Likewise, we define the *cost profile* of a job J_j to be an m-tuple $< \gamma_1, ..., \gamma_m >$ such that $c_{ij} = \frac{\varepsilon}{m} d_j (1 + \varepsilon)^{\gamma_i}$. Again, we adopt the convention that $\gamma_i = +\infty$ if $c_{ij} = +\infty$, and $\gamma_i = -\infty$ if $c_{ij} = 0$. Let us say that two jobs have the same profile iff they have the same execution and cost profile.

Lemma 2. *The number of distinct profiles is bounded by* ℓ $:=$ $\lceil 2 + 2 \log_{1+\varepsilon} \frac{m}{\varepsilon} \rceil^{2m}$.

Grouping jobs. Let $\delta := \frac{\varepsilon}{m}$, and partition the set of jobs in two subsets $L = \{J_j : d_j > \delta\}$ and $S = \{J_j : d_j \leq \delta\}$. Let us say that L is the set of *large* jobs, while S the set of *small* jobs. We further partition the set of small jobs into subsets S_i of jobs having the same profile, for $i = 1, ..., \ell$. Let J_a and J_b be two jobs from S_i such that $d_a, d_b \leq \delta/2$. We "group" together these two jobs to form a composed job J_c in which the processing time (and cost) on machine i is equal to the sum of the processing times (and costs) of J_a and J_b on machine i, and let $d_c = d_a + d_b$. We repeat this process, by using the modified set of jobs, until at most one job J_j from S_i has $d_j \leq \delta/2$. At the end, all jobs J_j in group S_i have $d_j \leq \delta$. The same procedure is performed for all other subsets S_i. At the end of this process, there are at most ℓ jobs, one for each subset S_i, having $d_j \leq \delta/2$. All the other jobs, have processing times larger than $\delta/2$. Therefore, the number of jobs in the transformed instance is bounded by $\frac{2D}{\delta} + \ell \leq \frac{2m}{\delta} + \ell = (\log m/\varepsilon)^{O(m)}$. Note that the procedure runs in linear time, and a feasible schedule for the original set of jobs can be easily obtained from a feasible schedule for the grouped jobs. We motivate the described technique with the following

Lemma 3. *With $1 + \varepsilon$ loss, the number of jobs can be reduced in linear time to be at most* $\min\{n, (\log m/\varepsilon)^{O(m)}\}$.

Proof. Consider the transformed instance I' according to Lemma 1 (small jobs are not yet grouped). Assume, w.l.o.g., that there exists a solution SOL' for I' of value OPT'. It is sufficient to show that, by using the small jobs grouped as described previously, there exists a schedule of value $(1 + \varepsilon)OPT'$.

Accordingly to SOL' we may assume that each machine executes the large jobs at the beginning of the schedule. Let c denote the total cost of large jobs when processed according to SOL', and let t_i denote the time at which machine i finishes to process large jobs, for $i = 1, ..., m$. Now, consider the following linear program LP_1:

$$\min T + c + \sum_{i=1}^{m} \sum_{J_j \in S} x_{ij} \frac{\varepsilon}{m} d_j (1 + \varepsilon)^{\gamma_i}$$
$$\text{s.t. } t_i + \sum_{J_j \in S} x_{ij} \frac{\varepsilon}{m} d_j (1 + \varepsilon)^{\pi_i} \leq T, \qquad i = 1, ..., m;$$
$$\sum_{i=1}^{m} x_{ij} = 1, \qquad J_j \in S;$$
$$x_{ij} \geq 0, \qquad i = 1, ..., m, J_j \in S.$$

Note that LP_1 formulates the integer relaxation of the original problem ILP for the subset of small jobs: we are assuming that machine i can start processing small jobs only at time t_i, and when the processing times and costs are structured as in Lemma 1.

For each S_ϕ, $\phi = 1, ..., \ell$, consider a set of decision variables $y_{\phi i} \in [0, 1]$ for $i = 1, ..., m$. The meaning of these variables is that $y_{\phi i}$ represents the fraction of jobs from S_ϕ processed on machine i. Consider the following linear program LP_2:

$$\min T + c + \sum_{i=1}^m \sum_{\phi=1}^\ell y_{\phi i} \sum_{J_j \in S_\phi} \tfrac{\varepsilon}{m} d_j (1 + \varepsilon)^{\gamma_i}$$
$$\text{s.t. } t_i + \sum_{\phi=1}^\ell y_{\phi i} \sum_{J_j \in S_\phi} \tfrac{\varepsilon}{m} d_j (1 + \varepsilon)^{\pi_i} \le T, \qquad i = 1, \dots, m;$$
$$\sum_{i=1}^m y_{\phi i} = 1, \qquad\qquad\qquad \phi = 1, \dots, \ell;$$
$$y_{\phi i} \ge 0, \qquad\qquad\qquad i = 1, \dots, m, \phi = 1, \dots, \ell.$$

By setting $y_{\phi i} = \dfrac{\sum_{J_j \in S_\phi} x_{ij} d_j}{\sum_{J_j \in S_\phi} d_j}$, it is easy to check that any feasible set of values (x_{ij}) for LP_1 gives a feasible set of values $(y_{\phi i})$ for LP_2; furthermore, by these settings, the objective function value of LP_2 is equal to that of LP_1. But LP_1 is a relaxation of the original problem. Therefore, if we were able, by using the grouped small jobs, to get a feasible schedule of length at most $1 + \varepsilon$ times the optimal value of LP_2, we would be done. In the remainder we show how to generate such a schedule. This solution is obtained by using the solution of LP_2 and the small grouped jobs.

Let $y_{\phi i}^*$ denote the values of variables $y_{\phi i}$ according to the optimal solution of LP_2. For every positive value $y_{\phi i}^*$, schedule a subset of grouped jobs from S_ϕ on machine i until either (a) the jobs from S_ϕ are exhausted or (b) the total fraction of jobs assigned to i is equal to $y_{\phi i}^*$ (if necessary fractionalize one job to use up $y_{\phi i}^*$ exactly). We repeat this for the not yet assigned grouped small jobs and for every positive value $y_{\phi i}^*$. Note that if $y_{\phi i}^*$ is not fractional, then the jobs from S_ϕ are not preempted by the previous algorithm. In general, the number of preempted jobs from S_ϕ is at most $f_\phi - 1$, where $f_\phi = |\left\{y_{\phi i}^* : y_{\phi i}^* > 0, i = 1, ..., m\right\}|$. Now remove all the preempted jobs J_j and schedule them at the end on machines m_j. This increases the makespan and the cost by at most $\Delta = \delta \cdot \sum_{\phi=1}^\ell (f_\phi - 1)$, since every grouped small job has cost plus processing time bounded by δ when processed on machine m_j. A basic feasible solution of LP_2 has the property that the number of positive variables is at most the number of rows in the constraint matrix, $m + \ell$, therefore $\sum_{\phi=1}^\ell (f_\phi - 1) \le m$. In order to bound the total increase Δ by ε we have to choose δ such that $\delta \le \tfrac{\varepsilon}{m}$, and the claim follows. $\square$

By using a dynamic programming similar to that used in [7] it is possible to compute a $(1 + \varepsilon)$-approximate schedule for the transformed instance in $(\log m/\varepsilon)^{O(m^2)}$ time. We omit the details in this extended abstract, and observe that by Lemma 3, and by adopting this dynamic programming approach, we have the following

Theorem 1. *For the problem of minimizing the weighted sum of the cost and the makespan in scheduling n jobs on m unrelated machines (m fixed), there exists a fully polynomial time approximation scheme that runs in $O(n) + (\log m/\varepsilon)^{O(m^2)}$ time.*

3   Makespan Minimization in Job Shops

Let $\varepsilon > 0$ be an arbitrary small rational number, and let $m \geq 2$ and $\mu \geq 1$ be integral values. Throughout this section, the values ε, m and μ are considered to be constants and not part of the input. For simplicity, we assume that $1/\varepsilon$ is integral. We begin by providing some lower and upper bounds of the minimum makespan. Then we show how to reduce the number of jobs to a constant: the reduced set of jobs is computed in linear time by structuring the input and grouping jobs that share the same structure. This directly gives a linear time approximation scheme for the makespan minimization of the job shop scheduling problem.

For a given instance of the job shop scheduling problem, the value of the optimum makespan will be denoted as OPT. Let $d_j = \sum_{i=1}^{\mu} p_{ij}$ be the total processing time of job $J_j \in \mathcal{J}$, and let $D = \sum_{J_j \in \mathcal{J}} d_j$. Clearly, $D/m \leq OPT$ and a schedule of length at most D can be obtained by scheduling one job after the other. Then, we get directly the following bounds: $\frac{D}{m} \leq OPT \leq D$. By dividing all execution times p_{ij} by D/m, we assume, without loss of generality, that $D/m = 1$ and

$$1 \leq OPT \leq m.$$

Negligible operations. Let us use $\mathcal{N}_j = \left\{ O_{ij} : i = 1, ..., \mu \text{ and } p_{ij} \leq \frac{\varepsilon}{\mu m} d_j \right\}$ to denote the set of *negligible* operations for job J_j, $j = 1, ..., n$. Then we have the following

Lemma 4. *With $1 + 2\varepsilon$ loss, we assume that for each job J_j the following holds:*

- $p_{ij} = 0$, *for* $O_{ij} \in \mathcal{N}_j$;
- $p_{ij} = \frac{\varepsilon}{\mu m} d_j (1 + \varepsilon)^{\pi_i}$, *where* $\pi_i \in \mathbb{N}$ *and for* $O_{ij} \notin \mathcal{N}_j$.

Proof. Set $p_{ij} = 0$, for every $O_{ij} \in \mathcal{N}_j$ and $j = 1, ..., n$. Clearly, the corresponding optimal makespan cannot be larger than OPT. Furthermore, if we replace the zero processing times of the negligible operations with the original processing times, we may potentially increase the makespan by at most $\sum_{O_{ij} \in \mathcal{N}} \frac{\varepsilon}{\mu m} d_j = \sum_{j=1}^{n} \sum_{O_{ij} \in \mathcal{N}_j} \frac{\varepsilon}{\mu m} d_j \leq \varepsilon \frac{D}{m} = \varepsilon$. Any non negligible operation O_{ij} has processing times p_{ij} greater than $\frac{\varepsilon}{\mu m} d_j$. Round each p_{ij} down to the nearest lower value of $\frac{\varepsilon}{\mu m} d_j (1 + \varepsilon)^h$, where $h \in \mathbb{N}$ and $O_{ij} \notin \mathcal{N}_j$. By replacing the rounded values with the original ones we may potentially increases the makespan by at most a factor of $1 + \varepsilon$. $\square$

Now we define the *profile* of a job J_j to be a $(\mu \cdot m)$-tuple $\langle \pi_{1,1}, \dots, \pi_{1,m}, \pi_{2,1}, \dots, \pi_{\mu,m} \rangle$ such that $p_{ij} = \frac{\varepsilon}{\mu m} d_j (1 + \varepsilon)^{\pi_{i,h}}$ if $m_{ij} = h$. We adopt the convention that $\pi_{i,h} = -\infty$ if $p_{ij} = 0$ or if $m_{ij} \neq h$.

Lemma 5. *The number of distinct profiles is bounded by $\ell := \lfloor 2 + \log_{1+\varepsilon} \frac{\mu m}{\varepsilon} \rfloor^{\mu} \cdot m^{\mu}$.*

Grouping jobs. Let $\delta := m(\frac{\varepsilon}{4\mu^4 m^2})^{m/\varepsilon}$, and partition the set of jobs in two subsets of large jobs $L = \{J_j : d_j > \delta\}$ and small jobs $S = \{J_j : d_j \le \delta\}$. Again, we further partition the set of small jobs into subsets S_i of jobs having the same profile, for $i = 1, ..., \ell$. For each subset S_i, we group jobs from S_i as described for the unrelated parallel machines scheduling problem, but with the following difference: here grouping two jobs, J_a and J_b, means to form a composed job for which the processing time of the i-th operation is equal to the sum of the processing times of the i-th operations of J_a and J_b.

Lemma 6. *With $1 + 2\varepsilon$ loss, the number of jobs can be reduced in linear time to be at most $\min\{n, \frac{2m}{\delta} + \ell\}$.*

Proof. Consider the transformed instance I' according to Lemma 4 (small jobs are not yet grouped). Assume, w.l.o.g., that there exists a solution SOL' for I' of value OPT'. It is sufficient to show that, by using the small jobs grouped as described previously, there exists a schedule of value $(1 + 2\varepsilon)OPT'$.

Partition the set of large jobs into three subsets L_1, L_2, L_3 as follows. Set $\rho = \frac{\varepsilon}{4\mu^4 m^2}$ and let α denote an integer defined later and such that $\alpha = 0, 1, ..., m/\varepsilon - 1$. Set L is partitioned in the following three subsets:

$$L_1 = \{J_j : m\rho^\alpha < d_j\},$$
$$L_2 = \{J_j : m\rho^{\alpha+1} < d_j \le m\rho^\alpha\},$$
$$L_3 = \{J_j : m\rho^{m/\varepsilon} < d_j \le m\rho^{\alpha+1}\}.$$

Note that each set size is bounded by a constant, and sets L_1, L_2, L_3 and S establish a partition of all jobs. The number α can be chosen so that

$$\sum_{J_j \in L_2} d_j \le \varepsilon. \tag{1}$$

This is done as follows. Starting from $\alpha := 0$, check each time whether the set L_2 corresponding to the current value of α satisfies inequality (1); if it is not the case, set $\alpha := \alpha + 1$ and repeat. Note that for different α-values, the corresponding L_2 sets are disjoint. The total length of all jobs is m, and so there exists a value $\alpha' \le m/\varepsilon - 1$ for which the corresponding set L_2 satisfies inequality (1). We set $\alpha := \alpha'$.

In the following we consider an artificial situation that we use as a tool. Focus on solution SOL', remove from SOL' all the jobs except those from L_1. Clearly, this creates gaps in the resulting schedule σ. The starting and finishing times of the operations from L_1 divide the time into intervals: $t_1, t_2, ..., t_g$, where t_1 is the interval whose right boundary is the starting time of the first operation according to σ, and t_g is the interval with left boundary defined by the finishing time of the jobs from L_1. Furthermore, let l_v denote the length of interval t_v, $1 \le v \le g$. It follows that $OPT' = \sum_{s=1}^{g} l_s$. Note that the number g of intervals is bounded by $2\mu|L_1| + 1 \le 2\mu/\rho^\alpha$. Let G be the set of pairs (gaps of σ) (v, i) such that no job from L_1 is processed in interval v and by machine i, for $v = 1, ..., g$ and $i = 1, ..., m$.

Since the total length of jobs from L_2 is at most ε, we can get rid of these jobs by assuming that they are processed at the end of the schedule one after the other; this increases the schedule by at most ε.

Consider the problem of placing jobs from $L_3 \cup S$ into the gaps of σ and such that the length of the resulting schedule is OPT'. In the following we first describe a linear program LP_1 which is a relaxation of this problem. Then we propose another linear program LP_2 which is a relaxation of LP_1. By using the solution of LP_2 we show that the jobs from L_3 and the grouped small jobs can be scheduled into the gaps by increasing the length of the gaps a little; we show that the total increase of the length can be bounded by ε, and the claim follows.

We formulate LP_1 as follows. Consider the set S of (not grouped) small jobs. For each job $J_j \in S \cup L_3$ we use a set of decision variables $x_{j,\tau} \in [0,1]$ for tuples $\tau = (\tau_1, \ldots, \tau_\mu) \in A$, where $A = \{(\tau_1, \ldots, \tau_\mu) | 1 \leq \tau_1 \leq \tau_2 \leq \ldots \leq \tau_\mu \leq g\}$. The meaning of these variables is that $x_{j,\tau}$ represents the fraction of job J_j whose operations are processed according to $\tau = (\tau_1, \ldots, \tau_\mu)$, i.e., the i-th operation is scheduled in interval τ_k for each $1 \leq k \leq \mu$. Note that by the way in which we numbered the operations, any tuple $(\tau_1, \ldots, \tau_\mu) \in A$ represents a valid ordering for the operations. Let the load $L_{v,h}$ on machine h in interval v be defined as the total processing time of operations from small jobs that are executed by machine h during interval s, i.e., $L_{v,h} = \sum_{J_j \in S \cup L_3} \sum_{\tau \in A} \sum_{k=1,\ldots,\mu | \tau_k = v, m_{kj} = h} x_{j,\tau} p_{kj}$. Let us write the load $L_{v,h}$ as the sum of $L^3_{v,h} + L^S_{v,h}$, where $L^3_{v,h}$ is the load of the jobs from L_3 while $L^S_{v,h}$ is the load of jobs from S. By Lemma 4, we have that $L^S_{v,h} = \sum_{\tau \in A} \sum_{k=1,\ldots,\mu | \tau_k = v} \sum_{J_j \in S} x_{j,\tau} \frac{\varepsilon}{\mu m} d_j (1+\varepsilon)^{\pi_{k,h}}$. Then LP_1 is the following

$$
\begin{aligned}
&(1)\ L^3_{v,h} + L^S_{v,h} \leq l_v, &&(v,h) \in G; \\
&(2)\ \sum_{\tau \in A} x_{j\tau} = 1, &&J_j \in S \cup L_3; \\
&(3)\ x_{j\tau} \geq 0, &&\tau \in A,\ J_j \in S \cup L_3.
\end{aligned}
$$

Constraint (1) ensures that the total length of operations assigned to gap (v,h) does not exceed the length of the interval, while constraint (2) ensures that job J_j is completely scheduled.

Let S_ϕ denote the set of small jobs having the same profile, where $\phi = 1, ..., \ell$. For each S_ϕ ($\phi = 1, ..., \ell$) we use a set of decision variables $y_{\phi\tau} \in [0,1]$ for tuples $\tau = (\tau_1, \ldots, \tau_\mu) \in A$. The meaning of these variables is that $y_{\phi\tau}$ represents the fraction of jobs from S_ϕ whose operations are processed according to $\tau = (\tau_1, \ldots, \tau_\mu)$, i.e., the i-th operation is scheduled in interval τ_k for each $1 \leq k \leq \mu$. Let $L^*_{v,h} = \sum_{\tau \in A} \sum_{k=1,\ldots,\mu | \tau_k = v} \sum_{\phi=1}^{\ell} y_{\phi\tau} \sum_{J_j \in S_\phi} x_{j,\tau} \frac{\varepsilon}{\mu m} d_j (1+\varepsilon)^{\pi_{k,h}}$. Then LP_2 is the following

$$
\begin{aligned}
&(1)\ L^3_{v,h} + L^*_{v,h} \leq t_s, &&(v,h) \in G; \\
&(2)\ \sum_{\tau \in A} x_{j\tau} = 1, &&J_j \in L_3; \\
&(3)\ \sum_{\tau \in A} y_{\phi\tau} = 1, &&\phi = 1, ..., \ell; \\
&(4)\ x_{j\tau} \geq 0, &&\tau \in A,\ J_j \in S; \\
&(5)\ y_{\phi\tau} \geq 0, &&\tau \in A,\ \phi = 1, ..., \ell.
\end{aligned}
$$

By setting $y_{\phi\tau} = \frac{\sum_{J_j \in S_\phi} x_{j,\tau} d_j}{\sum_{J_j \in S_\phi} d_j}$ it is easy to check that any feasible set of values $(x_{j,\tau})$ for LP_1 gives a feasible set of values $(y_{\phi\tau})$ for LP_2. Since by construction a feasible solution for LP_1 exists, a feasible solution for LP_2 exists as well. We show now that by using the optimal solution of LP_2 we can derive a schedule without increasing too much the makespan.

Let $y^*_{\phi\tau}$ $(x^*_{j\tau})$ denote the values of variables $y_{\phi\tau}$ $(x_{j\tau})$ according to the optimal solution of LP_2. For every positive value $y^*_{\phi\tau}$, schedule a subset $H_{\phi\tau}$ of grouped jobs from S_ϕ on machine i until either (a) the jobs from S_ϕ are exhausted or (b) the total fraction (i.e. $\frac{\sum_{J_j \in H_{\phi\tau}} d_j}{\sum_{J_j \in S_\phi} d_j}$) of jobs assigned to i is equal to $y^*_{\phi\tau}$ (if necessary fractionalize one job to use up $y^*_{\phi\tau}$ exactly). We repeat this for the not yet assigned grouped small jobs and for every positive value $y^*_{\phi\tau}$. Note that if $y^*_{\phi\tau}$ is not fractional, then the jobs from S_ϕ are not preempted by the previous algorithm. In general, the number of preempted jobs from S_ϕ is at most $f_\phi - 1$, where $f_\phi = |\left\{ y^*_{\phi\tau} : y^*_{\phi\tau} > 0, \tau \in A \right\}|$. According to the optimal solution of LP_2, let us say that job $J_j \in L_3$ is preempted if the corresponding $(x^*_{j\tau})$-values are fractional. Let $f_j = |\left\{ x^*_{j\tau} : x^*_{j\tau} > 0, \tau \in A \right\}|$, for $J_j \in L_3$, then we have that the number of preemptions of job $J_j \in L_3$ is $f_j - 1$. Therefore, the total number of preemptions is $f = \sum_{\phi=1}^\ell (f_\phi - 1) + \sum_{J_j \in L_3} (f_j - 1)$, and this gives also an upper bound on the number of preempted jobs. Now remove all the preempted jobs J_j from $S \cup L_3$, and schedule these set of jobs at the end of the schedule, one after the other. Since every job from S has a smaller total processing time than any job from L_3, we can bound the total increase of the schedule length by $\Delta = f \cdot m \rho^{\alpha+1}$. A basic feasible solution of LP_2 has the property that the number of positive variables is at most the number of rows in the constraint matrix, $mg + \ell + |L_3|$, therefore $f \le mg \le 2m\mu/\rho^\alpha$, and $\Delta = f \cdot m\rho^{\alpha+1} \le 2m^2\mu\rho$. By the previous algorithm, we have assigned all the jobs from $S \cup L_3$ to gaps with a total increase of the schedule length of $2m^2\mu\rho$. Now we consider the problem of schedule jobs from $S \cup L_3$ within each interval. This is simply a smaller instance of the job shop problem, and by using Sevastianov's algorithm [14] it is possible to find a feasible schedule for each interval t_v of length at most $l_v + \mu^3 m \cdot m\rho^{\alpha+1}$; this increases the total length by at most $\mu^3 mg \cdot m\rho^{\alpha+1} \le 2m^2\mu^4\rho$. Therefore, the total increase is $2m^2\mu\rho + 2m^2\mu^4\rho \le 4m^2\mu^4\rho$, and by setting $\rho = \frac{\varepsilon}{4m^2\mu^4}$ the claim follows. $\square$

By the previous lemma a $(1 + \varepsilon)$-approximate schedule can be obtained by finding the optimum schedule for the reduced set of jobs. It follows that:

Theorem 2. *There exists a linear time PTAS for the job shop scheduling problem whose multiplicative constant hidden in the $O(n)$ running time is reasonably small and does not depend on the error ε, whereas the additive constant is exponential in m, μ and $1/\varepsilon$.*

References

1. A.K. Amoura, E. Bampis, C. Kenyon, and Y. Manoussakis, *Scheduling indepen-dent multiprocessor tasks*, Proceedings of the 5th Annual European Symposium on Algorithms, vol. 1284, LNCS, 1997, pp. 1–12.
2. J. Chen and A. Miranda, *A polynomial time approximation scheme for general multiprocessor job scheduling*, Proceedings of the 31st Annual ACM Symposium on the Theory of Computing, 1999, pp. 418–427.
3. M. R. Garey and D. S. Johnson, *Computers and intractability; a guide to the theory of np-completeness*, W.H. Freeman, 1979.
4. M. D. Grigoriadis and L. G. Khachiyan, *Coordination complexity of parallel price-directive decomposition*, Mathematics of Operations Research **21** (1996), 321–340.
5. L.A. Goldberg, M. Paterson, A. Srinivasan, and E. Sweedyk, *Better approximation guarantees for job-shop scheduling*, SIAM Journal on Discrete Mathematics **14** (2001), no. 1, 67–92.
6. E. Horowitz and S. Sahni, *Exact and approximate algorithms for scheduling non-identical processors*, Journal of the ACM **23** (1976), 317–327.
7. K. Jansen and L. Porkolab, *Improved approximation schemes for scheduling unre-lated parallel machines*, Proceedings of the 31st Annual ACM Symposium on the Theory of Computing, 1999, pp. 408–417.
8. K. Jansen and L. Porkolab, *Linear-time approximation schemes for scheduling malleable parallel tasks*, Proceedings of the 10th Annual ACM-SIAM Symposium on Discrete Algorithms, 1999, pp. 490–498.
9. K. Jansen and L. Porkolab, *Polynomial time approximation schemes for general multiprocessor job shop scheduling*, ICALP'00, 2000, pp. 878–889.
10. K. Jansen, R. Solis-Oba, and M. Sviridenko, *Makespan minimization in job shops: a polynomial time approximation scheme*, Proceedings of the 31st Annual ACM Symposium on the Theory of Computing, 1999, pp. 394–399.
11. K. Jansen, R. Solis-Oba, and M. Sviridenko, *A linear time approximation scheme for the job shop scheduling problem*, APPROX'99, vol. 1671, 1999, pp. 177–188.
12. E.L. Lawler, J.K. Lenstra, A.H.G. Rinnooy Kan, and D.B. Shmoys, *Sequencing and scheduling: Algorithms and complexity*, Handbook in Operations Research and Management Science **4** (1993), 445–522.
13. J. K. Lenstra, D. B. Shmoys, and E. Tardos, *Approximation algorithms for schedul-ing unrelated parallel machines*, Mathematical Programming **46** (1990), 259–271.
14. S. V. Sevastianov, *On some geometric methods in scheduling theory: a survey*, Discrete Applied Mathematics **55** (1994), 59–82.
15. D.B. Shmoys and E. Tardos, *An approximation algorithm for the generalized as-signment problem*, Mathematical Programming (1993), 461–474.
16. D.B. Shmoys, C. Stein and J. Wein, *Improved approximation algorithms for shop scheduling problems*, SIAM Journal on Computing **23** (1994), 617–632.
17. D.P. Williamson, L.A. Hall, J.A. Hoogeveen, C.A.J. Hurkens, J.K. Lenstra, S.V. Sevastianov, and D.B. Shmoys, *Short shop schedules*, Operations Research **45** (1997), 288–294.

A 2-Approximation Algorithm for the Multi-vehicle Scheduling Problem on a Path with Release and Handling Times

Yoshiyuki Karuno[1] and Hiroshi Nagamochi[2]

[1] Kyoto Institute of Technology, Matsugasaki, Sakyo,
Kyoto 606-8585, Japan.
karuno@ipc.kit.ac.jp
[2] Toyohashi University of Technology, Tempaku-cho,
Toyohashi 441-8580, Japan.
naga@ics.tut.ac.jp

Abstract. In this paper, we consider a scheduling problem of vehicles on a path. Let $G = (V, E)$ be a path, where $V = \{v_1, v_2, \ldots, v_n\}$ is its set of n vertices and $E = \{\{v_j, v_{j+1}\} \mid j = 1, 2, \ldots, n - 1\}$ is its set of edges. There are m identical vehicles ($1 \leq m \leq n$). The travel times $w(v_j, v_{j+1})$ ($= w(v_{j+1}, v_j)$) are associated with edges $\{v_j, v_{j+1}\} \in E$. Each job j which is located at each vertex $v_j \in V$ has release time r_j and handling time h_j. Any job must be served by exactly one vehicle. The problem asks to find an optimal schedule of m vehicles that minimizes the maximum completion time of all the jobs. The problem is known to be NP-hard for any fixed $m \geq 2$. In this paper, we give an $O(mn^2)$ time 2-approximation algorithm to the problem, by using properties of optimal gapless schedules.

1 Introduction

In this paper, we consider a scheduling problem of vehicles on a path with release and handling times. The scheduling problem of vehicles, such as AGVs (automated guided vehicles), handling robots, buses, trucks and so forth, on a given road network is an important topic encountered in various applications. In particular, in FMS (flexible manufacturing system) environment, scheduling of the movement of AGVs, which carry materials and products between machining centers, has a vital effect on the system efficiency.

The single-vehicle scheduling problem (VSP, for short) contains the traveling salesman problem (TSP) and the delivery man problem (DMP) [3] as its special cases. In the TSP, a salesman (a vehicle) visits each of n customers (jobs) situated at different locations on a given network before returning to the initial location. The objective is to minimize the tour length. In the DMP, the same scenario is considered but the objective is to minimize the total completion time of all the jobs. The VSP usually takes into account the time constraints of jobs (i.e., release, handling and/or due times), and therefore another important objective

F. Meyer auf der Heide (Ed.): ESA 2001, LNCS 2161, pp. 218–229, 2001.

functions, such as the tour time, the maximum completion time of jobs, the maximum lateness from the due times and so forth, are also considered. Since path and tree are important network topologies from both practical and graph theoretical views, VSPs on these networks have been studied in several papers, e.g., Psaraftis, Solomon, Magnanti and Kim [14], Tsitsiklis [15], Averbakh and Berman [3], Karuno, Nagamochi and Ibaraki [8,9,10,11], Nagamochi, Mochizuki and Ibaraki [12,13], and Asano, Katoh and Kawashima [1].

The multi-vehicle scheduling problem (MVSP, for short) is a more general problem than the VSP. The MVSP on a general network can be viewed as a variant of the so-called vehicle routing problem with time windows (VRPTW) (e.g., see Desrosiers, Dumas, Solomon and Soumis [6]). The typical VRPTW uses vehicles with limited capacity, and all the vehicles start their routes from a common depot and return to the depot. The objective is to minimize the total tour length (or the total tour time) of all the vehicles under the minimum number of routes. Therefore, the MVSP may also appear in a variety of industrial and service sector application, e.g., bank deliveries, postal deliveries, industrial refuse collection, distribution of soft drinks, beer and gasoline, school bus routing, transportation of handicapped people and security patrol services.

The multi-vehicle problem on a path to be discussed here is called PATH-MVSP, and the number of vehicles is denoted by m ($1 \leq m \leq n$). Problem PATH-MVSP asks to find an optimal schedule of m vehicles (i.e., their optimal sequences of jobs) that minimizes the maximum completion time of all the jobs. Note that the objective is equivalent to minimizing the maximum workload of all the vehicles. In Averbakh and Berman [4,5], dealing with an MTSP (i.e., multiple traveling salesmen problem) on a tree to minimize the maximum tour length of all the vehicles, they mentioned that such a minmax objective may be motivated, first, by the desire to distribute the workload to the vehicles in a fair way, second, by natural restrictions such as limited working day of the vehicles. They presented an approximation algorithm with the worst-case performance ratio $2 - 2/(m + 1)$ and with the running time $O(n^{m-1})$ for the MTSP. They also considered an MDMP (i.e., multiple delivery men problem) on a path in a different paper by Averbakh and Berman [2] and showed that it can be solved in $O(n^4)$ time.

The PATH-MVSP is NP-hard for any fixed $m \geq 2$, and is NP-hard in the strong sense for m arbitrary, since PARTITION and 3-PARTITION (e.g., see Garey and Johnson [7]) are its special cases, respectively. The PATH-MVSP with $m = 1$ (i.e., the VSP on a path) was proved by Tsitsiklis [15] to be NP-hard if the initial location of a vehicle is specified. The PATH-MVSP with $m = 1$ is 2-approximable due to the results by Psaraftis et al. [14], and it was shown to be 1.5-approximable by Karuno et al. [11] if the initial and goal locations of a vehicle are specified as one end of the path.

In this paper, we consider the PATH-MVSP with symmetric edge weights. For a schedule of the problem, we refer to a maximal subpath of a given path which is traversed by a certain vehicle as its *zone*. A feasible schedule using m' vehicles ($m' \leq m$) is called a *zone schedule* if any two zones do not intersect and thus there are $m' - 1$ edges which are not traversed by any vehicle. Such an edge that is not traversed by any vehicle is called a *gap*. Conversely, a schedule is called *gapless* if each edge is traversed at least once by some vehicle. When the fleet of vehicles follows a zone schedule, any two vehicles do not interfere each other on the given path. As such non-interference between the vehicles is important in controlling them, the zone schedule is often required in practice. Hence it would be natural to ask how much the maximum completion time becomes larger if we restrict ourselves only on zone schedules. In this paper, for the PATH-MVSP of finding an optimal gapless schedule (with symmetric edge weights), we prove that there always exists a zone schedule with the maximum completion time at most twice the optimal. Based on this fact, we present a polynomial time 2-approximation algorithm to the problem.

The remainder of this paper is organized as follows. After providing the mathematical description of the PATH-MVSP in Section 2, we in Section 3 discuss how to construct a zone schedule whose maximum completion time is at most twice of the maximum completion time of any gapless schedule. In Section 4, by using a dynamic programming, we present an $O(mn^2)$ time 2-approximation algorithm for the PATH-MVSP. In Section 5, we give a conclusion.

2   Multi-vehicle Scheduling Problem on a Path

2.1   Problem Description

Problem PATH-MVSP is formulated as follows. Let $G = (V, E)$ be a path network, where $V = \{v_1, v_2, \ldots, v_n\}$ is a set of n vertices and $E = \{\{v_j, v_{j+1}\} \mid j = 1, 2, \ldots, n-1\}$ is a set of edges. In this paper, we assume that vertex v_1 is the *left end* of G, and v_n the *right end* of it. There is a job j at each vertex $v_j \in V$. The job set is denoted by $J = \{j \mid j = 1, 2, \ldots, n\}$. There are m vehicles on G $(1 \leq m \leq n)$, which are assumed to be identical. Each job must be served by exactly one vehicle.

The *travel time* of a vehicle is $w(v_j, v_{j+1}) \geq 0$ to traverse $\{v_j, v_{j+1}\} \in E$ from v_j to v_{j+1}, and is $w(v_{j+1}, v_j) \geq 0$ to traverse it in the opposite direction. Edge weight $w(v_j, v_{j+1})$ for $\{v_j, v_{j+1}\} \in E$ is called *symmetric* if $w(v_j, v_{j+1}) = w(v_{j+1}, v_j)$ holds. In this paper, we assume that all edge weights are symmetric. The travel time for a vehicle to move from vertex v_i to vertex v_j on G is the sum of edge weights belonging to the unique path from v_i to v_j. Each job $j \in J$ has its release time $r_j \geq 0$ and handling time $h_j \geq 0$: That is, a vehicle cannot start serving job j before time r_j, and it takes h_j time units to serve job j (no interruption of the service is allowed). A vehicle at vertex v_j may wait until time r_j to serve job j, or move to other vertices without serving job j if it is more advantageous (in this case, the vehicle must come back to v_j later to serve job j, or another vehicle must come to v_j to serve it). An instance of the problem PATH-MVSP is denoted by $(G(= (V, E)), r, h, w, m)$.

A motion schedule of the m vehicles is completely specified by m sequences of jobs $\pi^{[k]} = (j_1^{[k]}, j_2^{[k]}, \ldots, j_{n_k}^{[k]})$, $k = 1, 2, \ldots, m$, where n_k is the number of jobs to be served by vehicle k (hence, it holds that $\sum_{k=1}^{m} n_k = n$), and $j_i^{[k]}$ is its i-th job; i.e., vehicle k is initially situated at vertex $v_{j_1^{[k]}}$, starts serving job $j_1^{[k]}$ at time $\max\{0, r_{j_1^{[k]}}\}$, and takes $h_{j_1^{[k]}}$ time units to serve it. After completing job $j_1^{[k]}$, the vehicle immediately moves to $v_{j_2^{[k]}}$ along the unique path from $v_{j_1^{[k]}}$ to $v_{j_2^{[k]}}$, taking travel time of the path (i.e., $w(v_{j_1^{[k]}}, v_{j_1^{[k]}+1}) + \cdots + w(v_{j_2^{[k]}-1}, v_{j_2^{[k]}})$ or $w(v_{j_1^{[k]}}, v_{j_1^{[k]}-1}) + \cdots + w(v_{j_2^{[k]}+1}, v_{j_2^{[k]}}))$, and serves job $j_2^{[k]}$ after waiting until time $r_{j_2^{[k]}}$ if necessary, and so on, until it completes the last job $j_{n_k}^{[k]}$. A schedule is denoted by a set of m sequences of jobs $\pi = \{\pi^{[1]}, \pi^{[2]}, \ldots, \pi^{[m]}\}$. The completion time of vehicle k (i.e., the workload of it) is defined as the completion time of its last job $j_{n_k}^{[k]}$, which is denoted by $C(\pi^{[k]})$. The objective is to find a π that minimizes the maximum completion time of all the jobs, i.e.,

$$C_{max}(\pi) = \max_{1 \le k \le m} C(\pi^{[k]}). \tag{1}$$

In this paper, we denote by π^* an optimal schedule and by C_{max}^* the optimal value $C_{max}(\pi^*)$.

2.2 Subpath and Subinstance

Let $V(i,j) = \{v_i, v_{i+1}, \ldots, v_j\}$ ($\subseteq V$), where $i \le j$. Define $G(i,j) = (V(i,j), E(i,j))$ be a subpath of a given path $G = (V, E)$ induced by $V(i,j)$ and $E(i,j) = \{\{v_{j'}, v_{j'+1}\} \mid j' = i, i+1, \ldots, j-1\}$ ($\subseteq E$), and $J(i,j) = \{i, i+1, \ldots, j\}$ ($\subseteq J$) the corresponding subset of jobs to the subpath $G(i,j)$. This definition states that $G(1,n) = G$ and $J(1,n) = J$.

Next consider the scheduling problem of k ($\le m$) vehicles on $G(i,j)$. This is a *subinstance* of the original instance (G, r, h, w, m). We denote this subinstance by $(G(i,j), r, h, w, k)$; i.e., scheduling k vehicles on subpath $G(i,j) = (V(i,j), E(i,j))$ of the given path G with release times $r_{j'}$ and handling times $h_{j'}$ for $j' \in J(i,j)$ and with edge weights $w(v_{j'}, v_{j'+1})$ ($= w(v_{j'+1}, v_{j'})$) for $\{v_{j'}, v_{j'+1}\} \in E(i,j)$ (hence, the original instance is denoted by (G, r, h, w, m) as well as $(G(1,n), r, h, w, m)$).

2.3 Zone Schedule and Gapless Schedule

For a schedule π, assume that a vehicle *covers* a subpath $G(i,j) = (V(i,j), E(i,j))$: That is, all jobs served by the vehicle are on $G(i,j)$ and two jobs i and j located at the end vertices of $G(i,j)$ have to be served by it. But, there may be some jobs j' ($i < j' < j$) served by other vehicles. Then, the subpath $G(i,j)$ for the vehicle is referred to as its *zone*.

A feasible schedule π using m' vehicles ($m' \le m$) is referred to as a *zone schedule* if any two zones in π do not intersect and thus there are $m' - 1$ edges which are not traversed by any vehicle. Such an edge that is not traversed by any vehicle is called a *gap*. Moreover, a zone schedule is called a *1-way* zone schedule if any vehicle traverses an edge belonging to its zone exactly once (hence each vehicle traverses its zone from left to right or from right to left). A schedule π is called *gapless* if each edge $\{v_j, v_{j+1}\} \in E$ is traversed at least once (from v_j to v_{j+1} or from v_{j+1} to v_j) by some vehicle.

A zone schedule does not always attain an optimal value. Consider the following example:

Example 1. Let $G = (V, E)$ be a path with $m = 2$ vehicles and $n = 4$ jobs, where $V = \{v_1, v_2, v_3, v_4\}$ and $E = \{\{v_j, v_{j+1}\} \mid j = 1, 2, 3\}$. For a positive constant B, edge weights are given by $w(v_1, v_2) = w(v_3, v_4) = B$ and $w(v_2, v_3) = 0$, release times are given by $r_1 = r_2 = 0$, $r_3 = B$ and $r_4 = 2B$, and handling times are by $h_1 = h_4 = 0$ and $h_2 = h_3 = B$.   □

In this example, any zone schedule takes at least $3B$ time to complete all jobs, while an optimal schedule is given by $\pi^{[1]} = (v_1, v_3)$ and $\pi^{[2]} = (v_2, v_4)$ whose maximum completion time is $2B$.

3   A 2-Approximation Algorithm for Finding an Optimal Gapless Schedule

In this section, we show that the PATH-MVSP of finding an optimal gapless schedule is polynomially approximable within twice the optimal.

3.1   Notations

We use the following notations:

- $r_{max}(i, j) = \max\limits_{i \leq j' \leq j} r_{j'}$: the maximum release time for jobs in $J(i, j)$.
- $r_{max} = r_{max}(1, n)$: the maximum release time for all the jobs in J.
- $H(i, j) = \sum\limits_{j'=i}^{j} h_{j'}$: the sum of handling times for jobs in $J(i, j)$.
- $H = H(1, n)$: the sum of handling times for all the jobs in J.
- $W(i, j) = \sum\limits_{j'=i}^{j-1} w(v_{j'}, v_{j'+1})$: the sum of edge weights for edges in $E(i, j)$.
- $W = W(1, n)$: the sum of edge weights for all the edges in E.

The following lower bounds on the minimum of the maximum completion time of a gapless schedule in an instance (G, r, h, w, m) are immediately obtained:

$$LB1 = \max_{1 \leq j \leq n} \left\{ r_j + h_j \right\}, \quad LB2 = \frac{W + H}{m}, \quad LB3 = \max_{1 \leq j \leq n-1} w(v_j, v_{j+1}), \quad (2)$$

where $LB1$ is the maximum completion time to serve a single job, $LB2$ means the least time for a vehicle to serve and travel to complete the jobs assigned to the vehicle, and $LB3$ is the maximum time to travel an edge in the path. We denote the maximum of these lower bounds by:

$$\gamma = \max\{LB1, \ LB2, \ LB3\}. \tag{3}$$

3.2 Auxiliary Path and γ-Splitting Point Set

In this subsection, for a given path $G = (V, E)$, we introduce an *auxiliary path* $G_A = (V_A, E_A)$ and a set Γ of *points* on G_A.

The auxiliary path $G_A = (V_A, E_A)$ is defined as follows:

$$V_A = \{u_1, u_2, \ldots, u_{2n}\}, \quad E_A = E_V \cup E_E, \quad \text{where}$$

$$E_V = \{(u_{2j-1}, u_{2j}) \mid j = 1, 2, \ldots, n\}, \quad E_E = \{(u_{2j}, u_{2j+1}) \mid j = 1, 2, \ldots, n-1\},$$

and edge weights $w_A(u_i, u_{i+1})(= w_A(u_{i+1}, u_i))$ for $(u_i, u_{i+1}) \in E_A$ are given by $w_A(u_{2j-1}, u_{2j}) = h_j$ for $j = 1, 2, \ldots, n$ and $w_A(u_{2j}, u_{2j+1}) = w(v_j, v_{j+1})$ for $j = 1, 2, \ldots, n-1$. It is assumed that u_1 is the left end of G_A and u_{2n} the right end of it. We call edges in E_V and in E_E, *job-edges* and *travel-edges*, respectively. Each job-edge $(u_{2j-1}, u_{2j}) \in E_V$ corresponds to vertex $v_j \in V$ of a given path G, while each travel-edge $(u_{2j}, u_{2j+1}) \in E_E$ to edge $\{v_j, v_{j+1}\} \in E$ of G. Note that the total length of G_A is

$$\sum_{i=1}^{2n-1} w_A(u_i, u_{i+1}) = h_1 + w(v_1, v_2) + h_2 + w(v_2, v_3) + \cdots + h_n = W + H.$$

By viewing G_A as a set of $2n$ points in a 1-dimensional space, where $u_1, u_2, \ldots, u_{2n}$ are located in this order and the distance between u_i and u_{i+1} is $w_A(u_i, u_{i+1})$, we consider any point a (not necessarily a vertex of G_A) between u_1 and u_{2n} in the space.

For any two points a and b on G_A, let $S(a, b)$ denote the *segment* of G_A between a and b, and let $d(a, b)$ denote the length of $S(a, b)$ (i.e., the distance between a and b on G_A). For any edge $(u_i, u_{i+1}) \in E_A$, it is assumed that points u_i and u_{i+1} do not belong to the edge; $[u_i, u_{i+1})$ denotes edge (u_i, u_{i+1}) with its left end point u_i.

For the γ defined by Eq. (3), a set Γ of $m' - 1$ points on G_A is referred to as γ-*splitting point set*, if after deleting all points of Γ from G_A, the auxiliary path G_A will be divided into m' segments of each length equal to or less than γ; i.e., by letting $\tau_1, \tau_2, \ldots, \tau_{m'-1}$ be the points from set Γ, ordered according to their distances from the left end u_1 of G_A, then it holds that $d(\tau_k, \tau_{k+1}) = \gamma$ for $k = 0, 1, 2, \ldots, m' - 2$ and $d(\tau_{m'-1}, \tau_{m'}) \leq \gamma$, where we define the m' such that

$$(m' - 1)\gamma < W + H \leq m'\gamma \tag{4}$$

and we set $\tau_0 = u_1$ and $\tau_{m'} = u_{2n}$ for notational convenience.

3.3 A 2-Approximation for Optimal Gapless Schedule

By definition, for any instance (G, r, h, w, m), at most one point belonging in $\Gamma = \{\tau_1, \tau_2, \ldots, \tau_{m'-1}\}$ can fall on each edge of $G_A = (V_A, E_A)$. By assigning the jobs corresponding to job-edges fully contained in segment between τ_{k-1} and τ_k to the vehicle k, and by assigning the job corresponding to a job-edge containing τ_k to the vehicle k or $k + 1$, we obtain a zone schedule.

From Eqs. (3) and (4), if $(W+H)/m < \max\{\max_{1\le j\le n}(r_j+h_j), \max_{1\le j\le n-1} w(v_j, v_{j+1})\}$, then we reduce the number m of vehicles to m' such that $(m' - 1)\gamma < W+H \le m'\gamma$ (where $m-m'$ vehicles serve no jobs). Notice that this does not change the lower bound γ. In what follows, we assume that $\gamma = \frac{1}{m}(W+H)$ (which implies that $m' = m$).

Using γ-splitting point set Γ, the following algorithm obtains a zone schedule π^{ZONE} for an instance (G, r, h, w, m) of PATH-MVSP.

Algorithm ZONE($G, r, h, w, m; \pi^{\text{ZONE}}$)

Input: A path $G = (V, E)$, where $V = \{v_1, v_2, \ldots, v_n\}$ is its set of n vertices and $E = \{\{v_j, v_{j+1}\} \mid j = 1, 2, \ldots, n-1\}$ is its set of edges, release times r_j for $j \in J$, handling times h_j for $j \in J$, edge weights $w(v_j, v_{j+1})$ for $\{v_j, v_{j+1}\} \in E$, and the number of vehicles m.

Output: A zone schedule π^{ZONE} with $C_{max}(\pi^{\text{ZONE}}) \le 2 \cdot C_{max}(\pi_g^*)$ for an optimal gapless schedule π_g^*.

Step 1 (Initialization):
  Compute the auxiliary path G_A and its γ-splitting point set $\Gamma = \{\tau_1, \tau_2, \ldots, \tau_{m-1}\}$ from the given path G;
  $a(1) := 1$; $b(m) := n$; /* job 1 is served by the 1-st vehicle while job n is served by the m-th vehicle */

Step 2 (Allocating segments of G_A to m vehicles):
  for $k = 1, 2, \ldots, m-1$ **do**
    if $\tau_k \in [u_{2i-1}, u_{2i})$ for some job-edge (u_{2i-1}, u_{2i}) of G_A **then**
      if $d(u_{2i-1}, \tau_k) > \dfrac{w_A(u_{2i-1}, u_{2i})}{2}$ **then**
        $b(k) := i$; $a(k+1) := i+1$; /* job i is served by the k-th vehicle while job $i+1$ is served by the $(k+1)$-th vehicle */
      else
        $b(k) := i-1$; $a(k+1) := i$
      end; /* if */
    else
      /* $\tau_k \in [u_{2i}, u_{2i+1})$ for some travel-edge (u_{2i}, u_{2i+1}) of G_A */
      $b(k) := i$; $a(k+1) := i+1$ /* job i is served by the k-th vehicle while job $i+1$ is served by the $(k+1)$-th vehicle */
    end; /* if */
  end; /* for */

Step 3 (Scheduling m vehicles in each subpath $G(a(k), b(k))$, $k = 1, 2, \ldots, m$):
  for $k = 1, 2, \ldots, m$ **do**
    $\pi^{[k]} := (a(k), a(k)+1, \ldots, b(k))$; $\mu^{[k]} := (b(k), b(k)-1, \ldots, a(k))$;
    Compute the completion times $C(\pi^{[k]})$ and $C(\mu^{[k]})$ in $G(a(k), b(k))$;
    if $C(\pi^{[k]}) > C(\mu^{[k]})$ **then**
      $\pi^{[k]} := \mu^{[k]}$
    end; /* if */
  end; /* for */
  $\pi^{\text{ZONE}} := \{\pi^{[1]}, \pi^{[2]}, \ldots, \pi^{[m]}\}$.

Let $C_{max}(\pi_g^*)$ be the minimum of the maximum completion time of a gapless schedule, i.e., the maximum completion time of an optimal gapless schedule π_g^* in (G, r, h, w, m). Now the lower bound γ is equal to $\frac{W+H}{m}$.

Theorem 1. *For an instance (G, r, h, w, m) of PATH-MVSP, let $C_{max}(\pi_g^*)$ be the maximum completion time of an optimal gapless schedule π_g^*. Then there exists a 1-way zone schedule π^{ZONE} such that*

$$C_{max}(\pi^{ZONE}) \leq LB1 + LB2 \ (\leq 2 \cdot C_{max}(\pi_g^*)). \tag{5}$$

Moreover, such a 1-way zone schedule π^{ZONE} can be found in $O(n)$ time.

Proof. It is not difficult to see that algorithm ZONE runs in $O(n)$ time and outputs a feasible zone schedule π^{ZONE}. Hence it suffices to show that $C_{max}(\pi^{ZONE}) \leq 2 \cdot C_{max}(\pi_g^*)$ holds. For this, we show that $\min\{C(\pi^{[k]}), C(\mu^{[k]})\} \leq LB1 + LB2$ holds for each vehicle $k = 1, 2, \ldots, m$.

We consider the following three cases.

Case 1. $\tau_{k-1} \notin [u_{2a(k)-1}, u_{2a(k)})$ and $\tau_k \notin [u_{2b(k)-1}, u_{2b(k)})$: Consider the sequence of jobs $\pi^{[k]} = (a(k), a(k) + 1, \ldots, b(k))$ on subpath $G(a(k), b(k))$ of the given path G. Thus $W(a(k), b(k)) + H(a(k), b(k)) = d(u_{2a(k)-1}, u_{2b(k)}) \leq d(\tau_{k-1}, \tau_k) = \gamma$. By considering a schedule in which the vehicle starts the sequence of jobs after waiting at vertex $v_{a(k)}$ until time $r_{max}(a(k), b(k))$, we have

$$\begin{aligned} C(\pi^{[k]}) &\leq r_{max}(a(k), b(k)) + W(a(k), b(k)) + H(a(k), b(k)) \\ &\leq r_{max}(a(k), b(k)) + \gamma \leq LB1 + LB2. \end{aligned}$$

Case 2. $\tau_{k-1} \in [u_{2a(k)-1}, u_{2a(k)})$ and $\tau_k \notin [u_{2b(k)-1}, u_{2b(k)})$ (the case of $\tau_{k-1} \notin [u_{2a(k)-1}, u_{2a(k)})$ and $\tau_k \in [u_{2b(k)-1}, u_{2b(k)})$ can be treated symmetrically): In this case, τ_{k-1} is situated on $[u_{2a(k)-1}, u_{2a(k)})$ by $d(u_{2a(k)-1}, \tau_{k-1}) \leq w_A(u_{2a(k)-1}, u_{2a(k)})/2 \ (= h_{a(k)}/2)$. Thus $W(a(k), b(k)) + H(a(k), b(k)) = d(u_{2a(k)-1}, u_{2b(k)}) \leq d(u_{2a(k)-1}, \tau_{k-1}) + d(\tau_{k-1}, \tau_k) \leq h_{a(k)}/2 + \gamma$. Consider the sequence of jobs $\pi^{[k]} = (a(k), a(k) + 1, \ldots, b(k))$ on subpath $G(a(k), b(k))$ of G. If $r_{a(k)} + h_{a(k)} \geq r_{max}(a(k), b(k))$, then all jobs in $J(a(k), b(k)) \setminus \{a(k)\}$ do not require for the vehicle to wait at their vertices after it completes job $a(k)$, implying

$$\begin{aligned} C(\pi^{[k]}) &\leq r_{a(k)} + W(a(k), b(k)) + H(a(k), b(k)) \leq r_{a(k)} + \frac{h_{a(k)}}{2} + \gamma \\ &\leq \frac{r_{max}(a(k), b(k))}{2} + \frac{r_{a(k)} + h_{a(k)}}{2} + \gamma \leq \frac{LB1}{2} + \frac{LB1}{2} + LB2; \end{aligned}$$

On the other hand, if $r_{a(k)} + h_{a(k)} < r_{max}(a(k), b(k))$, then by considering a schedule such that the vehicle waits for $r_{max}(a(k), b(k)) - (r_{a(k)} + h_{a(k)})$ time after serving job $a(k)$, we have

$$\begin{aligned} C(\pi^{[k]}) &\leq r_{a(k)} + h_{a(k)} + \left(r_{max}(a(k), b(k)) - (r_{a(k)} + h_{a(k)}) \right) \\ &\quad + W(a(k), b(k)) + (H(a(k), b(k)) - h_{a(k)}) \\ &\leq r_{a(k)} + \frac{h_{a(k)}}{2} + \gamma + \left(r_{max}(a(k), b(k)) - (r_{a(k)} + h_{a(k)}) \right) \\ &\leq r_{max}(a(k), b(k)) + \gamma \leq LB1 + LB2. \end{aligned}$$

Case 3. $\tau_{k-1} \in [u_{2a(k)-1}, u_{2a(k)})$ and $\tau_k \in [u_{2b(k)-1}, u_{2b(k)})$: We can assume without loss of generality that $h_{a(k)} \geq h_{b(k)}$. In this case, τ_{k-1} is situated on $[u_{2a(k)-1}, u_{2a(k)})$ from $u_{2a(k)-1}$ with distance less than $w_A(u_{2a(k)-1}, u_{2a(k)})/2$ $(= h_{a(k)}/2)$, and τ_k is situated on $[u_{2b(k)-1}, u_{2b(k)})$ from $u_{2b(k)-1}$ with distance greater than $w_A(u_{2b(k)-1}, u_{2b(k)})/2$ $(= h_{b(k)}/2)$. Then $W(a(k), b(k)) + H(a(k), b(k)) = d(u_{2a(k)-1}, u_{2b(k)}) = d(u_{2a(k)-1}, \tau_{k-1}) + d(\tau_{k-1}, \tau_k) + d(\tau_k, u_{2b(k)}) \leq h_{a(k)}/2 + \gamma + h_{b(k)}/2$.

Consider the sequence of jobs $\pi^{[k]} = (a(k), a(k)+1, \ldots, b(k))$ on subpath $G(a(k), b(k))$ of G. If $r_{a(k)} + h_{a(k)} \geq r_{max}(a(k), b(k))$, then all jobs in $J(a(k), b(k)) \setminus \{a(k)\}$ do not require for the vehicle to wait at their vertices after it has completed job $a(k)$, indicating

$$C(\pi^{[k]}) \leq r_{a(k)} + W(a(k), b(k)) + H(a(k), b(k)) \leq r_{a(k)} + \frac{h_{a(k)}}{2} + \gamma + \frac{h_{b(k)}}{2}$$
$$\leq r_{a(k)} + h_{a(k)} + \gamma \leq LB1 + LB2;$$

Otherwise, if $r_{a(k)} + h_{a(k)} < r_{max}(a(k), b(k))$, then by considering a schedule such that the vehicle waits for $r_{max}(a(k), b(k)) - (r_{a(k)} + h_{a(k)})$ time after serving job $a(k)$, we have

$$C(\pi^{[k]}) \leq r_{a(k)} + h_{a(k)} + \left(r_{max}(a(k), b(k)) - (r_{a(k)} + h_{a(k)})\right)$$
$$+W(a(k), b(k)) + (H(a(k), b(k)) - h_{a(k)})$$
$$\leq r_{a(k)} + \left(r_{max}(a(k), b(k)) - (r_{a(k)} + h_{a(k)})\right) + \frac{h_{a(k)}}{2} + \gamma + \frac{h_{b(k)}}{2}$$
$$\leq r_{max}(a(k), b(k)) - \frac{h_{a(k)}}{2} + \frac{h_{b(k)}}{2} + \gamma$$
$$\leq r_{max}(a(k), b(k)) + \gamma \leq LB1 + LB2.$$

This case analysis shows that any vehicle has its completion time within twice the maximum completion time of the optimal gapless schedule. This completes the proof. □

4   A 2-Approximation Algorithm for General Schedules

Unfortunately, the optimal schedule π^* for a problem instance (G, r, h, w, m) is not always a gapless schedule, and hence $LB2$ (see Eq. (2)) cannot be used as a lower bound on the minimum of the maximum completion time C^*_{max} attained by general schedules. Thus, we need to take into account all configurations of gaps on G which are possible to be incurred by the optimal schedule. Based on this idea, we prove the next result.

Theorem 2. *For an instance (G, r, h, w, m) of PATH-MVSP, let C^*_{max} be the minimum of the maximum completion time of a schedule. Then there exists a 1-way zone schedule π^{ZONE} of $m - 1$ gaps such that*

$$C_{max}(\pi^{\text{ZONE}}) \leq 2 \cdot C^*_{max}. \tag{6}$$

Moreover, such a 1-way zone schedule can be found in $O(mn^2)$ time. □

The existence of such a 1-way schedule π^{ZONE} in the theorem can be shown as follows. Let π^* be an optimal schedule to the problem, which consists of several gapless schedules for subinstances of G. By $m \leq n$, we can assume without loss of generality that each of m vehicles serves at least one job in π^*. Let $e'_1, e'_2, \ldots, e'_\ell \in E$ be the gaps in π^*, and let $G_1, G_2, \ldots, G_{\ell+1}$ be the maximal subpaths of G induced by non-gap edges, where the jobs in each G_i is served by a gapless schedule $\pi_g^*(i)$ with m_i vehicles. For each gapless schedule $\pi_g^*(i)$, we see by Theorem 1 that there is a 1-way zone schedule π_i^{ZONE} of $m'_i(\leq m_i)$ vehicles which serves the jobs in G_i and has the completion time $C_{max}(\pi_i^{\text{ZONE}}) \leq 2C_{max}(\pi_g^*(i))$. We can assume $m'_i = m_i$, since G_i contains at least m_i vertices by the assumption on π^*, and hence if $m'_i < m_i$ we can modify π_i^{ZONE} into a zone schedule of $m_i - 1$ gaps without increasing the completion time. Since the completion time $C_{max}(\pi_g^*)$ is $\max\{C_{max}(\pi_g^*(1)), \ldots, C_{max}(\pi_g^*(\ell+1))\}$, the 1-way zone schedule π^{ZONE} consisting of these 1-way zone schedules $\pi_1^{\text{ZONE}}, \ldots, \pi_{\ell+1}^{\text{ZONE}}$ ensures the existence of a desired 1-way zone schedule in Theorem 2.

Now we show that such a 1-way zone schedule π^{ZONE} in Theorem 2 can be computed in $O(mn^2)$ time. For this, it suffices to prove that an optimal 1-way zone schedule π^{OPTZONE} with $m - 1$ gaps can be found in the same time complexity.

Lemma 1. *For an instance (G, r, h, w, m) of PATH-MVSP (where edge weights are not necessarily symmetric), an optimal 1-way zone schedule π^{OPTZONE} with $m - 1$ gaps can be found in $O(mn^2)$ time.*

Proof. We compute π^{OPTZONE} for (G, r, h, w, m) by the following dynamic programming. For $1 \leq i \leq j \leq n$, let $T(i, j)$ (resp., $T(j, i)$) denote the completion time of a 1-way zone schedule which serves the jobs in $J(i, j)$ from i to j (resp., from j to i). For $1 \leq i \leq j \leq n$ and $1 \leq \ell \leq m$, let $C(i, j, \ell)$ denote the maximum of the minimum completion times of 1-way zone schedules in subinstances $G(i', j')$ over all possible configurations of exactly $\ell - 1$ gaps on $G(i, j)$. Note that $C(1, n, m) = C_{max}(\pi^{\text{OPTZONE}})$. Hence, $C(1, n, m)$ can be computed in in $O(mn^2)$ time by the following dynamic programming.

Step 1: **for** $i = 1, 2, \ldots, n$ **do**
        $T(i, i) := r_i + h_i$
     end; /* for */
     for $k = 1, 2, \ldots, n - 1$ **do**
        for $i = 1, 2, \ldots, n - k$ **do**
           $j := i + k$;
           $T(i, j) := \max\{T(i, j - 1) + w(v_{j-1}, v_j), r_j\} + h_j$;
           $T(j, i) := \max\{T(j, i + 1) + w(v_{i+1}, v_i), r_i\} + h_i$;
           $C(i, j, 1) := \min\{T(i, j), T(j, i)\}$
        end; /* for */
     end; /* for */
Step 2: **for** $\ell = 2, 3, \ldots, m$ **do**
        for $j = \ell, \ell + 1, \ldots, n$ **do**

$$C(1, j, \ell) := \min_{1 \le j' \le j-1} \left\{ \max\{ C(1, j', \ell-1), C(j'+1, j, 1) \} \right\}$$

end; /* for */

end. /* for */

Notice that by backtracking the computation process, we can construct in $O(mn^2)$ time a 1-way zone scheduling π^{OPTZONE} which achieves the $C(1, n, m)$.

$\square$

The next example shows that the bound 2 in Theorem 2 is tight.

Example 2. Let $G = (V, E)$ be a path with m vehicles and $n = 3m$ jobs, where $V = \{v_{3k-2}, v_{3k-1}, v_{3k} \mid k = 1, 2, \ldots, m\}$ and $E = \{\{v_j, v_{j+1}\} \mid j = 1, 2, \ldots, 3m-1\}$. For positive constants $B \gg \delta > \varepsilon > 0$ (and $\delta \ge 1$), edge weights are given by $w(v_{3k-2}, v_{3k-1}) = \varepsilon$ and $w(v_{3k-1}, v_{3k}) = \varepsilon$ for $k = 1, 2, \ldots, m$, and by $w(v_{3k}, v_{3(k+1)-2}) = B\delta$ for $k = 1, 2, \ldots, m-1$. Release times are given by $r_{3k-2} = \delta + \varepsilon$, $r_{3k-1} = 0$, and $r_{3k} = \delta + 3\varepsilon$, and handling times are by $h_{3k-2} = 0$, $h_{3k-1} = \delta$ and $h_{3k} = 0$ for $k = 1, 2, \ldots, m$.

$\square$

In this example, the optimal schedule is $\pi^* = \{(2, 1, 3), (5, 4, 6), \ldots, (3m - 1, 3m - 2, 3m)\}$ with $C^*_{max} = \delta + 3\varepsilon = r_{max}$ (note that all the vehicles have the same completion time), while an optimal 1-way zone schedule is $\pi^{\text{OPTZONE}} = \{(1, 2, 3), (4, 5, 6), \ldots, (3m - 2, 3m - 1, 3m)\}$ with $C_{max}(\pi^{\text{OPTZONE}}) = 2\delta + 3\varepsilon$ (note that all the vehicles also have the same completion time with respect to π^{OPTZONE}). Therefore, we have $C_{max}(\pi^{\text{OPTZONE}})/C^*_{max} \to 2$ when $\varepsilon \to 0$.

5   Conclusion

In this paper, we discussed a scheduling problem of vehicles on a path with release and handling times, PATH-MVSP. The problem asks to find an optimal schedule of m vehicles serving n jobs that minimizes the maximum completion time of all the jobs. The PATH-MVSP is NP-hard for any fixed $m \ge 2$, and is NP-hard in the strong sense for m arbitrary. Even when there is a single vehicle (i.e., the case of $m = 1$), the problem is NP-hard if the initial location is specified. In this paper, for the PATH-MVSP with symmetric edge weights, we presented an approximation algorithm which delivers a zone schedule in $O(mn^2)$ time with the maximum completion time at most twice the optimal. It is left open whether or not the PATH-MVSP with asymmetric edge weights admits a good approximation algorithm.

References

1. Asano, T., Katoh, N. and Kawashima, K.: A new approximation algorithm for the capacitated vehicle routing problem on a tree. *Journal of Combinatorial Optimization*, **5** (2001) 213-231.
2. Averbakh, I. and Berman, O.: Routing and location routing p-delivery men problems on a path. *Transportation Science*, **28** (1994) 162-166.

3. Averbakh, I. and Berman, O.: Sales-delivery man problems on treelike networks. *Networks*, **25** (1995) 45-58.
4. Averbakh, I. and Berman, O.: A heuristic with worst-case analysis for minmax routing of two traveling salesmen on a tree. *Discrete Applied Mathematics*, **68** (1996) 17-32.
5. Averbakh, I. and Berman, O.: $(p-1)/(p+1)$-approximate algorithms for p-traveling salesmen problems on a tree with minmax objective. *Discrete Applied Mathematics*, **75** (1997) 201-216.
6. Desrosiers, J., Dumas, Y., Solomon, M. M. and Soumis, F.: Time constrained routing and scheduling. In Ball, M. O., Magnanti, T. L., Monma, C. L. and Nemhauser, G. L. (eds.): *Handbooks in Operations Research and Management Science Volume 8: Network Routing* (North-Holland, 1995), 35-139.
7. Garey, M. R. and Johnson, D. S.: *Computers and Intractability: A Guide to the Theory of NP-Completeness* (W. H. Freeman and Company, San Francisco, 1979).
8. Karuno, Y., Nagamochi, H. and Ibaraki, T.: Vehicle scheduling on a tree to minimize maximum lateness. *Journal of the Operations Research Society of Japan*, **39** (1996) 345-355.
9. Karuno, Y., Nagamochi, H. and Ibaraki, T.: Vehicle scheduling on a tree with release and handling times. *Annals of Operations Research*, **69** (1997) 193-207.
10. Karuno, Y., Nagamochi, H. and Ibaraki, T.: Computational complexity of the traveling salesman problem on a line with deadlines and general handling times. *Memoirs of the Faculty of Engineering and Design, Kyoto Institute of Technology*, **45** (1997) 19-22.
11. Karuno, Y., Nagamochi, H. and Ibaraki, T.: A 1.5-approximation for single-vehicle scheduling problem on a line with release and handling times. *Proceedings, ISCIE/ASME 1998 Japan-U.S.A. Symposium on Flexible Automation*, **3** (1998) 1363-1366.
12. Nagamochi, H., Mochizuki, K. and Ibaraki, T.: Complexity of the single vehicle scheduling problem on graphs. *Information Systems and Operations Research*, **35** (1997) 256-276.
13. Nagamochi, H., Mochizuki, K. and Ibaraki, T.: Solving the single-vehicle scheduling problems for all home locations under depth-first routing on a tree. *IEICE Transactions: Fundamentals*, **E84-A** (2001) 1135-1143.
14. Psaraftis, H., Solomon, M., Magnanti, T. and Kim, T.: Routing and scheduling on a shoreline with release times. *Management Science*, **36** (1990) 212-223.
15. Tsitsiklis, J. N.: Special cases of traveling salesman and repairman problems with time windows. *Networks*, **22** (1992) 263-282.

A Simple Shortest Path Algorithm
with Linear Average Time

Andrew V. Goldberg

STAR Laboratory, InterTrust Technologies Corp.
4750 Patrick Henry Dr., Santa Clara, CA 95054, USA
goldberg@intertrust.com

Abstract. We present a simple shortest path algorithm. If the input lengths are positive and uniformly distributed, the algorithm runs in linear time. The worst-case running time of the algorithm is $O(m + n \log C)$, where n and m are the number of vertices and arcs of the input graph, respectively, and C is the ratio of the largest and the smallest nonzero arc length.

1 Introduction

The shortest path problem with nonnegative arc lengths is very common in practice, and algorithms for this problem have been extensively studied, both from theoretical, e.g., [2,6,9,10,12,13,17,22,28,26,27,29,31,32], and computational, e.g., [4,5,11,19,21,20,33], viewpoints. Efficient implementations of Dijkstra's algorithm [12], in particular, received a lot of attention. At each step, Dijkstra's algorithm selects for processing a labeled vertex with the smallest distance label. For a nonnegative length function, this selection rule guarantees that each vertex is scanned at most once.

Suppose that the input graph has n vertices and m arcs. To state some of the previous results, we assume that the input arc lengths are integral. Let U denote the biggest arc length. We define C to be the ratio between U and the smallest nonzero arc length. Note that if the lengths are integral, then $C \leq U$. Modulo precision problems and arithmetic operation complexity, our results apply to real-valued arc lengths as well. To simplify comparing time bounds with and without U (or C), we make the similarity assumption [18]: $\log U = O(\log n)$.

Several algorithms for the problem have near-linear worst-case running times, although no algorithm has a linear running time if the graph is directed and the computational model is well-established. In the pointer model of computation, the Fibonacci heap data structure of Fredman and Tarjan [16] leads to an $O(m + n \log n)$ implementation of Dijkstra's algorithm. In a RAM model with word operations, the fastest currently known algorithms achieve the following bounds: $O(m+n(\log U \log \log U)^{1/3})$ [29], $O(m+n(\sqrt{\log n}))$ [28], $O(m \log \log U)$ [22], and $O(m \log \log n)$ [32].

For undirected graphs, Thorup's algorithm [31] has a linear running time in a word RAM model. A constant-time priority queue of [3] yields a linear-time

F. Meyer auf der Heide (Ed.): ESA 2001, LNCS 2161, pp. 230–241, 2001.

algorithm for directed graphs, but only in a non-standard computation model that is not supported by any currently existing computers.

In a recent paper [23,24], Meyer gives a shortest path algorithm with a linear average time for input arc lengths drawn independently from a uniform distribution on $[1, \ldots, M]$. He also proves that, under the same conditions, the running time is linear with high probability. Meyer's algorithm may scan some vertices more than once, and its worst-case time bound, $O(nm \log n)$, is far from linear. Both the algorithm and its analysis are complicated.

In this paper we show that an improvement of the multi-level bucket shortest path algorithm of [9] has an average running time that is linear, and a worst-case time of $O(m+n \log C)$. Our average-time bound holds for arc lengths distributed uniformly on $[1, \ldots, M]$. We also show that if the arc lengths are independent, the algorithm running time is linear with high probability. Both the algorithm and its analysis are natural and simple. Our algorithm is not an implementation of Dijkstra's algorithm: a vertex selected for scanning is not necessarily a minimum labeled vertex. However, the selected vertex distance label is equal to the correct distance, and each vertex is scanned at most once. This relaxation of Dijkstra's algorithm was originally introduced by Dinitz [13], used in its full strength by Thorup [31], and also used in [23]. Our technique can also be used to improve the worst-case running time of the above-mentioned $O(m+n(\log U \log \log U)^{1/3})$ algorithm of Raman [29] to $O(m + n(\log C \log \log C)^{1/3})$. (This new bound also applies if the input arc lengths are real-valued.)

Our results advance understanding of near-linear shortest path algorithms. Since many computational studies use graphs with uniform arc lengths, these results show that such problems are easy in a certain sense. Although we prove our results for the uniform arc length distribution, our algorithm achieves improved bounds on some other distributions as well. Our results may have practical implications in addition to theoretical ones. The multi-level bucket algorithm works well in practice [5,21] and our improvement of this algorithm is natural and easy to implement. It is possible that a variant of our algorithm is competitive with the current state-of-the-art implementations on all inputs while outperforming these implementations on some inputs. Although our algorithm looks attractive, the competing implementations are highly optimized and practicality of our results cannot be claimed without a careful implementation and experimentation.

2   Preliminaries

The input to the shortest path problem we consider is a directed graph $G = (V, A)$ with n vertices, m arcs, a source vertex s, and nonnegative arc lengths $\ell(a)$. The goal is to find shortest paths from the source to all vertices of the graph. We assume that arc lengths are integers in the interval $[1, \ldots, U]$, where U denotes the biggest arc length. Let δ be the smallest nonzero arc length and let C be the ratio of the biggest arc length to δ. If all arc lengths are zero or if $C < 2$, then the problem can be solved in linear time [13]; without loss of generality, we assume that $C \geq 2$ (and $\log C \geq 1$). This implies $\log U \geq 1$. We say that a statement holds *with high probability (w.h.p.)* if the probability that the statement is true approaches one as $m \to \infty$.

We assume the *word RAM* model of computation (see e.g., [1]). Our data structures need array addressing and the following unit-time word operations: addition, subtraction, comparison, and arbitrary shifts. To allow a higher-level description of our algorithm, we use a *strong RAM* computation model that also allows word operations including bitwise logical operations and the operation of finding the index of the most significant bit in which two words differ. The latter operation is in AC0; see [8] for a discussion of a closely related operation. The use of this more powerful model does not improve the amortized operation bounds, but simplifies the description.

Our shortest path algorithm uses a variant of the multi-level bucket data structure of Denardo and Fox [9]. Although we describe our result in the strong RAM model, following [6], one can also follow [9,21] and obtain an implementation of the algorithm in the weaker word RAM model. Although somewhat more complicated to describe formally, the latter implementation appears more practical.

3   Labeling Method and Related Results

The labeling method for the shortest path problem [14,15] works as follows (see e.g., [30]). The method maintains for every vertex v its distance label $d(v)$, parent $p(v)$, and status $S(v) \in \{\text{unreached}, \text{labeled}, \text{scanned}\}$. Initially $d(v) = \infty$, $p(v) = nil$, and $S(v) = \text{unreached}$. The method starts by setting $d(s) = 0$ and $S(s) = \text{labeled}$. While there are labeled vertices, the method picks such a vertex v, scans all arcs out of v, and sets $S(v) = \text{scanned}$. To scan an arc (v, w), one checks if $d(w) > d(v) + \ell(v, w)$ and, if true, sets $d(w) = d(v) + \ell(v, w)$, $p(w) = v$, and $S(w) = \text{labeled}$.

If the length function is nonnegative, the labeling method always terminates with correct shortest path distances and a shortest path tree. The efficiency of the method depends on the rule to chose a vertex to scan next. We say that $d(v)$ is *exact* if the distance from s to v is equal to $d(v)$. It is easy to see that if the method always selects a vertex v such that, at the selection time, $d(v)$ is exact, then each vertex is scanned at most once.

Dijkstra [12] observed that if ℓ is nonnegative and v is a labeled vertex with the smallest distance label, than $d(v)$ is exact. However, a linear-time implementation of Dijkstra's algorithm in the strong RAM model appears to be hard. Dinitz [13] and Thorup [31] use a relaxation of Dijkstra's selection rule to get linear-time algorithms for special cases of the shortest path problem. To describe this relaxation, we define the *caliber* of a vertex v, $c(v)$, to be the minimum length of an arc entering v, or infinity if no arc enters v. The following *caliber lemma* is implicit in [13,29].

Lemma 1. *Suppose ℓ is nonnegative and let μ be a lower bound on distance labels of labeled vertices. Let v be a vertex such that $\mu + c(v) \geq d(v)$. Then $d(v)$ is exact.*

4   Algorithm Description and Correctness

Our algorithm is based on the multi-level bucket implementation [9] of Dijkstra's algorithm, but we use Lemma 1 to detect and scan vertices with exact (but not necessarily minimum) distance labels. Our algorithm is a labeling algorithm. During the initialization, the algorithm also computes $c(v)$ for every vertex v. Our algorithm keeps labeled vertices in one of two places: a set F and a priority queue B. The former is implemented to allow constant time additions and deletions, for example as a doubly linked list. The latter is implemented using multi-level buckets as described below. The priority queue supports operations `insert`, `delete`, `decrease-key`, and `extract-min`. However, the `insert` operation may insert the vertex into B or F, and the `decrease-key` operation may move the vertex from B to F

At a high level, the algorithm works as follows. Vertices in F have exact distance labels and if F is nonempty, we remove and scan a vertex from F. If F is empty, we remove and scan a vertex from B with the minimum distance label. Suppose a distance label of a vertex u decreases. Note that u cannot belong to F. If u belongs to B, then we apply the `decrease-key` operation to u. This operation either relocates u within B or discovers that u's distance label is exact and moves u to F. If u was neither in B nor F, we apply the `insert` operation to u, and u is inserted either into B or, if $d(u)$ is determined to be exact, into F.

The bucket structure B contains $k + 1$ levels of buckets, where $k = \lceil \log U \rceil$. Except for the top level, a level contains two buckets. Conceptually, the top level contains infinitely many buckets. However, at most three consecutive top-level buckets can be nonempty at any given time (Lemma 2 below), and one can maintain only these buckets by wrapping around modulo three at the top level. (For low-level efficiency, one may want to have wrap around modulo four, which is a power of two.)

We denote bucket j at level i by $B(i,j)$; i ranges from 0 (bottom level) to k (top), and j ranges from 0 to 1, except at the top level discussed above. A bucket contains a set of vertices maintained in a way that allows constant-time insertion and deletion, e.g., in a doubly linked list. At each level i, we maintain the number of vertices at this level.

We maintain μ such that μ is a lower bound on the distance labels of labeled vertices. Initially $\mu = 0$. Every time an `extract-min` operation removes a vertex v from B, we set $\mu = d(v)$. Consider the binary representation of the distance labels and number bit positions starting from 0 for the least significant bit. Let $\mu_{i,j}$ denote the i-th through j-th least significant bit of μ and let μ_i denote the i-th least significant bit. Similarly, $d_i(u)$ denotes the i-th least significant bit of $d(u)$, and likewise for the other definitions. Note that μ and the $k + 1$ least significant bits of the binary representation of $d(u)$ uniquely determine $d(u)$: $d(u) = \mu + (u_{0,k} - \mu_{0,k})$ if $u_{0,k} > \mu_{0,k}$ and $d(u) = \mu + 2^k + (u_{0,k} - \mu_{0,k})$ otherwise.

For a given μ, let $\underline{\mu_i}$ and $\overline{\mu_i}$ be μ with the i least significant bits replaced by 0 or 1, respectively. Each level $i < k$ corresponds to the range of values $[\underline{\mu_{i+1}}, \overline{\mu_{i+1}}]$.

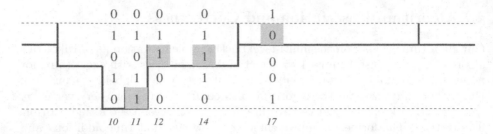

Fig. 1. Multi-level bucket example. $k = 3$, $\mu = 10$. Values on the bottom are in decimal. Values on top are in binary, with the least significant bit on the bottom. Shaded bits determine positions of the corresponding elements.

Each bucket $B(i, j)$ corresponds to the subrange containing all integers in the range with the i-th bit equal to j. At the top level, a bucket $B(k, j)$ corresponds to the range $[j \cdot 2^k, (j + 1) \cdot 2^k)$. The *width* of a bucket at level i is equal to 2^i: the bucket contains 2^i distinct values. We say that a vertex u *is in the range of* $B(i, j)$ if $d(u)$ belongs to the range corresponding to the bucket.

The position of a vertex u in B depends on μ: u belongs to the lowest-level bucket containing $d(u)$. More formally, let i be the index of the most significant bit in which $d(u)$ and $\mu_{0,k}$ differ, or 0 if they match. Note that $\underline{\mu_i} \leq d(u) \leq \overline{\mu_i}$. Given μ and u with $d(u) \geq \mu$, we define the *position of u* by $(i, d_i(\overline{u}))$ if $i < k$ and $B(k, \lfloor d(u) - \mu/2^k \rfloor)$ otherwise. If u is inserted into B, it is inserted into $B(i, j)$, where (i, j) is the position of u. For each vertex in B, we store its position.

Figure 1 gives an example of the bucket structure. In this example, $k = 3$ and $\mu = 10$. For instance, to find the position of a vertex v with $d(v) = 14$, we note that the binary representations of 10 and 14 differ in bit 2 (remember that we start counting from 0) and the bit value is 1. Thus v belongs to bucket 1 at level 2.

Our modification of the multi-level bucket algorithm uses Lemma 1 during the **insert** operation to put vertices into F whenever the lemma allows it. The details are as follows.

insert: Insert a vertex u into $B \cup F$ as follows. If $\mu + c(u) \geq d(u)$, put u into F. Otherwise compute u's position (i, j) in B and add u to $B(i, j)$.

decrease-key: Decrease the key of an element u in position (i, j) as follows. Remove u from $B(i, j)$. Set $d(u)$ to the new value and insert u as described above.

extract-min: Find the lowest nonempty level i. Find j, the first nonempty bucket at level i. If $i = 0$, delete a vertex u from $B(i, j)$. (In this case $\mu = d(u)$.) Return u. If $i > 0$, examine all elements of $B(i, j)$ and delete a minimum element u from $B(i, j)$. Note that in this case $\mu < d(u)$; set $\mu = d(u)$. Since μ increased, some vertex positions in B may have changed. We do *bucket expansion* of $B(i,j)$ and return u.

To understand bucket expansion, note that the vertices with changed positions are exactly those in $B(i,j)$. To see this, let μ' be the old value of μ and consider a vertex v in B. Let (i',j') be v's position with respect to μ'. By the choice of $B(i,j)$, if $(i,j) \neq (i',j')$, then either $i < i'$, or $i = i'$ and $j < j'$. In both cases, the common prefix of μ' and $d(v)$ is the same as the common prefix of $d(u)$ and $d(v)$, and the position of v does not change.

On the other hand, vertices in $B(i,j)$ have a longer common prefix with $d(u)$ than they have with μ' and these vertices need to move to a *lower* level. Bucket expansion deletes these vertices from $B(i)$ and uses the `insert` operation to add the vertices back into B or into F, as appropriate.

Although the formal description of the algorithm is a little complicated, the algorithm itself is relatively simple: At each step, remove a vertex from F; or, if F is empty, then remove the minimum-labeled vertex from B. In the latter case, expand the bucket from which the vertex has been removed, if necessary. Scan the vertex and update its neighbors if necessary. Terminate when both F and B are empty.

Note that we do bucket expansions only when F is empty and the expanded bucket contains a labeled vertex with the minimum distance. Thus μ is updated correctly.

In the original multi-level bucket algorithm, at any point of the execution all labeled vertices are contained in at most two consecutive top level buckets. A slightly weaker result holds for our algorithm.

Lemma 2. *At any point of the execution, all labeled vertices are in the range of at most three consecutive top level buckets.*

Proof. Let μ' be the current value of μ and let $B(k,j)$ be the top level bucket containing μ'. Except for s (for which the result holds trivially), a vertex v becomes labeled during a scan of another vertex u removed from either B or F. In the former case, at the time of the scan $d(u) = \mu \leq \mu'$, $d(v) = \mu + \ell(u,v) \leq \mu' + 2^k$, and therefore v is contained either in $B(k,j)$ or $B(k,j+1)$. In the latter case, when u has been added to F, the difference between $d(u)$ and μ was at most $c(u) \leq 2^k$, thus $d(u) \leq \mu' + 2^k$, $d(v) \leq d(u) + 2^k \leq \mu' + 2 \cdot 2^k$, and thus v belongs to $B(k,j)$, $B(k,j+1)$, or $B(k,j+2)$.

Algorithm correctness follows from Lemmas 1 and 2, and the observations that μ is always set to a minimum distance label of a labeled vertex, μ remains a lower bound on the labeled vertex labels (and therefore is monotonically non-decreasing), and F always contains vertices with exact distance labels.

5   Worst-Case Analysis

In this section we prove a worst-case bound on the running time of the algorithm. Some definitions and lemmas introduced in this section will be also used in the next section.

We start the analysis with the following lemmas.

Lemma 3. [6]

- *Given μ and u, we can compute the position of u with respect to μ in constant time.*
- *We can find the lowest nonempty level of B in $O(1)$ time.*

Lemma 4. *The algorithm runs in $O(m + n + \Phi)$ time, where Φ is the total number of times a vertex moves from a bucket of B to a lower level bucket.*

Proof. Since each vertex is scanned at most once, the total scan time is $O(m + n)$. A vertex is added to and deleted from F at most once, so the total time devoted to maintaining F is $O(n)$. An **insert** operation takes constant time, and these operations are caused by inserting vertices into B for the first time, by **decrease-key** operations, and by **extract-min** operations. The former take $O(n)$ time; we account for the remaining ones jointly with the other operations. A **decrease-key** operation takes constant time and is caused by a decrease of $d(v)$ due to a scan of an arc (u, v). Since an arc is scanned at most once, these operations take $O(m)$ total time. The work we accounted for so far is linear.

Next we consider the **extract-min** operations. Consider an **extract-min** operation that returns u. The operation takes $O(1)$ time plus an amount of time proportional to the number of vertices in the expanded bucket, excluding u. Each of these vertices moves to a lower level in B. Thus we get the desired time bound.

The $O(m + n \log U)$ worst-case time bound is easy to see. To show a better bound, we define $k' = \lfloor \log \delta \rfloor$.

Lemma 5. *Buckets at level k' and below are never used.*

Proof. Let (i, j) be the position of a vertex v of caliber $c(v) \geq \delta$. If $i \leq k'$, then $d(v) - \mu < 2^i \leq 2^{k'} \leq \delta \leq c(v)$ and the algorithm adds v to F, not B.

The above lemma implies the following bound.

Theorem 1. *The worst-case running time of the algorithm is $O(m + n \log C)$.*

Note that the lemma also implies that the algorithm needs only $O(\log C)$ words of memory to implement the bucket data structure.

Our optimization can also be used to improve other data structures based on multi-level buckets, such as radix heaps [2] and hot queues [6]. For these data structures, the equivalent of Lemma 5 allows one to replace time bound parameter U by C. In particular, the bound of the hot queue implementation of Raman [29] improves to $O(m + n(\log C \log \log C)^{1/3})$. The modification of Raman's algorithm to obtain this bound is straightforward given the results of the current section.

6   Average-Case Analysis

In this section we prove linear-time average and high-probability time bounds for our algorithm under the assumption that the input arc lengths are uniformly distributed on $[1, \ldots, M]$.[1]

A key lemma for our analysis is as follows.

Lemma 6. *The algorithm never inserts a vertex v into a bucket at a level less than or equal to $\log c(v) - 1$.*

Proof. Suppose during an **insert** operation v's position in B is (i, j) with $i \leq \log c(v) - 1$. Then the most significant bit in which $d(v)$ and μ differ is bit i and $d(v) - \mu < 2^{i+1} \leq c(v)$. Therefore **insert** puts v into F, not B.

The above lemma motivates the following definitions. The *weight of an arc* a, $w(a)$, is defined by $w(a) = k - \lfloor \log \ell(a) \rfloor$. The *weight of a vertex* v, $w(v)$, is defined to be the maximum weight of an incoming arc or zero if v has no incoming arcs. Lemma 6 implies that the number of times v can move to a lower level of B is at most $w(v) + 1$ and therefore $\Phi \leq m + \sum_V w(v)$. Note that k depends on the input, the weights are defined with respect to a given input.

For the probability distribution of arc weights defined above, we have $\mathbf{Pr}[\lfloor \log \ell(a) \rfloor = i] = 2^i / M$ for $i = 0, \ldots, k - 1$. The definition of w yields

$$\mathbf{Pr}[w(a) = t] = 2^{k-t}/M \quad \text{for } t = 1, \ldots, k. \tag{1}$$

Since $M \geq U$, we have $M \geq 2^{k-1}$, and therefore

$$\mathbf{Pr}[w(a) = t] \leq 2^{-t+1} \quad \text{for } t = 1, \ldots, k. \tag{2}$$

Theorem 2. *If arc lengths are uniformly distributed on $[1, \ldots, M]$, then the average running time of the algorithm is linear.*

Proof. Since $\Phi \leq m + \sum_V w(v)$, it is enough to show that $\mathbf{E}[\sum_V w(v)] = O(m)$. By the linearity of expectation and the definition of $w(v)$, we have $\mathbf{E}[\sum_V w(v)] \leq \sum_A \mathbf{E}[w(a)]$. The expected value of $w(a)$ is

$$\mathbf{E}[w(a)] = \sum_{i=1}^{k} i \mathbf{Pr}[w(a) = i] \leq \sum_{i=1}^{\infty} i 2^{-i+1} = 2 \sum_{i=1}^{\infty} i 2^{-i} = O(1).$$

Note that this bound holds for any k. Thus $\sum_A \mathbf{E}[w(a)] = O(m)$.

Remark The proof of the theorem works for any arc length distribution such that $\mathbf{E}[w(a)] = O(1)$. In particular, the theorem holds for real-valued arc lengths selected independently and uniformly from $[0, 1]$. In fact, for this distribution the

---

[1] As we shall see, if M is large enough then the result also applies to the range $[0, \ldots, M]$.

high-probability analysis below is simpler. However, the integer distribution is somewhat more interesting, for example because some test problem generators use this distribution.

Remark Note that Theorem 2 does not require arc lengths to be independent. Our proof of its high-probability variant, Theorem 3, requires the independence.

Next we show that the algorithm running time is linear w.h.p. by showing that $\sum_A w(a) = O(m)$ w.h.p. First, we show that w.h.p. U is not much smaller than M and δ is close to Mm^{-1} (Lemmas 7 and 8). Let S_t be the set of all arcs of weight t and note that $\sum_A w(a) = \sum_t t|S_t|$. We show that as t increases, the expected value of $|S_t|$ goes down exponentially. For small values of t, this is also true w.h.p. To deal with large values of t, we show that the total number of arcs with large weights is small, and so is the contribution of these arcs to the sum of arc weights.

Because of the space limit, we omit proofs of the following two lemmas.

Lemma 7. *W.h.p., $U \geq M/2$.*

Lemma 8. *W.h.p., $\delta \geq Mm^{-4/3}$. If $M \geq m^{2/3}$, then w.h.p. $\delta \leq Mm^{-2/3}$.*

From (1) and the independence of arc weights, we have $\mathbf{E}[|S_t|] = m2^{k-t}/M$. By the Chernoff bound (see e.g. [7,25]), $\mathbf{Pr}\big[|S_t| \geq 2m2^{k-t}/M\big] < \left(\frac{e}{4}\right)^{m2^{k-t}/M}$. Since $M \geq 2^{k-1}$, we have

$$\mathbf{Pr}\big[|S_t| \geq 4m2^{-t}\big] < \left(\frac{e}{4}\right)^{2m2^{-t}}.$$

As mentioned above, we bound the contributions of arcs with large and small weights to $\sum_A w(a)$ differently. We define $\beta = \log(m^{2/3})$ and partition A into two sets, A_1 containing the arcs with $w(a) \leq \beta$ and A_2 containing the arcs with $w(a) > \beta$.

Lemma 9. $\sum_{A_1} w(a) = O(m)$ *w.h.p.*

Proof. Assume that $\delta \geq Mm^{-4/3}$ and $U \geq M/2$; by Lemmas 7 and 8 this happens w.h.p. This assumption implies $C \leq m^{4/3}$. The probability that for some $t : 1 \leq t \leq \beta$, $|S_t| \geq 4m2^{-t}$ is, by the union bound and the fact that the probability is maximized for $t = \beta$, less than

$$\beta \left(\frac{e}{4}\right)^{m2^{-\beta}} \leq \log(m^{2/3}) \left(\frac{e}{4}\right)^{mm^{-2/3}} \leq \log m \left(\frac{e}{4}\right)^{m^{1/3}} \to 0 \quad \text{as } m \to \infty.$$

Thus w.h.p., for all $t : 1 \leq t \leq \beta$, we have $|S_t| < 4m2^{-t}$ and

$$\sum_{A_1} w(a) = \sum_{t=1}^{t \leq \beta} t|S_t| \leq 4m \sum_{t=1}^{\infty} t2^{-t} = O(m).$$

Lemma 10. $\sum_{A_2} w(a) = O(m)$ w.h.p.

Proof. If $M < m^{2/3}$, then $k \leq \beta$ and A_2 is empty, so the lemma holds trivially.

Now consider the case $M \geq m^{2/3}$. By Lemmas 7 and 8, w.h.p. $Mm^{-4/3} \leq \delta \leq Mm^{-2/3}$ and $U \geq M/2$; assume that this is the case. The assumption implies $m^{2/3}/2 \leq C \leq m^{4/3}$. Under this assumption, we also have $2^{k-1} \leq M \leq 2^{k+1}$. Combining this with (1) we get $2^{-2-t} \leq \mathbf{Pr}[w(a) = t] \leq 2^{1-t}$. This implies that

$$2^{-2-\beta} \leq \mathbf{Pr}[w(a) > \beta] \leq 2^{2-\beta},$$

therefore

$$\frac{m^{-2/3}}{8} \leq \mathbf{Pr}[w(a) > \beta] \leq 4m^{-1/3}$$

and by the independence of arc weights,

$$\frac{m^{1/3}}{8} \leq \mathbf{E}[|A_2|] \leq 4m^{2/3}$$

By the Chernoff bound,

$$\mathbf{Pr}[|A_2| > 2\mathbf{E}[|A_2|]] < \left(\frac{e}{4}\right)^{\mathbf{E}[|A_2|]}.$$

Replacing the first occurrence of $\mathbf{E}[|A_2|]$ by the upper bound on its value and the second occurrence by the lower bound (since $e/4 < 1$), we get

$$\mathbf{Pr}\left[|A_2| > 8m^{2/3}\right] < \left(\frac{e}{4}\right)^{m^{1/3}/8} \to 0 \quad \text{as } m \to \infty.$$

For all arcs a, $\ell(a) \geq \delta$, and thus

$$w(a) = k - \lfloor \ell(a) \rfloor \leq 1 + \log U + 1 - \log \delta = 2 + \log C \leq 2 + (4/3) \log m.$$

Therefore w.h.p.,

$$\sum_{A_2} w(a) \leq 8m^{2/3}(2 + (4/3) \log m) = o(m).$$

Thus we have the following theorem.

Theorem 3. *If arc lengths are independent and uniformly distributed on $[1, \ldots, M]$, then with high probability, the algorithm runs in linear time.*

Remark The expected and high probability bounds also apply if the arc lengths come from $[0, \ldots, U]$ and $U = \omega(m)$, as in this case with high probability no arc has zero length.

7   Concluding Remarks

We described our algorithm for the binary multi-level buckets, with two buckets at each level except for the top level. One can easily extend the algorithm for base-Δ buckets, for any integer $\Delta \geq 2$. One gets a worst-case bound for $\Delta = \theta(\frac{\log C}{\log \log C})$ [9] when the work of moving vertices to lower levels balances with the work of scanning empty buckets during bucket expansion. Our average-case analysis reduces the former but not the latter. We get a linear running time when Δ is constant and the empty bucket scans can be charged to vertices in nonempty buckets. An interesting open question is if one can get a linear average running time and a better worst-case running time, for example using techniques from [2,6,9], without running several algorithms "in parallel."

Our optimization is to detect vertices with exact distance labels before these vertices reach the bottom level of buckets and place them into F. This technique can be used not only in the context of multi-level buckets, but in the context of radix heaps [2] and hot queues [6].

Acknowledgments. The author would like to thank Jim Horning, Anna Karlin, Rajeev Raman, Bob Tarjan, and Eva Tardos for useful discussion and comments on a draft of this paper. We are also greatful to an anonymous referee for pointing out that Theorem 2 does not need arc lengths to be independent.

References

1. A. V. Aho, J. E. Hopcroft, and J. D. Ullman. *The Design and Analysis of Computer Algorithms*. Addison-Wesley, 1974.
2. R. K. Ahuja, K. Mehlhorn, J. B. Orlin, and R. E. Tarjan. Faster Algorithms for the Shortest Path Problem. *J. Assoc. Comput. Mach.*, 37(2):213–223, April 1990.
3. A. Brodnik, S. Carlsson, J. Karlsson, and J. I. Munro. Worst case constant time priority queues. In *Proc. 12th ACM-SIAM Symposium on Discrete Algorithms*, pages 523–528, 2001.
4. B. V. Cherkassky and A. V. Goldberg. Negative-Cycle Detection Algorithms. *Math. Prog.*, 85:277–311, 1999.
5. B. V. Cherkassky, A. V. Goldberg, and T. Radzik. Shortest Paths Algorithms: Theory and Experimental Evaluation. *Math. Prog.*, 73:129–174, 1996.
6. B. V. Cherkassky, A. V. Goldberg, and C. Silverstein. Buckets, Heaps, Lists, and Monotone Priority Queues. *SIAM J. Comput.*, 28:1326–1346, 1999.
7. H. Chernoff. A Measure of Asymptotic Efficiency for Test of a Hypothesis Based on the Sum of Observations. *Anals of Math. Stat.*, 23:493–509, 1952.
8. R. Cole and U. Vishkin. Deterministic Coin Tossing with Applications to Optimal Parallel List Ranking. *Information and Control*, 70:32–53, 1986.
9. E. V. Denardo and B. L. Fox. Shortest–Route Methods: 1. Reaching, Pruning, and Buckets. *Oper. Res.*, 27:161–186, 1979.
10. R. B. Dial. Algorithm 360: Shortest Path Forest with Topological Ordering. *Comm. ACM*, 12:632–633, 1969.
11. R. B. Dial, F. Glover, D. Karney, and D. Klingman. A Computational Analysis of Alternative Algorithms and Labeling Techniques for Finding Shortest Path Trees. *Networks*, 9:215–248, 1979.

12. E. W. Dijkstra. A Note on Two Problems in Connexion with Graphs. *Numer. Math.*, 1:269–271, 1959.
13. E. A. Dinic. Economical algorithms for finding shortest paths in a network. In Yu.S. Popkov and B.L. Shmulyian, editors, *Transportation Modeling Systems*, pages 36–44. Institute for System Studies, Moscow, 1978. In Russian.
14. L. Ford. Network Flow Theory. Technical Report P-932, The Rand Corporation, 1956.
15. L. R. Ford, Jr. and D. R. Fulkerson. *Flows in Networks*. Princeton Univ. Press, Princeton, NJ, 1962.
16. M. L. Fredman and R. E. Tarjan. Fibonacci Heaps and Their Uses in Improved Network Optimization Algorithms. *J. Assoc. Comput. Mach.*, 34:596–615, 1987.
17. M. L. Fredman and D. E. Willard. Trans-dichotomous Algorithms for Minimum Spanning Trees and Shortest Paths. *J. Comp. and Syst. Sci.*, 48:533–551, 1994.
18. H. N. Gabow. Scaling Algorithms for Network Problems. *J. of Comp. and Sys. Sci.*, 31:148–168, 1985.
19. G. Gallo and S. Pallottino. Shortest Paths Algorithms. *Annals of Oper. Res.*, 13:3–79, 1988.
20. F. Glover, R. Glover, and D. Klingman. Computational Study of an Improved Shortest Path Algorithm. *Networks*, 14:25–37, 1984.
21. A. V. Goldberg and C. Silverstein. Implementations of Dijkstra's Algorithm Based on Multi-Level Buckets. In P. M. Pardalos, D. W. Hearn, and W. W. Hages, editors, *Lecture Notes in Economics and Mathematical System 450 (Refereed Proceedings)*, pages 292–327. Springer Verlag, 1997.
22. T. Hagerup. Improved Shortest Paths in the Word RAM. In *27th Int. Colloq. on Automata, Languages and Programming, Geneva, Switzerland*, pages 61–72, 2000.
23. U. Meyer. Single-Source Shortest Paths on Arbitrary Directed Graphs in Linear Average Time. In *Proc. 12th ACM-SIAM Symposium on Discrete Algorithms*, pages 797–806, 2001.
24. U. Meyer. Single-Source Shortest Paths on Arbitrary Directed Graphs in Linear Average Time. Technical Report MPI-I-2001-1-002, Max-Planck-Institut für Informatik, Saarbrüken, Germany, 2001.
25. R. Motwani and P. Raghavan. *Randomized Algorithms*. Cambridge University Press, 1995.
26. K. Noshita. A Theorem on the Expected Complexity of Dijkstra's Shortest Path Algorithm. *J. Algorithms*, 6:400–408, 1985.
27. R. Raman. Fast Algorithms for Shortest Paths and Sorting. Technical Report TR 96-06, King's Colledge, London, 1996.
28. R. Raman. Priority Queues: Small, Monotone and Trans-Dichotomous. In *Proc. 4th Annual European Symposium Algorithms*, pages 121–137. Springer-Verlag, Lect. Notes in CS 1136, 1996.
29. R. Raman. Recent Results on Single-Source Shortest Paths Problem. *SIGACT News*, 28:81–87, 1997.
30. R. E. Tarjan. *Data Structures and Network Algorithms*. Society for Industrial and Applied Mathematics, Philadelphia, PA, 1983.
31. M. Thorup. Undirected Single-Source Shortest Paths with Positive Integer Weights in Linear Time. *J. Assoc. Comput. Mach.*, 46:362–394, 1999.
32. M. Thorup. On RAM Priority Queues. *SIAM Journal on Computing*, 30:86–109, 2000.
33. F. B. Zhan and C. E. Noon. Shortest Path Algorithms: An Evaluation using Real Road Networks. *Transp. Sci.*, 32:65–73, 1998.

A Heuristic for Dijkstra's Algorithm with Many Targets and Its Use in Weighted Matching Algorithms*

Kurt Mehlhorn and Guido Schäfer**

Max-Planck-Institut für Informatik
Stuhlsatzenhausweg 85, 66123 Saarbrücken, Germany

http://www.mpi-sb.mpg.de/~{mehlhorn|schaefer}

Abstract. We consider the single-source many-targets shortest-path (SSMTSP) problem in directed graphs with non-negative edge weights. A source node s and a target set T is specified and the goal is to compute a shortest path from s to a node in T. Our interest in the shortest path problem with many targets stems from its use in weighted bipartite matching algorithms. A weighted bipartite matching in a graph with n nodes on each side reduces to n SSMTSP problems, where the number of targets varies between n and 1.

The SSMTSP problem can be solved by Dijkstra's algorithm. We describe a heuristic that leads to a significant improvement in running time for the weighted matching problem; in our experiments a speed-up by up to a factor of 10 was achieved. We also present a partial analysis that gives some theoretical support for our experimental findings.

1   Introduction and Statement of Results

A matching in a graph is a subset of the edges no two of which share an endpoint. The weighted bipartite matching problem asks for the computation of a maximum weight matching in an edge-weighted bipartite graph $G = (A \cup B, E, w)$ where the cost function $w : E \to I\!R$ assigns a real weight to every edge. The weight of a matching M is simply the sum of the weights of the edges in the matching, i.e., $w(M) = \sum_{e \in M} w(e)$. One may either ask for a perfect matching of maximal weight (the weighted perfect matching problem or the assignment problem) or simply for a matching of maximal weight. Both versions of the problem can be solved by solving n, $n = \max(|A|, |B|)$, single-source many-targets shortest-path (SSMTSP) problems in a derived graph, see Sec. 4. We describe and analyse a heuristic improvement for the SSMTSP problem which leads to

---

*   Partially supported by the IST Programme of the EU under contract number IST-1999-14186 (ALCOM-FT).
** Funded by the Deutsche Forschungsgemeinschaft (DFG), Graduiertenkolleg (Graduate Studies Program) 'Quality Guarantees for Computer Systems', Department of Computer Science, University of the Saarland, Germany.

F. Meyer auf der Heide (Ed.): ESA 2001, LNCS 2161, pp. 242–253, 2001.

a significant speed-up in LEDA's weighted bipartite matching implementation, see Tab. 3.

In the SSMTSP problem we are given a directed graph $G = (V, E)$ whose edges carry a non-negative cost. We use $cost(e)$ to denote the cost of an edge $e \in E$. We are also given a source node s. Every node in V is designated as either free or non-free. We are interested in finding the shortest path from s to a free node.

The SSMTSP problem is easily solved by Dijkstra's algorithm. Dijkstra's algorithm (see Sec. 2) maintains a tentative distance for each node and a partition of the nodes into settled and unsettled. At the beginning all nodes are unsettled. The algorithm operates in phases. In each phase, the unsettled node with smallest tentative distance is declared settled and its outgoing edges are relaxed in order to improve tentative distances of other unsettled nodes. The unsettled nodes are kept in a priority queue. The algorithm can be stopped once the first free node becomes settled.

We describe a heuristic improvement. The improvement maintains an upper bound for the tentative distance of free nodes and performs only queue operations with values smaller than the bound. All other queue operations are suppressed. The heuristic significantly reduces the number of queue operations and the running time of the bipartite matching algorithm, see Tab. 2 and Tab. 3.

This paper is structured as follows. In Sec. 2 we discuss Dijkstra's algorithm for many targets and describe our heuristic. In Sec. 3 we give an analysis of the heuristic for random graphs and report about experiments on random graphs. In Sec. 4 we discuss the application to weighted bipartite matching algorithms and present our experimental findings for the matching problem.

The heuristic was first used by the second author in his jump-start routine for the general weighted matching algorithm [6,5]. When applied to bipartite graphs, the jump-start routine computes a maximum weight matching. When we compared the running time of the jump-start routine with LEDA's bipartite matching code [4, Sec. 7.8], we found that the jump-start routine is consistently faster. We traced the superiority to the heuristic described in this paper.

2   Dijkstra's Algorithm with Many Targets

It is useful to introduce some more notation. For a node $v \in V$, let $d(v)$ be the shortest path distance from s to v, and let $d_0 = \min \{d(v) \ ; \ v \text{ is free}\}$. If there is no free node reachable from s, $d_0 = +\infty$. Our goal is to compute (1) a node v_0 with $d(v_0) = d_0$ (or an indication that there is no such node), (2) the subset V' of nodes with $d(v) < d_0$, more precisely, $v \in V'$ if $d(v) < d_0$ and $d(v) \geq d_0$ if $v \notin V'$, and (3) the value $d(v)$ for every node $v \in \{v_0\} \cup V'$, i.e., a partial function $\tilde{d}$ with $\tilde{d}(v) = d(v)$ for any $v \in \{v_0\} \cup V'$. (Observe that nodes v with $d(v) = d_0$ may or may not be in V'.) We refer to the problem just described as the single-source many-targets shortest-path (SSMTSP) problem. It is easily solved by an adapted version of Dijkstra's algorithm as shown in Fig. 1.

DIJKSTRA'S ALGORITHM (ADAPTED VERSION):
$dist(s) = 0$ and $dist(u) = +\infty$ for all $u \in V, u \neq s$
$PQ.insert(s, 0)$ (insert $\langle s, 0 \rangle$ into PQ)
while not $PQ.empty()$ **do**
   $u = PQ.del\_min()$ (remove node u from PQ with minimal priority)
   if u is free **then STOP fi**
   RELAX ALL OUTGOING EDGES OF u
od

RELAX ALL OUTGOING EDGES OF u:
forall $e = (u, v) \in E$ **do**
   $c = dist(u) + cost(e)$
   if $c < dist(v)$ **then**
      if $dist(v) = +\infty$ (v is not contained in PQ)
      then $PQ.insert(v, c)$ (insert $\langle v, c \rangle$ into PQ)
      else $PQ.decrease\_p(v, c)$ (decrease priority of v in PQ to c)
      fi
      $dist(v) = c$
   fi
od

Fig. 1. Dijkstra's algorithm adapted for many targets. When the first free node is removed from the queue, the algorithm is stopped: v_0 is the node removed last and V' consists of all non-free nodes removed from the queue.

We maintain a priority queue PQ for the nodes of G. The queue is empty initially. For each node $u \in V$ we compute a tentative distance $dist(u)$ of a shortest path from s to u. Initially, we set $dist(s)$ to zero and put the item $\langle s, 0 \rangle$ into the priority queue. For each $u \in V, u \neq s$, we set $dist(u)$ to $+\infty$ (no path from s to u has been encountered yet). In the main loop, we delete a node u with minimum $dist$-value from the priority queue. If u is free, we are done: $v_0 = u$ and V' is the set of nodes removed in preceding iterations. Otherwise, we relax all edges out of u. Consider an edge $e = (u, v)$ and let $c = dist(u) + cost(e)$. We check whether c is smaller than the current tentative distance of v. If so, we distinguish two cases. (1) If e is the first edge into v that is relaxed (this is the case iff $dist(v)$ equals $+\infty$) we insert an item $\langle v, c \rangle$ into PQ. (2) Otherwise, we decrease the priority of v in PQ to c. If a queue operation is performed, we also update $dist(v)$.

We next describe a heuristic improvement of the scheme above. Let B be the smallest $dist$-value of a free node encountered by the algorithm; $B = +\infty$ initially. We claim that queue operations $PQ.op(\cdot, c)$ with $c \geq B$ may be skipped without affecting correctness. This is clear, since the algorithm stops when the first free node is removed from the queue and since the $dist$-value of this node is certainly at least as small as B. Thus all $dist$-values less than $d(v_0)$ will be computed correctly. The modified algorithm may output a different node v_0 and a different set V'. However, if all distances are pairwise distinct the same node v_0 and the same set V' as in the basic algorithm are computed.

DIJKSTRA'S ALGORITHM WITH PRUNING HEURISTIC:
$dist(s) = 0$ and $dist(u) = +\infty$ for all $u \in V, u \neq s$
$B = +\infty$ (initialize upper bound to $+\infty$)
$PQ.insert(s, 0)$ (insert $\langle s, 0\rangle$ into PQ)
while not $PQ.empty()$ **do**
   $u = PQ.del\_min()$ (remove node u from PQ with minimal priority)
   if u is free **then STOP fi**
   RELAX ALL OUTGOING EDGES OF u
od

RELAX ALL OUTGOING EDGES OF u:
forall $e = (u, v) \in E$ **do**
  $c = dist(u) + cost(e)$
  if $c \geq B$ **then continue fi**   (prune edge if bound is exceeded)
  if v is free **then** $B = \min\{c, B\}$ **fi**   (try to improve bound)
  if $c < dist(v)$ **then**
    if $dist(v) = +\infty$ (v is not contained in PQ)
    then $PQ.insert(v, c)$ (insert $\langle v, c\rangle$ into PQ)
    else $PQ.decrease\_p(v, c)$ (decrease priority of v in PQ to c)
    fi
    $dist(v) = c$
  fi
od

Fig. 2. Dijkstra's algorithm for many targets with a pruning heuristic. An upper bound B for $d(v_0)$ is maintained and queue operations $PQ.op(\cdot, c)$ with $c \geq B$ are not performed.

The pruning heuristic can conceivably save on queue operations, since fewer insert and decrease priority operations may be performed. Figure 2 shows the algorithm with the heuristic added.

3   Analysis

We perform a partial analysis of the basic and the modified version of Dijkstra's algorithm for many targets. We use n for the number of nodes, m for the expected number of edges and f for the expected number of free nodes. We assume that our graphs are random graphs in the $B(n, p)$ model with $p = m/n^2$, i.e., each of the n^2 possible edges is picked independently and uniformly at random with probability p. We use c to denote $pn = m/n$. We also assume that a node is free with probability $q = f/n$ and that edge costs are random reals between 0 and 1. We could alternatively use the model in which all graphs with m edges are equally likely and in which the free nodes form a random subset of f nodes. The results would be similar. We are mainly interested in the case, where $p = c/n$ for a small constant c, say $2 \leq c \leq 10$, and q a constant, i.e., the expected number of free nodes is a fixed fraction of the nodes.

Deletions from the Queue: We first analyze the number of nodes removed from the queue. If our graph were infinite and all nodes were reachable from s, the expected number would be $1/q$, namely the expected number of trials until the first head occurs in a sequence of coin tosses with success probability q. However, our graph is finite (not really a serious difference if n is large) and only a subset of the nodes is reachable from s. Observe, that the probability that s has no outgoing edge is $(1-p)^n \approx e^{-c}$. This probability is non-negligible. We proceed in two steps. We first analyze the number of nodes removed from the queue given the number R of nodes reachable from s and in a second step review results about the number R of reachable nodes.

Lemma 1. *Let R be the number of nodes reachable from s in G and let T be the number of iterations, i.e., in iteration T the first free node is removed from the queue or there is no free node reachable from s and $T = R$. Then, $\mathbf{Pr}\,(T = t \,|\, R = r) = (1-q)^{t-1}q$, for $1 \leq t < r$, and $\mathbf{Pr}\,(T = t \,|\, R = r) = (1-q)^{t-1}$, for $t = r$. Moreover, for the expected number of iterations we have: $\mathbf{E}\,[T \,|\, R = r] = 1/q - (1-q)^r/q$.*

The proof is given in Appendix A.

The preceding Lemma gives us information about the number of deletions from the queue. The expected number of edges relaxed is $c\mathbf{E}\,[(T-1) \,|\, R = r]$ since $T - 1$ non-free nodes are removed from the queue and since the expected out-degree of every node is $c = m/n$. We conclude that the number of edges relaxed is about $((1/q) - 1)(m/n)$.

Now, how many nodes are reachable from s? This quantity is analyzed in [2, pages 149–155]. Let $\alpha > 0$ be such that $\alpha = 1 - \exp(-c\alpha)$, and let R be the number of nodes reachable from s. Then R is bounded by a constant with probability about $1 - \alpha$ and is approximately αn with probability about α. More precisely, for every $\epsilon > 0$ and $\delta > 0$, there is a t_0 such that for all sufficiently large n, we have

$$1 - \alpha - 2\epsilon \leq \mathbf{Pr}\,(R \leq t_0) \leq 1 - \alpha + \epsilon$$

and

$$\alpha - 2\epsilon \leq \mathbf{Pr}\,((1-\delta)\alpha n < R < (1+\delta)\alpha n) \leq \alpha + 3\epsilon\,.$$

Table 1 indicates that small values of ϵ and δ work even for moderate n. For $c = 2$, we have $\alpha \approx 0.79681$. We generated 10000 graphs with $n = 1000$ nodes and 2000 edges and determined the number of nodes reachable from a given source node s. This number was either smaller than 15 or larger than 714. The latter case occurred in $7958 \approx \alpha \cdot 10000$ trials. Moreover, the average number of nodes reachable from s in the latter case was $796.5 \approx \alpha \cdot 1000 = \alpha n$.

For the sequel we concentrate on the case that $(1-\delta)\alpha n$ nodes are reachable from s. In this situation, the probability that all reachable nodes are removed from the queue is about

$$(1-q)^{\alpha n} = \exp(\alpha n \ln(1-q)) \approx \exp(-\alpha n q) = \exp(-\alpha f)\,.$$

Table 1. For all experiments (except the one in the last column) we used random graphs with $n = 1000$ nodes and $m = cn$ edges. For the last column we chose $n = 2000$ in order to illustrate that the dependency on n is weak. Nodes were free with probability q. The following quantities are shown; for each value of q and c we performed 10^4 trials.

α: the solution of the equation $\alpha = 1 - \exp(-c\alpha)$.

MS: the maximal number of nodes reachable from s when few nodes are reachable.

ML: the minimal number of nodes reachable from s when many nodes are reachable.

R: the average number of nodes reachable from s when many nodes are reachable.

F: the number of times many nodes are reachable from s.

c	2	5	8	8
α	0.7968	0.993	0.9997	0.9997
MS	15	2	1	1
ML	714	981	996	1995
R	796.5	993	999.7	1999.3
F	7958	9931	9997	9995

This is less than $1/n^2$, if $c \geq 2$ and $f \geq 4 \ln n$,[1] an assumption which we are going to make. We use the phrase "*R* and f are large" to refer to this assumption.

Insertions into the Queue: We next analyze the number of insertions into the queue, first for the standard scheme.

Lemma 2. *Let IS be the number of insertions into the queue in the standard scheme. Then* $\mathbf{E}\left[IS \mid T = t\right] = n - (n - 1)(1 - p)^{t-1}$ *and*

$$\mathbf{E}\left[IS \mid R \text{ and } f \text{ are large}\right] \approx \frac{c}{q} - c + 1 + o(1) .$$

The proof is given in Appendix B.

Observe that the standard scheme makes about c/q insertions into but only $1/q$ removals from the queue. This is where the refined scheme saves. Let *INRS* be the number of nodes which are inserted into the queue but never removed in the standard scheme. Then, by the above,

$$\mathbf{E}\left[INRS \mid R \text{ and } f \text{ are large}\right] \approx \frac{c}{q} - c + 1 - \frac{1}{q} \approx \frac{c-1}{q} .$$

The standard scheme also performs some *decrease_p* operations on the nodes inserted but never removed. This number is small since the average number of incoming edges scanned per node is small.

We turn to the refined scheme. We have three kinds of savings.

- Nodes that are removed from the queue may incur fewer queue operations because they are inserted later or because some distance decreases do not lead to a queue operation. This saving is small since the number of distance decreases is small (recall that only few incoming edges per node are scanned)

---

[1] For $c \geq 2$, we have $\alpha > 1/2$ and thus $\exp(-\alpha f) < \exp(-\frac{1}{2}f)$. Choosing $f \geq 4 \ln n$, we obtain: $\exp(-\alpha f) < 1/n^2$.

- Nodes that are never removed from the queue in the standard scheme are not inserted in the refined scheme. This saving is significant and we will estimate it below.
- Nodes that are never removed from the queue in the standard scheme are inserted in the refined scheme but fewer decreases of their distance labels lead to a queue operation. This saving is small for the same reason as in the first item.

We concentrate on the set of nodes that are inserted into but never removed from the queue in the standard scheme. How many of these *INRS* insertions are also performed in the refined scheme? We use *INRR* to denote their number. We compute the expectation of *INRR* conditioned on the event E_l, $l \in \mathbb{N}$, that in the standard scheme there are exactly l nodes which are inserted into the queue but not removed.

Let $e_1 = (u_1, v_1), \ldots, e_l = (u_l, v_l)$ be the edges whose relaxations lead to the insertions of nodes that are not removed, labeled in the order of their relaxations. Then, $d(u_i) \leq d(u_{i+1})$, $1 \leq i \leq l - 1$, since nodes are removed from the queue in non-decreasing order of their distance values.

Node v_i is inserted with value $d(u_i) + w(e_i)$; $d(u_i) + w(e_i)$ is a random number in the interval $[d(t), d(u_i) + 1]$, where t is the target node closest to s, since the fact that v_i is never removed from the queue implies $d(u_i) + w(e_i) \geq d(t)$ but reveals nothing else about the value of $d(u_i) + w(e_i)$.

In the refined scheme e_i leads to an insertion only if $d(u_i) + w(e_i)$ is smaller than $d(u_j) + w(e_j)$ for every free v_j with $j < i$. The probability for this event is at most $1/(k+1)$, where k is the number of free v_j preceding v_i. The probability would be exactly $1/(k + 1)$ if the values $d(u_h) + w(e_h)$, $1 \leq h \leq i$, were all contained in the same interval. Since the upper bound of the interval containing $d(u_h) + w(e_h)$ increases with h, the probability is at most $1/(k + 1)$.
Thus (the expectation is conditioned on the event E_l)

$$\mathbf{E}\left[INRR \mid E_l\right] \leq \sum_{1 \leq i \leq l} \sum_{0 \leq k < i} \binom{i - 1}{k} q^k (1 - q)^{i-1-k} \frac{1}{k + 1} \ .$$

In Appendix C, we show that

$$\mathbf{E}\left[INRR \mid E_l\right] \leq \frac{1}{q} \cdot (1 + \ln(lq)) \ .$$

Since $\ln(lq)$ is a convex function of l (its first derivative is positive and its second derivative is negative), we obtain an upper bound on the expectation of *INRR* conditioned on R and f being large, if we replace *INRS* by its expectation. We obtain

$$\mathbf{E}\left[INRR \mid R \text{ and } f \text{ are large}\right] \leq \frac{1}{q} \cdot (1 + \ln(q\mathbf{E}\left[INRS \mid R \text{ and } f \text{ are large}\right]))$$

$$\approx \frac{1}{q} \cdot \left(1 + \ln\left(q\frac{c - 1}{q}\right)\right) = \frac{1}{q} \cdot (1 + \ln(c - 1)) \ .$$

Table 2. For all experiments (except the one in the last column) we used random graphs with $n = 1000$ nodes and $m = cn$ edges. For the last column we chose $n = 2000$ in order to illustrate that the dependency on n is weak. Nodes were free with probability q. The following quantities are shown; for each value of q and c we performed 10^4 trials. Trials where only a small number of nodes were reachable from s were ignored, i.e., about $(1 - \alpha)n$ trials were ignored.

D: the number of deletions from the queue.

$D^* = 1/q$: the predicted number of deletions from the queue.

IS: the number of insertions into the queue in the standard scheme.

$IS^* = \frac{c(1-q)}{q+(1-q)c/n} - \frac{(1-q)c/n}{q+(1-q)c/n} + 1$: the predicted number of insertions into the queue.

$INRS$: the number of nodes inserted but never removed.

$INRS^* = IS^* - D^*$: the predicted number.

$INRR$: the number of extra nodes inserted by the refined scheme.

$INRR^* = \frac{1}{q} \cdot (1 + \ln(qN^*))$: the predicted number.

DP_s: the number of decrease priority operations in the standard scheme.

DP_r: the number of decrease priority operations in the refined scheme.

Q_s: the total number of queue operations in the standard scheme.

Q_r: the total number of queue operations in the refined scheme.

$S = Q_s - Q_r$: the number of saved queue operations.

S^*: the lower bound on the number of saved queue operations.

$P = S/Q_s$: the percentage of queue operations saved.

c	2	2	2	5	5	5	8	8	8	8
q	0.02	0.06	0.18	0.02	0.06	0.18	0.02	0.06	0.18	0.18
D	49.60	16.40	5.51	49.33	16.72	5.50	50.22	16.79	5.61	5.53
D^*	50.00	16.67	5.56	50.00	16.67	5.56	50.00	16.67	5.56	5.56
IS	90.01	31.40	10.41	195.20	73.71	22.98	281.30	112.90	36.45	36.52
IS^*	90.16	31.35	10.02	197.60	73.57	23.25	282.30	112.30	36.13	36.77
$INRS$	40.41	15.00	4.89	145.80	56.99	17.49	231.00	96.07	30.85	30.99
$INRS^*$	40.16	14.68	4.46	147.60	56.90	17.69	232.30	95.60	30.57	31.22
$INRR$	11.00	4.00	1.00	35.00	12.00	4.00	51.00	18.00	5.00	5.00
$INRR^*$	39.05	14.56	4.34	104.10	37.13	11.99	126.80	45.78	15.03	15.15
DP_s	1.42	0.19	0.02	13.78	1.90	0.19	36.55	5.28	0.56	0.28
DP_r	0.71	0.09	0.01	2.63	0.31	0.03	4.60	0.50	0.05	0.03
Q_s	140.00	46.98	14.94	257.30	91.33	27.67	367.00	133.90	41.62	41.34
Q_r	110.40	36.12	11.52	134.50	45.33	13.97	154.40	50.85	16.00	15.77
S	29.58	10.86	3.42	122.80	46.00	13.69	212.70	83.08	25.62	25.57
S^*	1.12	0.13	0.12	43.47	19.77	5.70	105.50	49.82	15.54	16.07
P	21.12	23.11	22.87	47.74	50.37	49.50	57.94	62.03	61.55	61.85

We can now finally lower bound the number S of queue operations saved. By the above the saving is at least $INRS - INRR$. Thus

$$\mathbf{E}\,[S \mid R \text{ and } f \text{ are large}] \geq \frac{c-1}{q} - \frac{1}{q}(1 + \ln(c-1)) \approx \frac{c}{q}\left(1 - \frac{2 + \ln c}{c}\right).$$

We have a guaranteed saving of $\frac{2+\ln c}{c}$. Moreover, if $\frac{2+\ln c}{c} < 1$ we are guaranteed to save a constant fraction of the queue operations. For example, if $c = 8$, we will save at least a fraction of $1 - \frac{2+\ln 8}{8} \approx 0.49$ of the queue operations. The actual savings are higher, see Tab. 2. Also, there are substantial savings, even if the assumption of R and f being large does not hold (e.g., for $c = 2$ and $q = 0.02$).

It is interesting to observe how our randomness assumptions were used in the argument above. G is a random graph and hence the number of nodes reachable from s is either bounded or very large. Also, the expected number of nodes reached after t removals from the queue has a simple formula. The fact that a node is free with fixed probability gives us the distribution of the number of deletions from the queue. In order to estimate the savings resulting from the refined scheme we use that every node has the same chance of being free and that edge weights are random. For this part of the argument we do not need that our graph is random.

4   Bipartite Matching Problems

Both versions of the weighted bipartite matching problem, i.e., the assignment problem and the maximum weight matching problem, can be reduced to solving n, $n = \max(|A|, |B|)$, SSMTSP problems; we discuss the reduction for the assignment problem.

A popular algorithm for the assignment problem follows the primal dual paradigm [1, Sec. 12.4], [4, Sec. 7.8], [3]. The algorithm constructs a perfect matching and a dual solution simultaneously. A dual solution is simply a function $\pi : V \to \mathbb{R}$ that assigns a real potential to every node. We use V to denote $A \cup B$. The algorithm maintains a matching M and a potential function π with the property that

(a) $w(e) \leq \pi(a) + \pi(b)$ for every edge $e = (a, b)$,
(b) $w(e) = \pi(a) + \pi(b)$ for every edge $e = (a, b) \in M$ and
(c) $\pi(b) = 0$ for every free[2] node $b \in B$.

Initially, $M = \emptyset$, $\pi(a) = \max_{e \in E} w(e)$ for every $a \in A$ and $\pi(b) = 0$ for every $b \in B$. The algorithm stops when M is a perfect matching[3] or when it discovers that there is no perfect matching. The algorithm works in phases. In each phase the size of the matching is increased by one (or it is determined that there is no perfect matching).

A phase consists of the search for an augmenting path of minimum reduced cost. An augmenting path is a path starting at a free node in A, ending at a free node in B and using alternately edges not in M and in M. The reduced cost of an edge $e = (a, b)$ is defined as $\overline{w}(e) = \pi(a) + \pi(b) - w(e)$; observe that edges in M have reduced cost zero and that all edges have non-negative reduced cost. The reduced cost of a path is simply the sum of the reduced costs of the edges

---

[2] A node is free if no edge in M is incident to it.

[3] It is easy to see that M has maximal weight among all perfect matchings. Observe that if M' is any perfect matching and π is any potential function such that (a) holds then $w(M') \leq \sum_{v \in V} \pi(v)$. If (b) also holds, we have a pair (M', π) with equality and hence the matching has maximal weight (and the node potential has minimal weight among all potentials satisfying (a)).

contained in it. There is no need to search for augmenting paths from all free nodes in A; it suffices to search for augmenting paths from a single arbitrarily chosen free node $a_0 \in A$.

If no augmenting path starting in a_0 exists, there is no perfect matching in G and the algorithm stops. Otherwise, for every $v \in V$, let $d(v)$ be the minimal reduced cost of an alternating path from a_0 to v. Let $b_0 \in B$ be a free node in B which minimizes $d(b)$ among all free nodes b in B. We update the potential function according to the rules (we use π' to denote the new potential function):

(d) $\pi'(a) = \pi(a) - \max(d(b_0) - d(a), 0)$ for all $a \in A$,
(e) $\pi'(b) = \pi(b) + \max(d(b_0) - d(b), 0)$ for all $b \in B$.

It is easy to see that this change maintains (a), (b), and (c) and that all edges on the least cost alternating path p from a_0 to b_0 become tight[4]. We complete the phase by switching the edges on p: matching edges on p become non-matching and non-matching edges become matching edges. This increases the size of the matching by one.[5]

A phase is tantamount to a SSMTSP problem: a_0 is the source and the free nodes are the targets. We want to determine a target (= free node) b_0 with minimal distance from a_0 and the distance values of all nodes v with $d(v) < d(b_0)$. For nodes v with $d(v) \geq d(b_0)$, there is no need to know the exact distance. It suffices to know that the distance is at least $d(b_0)$.

Table 3 shows the effect of the pruning heuristic for the bipartite matching algorithm. (The improved code will be part of LEDA Version 4.3.)

A   Proof of Lemma 1

Proof (Lemma 1). Since each node is free with probability $q = f/n$ and since the property of being free is independent from the order in which nodes are removed from the queue, we have $\mathbf{Pr}\,(T = t \,|\, R = r) = (1 - q)^{t-1}q$ and $\mathbf{Pr}\,(T \geq t \,|\, R = r) = (1 - q)^{t-1}$, for $1 \leq t < r$. If $t = r$, $\mathbf{Pr}\,(T = r \,|\, R = r) = (1 - q)^{r-1} = \mathbf{Pr}\,(T \geq r \,|\, R = r)$.

The expected number of iterations is

$$\mathbf{E}\,[T \,|\, R = r] = \sum_{t \geq 1} \mathbf{Pr}\,(T \geq t \,|\, R = r) = \sum_{1 \leq t < r}(1 - q)^{t-1} + (1 - q)^{r-1}$$
$$= \frac{1-(1-q)^r}{1-(1-q)} = \frac{1}{q} - \frac{(1-q)^r}{q} \,.$$

$\square$

---

[4] An edge is called tight if its reduced cost is zero.
[5] The correctness of the algorithm can be seen as follows. The algorithm maintains properties (a), (b), and (c) and hence the current matching M is optimal in the following sense. Let A_m be the nodes in A that are matched. Then M is a maximal weight matching among the matchings that match the nodes in A_m and leave the nodes in $A \setminus A_m$ unmatched. Indeed if M' is any such matching then $w(M') \leq \sum_{a \in A_m} \pi(a) + \sum_{b \in B} \pi(b) = w(M)$, where the inequality follows from (a) and (c) and the equality follows from (b) and (c).

Table 3. Effect of the pruning heuristic. LEDA stands for LEDA's bipartite matching algorithm (up to version LEDA-4.2) as described in [4, Sec. 7.8] and MS stands for a modified implementation with the pruning heuristic. We created random graphs with n nodes on each side of the bipartition and cn edges inbetween. The running time is stated in CPU-seconds and is an average of 10 trials.

Unit Weights

n	c	LEDA	MS	c	LEDA	MS	c	LEDA	MS
10000	2	24.14	8.84	3	31.56	6.09	4	34.64	4.17
20000	2	83.95	30.77	3	113.14	21.73	4	125.60	13.94
40000	2	300.38	107.40	3	426.43	75.12	4	477.63	44.91

Random Weights [0 ... 1000]

n	c	LEDA	MS	c	LEDA	MS	c	LEDA	MS
10000	2	1.20	0.88	3	4.94	2.47	4	15.07	6.41
20000	2	2.61	1.86	3	10.35	5.09	4	35.76	14.34
40000	2	6.08	4.17	3	23.85	11.51	4	84.57	33.98

Random Weights [1000 ... 1005]

n	c	LEDA	MS	c	LEDA	MS	c	LEDA	MS
10000	2	9.80	6.55	3	14.77	9.12	4	17.57	9.44
20000	2	26.27	19.15	3	46.13	27.02	4	62.45	30.24
40000	2	86.32	59.10	3	155.98	86.65	4	166.04	92.63

B   Proof of Lemma 2

Proof (Lemma 2). In the standard scheme every node that is reached by the search is inserted into the queue. If we remove a total of t elements from the queue, the edges out of $t - 1$ elements are scanned. A node v, $v \neq s$, is not reached if none of these $t - 1$ nodes has an edge into v. The probability for this to happen is $(1 - p)^{t-1}$ and hence the expected number $\mathbf{E}\,[IS \,|\, T = t]$ of nodes reached is $n - (n - 1)(1 - p)^{t-1}$. This is also the number of insertions into the queue under the standard scheme.

If R and f are large, we have

$$
\mathbf{E}\,[IS \,|\, R \text{ and } f \text{ are large}]
$$
$$
= \sum_{t=1}^{R} \mathbf{E}\,[IS \,|\, T = t \text{ and } R \text{ and } f \text{ are large}]\,\mathbf{Pr}\,(T = t \,|\, R \text{ and } f \text{ are large})
$$
$$
= \sum_{t \geq 1} \left(n - (n-1)(1-p)^{t-1}\right)(1-q)^{t-1}q + \left(n - (n-1)(1-p)^{R-1}\right)(1-q)^{R-1}
$$
$$
\quad - \sum_{t \geq R} \left(n - (n-1)(1-p)^{t-1}\right)(1-q)^{t-1}q
$$
$$
= \sum_{t \geq 1} \left(n - (n-1)(1-p)^{t-1}\right)(1-q)^{t-1}q + o(1)
$$
$$
= n - q(n-1)\sum_{t \geq 0}(1-q)^t(1-p)^t + o(1) = n - q(n-1)\frac{1}{1-(1-p)(1-q)} + o(1)
$$
$$
= n - 1 - (n-1)\frac{q}{p+q-pq} + 1 + o(1) = (n-1)\frac{p-pq}{p+q-pq} + 1 + o(1)
$$
$$
= \frac{c(1-q)}{q+(1-q)c/n} - \frac{(1-q)c/n}{q+(1-q)c/n} + 1 + o(1) \approx \frac{c}{q} - c + 1 + o(1) .
$$

$\square$

The final approximation is valid if $c/n \ll q$. The approximation makes sense intuitively. We relax the edges out of $1/q - 1$ nodes and hence relax about c times as many edges. There is hardly any sharing of targets between these edges, if n is large. We conclude that the number of insertions into the queue is $\frac{c}{q} - c + 1$.

C   Estimation of $\mathbf{E}\left[INRR \mid E_l\right]$

We have:

$$
\begin{aligned}
\mathbf{E}\left[INRR \mid E_l\right] &\leq \sum_{1 \leq i \leq l} \sum_{0 \leq k < i} \binom{i-1}{k} q^k (1-q)^{i-1-k} \frac{1}{k+1} \\
&= \sum_{1 \leq i \leq l} \frac{1}{iq} \sum_{0 \leq k < i} \binom{i}{k+1} q^{k+1} (1-q)^{i-(k+1)} = \sum_{1 \leq i \leq l} \frac{1}{iq} \sum_{1 \leq k \leq i} \binom{i}{k} q^k (1-q)^{i-k} \\
&= \sum_{1 \leq i \leq l} \frac{1}{iq} \left(1 - (1-q)^i\right) ,
\end{aligned}
$$

where the first equality follows from $\binom{i-1}{k} \frac{1}{k+1} = \frac{1}{i} \binom{i}{k+1}$. The final formula can also be interpreted intuitively. There are about iq free nodes preceding v_i and hence v_i is inserted with probability about $1/(iq)$.

In order to estimate the final sum we split the sum at a yet to be determined index i_0. For $i < i_0$, we estimate $(1 - (1-q)^i) \leq iq$, and for $i \geq i_0$, we use $(1 - (1-q)^i) \leq 1$. We obtain

$$
\mathbf{E}\left[INRR \mid E_l\right] \leq i_0 + \frac{1}{q} \sum_{i_0 \leq i \leq l} \frac{1}{i} \approx i_0 + \frac{1}{q} \ln \frac{l}{i_0} .
$$

For $i_0 = 1/q$ (which minimizes the final expression[6]) we have

$$
\mathbf{E}\left[INRR \mid E_l\right] \leq \frac{1}{q} \cdot (1 + \ln(lq)) .
$$

References

[1] R.K. Ahuja, T.L. Magnanti, and J.B. Orlin. *Network Flows*. Prentice Hall, 1993.

[2] N. Alon, J.H. Spencer, and P. Erdös. *The Probabilistic Method*. John Wiley & Sons, 1992.

[3] Z. Galil. Efficient algorithms for finding maximum matching in graphs. *ACM Computing Surveys*, 18(1):23–37, 1986.

[4] K. Mehlhorn and S. Näher. *The LEDA Platform for Combinatorial and Geometric Computing*. Cambridge University Press, 1999. 1018 pages.

[5] K. Mehlhorn and G. Schäfer. Implementation of $O(nm\log n)$ weighted matchings in general graphs: The power of data structures. In *Workshop on Algorithm Engineering (WAE)*, Lecture Notes in Computer Science, to appear, 2000. www.mpi-sb.mpg.de/~mehlhorn/ftp/WAE00.ps.gz.

[6] G. Schäfer. Weighted matchings in general graphs. Master's thesis, Fachbereich Informatik, Universität des Saarlandes, Saarbrücken, Germany, 2000.

---

[6] Take the derivative with respect to i_0 ... .

A Separation Bound for Real Algebraic Expressions*

Christoph Burnikel[1], Stefan Funke[2], Kurt Mehlhorn[2],
Stefan Schirra[3], and Susanne Schmitt[2]

[1] ENCOM GmbH, 66740 Saarlouis,
burnikel@mpi-sb.mpg.de,
[2] MPI für Informatik, 66123 Saarbrücken,
{funke,mehlhorn,sschmitt}@mpi-sb.mpg.de,
[3] think & solve Beratungsgesellschaft, 66111 Saarbrücken,
stschirr@mpi-sb.mpg.de

Abstract. Real algebraic expressions are expressions whose leaves are integers and whose internal nodes are additions, subtractions, multiplications, divisions, k-th root operations for integral k, and taking roots of polynomials whose coefficients are given by the values of subexpressions. We consider the sign computation of real algebraic expressions, a task vital for the implementation of geometric algorithms. We prove a new separation bound for real algebraic expressions and compare it analytically and experimentally with previous bounds. The bound is used in the sign test of the number type leda_real.

1 Introduction

Real algebraic expressions are expressions whose leaves are integers and whose internal nodes are additions, subtractions, multiplications, divisions, k-th root operations for integral k, and taking roots of polynomials whose coefficients are given by the values of subexpressions; the exact definition is given below. Examples are $\sqrt{17} + \sqrt{21} - \sqrt{\sqrt{17} + \sqrt{21} + 2\sqrt{357}}$ and $\frac{17+\sqrt{21}}{19} - \frac{18+\sqrt{22}}{20}$. We consider the sign computation of real algebraic expressions.

Our main motivation is the implementation of geometric algorithms. The evaluation of geometric predicates, such as the incircle or the side-of predicate, amounts to the computation of the sign of an expression. Non-linear objects (circles, ellipses, ...) lead to expressions involving roots and hence an efficient method for computing signs of algebraic expressions is an essential basis for the robust implementation of geometric algorithms dealing with non-linear objects.

The separation bound approach is the most successful approach to sign computation; it is, for example, used in the number type leda_real [4,1,12] and the number type Expr of the CORE package [7]. A *separation bound* is an easily computable function sep mapping expressions into positive real numbers such that the value ξ of any non-zero expression E is lower bounded by $sep(E)$, i.e., either $\xi = 0$ or $|\xi| \geq sep(E)$. Separation bounds allow one to determine the sign of an expression by numerical computation. An error bound Δ is initialized to some positive value, say $\Delta = 1$, and an approximation $\widetilde{\xi}$ of ξ with $|\xi - \widetilde{\xi}| \leq \Delta$ is computed using approximate arithmetic, say floating point

---

* Partially supported by ESPRIT LTR project (Effective Computational Geometry for Curves and Surfaces).

F. Meyer auf der Heide (Ed.): ESA 2001, LNCS 2161, pp. 254–265, 2001.

arithmetic with arbitrary-length mantissa. If $|\widetilde{\xi}| > \Delta$, the sign of ξ is equal to the sign of $\widetilde{\xi}$. Otherwise, $|\widetilde{\xi}| \leq \Delta$ and hence $|\xi| < 2\Delta$. If $2\Delta \leq sep(E)$, we have $\xi = 0$. If $2\Delta > sep(E)$, we halve Δ and repeat. The worst case complexity of the procedure just outlined is determined by the separation bound; $\log(1/sep(E))$ determines the maximal precision needed for the computation of $\widetilde{\xi}$ and we refer to $\log(1/sep(E))$ as the *bit bound*. If $\xi \neq 0$, the actual precisions required is $\log \xi$ and hence "easy sign tests" are much faster than the worst case. This feature distinguishes the separation bound approach to sign computation from approaches that explicitly compute a defining polynomial.

Separation bounds have been studied extensively in computer algebra [5,10,15,11, 17], as well as in computational geometry [3,2,16,9,13]. We prove a new separation bound for the following class of *real algebraic expressions*. The value of a real algebraic expression is either a real algebraic number or undefined (at the end of Section 3 we show how to test whether the value of an expression is defined).

(1) Any integer v is a real algebraic expression. The integer is also its value.
(2) If E_1 and E_2 are real algebraic expressions, so are $E_1 + E_2$, $E_1 - E_2$, $E_1 \cdot E_2$, E_1/E_2, and $\sqrt[k]{E_1}$, where $k \geq 2$ is an integer. The value of $\sqrt[k]{E_1}$ is undefined if k is even and the value of E_1 is negative. The value of E_1/E_2 is undefined, if the value of E_2 is zero. The value of E_1 op E_2 or $\sqrt[k]{E_1}$ is undefined, if the value of E_1 or the value of E_2 is undefined. Otherwise the value of $E_1 + E_2$, $E_1 - E_2$, $E_1 \cdot E_2$, and E_1/E_2 is the sum, the difference, the product and the quotient of the values of E_1 and E_2 respectively and the value of $\sqrt[k]{E_1}$ is the k-th root of the value of E_1.
(3) If E_d, E_{d-1}, ... , E_1, E_0 are real algebraic expressions and j is a positive integer with $0 \leq j < d$, then $\diamond(j, E_d, E_{d-1}, \ldots, E_1, E_0)$ is an expression. If the values of the E_i are defined and ξ_i is the value of E_i, the value of the expression is the j-th smallest real root of the polynomial $\xi_d X^d + \xi_{d-1} X^{d-1} + \ldots + \xi_0$, if the polynomial has at least j real roots. Otherwise, the value is undefined.

Below, expression always means real algebraic expression. An expression is given as a directed acyclic graph (dag) whose source nodes are labeled by the operands and whose internal nodes are labeled by operators. We call an expression *simple* if only items (1) and (2) are used in its definition and we call it *simple and division-free* if, in addition, no division operator occurs in the expression.

The starting point for the present work is the bound given by Burnikel et al. [2] for simple expressions. We refer to this bound as the BFMS bound in the sequel.

Lemma 1 ([2]). *Let E be an expression with integral operands and operations $+$, $-$, $\cdot$, $/$, $\sqrt[k]{}$ for integral $k \geq 2$. Let ξ be the value of E, let the weight $D(E)$ of E be the product of the indices (the index of a $\sqrt[k]{}$ operation is k) of the radical operations in E, and let $u(E)$ and $l(E)$ be defined inductively on the structure of E by the rules shown in the table below.*

	$u(E)$	$l(E)$
integer N	$|N|$	1
$E_1 \pm E_2$	$u(E_1) \cdot l(E_2) + l(E_1) \cdot u(E_2)$	$l(E_1) \cdot l(E_2)$
$E_1 \cdot E_2$	$u(E_1) \cdot u(E_2)$	$l(E_1) \cdot l(E_2)$
E_1/E_2	$u(E_1) \cdot l(E_2)$	$l(E_1) \cdot u(E_2)$
$\sqrt[k]{E_1}$	$\sqrt[k]{u(E_1)}$	$\sqrt[k]{l(E_1)}$

Then $\xi = 0$ *or* $\left(l(E)u(E)^{D(E)^2-1} \right)^{-1} \leq |\xi| \leq u(E)l(E)^{D(E)^2-1}$. *If E is division-free,* $l(E) = 1$, *and the above bound holds with $D(E)^2$ replaced by $D(E)$.*

Observe the difference between the division-free case and the general case. For simple division-free expressions, the BFMS-bound is the best bound known. Expressions with divisions arise naturally in geometric applications. Inputs to expressions are frequently fractions and, e.g., normalizing a line equation amounts to a division. For expressions with divisions, the BFMS-bound is much weaker than for expressions without divisions. We give an example. Consider the expression $\dfrac{2^{10} \sqrt[8]{2^8 - (2^8-1)} - 2^6}{1}$. Here $u(E) \approx 2^{10}$, $l(E) = 1$ and $D(E) = 8$. So the BFMS bound is $2^{-10 \cdot 63} = 2^{-630}$, since E is not division-free and hence the dependence (of the logarithm of the bound) on D is quadratic. Without the final redundant division, the expression is division-free and the bound becomes $2^{-10 \cdot 7} = 2^{-70}$. Our new bound handles divisions much better and also applies to a wider class of expressions than the BFMS bound.

This paper is structured as follows. In Section 2, we review the proof of the BFMS bound and motivate our new way of dealing with divisions. In Section 3, we prove our main theorem, a separation bound for expressions defined by (1), (2), and (3). In Sections 4 and 5 we compare our bound analytically and experimentally to previous bounds.

2 A Review of the BFMS Bound

An algebraic integer is the root of a polynomial with integer coefficients and leading coefficient one. The following three Lemmas were already used in [2] and [9].

Lemma 2. *Let α be an algebraic integer and let $\deg(\alpha)$ be the algebraic degree of α. If U is an upper bound on the absolute value of all conjugates of α, then $|\alpha| \geq U^{1-\deg(\alpha)}$.*

Proof. The proof is simple. Let d be the degree of α and let $\alpha_1 = \alpha, \alpha_2, \ldots, \alpha_d$ be the conjugates of α. The product of the conjugates is equal to the constant coefficient of the defining polynomial and hence in $\mathbb{Z}$. Thus $|\alpha| \cdot U^{d-1} \geq 1$.   □

Lemma 3 ([6,8] or [2, Theorem 4]). *Let α and β be algebraic integers. Then $\alpha \pm \beta$, $\alpha\beta$ and $\sqrt[k]{\alpha}$ are algebraic integers.*

We also need to cover item (3) in the definition of algebraic expressions.

Lemma 4. *Let ϱ be the root of a monic polynomial $P(X) = X^n + \alpha_{n-1}X^{n-1} + \cdots + \alpha_0$ of degree n where the coefficients $\alpha_{n-1}, \ldots, \alpha_0$ are algebraic integers. Then ϱ is an algebraic integer.*

Proof. This fact is well-known, a proof can, for example, be found in [14, Theorem 2.4]. We include a proof for completeness. The proof uses an argument similar to the proof of Lemma 3. Let $\alpha_j^{(i_j)}$, $1 \leq i_j \leq \deg(\alpha_j)$, be the conjugates of α_j for $0 \leq j \leq n - 1$ and let $\widehat{\alpha}_j$ be the vector formed by the conjugates of α_j. Consider the polynomial

$$Q(X) = \prod_{i_0} \prod_{i_1} \cdots \prod_{i_{n-1}} \left(X^n + \alpha_{n-1}^{(i_{n-1})} X^{n-1} + \alpha_{n-2}^{(i_{n-2})} X^{n-2} + \cdots + \alpha_0^{(i_0)} \right) .$$

ϱ is a root of $Q(X)$ and $Q(X)$ is symmetric in the $\alpha_j^{(i_j)}$ for all j. The theorem on elementary symmetric function implies that $Q(X)$ is a polynomial in X and the elementary symmetric functions $\sigma_1(\widehat{\alpha}_j), \ldots, \sigma_{\deg(\alpha_j)}(\widehat{\alpha}_j)$. The elementary symmetric function $\sigma_l(\widehat{\alpha}_j)$ is the coefficient of $X^{\deg(\alpha_j)-l}$ in the minimal polynomial of α_j and hence in $\mathbb{Z}$ (since α_j is an algebraic integer). Thus $Q(X)$ is a monic polynomial in $\mathbb{Z}[X]$ and ϱ is an algebraic integer.   □

Lemma 5 ([2, Lemma 6]). *Let α and β be algebraic integers and let U_α and U_β be upper bounds on the absolute size of the conjugates of α and β, respectively. Then $U_\alpha + U_\beta$ is an upper bound on the absolute size of the conjugates of $\alpha \pm \beta$, $U_\alpha U_\beta$ is an upper bound on the absolute size of the conjugates of $\alpha\beta$, and $\sqrt[k]{U_\alpha}$ is an upper bound on the absolute size of the conjugates of $\sqrt[k]{\alpha}$.*

We also need bounds for the absolute size of roots of monic polynomials. Let $P(X) = X^n + a_{n-1}X^{n-1} + a_{n-2}X^{n-2} + \cdots + a_0$ be a monic polynomial with arbitrary real coefficients, not necessarily integral, and let α be a root of $P(X)$. A *root bound* Φ is any function of the coefficients of P that bounds the absolute value of α, i.e., $|\alpha| \leq \Phi(a_{n-1}, a_{n-2}, \ldots, a_0)$. We require that Φ is monotone, i.e., if $|a_i| \leq b_i$ for $0 \leq i \leq n-1$, then $\Phi(a_{n-1}, a_{n-2}, \ldots, a_0) \leq \Phi(b_{n-1}, b_{n-2}, \ldots, b_0)$.

Examples of root bounds are:

$$|\alpha| \leq 2 \max\left(|a_{n-1}|, \sqrt{|a_{n-2}|}, \sqrt[3]{|a_{n-3}|}, \ldots, \sqrt[n]{|a_0|}\right)$$

$$|\alpha| \leq 1 + \max\left(|a_{n-1}|, |a_{n-2}|, \ldots, |a_0|\right)$$

$$|\alpha| \leq \max\left(n|a_{n-1}|, \sqrt{n|a_{n-2}|}, \sqrt[3]{n|a_{n-3}|}, \ldots, \sqrt[n]{n|a_0|}\right)$$

A proof of all bounds can be found in [17]. The first bound is called the Lagrange-Zassenhaus bound and the last two bounds are called the Cauchy bounds.

We next briefly review the proof of the BFMS bound. For a division-free simple expression E one observes that the value ξ of E is an algebraic integer (by Lemma 3) and that $u(E)$ is an upper bound on ξ and all its conjugates (by Lemma 5). Furthermore $D(E)$ is an upper bound for the algebraic degree of E. Thus $|\xi| \leq u(E)$ and $|\xi| \geq 1/(u(E)^{D(E)-1})$ by Lemma 2.

Expressions with divisions are handled by reduction to the division-free case. Let E be a simple expression and let ξ be its value. We construct a new expression dag, also with value $\xi = val(E)$, containing only a single division. Moreover, the division is the final operation in the dag and hence $val(E) = val(E_1)/val(E_2)$, where E_1 and E_2 are the inputs to the division. The bounds for the division free case apply to E_1 and E_2 and $D(E_1)$ and $D(E_2)$ are at most $D(E)^2$. The construction of the new dag is straightforward. For every node A in the original dag there are two nodes A_1 and A_2 in the new dag such that $val(A) = val(A_1)/val(A_2)$. For the leaves (which stand for integers) the replacement is trivial (we take $A_1 = A$ and $A_2 = 1$) and for interior nodes we use the rules

$$\frac{A_1}{A_2} \pm \frac{B_1}{B_2} \Rightarrow \frac{A_1 B_2 \pm A_2 B_1}{A_2 B_2} \quad \frac{A_1}{A_2} \cdot \frac{B_1}{B_2} \Rightarrow \frac{A_1 A_2}{B_1 B_2} \quad \frac{A_1}{A_2} / \frac{B_1}{B_2} \Rightarrow \frac{A_1 B_2}{A_2 B_1} \quad \sqrt[k]{\frac{A_1}{A_2}} \Rightarrow \frac{\sqrt[k]{A_1}}{\sqrt[k]{A_2}} \; .$$

In this way, *each root operation in the original dag gives rise to two root operations in the new dag*. This may square the D-value of the expression.

The starting point for the present paper was a simple but powerful observation. Although the transformation rules above are natural, they are not the only way of obtaining division free expressions E_1 and E_2 with $val(E) = val(E_1)/val(E_2)$. Instead of the last rule we may also use $\sqrt[k]{\frac{A_1}{A_2}} \Rightarrow \frac{\sqrt[k]{A_1 A_2^{k-1}}}{A_2}$ or $\frac{A_1}{\sqrt[k]{A_1^{k-1} A_2}}$. The new rule does not increase the total degree of the expression and hence $D(E_1)$ and $D(E_2)$ are at most $D(E)$. In an earlier version of the paper, we only used the first alternative of the new rule. Chee Yap (personal communication, January 2001) pointed out to us that it is advantageous to have both rules (see the proof of Lemma 6).

3   The New Bound

We derive a separation bound for the expressions defined by items (1) to (3). For items (1) and (2), we use the BFMS rules with the modification proposed in the previous paragraph.

The diamond operation allows one to take the root of a polynomial $P(X) = \alpha_d X^d + \alpha_{d-1} X^{d-1} + \cdots + \alpha_1 X + \alpha_0$ where the α_i are arbitrary real algebraic numbers. Every real algebraic number can be written as the quotient of two algebraic integers; this is well-known, but will be reproved below as part of the proof of our main theorem. Let $\alpha_i = \nu_i/\delta_i$ where ν_i and δ_i are algebraic integers. Then $P(X) = \frac{\nu_d}{\delta_d} X^d + \frac{\nu_{d-1}}{\delta_{d-1}} X^{d-1} + \cdots + \frac{\nu_1}{\delta_1} X + \frac{\nu_0}{\delta_0}$. Let $D = \prod \delta_i$. By multiplication with D we obtain $D \cdot P(X) = (\nu_d D/\delta_d) X^d + (\nu_{d-1} D/\delta_{d-1}) X^{d-1} + \cdots + (\nu_0 D/\delta_0)$, a polynomial with algebraic integral coefficients.

We next derive a monic polynomial. To get rid of the leading coefficient $(\nu_d D/\delta_d)$, we multiply by $(\nu_d D/\delta_d)^{d-1}$ and substitute $X/(\nu_d D/\delta_d)$ for X. We obtain $Q(X) = D \cdot (\nu_d D/\delta_d)^{d-1} \cdot P\left(\frac{X}{\nu_d D/\delta_d}\right)$ with

$$Q(X) = X^d + (\nu_d D/\delta_d)(\nu_{d-1} D/\delta_{d-1}) X^{d-1} + \cdots + (\nu_d D/\delta_d)^d (\nu_0 D/\delta_0)$$

which is monic and has algebraic integer coefficients. The root bounds of Section 2 provide us with an upper bound on the size of the roots of $Q(X)$: the size of any root of $Q(X)$ is bounded by $u = \Phi((\nu_d D/\delta_d)(\nu_{d-1} D/\delta_{d-1}), \ldots, (\nu_d D/\delta_d)^d (\nu_0 D/\delta_0))$. Since the roots of P are simply the roots of Q divided by $\nu_d D/\delta_d$, this suggests to extend the definitions of u and l as follows: For an expression E denoting a root of a polynomial of degree d with coefficients given by $E_d, E_{d-1}, E_{d-2}, \ldots, E_0$ we define

$$u(E) = \Phi\left(\ldots, \left(u(E_d) \prod_{k \neq d} l(E_k) \right)^{d-i} u(E_i) \prod_{k \neq i} l(E_k), \ldots \right), \quad l(E) = u(E_d) \prod_{k \neq d} l(E_k) .$$

We still need to define the weight $D(E)$ of an expression. We do so in the obvious way. The weight $D(E)$ of an expression dag E is the product of the weights of the nodes and leaves of the dag. Leaves and $+$, $-$, $\cdot$ and $/$-operations have weight 1, a $\sqrt[k]{}$-node has weight k, and a $\diamond(j, E_d, \ldots)$-operation has weight d.

We can now state our main theorem.

Theorem 1. *Let E be an expression with integer operands and operations $+, -, \cdot, \sqrt[k]{}$ for integral k and $\diamond(j, \dots)$ operations and let $D(E)$ be the weight of E. Let ξ be the value of E. Let $u(E)$ and $l(E)$ be defined inductively on the structure of E according to the following rules:*

	$u(E)$	$l(E)$
integer N	$\|N\|$	1
$E_1 \pm E_2$	$u(E_1) \cdot l(E_2) + l(E_1) \cdot u(E_2)$	$l(E_1) \cdot l(E_2)$
$E_1 \cdot E_2$	$u(E_1) \cdot u(E_2)$	$l(E_1) \cdot l(E_2)$
E_1/E_2	$u(E_1) \cdot l(E_2)$	$l(E_1) \cdot u(E_2)$
$\sqrt[k]{E_1}$ *and* $u(E_1) \geq l(E_1)$	$\sqrt[k]{u(E_1)l(E_1)^{k-1}}$	$l(E_1)$
$\sqrt[k]{E_1}$ *and* $u(E_1) < l(E_1)$	$u(E_1)$	$\sqrt[k]{(u(E_1)^{k-1}l(E_1)}$
$\diamond(j, E_d, \dots, E_0)$	$\Phi(\dots, \left(l(E)^{d-i}u(E_i) \prod_{k \neq i} l(E_k)\right), \dots)$	$u(E_d) \prod_{k \neq d} l(E_k)$

Then either $\xi = 0$ or $\left(l(E)u(E)^{D(E)-1}\right)^{-1} \leq |\xi| \leq u(E)l(E)^{D(E)-1}$.

Proof. We show that the rules for u and l keep the invariant that there are algebraic integers β and γ such that $\xi = \beta/\gamma$ and $u(E)$ is an upper bound on the absolute size of the conjugates of β and $l(E)$ is an upper bound on the absolute size of the conjugates of γ.

We prove this by induction on the structure of E. The base case is trivial. If E is an integer N, we take $\beta = N$ and $\alpha = 1$; β is the root of the polynomial $X - N$ and α is a root of $X - 1$.

Now let $E = E_1 \pm E_2$. By induction hypothesis we have $\xi_j = \beta_j/\gamma_j$ for $j = 1, 2$. We set $\beta = \beta_1\gamma_2 \pm \beta_2\gamma_1$ and $\gamma = \gamma_1\gamma_2$. Since algebraic integers are closed under additions, subtractions and multiplications, β and γ are algebraic integers. By Lemma 5, $u(E) = u(E_1) \cdot l(E_2) + l(E_1) \cdot u(E_2)$ is an upper bound on the absolute size of the conjugates of β. Similarly, $l(E)$ is an upper bound on the absolute size of the conjugates of γ.

If $E = E_1 \cdot E_2$, we set $\beta = \beta_1\beta_2$ and $\gamma = \gamma_1\gamma_2$. The claim follows analogously to the previous case by Lemma 5.

If $E = E_1/E_2$, we set $\beta = \beta_1\gamma_2$ and $\gamma = \beta_2\gamma_2$. Again, the claim follows by Lemma 5.

If $E = \sqrt[k]{E_1}$ and $\beta_1 \geq \gamma_1$, we set $\beta = \sqrt[k]{\beta_1\gamma_1^{k-1}}$ and $\gamma = \gamma_1$. Since algebraic integers are closed under $\sqrt[k]{}$-operations, β is an algebraic integer. By Lemma 5, $u(E)$ is an upper bound on the absolute size of the conjugates of β. There is nothing to show for $\gamma = \gamma_1$.

If $E = \sqrt[k]{E_1}$ and $\beta_1 < \gamma_1$, we set $\beta = \beta_1$ and $\gamma = \sqrt[k]{\beta_1^{k-1}\gamma_1}$. Since algebraic integers are closed under $\sqrt[k]{}$-operations, γ is an algebraic integer. By Lemma 5, $l(E)$ is an upper bound on the absolute size of the conjugates of γ. There is nothing to show for $\beta = \beta_1$.

Finally, let E be defined by a $\diamond(j, E_d, \dots, E_0)$-operation. We set

$$\beta = \diamond(j, 1, \beta_{d-1}\gamma_d\gamma_{d-2} \cdots \gamma_0, \gamma\beta_{d-2}\gamma_d\gamma_{d-1}\gamma_{d-3} \cdots \gamma_0, \dots, \gamma^{n-1}\gamma_d\gamma_{d-1}\gamma_{d-2} \cdots \gamma_1\beta_0)$$

and $\gamma = \beta_d\gamma_{d-1}\gamma_{d-2} \cdots \gamma_0$. By the discussion preceding the statement of our main theorem, $\xi = \beta/\gamma$, β and γ are algebraic integers, $l(E)$ is an upper bound on the absolute size of the conjugates of γ, and $u(E)$ is an upper bound on the absolute value of the conjugates of β. This completes the induction step.

Rewriting ξ as β/γ corresponds to a restructuring of the expression dag defining E into an expression dag E' with a single division-operation. We have $D(E') = D(E)$.

We still need to argue that $D(E)$ is an upper bound on the algebraic degree of β. This follows from the fact that every operation leads to a field extensions whose degree is bounded by the weight of the operation.

We now have collected all ingredients to bound the absolute value of ξ from below. If $\xi \neq 0$, we have $\beta \neq 0$. The absolute value of β and all its conjugates is bounded by $u(E)$. Thus $|\beta| \geq (u(E)^{\deg(\beta)-1})^{-1}$ by Lemma 2. Also $|\gamma| \leq l(E)$. Thus

$$|\xi| = \frac{\beta}{\gamma} \geq \frac{1}{u(E)^{\deg(\beta)-1}} \cdot \frac{1}{l(E)} \geq \frac{1}{u(E)^{D(E)-1} \cdot l(E)} \ .$$

$\square$

The value of an algebraic expression may be undefined. Divisions by zero and taking a root of even degree of a negative number are easily caught by the sign test. We next argue that the sign test also allows us to test whether the diamond-operation is well defined. For this matter, we need to determine the number of zeros of a polynomial. Sturm sequences, see [11, chapter 5] or [17, Chapter 7] are the appropriate tool. The computation of Sturm sequences amounts to a gcd computation between a polynomial and its derivative. Our sign test is sufficient to implement a gcd computation.

4   Comparison to Other Constructive Root Bounds

We compare our new bound to previous root bounds provided by Mignotte [11], Canny [5], Dubé/Yap [18], BFMS [2,13], Scheinermann [15], Li/Yap [9]. We refer to the bound presented in this paper as BFMSS. Root bounds can be compared along two axees: according to the class of expressions to which they apply and according to their value.

The bounds by Mignotte, Dubé/Yap and Scheinerman apply to division-free simple expressions, BFMS applies to simple expressions. The bounds in [9] and [13] apply to expressions defined by items (1) to (3) with the restriction that the E_d to E_0 in (3) must be integers. Canny's bound is most general. It applies to algebraic numbers defined by systems of multi-variate polynomial equations with integer coefficients.

We next discuss the quality of the bounds. In [2,9] it was already shown that the BFMS-bound it never worse than the bounds by Mignotte, Canny, Dubé/Yap, and Scheinermann. In [9] it was also shown that the BFMS bound and the Li/Yap bound are incomparable.

Lemma 6 (C. Yap, personal communication). *Let E be an arbitrary simple expression, let u and l be defined as in the original BFMS-bound, let u' and l' be defined as in Theorem 1, and let $D = D(E)$ be the degree bound of E. Then*

$$l(E)u(E)^{D^2-1} \geq l'(E)u'(E)^{D-1} \ ,$$

i.e., the improved bound is always as least as strong as the orginal BFMS-bound

Proof. We show $\frac{u(E)}{l(E)} = \frac{u'(E)}{l'(E)}$ and $u'(E) \leq u(E)^{D(E)}$ and $l'(E) \leq l(E)^{D(E)}$ by induction on the structure of E. Assume that these relations hold. Then

$$l(E)u(E)^{D^2-1} = \frac{l(E)}{u(E)}(u(E)^D)^D \geq \frac{l'(E)}{u'(E)}u'(E)^D = l'(E)u'(E)^{D-1}$$

and we are done.

The proof of the equality is a simple induction on the structure of E. The base case is clear. In the inductive step we write u_1 instead of $u(E_1)$ and similarly for E_2, l, u' and l'. If $E = E_1+E_2$, we have $\frac{u(E)}{l(E)} = \frac{u_1l_2+u_2l_1}{l_1l_2} = \frac{u_1}{l_1}+\frac{u_2}{l_2} = \frac{u'_1}{l'_1}+\frac{u'_2}{l'_2} = \frac{u'(E)}{l'(E)}$. Multiplication and division are handled similarly. If $E = \sqrt[k]{E_1}$, we have (assuming $u'_1 \geq l'_1$, the case $u'_1 < l'_1$ is handled similarly) $\frac{u(E)}{l(E)} = \frac{\sqrt[k]{u_1}}{\sqrt[k]{l_1}} = \sqrt[k]{\frac{u_1}{l_1}} = \sqrt[k]{\frac{u'_1}{l'_1}} = \frac{\sqrt[k]{u'_1(l'_1)^{k-1}}}{l'_1} = \frac{u'(E)}{l'(E)}$.

For the inequalities we have to work slightly harder. The base case is again clear; observe that $D = 1$ in the base case. It is also clear that $u(E) \geq 1$ (or $u(E) = 0$) and $l(E) \geq 1$ for all E. If $E = E_1 \pm E_2$, we have (using $D \geq D_1$ and $D \geq D_2$)

$$u(E)^D = (u_1l_2 + u_2l_1)^D \geq (u_1l_2)^D + (u_2l_1)^D \geq u_1^{D_1}l_2^{D_2} + u_2^{D_2}l_1^{D_1} \geq u'_1l'_2 + u'_2l'_1 = u'(E)$$

and $l(E)^D = (l_1l_2)^D \geq l_1^{D_1}l_2^{D_2} \geq l'_1l'_2 = l'(E)$. Multiplication and division are handled similarly. If $E = \sqrt[k]{E_1}$, we have $D(E_1) = D(E)/k$ and hence (assuming $u'_1 \geq l'_1$, the case $u'_1 < l'_1$ is handled similarly) $u(E)^D = u_1^{D/k} = u_1^{D_1} \geq u'_1 \geq \sqrt[k]{u'_1(l'_1)^{k-1}} = u'(E)$ and $l(E)^D = l_1^{D/k} = l_1^{D_1} \geq l'_1 = l'(E)$.   □

We next show that the new bound can be significantly better than the old bound. Consider the expression $F = \sqrt[k]{x/a}$ and $E = F - F$ where x is a ck-bit integer for some constant c and a is a d-bit integer for some constant d. Then $D(E) = k$. We evaluate both bounds as functions of k.

For the BFMS-bound we have $\log u(F) = (1/k)ck = c$, $\log l(F) = d/k$, $\log u(E) = 1 + c + d/k$, $\log l(E) = 2d/k$ and hence the BFMS bit bound is $(k^2 - 1)\log u(E) + \log l(E) = \Theta(k^2)$.

For the BFMSS-bound we have $\log u(F) = (1/k)(ck+d) = c+d/k$, $\log l(F) = d$, $\log u(E) = 1 + c + d/k + d$, $\log l(E) = 2d$ and hence the BFMSS bit bound is $(k - 1)\log u(E) + \log l(E) = \Theta(k)$.

It remains to compare the BFMSS and the Li/Yap bound. For division-free simple expressions, the bounds are identical. For expressions with divisions, the bounds are incomparable.

We start with an example, where the BFMSS-bound is significantly better. Let[1] $E_0 = 17/3$, let $F_i = \sqrt{E_{i-1}}$ and $E_i = F_i + F_i$ for $1 \leq i \leq k$, and let $E = E_k - E_k$. Then $\deg(E_i) = \deg(F_i) = 2^i$. We evaluate both bounds as functions of k.

For the BFMSS bound, we have $\log u(E_0) = \log 17$, $\log l(E_0) = \log 3$, $\log l(F_i) = \log l(E_{i-1})$, $\log u(F_i) = \frac{1}{2}(\log u(E_{i-1}) + \log l(E_{i-1}))$, $\log l(E_i) = 2\log l(F_i) = 2^i \log 3$, $\log u(E_i) = 1 + \log u(F_i) + \log l(F_i) = 1 + \frac{1}{2}\log u(E_{i-1}) + \frac{3}{2}2^{i-1}\log 3 \leq 2 + 2^{-i}\log 17 + 2^i \log 3$ and hence $\log u(E) = 1 + \log u(E_k) + \log l(E_k) \leq 3 + 4 \cdot 2^k$

and

$\log l(E) = 2\log l(E_k) \leq 4 \cdot 2^k$. We conclude that the BFMSS bit bound is equal to

---

[1] Any other fraction will also work as the initial value.

$(2^k - 1)(3 + 4 \cdot 2^k) + 4 \cdot 2^k = \Theta(4^k)$. Increasing k by one, quadruples the numbers of bits.

The Li/Yap bound involves the lead coefficient of the minimal polynomial and is at least the logarithm of the lead coefficient. Li and Yap compute the following estimates lc for the lead coefficients. Let $d_i = D(E_i) = D(F_i) = 2^i$. Then $\log lc(E_0) = \log 3 \geq 1$, $\log lc(F_i) = \log lc(E_{i-1})$, $\log l(E_i) = 2 \cdot d_i \cdot \log lc(F_i) = 2 \cdot 2^i \log lc(E_{i-1}) = 2^i \prod_{1 \leq j \leq i} 2^j \cdot \log 3 \geq 2^{i(i+3)/2}$, $\log lc(E) = 2 \cdot 2^k \log lc(E_k)$ and hence the Li-Yap bit bound is $\Omega(2^{k^2/2})$. Increasing n by one multiplies the required number of bits by more than 2^k.

We next give an example where the Li/Yap bound is better. We start with the fraction $17/3$, square k times and then take roots k times. The weight of the expression is 2^k and $\log u(E) \geq 2^k$. The BFMSS bit bound is therefore at least $\Omega(4^k)$. On the other hand, the Li/Yap bound is $O(2^k)$.

An implementation should compute the Li/Yap and BFMSS bounds and use the better of the bounds.

5   Experimental Evaluation

The separation bound approach to sign determination of algebraic numbers is used in the number types `real` of LEDA [12] and `Expr` of CORE [7]. We report about the improvements in running time due to the new separation bounds and due to a recent reimplementation of `leda_real`. We also compare CORE and `leda_real`.

All tests are based on LEDA 4.2.1 with the most recent arithmetic module incorporated. For the tests with the CORE library we used CORE v1.3 available from [9]; it uses the Li/Yap-bound. All benchmarks are performed on a Sun Ultra 5 with 333 MHz, 128 MB RAM, running Solaris 2.7. We used g++ 2.95.2.1 as a compiler, times are always stated in seconds.

We briefly review the implementation of `leda_real`, a detailed description is available in [4]. The number type supports the sign determination of simple algebraic expressions. Expressions are represented by their expression dag $\mathcal{G}(E)$. The input values of E are contained in the leaves of the dag, every inner node corresponds to an arithmetical operation, and the root corresponds to E.

When the sign of an algebraic number E needs to be dermined, the datatype first computes a separation bound q_E. Using `leda_bigfloat` arithmetic (= floating-point numbers with exponent and mantissa of arbitrary length), the datatype computes successively intervals of decreasing length that include E, until the interval does not contain zero or the length of the interval is less than q_E.

Several shortcuts are used to speed up the computation of the sign. First, a double approximation $\widetilde{E}$ and an error bound err such that $|E - \widetilde{E}| \leq err$ is stored with every node of the expression dag. As long as the double approximation $\widetilde{E}$ is known to be exact, i.e. $err = 0$, no expression graph is constructed and $\widetilde{E}$ represents E.

Secondly, if the double approximation $\widetilde{E}$ suffices to determine the sign of E, i.e. $0 \notin [\widetilde{E} - err, \widetilde{E} + err]$, no bigfloat computation is triggered. This technique is called a *floating-point filter*.

In the reimplementation, we made the following improvements:

(1) the separation bound is the better of the Li/Yap and the BFMSS bound.
(2) the implementation of the underlying bigfloat arithmetic has been improved; at the beginning it was based on number type `leda_integer` for integer numbers of arbitrary size, now it directly operates on vectors of `long` integers.
(3) memory management within the `real` datatype has been improved; in particular, space for the bigfloat approximations is now only allocated if bigfloat computation is necessary for a sign determination.
(4) the built-in floating-point filters have been improved, both with respect to running time as well as precision.

Overall, the efficiency has improved for 'easy instances' (i.e. instances that do not need the bigfloat computation) due to improved floating-point filter techniques as well as for 'difficult instances' due to the improved separation bounds and bigfloat implementation.

We turn to our experiments. The source code of all experiments is available at `http://www.mpi-sb.mpg.de/~funke/SepBoundESA01.html`. Many of the experiments make use of L-bit random integers. We generated them outside the `leda_real` number type and used them as inputs for our expressions.

(1) The first test is a simple check of a binomial expression. Let $x = \frac{a}{b}, y = \frac{c}{d}$ where a, b, c, and d are L-bit integers and let $E = (\sqrt{x} + \sqrt{y}) - \sqrt{x + y + 2\sqrt{xy}}$. For the old BFMS-bound we get a $\text{sep}_{BFMS} = 160L + 381$, for our improved bound $\text{sep}_{improv} = 96L + 60$, whereas the LiYap-bound gives $\text{sep}_{LiYap} = 28L + 60$. This is of course reflected in the running times in Table 1, left half.

Table 1. Running time for experiments (1) and (2)

L	Experiment (1)							Experiment (2)					
	25	50	100	200	400	800	1600	500	1000	2000	4000	8000	16000
BFMS	0.04	0.10	0.27	0.77	2.21	6.55	20.73	0.01	0.03	0.08	0.25	0.72	2.33
Improv	0.01	0.04	0.10	0.27	0.77	2.26	7.07	0.01	0.03	0.08	0.24	0.73	2.32
LiYAP	0.00	0.01	0.02	0.04	0.11	0.29	0.91	0.36	1.05	3.17	9.47	28.5	85.6

(2) Let x and y be L-bit integers, $C = (\sqrt{x} - \sqrt{y})/(x - y)$ and $E = C - C$. For both our old and improved bound we get $\text{sep}_{BFMS} = \text{sep}_{Improv} = 6L + 64$, whereas the LiYap-bound gives $\text{sep}_{LiYap} = 65L + 91$. See Table 1, right half.

We now turn to examples for which we have already proved differing asymptotic behaviour of the bounds in Section 4.

(3) First consider $F = \sqrt[k]{x/y}$ and $E = F - F$ where x is a $100k$-bit integer and y a 32-bit integer. The BFMS bound is $\Theta(k^2)$, whereas the new bound is $\Theta(k)$. The Li/Yap bound is also $\Theta(k)$ and even better than our new bound.

The running time of multiplication, division, and the root operation for L-bit numbers in `leda_bigfloat` is $L^{\log 3}$. Doubling k in case of the BFMS bound quadruples the separation bound and hence multiplies the running time by about[2] 9, whereas in case of the improved bound, the separation bound doubles and the running time roughly triples. See Table 2, left half, for the results.

---

[2] As machines get slower as they use more memory, we see a factor of slightly more than 9.

Table 2. Running times and separation bounds for experiments (3) and (4)

k		Experiment (3)					Experiment (4)				
		2	4	8	16	32	64	2	3	4	5	6
BFMS time		0.01	0.02	0.20	2.22	24.36	86.6	0.01	0.01	0.02	0.80	8.86
BFMS bound		391	1683	6751	26781	106396	421787	237	1213	5885	27645	126973
Improv. time		0.01	0.01	0.01	0.04	0.11	0.36	0.01	0.01	0.01	0.04	0.32
Improv. bound		214	538	1198	2524	5179	10459	76	284	1084	4220	16636
LiYap time		0.01	0.01	0.01	0.03	0.04	0.12	0.01	0.01	1.98	1781	(too long)
LiYap bound		150	346	750	1564	3195	6427	140	2076	65596	4194428	536871164

(4) For $E_0 = 17/3, F_i = \sqrt{E_{i-1}}, E_i = F_i + F_i, E = E_k - E_k$, our bounds for E are $\Theta(4^k)$ (but with different constant factors), whereas the Li-Yap bound is $\Theta(2^{k^2})$. See Table 2, right half, for the results.

In the following test we compare different implementations: `real(1)` denotes our old implementation and `real(2)` the new implementation.

(5) As in our first example we take $x = \frac{a}{b}, y = \frac{c}{d}$, where a, b, c, d are L-bit integers, and $E = (\sqrt{x} + \sqrt{y}) - \sqrt{x + y + 2\sqrt{xy}}$. As we can see in Table 3, the improved implementation of the LEDA `real` datatype already leads to a speedup of factor 4, even with the same separation bound. The new separation bound gives another speedup of factor 3. We did not expect the currently available CORE/Expr implementation that far behind, since it uses the Li-Yap bound which is superior to our bounds in this example. We neither understand why there is no difference in running time for $L = 100$ and $L = 200$, nor the change in running time when doubling the bitlength of the input values.

Table 3. Experiment (5)

L		25	50	100	200	400	800
real(1)	sep_{BFMS}	0.12	0.30	0.98	2.63	8.48	23.97
real(2)	sep_{BFMS}	0.04	0.10	0.27	0.77	2.21	6.55
real(2)	sep_{improv}	0.01	0.04	0.10	0.27	0.77	2.26
CORE/Expr	$sep_{OldLiYap}$	2.32	15.7	116.9	116.84	692	3973

Table 4. Experiment (6)

L		50	100	200
double		0.08	0.08	0.08
real(1)		1.64	1.65	194
real(2)		1.22	1.23	120
CORE/Expr		568	555	672

(6) The final comparison concerns easy sign tests. The following expression arises during Fortune's sweep-line algorithm for Voronoi diagrams: $E = \frac{a+\sqrt{b}}{c} - \frac{a'+\sqrt{b'}}{c'}$ where $a, a', b, b', c,$ and c' are random $3L$-, $6L$-, and $2L$-bit integers. The root bounds do not play a role here, only the efficiency of the implementation, in particular the floating-point filters comes into play. To get meaningful results we measured the time of 200000 sign computations, see Table 4.

Clearly, pure `double` arithmetic is the fastest, creating the expression dag does not come without cost. But as you can see, our new implementation gains about 25% compared to the old one. The huge increase in running time for $L = 200$ can be explained by the fact that in this case, the numbers get too large to be representable by a double (remember that we create integers of length $6L$). Therefore the floating-point filters will always fail and `bigfloat` arithmetic has to be used. CORE does not have built-in floating-point filters so it is much slower then `leda_real`.

6 Conclusions

We presented a new separation bound which applies to a wide class of algebraic expressions and is easily computable. For many expressions it gives much better bounds than previous bounds resulting in significant gains in running time. We see two main challenges: (1) For algebraic numbers defined by systems of polynomials, Canny's bound is the best bound known. Provide a better bound. (2) Our bound as well as the Li/Yap bound is very easy to compute. In the context of expensive sign computations it is worthwile to investigate more expensive methods for computing separation bounds.

References

1. C. Burnikel, R. Fleischer, K. Mehlhorn, and S. Schirra. Exact efficient computational geometry made easy. In *Proceedings of the 15th Annual Symposium on Computational Geometry (SCG'99)*, pages 341–350, 1999. www.mpi-sb.mpg.de/~mehlhorn/ftp/egcme.ps.
2. C. Burnikel, R. Fleischer, K. Mehlhorn, S. Schirra. A strong and easily computable separation bound for arithmetic expressions involving radicals. *Algorithmica*, 27:87–99, 2000.
3. C. Burnikel, K. Mehlhorn, and S. Schirra. How to compute the Voronoi diagram of line segments: Theoretical and experimental results. In Springer, editor, *Proceedings of the 2nd Annual European Symposium on Algorithms - ESA'94*, volume 855 of *Lecture Notes in Computer Science*, pages 227–239, 1994.
4. C. Burnikel, K. Mehlhorn, and S. Schirra. The LEDA class *real* number. Technical Report MPI-I-96-1-001, Max-Planck-Institut für Informatik, Saarbrücken, 1996.
5. J.F. Canny. *The Complexity of Robot Motion Planning*. The MIT Press, 1987.
6. E. Hecke. *Vorlesungen über die Theorie der algebraischen Zahlen*. Chelsea, New York, 1970.
7. V. Karamcheti, C. Li, I. Pechtchanski, and Chee Yap. A core library for robust numeric and geometric computation. In *Proceedings of the 15th Annual ACM Symposium on Computational Geometry*, pages 351–359, Miami, Florida, 1999.
8. R. Loos. Computing in algebraic extensions. In B. Buchberger, G. E. Collins, and R. Loos, editors, *Computer Algebra. Symbolic and Algebraic Computation*, volume 4 of *Computing Supplementum*, pages 173–188. Springer-Verlag, 1982.
9. C. Li and C. Yap. A new constructive root bound for algebraic expressions. In *Proceedings of the 12th Annual ACM-SIAM Symposium on Discrete Algorithms*, pages 496–505, 2001.
10. M. Mignotte. Identification of Algebraic Numbers. *Journal of Algorithms*, 3(3):197–204, September 1982.
11. M. Mignotte. *Mathematics for Computer Algebra*. Springer, 1992.
12. K. Mehlhorn and S. Näher. *The LEDA Platform for Combinatorial and Geometric Computing*. Cambridge University Press, 1999. http://www.mpi-sb.mpg.de/LEDA/leda.html.
13. K. Mehlhorn and St. Schirra. Exact computation with leda_real - theory and geometric applications. In G. Alefeld, J. Rohn, S. Rumpf, and T. Yamamoto, editors, *Symbolic Algebraic Methods and Verification Methods*. Springer Verlag, Vienna, 2001.
14. J. Neukirch. *Algebraische Zahlentheorie*. Springer-Verlag, 1990.
15. E. R. Scheinerman. When close enough is close enough. *American Mathematical Monthly*, 107:489–499, 2000.
16. C.K. Yap. Towards exact geometric computation. *CGTA: Computational Geometry: Theory and Applications*, 7, 1997.
17. C.K. Yap. *Fundamental Problems in Algorithmic Algebra*. Oxford University Press, 1999.
18. C.K. Yap and T. Dube. The exact computation paradigm. In *Computing in Euclidean Geometry II*. World Scientific Press, 1995.

Property Testing with Geometric Queries

(Extended Abstract)*

Artur Czumaj[1] and Christian Sohler[2]

[1] Department of Computer and Information Science, New Jersey Institute of Technology,
University Heights, Newark, NJ 07102-1982, USA. czumaj@cis.njit.edu
[2] Heinz Nixdorf Institute and Department of Mathematics & Computer Science, University of
Paderborn, D-33095 Paderborn, Germany. csohler@uni-paderborn.de

Abstract. This paper investigates geometric problems in the context of property testing algorithms. Property testing is an emerging area in computer science in which one is aiming at verifying whether a given object has a predetermined property or is "far" from any object having the property. Although there has been some research previously done in testing geometric properties, prior works have been mostly dealing with the study of *combinatorial* notion of the distance defining whether an object is "far" or it is "close"; very little research has been done for *geometric* notion of distance measures, that is, distance measures that are based on the geometry underlying input objects.

The main objective of this work is to develop sound models to study geometric problems in the context of property testing. Comparing to the previous work in property testing, there are two novel aspects developed in this paper: geometric measures of being close to an object having the predetermined property, and the use of geometric data structures as basic primitives to design the testers. We believe that the second aspect is of special importance in the context of property testing and that the use of specialized data structures as basic primitives in the testers can be applied to other important problems in this area.

We shall discuss a number of models that in our opinion fit best geometric problems and apply them to study geometric properties for three very fundamental and representative problems in the area: testing convex position, testing map labeling, and testing clusterability.

1   Introduction

A classical problem in computer science is to verify if a given object possesses a certain property. For example, we want to determine if a boolean formula is satisfiable, or if a set of polygons in the Euclidean plane is intersection free. In its very standard formulation, the goal is to give an exact solution to the problem, that is, to provide an algorithm that always returns a correct answer. In many situation, however, this formulation is too restrictive, for example, because there is no fast (or just fast enough) algorithm that gives the exact solution. Very recently, many researchers started studying a relaxation of the "exact decision task" and consider various forms of approximation algorithms

---

* Research supported in part by an SBR grant No. 421090, by DFG grant Me872/7-1, and by the IST Programme of the EU under contract number IST-1999-14186 (ALCOM-FT).

F. Meyer auf der Heide (Ed.): ESA 2001, LNCS 2161, pp. 266–277, 2001.

for decision problems. In *property testing* (see, e.g., [3,4,7,10,12,13,14,15,18]), one considers the following class of problems:

> Let $\mathfrak{C}$ be a class of objects, $\mathfrak{O}$ be an unknown object from $\mathfrak{C}$, and Q be a fixed property of objects from $\mathfrak{C}$. The goal is to determine (possibly probabilistically) if $\mathfrak{O}$ has property Q or if it is *far* from any object in $\mathfrak{C}$ which has property Q, where distance between two objects is measured with respect to some distribution $\mathcal{D}$ on $\mathfrak{C}$.

The motivation behind this notion of property testing, is that while relaxing the exact decision task, we expect the testing algorithm to be significantly more efficient than any exact decision algorithm, and in many cases, we achieve this goal by exploring only a small part of the input. And so, for example, in [4] it is shown that all first order graph properties of the type "∃∀" can be tested in time independent of the input size (see also, [13,14,18] for some other most striking results).

In the standard context of property testing, the first general study of geometric properties appeared in [7]. In this paper the authors studied property testing for classical geometric problems like being in convex position, for disjointness of geometric objects, for Euclidean minimum spanning tree, etc. Roughly at the same time, in [3], property testing for some clustering problems has been investigated. In [10], the problem of testing if a given list of points in $\mathbb{R}^2$ represents a convex polygon is investigated. In all these papers, the common measure of being close to having the predetermined property was the *Hamming distance*. That is, for an object $\mathfrak{O}$ from a class $\mathfrak{C}$, a property Q, and a real ε, $0 \le \varepsilon \le 1$, we say $\mathfrak{O}$ is ε-far from having property Q, if any object $\mathfrak{O}$ from $\mathfrak{C}$ that has property Q has the Hamming distance at least $\varepsilon \cdot |\mathfrak{O}|$ from $\mathfrak{O}$. The Hamming distance is a standard measure to analyze combinatorial problems, but in the opinion of the authors, other more geometric distance measures should also be considered in the context of Computational Geometry. The reason is that this measure does not explore geometry underlying investigated problems, but only their combinatorial structure (how many "atom" objects must be modified to transform the object into one possessing the requiring property). This issue has been partly explored in the context of the *metrology of geometric tolerancing* [5,6,8,17,19,20]. In this area (motivated by manufacturing processes) one considers the problems of verifying if a geometric object is within some given tolerance from having certain property $\mathfrak{Q}$. In geometric tolerancing, the researchers have been studying among others, the "roundness property," the "flatness property," etc. [5,6,8,17]. We emphasize, however, that there is a major difference between the notion of geometric property testing and geometric tolerancing in that in the former one allows to reject (as well as accept) any object that does not satisfy the property $\mathfrak{Q}$, while in geometric tolerancing one should accept such an object if it is within given tolerance.

Our contribution. This paper is partly of a methodological character. The main objective of this paper is to develop proper models to study geometric problems in the context of property testing. We shall discuss a number of models that in our opinion best fit geometric problems and apply them to study geometric properties for the most fundamental problems in the area. Comparing to the previous work in property testing, in the current paper we develop two main novel ideas:

☞ geometric measures of being close to an object having the predetermined property, and

☞ the use of geometric data structures to develop the testers.

We discuss these two issues in details in Sections 2 and 3 while testing *convex position*. We demonstrate the need of geometric distance measures for geometric problems and propose three new models that in our opinion suit best to study geometric properties. We show also that the complexity measures used in standard property testing have to be modified in order to achieve something non-trivial using geometric distance measures. We propose a model of computation that uses *queries for geometric primitives* (in this case range queries) as its basis and discuss its use and practical justifications. Finally, we illustrate all these issues by designing property testing algorithms for convex position. Unlike in the model investigated in [7], our testing algorithms run in time either completely independent of the input size or only with a polylogarithmic dependency, and we believe that they fit much better the geometry underlying the problem of testing convex position.

In Section 4, we investigate the *map labeling* problem. We first show that in the classical property testing setting (that uses uniform sampling of the input points) this problem does not have fast testing algorithms. Next, we show that by using geometric queries as basic operations one can obtain very efficient testing algorithms. We present an ε-tester for map labeling that requires only $poly(1/\varepsilon)$ range queries of the form: *What is the i-th point in the orthogonal range R ?*

Then, in Section 5, we consider *clustering problems* in our context and provide efficient testers for clustering problems in most reasonable geometric models. The goal of a clustering problem is to partition a point set in $\mathbb{R}^d$ into k different clusters such that the cost of each cluster is at most b. We consider three different variants of the clustering problem (see, e.g., [3]): *radius clustering*, *discrete radius clustering*, and *diameter clustering*. We say that a set of points is ε-*far from clusterable* with k clusters of size b, if there is no clustering into k clusters of size $(1 + \varepsilon) b$. We show that it is possible to test clusterability using $\mathcal{O}(k/\varepsilon^d)$ oracle range queries.

Comparing our results to those in [3], we use a more powerful oracle but we also have a more restrictive distance measure. Using our distance measure and the classical oracle from [3], it is impossible to design a sublinear property tester for this problem.

Further, we show how to use our tester to maintain (under insertion and deletion of points) an approximate k-clustering in $\mathbb{R}^d$ of size at most $(1 + \varepsilon)$ times the optimum in time $polylog(n)$ for any constants k, d, and ε. Here, n denotes the current number of points.

2   Testing Convex Position

Let us first consider the classical problem of testing if a point set P in the plane is in *convex position* (that is, the interior of the convex hull $conv(P)$ contains no point from P, or equivalently, all points in P are *extreme*). Our goal is to consider a practical situation in which we allow some relaxation of the exact decision test and we consider the following type of testers:

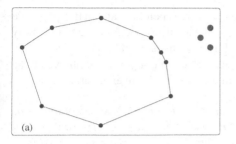
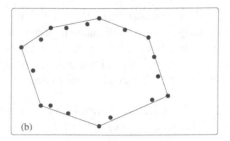

Fig. 1. Which of the two point sets is "more convex?" In Figure (a) it is enough to delete only 3 points (those in the top right corner) to obtain a point set in convex position; in Figure (b) one has to remove much more points to do so. On the other hand, the points from Figure (a) are visually far from convex position, while points in Figure (b) look similar as they were in convex position; it is enough to perturbate them very little to obtain a point set in convex position.

☞ If P is in convex position, then the tester must accept the input.

☞ If P is "far" from convex position, then the tester "typically" rejects the input.

☞ If P is not in convex position, but it is close to being so, then the answer may be arbitrary.

In order to use this concept we must formalize some of the notions used above. First of all, we assume a tester is a possibly randomized algorithm and, following standard literature in this area, by "typically" we shall mean that the required answer is output with probability at least $\frac{2}{3}$, where the probability is over the random choices made by the tester (and thus, this lower bound of $\frac{2}{3}$ is independent of the input).

2.1   Distance Measures — Far or Close

A more subtle issue is what do we mean by saying that P is "far" from convex position. We pick a parameter ε, $0 \le \varepsilon \le 1$, which will measure the quality of how "close" is P to convex position In the standard terminology used in the property testing literature (see, e.g., [13,18]), one uses the following definition:

Definition 2.1. (Hamming distance) *A point set P in the Euclidean space $\mathbb{R}^d$ is ε-far from convex position (according to the Hamming distance), if for any subset $S \subseteq P$, $|S| \le \varepsilon \cdot |P|$, set $P \setminus S$ is not in convex position.*

We found, however, that this measure often does not correspond to notions of the distance used in geometry (see, e.g., Figure 1). It tells only about combinatorial properties of the object at hand, but it tells very little about geometry behind the object. For example, do we want to accept an n-point set P if it contains $\frac{1}{2}\varepsilon n$ points that are very far away from the remaining points that are in convex position (as, for example, in Figure 1 (a))? Or perhaps, we consider such a set P as far from convex position? On the other hand, if P contains an ε fraction of points which make P non-convex, but after a very small perturbation of these points, the obtained set will be in convex position (see, e.g., Figure 1 (b)). Do we want to call such a point set ε-far from convex position or not?

It is clear that the distance notion is very application dependent, and in this paper we investigate various distance measures which should be of practical interest, and study basic problems from computational geometry for these distance measures.

We begin with a distance that measures how much the input points are allowed to be moved (perturbated) in order to transform them into being in convex position.

Definition 2.2. (Perturbation measure) *A point set P in the d-dimensional unit cube is ε-far from convex position (according to the Perturbation measure), if for any perturbation of points in P that moves any point by distance at most ε, the resulting point set is not in convex position.*

Because of scaling, the fact that P is enclosed by the unit cube is assumed without loss of generality.

We introduce also another measure that although very similar to the Perturbation measure, will be more useful for our applications.

Definition 2.3. (Neighborhood measure) *A point set P in the d-dimensional unit cube is ε-far from convex position (according to the Neighborhood measure), if there exists a point $p \in P$ for which the d-dimensional ball of radius ε with center at p does not intersect the boundary of $conv(P)$.*

The next measure is more related to the volume discrepancy of the convex hull of the input points. It differs significantly from the Perturbation and Neighborhood measures, because this measure is relative to the volume of $conv(P)$.

If a point set P is in convex position, then all points in P lie on the boundary of convex hull $conv(P)$ and therefore $conv(P)$ is also the maximal convex hull defined by any (non-trivial) subset of P whose interior contains no point from P. In view of this, we may want to consider P to be close to convex position, if a maximum (with respect to the volume) convex hull defined by a subset of P that contains no point from P in its interior[1] is almost the same as $conv(P)$. If we use the volume measures for these two objects, then we get the following definition:

Definition 2.4. (Volume measure) *A point set P in $\mathbb{R}^d$ is ε-far from convex position (according to the Volume measure), if $\frac{vol(EmpInt(P))}{vol(conv(P))} \leq 1 - \varepsilon$, where $vol(X)$ denotes the volume of object X and $EmpInt(P)$ is a maximum volume convex hull defined by a subset of P that contains no point from P in its interior.*

Now, we are ready to formally define property testing algorithms.

Definition 2.5. (ε-Testers) *An algorithm is called an ε-tester for a property Q, if it always accepts any input satisfying property Q and with probability at least $\frac{2}{3}$, rejects any input that is ε-far from satisfying property Q.*

Throughout the paper, we say P is ε-close to convex position if it is not ε-far from convex position.

---

[1] Observe that in general that may be many such maximum convex hulls.

Relations between different measures of closeness. How can we relate the four measures defined in Definitions 2.1–2.4? As we observed above, a point set P can be close to convex position according to the Hamming distance, even if it is far (very far!) from convex position according to the Perturbation, the Neighborhood, and the Volume measures. Similarly, the opposite is also true: P may be very close to convex position according to the Perturbation, the Neighborhood, and the Volume measures even it it fails for the Hamming distance. But how about any relationship between the Perturbation, the Neighborhood, and the Volume measures?

The first lemma shows that the first two measures are somehow equivalent for asymptotic complexity of the testers. It holds for any cost measure of query complexity, because exactly the same tester is to be used.

Lemma 2.1. *There is an ε-tester for convex position according to the Perturbation measure with query complexity $T(n, \varepsilon)$ if and only if there is an $\Theta(\varepsilon)$-tester for convex position according to the Neighborhood measure with query complexity $T(n, \Theta(\varepsilon))$.* □

Unfortunately, we were unable to provide a similar relationship between the Neighborhood and the Volume measures. It seems to us that the latter one is more complicated. We can only prove some partial results about similarity of these two measures, for example:

Lemma 2.2. *An ε^d-tester for the Volume measure is an $\mathcal{O}(\varepsilon)$-tester for the Neighborhood measure.* □

3   A New Model Using Geometric Queries

In the previous works on property testing, the complexity of a tester has been typically measured as the number of input "atom objects" inspected, that is, as the number of *queries* to the input. The form of the queries allowed for the algorithm depended on the input representation. And so, for example, if an input consists of a set of n points (as it is the case for testing if the points are in convex position), then it has been typically assumed that one can use queries of the form: "*what is the position of the kth point in the input.*" In the standard query complexity additional computational work is not counted (for example, if we know positions of points in S, $S \subseteq P$, then the cost of computing a convex hull of S is not counted in the query complexity[2]). Our main observation is that this notion of query complexity often does not suit well to study geometric properties, or actually, to distance measures different than the Hamming distance. Indeed, if we want to check if a point set P is ε-close to convex position according to the Perturbation

---

[2] For example, in [7], it is shown that the query complexity for testing (according to Hamming distance) if a point set in $\mathbb{R}^d$ is in convex position is $\Theta(n^d/\varepsilon)^{1/(d+1)}$, while for $d \geq 4$, the "running time complexity" (which measures also the time required for all computations used by the tester) is $\mathcal{O}(n\,polylog(1/\varepsilon) + (n/\varepsilon)^{(\lceil d/2 \rceil)/(1+\lceil d/2 \rceil)}\,polylog(n))$, and it is quite possible that it is optimal. Thus, in the most basic case, for $d = 4$ and constant ε^{-1}, the query complexity is $\Theta(n^{3/4})$ while the "running time complexity" is $\mathcal{O}(n)$. This difference vanishes for $d = 2, 3$, because in this case very efficient (almost linear-time) algorithms for testing convex position are available.

measure, then even a single point might be far away from the remaining points to make this property false. Therefore, the algorithm must find this point with probability at least $\frac{2}{3}$, and this clearly requires $\Theta(n)$ query complexity. Similar phenomenon holds also for the Neighborhood measure and the Volume measure.

This observation shows that in order to model property testing for geometric properties and in order to obtain very efficient (sublinear-time) algorithms one has to reconsider and change the notion of query complexity. Unlike in the very standard model, here we want to allow more complex queries: those using certain geometric properties of the input.

In most of geometric models (and in applications) even if the input is represented by positions of the points (or other geometric objects), very often one maintains some additional data structures for efficient and structured access to the input. One of the most fundamental abstract data structure maintained by many algorithms working with points are data structures for efficient answering *range queries* (cf. [1]). For the purposes of this paper we adopt a model of computation in which the basis operation is a range query to the input, and the *query complexity* is the number of range queries to the input.

[3]Formally, we are given an unknown set P of n points in $\mathbb{R}^d$ that is defined by an unknown function $\mathcal{F}_P : \mathbb{R}^d \times \mathbb{N} \to \mathbb{R}^d \cup \{empty\}$ such that $\mathcal{F}_P(R, i)$ returns the ith point in a query range R (according to some unknown fixed order) or the symbol $empty$, if there are less than i points in the query range R.

The model defined above uses a very powerful oracle since we are allowed to specify an arbitrary range when we query the oracle. To make our consideration of practical value, it seems reasonable to require that such an oracle must be efficiently implemented. Therefore, in this model we will restrict ourselves to the case that R is a (possibly unbounded) simplex. Most of the results presented in this paper hold even for *orthogonal range queries*. Such queries are supported by many well known data structures such as partition trees and cutting trees, as well as practical structures based on quad-trees or R-trees (see, e.g., [1] for a more detailed discussion). There are efficient data structures (see, e.g., [1]) to support our queries and a single query to such a data structure is usually performed very fast (i.e., much faster than processing the whole point set). We believe that the use of such range queries is very natural in our context, since many applications (such as GISs) use data structures for range queries (e.g., R-trees) to answer other kinds of queries anyway.

In a similar way, depending on the problem at hand, one could assume that some other very basic *geometric queries* are available; we do not discuss this issue in more details however.

3.1   Property Testing Algorithms for Convex Position in the New Model

In this section we present our first ε-tester for convex position. It works for the Neighborhood measure, and it shows that the use of geometric queries (orthogonal range queries) allows to beat the lower bounds discussed in Section 3 and obtain the query complexity of $polylog(1/\varepsilon)$.

---

[3] We formalize our model of computation only to inputs that are in the form of point sets; for other input types the model can be defined accordingly in a similar way.

The input for our tester consists of a point set P in the d-dimensional unit cube and a real number ε, $0 < \varepsilon < 1$, which defines the quality of the tester.

CONVEXITY-TEST I (P, ε):

$\quad S = \emptyset$

$\quad$ partition the unit cube into $(2\sqrt{d}/\varepsilon)^d$ sub-cubes of side-length $\frac{\varepsilon}{2\sqrt{d}}$

$\quad$ **for each** such sub-cube c **do**

$\quad\quad$ **if** c contains a point from P **then** add any such a point to S

$\quad$ **if** S is in convex position **then** *accept*

$\quad\quad\quad\quad\quad$ **else** *reject*

In algorithm CONVEXITY-TEST I (P, ε) the operation of verifying if c contains a point from P as well as the operation of returning a point from $P \cap c$ is performed using orthogonal range queries.

Theorem 3.1. *Let P be a point set in the d-dimensional unit cube and let ε be a real number, $0 < \varepsilon < 1$. Algorithm CONVEXITY-TEST I (P, ε) is a property tester that accepts P only if P is ε-close (according to the Neighborhood measure) to convex position. It uses $\mathcal{O}((\sqrt{d}/\varepsilon)^d)$ orthogonal range queries.* $\square$

Actually, we can slightly improve the complexity of CONVEXITY-TEST I (P, ε) and design an ε-tester that uses only $\mathcal{O}((\sqrt{d}/\varepsilon)^{d-1})$ orthogonal range queries.

In the previous sections we discussed testing convexity properties in geometric setting. Now, we give a tester for testing convexity properties of planar point sets using a distance measure that is related to the Hamming distance (in fact, the distance measure below is stronger). A tester for Hamming distance is presented in [7]. The main difference in our approach here is the use of geometric queries that leads a to substantial speed up.

Definition 3.1. *A set P of n points in the plane is ε-far from being in convex position, if at least εn points in P are not extreme. (A point is extreme if it belongs to the boundary of the convex hull of P.)*

It immediately follows.

Lemma 3.1. *A tester for the distance measure in Definition 3.1 is also a tester for the Hamming distance.* $\square$

We can also prove the following lemma.

Lemma 3.2. *In the standard property testing model (see, e.g., [7,13,18]), there is no testing algorithm for the distance measure from Definition 3.1 that has $o(n)$ query complexity.* $\square$

We can prove that the use of appropriate data structures for the geometric queries allows us to design a tester with logarithmic query complexity. We assume the input point set P is in general position.

Theorem 3.2. *There is a tester for convex position in the plane with query complexity $\mathcal{O}(\log n/\varepsilon)$ that uses only triangular range queries.* $\square$

4 Map Labeling

In this section we consider the following basic *map labeling* problem:

Let P be a set of n points in the plane. Decide whether it is possible to place n axis-parallel unit squares such that

- all squares are pairwise disjoint (labels do not overlap),
- each point is a corner of exactly one square (each point is labeled), and
- each square has exactly one point on its corners (each point has a unique label).

If a set S of n squares satisfies the conditions above, then S is called a *valid labeling for P*. The map labeling problem is known to be $\mathcal{NP}$-complete and the corresponding optimization problem is known to have no approximation algorithm with ratio better than 2, unless $\mathcal{P} = \mathcal{NP}$ [11].

In this section we develop a property tester for the map labeling problem. We use the following Hamming distance measure:

Definition 4.1. *A set P of n points in the plane is ε-far from having a valid labeling, if we have to delete at least $\varepsilon\, n$ points to obtain a set of points that has a valid labeling.*

When we consider the standard property testing model [13,18] that allows only to sample random subsets of P with a uniform distribution we can prove the following result.

Theorem 4.1. *For any constant δ, $0 < \delta < 1$, there is a positive constant ε such that there is no ε-tester for the labeling problem with $o(n^{1-\frac{1}{2\delta+1}})$ query complexity in the standard testing model.*   □

We show now that if we use the computational model that allows/supports geometric queries we can design a tester with $\mathcal{O}(1/\varepsilon^3)$ query complexity. It is based on the approach developed in [9] and [16].

LABELTEST(P):

    choose a sample set S of size $\mathcal{O}(1/\varepsilon)$ uniformly at random from P
    for each $p \in S$ **do**
        $i = 0, \mathcal{T} = \emptyset$
        Let S be the axis parallel square with center p and side length $16\lceil 1/\varepsilon \rceil$
        while $i \leq (16\lceil 1/\varepsilon \rceil + 2)^2$ **do**
            Let q be the i-th point in the query range S
            if $q \neq \emptyset$ **then** $\mathcal{T} = \mathcal{T} \cup \{q\}$
        if $\mathcal{T}$ does not have a valid labeling **then** *reject*
    accept

Theorem 4.2. *Algorithm* LABELTEST *is a tester for the labeling problem that has query complexity $\mathcal{O}(1/\varepsilon^3)$ and running time $\exp(\mathcal{O}(1/\varepsilon^2))$.*   □

5   Clustering Problems

In this section we design testing algorithms for three geometric clustering problems. The goal of a clustering problem is to decide whether a set of n points P in $\mathbb{R}^d$ can be partitioned into k subsets (called *clusters*) $S_1, \ldots, S_k$ such that the *cost* of each cluster is at most b. There are several different ways to define the cost of a cluster. Let S be a set of points in $\mathbb{R}^d$. We consider the following variants:

Radius Clustering: The cost $cost_R(S)$ of a cluster S is twice the minimum radius of a ball containing all points of the cluster.

Discrete Radius Clustering: The cost $cost_{DR}(S)$ of a cluster S is the minimum radius of a ball containing all points of the cluster and having its center among the points from P.

Diameter Clustering: The cost $cost_D(S)$ of a cluster S is the maximum distance between a pair of points of the cluster.

The goal of our property tester is to accept all instances that admit a clustering into k subsets of cost b and to reject with high probability those instances that cannot be clustered into k subsets of cost $(1 + \varepsilon) b$.

Definition 5.1. *A point set P is (b, k)-clusterable for a cost measure cost(), if there is a partition of P into sets $S_1, \ldots S_k$ such that $cost(S_i) \le b$ for all $1 \le i \le k$. A point set P is ε-far from being (b, k)-clusterable, if for any partition of P into sets $S_1, \ldots S_k$ at least one set S_i has cost larger than $(1 + \varepsilon) b$.*

In the standard context of property testing, Hamming distance (that is, a point set is ε-far from clusterable, if we have to remove at least εn points to make it clusterable) has been used before [3]. For the diameter clustering problem the distance measure used in [3] has the additional relaxation that a point set is ε-far from (b, k)-clusterable, if one has to remove εn points to make the set $((1+\varepsilon) b, k)$-clusterable. Thus, this definition assumes a geometric and a combinatorial relaxation of the corresponding decision problem. We require only the geometric relaxation.

Let us assume, without loss of generality, that $b = 1$ and thus we want to design a tester for the problem whether a point set P is $(1, k)$-clusterable for the three cost measures above. We partition $\mathbb{R}^d$ into grid cells of side length $\varepsilon/(3 \sqrt{d})$. For each cell containing an input point, we choose arbitrary input point from the cell as its representative. Then, we compute whether the set of representatives is $(1, k)$-clusterable. If it is so, then we accept it, if it is not so, then we reject it. Clearly, any set of points that is $(1, k)$-clusterable is accepted by the algorithm. On the other hand, any instance that is ε-far from $(1, k)$-clusterable will be rejected. (This approach has been introduced in [2] to obtain a $(1 + \varepsilon)$-approximation algorithm for the radius clustering problem.)

Our algorithms starts with an empty box with endpoints at infinity. Then we query for a point in this box. We allocate the corresponding grid cell and partition the box into the 3^d sub-boxes induced by the hyperplanes bounding the grid cell. Then we continue with one of these sub-boxes. If we find an empty sub-box, it will be marked. If there are only marked boxes the algorithm terminates.

So far, our partition into grid cells works fine, if there are many points in a single cell. On the other hand, if no two points are in the same grid cell, the algorithm has $\Omega(n)$

query complexity. Thus we need an upper bound on the number of grid cells, whose representative may form a cluster.

Lemma 5.1. *Let S be a set of points in $\mathbb{R}^d$ no two points of which belonging to the same cell of a grid of size $\varepsilon/(3\sqrt{d}) < 1$. If $cost(S) \leq 1$ for any of the three cost measures described above, then $|S| \leq (6\sqrt{d}/\varepsilon)^d$, where $cost(\cdot) \in \{cost_R, cost_{DR}, cost_D\}$.*   □

Let $V = k \cdot (6\sqrt{d}/\varepsilon)^d$ the maximum number of cells that can contain points that belong to one of the k clusters. We observe that we can stop our procedure if the number of representatives is V. Thus, we can guarantee that the algorithm requires at most $V \cdot 3^d$ range queries.

Theorem 5.1. *There is an ε-tester for the radius clustering and diameter clustering problem that uses at most $k \cdot (18\sqrt{d}/\varepsilon)^d$ orthogonal range queries. There is an ε-tester for the discrete radius clustering problem that uses $k \cdot (162\sqrt{d}/\varepsilon)^d$ orthogonal range queries.*   □

5.1 Dynamic Clustering

In this section we consider the problem of maintaining an approximate clustering of points in $\mathbb{R}^d$ under the operations *insert* and *delete*. Obviously, we can call the decision procedure from the previous section $\mathcal{O}(\log_{1+\varepsilon} B)$ times to find a clustering of size at most $(1+\varepsilon)B$ where B is the size of an optimal clustering and $B \geq 1$. When we combine this with a dynamic data structure that supports orthogonal range queries in time $A(n)$ (to report a single point in the query range) and update time $U(n)$ we immediately obtain the following result.

Corollary 5.1. *We can maintain an $(1 + 5\varepsilon)$ approximate radius/diameter clustering of a point set P in $\mathbb{R}^d$ (d constant) under the operations insert and delete in time $\mathcal{O}(U(n) + \log_{1+\varepsilon} B \cdot (A(n) \cdot k/\varepsilon^d + \exp(\mathcal{O}(k/\varepsilon^d))))$. If the parameters ε, d, and k are constants this is $\mathcal{O}(U(n) + A(n) + \log_{1+\varepsilon} B)$ time.*   □

Now, we want to obtain a time bound that is independent of the size of the clustering. We shall require an additional kind of oracle access: we allow the tester to query the oracle for the number of points within a certain range (this procedure could be also performed in our prior model in a logarithmic cost). We also need a procedure to compute a minimum (axis parallel) bounding box for the points inside a given cell. This can be easily done with $\mathcal{O}(d \log n)$ expected oracle accesses.

To avoid a simple binary search we use this bounding box. The size of the bounding box will always be the length of its longest side. Then we compute a clustering C for the current grid size (using the representatives for each cell). If l is the size of the largest bounding box of all grid cells, then we know that P can be clustered at cost at most $cost(C) + 2 \cdot \sqrt{d} \cdot l$. Note that we can stop our process if $\frac{s}{cost(C)} \leq \varepsilon/(3\sqrt{d})$ where s is the current size of the grid.

If we cannot stop we continue with a grid of size $l/2$. This way the number of grid cells with representatives is at most 3^d times the previous number of grid cells and there is at least 1 more such cell. We continue this procedure until we get a lower bound on the size of the current clustering. Then we have to do a logarithmic number of further steps and we are done.

Theorem 5.2. *We can maintain an* $(1 + \varepsilon)$ *approximate radius/diameter clustering of a point set* P *in* $\mathbb{R}^d$ *(d constant) under the operations* insert *and* delete *in time* $U(n) + \mathcal{O}(\exp(\mathcal{O}(V)) \cdot (k + \log(1/\varepsilon) + A(n) \cdot V \cdot \log n \cdot (k + \log(1/\varepsilon)))$. *If k, d, and* ε *are constants then this is* $\mathcal{O}(U(n) + A(n) \cdot \log n)$. $\square$

References

1. P. K. Agarwal. Range searching. Chapter 31 in J. E. Goodman and J. O'Rourke, editors, *Handbook of Discrete and Computational Geometry*, CRC Press, 1997.
2. P. K. Agarwal, C. M. Procopiuc. Exact and approximation algorithms for clustering. *Proc. 9th ACM-SIAM SODA*, pp. 658–667, 1998.
3. N. Alon, S. Dar, M. Parnas, and D. Ron. Testing of clustering. *Proc. 41st IEEE FOCS*, pp. 240–250, 2000.
4. N. Alon, E. Fischer, M. Krivelevich, and M. Szegedy. Efficient testing of large graphs. *Proc. 40th IEEE FOCS*, pp. 656–666, 1999.
5. P. Bose and P. Morin. Testing the quality of manufactured disks and cylinders. *Proc. 9th ISAAC*, pp. 129–138, 1998.
6. P. Bose and P. Morin. Testing the quality of manufactures balls. *Proc. 6th WADS*, pp. 145–156, 1999.
7. A. Czumaj, C. Sohler, and M. Ziegler. Property testing in computational geometry. *Proc. 8th ESA*, pp. 155–166, 2000.
8. C. A. Duncan, M. T. Goodrich, and E. A. Ramos. Efficient approximation and optimization algorithms for computational metrology. *Proc. 8th ACM-SIAM SODA*, pp. 121–130, 1997.
9. S. Doddi, M. Marathe, A. Mirzaian, B. Moret, and B. Zhu. Map labeling and its generalizations. *Proc. 8th ACM-SIAM SODA*, pp. 148-157, 1997.
10. F. Ergün, S. Kannan, S. R. Kumar, R. Rubinfeld, and M. Viswanathan. Spot-checkers. *JCSS*, 60:717–751, 2000.
11. M. Formann and F. Wagner. A packing problem with applications to lettering of maps. *Proc. 7th ACM SoCG*, pp. 281-288, 1991.
12. A. Frieze and R. Kannan. Quick approximation to matrices and applications. *Combinatorica*, 19:175–220, 1999.
13. O. Goldreich. Combinatorial property testing (A survey). *Proc. DIMACS Workshop on Randomization Methods in Algorithm Design*, pp. 45–59, 1997.
14. O. Goldreich, S. Goldwasser, and D. Ron. Property testing and its connection to learning and approximation. *JACM*, 45(4):653–750, 1998.
15. O. Goldreich and D. Ron. A sublinear bipartiteness tester for bounded degree graphs. *Combinatorica*, 19(3):335–373, 1999.
16. H. B. Hunt III, M. V. Marather, V. Radhakrishnan, S. S. Ravi, D. J. Rosenkrantz, and R. E. Stearns. NC-approximation schemes for NP- and PSPACE-hard problems for geometric graphs. *J. Algorithms*, 26(2):238–274, 1998.
17. K. Mehlhorn, T. C. Shermer, and C. K. Yap. A complete roundness classification procedure. *Proc. 13th ACM SoCG*, pp. 129–138, 1997.
18. D. Ron. Property testing. To appear in *Handobook of Randomized Algorithms*, P. M. Pardalos, S. Rajasekaran, J. Reif, and J. D. P. Rolim, eds., Kluwer Academic Publishers.
19. C. Yap. Exact computational geometry and tolerancing metrology. In D. Avis and J. Bose, editors, *Snapshots of Computational and Discrete Geometry*, volume 3. McGill School of Computer Science, 1995. Technical Report SOCS-94.50.
20. C. Yap and E.-C. Chang. Issues in the metrology of geometric tolerancing. *Proc. 2nd Workshop on Algorithmic Foundations of Robotics*, pp. 393–400, 1996.

Smallest Color-Spanning Objects*

Manuel Abellanas[1], Ferran Hurtado[2], Christian Icking[3], Rolf Klein[4],
Elmar Langetepe[4], Lihong Ma[4], Belén Palop[5], and Vera Sacristán[2]

[1] Dept. de Matemática Aplicada, Universidad Politécnica de Madrid, Spain
[2] Dept. de Matemàtica Aplicada II, Universitat Politècnica de Catalunya, Spain
[3] Praktische Informatik VI, FernUniversität Hagen, Germany
[4] Institut für Informatik I, Universität Bonn, Germany
[5] Dept. de Ciencias de la Computación, Universidad Alcalá de Henares, Spain

Abstract. Motivated by questions in location planning, we show for a
set of colored point sites in the plane how to compute the smallest—
by perimeter or area—axis-parallel rectangle and the narrowest strip
enclosing at least one site of each color.

1 Introduction

We are given a set of n point sites in the plane and $k \leq n$ colors, each site is
associated one color. A region of the plane is called *color-spanning* if it contains
at least one point of each color. For different kinds of regions we are interested
in the smallest color-spanning one of that kind.

The original motivation for our questions comes from location planning. Suppose there are k types of facilities, e. g. schools, post offices, supermarkets, modeled by n colored points in the plane, each type by its own color. One basic
goal in choosing a residence location is in having at least one representative of
each facility type in the neighborhood, where there are various specifications of
the term "neighborhood". A natural question is to ask for the smallest color-
spanning *circle*. It can be found using the upper envelope of *Voronoi surfaces*, as
described by Huttenlocher et al. [11] and Sharir and Aggarwal [19, Section 8.7];
their algorithm for computing the solution runs in time $O(kn \log n)$. Similarly
one can determine the smallest color-spanning axis-parallel *square* and other objects with fixed orientation which are unit circles of a convex distance function.

In this paper, we propose to solve more complicated problems of this context:
we give algorithms to compute the smallest color-spanning axis-parallel rectangle
in Sect. 2 and the narrowest color-spanning strip in Sect. 3.

A couple of related optimization problems for a set S of n points have already
been studied in the literature, with motivations from statistical clustering or
pattern recognition. For example, the convex polygon with minimum perimeter

* The Spanish authors acknowledge partial support from Acción integrada HA1999-
0094, MEC-DGES-SEUID PB98-0933, and Gen. Cat. 1999SGR000356, while the
German team was supported by DAAD grant 314-AI-e-dr.

F. Meyer auf der Heide (Ed.): ESA 2001, LNCS 2161, pp. 278–289, 2001.
© Springer-Verlag Berlin Heidelberg 2001

containing k points of S can be found by using the methods of Dobkin et al. [4], Aggarwal et al. [2], or finally Eppstein and Erickson [6], the last one in time $O(n \log n + k^3 n)$. The minimum *area* convex polygon containing k points of S can be determined in time $O(n^2 \log n + kn^2 \min(k^2, n))$ combining results of [6] and Eppstein et al. [7]. Similar problems for selecting k points out of n use as optimization criterion the diameter or the variance of the k-set, or they ask for the smallest circle with respect to a certain metric containing at least k points [2, 4,5,7,13,20], this latter problem is of course very closely related to the Voronoi diagram of order k.

Other very natural optimization criteria are the perimeter and the area of the axis-parallel rectangle enclosing a k-point set, these criteria are sometimes briefly called the L_∞ perimeter and L_∞ area. For computing the smallest perimeter, the best known running time is $O(n \log n + k^2 n)$ for algorithms by Datta et al. [3] and by Eppstein and Erickson [6]. The algorithm of Aggarwal et al. [2] can be used for both variants of the problem, the area and the perimeter, and takes time $O(\min(k^2 n \log n, n^3))$, while Segal and Kedem's solution [18] for both variants runs in $O(n + k(n - k)^2)$ time and is applicable only for $k > n/2$.

Not much seems to be known about lower bounds for these problems. Matoušek [12] reports that at least some of them are known to be n^2-hard, a notion introduced by Gajentaan and Overmars [9]; compare also Erickson and Seidel [8].

Interestingly, the approach in [2] for the smallest rectangle, like some approaches for the smallest circle, is also based on the Voronoi diagram of higher order, in this case of order $6k - 6$. This is because the optimal k-point set can be shown to be contained in a circle centered at such a Voronoi vertex which passes through the corresponding sites. Eppstein and Erickson's approach [6] uses the fact that the members of the optimal k-point set are always among the $16k$ nearest rectilinear neighbors of each of them. But neither of these properties based on proximity seem to be extensible to our new problem that involves colors.

For multicolored point sets, there are solutions to several problems, such as the *bichromatic closest pair*, see e.g. Preparata and Shamos [17, Section 5.7], Agarwal et al. [1], and Graf and Hinrichs [10], the *group Steiner tree* where, for a graph with colored vertices, the objective is to find a minimum-weight subtree that covers all colors, see Mitchell [14, Section 7.1], or the *chromatic nearest neighbor search*, see Mount et al. [15].

2   The Smallest Color-Spanning Rectangle

Among all rectangles whose sides are parallel to the x- and y-axis and which contain at least one site of each color we are looking for the smallest one, by perimeter or by area.

Some special cases are immediately solved. For $k = 1$ the problem is trivial, and for $k = n$, i.e., we have exactly one point for each color, the solution is the bounding box of the point set. And also the case $k = n - C$ for some constant C can be solved in time $O(n)$ because in this case there is only a constant ($\leq 2C$) number of sites with colors that have more than one site. Finally for $k = 2$ in the perimeter case, we can make use of an algorithm that computes the bichromatic L_1-closest pair in time $O(n \log n)$, see e. g. Graf and Hinrichs [10]. For the general case however, new ideas are necessary.

In Sect. 2.1 we show that the optimal rectangle must fulfill the so-called *non-shrinkable* property, we present a simple algorithm with running time in $O(n(n-k)^2)$, and we prove the tight bound of $\Theta((n-k)^2)$ for the number of non-shrinkable rectangles. In Sect. 2.2 we give necessary and sufficient conditions for non-shrinkable rectangles, and we refine the algorithm to a running time of $O(nk(n-k))$. Finally, in Sect. 2.3 we use a result by Overmars and van Leeuwen [16] for dynamically maintaining the maximal elements of a point set to further improve the running time to a near-optimal $O(n(n-k) \log^2 k)$.

2.1   Non-shrinkable Rectangles and a First Algorithm

Let p_x and p_y denote the coordinates of a site and p_{col} denote its color. For the sake of simplicity of the presentation, we make the following assumption on general position. No two x- or y-coordinates are equal, i. e., there is no horizontal or vertical line passing through two points. We exclude the trivial cases and assume for the remaining part of the paper that $1 < k < n$. It is clear that the smallest color-spanning rectangle, by perimeter or by area, must be non-shrinkable in the following sense.

Definition 1. *An axis-parallel rectangle is called* non-shrinkable *if it contains sites of all k colors and it does not properly contain another axis-parallel rectangle that contains all colors.*

Therefore, each non-shrinkable rectangle must touch a site with each of its four edges, such that there are two, three, or four sites on its boundary, among them no two of the same color. The colors on its boundary do not appear at sites in its interior.

Our algorithms will systematically find all non-shrinkable rectangles and compare their perimeters or areas to determine the smallest one, thereby solving the two variants of the problem at the same time. A first and quite simple idea to do this is shown next, this is similar to the procedure of [2].

Algorithm 1 The lower-left corner of a candidate is either determined by one site or by a pair of sites of different colors such that these two sites lie on the candidate's bottom and left edges.
For each such lower-left corner, we proceed as follows. Let U be the set of sites which lie above and to the right of the corner.
1. Initially the top edge of the rectangle starts at infinity. The right edge starts at the x-coordinate of the corner, it is moved right over the sites of U until all colors are contained in the actual rectangle.
2. Then, in decreasing y-order, the top edge steps through the sites of the rectangle as long as it still contains all colors; when this stops, we have found a candidate.

3. The right edge advances until it reaches a site with the color of the site at the actual top edge.
4. As long as the right edge has not stepped over all points of U, we repeat from step 2.

It is clear that all non-shrinkable rectangles are checked as candidates by Algorithm 1, but also some more rectangles that may contain sites of the same color as the left or bottom edges. For each corner the algorithm spends time $O(n)$ if the sites are previously sorted by their x-coordinates and by their y-coordinates.

Remark that for one fixed corner we cannot have more than $n - k + 1$ candidates because the right edge has stepped over at least k sites in step 1. Furthermore, the left edge of a candidate can be only at the first $n - k + 1$ sites in x-order and the lower edge only at the first $n - k + 1$ sites in y-order, so we obtain a $O(n(n - k)^2)$ bound for the running time of Algorithm 1.

Before trying to improve on this time bound, we are interested in determining the exact number of non-shrinkable rectangles, since in the worst case it seems unavoidable to check (nearly) all of them.

Lemma 1.   *There are $\Theta((n - k)^2)$ non-shrinkable rectangles.*

Proof. We start by proving the upper bound. As we have remarked earlier, each edge of a non-shrinkable rectangle N must contain a site of a color that occurs only once in N. First consider the case that a site is a corner of the rectangle, i. e., the site touches two edges. A site can be the, e. g., lower left corner of a non-shrinkable rectangle only if it belongs to the $n - k + 1$ leftmost sites because it must have $k - 1$ sites to its right. Also, in the analysis of Algorithm 1 we have seen that there are at most $n - k + 1$ non-shrinkable rectangles for one fixed corner. Thus, the upper bound holds for all non-shrinkable rectangles that have at least one site at a corner. In particular, this also settles the cases $k = 2, 3$.

So we can assume that $k \geq 4$ and that each edge contains exactly one site in its interior. Let l, b, and r denote the colors of the singular points on the left, bottom, and right edge of N, correspondingly. We enlarge N by moving its upper edge upwards until one of the following events occurs. Either, the upper edge hits a point of color b; then we have obtained a so-called *enlarged candidate* with singular points on its left and on its right edge that contains points of color b on its top and bottom edges, but no further b-colored points (type 1). Or the upper edge hits a point of color l or r, say l; then we have an enlarged candidate containing two points of color l on its top and left edges, but no further points of color l, and singular points on its right and bottom edges (type 2). If the upper edge does not hit a point of color b, l, or r then we obtain an enlarged candidate with upper edge at infinity and singular points on its other three edges (type 3).

This way we have mapped each non-shrinkable rectangle on an enlarged candidate of type 1, 2, or 3. The mapping is one-to-one; given an enlarged candidate of any type we can just lower its top edge until, for the first time, the lowest point of some color is hit, and obtain a non-shrinkable rectangle. Thus, it suffices to show that there are only $O((n - k)^2)$ enlarged candidates of each type.

In order to bound the number of type 1 rectangles we fix an arbitrary point, p, of color p_{col} and show that there are at most $O(n-k)$ type 1 rectangles $R_{i,j}$ that have p on their bottom edge and another p_{col}-colored point q_i on their top edge to the right of p. Indexing is such that $q_1, \ldots, q_m$ have increasing x-coordinates. Let $r_{i,j}$ be the singular point on the right edge of rectangle $R_{i,j}$, for $1 \leq j \leq m_i$. Clearly, different rectangles $R_{i,j}$ with the same index i must have different right points $r_{i,j}$. Since none of these rectangles can contain a third point of color p_{col}, point q_{i+1} must be below q_i and to the right of all points $r_{i,j}, 1 \leq j \leq m_i$. Trivially, all points $r_{i+1,j}, 1 \leq j \leq m_{i+1}$, are to the right of q_{i+1}. A continuation of this argument shows that *all* points $r_{i,j}$ are pairwise different. This proves the claim since only $n - k + 1$ points can have the right edge, and the same for the bottom edge.

The argument for the type 2 rectangles is quite similar. We fix the point p on the left edge and consider all rectangles $R_{i,j}$ of type 2 that have a point q_i of the same color on their top edges. Again, all singular points $r_{i,j}$ on the right edges are pairwise different.

The unbounded enlarged candidates of type 3 are even easier to count: for a fixed point on the bottom edge there can be only $n-k+1$ of them, since for each possible left edge there is at most one right edge, if any, and only $n - k + 1$ sites can have the left edge. (By the way, k is also an upper bound on the number of rectangles of this type for a fixed bottom edge, as will follow from Lemma 2.)

It remains to show the lower bound, i.e., we are given numbers n and k, and we want to place n sites with k colors such that there are $\Omega((n-k)^2)$ different non-shrinkable rectangles. To this end, we make a construction as sketched in Fig.1.

We construct three groups of sites. The first group consists of $\lfloor (n-k)/2 \rfloor + 1$ sites of color 1 and is placed on the line $y = -1 - x$ at positions with negative x- and y-coordinates, the second group has $\lceil (n-k)/2 \rceil + 1$ sites but of color 2 and is placed on the line $y = 1 - x$ at positions with positive coordinates, and the third group contains one site of each of the other $k - 2$ colors and is placed very close to the origin.

Now each rectangle spanned by a site of color 1 as the lower left corner and by a site of color 2 as the upper right corner contains all colors and is one of $\Omega((n-k)^2)$ non-shrinkable rectangles. $\qquad\qquad\square$

2.2   An Improved Approach

The question arises if the proof method for the $O((n-k)^2)$ upper bound can be used for efficiently constructing all non-shrinkable rectangles. In fact, we are able to enumerate all enlarged candidates of types 1, 2, and 3 within time $O(n^2 \log k)$. The difficulty is in efficiently moving down the upper edges of these rectangles, in order to obtain non-shrinkable rectangles. This can be done within the same time bound for the types 2 and 3, but seems quite hard to do for the type 1 rectangles.

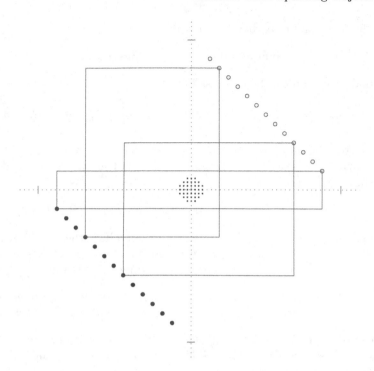

Fig. 1. Each rectangle spanned by a site of color 1 and a site of color 2 is non-shrinkable.

Therefore, we resort to a more direct method that is a refinement of Algorithm 1. Instead of fixing the lower left corner, let us try to fix the upper and lower edges, i. e., for each pair of sites a and b with $a_y < b_y$ we check all non-shrinkable rectangles with lower y-coordinate a_y and upper y-coordinate b_y.

We consider conditions that must be fulfilled by such a non-shrinkable rectangle with left edge at l and right edge at r, l and r may coincide with a or b. First, it is clear that a and b must be contained in the rectangle. Second, the interior of the rectangle must not contain sites of the colors of a and b. Third, the colors of l and r are not contained in the interior either.

More formally, for a given color c we define the following numbers.

$$L_c(a,b) = \max_{p \in S,\, p_{col}=c} \{\, p_x \mid a_y < p_y < b_y \text{ and } p_x < a_x \,\}$$

$$R_c(a,b) = \min_{p \in S,\, p_{col}=c} \{\, p_x \mid a_y < p_y < b_y \text{ and } p_x > a_x \,\}$$

In other words, $L_c(a,b)$ is the maximum x-coordinate of all sites of color c in the horizontal strip between a and b and to the left of a_x, and $R_c(a,b)$ the analogous minimum to the right of a_x; they take on the values of $-\infty$ resp. $+\infty$ if no such site exists.

Now the first condition above means that

$$l_x \leq \min(a_x, b_x) \quad \text{and} \quad r_x \geq \max(a_x, b_x). \tag{1}$$

The second condition can be expressed as

$$l_x > \max\left(L_{a_{col}}(a, b), L_{b_{col}}(a, b)\right) \quad \text{and} \quad r_x < \min\left(R_{a_{col}}(a, b), R_{b_{col}}(a, b)\right). \tag{2}$$

In other words, we have an x-interval for the possible positions of the left edge of a non-shrinkable rectangle from a_y to b_y, and another one for the right edge.

The third condition transforms to

$$l_x = L_{l_{col}}(a, b) \text{ if } l \neq a, b \quad \text{and} \quad r_x = R_{r_{col}}(a, b) \text{ if } r \neq a, b, \tag{3}$$

i. e., the site l on the left edge, if it is not a or b itself, is the x-maximal site of its color in the horizontal strip between a and b and to the left of $\min(a_x, b_x)$, and correspondingly with r. Therefore the following assertion holds.

Lemma 2. *Let a and b be two sites of S. Independently of n, there are at most $k - 1$ non-shrinkable rectangles with lower edge at a and upper edge at b.*

Proof. According to (3), the left edge of such a non-shrinkable rectangle has only $k - 2$ possible positions if its color is different from a_{col} and b_{col}, and $\min(a_x, b_x)$ as one additional possibility. □

For fixed a, the quantities $L_c(a, b)$ and $R_c(a, b)$ can easily be updated if b steps through the sites in y-order. For each b it remains to match the correct pairs of sites at the left and at the right edges, this is done by the following algorithm.

Algorithm 2 The sites are sorted in y-order. For each site a we do the following to find all candidate rectangles with lower y-coordinate a_y.

1. Let L and R be arrays over all colors, initialized to $-\infty$ resp. $+\infty$; they will contain the values $L_c(a, b)$ and $R_c(a, b)$ for the actual a and b and for all colors c.
   The lists *SortL* and *SortR* will contain all sites that actually contribute an entry to L resp. R, sorted in x-direction.
2. For all sites b with $b_y > a_y$ in y-order we do. Perform steps 2a to 2c only if $b_{col} \neq a_{col}$, in any case perform step 2d.
   a) *InclL* := $\min(a_x, b_x)$;   *ExclL* := $\max(L_{a_{col}}, L_{b_{col}})$;
      InclR := $\max(a_x, b_x)$;   *ExclR* := $\min(R_{a_{col}}, R_{b_{col}})$;
      In list *SortL* we mark the sites with x-coordinates greater than *ExclL* and smaller than *InclL*, and correspondingly with *SortR* from *InclR* to *ExclR*.
   b) The left edge starts at the first marked element of *SortL*. The right edge starts at *InclR* and if necessary steps over the marked sites in *SortR* until all colors are contained in the actual rectangle.

c) As long as the marked parts of *SortL* and *SortR* are not exhausted, we repeat the following steps.

    i. The left edge advances over the marked sites in *SortL* and finally *InclL* as long as the rectangle contains all colors; when this stops, we have found a candidate.

    ii. The right edge advances over the marked sites in *SortR* until it reaches the color of the site at the actual left edge.

d) If $b_x < a_x$

    then $L_{b_{col}} := \max(L_{b_{col}}, b_x)$, also unmark and update *SortL*

    else $R_{b_{col}} := \min(R_{b_{col}}, b_x)$, also unmark and update *SortR*.

Lemma 3. *The candidates reported by Algorithm 2 are precisely all non-shrinkable rectangles. Its running time is in $O(nk(n - k))$.*

Proof. Let us consider what happens if steps 2a to 2d are executed for certain sites a and b.

Step 2d has been performed for all previous values of b, so L and R contain the correct $L_c(a, b)$ and $R_c(a, b)$ for all colors. Remark that this also holds for b_{col} because the update of L and R concerning b is done at the end of the loop.

SortL and *SortR* contain the sites corresponding to the values of L resp. R, and only these values are possible left resp. right edges of the rectangle, as we have seen earlier. The marked parts correspond to the intervals (*ExclL*, *InclL*) resp. (*InclR*, *ExclR*) and reflect conditions (1) and (2); sites a or b can also be at the left or right edges of a non-shrinkable rectangle, this is taken into account by starting the right edge at *InclR* in step 2b and finishing with the left edge at *InclL* in step 2(c)i.

For each possible left edge the matching right edge, if any, is found in steps 2(c)i and 2(c)ii, these steps are very similar to steps 2 to 3 of Algorithm 1.

The case in which there is no non-shrinkable rectangle for a at the bottom edge and b at the top is quickly detected: If some colors are missing in the horizontal strip between a and b then step 2b already does not succeed. If another site of color a_{col} or b_{col} is contained in the rectangle spanned by a and b then *ExclL* > *InclL* or *InclR* > *ExclR* and one of the lists is not marked at all. Finally, the case $b_{col} = a_{col}$ is explicitly excluded in step 2.

The running time can be estimated as follows. Site a at the bottom edge needs to be iterated only over the first $n - k + 1$ sites in y-order, so this factor is contributed. Factor n is for the loop over all b (not $n - k$ because the updates in step 2d need to be executed for all b above a). Finally, the repetition of steps 2(c)i and 2(c)ii results in a factor k, as well as the (un)marking and the updates of the sorted lists. □

For small and in particular for constant k Algorithm 2 is the method of choice, because it is very simple and can be implemented using just a few lines of code. On the other hand, for large k the $O(n(n - k)^2)$ method of Algorithm 1 is preferable. In the general case however, there is still room for improvements.

2.3   Maximal Elements

Definition 2. *A maximal element of a set, T, of points in the plane is a point $p \in T$ such that there is no $q \in T$ with $p_x > q_x$ and $p_y < q_y$.*

Remark that for our special purpose we have slightly deviated from the usual definition, see [16], we are interested in maximal elements in the upper *left* (instead of right) direction, see Fig.2. Our maximal elements are those that are not *dominated* from left and above by another point of the set.

Fig. 2. Maximal elements of a point set in the upper left direction.

Now consider, for given a and b, the values of $L_c(a, b)$ and $R_c(a, b)$. We transform these values to points in 2D, using L for the x-coordinates and R for the y-coordinates.

$$T_c(a, b) = \Big(L_c(a, b), R_c(a, b)\Big) \text{ for all colors } c \neq a_{col}, b_{col}$$

Some of the coordinates of these points may be $\pm\infty$. With $T(a, b)$ we denote the set of all points $T_c(a, b)$. The next lemma shows that maximal elements are closely related to spanning colors.

Lemma 4. *Assume that the horizontal strip between a and b contains all colors. The point $T_c(a, b)$ for some color c is a maximal element of $T(a, b)$ if and only if the rectangle with a and b at the bottom and top edges and with $L_c(a, b)$ as left and $R_c(a, b)$ as right edge contains all colors with the possible exception of b_{col}.*

Proof. Let $T_c(a, b)$ be a maximal element of $T(a, b)$. Suppose there is a color, c', which is not contained in the rectangle between a, b, $L_c(a, b)$, and $R_c(a, b)$. Then $L_{c'}(a, b) < L_c(a, b)$ and $R_{c'}(a, b) > R_c(a, b)$, and $T_{c'}$ dominates $T_c(a, b)$, a contradiction. Conversely, if all colors are contained in the rectangle then $T_c(a, b)$ must be a maximal element because it can't be dominated by any other color. □

Now we have an interesting relation between non-shrinkable rectangles and maximal elements.

Lemma 5. *For a non-shrinkable rectangle with sites a, b, l, r at the bottom, top, left, and right edges with $l \neq a, b$ and $r \neq a, b$, $T_{l_{col}}(a, b)$ and $T_{r_{col}}(a, b)$ are two successive maximal elements of the set of points in $T(a, b)$.*

Proof. Assume that $T_{l_{col}}(a, b)$ is dominated by some $T_c(a, b)$. Clearly we have $L_c(a, b) < L_{l_{col}}(a, b) = l_x$, but also $R_c(a, b) > R_{l_{col}}(a, b) > r_x$ holds because l_{col} cannot appear a second time in the rectangle. This means that color c is not contained in the rectangle, a contradiction, and analogously for $T_{r_{col}}(a, b)$.

Now assume some $T_c(a, b)$ is maximal element between the two maximal elements $T_{r_{col}}(a, b)$ and $T_{l_{col}}(a, b)$. Then we have $L_{r_{col}}(a, b) < L_c(a, b) < L_{l_{col}}(a, b) = l_x$ and $r_x = R_{r_{col}}(a, b) < R_c(a, b) < R_{l_{col}}(a, b)$, and again c is not contained in the rectangle. $\qquad\square$

And the converse is also true, in some sense.

Lemma 6. *Consider two sites a and b and two colors $c, c' \neq a_{col}, b_{col}$ such that $T_c(a, b)$ and $T_{c'}(a, b)$ are two successive maximal elements of $T(a, b)$ and assume that the horizontal strip between a and b contains all colors. Then the rectangle between a, b, l with $l_x = L_{c'}(a, b)$ as left edge, and r with $r_x = R_c(a, b)$ as right edge is non-shrinkable if additionally conditions (1) and (2) hold.*

Proof. From Lemma 4 we know that the rectangle between $a, b, L_c(a, b)$ and $R_c(a, b)$ contains all colors. Now let the left edge move right until the rectangle is non-shrinkable. The left edge must now be situated at $L_{c'}(a, b)$, otherwise there would be another maximal element between $T_c(a, b)$ and $T_{c'}(a, b)$. Conditions (1) and (2) are necessary to guarantee that no other sites of color a_{col} or b_{col} are contained in the rectangle. $\qquad\square$

Theorem 1. *Given n sites and k colors, the smallest color-spanning rectangle can be found in time $O(n(n - k) \log^2 k)$.*

Proof. We modify Algorithm 2. The main difference is in maintaining a dynamic tree *MaxElem* of maximal elements instead of the lists *SortL* and *SortR*.

In step 2d now *MaxElem* is updated if the value of $L_{b_{col}}$ or $R_{b_{col}}$ has changed; this can be done in time $O(\log^2 k)$ using the method of Overmars and van Leeuwen [16].

The marking of the lists is replaced in the following way. The values *ExclL*, *InclL*, *InclR*, and *ExclR* are computed as before. Then the subsequence of elements in *MaxElem* is extracted that is included in (*ExclL, InclL*) in x-direction as well as in (*InclR, ExclR*) in y-direction. This can be done in time $O(\log k)$ plus the length of this subsequence which in turn is essentially the same as the number of non-shrinkable rectangles reported. It remains to report the matchings between left and right edges as described in Lemma 6.

So the running time of this method is $O(n(n - k) \log^2 k)$ plus the total number of reported non-shrinkable rectangles but which is fortunately bounded by $O((n - k)^2)$, see Lemma 1. $\qquad\square$

3   The Narrowest Color-Spanning Strip

The narrowest color-spanning strip problem asks for two parallel lines in the plane that contain all colors in between them such that their distance is minimized.

Notice that the solution strip must have three sites of three different colors on its boundary, because if they were only two or they had a coincident color, the strip could be shrunk by rotation.

A brute force approach could work as follows. Consider the $O(n^2)$ lines defined by two sites of different colors, and sort them by slopes in $O(n^2 \log n)$ time. Start with one of them and project all the other sites, following the direction of the line, onto any perpendicular line. By sorting the projected points and walking along them we can find the solution in that direction in $O(n \log n)$ time. Now, at each change of direction, we only need to update the order of the projected points in $O(1)$ time and explore the points, again walking along them, in $O(n)$ time to find the solution for the new direction. Hence, the algorithm works in $O(n^3)$ time.

When the direction changes, the cluster of points that gives the optimal solution may completely change. This is the reason why we don't envisage a more clever updating. Using techniques of inversion and outer envelopes we obtain a much better algorithm, as the following theorem states.

Theorem 2. *Given n sites and k colors, the narrowest color-spanning strip can be found in $O(n^2 \alpha(k) \log k)$ time.*

The proof is omitted here because of lack of space.

4   Conclusions

We have solved two optimization problems by giving algorithms that are likely to be close to optimal. The narrowest color-spanning strip problem can be solved in time $O(n^2 \alpha(k) \log k)$, and it is n^2-hard in the sense of [9], this proof is not very difficult but omitted here for brevity.

On the other hand, the smallest color-spanning rectangle problem can be solved in time $O(n(n - k) \log^2 k)$ while we have the tight quadratic bound for the number of non-shrinkable rectangles. It would be interesting to have a formal proof (or a refutation) that this number is also a lower bound for this problem.

We must admit though that we do not have a lower bound better than $\Omega(n \log n)$ for the problem; this one at least can be obtained in several ways. For example, the problem of finding the bichromatic closest pair can be transformed to our problem with $k = 2$, or the maximum gap problem [17] can be transformed to it with $k = n/2$. Of course this is not really satifying, because it seems that a quadratic lower bound or a n^2-hardness result is possible.

The smallest color-spanning rectangle with an arbitrary orientation would be the next natural generalization.

References

1. P. K. Agarwal, H. Edelsbrunner, O. Schwarzkopf, and E. Welzl. Euclidean minimum spanning trees and bichromatic closest pairs. *Discrete Comput. Geom.*, 6(5):407–422, 1991.
2. A. Aggarwal, H. Imai, N. Katoh, and S. Suri. Finding k points with minimum diameter and related problems. *J. Algorithms*, 12:38–56, 1991.
3. A. Datta, H.-P. Lenhof, C. Schwarz, and M. Smid. Static and dynamic algorithms for k-point clustering problems. *J. Algorithms*, 19:474–503, 1995.
4. D. P. Dobkin, R. L. Drysdale, III, and L. J. Guibas. Finding smallest polygons. In F. P. Preparata, editor, *Computational Geometry*, volume 1 of *Adv. Comput. Res.*, pages 181–214. JAI Press, Greenwich, Conn., 1983.
5. A. Efrat, M. Sharir, and A. Ziv. Computing the smallest k-enclosing circle and related problems. *Comput. Geom. Theory Appl.*, 4:119–136, 1994.
6. D. Eppstein and J. Erickson. Iterated nearest neighbors and finding minimal polytopes. *Discrete Comput. Geom.*, 11:321–350, 1994.
7. D. Eppstein, M. H. Overmars, G. Rote, and G. Woeginger. Finding minimum area k-gons. *Discrete Comput. Geom.*, 7:45–58, 1992.
8. J. Erickson and R. Seidel. Better lower bounds on detecting affine and spherical degeneracies. *Discrete Comput. Geom.*, 13:41–57, 1995.
9. A. Gajentaan and M. H. Overmars. On a class of $O(n^2)$ problems in computational geometry. *Comput. Geom. Theory Appl.*, 5:165–185, 1995.
10. T. Graf and K. Hinrichs. Algorithms for proximity problems on colored point sets. In *Proc. 5th Canad. Conf. Comput. Geom.*, pages 420–425, 1993.
11. D. P. Huttenlocher, K. Kedem, and M. Sharir. The upper envelope of Voronoi surfaces and its applications. *Discrete Comput. Geom.*, 9:267–291, 1993.
12. J. Matoušek. On geometric optimization with few violated constraints. In *Proc. 10th Annu. ACM Sympos. Comput. Geom.*, pages 312–321, 1994.
13. J. Matoušek. On enclosing k points by a circle. *Inform. Process. Lett.*, 53:217–221, 1995.
14. J. S. B. Mitchell. Geometric shortest paths and network optimization. In J.-R. Sack and J. Urrutia, editors, *Handbook of Computational Geometry*, pages 633–701. Elsevier Science Publishers B.V. North-Holland, Amsterdam, 2000.
15. D. M. Mount, N. S. Netanyahu, R. Silverman, and A. Y. Wu. Chromatic nearest neighbour searching: a query sensitive approach. *Comput. Geom. Theory Appl.*, 17:97–119, 2000.
16. M. H. Overmars and J. van Leeuwen. Maintenance of configurations in the plane. *J. Comput. Syst. Sci.*, 23:166–204, 1981.
17. F. P. Preparata and M. I. Shamos. *Computational Geometry: An Introduction.* Springer-Verlag, New York, NY, 1985.
18. M. Segal and K. Kedem. Enclosing k points in the smallest axis parallel rectangle. *Inform. Process. Lett.*, 65:95–99, 1998.
19. M. Sharir and P. K. Agarwal. *Davenport-Schinzel Sequences and Their Geometric Applications.* Cambridge University Press, New York, 1995.
20. M. Smid. Finding k points with a smallest enclosing square. Report MPI-I-92-152, Max-Planck-Institut Inform., Saarbrücken, Germany, 1992.

Explicit Deterministic Constructions for Membership in the Bitprobe Model

Jaikumar Radhakrishnan[1], Venkatesh Raman[2], and S. Srinivasa Rao[2]

[1] Tata Institute of Fundamental Research, Mumbai.
jaikumar@tcs.tifr.res.in
[2] Institute of Mathematical Sciences, Chennai, India 600 113.
{vraman,ssrao}@imsc.ernet.in

Abstract. We look at time-space tradeoffs for the static membership problem in the bit-probe model. The problem is to represent a set of size up to n from a universe of size m using a small number of bits so that given an element of the universe, its membership in the set can be determined with as few bit probes to the representation as possible.

We show several deterministic upper bounds for the case when the number of bit probes, is small, by explicit constructions, culminating in one that uses $o(m)$ bits of space where membership can be determined with $\lceil \lg \lg n \rceil + 2$ adaptive bit probes. We also show two tight lower bounds on space for a restricted two probe adaptive scheme.

1 Introduction

We look at the static membership problem: Given a subset S of up to n keys drawn from a universe of size m, store it so that queries of the form "Is x in S?" can be answered quickly. We study this problem in the bit-probe model where space is counted as the number of bits used to store the data structure and time as the number of bits of the data structure looked at in answering a query.

A simple characteristic bit vector gives a solution to the problem using m bits of space in which membership queries can be answered using one bit probe. On the other hand, the structures given by Fredman et al.[4], Brodnik and Munro [1] and Pagh [5] can be used to get a scheme that uses $O(n \lg m)$ bits of space in which membership queries can be answered using $O(\lg m)$ bit probes. Recently Pagh [6] has given a structure that requires $O(s_{m,n})$ bits of space and supports membership queries using $O(\lg(m/n))$ bit probes to the structure, where $s_{m,n} = \Theta(n \lg(m/n))$ is the information theoretic lower bound on space for any structure storing an n element subset of an m element universe.

Buhrman et al.[2] have shown that both the above schemes are optimal. In particular they have shown that any deterministic scheme that answers membership queries using one bit probe requires at least m bits of space and any deterministic scheme using $O(s_{m,n})$ bits of space requires at least $\Omega(\lg(m/n))$ probes to answer membership queries. They have considered the intermediate ranges and have given some upper and lower bounds for randomized as well as deterministic versions. Their main result is that the optimal $O(n \lg m)$ bits (for

F. Meyer auf der Heide (Ed.): ESA 2001, LNCS 2161, pp. 290–299, 2001.

$n \leq m^{1-\Omega(1)}$) and one bit probe per query are sufficient, if the query algorithm is allowed to make errors (both sided) with a small probability. For the deterministic case, however, they have given some non-constructive upper bounds. They have also given some explicit structures for the case when t is large ($t \geq \lg n$).

Our main contribution in this paper, is some improved deterministic upper bounds for the problem using explicit constructions, particularly for small values of t. For sets of size at most 2, we give a scheme that uses $O(m^{2/3})$ bits of space and answers queries using 2 probes. This improves the $O(m^{3/4})$ bit scheme in [2] shown using probabilistic arguments. We also show that the space bound is optimal for a restricted two probe scheme. We then generalize this to a $\lceil \lg \lg n \rceil + 2$ probe scheme for storing sets of size at most n, which uses $o(m)$ bits of space. This is the best known constructive scheme (in terms of the number of bit probes used) for general n that uses $o(m)$ bits of space, though it is known [2] (using probabilistic arguments) that there exists a scheme using $o(m)$ bits of space where queries can be answered using a constant number of bit probes.

The next section introduces some definitions. The following section gives improved upper bounds for deterministic schemes. In section 4, we give some space lower bounds for a restricted class of two probe schemes, matching our upper bound. Finally, Section 5 concludes with some remarks and open problems.

2   Definitions

We reproduce the definition of a storing scheme, introduced in [2]. An (n, m, s)-storing scheme, is a method for representing any subset of size at most n over a universe of size m as an s-bit string. Formally, an (n, m, s)-storing scheme is a map ϕ from the subsets of size at most n of $\{1, 2, \ldots, m\}$ to $\{0, 1\}^s$. A deterministic (m, s, t)-query scheme is a family of m boolean decision trees $\{T_1, T_2, \ldots, T_m\}$, of depth at most t. Each internal node in a decision tree is marked with an index between 1 and s, indicating an address of a bit in an s-bit data structure. All the edges are labeled by "0" or "1" indicating the bit stored in the parent node. The leaf nodes are marked "Yes" or "No". Each tree T_i induces a map from $\{0, 1\}^s \rightarrow \{\text{Yes}, \text{No}\}$. An (n, m, s)-storing scheme and an (m, s, t)-query scheme T_i together form an (n, m, s, t)-scheme which solves the (n, m)-membership problem if $\forall S, x$ s.t. $|S| \leq n, x \in U : T_x(\phi(S)) = Yes$ if and only if $x \in S$. A non-adaptive query scheme is a deterministic scheme where in each decision tree, all nodes on a particular level are marked with the same index.

We follow the convention that whenever the universe $\{1, \ldots, m\}$ is divided into blocks of size b (or m/b blocks), the elements $\{(i-1)b+1, \ldots, ib\}$ from the universe belong to the ith block, for $1 \leq i \leq \lfloor m/b \rfloor$ and the remaining (at most b) elements belong to the last block. For integers x and a we define, $div(x, a) = \lfloor x/a \rfloor$ and $mod(x, a) = x - a\ div(x, a)$. To simplify the notation, we ignore integer rounding ups and downs at some places where they do not affect the asymptotic analysis.

3   Upper Bounds for Deterministic Schemes

As observed in [2], the static dictionary structure given by Fredman, Komlos and Szemeredi [4] can be modified to give an adaptive (n, m, s, t)-scheme with $s = O(nkm^{1/k})$ and $t = O(\lg n + \lg \lg m) + k$, for any parameter $k \geq 1$. This gives a scheme when the number of probes is larger than $\lg n$. In this section, we look at schemes which require fewer number of probes albeit requiring more space.

For two element sets, Buhrman et al.[2] have given a non-adaptive scheme that uses $O(\sqrt{m})$ bits of space and answers queries using 3 probes. If the query scheme is adaptive, there is even a simpler structure. Our starting point is a generalization of this scheme for larger n.

Theorem 1. *There is an explicit adaptive (n, m, s, t)-scheme with $t = \lceil \lg(n + 1) \rceil + 1$ and $s = (n + \lceil \lg(n+1) \rceil)m^{1/2}$.*

Proof. The structure consists of two parts. We divide the universe into blocks of size $m^{1/2}$. The first part consists of a table T of size $m^{1/2}$, each entry corresponding to a block. We call a block non-empty if at least one element from the given set falls into that block and empty otherwise. For each non-empty block, we store its rank (the number of non-empty blocks appearing before and including it) in the table entry of that block and store a string of zeroes for each empty block. Since the rank can be any number in the range $[1, \ldots, n]$ (and we store a zero for the empty blocks), we need $\lceil \lg(n+1) \rceil$ bits for storing each entry of the table T.

In the second part, we store the bit vectors corresponding to each non-empty block in the order in which they appear in the first part. For convenience, we call the jth bit vector as table T_j. Thus the total space required for the structure is at most $(n + \lceil \lg(n + 1) \rceil)m^{1/2}$ bits.

Every element $x \in [m]$ is associated with $l+1$ locations, where l is the number of non-empty blocks: $t(x) = div(x, m^{1/2})$ in table T and $t_j(x) = mod(x, m^{1/2})$ in table T_j for $1 \leq j \leq l$. Given an element x, the query scheme first reads the entry j at location $t(x)$ in table T. If $j = 0$, the scheme answers 'No'. Otherwise it looks at the bit $T_j(t_j(x))$ in the second part and answers 'Yes' if and only if it is a one. $\qquad \square$

If only two probes are allowed, Buhrman et al.[2] have shown that, any non-adaptive scheme must use m bits of space. For sets of size at most 2, they have also proved the existence of an adaptive scheme using 2 probes and $O(m^{3/4})$ bits of space. We improve it to the following:

Theorem 2. *There is an explicit adaptive scheme that stores sets of size at most 2 from a universe of size m using $O(m^{2/3})$ bits and answers queries using 2 bit-probes.*

Proof. Divide the universe into blocks of size $m^{1/3}$ each. There are $m^{2/3}$ blocks. Group $m^{1/3}$ consecutive blocks into a superblock. There are $m^{1/3}$ superblocks of size $m^{2/3}$ each.

The storage scheme consists of three tables T, T_0 and T_1, each of size $m^{2/3}$ bits. Each element $x \in [m]$ is associated with three locations, $t(x)$, $t_0(x)$ and $t_1(x)$, one in each of the three tables, as defined below. Let $b = m^{2/3}$ and $b_1 = m^{1/3}$. Then, $t(x) = div(x, b_1)$, $t_0(x) = mod(x, b)$ and $t_1(x) = div(x, b) \, b_1 + mod(x, b_1)$. Given an element $x \in [m]$, the query scheme first looks at $T(t(x))$. If $T(t(x)) = j$, it looks at $T_j(t_j(x))$ and answers 'Yes' if and only if it is 1, for $j \in \{0, 1\}$.

To represent a set $\{x, y\}$, if both the elements belong to the same superblock (i.e. if $div(x, b) = div(y, b)$), then we set the bits $T(t(x))$ and $T(t(y))$ to 0, all other bits in T to 1; $T_0(t_0(x))$ and $T_0(t_0(y))$ to 1 and all other bits in T_0 and T_1 to 0. In other words, we represent the characteristic vector of the superblock containing both the elements, in T_0, in this case.

Otherwise, if both the elements belong to different superblocks, we set $T(t(x))$, $T(t(y))$, $T_1(t_1(x))$ and $T_1(t_1(y))$ to 1 and all other bits in T, T_0 and T_1 to 0. In this case, each superblock has at most one non-empty block containing one element. So in T_1, for each superblock, we store the characteristic vector of the only non-empty block in it (if it exists) or any one block in it (which is a sequence of zeroes) otherwise. One can easily verify that the storage scheme is valid and that the query scheme answers membership queries correctly.   □

One can immediately generalize this scheme for larger n to prove the following. Notice that the number of probes is slightly smaller than that used in Theorem 1, though the space used is larger.

Theorem 3. *There is an explicit adaptive (n, m, s, t)-scheme with $t = 1 + \lceil \lg(\lfloor n/2 \rfloor + 2) \rceil$ and $s = O(m^{2/3}(n/2 + \lg(n/2 + 2) + 1))$.*

Proof Sketch: The idea is to distinguish superblocks containing at least 2 elements from those containing at most one element.

In the first level, if a superblock contains at least 2 elements, we store its rank among all superblocks containing at least 2 elements, with all its blocks. Since there can be at most $\lfloor n/2 \rfloor$ superblocks containing at least 2 elements, the rank can be any number in the range $\{1, \ldots, \lfloor n/2 \rfloor\}$. For blocks which fall into superblocks containing at most one element, we store the number $\lfloor n/2 \rfloor + 1$, if the block is non-empty and a sequence of $\lceil \lg(\lfloor n/2 \rfloor + 2) \rceil$ zeroes, otherwise.

The second level consists of $\lfloor n/2 \rfloor + 1$ bit vectors of size $m^{2/3}$ each. We will store the characteristic vector of the jth superblock in the jth bit vector for $1 \leq j \leq l$, where l is the number of superblocks containing at least 2 elements. We will store all zeroes in the bit vectors numbered $l + 1$ to $\lfloor n/2 \rfloor$. In the $(\lfloor n/2 \rfloor + 1)$st bit vector, for each superblock we store the characteristic vector of the only non-empty block in it, if it has exactly one non-empty block or a sequence of zeroes otherwise.

On query x, we look at the first level entry of the block corresponding to x. We answer that the element is not present, if the entry is a sequence of zeroes. Otherwise, if it is a number k in the range $[1, \ldots, \lfloor n/2 \rfloor]$, we look at the corresponding location of x in the kth bit vector in the second level (which stores the bit vector corresponding to the superblock containing x). Otherwise (if the number is $\lfloor n/2 \rfloor + 1$), we look at the corresponding location of x in the last bit vector and answer accordingly.   □

This can be further generalized as follows. In the first level, we will distinguish the superblocks having at least k elements (for some integer k) from those with at most $k-1$ elements in them. For superblocks having at least k elements, we store the rank of that superblock among all such superblocks, in all the blocks of that superblock. For the other superblocks, we store the rank of the block among all non-empty blocks in that superblock, if the block is non-empty and a sequence of zeroes otherwise. The second level will have $\lfloor n/k \rfloor + k - 1$ bit vectors of length $m^{2/3}$ each where in the first $\lfloor n/k \rfloor$ bit vectors, we store the characteristic vectors of the at most $\lfloor n/k \rfloor$ superblocks containing at least k elements in them (in the order of increasing rank) and pad the rest of them with zeroes. Each of the $(\lfloor n/k \rfloor + j)$th bit vectors, for $1 \le j \le k - 1$, stores one block for every superblock. This block is the jth non-empty block in that superblock, if that superblock contains at least j non-empty blocks and at most $k-1$ elements; we store a sequence of zeroes otherwise. The query scheme is straightforward. This results in the following.

Corollary 1. *There is an explicit adaptive (n, m, s, t)-scheme with $t = 1 + \lceil \lg(\lfloor n/k \rfloor + k) \rceil$ and $s = O(m^{2/3}(n/k + \lg(n/k + k) + k))$.*

Choosing $k = \lceil \sqrt{n} \rceil$, we get an explicit adaptive (n, m, s, t)-scheme with $t = 2 + \lceil \frac{1}{2} \lg n \rceil$ and $s = O(m^{2/3}\sqrt{n})$.

Actually, by choosing the block sizes to be $\frac{m^{1/3}(\lg n)^{2/3}}{n^{1/3}}$ and the sizes of the superblocks to be $\frac{m^{2/3}(\lg n)^{1/3}}{n^{1/6}}$ we get the following improved scheme:

Corollary 2. *There is an explicit adaptive (n, m, s, t)-scheme with $t = 2 + \lceil \frac{1}{2} \lg n \rceil$ and $s = O(m^{2/3}(n \lg n)^{1/3})$.*

We generalize this to the following:

Theorem 4. *There is an explicit adaptive (n, m, s, t)-scheme with $t = \lceil \lg k \rceil + \lceil \frac{1}{k} \lg n \rceil + 1$ and $s = m^{k/(k+1)} \left(\lg k + \frac{1}{k} \lg n + k n^{1/k} \right)$, for $k \ge 1$.*

Proof. We divide the universe into blocks of size b (to be determined later) and construct a complete b-ary tree with these blocks at the leaves. Let the height of this tree be k. Thus, we have $m = b^{k+1}$ or $b = m^{1/(k+1)}$. Given a set S of n elements from the universe, we store it using a three level structure. We call a block non-empty if at least one element of the given set S belongs to that block and call it empty otherwise. We define the height of a node in the tree to be the length of the path (the number of nodes in the path) from that node to any leaf in the subtree rooted at that node. Note that the height of the root is $k+1$ and that of any leaf is one.

In the first level we store an index in the range $[0, \ldots, k-1]$ corresponding to each block. Thus the first level consists of a table B of size b^k where each entry is a $\lceil \lg k \rceil$ bit number. The index stored for an empty block is 0. For a non-empty block, we store the height $h \leq k - 1$ of its ancestor (excluding the root and the first level nodes of the tree) x of maximum height such that the total number of elements falling into all the blocks in the subtree rooted at node x is more than $\lfloor n^{h/k} \rfloor$. This will be a number in the range $[0, \ldots, k - 1]$.

In the second level we store a number in the range $[1, \ldots, \lceil n^{1/k} \rceil - 1]$ corresponding to each block. Thus this level consists of a table T of size b^k, each entry of which is a $\lceil \lg n^{1/k} \rceil$ bit number. The number stored for an empty block is 0. For a non-empty block, we store the following:

Observe that given any node x at height h which has at most $\lfloor n^{h/k} \rfloor$ elements from the set, the number of its children which have more than $\lfloor n^{(h-1)/k} \rfloor$ elements from the set is less than $\lceil n^{1/k} \rceil$. Suppose the index stored for a block is l. It means that the ancestor x of that block at height l has more than $\lfloor n^{l/k} \rfloor$ elements and the ancestor y at height $l+1$ has at most $\lfloor n^{(l+1)/k} \rfloor$ elements. Hence y can have less than $\lceil n^{1/k} \rceil$ children which have more than $\lfloor n^{l/k} \rfloor$ elements. Call these the 'large' children. With all the leaves rooted at each large child of y, we store the rank of that child among all large children (from left to right) in the second level.

In the third level, we have k tables, each of size $\lceil n^{1/k} \rceil m/b$ bits. The ith table stores the representations of all blocks whose first level entry (in table B) is i. We think of the ith table as a set of $\lceil n^{1/k} \rceil$ bit vectors, each of length m/b. Each of these bit vectors in the ith level stores the characteristic vector of a particular child for each node at height i of the tree, in the left to right order. For each block (of size b) with first level entry i and second level entry j, we store the characteristic vector of that block in the jth bit vector of the ith table at the location corresponding to its block of size b^{k-i}. We store zeroes (i.e. the characteristic vector of an empty block of appropriate size) at all other locations not specified above.

Every element $x \in [m]$ is associated with $k + 2$ locations $b(x)$, $t(x)$ and $t_i(x)$ for $0 \leq i \leq k - 1$, as defined below: $b(x) = t(x) = div(x, b)$, $t_i(x) = mod(div(x, b^{k-i})b^i + mod(x, b^i), b^k)$.

Given an element x, the query scheme first reads $i = B(b(x))$ and $j = T(t(x))$ from the first two levels of the structure. If $j = 0$, it answers 'No'. Otherwise, it reads the jth bit in the table entry at location $t_i(x)$ in table T_i and answers 'Yes' if and only if it is 1.

The space required for the structure is $s = b^k(\lceil \lg k \rceil + \lceil \frac{1}{k} \lg n \rceil + \frac{m}{b} k \lfloor n^{1/k} \rfloor)$ bits. Substituting $b = m^{1/(k+1)}$ makes the space complexity to be $m^{k/(k+1)}(\lceil \lg k \rceil + \lceil \frac{1}{k} \lg n \rceil + kn^{1/k})$. The number of probes required to answer a query is $t = \lceil \lg k \rceil + \lceil \frac{1}{k} \lg n \rceil + 1$.   □

One can slightly improve the space complexity of the above structure by choosing non-uniform block sizes and making the block sizes (branching factors at each level, in the above tree structure) to be a function of n. More precisely, by choosing the branching factor of all the nodes at level i in the above tree structure to be b_i, where $b_i = m^{1 - \frac{i}{k+1}} \left(\frac{\lceil \lg k \rceil + \lceil \frac{1}{k} \lg n \rceil}{\lceil n^{1/k} \rceil} \right)^{i/(k+1)}$, we get

Corollary 3. *There is an explicit adaptive (n, m, s, t)-scheme with $t = \lceil \lg k \rceil + \lceil \frac{1}{k} \lg n \rceil + 1$ and $s = (k+1)m^{k/(k+1)} \left(n(\lceil \lg k \rceil + \lceil \lg n^{1/k} \rceil) \right)^{1/(k+1)}$, for $k \geq 1$.*

By setting $k = \lg n$, we get

Corollary 4. *There is an explicit adaptive (n, m, s, t)-scheme with $t = \lceil \lg \lg n \rceil + 2$ and $s = o(m)$ when n is $O(m^{1/\lg \lg m})$.*

In the above adaptive scheme we first read $\lceil \lg k \rceil + \lceil \frac{1}{k} \lg n \rceil$ bits from the structure, and depending on these bits we look at one more bit in the next level to determine whether the query element is present. An obvious way to make this scheme non-adaptive is to read the $\lceil \lg k \rceil + \lceil \frac{1}{k} \lg n \rceil$ bits and all possible $k \lceil n^{1/k} \rceil$ bits (in the next level) and determine the membership accordingly. Thus we get an explicit non-adaptive (n, m, s, t)-scheme with $t = \lceil \lg k \rceil + \lceil \frac{1}{k} \lg n \rceil + k \lceil n^{1/k} \rceil$ and $s = tm^{k/(k+1)}$. By setting $k = \lceil \lg n \rceil$ in this, we get a non-adaptive scheme with $t = O(\lg n)$ and $s = o(m)$.

These schemes give the best known explicit adaptive and non-adaptive schemes respectively for general n using $o(m)$ bits.

4   Lower Bounds

Buhrman et al.[2] have shown that for any (n, m, s, t) scheme s is $\Omega(ntm^{1/t})$. One can achieve this bound easily for $n = 1$. They have also shown that for $n \geq 2$ any two probe non-adaptive scheme must use at least m bits of space. In this section, we show a space lower bound of $\Omega(m^{2/3})$ bits for a restricted class of adaptive schemes using two probes, for $n \geq 2$. Combining this with the upper bound of Theorem 2, this gives a tight lower bound for this class of restricted schemes. We conjecture that the lower bound applies even for unrestricted schemes. We also show a lower bound of $\Omega(m)$ bits for this restricted class of schemes for $n \geq 3$.

Any two-probe $O(s)$ bit adaptive scheme to represent sets of size at most 2 from a universe U of size m, can be assumed to satisfy the following conditions (without loss of generality):

1. It has three tables A, B and C each of size s bits.
2. Each $x \in U$ is associated with three locations $a(x)$, $b(x)$ and $c(x)$.
3. On query x, the query scheme first looks at $A(a(x))$. If $A(a(x)) = 0$ then it answers 'Yes' if and only if $B(b(x)) = 1$ else if $A(a(x)) = 1$ then it answers 'Yes' if and only if $C(c(x)) = 1$.
4. Let $A_i = \{x \in [m] : a(x) = i\}$, $B_i = \{b(x) : x \in A_i\}$ and $C_i = \{c(x) : x \in A_i\}$ for $1 \leq i \leq s$. For all $1 \leq i \leq s$, $|B_i| = |A_i|$ or $|A_i| = |C_i|$. I.e. the set of elements looking at a particular location in table A will all look at a distinct locations in one of the tables, B and C. (Otherwise, let $x, y, x', y' \in A_i$, $x \neq y$ and $x' \neq y'$ be such that $b(x) = b(y)$ and $c(x') = c(y')$. Then we can not represent the set $\{x, x'\}$.)

5. Each location of A, B and C is looked at by at least two elements of the universe, unless $s \geq m$. (If a location is looked at by only one element, then set that location to 1 or 0 depending on whether the corresponding element is present or not; we can remove that location and the element out of our scheme.)

6. There are at most two ones in B and C put together.

Define the following restrictions:

- R1. For $x, y \in [m], x \neq y$, $a(x) = a(y) \Rightarrow b(x) \neq b(y)$ and $c(x) \neq c(y)$.
- R2. For $i, j \in [s], i \neq j$, $B_i \cap B_j \neq \phi \Rightarrow C_i \cap C_j = \phi$.
- R3. Either B or C is all zeroes.

We show that if an adaptive $(2, m, s, 2)$ scheme satisfies R3 (or equivalently R1 and R2, as we will show), then s is $\Omega(m^{2/3})$. Note that the scheme given in Theorem 2 satisfies all these three conditions. We then show that if an adaptive $(n, m, s, 2)$ scheme for $n \geq 3$ satisfies R3, then $s \geq m$.

Theorem 5. *If an adaptive $(2, m, s, 2)$ scheme satisfies condition R3, then s is $\Omega(m^{2/3})$.*

Proof. We first show that (R3 $\Rightarrow$ R1 and R2) and then show that (R1 and R2 $\Rightarrow$ s is $\Omega(m^{2/3})$).

Let $a(x) = a(y)$ and $b(x) = b(y)$ for $x, y \in [m], x \neq y$. Consider an element $z \neq x$ such that $c(x) = c(z)$ (such an element exists by condition 5 above). Now, the set $\{y, z\}$ cannot be represented satisfying R3. Thus we have, R3 $\Rightarrow$ R1.

Again, let $a(x_1) = a(x_2) = i$, $a(y_1) = a(y_2) = j$, $b(x_1) = b(y_1)$ and $c(x_2) = c(y_2)$ (so that R2 is violated). Then, the set $\{x_2, y_1\}$ cannot be represented satisfying R3. Thus we have, R3 $\Rightarrow$ R2.

Observe that R1 implies

$$|A_i| = |B_i| = |C_i|, \ \forall i, 1 \leq i \leq s. \tag{1}$$

Hence

$$\sum_{i=1}^{s} |B_i| = \sum_{i=1}^{s} |A_i| = m. \tag{2}$$

By R2, the sets $B_i \times C_i$ are disjoint (no pair occurs in two of these Cartesian products). Thus, by Equation (1), $\sum_{i=1}^{s} |B_i|^2 \leq s^2$. By Cauchy-Schwarz, $s(\sum_{i=1}^{s} |B_i|/s)^2 \leq \sum_{i=1}^{s} |B_i|^2 \leq s^2$. By Equation (2), $\sum_i |B_i| = m$. Thus, $m^2/s \leq s^2$ or $s \geq m^{2/3}$. $\square$

Remark: We observe that, in fact the condition R3 is equivalent to R1 and R2. To show this, it is enough to prove that R1 and R2 $\Rightarrow$ R3. We argue that any scheme that satisfies R1 and R2 can be converted into a scheme that satisfies R3 also.

Consider any scheme which satisfies R1 and R2 but not R3. So, there exists a set $\{x, y\}$ such that $a(x) \neq a(y)$ for which the scheme stores this set as follows (without loss of generality): $A(a(x)) = 0, A(a(y)) = 1, B(b(x)) = 1, C(c(y)) = 1, A(a(z)) = 1$ for all z for which $b(z) = b(x), A(a(z)) = 0$ for all z for which $c(z) = c(y)$ and all other locations as zeroes.

Let $a(x) = i$ and $a(y) = j$. If $B_i \cap B_j = \phi$ then we can store this set as follows: $A(a(x)) = A(a(y)) = 0, B(b(x)) = B(b(y)) = 1$ and all other entries in A as 1s, and all entries in B and C as zeroes, satisfying R3. Condition R1 (and the fact that $B_i \cap B_j = \phi$) ensures that this is a valid scheme to represent the set $\{x, y\}$.

If $B_i \cap B_j \neq \phi$, then R2 ensures that $C_i \cap C_j = \phi$. In this case, to store the set $\{x, y\}$ we can set $A(a(x)) = A(a(y)) = 1, C(c(x)) = C(c(y)) = 1$ and all other entries as zeroes, satisfying R3.

We now show the following.

Theorem 6. *If an adaptive* $(n, m, s, 2)$ *scheme, for* $n \geq 3$ *satisfies condition R3, then* $s \geq m$.

Proof. We first observe that any two probe adaptive scheme satisfies conditions 1 to 5 of the adaptive schemes for sets of size at most 2. Consider an adaptive $(3, m, s, 2)$ scheme with $s < m$. One can find five elements x, y, y', z and z' from the universe such that $a(y) = a(y'), a(z) = a(z'), b(x) = b(y)$ and $c(x) = c(z)$. (Start by fixing x, y, z and then fix x' and y'.) Existence of such a situation is guaranteed by condition 5, as $s < m$. Then we can not represent the set $\{x, y', z'\}$ satisfying R3, contradicting the assumption. Hence, $s \geq m$.   □

5   Conclusions

We have given several deterministic explicit schemes for the membership problem in the bit probe model for small values of t. Our main goal is to achieve $o(m)$ bit space and answer queries using as few probes as possible. We could achieve $\lceil \lg \lg n \rceil + 2$ adaptive probes through an explicit scheme, though it is known (probabilistically) that one can get a $o(m)$ bit structure which uses only 5 probes to answer queries. It is a challenging open problem to come up with explicit scheme achieving this bound. We conjecture that one can not get a three probe $o(m)$ bit structure.

One can also fix some space bound and ask for the least number of probes required to answer the queries. For example, if $s = O(n\sqrt{m})$, Theorem 1 gives a $\lg(n + 1) + 1$ probe adaptive scheme. It would be interesting to see if this can be improved. Also this scheme immediately gives an $n + O(\lg n)$ probe non-adaptive scheme, with the same space bound. Demaine et al.[3] have improved this to an $O(\sqrt{n \lg n})$ probe non-adaptive scheme with $s = O(\sqrt{mn \lg n})$.

Acknowledgment. Part of the work was done while the second author was visiting the University of Waterloo, Canada. He thanks Ian Munro and Erik Demaine for useful discussions.

References

1. A. Brodnik and J. I. Munro, "Membership in constant time and almost minimum space", *SIAM Journal on Computing*, **28(5)**, 1628-1640 (1999).
2. H. Buhrman, P. B. Miltersen, J. Radhakrishnan and S. Venkatesh, "Are Bitvectors Optimal?", *Proceedings of Symposium on Theory of Computing* (2000) 449-458.
3. E. D. Demaine, J. I. Munro, V. Raman and S. S. Rao, "Beating Bitvectors with Oblivious Bitprobes", *I.M.Sc. Technical Report* (2001).
4. M. L. Fredman, J. Komlós and E. Szemerédi, "Storing a sparse table with $O(1)$ access time", *Journal of the Association for Computing Machinery*, **31** (1984) 538-544.
5. Rasmus Pagh, "Low redundancy in dictionaries with O(1) worst case lookup time", *Proceedings of the International Colloquium on Automata, Languages and Programming, LNCS* **1644** (1999) 595-604.
6. Rasmus Pagh, "On the Cell Probe Complexity of Membership and Perfect Hashing", Proceedings of Symposium on Theory of Computing (2001).

Lossy Dictionaries

Rasmus Pagh* and Flemming Friche Rodler

BRICS**
Department of Computer Science
University of Aarhus, Denmark
{pagh,ffr}@brics.dk

Abstract. Bloom filtering is an important technique for space efficient storage of a conservative approximation of a set S. The set stored may have up to some specified number of "false positive" members, but all elements of S are included. In this paper we consider *lossy dictionaries* that are also allowed to have "false negatives". The aim is to maximize the weight of included keys within a given space constraint. This relaxation allows a very fast and simple data structure making almost optimal use of memory. Being more time efficient than Bloom filters, we believe our data structure to be well suited for replacing Bloom filters in some applications. Also, the fact that our data structure supports information associated to keys paves the way for new uses, as illustrated by an application in lossy image compression.

1   Introduction

Dictionaries are part of many algorithms and data structures. A dictionary provides access to information indexed by a set S of *keys*: Given a key, it returns the associated information or reports that the key is not in the set. In this paper we will not be concerned with updates, i.e., we consider the *static* dictionary problem. The main parameters of interest are of course the space used by the dictionary and the time for looking up information. We will assume keys as well as the information associated with keys to have a fixed size.

A large literature has grown around the problem of constructing efficient dictionaries, and theoretically satisfying solutions have been found. Often a slightly easier problem has been considered, namely the *membership* problem, which is the dictionary problem without associated information. It is usually easy to derive a dictionary from a solution to the membership problem, using extra space corresponding to the associated information. In this paper we are particularly interested in dictionary and membership schemes using little memory. Let n denote the size of the key set S. It has been shown that when keys are w-bit machine words, lookups can be performed in constant time in a membership

---

* Partially supported by the IST Programme of the EU under contract number IST-1999-14186 (ALCOM-FT).

** Basic Research in Computer Science (www.brics.dk), funded by the Danish National Research Foundation.

F. Meyer auf der Heide (Ed.): ESA 2001, LNCS 2161, pp. 300–311, 2001.

data structure occupying $B + o(B)$ bits of memory, where $B = \log\binom{2^w}{n}$ is the minimum amount of memory needed to be able to represent any subset of size n [2] (logarithms in this paper are base 2). However, constant factors in the lower order term and lookup time make this and similar schemes less than one could hope for from an applied point of view. Also, difficulty of implementation is an obstacle to practical use. In total, current schemes with asymptotically optimal space usage appear to be mainly of theoretical interest.

If one relaxes the requirements to the membership data structure, allowing it to store a slightly different key set than intended, new possibilities arise. A technique finding many applications in practice is *Bloom filtering* [1]. This technique allows space-efficient storage of a superset S' of the key set S, such that $S' \setminus S$ is no more than an ϵ fraction of $\{0,1\}^w$. For $n \ll 2^w$, about $\log(1/\epsilon)$ bits per key in S are necessary and sufficient for this [4]. This is a significant savings compared to a membership data structure using $B \approx n\log(\frac{2^w \epsilon}{n})$ bits. Lookup of a key using Bloom filtering requires $O(\log(1/\epsilon))$ memory accesses and is thus relatively slow compared to other hashing schemes when ϵ is small. Bloom filtering has applications in, for example, cooperative caching and differential files, where one wants no more than a small chance that an expensive operation is performed in vain. Bloom filtering differs from most other hashing techniques in that is does *not* yield a solution to the dictionary problem.

1.1   This Paper

In this paper we introduce the concept of *lossy dictionaries* that can have not only false positives (like Bloom filters), but also false negatives. That is, some keys in S (with associated information) are thrown away when constructing the dictionary. For false positives there is no guarantee on the associated information returned. We let each key in S have a weight, and try to maximize the sum of weights of keys in the dictionary under a given space constraint. This problem, with no false positives allowed, arises naturally in lossy image compression, and is potentially interesting for caching applications. Also, a dictionary with two-sided errors could take the place of Bloom filters in cooperative caching.

We study this problem on a unit cost RAM, in the case where keys arc machine words of w bits. We examine a very simple and efficient data structure from a theoretical as well as an experimental point of view. Experimentally we find that our data structure has surprisingly good behavior with respect to keeping the keys of largest weight. The experimental results are partially explained by our theoretical considerations, under strong assumptions on the hash functions involved. Specifically, we assume in our RAM model that for a number of random functions, arbitrary function values can be returned in constant time by an oracle.

1.2   Related Work

Most previous work related to static dictionaries has considered the membership problem on a unit cost RAM with word size w. The first membership data structure with worst case constant lookup time using $O(n)$ words of space was constructed by Fredman et al. [7]. For constant $\delta > 0$, the space usage is $O(B)$ when $2^w > n^{1+\delta}$, but in general the data structure may use space $\Omega(Bw)$. The space usage has been lowered to $B + o(B)$ bits by Brodnik and Munro [2]. The lower order term was subsequently improved to $o(n) + O(\log w)$ bits by the first author [11]. The main concept used in the latter paper is that of a *quotient function* q of a hash function h, defined simply to be a function such that the mapping $k \mapsto (h(k), q(k))$ is injective.

The membership problem with false positives was first considered by Bloom [1]. Apart from Bloom filtering the paper presents a less space efficient data structure that is readily turned into a lossy dictionary with only false positives. However, the space usage of the derived lossy dictionary is not optimal. Carter et al. [4] provided a lower bound of $n \log(1/\epsilon)$ bits on the space needed to solve membership with an ϵ fraction false positives, for $n \ll 2^w$, and gave data structures with various lookup times matching or nearly matching this bound. Though none of the membership data structures have constant lookup time, such a data structure follows by plugging the abovementioned results on space optimal membership data structures [2,11] into a general reduction provided in [4]. In fact, the dictionary of [11] can be easily modified to a lossy dictionary with false positives, thus also supporting associated information, using $O(n + \log w)$ bits more than the lower bound.

Another relaxation of the membership problem was recently considered by Buhrman et al. [3]. They store the set S exactly, but allow the lookup procedure to use randomization and to have some probability of error. For two-sided error ϵ they show that there exists a data structure of $O(nw/\epsilon^2)$ bits in which lookups can be done using just *one* bit probe. To do the same without false negatives it is shown that $O(n^2w/\epsilon^2)$ bits suffice and that this is essentially optimal. Schemes using more bit probes and less space are also investigated. If one fixes the random bits of the lookup procedure appropriately, the result is a lossy dictionary with error ϵ. However, it is not clear how to fix the parameters in a reasonable model of computation.

2   Lossy Dictionaries

Consider a set S containing keys $k_1, \ldots, k_n$ with associated information $a_1, \ldots, a_n$ and positive weights $v_1, \ldots, v_n$. Suppose we are given an upper bound m on available space and an error parameter ϵ. The *lossy dictionary problem* for $\epsilon = 0$ is to store a subset of the keys in S and corresponding associated information in a data structure of m bits, trying to optimize the sum of weights of included keys, referred to as the *value*. For general ϵ we allow the dictionary to contain also $2^w \epsilon$ keys from the complement of S. In this section we show the following theorem.

Theorem 1. *Let a sequence of keys $k_1, \ldots, k_n \in \{0,1\}^w$, associated information $a_1, \ldots, a_n \in \{0,1\}^l$, and weights $v_1 \geq \cdots \geq v_n > 0$ be given. Let $r > 0$ be an* *even integer, and $b \geq 0$ an integer. Suppose we have oracle access to random functions $h_1, h_2 : \{0,1\}^w \to \{1, \ldots, r/2\}$ and corresponding quotient functions $q_1, q_2 : \{0,1\}^w \to \{0,1\}^s \setminus 0^s$. There is a lossy dictionary with the following properties:*

1. *The space usage is $r(s - b + l)$ bits (two tables with $r/2$ cells of $s - b + l$ bits).*
2. *The fraction of false positives is bounded by $\epsilon \leq (2^b - 1)r/2^w$.*
3. *The expected weight of the keys in the set stored is $\sum_{i=1}^n p_{r,i} v_i$ where*

$$p_{r,i} \geq \begin{cases} 1 - 52\, r^{-1}/(\frac{r/2}{i} - 1), & \text{for } i < r/2 \\ 2\,(1 - 2/r)^{i-1} - (1 - 2/r)^{2(i-1)}, & \text{for } i \geq r/2 \end{cases}$$

    is the probability that k_i is included in the set.
4. *Lookups are done using at most two (independent) accesses to the tables.*
5. *It can be constructed in time $O(n \log^* n + rl/w)$.*

As discussed in Sect. 2.1 there exist quotient functions for $s = w - \log r + O(1)$ if the hash functions map approximately the same number of elements to each value in $\{1, \ldots, r/2\}$. The inequality in item 2 is satisfied for $b = \lfloor \log(2^w \epsilon/r + 1) \rfloor$, so for $s = w - \log r + O(1)$ an ϵ fraction of false positives can be achieved using space $r\,(\log(\frac{1}{\epsilon + r/2^w}) + l + O(1))$. As can be seen from item 3, almost all of the keys $\{k_1, \ldots, k_{r/2}\}$ are expected to be included in the set represented by the lossy dictionary. For $i \geq r/2$ our bound on $p_{i,r}$ is shown in Fig. 1 of Sect. 3, together with experimentally observed probabilities. If $n \geq r$ and r is large enough it can be shown by integration that, in the expected sense, more than 70% of the keys from $\{k_1, \ldots, k_r\}$ are included in the set (our experiments indicate 84%). We show in Sect. 2.5 that the amount of space we use to achieve this is within a small constant factor of optimal.

Note that by setting $b = 0$ we obtain a lossy dictionary with no false positives. Another point is that given a desired maximum space usage m and false positive fraction ϵ, the largest possible size r of the tables can be chosen efficiently.

2.1 Preliminaries

The starting point for the design of our data structure is a static dictionary recently described in [12]. In this dictionary, two hash tables T_1 and T_2 are used together with two hash functions $h_1, h_2 : \{0,1\}^w \to \{1, \ldots, r/2\}$, where r denotes the combined size of the hash tables, assumed to be even. A key $x \in S$ is stored in either cell $h_1(x)$ of T_1 or cell $h_2(x)$ of T_2. It was shown that if $r \geq (2 + \delta)\,n$, for $\delta > 0$, and h_1, h_2 are random functions, there exists a way of arranging the keys in the tables according to the hash functions with probability at least $1 - \frac{52}{\delta r}$. For small δ this gives a dictionary utilizing about 50% of the hash table cells. The arrangement of keys was shown to be computable in expected linear time.

Another central concept is that of *quotient functions*. Recall that a quotient function q of a hash function h is a function such that the mapping $k \mapsto (h(k), q(k))$ is injective [11]. When storing a key k in cell $h(k)$ of a hash table, it is sufficient to store $q(k)$ to uniquely identify k among all other elements hashing to $h(k)$. To mark empty cells one needs a bit string not mapped to by the quotient function, e.g., 0^s for the quotient functions of Theorem 1. The idea of using quotient functions is, of course, that storing $q(k)$ may require fewer bits than storing k itself. If a fraction $O(1/r)$ of all possible keys hashes to each of r hash table cells, there is a quotient function whose function values can be stored in $w - \log r + O(1)$ bits. This approach was used in [11] to construct a dictionary using space close to the information theoretical minimum. As an example, we look at a hash function family from [6] mapping from $\{0,1\}^w$ to $\{0,1\}^t$. It contains functions of the form $h_a(k) = (ak \mod 2^w) \operatorname{div} 2^{w-t}$ for a odd and $0 \le a < 2^w$. Letting bit masks and shifts replace modulo and division these hash functions can be evaluated very efficiently. A corresponding family of quotient functions is given by $q_a(k) = (ak \mod 2^w) \mod 2^{w-t}$, whose function values can be stored in $w - \log r$ bits.

2.2   Our Data Structure

The idea behind our lossy dictionary, compared to the static dictionary of [12] described above, is to try to fill the hash tables almost completely, working with key sets of size similar to or larger than r. Each key has two hash table cells to which it can be matched. Thus, given a pair of hash functions, the problem of finding a maximum value subset of S that can be arranged into the hash tables is a maximum weight matching problem that can be solved in polynomial time, see e.g. [5]. In Sect. 2.3 we will present an algorithm that finds such an optimal solution in time $O(n \log^* n)$, exploiting structural properties. The term $O(rl/w)$ in the time bound of Theorem 1 is the time needed to copy associated information to the tables. Assume for now that we know which keys are to be represented in which hash table cells.

For $b = 0$ we simply store quotient function values in nonempty hash table cells and 0^s in empty hash table cells, using s bits per cell. For general b we store only the first $s - b$ bits. Observe that no more than 2^b keys with the same hash function value can share the first $s - b$ bits of the quotient function value. This means that there are at most $2^b - 1$ false positives for each nonempty cell. Since 0^s is not in the range, this is also true for empty cells. In addition to the $s - b$ bits, we use l bits per cell to store associated information.

We now proceed to fill in the remaining details on items 3 and 5 of Theorem 1.

2.3   Construction Algorithm

First note that it may be assumed without loss of generality that weights are distinct. If there are consecutive equal weights $v_j = \cdots = v_k$, we can imagine making them distinct by adding positive integer multiples of some quantity δ much smaller than the difference between any pair of achievable values. For sufficiently small δ, the relative ordering of the values of any two solutions with distinct values will not change.

When weights are distinct, the set of keys in an optimal solution is unique, as shown in the following lemma:

Lemma 2. *Suppose that weights are distinct. For $1 \leq i \leq n$, an optimal solution includes key k_i if and only if there is an optimal solution for the set $\{k_1, \ldots, k_{i-1}\}$ such that cell $h_1(k_i)$ of T_1 or cell $h_2(k_i)$ of T_2 is empty. In particular, the set of keys included in an optimal solution is unique.*

Proof. We proceed by induction on n. For $n = 1$ the claim is obvious. In general, the claim follows for $i < n$ by using the induction hypothesis on the set $\{k_1, \ldots, k_i\}$. For $i = n$ consider the unique set K of keys included in an optimal solution for $\{k_1, \ldots, k_{n-1}\}$. If there is an arrangement of K not occupying both cell $h_1(k_n)$ of T_1 and cell $h_2(k_n)$ of T_2, we may add the key k_n to the arrangement which must yield an optimal arrangement. On the other hand, if all arrangements of K occupy both cell $h_1(k_n)$ of T_1 and cell $h_2(k_n)$ of T_2, there is no way of including k_n without discarding a key in K. However, this would yield a lower value and hence cannot be optimal. $\square$

Given hash functions h_1 and h_2 and a key set K, we define the bipartite graph $G(K)$ with vertex set $\{1, 2\} \times \{1, \ldots, r/2\}$, corresponding in a natural way to hash table cells, and the multiset of edges $\{\{(1, h_1(k)), (2, h_2(k))\} \mid k \in K\}$, corresponding to keys. Note that there may be parallel edges if several keys have the same pair of hash function values. We will use the terms keys/edges and cells/vertices synonymously. Define a connected component of $G(K)$ to be *saturated* if the number of edges is greater than or equal to the number of vertices, i.e., if it is not a tree. We have the following characterization of the key set of the optimal solution:

Lemma 3. *Suppose that weights are distinct. The optimal solution includes key k_i if and only if at least one of $(1, h_1(k_i))$ and $(2, h_2(k_i))$ is in a non-saturated connected component of $G(\{k_1, \ldots, k_{i-1}\})$.*

Proof. By Lemma 2 it is enough to show that the keys included in an optimal solution for a set of keys K can be arranged such that cell z is empty if and only if the connected component of z in $G(K)$ is not saturated. Consider the key set K' in an optimal solution for K. For every subset $H \subseteq K'$ it must hold that $|h_1(H)| + |h_2(H)| \geq |H|$ since otherwise not all keys could have been placed. Thus, a connected component with key set C is saturated if and only if $|h_1(C \cap K')| + |h_2(C \cap K')| = |C \cap K'|$. In particular, when arranging the keys of $C \cap K'$, where C is the set of keys of a saturated component, every cell in the connected component must be occupied. On the other hand, suppose there is no arrangement of K' such that z, say cell number i of T_1, is empty. Hall's theorem says that there must be a set $H \subseteq K'$ such that $|h_1(H) \backslash \{i\}| + |h_2(H)| < |H|$. In fact, as no other connected components are affected by blocking z, we may chose H as a subset of the keys in the connected component of z. But then the graph of edges in H must contain a cycle, meaning that the connected component is saturated. The case of z being in T_2 is symmetrical. $\square$

The lemma implies that the following algorithm finds the key set S' of the optimal solution, given keys sorted according to decreasing weight.

1. Initialize a union-find data structure for the cells of the hash tables.
2. For each equivalence class, set a "saturated" flag to `false`.
3. For $i = 1, \ldots, n$:
   a) Find the equivalence class c_b of cell $h_b(k_i)$ in T_b, for $b = 1, 2$.
   b) If c_1 or c_2 is not saturated:
      i. Include k_i in the solution.
      ii. Join c_1 and c_2 to form an equivalence class c.
      iii. Set the saturated flag of c if $c_1 = c_2$ or if the flag is set for c_1 or c_2.

In the loop, equivalence classes correspond to the connected components of the graph $G(\{k_1, \ldots, k_{i-1}\})$. There is a simple implementation of a union-find data structure for which operations take $O(\log^* n)$ amortized time; see [16] which actually gives an even better time bound.

What remains is arranging the optimal key set S' in the tables. Consider a vertex in $G(S')$ of degree one. It is clear that there must be an arrangement such that the corresponding cell contains the key of the incident edge. Thus, one can iteratively handle edges incident to vertices of degree one and delete them. As we remove the same number of edges and vertices from each connected component, the remaining graph consists of connected components with no more edges than vertices and no vertices of degree one, i.e., cycles. The arrangement of edges in a cycle follows as soon as one key has been put (arbitrarily) into one of the tables. The above steps are easily implemented to run in linear time. This establishes item 5 of Theorem 1.

2.4 Quality of Solution

We now turn to the problem of estimating the quality of the solution. Again we will use the fact that weights can be assumed to be unique. A consequence of Lemma 2 is that the set of keys in an optimal solution does not depend on the actual weights, but only on the sequence of hash function values. Thus, the expected value of the optimal solution is $\sum_{i=1}^{n} p_{r,i} v_i$, where $p_{r,i}$ is the probability that the ith key is included in the optimal set of keys.

Lemma 2 says that if $\{k_1, \ldots, k_i\}$ can be accommodated under the given hash functions, they are included in an optimal solution. Using the earlier mentioned result of [12] on $\{k_1, \ldots, k_i\}$ with $\delta = \frac{r/2}{i} - 1$, we have that for $i < r/2$ this happens with probability at least $1 - 52\, r^{-1}/(\frac{r/2}{i} - 1)$. In particular, $p_{r,i}$ is at least this big.

For $i \geq r/2$ we derive a lower bound on $p_{r,i}$ as follows. If one of the vertices $(1, h_1(k_i))$ and $(2, h_2(k_i))$ in $G(\{k_1, \ldots, k_{i-1}\})$ is isolated, it follows by Lemma 3 that k_i is in the optimal solution. Under the assumption that hash function values are truly random, $G(\{k_1, \ldots, k_{i-1}\})$ has $i-1$ randomly and independently chosen edges. Thus, we have the bound $p_{r,i} \geq 1 - (1 - (1 - 2/r)^{i-1}))^2 = 2(1 - 2/r)^{i-1} - (1 - 2/r)^{2(i-1)} \approx 2e^{-i/r} - e^{-2i/r}$. This concludes the proof of Theorem 1.

2.5   A Lower Bound

This section gives a lower bound on the amount of memory needed by a lossy dictionary with an ϵ fraction of false positives and γn false negatives. Our proof technique is similar to that used for the lower bound in [4] for the case $\gamma = 0$.

Proposition 4. *For* $0 < \epsilon < 1/2$ *and* $0 < \gamma < 1$, *a lossy dictionary representing a set* $S \subseteq \{0,1\}^w$ *of* $n \leq 2^{w-1}$ *keys with at most* $2^w \epsilon$ *false positives and at most* γn *false negatives must use space at least*

$$(1 - \gamma)\, n \log \left(\frac{1}{\epsilon + n/2^w} \right) - O(n) \ bits.$$

Proof. We can assume without loss of generality that γn is integer. Consider the set of all data structures used for the various subsets of n elements from $\{0,1\}^w$. Any of these data structures must represent a set of at most $2^w \epsilon + n$ keys, in order to meet the requirement on the number of false positives. Thus, the number of n-element sets having up to γn keys not in the set represented by a given data structure is at most $\sum_{i=0}^{\gamma n} \binom{2^w \epsilon + n}{n-i} \binom{2^w}{i} \leq n \binom{2^w \epsilon + n}{n - \gamma n} \binom{2^w}{\gamma n}$. To represent all $\binom{2^w}{n}$ key sets one therefore needs space at least

$$\log \binom{2^w}{n} - \log \left(n \binom{2^w \epsilon + n}{(1-\gamma)n} \binom{2^w}{\gamma n} \right)$$

$$\geq \log \left(\frac{2^w}{n} \right)^n - \log \left(n \left(\frac{(2^w \epsilon + n)e}{(1-\gamma)n} \right)^{(1-\gamma)n} \left(\frac{2^w e}{\gamma n} \right)^{\gamma n} \right)$$

$$= n \log \left(\frac{(1-\gamma)/e}{\epsilon + n/2^w} \right) - \gamma n \log \left(\frac{(1-\epsilon)(1-\gamma)}{\gamma(\epsilon + n/2^w)} \right) - \log n \ .$$

The argument is concluded by first using $\gamma n \log(1/\gamma) = O(n)$, then merging the two first terms, and finally using $(1 - \gamma)n \log(1 - \gamma) = O(n)$.   □

In the discussion following Theorem 1 we noted that if there are quotient functions with optimal range, the space usage of our scheme is $n \log(\frac{1}{\epsilon + n/2^w}) + O(n)$ when tables of combined size n are used. The expected fraction γ of false negatives is less than $3/10$ by Theorem 1. This means that our data structure uses within $O(n)$ bits of $10/7$ times the lower bound. The experiments described in Sect. 3 indicate that the true factor is less than $6/5$.

2.6   Using More Tables

We now briefly look at a generalization of the two table scheme to schemes with more tables. Unfortunately the algorithm described in Sect. 2.3 does not seem to generalize to more than two tables. An optimal solution can again be found using maximum weight matching, but the time complexity of this solution is not attractive. Instead we can use a variant of the *cuckoo* scheme described by the authors in [13], attempting to insert keys in order $k_1, \ldots, k_n$. For two tables an insertion attempt for k_i works as follows: We store k_i in cell $h_1(k_i)$ of T_1 pushing the previous occupant, if any, away and thus making it *nestless*. If cell $h_1(k_i)$ was free we are done. Otherwise we insert the new nestless element in T_2, possibly pushing out another element. This continues until we either find a free cell or loop around unable to find a free cell, in which case k_i is discarded. It follows from [13] and the analysis in Sect. 2.3 that this algorithm finds an optimal solution, though not as efficiently as the algorithm given in Sect. 2.2. When using three or more tables it is not obvious in which of the tables one should attempt placing the "nestless" key. One heuristic that works well is to simply pick one of the two possible tables at random. It is interesting to compare this heuristic to a random walk on an expander graph, which will provably cross any large subset of the vertices with high probability.

The main drawback of using three tables is, of course, that another memory probe is needed for lookups. Furthermore, as the range of the hash functions must be smaller than when using two tables, the smallest possible range of quotient functions is larger, so more space may be needed for each cell.

3   Experiments

An important performance parameter of our lossy dictionaries is the ability to store many keys with high weight. Our first experiment tests this ability for lossy dictionaries using two and three tables. For comparison, we also test the simple one table scheme that stores in each cell the key of greatest weight hashing to it. The tests were done using truly random hash function values, obtained from a high quality collection of random bits freely available on the Internet [10]. Figure 1 shows experimentally determined values of $p_{r,\alpha r}$, the probability that key with index $i = \alpha r$ is stored in the dictionary, determined from 10^4 trials. For the experiments with one and two tables we used table size $r = 2048$ while for the experiment with three tables we used $r = 1536$. We also tried various other table sizes, but the graphs were almost indistinguishable from the ones shown. From the figure we see the significant improvement of moving from one to more tables. As predicted, nearly all of the $r/2$ heaviest keys are stored when using two tables. For three tables this number increases to about $.88r$. Of the r heaviest keys, about 84% are stored when using two tables, and 95% are stored when using three tables.

Apart from asymptotically vanishing differences around the point where the curves start falling from 1, the graphs of Fig. 1 seem independent of r. For two tables the observed value of $p_{r,\alpha r}$ for $\alpha > 1/2$ is approximately $3.5/9.6^\alpha$.

3.1   Application

To examine the practicality of our dictionaries we turn to the real world example of lossy image compression using wavelets. Today most state-of-the-art image coders, such as JPEG2000, are based on wavelets. The wavelet transform has the

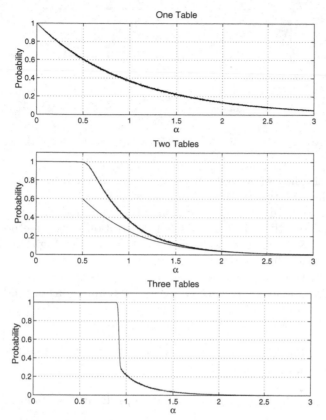

Fig. 1. The observed probability that the element with (αr)th highest weight is stored. For two tables our lower bound is shown.

ability to efficiently approximate nonlinear and nonstationary signals with coefficients whose magnitudes, in sorted order, decay rapidly towards zero. This is illustrated in Fig. 2. The figure shows the sorted magnitudes of the wavelet coefficients for the Lena image, a standard benchmark in image processing, computed using Daubechies second order wavelets. Thresholding the wavelet coefficients by a small threshold, i.e., setting small valued coefficients to zero, introduces only a small *mean squared error* (MSE) while leading to a sparse representation that can be exploited for compression purposes. The main idea of most wavelet based compression schemes is to keep the value and position of the r coefficients of largest magnitude. To this end many advanced schemes, such as zero-tree coding, have been developed. None of these schemes support access to a single pixel without decoding significant portions of the image.

Recently, interest in fast random access to decoded data, accessing only a few wavelet coefficients, has arisen [8,9,14,15]. In [15] we show that lossy dictionaries are well suited for this purpose. Based on our data structure for lossy dictionaries, we present a new approach to lossy storage of the coefficients of wavelet transformed data. The approach supports fast random access to individual data

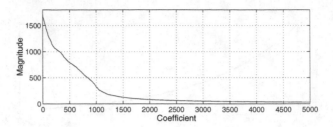

Fig. 2. Largest 5000 magnitudes of 67615 wavelet coefficients of the Lena image.

elements within the compressed representation. Compared to the previously best methods in the literature [8,14] our lossy dictionary based scheme performs about 50%-80% better in terms of compression ratio, while reducing the random access time by more than 60%. A detailed description of the method is outside the scope of this paper. Instead we use the Lena image to give a flavor of the usefulness of lossy dictionaries on real world data. We store the coefficients of Fig. 2 in a two table lossy dictionary of total table size $r = 2^{11}$, using a simple family of hash functions. Specifically, we use hash functions of the form

$$h(k) = ((a_2 k^2 + a_1 a_2 k + a_0) \bmod p) \bmod r/2,$$

where p is a prime larger than any key, $0 < a_0, a_1, a_2 < p$ and a_1 is even. A corresponding quotient function is

$$q(k) = 2(((a_2 k^2 + a_1 a_2 k + a_0) \bmod p) \operatorname{div} r/2) + k \bmod 2 \ .$$

Again, 10^4 iterations were made, selecting random functions from the above family using C's **rand** function. The graph of $p_{r,\alpha r}$ is indistinguishable from that in Fig. 1. For our application, we obtain an MSE of 200, which is 27% more than the MSE when storing the r coefficients of largest magnitude. This difference would be difficult at best to detect in the reconstructed image. The previously mentioned family of [6] had somewhat worse performance. Using three tables reduces the MSE increase to a mere 1%.

4   Conclusion

We have introduced the concept of lossy dictionaries and presented a simple and efficient data structure implementing a lossy dictionary. Our data structure combines very efficient lookups and near-optimal space utilization, and thus seems a promising alternative to previously known data structures when a small percentage of false negatives is tolerable.

Though simple and efficient hash functions seem to work well in practice with our data structure, the challenge of finding such families that provably work well remains. Furthermore, the last two graphs in Fig. 1 are not completely understood. The same is true for the insertion heuristic for three or more tables.

Acknowledgment. We thank Stephen Alstrup and Theis Rauhe for helpful discussions on the construction of our two table data structure.

References

[1] Burton H. Bloom. Space/time trade-offs in hash coding with allowable errors. *Communications of the ACM*, 13(7):422–426, July 1970.

[2] Andrej Brodnik and J. Ian Munro. Membership in constant time and almost-minimum space. *SIAM J. Comput.*, 28(5):1627–1640 (electronic), 1999.

[3] Harry Buhrman, Peter Bro Miltersen, Jaikumar Radhakrishnan, and S. Venkatesh. Are bitvectors optimal? In *Proceedings of the 32nd Annual ACM Symposium on Theory of Computing (STOC '00)*, pages 449–458. ACM Press, New York, 2000.

[4] Larry Carter, Robert Floyd, John Gill, George Markowsky, and Mark Wegman. Exact and approximate membership testers. In *Proceedings of the 10th Annual ACM Symposium on Theory of Computing (STOC '78)*, pages 59–65. ACM Press, New York, 1978.

[5] William J. Cook, William H. Cunningham, William R. Pulleyblank, and Alexander Schrijver. *Combinatorial optimization*. John Wiley & Sons Inc., New York, 1998. A Wiley-Interscience Publication.

[6] Martin Dietzfelbinger, Torben Hagerup, Jyrki Katajainen, and Martti Penttonen. A reliable randomized algorithm for the closest-pair problem. *Journal of Algorithms*, 25(1):19–51, 1997. doi:10.1006/jagm.1997.0873.

[7] Michael L. Fredman, János Komlós, and Endre Szemerédi. Storing a sparse table with $O(1)$ worst case access time. *J. Assoc. Comput. Mach.*, 31(3):538–544, 1984.

[8] Insung Ihm and Sanghun Park. Wavelet-based 3D compression scheme for very large volume data. *Graphics Interface*, pages 107–116, 1998.

[9] Tae-Young Kim and Yeong Gil Shin. An efficient wavelet-based compression method for volume rendering. In *Seventh Pacific Conference on Computer Graphics and Applications*, pages 147–156, 1999.

[10] George Marsaglia. The Marsaglia random number CDROM including the diehard battery of tests of randomness. http://stat.fsu.edu/pub/diehard/.

[11] Rasmus Pagh. Low Redundancy in Static Dictionaries with $O(1)$ Lookup Time. In *Proceedings of the 26th International Colloquium on Automata, Languages and Programming (ICALP '99)*, volume 1644 of *Lecture Notes in Computer Science*, pages 595–604. Springer-Verlag, Berlin, 1999.

[12] Rasmus Pagh. On the Cell Probe Complexity of Membership and Perfect Hashing. In *Proceedings of the 33rd Annual ACM Symposium on Theory of Computing (STOC '01)*. ACM Press, New York, 2001.

[13] Rasmus Pagh and Flemming Friche Rodler. Cuckoo hashing. To appear in Proceedings of ESA 2001, 2001.

[14] Flemming Friche Rodler. Wavelet based 3D compression with fast random access for very large volume data. In *Seventh Pacific Conference on Computer Graphics and Applications*, pages 108–117, Seoul, Korea, 1999.

[15] Flemming Friche Rodler and Rasmus Pagh. Fast random access to wavelet compressed volumetric data using hashing. Manuscript.

[16] Robert Endre Tarjan. Efficiency of a good but not linear set union algorithm. *J. Assoc. Comput. Mach.*, 22:215–225, 1975.

Splitting a Delaunay Triangulation in Linear Time[*]

Bernard Chazelle[1], Olivier Devillers[2], Ferran Hurtado[3], Mercè Mora[3], Vera Sacristán[3], and Monique Teillaud[2]

[1] Princeton University, Computer Science Department, 35 Olden Street,Princeton, NJ 08544,USA. chazelle@cs.princeton.edu
[2] INRIA, BP93, 06902 Sophia-Antipolis, France. {Olivier.Devillers,Monique.Teillaud}@sophia.inria.fr
[3] Dept. Matemàtica Aplicada II, Univ. Politècnica de Catalunya, Pau Gargallo, 5, 08028 Barcelona, Spain. {hurtado,mora,vera}@ma2.upc.es

1 Introduction

Computing the Delaunay triangulation of n points is well known to have an $\Omega(n \log n)$ lower bound. Researchers have attempted to break that bound in special cases where additional information is known.

The Delaunay triangulation of the vertices of a convex polygon is such a case where the lower bound of $\Omega(n \log n)$ does not hold. This problem has been solved in linear time with a deterministic algorithm of Agarwal, Guibas, Saxe and Shor [1]. Chew has also proposed a very simple randomized algorithm [8] for the same problem, which we sketch in Sect.2.2. These two algorithms can also compute the skeleton of a convex polygon in linear time and support the deletion of a point from a Delaunay triangulation in time linear in its degree.

Another result is that if a spanning subgraph of maximal degree d of the Delaunay triangulation is known, then the remaining part of the Delaunay triangulation can be computed in $O(nd \log^\star n)$ expected randomized time [14]. The Euclidean minimum spanning tree is an example of such a graph of bounded degree 6. This $O(n \log^\star n)$ result applies also if the points are the vertices of a chain monotone in both x and y directions but, in this special case, linear complexity has been achieved by Djidjev and Lingas [15] generalizing the result of Agarwal et al. for convex polygons.

Beside these results, where knowing some information on the points helps to construct the Delaunay triangulation, it has been proven that knowing the order of the points along any one given direction does not help [15].

Breaking a lower bound by using some additional information arises similarly in some other problems. One of the most famous is the triangulation of a simple polygon in linear time [6,18,2]. Other related problems are the constrained Delaunay triangulation of a simple polygon in $O(n)$ time [17]; the medial axis of a

[*] The French team was partially supported by Picasso French-Spanish collaboration program. The Spanish team was partially supported by CUR Gen. Cat. 1999SGR00356, Proyecto DGES-MEC PB98-0933 and Acción Integrada Francia-España HF99-112.

F. Meyer auf der Heide (Ed.): ESA 2001, LNCS 2161, pp. 312–320, 2001.

simple polygon in linear time [10]; the computation of one cell in the intersection of two polygons in $O(n \log^{*2} n)$ time [12]; the L^∞ Delaunay triangulation of points sorted along x and y axes in $O(n \log \log n)$ time [9]. Also, given the 3D convex hull of a set of blue points and the convex hull of the set of red points, the convex hull of all points can be computed in linear time [7].

The problem we address in this paper is the following: given the Delaunay triangulation $\mathcal{DT}(S)$ of a point set S in E^2, a partition of S into S_1, S_2, can we compute both $\mathcal{DT}(S_i)$ in $o(n \log n)$ time?

The reverse problem, given a partition of S into S_1, S_2, reconstruct $\mathcal{DT}(S)$ from $\mathcal{DT}(S_i)$, is known to be doable in linear time [7]. Indeed, the 3D convex hull of the vertices of two convex polyhedra can be computed in linear time [7] and, by standard transformation of the Delaunay triangulation to the convex hull, we get the result. This reverse operation can be used as the merging step of a divide and conquer algorithm.

In this paper, we propose an $O(n)$ randomized algorithm in the spirit of Chew's algorithm for the Delaunay triangulation of a convex polygon.

2   Preliminaries

We assume in the sequel that a triangulation allows constant time access from a triangle to its three neighbors and to its three vertices, and from a vertex to one incident triangle. This is provided by any reasonable representation of a triangulation, either based on triangles [4] or as in the $DCEL$ or winged-edge structure [13, pp. 31-33].

2.1   Classical Randomized Incremental Constructions

Randomized incremental constructions have been widely used for geometric problems [11,3] and specifically for the Delaunay triangulation [5,16,14]. These algorithms insert the points one by one in a random order in some data structure to locate the new point and update the triangulation. The location step has an $O(\log n)$ expected complexity. The update step has constant expected complexity as can be easily proved by backwards analysis [19]. Indeed, the update cost of inserting the last point in the triangulation is its degree in the final triangulation. Since the last point is chosen randomly, its insertion cost is the average degree of a planar graph, which is less than 6.

2.2   Chew's Algorithm for the Delaunay Triangulation of a Convex Polygon

Chew's algorithm [8] for the Delaunay triangulation of a convex polygon uses the ideas above for the analysis of the insertion of the last point. The main idea is to avoid the location cost using the additional information of the convex polygon.

As noticed earlier, for any vertex v we know one of its incident triangles. In the case of Chew's algorithm, it is required that the triangle in question be incident to the convex hull edge following v in counterclockwise order.

The algorithm can be stated as follows:
1. Choose a random vertex p of the polygon $\mathcal{P}$.
2. Store the point q before p on the convex hull.
3. Compute the convex polygon $\mathcal{P} \setminus \{p\}$.
4. Compute recursively $\mathcal{DT}(S \setminus \{p\})$.
5. Let t be the triangle pointed to by q.
6. Create a triangle neighbor of t with p as vertex, flip diagonals if necessary using the standard Delaunay criterion and update links from vertices to incident triangles.

By standard backwards analysis, the flipping step has expected constant cost. Other operations, except the recursive call, require constant time. Thus we get a linear expected complexity.

The important thing is that we avoid the location step. Thus Chew's algorithm applies to other cases where the location step can be avoided, e.g. deletion of a point in a Delaunay triangulation.

3   Algorithm

3.1   General Scheme

The main idea is similar to Chew's algorithm, that is to delete a random point $p \in S_i$ from $\mathcal{DT}(S)$, to split the triangulation and then to insert p in the triangulation $\mathcal{DT}(S_i \setminus \{p\})$ avoiding the usual location step. The location of p can be done by computing the nearest neighbor of p in S_i, which can be done in time $T(p) \log T(p)$ for some number $T(p)$ depending on p, whose expectation is $O(1)$ (details will be given at Step 2 of the algorithm of Section 3.3). However, it is possible for example to have one point p, chosen with probability $\frac{1}{n}$ such that $T(p) = n$, which brings the expected cost to $E\left(T(p) \log T(p)\right) = \Omega(n \log n)$. The idea is to choose two points p_α, p_β and to take for p the better of the two, in order to concentrate the distribution around its mean. Here is the algorithm:

Given $\mathcal{DT}(S)$,
1. Choose two random points $p_\alpha, p_\beta \in S$. Let $i, j \in \{1, 2\}$ such that $p_\alpha \in S_i$ and $p_\beta \in S_j$ (i and j do not need to be different).
2. Look simultaneously for the nearest neighbor of p_α in S_i and the nearest neighbor of p_β in S_j. As soon as one of the two is found, say the neighbor q of p_α in S_i, stop all searching and let p be p_α.
3. Remove p from $\mathcal{DT}(S)$ to get $\mathcal{DT}(S \setminus \{p\})$.
4. Recursively compute $\mathcal{DT}(S_1 \setminus \{p\})$ and $\mathcal{DT}(S_2 \setminus \{p\})$ from $\mathcal{DT}(S \setminus \{p\})$.
5. Determine the triangle of $\mathcal{DT}(S_i \setminus \{p\})$ incident to q that is traversed by the segment pq.
6. Apply the usual Delaunay flip procedure to obtain $\mathcal{DT}(S_i)$ from $\mathcal{DT}(S_i \setminus \{p\})$.

3.2   Combination Lemmas

Note that in the algorithm, p is not a random point uniformly distributed among S, but one chosen among two random points. In this section, we investigate how this choice influences the mean value of some variable depending on p.

Let $X(p)$ be a positive random variable depending on a point p, bounded above by n and of constant expected value $E(X) = c$. Since p is one point among n, $X(p)$ can take n values $n \geq X_1 \geq X_2 \geq \ldots \geq X_n \geq 0$ and $\sum X_i = n \cdot c$.

Pick i, j uniformly in $\{1, \ldots, n\}$ independently and let $Y = \max\{X_i, X_j\}$ and $Z = \min\{X_i, X_j\}$.

Lemma 1.
$$E(Y) \leq 2c.$$

Proof.

$$E(Y) = \frac{1}{n^2} \sum_{1 \leq i,j \leq n} \max(X_i, X_j)$$

$$= \frac{1}{n^2} \left(2 \sum_{1 \leq i < j \leq n} X_i + \sum_{1 \leq i \leq n} X_i \right)$$

$$= \frac{2}{n^2} \sum_{1 \leq i \leq n} (n - i + \frac{1}{2}) X_i$$

$$\leq \frac{2n}{n^2} \sum_{1 \leq i \leq n} X_i = 2c. \square$$

Lemma 2. *If f is a concave non decreasing function,*
$$E(Z \cdot f(Z)) \leq 2c \cdot f(c).$$

Proof.

$$E(Z \cdot f(Z)) = \frac{1}{n^2} \sum_{1 \leq i,j \leq n} \min(X_i, X_j) f(\min(X_i, X_j))$$

$$= \frac{1}{n^2} \left(2 \sum_{1 \leq i < j \leq n} X_j f(X_j) + \sum_{1 \leq j \leq n} X_j f(X_j) \right)$$

$$= \frac{2}{n^2} \sum_{1 \leq j \leq n} (j - \frac{1}{2}) X_j f(X_j)$$

$$\leq \frac{2}{n^2} \sum_{1 \leq j \leq n} j X_j f(X_j).$$

Clearly $j X_j \leq \sum_{1 \leq i \leq j} X_i \leq \sum_{1 \leq i \leq n} X_i \leq cn$;

$$E(Z \cdot f(Z)) \leq \frac{2c}{n} \sum_{1 \leq j \leq n} f(X_j)$$

$$\leq \frac{2c}{n} n \cdot f(c) \quad \text{by concavity of } f. \square$$

3.3   Algorithmic Details and Randomized Analysis

Referring to the six different steps of the algorithm, here is a detailed cost analysis:

1. Done in time $O(1)$.
2. The nearest neighbor in S_i of a point $p \in S_i$ can be found in the following way. Start considering all the Delaunay edges incident to p. Put them in a priority queue by increasing order of their distance to p. Explore the queue in the following way: each time that we consider a point q, there are two possibilities:
   - If $q \in S_i$, we are done: q is p's nearest neighbor in S_i.
   - If $q \notin S_i$, insert in the queue all its Delaunay neighbors, delete q and proceed to the following point in the queue.

   The correctness of this process is based on the fact that it simulates the way in which a circle centered in p would grow. In other words, if $q \in S_i$ is the point we are looking for, the algorithm computes and orders all the points that are closer to p than q (obviously, none of them belongs to S_i). The proof is based on the following observation.

Fact. *Let S be a set of points. Let C be any disk in the plane that contains a point $s \in S$ on its boundary. Let $p_1, \ldots, p_k$ be all the points of S contained in C. Then s must have a Delaunay neighbor among $p_1, \ldots, p_k$.*

Proof. Grow a circle C_s through s, tangent to C and interior to C, until it reaches the first point p_i (see Fig. 1).

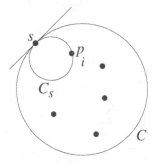

Fig. 1. The points s and p_i are Delaunay neighbors.

The emptiness of C_s is obvious, and therefore sp_i is a Delaunay edge.   □
In this procedure, we have explored and ordered all the points that lie closer to p than q, together with all their neighbors. Can $T(p)$, the number of such points, be too big on average? As the randomly chosen point can belong either to S_1 or to S_2, we want to bound the following amount:

$$E(T) = \frac{1}{n} \left(\sum_{p \in S_1} \sum_{q \in D(p, NN_1(p))} \deg(q) + \sum_{p \in S_2} \sum_{q \in D(p, NN_2(p))} \deg(q) \right)$$

where $NN_i(p)$ denotes the nearest neighbor of p in S_i, $D(p,s)$ is the disk of center p passing through s and $\deg(q)$ denotes the degree of q in $DT(S)$. We bound the summands in the following way:

$$\sum_{p \in S_1} \sum_{q \in D(p,NN_1(p))} \deg(q) = \sum_{q \in S_2} \sum_{p \text{ s.t. } q \in D(p,NN_1(p))} \deg(q)$$

$$= \sum_{q \in S_2} \deg(q) \ \text{number}\{p \text{ s.t. } q \in D(p, NN_1(p))\}$$

$$\leq 6 \sum_{q \in S_2} \deg(q).$$

The last inequality is due to the fact that the number of disks of the kind $D(p, NN_1(p))$ that can contain a point $q \in S_2$ is at most 6, because in the set $S_1 \cup \{q\}$ such a point p would have q as closest neighbor, and the maximum degree of q in the nearest neighbor graph of $S_1 \cup \{q\}$ is 6. Thus we get

$$E(T) \leq \frac{6}{n} \left(\sum_{q \in S_2} \deg(q) + \sum_{q \in S_1} \deg(q) \right) \leq 36.$$

Since the algorithm requires a priority queue, the cost of searching for q is $O(T \log T)$ if we use a balanced priority queue or even $O(T^2)$ if we use a simple list to implement the queue and $E(T^2)$ cannot be bounded by a constant. But the time for deciding which of p_α and p_β will be p is the minimum of the times for finding the neighbors of p_α and p_β and thus expected to be constant by Lemma 2. This step has expected cost $O(1)$.

3. It is known that it can be done in time proportional to the degree of p with Chew's algorithm. Since for a random point, the expected degree is 6, the expected degree of p is smaller than 12 by Lemma 1. Hence, this step has expected cost $O(1)$.

4. If the cost of the algorithm is $T(n)$, this step can be done in $T(n-1)$.

5. Exploring all the triangles incident to q takes time proportional to the degree of q in $DT(S_i \setminus \{p\})$. But q is not a random point, but the nearest neighbor of p, itself chosen among two random points. We will prove below that the degree of the nearest neighbor in S_i of a random point $p \in S_i$ is at most 42, and thus by Lemma 1 the expected degree of q is less than 84 and this step can be done in time $O(1)$.

Fact. *Given a random point p in a set of points R, the expected degree in $DT(R \setminus \{p\})$ of the nearest neighbor of p in R is at most 42.*

Proof. We have to consider the degree of a point in several graphs. Let $\deg_{NN}(q)$ be the degree of q in the nearest neighbor graph of R, $\deg(q)$ be the degree of q in $DT(R)$ and $\deg_p(q)$ be the degree of q in $DT(R \setminus \{p\})$. It is known that $\deg_{NN}(q)$ is at most 6. When p is removed from $DT(R)$ the new neighbors of q are former neighbors of p , thus $\deg_p(q) \leq \deg(p) + \deg(q)$.

The expected value of $deg_p(NN(p))$ is:

$$E\left(\deg_p(NN(p))\right) = \frac{1}{n} \sum_{p \in R} \deg_p(NN(p))$$

$$\leq \frac{1}{n} \sum_{p \in R} (\deg(p) + \deg(NN(p)))$$

$$\leq 6 + \frac{1}{n} \sum_{\substack{p,q \in R \ q=NN(p)}} \deg(q)$$

$$\leq 6 + \frac{1}{n} \sum_{q \in R} (\deg_{NN}(q) \deg(q))$$

$$\leq 6 + \frac{1}{n} \sum_{q \in R} (6 \deg(q)) \leq 6 + 36 = 42.\square$$

6. It is known that this step can be done in time proportional to the degree of p, that is, in expected time $O(1)$ by Lemma 1.

As a conclusion, we have proved the following

Theorem 3. *Given a set of n points S and its Delaunay triangulation, for any partition of S into two disjoint subsets, S_1 and S_2, the Delaunay triangulations $DT(S_1)$ and $DT(S_2)$ can be computed in $O(n)$ expected time.*

4   Concluding Remarks

4.1   Alternative Ideas

We should mention several simpler ideas that do not work. A first idea consists in deleting all the points of S_2 from $DT(S)$ in a random order, but the degree of a random point in S_2 cannot be controlled; in fact if we take points on the part of the unit parabola with positive abscissa, the Delaunay triangulation links the point of highest curvature to all others (see Fig. 2). If we split the set into two parts along the parabola and we remove the highest-curvature half of the point set in a random order, then the probability of removing the highest curvature point increases as the set of point decreases and the expected time to remove half the points is $O(n \log n)$.

Another idea is to remove the points not at random, but by increasing degree, but in that case the set of points to remove must be kept sorted by degree, although the degrees change during the algorithm.

4.2   Convex Hull in 3D

Through the projection of the plane on a paraboloid in 3D, Delaunay triangulations are closely related to convex hulls in three dimensions.

Unfortunately, our algorithm, or more precisely its complexity analysis, does not generalize to 3D convex hulls. In this paper we use the fact that the nearest

Fig. 2. Points on a parabola

neighbor graph is a subgraph of the Delaunay triangulation having bounded degree, and to generalize the algorithm we would need to define a neighboring relation which is a subgraph of the convex hull; several possibilities for such a subgraph exist but they do not provide bounded degree and thus the analysis does not generalize.

Acknowledgments. The authors thank Oswin Aichholzer for various discussions about this problem.

References

1. A. Aggarwal, L. J. Guibas, J. Saxe, and P. W. Shor. A linear-time algorithm for computing the Voronoi diagram of a convex polygon. *Discrete Comput. Geom.*, 4(6):591–604, 1989.
2. N. M. Amato, M. T. Goodrich, and E. A. Ramos. Linear-time triangulation of a simple polygon made easier via randomization. In *Proc. 16th Annu. ACM Sympos. Comput. Geom.*, pages 201–212, 2000.
3. J.-D. Boissonnat, O. Devillers, R. Schott, M. Teillaud, and M. Yvinec. Applications of random sampling to on-line algorithms in computational geometry. *Discrete Comput. Geom.*, 8:51–71, 1992.
4. J. D. Boissonnat, O. Devillers, M. Teillaud, and M. Yvinec. Triangulations in CGAL. In *Proc. 16th Annu. ACM Sympos. Comput. Geom.*, pages 11–18, 2000.
5. J.-D. Boissonnat and M. Teillaud. On the randomized construction of the Delaunay tree. *Theoret. Comput. Sci.*, 112:339–354, 1993.
6. B. Chazelle. Triangulating a simple polygon in linear time. *Discrete Comput. Geom.*, 6(5):485–524, 1991.
7. B. Chazelle. An optimal algorithm for intersecting three-dimensional convex polyhedra. *SIAM J. Comput.*, 21(4):671–696, 1992.
8. L. P. Chew. Building Voronoi diagrams for convex polygons in linear expected time. Technical Report PCS-TR90-147, Dept. Math. Comput. Sci., Dartmouth College, Hanover, NH, 1986.
9. L. P. Chew and S. Fortune. Sorting helps for Voronoi diagrams. *Algorithmica*, 18:217–228, 1997.
10. F. Chin, J. Snoeyink, and C. A. Wang. Finding the medial axis of a simple polygon in linear time. *Discrete Comput. Geom.*, 21(3):405–420, 1999.

11. K. L. Clarkson and P. W. Shor. Applications of random sampling in computational geometry, II. *Discrete Comput. Geom.*, 4:387–421, 1989.

12. M. de Berg, O. Devillers, K. Dobrindt, and O. Schwarzkopf. Computing a single cell in the overlay of two simple polygons. *Inform. Process. Lett.*, 63(4):215–219, Aug. 1997.

13. M. de Berg, M. van Kreveld, M. Overmars, and O. Schwarzkopf. *Computational Geometry: Algorithms and Applications*. Springer-Verlag, Berlin, 1997.

14. O. Devillers. Randomization yields simple $O(n \log^* n)$ algorithms for difficult $\Omega(n)$ problems. *Internat. J. Comput. Geom. Appl.*, 2(1):97–111, 1992.

15. H. Djidjev and A. Lingas. On computing Voronoi diagrams for sorted point sets. *Internat. J. Comput. Geom. Appl.*, 5:327–337, 1995.

16. L. J. Guibas, D. E. Knuth, and M. Sharir. Randomized incremental construction of Delaunay and Voronoi diagrams. *Algorithmica*, 7:381–413, 1992.

17. R. Klein and A. Lingas. A linear-time randomized algorithm for the bounded Voronoi diagram of a simple polygon. *Internat. J. Comput. Geom. Appl.*, 6:263–278, 1996.

18. R. Seidel. A simple and fast incremental randomized algorithm for computing trapezoidal decompositions and for triangulating polygons. *Comput. Geom. Theory Appl.*, 1(1):51–64, 1991.

19. R. Seidel. Backwards analysis of randomized geometric algorithms. In J. Pach, editor, *New Trends in Discrete and Computational Geometry*, volume 10 of *Algorithms and Combinatorics*, pages 37–68. Springer-Verlag, 1993.

A Fast Algorithm for Approximating the Detour of a Polygonal Chain

Annette Ebbers-Baumann[1], Rolf Klein[1], Elmar Langetepe[1], and
Andrzej Lingas[2]

[1] Universität Bonn, Institut für Informatik I, D-53117 Bonn, Germany.
[2] Department of Computer Science, Lund University, 22100 Lund, Sweden.

Abstract. Let C be a simple[1] polygonal chain of n edges in the plane, and let p and q be two points on C. The *detour* of C on (p, q) is defined to be the length of the segment of C that connects p with q, divided by the Euclidean distance between p and q. Given an $\epsilon > 0$, we compute in time $O(n \log n)$ a pair of points on which the chain makes a detour at least $1/(1 + \epsilon)$ times the maximum detour.

1   Introduction

A transportation route like a highway is supposed to provide a reasonably short connection between the points it passes through. More precisely, for any two points, p and q, on an open curve C in the plane that consists of a bounded number of smooth pieces, we call the value

$$d_C(p, q) = \frac{|C_p^q|}{|pq|}$$

the *detour of C on the pair (p, q)*; here C_p^q denotes the unique segment of C that connects p with q, $|C_p^q|$ denotes its length, and $|pq|$ is the Euclidean distance between p and q.

Clearly, $d_C(p, q) \geq 1$ holds. We assume that C does not self-intersect. Then $d_C(p, q)$ is bounded, and it tends to 1 as q tends to p along C.[2] We are interested in the value

$$d_C = \max_{p, q \in C} d_C(p, q)$$

called the *maximum detour* of curve C.

The detour of a curve is an important notion in analyzing on-line navigation strategies, where the length of a path created by some robot must be compared to the shortest path connecting two points; see, e. g., Icking and Klein [6]. A different type of application is in comparing the Fréchet distance between two curves with their Hausdorff distance; while the first is always greater or equal

---

[1] C has no self-intersections.
[2] If C passes through the same point, p, twice then $d_C(p, q)$ is unbounded for certain points q close to p.

F. Meyer auf der Heide (Ed.): ESA 2001, LNCS 2161, pp. 321–332, 2001.

than the latter, there is, in general, no bound for the other direction. Only for curves of bounded maximum detour can such a bound be shown, see Alt et al. [2]. The vertex-to-vertex maximum detour of graphs was already considered as the dual t-spanner problem by Narasimhan and Smid [9]. They provide an approximation of the vertex-to-vertex *stretch factor* in a more general setting.

Intuitively, a curve that does not meander wildly should have a small maximum detour. This idea can be made precise in several ways. An oriented curve running from s to t is called self-approaching if, for each point p, the curve segment C_p^t fits in a 90° wedge with apex in p. The maximum detour of self-approaching curves has been tightly bounded by 5.3331...; see Icking et al. [7]. This result can be generalized to wedges of arbitrary angles, see Aichholzer et. al. [1]. Rote [10] has shown a tight upper bound of $2/3\pi$ for the detour of curves of increasing chords, i. e., curves that are self-approaching in both directions.[3]

These results were all obtained in an indirect way, by bounding the curve's length by the perimeter of some simple, convex container.

In this paper we present an $O(n \log n)$ algorithm for computing directly the maximum detour of an arbitrary polygonal chain of n edges, up to an $1+\epsilon$ factor. The paper is organized as follows. In Sect. 2 we first state a local criterion that is necessary for a pair of points (p, q) on which the chain takes on its maximum detour. As a consequence, one of p, q can be assumed to be a vertex of the chain.

In Sect. 3 we prove some global properties. It turns out that the maximum detour is always attained by a vertex-edge cut of the chain C, that is by a pair (p, q) of *co-visible* points of the chain, one of which is a vertex whereas the other may be an interior point of an edge.

Moreover, we prove a certain property of the detours related to cuts that cross each other. While this property is weaker than the Monge property (see Burkard et al. [3]) it does imply that cuts attaining the maximum detour do not cross and must be, therefore, linear in number.

In Sect. 4 we present an algorithm that computes, for a given real number $\epsilon > 0$, within time $O(n \log n)$ a vertex-edge cut (p, q) such that the maximum detour of chain C is at most $1 + \epsilon$ times the detour of C on (p, q). Our algorithm uses the result by Gutwin and Keil [8] on constructing sparse spanners for finding a pair of *vertices* of C whose detour is close to the maximum detour chain C makes on all vertex pairs.

Finally, in Sect. 5 we mention some open problems that naturally arise from this work.

---

[3] The relationship between wedge containment and detour seems to work in only one way, because there are curves of arbitrary small maximum detour that do not fit in small wedges.

2 Local Properties

Throughout this paper, let C be a simple, planar, polygonal chain of n edges. Simple means that C does not have self-intersections. That is, if any two edges intersect at all they must be neighbors in the chain, and their intersection is just a common vertex.

Now let p and q denote two points on C, and let

$$d_C(p, q) = \frac{|C_p^q|}{|pq|}$$

be the detour of C on (p, q). We want to analyze how the detour changes at points close to p on the same edge, e, while q remains fixed. Let the positive direction of e be such that the length of the chain segment C_p^q increases, and let β denote the angle between the positive part of e and the line segment pq; see Fig. 1. Excluding trivial cases we assume that $0 < \beta < \pi$ holds.

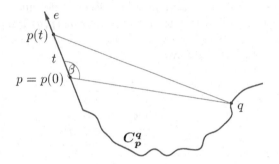

Fig. 1. When does the detour increase as $p(t)$ moves?

Lemma 1. *For a fixed point q, the function $d_C(., q)$ takes on a unique maximum on edge e. If*

$$\cos \beta = -\frac{|pq|}{|C_p^q|}$$

holds then this maximum is attained at p. If $\cos \beta$ is bigger (resp. smaller) than the right hand side, the detour can be increased by moving p forwards (resp. backwards).

Proof. Let $p(t)$ be the point on edge e that lies in distance $|t|$ behind or before $p = p(0)$, depending on the sign of t. By the law of cosine we have

$$d_C(p(t), q) = \frac{t + |C_p^q|}{\sqrt{t^2 + |pq|^2 - 2t|pq|\cos \beta}}$$

for positive and negative values of t. The derivative with respect to t has a positive denominator. Its numerator is of the same sign as

$$|pq|\frac{|pq| + |C_p^q| \cos \beta}{|pq| \cos \beta + C_p^q} - t$$

because the term $|pq| \cos \beta + C_p^q$ is positive. This implies the claims.   □

We obtain the following important consequence:

Lemma 2. *Any polygonal chain makes its maximum detour on a pair of points at least one of which is a vertex.*

Proof. Let (p, q) be a pair of points on which the chain C attains its maximum detour, and assume that neither of p, q is a vertex of C. By Lemma 1, the line segment pq must form the same angle,

$$\beta = \arccos(-\frac{|pq|}{|C_p^q|}),$$

with the two edges containing p and q. Otherwise, the detour could be increased by moving one of the points. But then the detour $d_C(p, q)$ remains constant as we move both points simultaneously until one of them reaches the endpoint of its edge; see Fig. 2. In fact, we have

$$d_C(p', q') = \frac{|C_p^q| + 2t}{|pq| - 2t \cos \beta}$$

$$= \frac{|C_p^q|}{|pq|} = d_C(p, q).$$

□

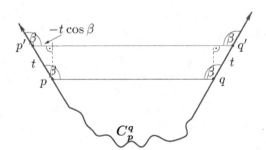

Fig. 2. Chain C attains its maximum detour on both pairs, (p, q) and (p', q').

If we are given a point p and an edge e of the chain we can apply Lemma 1 to determine, in time $O(1)$, the unique point q on e that maximizes the detour

$d_C(p, .)$ on this edge. If we do this for all vertex-edge pairs (p, e) of the chain C, then the maximum value encountered will be the maximum detour of C, by Lemma 2.

This approach yields an $O(n^2)$ algorithm for computing the maximum detour. In the next section we will discuss how to improve on this bound using global properties.

3   Global Properties

The first observation is that we do not need consider pairs (p, q) of points that cannot see each other because of C.

Lemma 3. *The maximum detour of C is attained by a vertex-edge cut (p, q), where p is a vertex, q is a point on some edge, and p, q are co-visible.*[4]

Proof. Let p, q be two arbitrary points of C, and let $p = p_0, p_1, \ldots, p_k = q$ be the points of C intersected by the line segment pq, ordered by their appearance on pq. For each pair (p_i, p_{i+1}) of consecutive points let C_i denote the segment of C that connects them. These segments need not be disjoint, so the sum of their lengths is at least as large as $|C_p^q|$. Hence,

$$
\begin{aligned}
d_C(p, q) &= \frac{|C_p^q|}{|pq|} \\
&\leq \frac{\sum_{i=0}^{k-1} |C_i|}{\sum_{i=0}^{k-1} |p_i p_{i+1}|} \\
&\leq \max_{0 \leq i \leq k-1} \frac{|C_i|}{|p_i p_{i+1}|} \\
&= \max_{0 \leq i \leq k-1} d_C(p_i, p_{i+1}).
\end{aligned}
$$

To prove the last inequality we note that if $a_i/b_i \leq q$ holds for all i, then $\sum_i a_i / \sum_i b_i \leq q$ follows.   ⊔

Hershberger [5] has shown how to compute all co-visible vertex-vertex pairs of a simple polygon in time proportional to their number. Lemma 3 would invite us to generalize this algorithm to the m many co-visible vertex-edge pairs of a chain, and obtain an $O(m)$ algorithm for computing the maximum detour. Unfortunately, m can still be quadratic in n.

An interesting example is the case of a convex chain, C, whose total turning angle is less than π. There are $\Omega(n^2)$ many co-visible pairs of vertices, but one can show that the maximum detour is always attained at an end point of C. Thus, there are only $O(n)$ many vertex-edge candidate pairs to be checked. One

---

[4] Two points, p and q, are called *co-visible* if the line segment connecting them contains no points of the chain C in its interior.

end point of C can attain the maximum detour with a point on each edge of C; such a chain can be constructed by approximating an exponential spiral, which is defined by the fact that all tangents form the same angle with the radii to the spiral's center.

Now we show that even for general chains there is at most a linear number of vertex-edge cuts that can attain the maximum detour. To this end, we prove the following fact illustrated by Fig. 3.

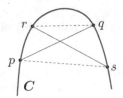 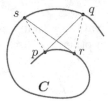

Fig. 3. The minimum of $d_C(p,q)$ and $d_C(r,s)$ is less than the maximum of $d_C(r,q)$ and $d_C(p,s)$, in either case.

Lemma 4. *Let p, r, q, s be consecutive points on C, and assume that (p, q) and (r, s) are two cuts of C that cross each other. Then*

$$\min(d_C(p,q), d_C(r,s)) < \max(d_C(r,q), d_C(p,s)).$$

The same statement holds if the points appear in order p, r, s, q on C.

Proof. Let us assume that $d_C(p,q) \leq d_C(r,s)$ holds, and $d_C(p,q) \geq d_C(r,q)$. We have to show $d_C(p,q) < d_C(p,s)$. By the triangle inequality, we have

$$|ps| + |rq| < |pq| + |rs|,$$

therefore

$$
\begin{aligned}
|C_p^q|(|ps| + |rq|) &< |C_p^q|(|pq| + |rs|) \\
&\leq |C_p^q||pq| + |C_r^s||pq| \\
&= |C_p^s||pq| + |C_r^q||pq| \\
&\leq |C_p^s||pq| + |C_p^q||rq|,
\end{aligned}
$$

and the claim follows.   □

This property is weaker than the Monge property (see Burkard et al. [3])

$$d_C(p,q) + d_C(r,s) \leq d_C(p,s) + d_C(r,q),$$

which is not always fulfilled here. But we can draw the following consequence.

Lemma 5. *Let (p, q) and (r, s) be two vertex-edge cuts that attain the maximum detour, d_C. Then the cuts do not cross. Consequently, there are only $O(n)$ many such cuts altogether.*

Proof. If the line segments pq and rs crossed, then chain C would visit the points p, q, r, s in one of the two ways depicted in Fig. 3, and from Lemma 4 we would obtain a contradiction to the maximality of the cuts. By Euler's formula for planar graphs, there can be only $O(n)$ many non-crossing vertex-edge cuts.[5]   □

The non-crossing property shown in Lemma 4 and Lemma 5 need not be fulfilled for locally optimal cuts. For example, there can be cuts $(p, q), (r, s)$ satisfying

$$d_C(p, q) = \max_{q'} d_C(p, q')$$
$$d_C(r, s) = \max_{s'} d_C(r, s')$$

that cross each other, see Fig. 4.

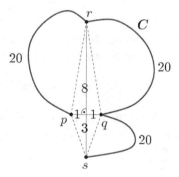

Fig. 4. Locally optimal cuts can cross.

4   An Efficient Algorithm

First, we have to solve two restricted versions of the maximum detour problem.

Lemma 6. *Let ρ be a given angle. A cut attaining the maximum detour among all vertex-edge cuts in direction ρ can be found in time $O(n \log n)$.*

Proof. Let us assume that $\rho = \pi/2$ holds, so that we are interested only in vertical vertex-edge cuts. For each vertex p of the chain C we construct its upper

---

[5] Formally, we can identify the non-vertex endpoints of all cuts hitting the same edge with one extra vertex.

and lower vertical extension, i. e., the vertical line segments that connect p to the first points of C above and below; see Fig. 5. This vertical decomposition was used by Seidel [11] for the purpose of point location. Using a sweep algorithm, we can construct it in time $O(n \log n)$.[6] Once all vertical extensions are available we traverse the chain and mark each endpoint of each vertical segment with the current odometer reading. When a segment is encountered for the second time, we can thus compute its detour, and determine the maximum of all.   □

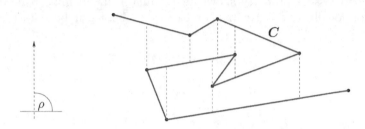

Fig. 5. The vertical extensions of the vertices of C.

Another version of the maximum detour problem results if we restrict ourselves to vertex-vertex pairs. That is, we are interested in the value

$$d_C^V = \max\{d_C(p,q); p, q \text{ vertices of } C\}.$$

One should note that the claim of Lemma 3 does not hold in this case: two vertices attaining maximum detour need not be co-visible. We can prove the following approximation result.

Theorem 1. *Let C be a simple polygonal chain of n edges in the plane, and let $\eta > 0$. In time $O(n \log n)$ we can compute a pair (p, q) of vertices of C satisfying*

$$d_C^V \le (1 + \eta)d_C(p,q).$$

Proof. Let V denote the set of all vertices of C, and let S be a sparse $1 + \eta$-spanner of V. That means, S is a graph of $O(n)$ edges over V, and for any two points p, q of V there exists a path in S whose length is at most $1 + \eta$ times the Euclidean distance $|pq|$.[7]

Now let (p, q) denote a vertex pair for which $d_C^V = d_C(p,q)$ holds, and let $p = p_0, p_1, \ldots, p_k = q$ be the approximating path in S. Moreover, let C_i denote the segment of C that connects vertex p_i to p_{i+1}. Similar to the proof of Lemma 3, we argue as follows:

---

[6] Faster algorithms are known, but not necessary for our purpose.
[7] Path length in S is defined as the sum of the Euclidean distances between consecutive vertices on the path.

$$d_C^V = d_C(p, q) = \frac{|C_p^q|}{|pq|} \leq \frac{\sum_{i=0}^{k-1} |C_i|}{|pq|}$$

$$\leq (1 + \eta) \frac{\sum_{i=0}^{k-1} |C_i|}{\sum_{i=0}^{k-1} |p_i p_{i+1}|}$$

$$\leq (1 + \eta) \max_{0 \leq i \leq k-1} \frac{|C_i|}{|p_i p_{i+1}|}$$

$$= (1 + \eta) \max_{0 \leq i \leq k-1} d_C(p_i, p_{i+1}).$$

A similar result was also obtained by Narasimhan and Smid [9] in a more general setting, approximating the stretch factor of an Euclidean path in time $O(n \log n)$. Theorem 1 suggests the following algorithm.

Algorithm 1
Input: A polygonal chain C on n edges and a real number $\eta > 0$.
Output: A vertex pair of C whose detour is within $1 + \eta$ of the maximum vertex detour, d_C^V, of C.

1. Construct a sparse $(1 + \eta)$-spanner of the vertices of C.
2. For each edge of the spanner, compute its detour.
3. Output an edge of the spanner having the largest detour.

By a result of Gutwin and Keil [8], step (1) can be carried out in time $O(n \log n)$, see also the spanner survey of Eppstein [4] for alternate methods. Because of the spanner's sparseness, step (2) takes only linear time; it can be implemented by traversing C, as in the proof of Lemma 6. Hence, the claim follows.   □

Now we can prove our main result.

Theorem 2. *Let C be a simple polygonal chain of n edges in the plane, and let $\epsilon > 0$. In time $O(n \log n)$ we can compute a vertex-edge cut (p, q) of C that approximates the maximum detour of C, i. e., such that*

$$d_C \leq (1 + \epsilon) d_C(p, q).$$

Proof. Let $\eta < \epsilon$ be small enough to satisfy

$$\frac{1 - \eta}{1 + \eta} \geq \frac{1}{1 + \epsilon},$$

choose an angle $\overline{\beta}$ in $(\pi/2, \pi)$ so large that for all angles $\beta \in [\overline{\beta}, \pi)$

$$- \cos \beta \geq \frac{1}{1 + \epsilon}$$

holds, and let finally the angle ρ so small that it satisfies

$$\frac{\sin \rho}{\sin \overline{\beta}} \leq \eta.$$

We run the following algorithm.

Algorithm 2

Input: A polygonal chain C on n edges and a real number $\epsilon > 0$.
Output: A vertex-edge cut of C whose detour is within $1 + \epsilon$ of the maximum detour, d_C, of C.

1. Let ρ be defined as above;
   for each integer m between 0 and $2\pi/\rho$, and for each vertex p,
   compute the first point of C hit by a ray from p in direction $m\rho$;
   move this hit point along its edge such as to maximize the detour locally.
2. Let (p_1, q_1) be the maximum detour cut thus obtained.
3. Let η be as defined above;
   compute a pair (p_2, q_2) of vertices satisfying $d_C^V \leq (1+\eta)d_C(p_2, q_2)$.
4. Compute the maximum of $d_C(p_1, q_1)$ and $d_C(p_2, q_2)$, and output the corresponding pair of points.

First, we address the correctness of Algorithm 2. Lemma 2 ensures that d_C is attained by some cut (p, q), where p is a vertex of C. If q is a vertex, too, then

$$d_C = d_C^V \leq (1 + \eta)d_C(p_2, q_2) \leq (1 + \epsilon)d_C(p_2, q_2)$$

holds for the vertex pair (p_2, q_2) computed in (3), and we are done. Let us assume that q is an interior point of some edge, e. By Lemma 1, the outgoing part of e forms an angle $\beta > \pi/2$ with the line segment pq, where

$$-\cos \beta = \frac{|pq|}{|C_p^q|}.$$

If $\beta \geq \overline{\beta}$ then, by definition of $\overline{\beta}$, $d_C = -1/\cos \beta \leq 1 + \epsilon$ holds, and our output is certainly correct. So, let us assume that $\beta \in (\pi/2, \overline{\beta})$. We distinguish two cases illustrated by Fig. 6.

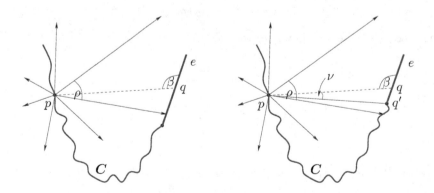

Fig. 6. Two cases in the proof of Theorem 2.

In case (i), there exists an integer m such that the ray emanating from vertex p in direction $m\rho$ hits the edge e that contains point q. This ray will be encountered in step (1), and by local maximization, point q on e will be discovered. Hence, the cut (p_1, q_1) computed in step (2) attains maximum detour, d_C.

In case (ii), all rays emanating from vertex p are missing edge e, i. e., this edge is fully contained in a wedge of angle ρ. Let q' be the endpoint of e first reached from p, and let $\nu < \rho$ be the angle between pq and pq'. By the law of sines we have

$$\frac{|pq'|}{\sin \beta} = \frac{|qq'|}{\sin \nu} = \frac{|pq|}{\sin(\beta - \nu)},$$

hence

$$
\begin{aligned}
d_C(p, q') &= \frac{|C_p^q| - |qq'|}{|pq'|} \\
&= \frac{\sin(\beta - \nu)}{\sin \beta} \frac{|C_p^q|}{|pq|} - \frac{\sin \nu}{\sin \beta} \\
&> d_C - \frac{\sin \rho}{\sin \overline{\beta}} \\
&\geq d_C - \eta \\
&\geq (1 - \eta)d_C,
\end{aligned}
$$

because of $\sin(\beta - \nu) > \sin \beta$, $\sin \nu < \sin \rho$, and $\sin \beta > \sin \overline{\beta}$. Since both p and q' are vertices of C, we obtain for the vertex pair (p_2, q_2) computed in step (3)

$$
\begin{aligned}
d_C(p_2, q_2) &\geq \frac{1}{1 + \eta} d_C^V \\
&\geq \frac{1}{1 + \eta} d_C(p, q') \\
&\geq \frac{1 - \eta}{1 + \eta} d_C \\
&\geq \frac{1}{1 + \epsilon} d_C.
\end{aligned}
$$

It remains to account for the running time of Algorithm 2. For each fixed direction $m\rho$, step (1) can be implemented to run in time $O(n \log n)$, by combining Lemma 6 and Lemma 1. The number of directions to be dealt with is a constant dependent only on ϵ. Step (3) runs in time $O(n \log n)$, by Theorem 1. This completes the proof.[8]   □

---

[8] Taking the result of Gutwin and Keil [8] into account, the overall dependency on $1/\epsilon$ is not worse than quadratic for small ϵ.

5   Conclusions

We have presented the first $O(n \log n)$ algorithm for approximating the maximum detour of a planar polygonal chain over n edges. This result gives rise to a number of interesting questions. Is the true complexity of this problem less than quadratic? How fast can we compute the maximum detour attained by a pair of co-visible vertices? How can smooth curves be handled? And finally, coming back to the evaluation of transportation routes, if a certain amount of money is available for building shortcuts of total length at most c, how far can the maximum detour be reduced?

Acknowledgements. We would like to thank Helmut Alt for pointing out the detour problem, and Pankaj Agarwal, Christian Knauer, Micha Sharir and Günter Rote for interesting discussions.

Additionally we would like to thank the anonymous referees for their valuable remarks and comments.

References

1. O. Aichholzer, F. Aurenhammer, C. Icking, R. Klein, E. Langetepe, and G. Rote. Generalized self-approaching curves. *Discrete Appl. Math.*, 109:3–24, 2001.
2. H. Alt, C. Knauer, and C. Wenk. Bounding the Fréchet distance by the Hausdorff distance. In *Abstracts 17th European Workshop Comput. Geom.*, pages 166–169. Freie Universität Berlin, 2001.
3. R. E. Burkard, B. Klinz, and R. Rudolf. Perspectives of Monge properties in optimization. *Discrete Appl. Math.*, 70:95–161, 1996.
4. D. Eppstein. Spanning trees and spanners. In J.-R. Sack and J. Urrutia, editors, *Handbook of Computational Geometry*, pages 425–461. Elsevier Science Publishers B.V. North-Holland, Amsterdam, 2000.
5. J. Hershberger. An optimal visibility graph algorithm for triangulated simple polygons. *Algorithmica*, 4:141–155, 1989.
6. C. Icking and R. Klein. Searching for the kernel of a polygon: A competitive strategy. In *Proc. 11th Annu. ACM Sympos. Comput. Geom.*, pages 258–266, 1995.
7. C. Icking, R. Klein, and E. Langetepe. Self-approaching curves. *Math. Proc. Camb. Phil. Soc.*, 125:441–453, 1999.
8. J. M. Keil and C. A. Gutwin. Classes of graphs which approximate the complete Euclidean graph. *Discrete Comput. Geom.*, 7:13–28, 1992.
9. G. Narasimhan and M. Smid. Approximating the stretch factor of Euclidean graphs. *SIAM J. Comput.*, 30:978–989, 2000.
10. G. Rote. Curves with increasing chords. *Math. Proc. Camb. Phil. Soc.*, 115:1–12, 1994.
11. R. Seidel. A simple and fast incremental randomized algorithm for computing trapezoidal decompositions and for triangulating polygons. *Comput. Geom. Theory Appl.*, 1(1):51–64, 1991.

An Approximation Algorithm for Minimum Convex Cover with Logarithmic Performance Guarantee

Stephan Eidenbenz and Peter Widmayer

Institut für Theoretische Informatik, ETH Zentrum, CLW C 2, Clausiusstrasse 49, 8092 Zürich, Switzerland. {eidenben,widmayer}@inf.ethz.ch

Abstract. The problem Minimum Convex Cover of covering a given polygon with a minimum number of (possibly overlapping) convex polygons is known to be NP-hard, even for polygons without holes [3]. We propose a polynomial-time approximation algorithm for this problem for polygons with or without holes that achieves an approximation ratio of $O(\log n)$, where n is the number of vertices in the input polygon. To obtain this result, we first show that an optimum solution of a restricted version of this problem, where the vertices of the convex polygons may only lie on a certain grid, contains at most three times as many convex polygons as the optimum solution of the unrestricted problem. As a second step, we use dynamic programming to obtain a convex polygon which is maximum with respect to the number of "basic triangles" that are not yet covered by another convex polygon. We obtain a solution that is at most a logarithmic factor off the optimum by iteratively applying our dynamic programming algorithm. Furthermore, we show that Minimum Convex Cover is APX-hard, i.e., there exists a constant $\delta > 0$ such that no polynomial-time algorithm can achieve an approximation ratio of $1 + \delta$. We obtain this result by analyzing and slightly modifying an already existing reduction [3].

1 Introduction and Problem Definition

The problem Minimum Convex Cover is the problem of covering a given polygon T with a minimum number of (possibly overlapping) convex polygons that lie in T. This problem belongs to the family of classic art gallery problems; it is known to be NP-hard for input polygons with holes [14] and without holes [3]. The study of approximations for hard art gallery problems has rarely led to good algorithms or good lower bounds; we discuss a few exceptions below. In this paper, we propose the first non-trivial approximation algorithm for Minimum Convex Cover. Our algorithm works for both, polygons with and without holes. It relies on a strong relationship between the continuous, original problem version and a particular discrete version in which all relevant points are restricted to lie on a kind of grid that we call a quasi-grid. The quasi-grid is the set of intersection points of all lines connecting two vertices of the input polygon. Now, in the Restricted Minimum Convex Cover problem, the vertices of

F. Meyer auf der Heide (Ed.): ESA 2001, LNCS 2161, pp. 333–344, 2001.
© Springer-Verlag Berlin Heidelberg 2001

the convex polygons that cover the input polygon may only lie on the quasi-grid. We prove that an optimum solution of the RESTRICTED MINIMUM CONVEX COVER problem needs at most three times the number of convex polygons that the MINIMUM CONVEX COVER solution needs. To find an optimum solution for the RESTRICTED MINIMUM CONVEX COVER problem, we propose a greedy approach: We compute one convex polygon of the solution after the other, and we pick as the next convex polygon one that covers a maximum number of triangles defined on an even finer quasi-grid, where these triangles are not yet covered by previously chosen convex polygons. We propose an algorithm for finding such a maximum convex polygon by means of dynamic programming. To obtain an upper bound on the quality of the solution, we interpret our covering problem on triangles as a special case of the general MINIMUM SET COVER problem that gives as input a base set of elements and a collection of subsets of the base set, and that asks for a smallest number of subsets in the collection whose union contains all elements of the base set. In our special case, each triangle is an element, and each possible convex polygon is a possible subset in the collection, but not all of these subsets are represented explicitly (there could be an exponential number of subsets). This construction translates the logarithmic quality of the approximation from MINIMUM SET COVER to MINIMUM CONVEX COVER [10].

On the negative side, we show that MINIMUM CONVEX COVER is APX-hard, i.e., there exists a constant $\delta > 0$ such that no polynomial-time algorithm can achieve an approximation ratio of $1 + \delta$. This inapproximability result is based on a known problem transformation [3]; we modify this transformation slightly and show that it is gap-preserving (as defined in [1]).

As for previous work, the related problem of partitioning a given polygon into a minimum number of non-overlapping convex polygons is polynomially solvable for input polygons without holes [2]; it is NP-hard for input polygons with holes [12], even if the convex partition must be created by cuts from a given family of (at least three) directions [13]. Other related results for art gallery problems include approximation algorithms with logarithmic approximation ratios for MINIMUM VERTEX GUARD and MINIMUM EDGE GUARD [8], as well as for the problem of covering a polygon with rectangles in any orientation [9]. Furthermore, logarithmic inapproximability results are known for MINIMUM POINT/VERTEX/EDGE GUARD for polygons with holes [5], and APX-hardness results for the same problems for polygons without holes [6]. The related problem RECTANGLE COVER of covering a given orthogonal polygon with a minimum number of rectangles can be approximated with a constant ratio for polygons without holes [7] and with an approximation ratio of $O(\sqrt{\log n})$ for polygons with holes [11]. For additional results see the surveys on art galleries [15,16]. The general idea of using dynamic programming to find maximum convex structures has been used before to solve the problem of finding a maximum (with respect to the number of vertices) empty convex polygon, given a set of vertices in the plane [4], and for the problem of covering a polygon with rectangles in any orientation [9].

This paper is organized as follows: In Sect. 2, we define the quasi-grid and its refinement into triangles. Section 3 contains the proof of the linear relationship between the sizes of the optimum solutions of the unrestricted and restricted

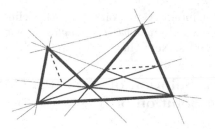

Fig. 1. Construction of first-order basic triangles

convex cover problems. We propose a dynamic programming algorithm to find a maximum convex polygon in Sect. 4, before showing how to iteratively apply this algorithm to find a convex cover in Sect. 5. In Sect. 6, we present the outline of our proof of the APX-hardness of MINIMUM CONVEX COVER. A few concluding remarks can be found in Sect. 7.

2   From the Continuous to the Discrete

Consider simple input polygons with and without holes, where a polygon T is given as an ordered list of vertices in the plane. If T contains holes, each hole is also given as an ordered list of vertices. Let V_T denote the set of vertices (including the vertices of holes, if any) of a given polygon T. While, in the general MINIMUM CONVEX COVER problem, the vertices of the convex polygons that cover the input polygon can be positioned anywhere in the interior or on the boundary of the input polygon, we restrict their positions in an intermediate step: They may only be positioned on a quasi-grid in the RESTRICTED MINIMUM CONVEX COVER problem.

In order to define the RESTRICTED MINIMUM CONVEX COVER problem more precisely, we partition the interior of a polygon T into *convex components* (as proposed in [8] for a different purpose) by drawing a line through each pair of vertices of T. We then triangulate each convex component arbitrarily. We call the triangles thus obtained *first-order basic triangles*. Figure 1 shows in an example the first-order basic triangles of a polygon (thick solid lines) with an arbitrary triangulation (fine solid lines and dashed lines). If a polygon T consists of n vertices, drawing a line through each pair of vertices of T will yield less than $\binom{n}{2} \cdot \binom{n}{2} \in O(n^4)$ intersection points. Let V_T^1 be the set of these intersection points that lie in T (in the interior or on the boundary). Note that $V_T \subseteq V_T^1$. The first-order basic triangles are a triangulation of V_T^1 inside T, therefore the number of first-order basic triangles is also $O(n^4)$. The RESTRICTED MINIMUM CONVEX COVER problem asks for a minimum number of convex polygons, with vertices restricted to V_T^1, that together cover the input polygon T. We call V_T^1 a quasi-grid that is imposed on T. For solving the RESTRICTED MINIMUM CONVEX COVER problem, we make use of a finer quasi-grid: Simply partition T by drawing lines through each pair of points from V_T^1. This yields again convex components, and we triangulate them again arbitrarily. This higher resolution partition yields $O(n^{16})$ intersection points, which define the set V_T^2. We call the resulting triangles *second-order basic triangles*. Obviously, there are $O(n^{16})$ second-order basic triangles. Note that $V_T \subseteq V_T^1 \subseteq V_T^2$.

3   The Optimum Solution of MINIMUM CONVEX COVER vs. the Optimum Solution of RESTRICTED MINIMUM CONVEX COVER

The quasi-grids V_T^1 and V_T^2 serve the purpose of making a convex cover computationally efficient while at the same time guaranteeing that the cover on the discrete quasi-grid is not much worse than the desired cover in continuous space. The following theorem proves the latter.

Theorem 1. *Let T be an arbitrary simple input polygon with n vertices. Let OPT denote the size of an optimum solution of* MINIMUM CONVEX COVER *with input polygon T and let OPT' denote the size of an optimum solution of* RESTRICTED MINIMUM CONVEX COVER *with input polygon T. Then:*

$$OPT' \leq 3 \cdot OPT$$

Proof. We proceed as follows: We show how to *expand* a given, arbitrary convex polygon C to another convex polygon C' with $C \subseteq C'$ by iteratively expanding edges. We then replace the vertices in C' by vertices from V_T^1, which results in a (possibly) non-convex polygon C'' with $C' \subseteq C''$. Finally, we describe how to obtain three convex polygons C_1'', C_2'', C_3'' with $C'' = C_1'' \cup C_2'' \cup C_3''$ that only contain vertices from V_T^1. This will complete the proof, since each convex polygon from an optimum solution of MINIMUM CONVEX COVER can be replaced by at most 3 convex polygons that are in a solution of RESTRICTED MINIMUM CONVEX COVER. Following this outline, let us present the proof details.

Let C be an arbitrary convex polygon inside polygon T. Let the vertices of C be given in in clockwise order. We obtain a series of convex polygons $C^1, C^2, \ldots, C'$ with $C = C^0 \subseteq C^1 \subseteq C^2 \subseteq \ldots \subseteq C'$, where C^{i+1} is obtained from C^i as follows (see Fig. 2):

Let a, b, c, d be consecutive vertices (in clockwise order) in the convex polygon C^i that lies inside polygon T. Let vertices $b, c \notin V_T$, with b and c not on the same edge of T. Then, the edge (b, c) is called *expandable*. If there exists no expandable edge in C^i, then $C' = C^i$, which means we have found the end of the series of convex polygons. If (b, c) is an expandable edge, we *expand* the edge from vertex b to vertex c as follows:

– If b does not lie on the boundary of T, then we let a point p start on b and move on the halfline through a and b away from b until either one of two events happens: p lies on the line through c and d, or the triangle p, c, b touches the boundary of T. Fix p as soon as the first of these events happens. Figure 2 shows a list of all possible cases, where the edges from polygon T are drawn as thick edges: Point p either lies on the intersection point of the lines from a through b and from c through d as in case (a), or there is a vertex v_l on the line segment from p to c as in case (b), or p lies on an edge of T as in case (c).

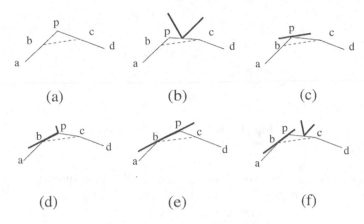

Fig. 2. Expansion of edge (b,c)

- If b lies on the boundary of T, i.e. on some edge of T, say from v_k to v_{k+1}, then let p move as before, except that the direction of the move is now on the way from v_k through b up until v_{k+1} at most (instead of the ray from a through b).

  Figure 2 shows a list of all possible cases: Point p either lies at vertex v_{k+1} as in case (d), or on the intersection point of the lines from b to v_{k+1} and from d through c as in case (e), or there is a vertex v_l on the line segment from p to c as in case (f).

A new convex polygon C_p^i is obtained by simply adding point p as a vertex in the ordered set of vertices of C^i between the two vertices b and c. Furthermore, eliminate all vertices in C_p^i that have collinear neighbors and that are not vertices in V_T.

Note that an edge from two consecutive vertices b and c with $b, c \notin V_T$ can always be expanded in such a way that the triangle b, p, c that is added to the convex polygon is non-degenerate, i.e., has non-zero area, unless b and c both lie on the same edge of polygon T. This follows from the cases (a) - (f) of Fig. 2.

Now, let $C^{i+1} = C_p^i$, if either a new vertex of T has been added to C_p^i in the expansion of the edge, which is true in cases (b), (d), and (f), or the number of vertices of C^i that are not vertices of T has decreased, which is true in case (a). If p is as in case (c), we expand the edge (p, c), which will result in either case (d), (e), or (f). Note that in cases (d) and (f), we have found C^{i+1}. If p is as in case (e), we expand the edge (p, d), which will result in either case (d), (e), or (f). If it is case (e) again, we repeat the procedure by expanding the edge from p and the successor (clockwise) of d. This needs to be done at most $|C^i|$ times, since the procedure will definitely stop once it gets to vertex a. Therefore, we obtain C^{i+1} from C^i in a finite number of steps. Let τ_i denote the number of vertices in C^i that are also vertices in T and let $\hat{\tau}_i$ be the number of vertices in C^i that are not vertices in T. Now note that $\phi(i) = \hat{\tau}_i - 2\tau_i + 2n$ is a function that bounds the number of remaining steps, i.e., it strictly decreases with every increase in

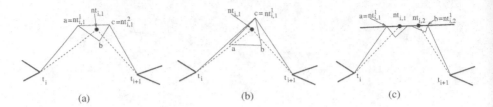

Fig. 3. Replacing non-T-vertices

i and cannot become negative. The existence of this bounding function implies the finiteness of the series $C^1, C^2, \ldots, C'$ of convex polygons.

By definition, there are no expandable edges left in C'. Call a vertex of C' a T-vertex, if it is a vertex in T. From the definition of expandable edges, it is clear that there can be at most two non-T-vertices between any two consecutive T-vertices in C', and if there are two non-T-vertices between two consecutive T-vertices, they must both lie on the same edge in T. Let the T-vertices in C' be $t_1, \ldots, t_l$ in clockwise order, and let the non-T-vertices between t_i and t_{i+1} be $nt_{i,1}$ and $nt_{i,2}$ if they exist. We now replace each non-T-vertex $nt_{i,j}$ in C' by one or two vertices $nt^1_{i,j}$ and $nt^2_{i,j}$ that are both elements of V^1_T. This will transform the convex polygon C' into a non-convex polygon C'' (we will show later how C'' can be covered by at most three convex polygons C''_1, C''_2, C''_3).

To this end, let a, b, c be the first-order basic triangle in which non-T-vertex $nt_{i,j}$ lies, as illustrated in Fig. 3. Points a, b, c are all visible from both vertices t_i and t_{i+1}. To see this, assume by contradiction that the view from, say, t_i to a is blocked by an edge e of T. Since $nt_{i,j}$ must see t_i, the edge e must contain a vertex e' in the triangle $t_i, a, nt_{i,j}$, but then a cannot be a vertex of the first-order basic triangle in which $nt_{i,j}$ lies, since the line from vertex t_i through vertex e' would cut through the first-order basic triangle, an impossibility. Now, let d_i be the intersection point of the line from t_{i-1} through t_i and the line from t_{i+1} through t_{i+2}. With similar arguments, the triangle t_i, d_i, t_{i+1} completely contains triangle a, b, c.

Assume that only one non-T-vertex $nt_{i,1}$ exists between t_i and t_{i+1}. If the triangle formed by t_i, t_{i+1} and a completely contains the triangle $t_i, nt_{i,1}, t_{i+1}$, we let $nt^1_{i,1} = a$, likewise for b and c (see Fig. 3 (b)). Otherwise, we let $(nt^1_{i,1}, nt^2_{i,1})$ be $(a, b), (a, c)$, or (b, c) as in Fig. 3 (a), such that the polygon $t_i, nt^1_{i,1}, nt^2_{i,1}, t_{i+1}$ is convex and completely contains the triangle $t_i, nt_{i,1}, t_{i+1}$. This is always possible by the definition of points a, b, c.

Now, assume that two non-T-vertices $nt_{i,1}$ and $nt_{i,2}$ exist between t_i and t_{i+1}. From the definition of C', we know that $nt_{i,1}$ and $nt_{i,2}$ must lie on the same edge e of T. Therefore, the basic triangle in which $nt_{i,1}$ lies must contain a vertex a either at $nt_{i,1}$ or preceeding $nt_{i,1}$ on edge e along T in clockwise order.

Let $nt^1_{i,1} = a$. The basic triangle in which $nt_{i,2}$ lies must contain a vertex b either at $nt_{i,2}$ or succeeding $nt_{i,2}$ on edge e. Let $nt^1_{i,2} = b$. See Fig. 3 (c). Note that the convex polygon $t_i, nt^1_{i,1}, nt^1_{i,2}, t_{i+1}$ completely contains the polygon $t_i, nt_{i,1}, nt_{i,2}, t_{i+1}$.

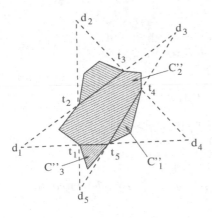

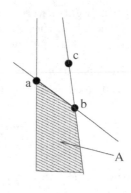

Fig. 4. Covering C'' with three convex polygons

Fig. 5. Dynamic Programming

After applying this change to all non-T-vertices in C', we obtain a (possibly) non-convex polygon C''. First, assume that C'' contains an odd number f of T-vertices. We let C_1'' be the polygon defined by vertices $t_i, nt_{i,j}^k$ and t_{i+1} for all j, k and for all odd i, but $i \neq f$. By construction, C_1'' is convex. Let C_2'' be the polygon defined by vertices $t_i, nt_{i,j}^k$ and t_{i+1} for all j, k and for all even i. Finally, let C_3'' be the polygon defined by vertices $t_f, nt_{f,j}^k$ and t_1 for all j, k. Figure 4 shows an example. Obviously, C_1'', C_2'', and C_3'' are convex and together cover all of C''. Second, assume that C'' contains an even number of T-vertices, and cover it with only two convex polygons using the same concept. This completes the proof.

4   Finding Maximum Convex Polygons

Assume that each second-order basic triangle from a polygon T is assigned a weight value of either 1 or 0. In this section, we present an algorithm using dynamic programming that computes the convex polygon M in a polygon T that contains a maximum number of second-order basic triangles with weight 1 and that only has vertices from V_T^1. For simplicity, we call such a polygon a *maximum convex polygon*. The weight of a polygon M is defined as the sum of the weights of the second-order basic triangles in the polygon and is denoted by $|M|$. We will later use the algorithm described below to iteratively compute a maximum convex polygon with respect to the triangles that are not yet covered, to eventually obtain a convex cover for T.

Let $a, b, c \in V_T^1$. Let $P_{a,b,c}$ denote the maximum convex polygon that:

- contains only vertices from V_T^1, and
- contains vertices a, b, c in counterclockwise order, and
- has a as its left-most vertex, and
- contains additional vertices only between vertices a and b.

Given three vertices $a, b, c \in V_T^1$, let A be the (possibly infinite) area of points that are:

- to the right of vertex a, and
- to the left of the line oriented from b through a, and
- to the left of the line oriented from b through c.

For an illustration, see Fig. 5. Let $P'_{a,b,c} = \max_{d \in V_T^1 \cap A} P_{a,d,b} \cup \Delta a, b, c$, where max is defined as follows (to simplify notation):

$$\max\{P_1, P_2\} = \begin{cases} P_1 \text{ if } |P_1| \geq |P_2| \\ P_2 \text{ otherwise} \end{cases}.$$

Lemma 1. $P_{a,b,c} = P'_{a,b,c}$, if the triangle a, b, c is completely contained in the polygon T.

Proof. Consider $P_{a,b,c}$, which is maximum by definition. $P_{a,b,c}$ must contain additional vertices between a and b (otherwise the lemma is trivially true). Let d' be the predecessor of b in the counterclockwise order of $P_{a,b,c}$. Vertex d' must lie in area A as defined above, otherwise the polygon a, d', b, c would either be non-convex, not have a as its left-most vertex, or not be in the required counterclockwise order. Now consider $P'' = P_{a,b,c} - \Delta a, b, c$. From the definition of area A it is clear the P'' can only contain vertices that lie in A. Now $P_{a,d',b}$ is maximum by definition, and it is considered when computing $P'_{a,b,c}$.

Let M be a maximum convex polygon for a polygon T with weights assigned to the second-order basic triangles. Let a be the left-most vertex of M, let c be the predecessor of a in M in counter clockwise order, and let b be the predecessor of c. Then $|P_{a,b,c}| = |M|$ by definition.

We will now use Lemma 1 to construct an algorithm, which takes as input a polygon T and an assignment of weight 0 or 1 to each second-order basic triangle of T and computes the maximum convex polygon. To this end, we fix vertex $a \in V_T^1$. Let a' be a point with the same x-coordinate and smaller y-coordinate than a. Now, order all other vertices $b \in V_T^1$ to the right of a according to the angle formed by b, a, a'. Let the resulting ordered set be B and let B' be the empty set. Take the smallest element b from B, remove it from B and add it to set B', then for all $c \in V_T^1 \backslash B'$ and to the right of a, compute weight $|\Delta a, b, c|$ of the triangle a, b, c and compute $P_{a,b,c}$ according to Lemma 1. Compute weight $|P_{a,b,c}|$ by adding $|\Delta a, b, c|$ to $|P_{a,d,b}|$, where d is the maximizing argument. Note that the computation of $P_{a,b,c}$ according to Lemma 1 is always possible, since all possible vertices d in $P_{a,d,b}$ lie to the left of the line from b to a (see also definition of area A) and have therefore smaller angles d, a, a' than b, a, a', and have therefore already been computed. The algorithm is executed for every $a \in V_T^1$, and the maximum convex polygon found is returned.

Note that $|T| = n$, $|V_T^1| = O(n^4)$, and $|V_T^2| = O(n^{16})$. Ordering $O(n^4)$ vertices takes $O(n^4 \log n)$ time. Computing the weight of a triangle takes $O(n^{16})$ time. Computing $P_{a,b,c}$ takes $O(n^4)$ time. We have to compute the weight of $O(n^8)$ triangles, which gives a total time of $O(n^{24})$. Finally, we have to execute our algorithm for each $a \in V_T^1$, which gives a total running time of $O(n^{28})$. Space requirements are $O(n^{12})$ by using pointers.

5   An Approximation Algorithm for MINIMUM CONVEX COVER

Given a polygon T, we obtain a convex cover by iteratively applying the algorithm for computing a maximum convex polygon from Sect. 4. It works as follows for an input polygon T.

1. Let all second-order basic triangles have weight 1. Let $S = \emptyset$.
2. Find the maximum convex polygon M of polygon T using the algorithm from Sect. 4, and add M to the solution S. Decrease the weight of all second-order basic triangles that are contained in M to 0.[1]
3. Repeat step 2 until there are no second-order basic units with weight 1 left. Return S.

To obtain a performance guarantee for this algorithm, consider the MINIMUM SET COVER instance I, which has all second-order basic triangles as elements and where the second-order basic triangles with weight 1 of each convex polygon in T, which only contains vertices from V_T^1, form a set in I. The greedy heuristic for MINIMUM SET COVER achieves an approximation ratio of $1 + \ln n'$, where n' is the number of elements in I [10] and it works in exactly the same way as our algorithm. However, we do not have to (and could not afford to) compute all the sets of the MINIMUM SET COVER instance I (which would be a number exponential in n'): It suffices to always compute a set, which contains a maximum number of elements not yet covered by the solution thus far. This is achieved by reducing the weights of the second-order basic triangles already in the solution to 0; i.e. a convex polygon with maximum weight is such a set.

Note that $n' = O(n^{16})$. Therefore, our algorithm achieves an approximation ratio of $O(\log n)$ for RESTRICTED MINIMUM CONVEX COVER on input polygon T. Because of Theorem 1, we know that the solution found for RESTRICTED MINIMUM CONVEX COVER is also a solution for the unrestricted MINIMUM CONVEX COVER that is at most a factor of $O(\log n)$ off the optimum solution.

As for the running time of this algorithm, observe that the algorithm adds to the solution in each round a convex polygon with non-zero weight. Therefore, there can be at most $O(n^{16})$ rounds, which yields a total running time of $O(n^{44})$. This completes the proof of our main theorem:

Theorem 2. MINIMUM CONVEX COVER *for input polygons with or without holes can be approximated by a polynomial time algorithm with an approximation ratio of $O(\log n)$, where n is the number of polygon vertices.*

6   *APX*-Hardness of MINIMUM CONVEX COVER

The upper bound of $O(\log n)$ on the approximation ratio for MINIMUM CONVEX COVER is not tight: We will now prove that there is a constant lower bound on the approximation ratio, and hence a gap remains. More precisely, we prove

---

[1] Note that by the definition of second-order basic triangles, a second-order basic triangle is either completely contained in M or completely outside M.

MINIMUM CONVEX COVER to be APX-hard. Our proof of the APX-hardness of MINIMUM CONVEX COVER for input polygons with or without holes uses the construction that is used to prove the NP-hardness of this problem for input polygons without holes[2] [3]. However, we reduce the problem MAXIMUM 5-OCCURRENCE-3-SAT rather than SAT (as done in the original reduction [3]) to MINIMUM CONVEX COVER, and we design the reduction to be gap-preserving [1]. MAXIMUM 5-OCCURRENCE-3-SAT is the variant of SAT, where each variable may appear at most 5 times in clauses and each clause contains at most 3 literals. MAXIMUM 5-OCCURRENCE-3-SAT is APX-complete [1].

Theorem 3. *Let I be an instance of* MAXIMUM 5-OCCURRENCE-3-SAT *consisting of n variables, m clauses with a total of l literals, and let I' be the corresponding instance of* MINIMUM CONVEX COVER. *Let OPT be the maximum number of satisfied clauses of I by any assignment of the variables. Let OPT' be the minimum number of convex polygons needed to cover the polygon of I'. Then:*

$$OPT = m \Longrightarrow OPT' = 5l + n + 1$$
$$OPT \leq (1 - 15\epsilon)m \Longrightarrow OPT' \geq 5l + n + 1 + \epsilon n$$

Proof. Theorem 3 is proved by showing how to transform the convex polygons of a solution of the MINIMUM CONVEX COVER I' in such a way that their total number does not increase and in such a way that a truth assignment of the variables can be "inferred" from the convex polygons that satisfies the desired number of clauses. The proof employs concepts similar to those used in [6]; we do not include details, due to space limitation.

In the promise problem of MAXIMUM 5-OCCURRENCE-3-SAT as described above, we are promised that either all clauses are satisfiable or at most a fraction of $1 - 15\epsilon$ of the clauses is satisfiable, and we are to find out, which of the two possibilities is true. This problem is NP-hard for small enough values of $\epsilon > 0$ [1]. Therefore, Theorem 3 implies that the promise problem for MINIMUM CONVEX COVER, where we are promised that the minimum solution contains either $5l + n + 1$ convex polygons or $5l + n + 1 + \epsilon n$ convex polygons, is NP-hard as well, for small enough values of $\epsilon > 0$. Therefore, MINIMUM CONVEX COVER cannot be approximated with a ratio of: $\frac{5l+n+1+\epsilon n}{5l+n+1} \geq 1 + \frac{\epsilon n}{25n+n+1} \geq 1 + \frac{\epsilon}{27}$, where we have used that $l \leq 5n$ and $n \geq 1$. This establishes the following:

Theorem 4. MINIMUM CONVEX COVER *on input polygons with or without holes is APX-hard.*

7   Conclusion

We have proposed a polynomial time approximation algorithm for MINIMUM CONVEX COVER that achieves an approximation ratio that is logarithmic in the

---

[2] APX-hardness for MINIMUM CONVEX COVER for input polygons without holes implies APX-hardness for the same problem for input polygons with holes.

number of vertices of the input polygon. This has been achieved by showing that there is a discretized version of the problem using no more than three times the number of cover polygons. The discretization may shed some light on the long-standing open question of whether the decision version of the MINIMUM CONVEX COVER problem is in NP [15]. We know now that convex polygons of optimum solutions only contain a polynomial number of vertices and that a considerable fraction of these vertices are actually vertices from the input polygon. Apart from the discretization, our algorithm applies a MINIMUM SET COVER approximation algorithm to a MINIMUM SET COVER instance with an exponential number of sets that are represented only implicitly, through the geometry. We propose an algorithm that picks a best of the implicitly represented sets with a dynamic programming approach, and hence runs in polynomial time. This technique may prove to be of interest for other problems as well. Moreover, by showing APX-hardness, we have eliminated the possibility of the existence of a polynomial-time approximation scheme for this problem. However, polynomial time algorithms could still achieve constant approximation ratios. Whether our algorithm is the best asymptotically possible, is therefore an open problem. Furthermore, our algorithm has a rather excessive running time of $O(n^{44})$, and it is by no means clear whether this can be improved substantially.

Acknowledgement. We want to thank anonymous referees for pointing us to [9,13] and for additional suggestions.

References

1. S. Arora and C. Lund; *Hardness of Approximations*; in: Approximation Algorithms for NP-Hard Problems (ed. Dorit Hochbaum), PWS Publishing Company, pp. 399-446, 1996.
2. B. Chazelle, D.P. Dobkin; *Optimal Convex Decompositions*; Computational Geometry, pp. 63-133, Elsevier Science B. V., 1985.
3. J. C. Culberson and R. A. Reckhow; *Covering Polygons is Hard*; Journal of Algorithms, Vol. 17, No. 1, pp. 2 – 44, July 1994.
4. D. P. Dobkin, H. Edelsbrunner, and M. H. Overmars; *Searching for Empty Convex Polygons*; Algorithmica, 5, pp. 561 – 571, 1990.
5. S. Eidenbenz, C. Stamm, and P. Widmayer; *Inapproximability of some Art Gallery Problems*; Proc. 10th Canadian Conf. Computational Geometry, pp. 64-65, 1998.
6. S. Eidenbenz; *Inapproximability Results for Guarding Polygons without Holes*; Lecture Notes in Computer Science, Vol. 1533 (ISAAC'98), p. 427-436, 1998.
7. D.S. Franzblau; *Performance Guarantees on a Sweep Line Heuristic for Covering Rectilinear Polygons with Rectangles*; SIAM J. Discrete Math., Vol 2,3, pp. 307 – 321, 1989.
8. S. Ghosh; *Approximation Algorithms for Art Gallery Problems*; Proc. of the Canadian Information Processing Society Congress, 1987.
9. J. Gudmundsson, C. Levcopoulos; *Close Approximations of Minimum Rectangular Coverings*; J. Comb. Optimization, Vol. 4, No. 4, pp. 437 – 452, 1999.
10. D. Hochbaum; *Approximating Covering and Packing Problems: Set Cover, Vertex Cover, Independent Set, and Related Problems*; in: Approximation Algorithms for NP-Hard Problems (ed. Dorit Hochbaum), PWS Publishing Company, pp. 94-143, 1996.

11. V.S.A. Kumar, H. Ramesh; *Covering Rectilinear Polygons with Axis-parallel Rectangles*; Proc. STOC'99, pp. 445 – 454, 1999.
12. A. Lingas; *The Power of Non-Rectilinear Holes*; Proc. 9th Colloquium on Automata, Languages, and Programming, pp. 369 – 383, 1982.
13. A. Lingas, V. Soltan; *Minimum Convex Partition of a Polygon with Holes by Cuts in Given Directions*; Theory of Comuting Systems, Vol. 31, pp. 507 – 538, 1998.
14. J. O'Rourke and K. J. Supowit; *Some NP-hard Polygon Decomposition Problems*; IEEE Transactions on Information Theory, Vol IT-29, No. 2, 1983.
15. T. Shermer; *Recent results in Art Galleries*; Proc. of the IEEE, 1992.
16. J. Urrutia; *Art Gallery and Illumination Problems*; in Handbook on Computational Geometry, ed. J.-R. Sack and J. Urrutia, Elsevier, Chapter 22, pp. 973 – 1027, 2000.

Distributed $O(\Delta \log n)$-Edge-Coloring Algorithm[*]

A. Czygrinow[1], M. Hańćkowiak[2], and M. Karoński[3]

[1] Faculty of Mathematics and Computer Science
Adam Mickiewicz University, Poznań, Poland
and Arizona State University, Tempe, Az 85287-1804, USA
andrzej@math.la.asu.edu
[2] Faculty of Mathematics and Computer Science
Adam Mickiewicz University, Poznań, Poland
mhanckow@main.amu.edu.pl
[3] Faculty of Mathematics and Computer Science
Adam Mickiewicz University, Poznań, Poland
karonski@main.amu.edu.pl
and Emory University, Atlanta, Ga 30322, USA
michal@mathcs.emory.edu

Abstract. Let $G = (V, E)$ be a graph on n vertices and let Δ denote the maximum degree in G. We present a distributed algorithm that finds a $O(\Delta \log n)$ -edge-coloring of G in time $O(\log^4 n)$.

1 Introduction

In this paper, we consider a problem of edge-coloring of a graph in a distributed model of computations. In our model a network is represented by an undirected graph $G = (V, E)$ where each vertex represents a processor of the network and an edge corresponds to a connections between processors. We assume full synchronization of the network: in every step, each processor sends messages to all its neighbors, receives messages from all of its neighbors, and can perform some local computations. The number of steps should be polylogarithmic in the size of the graph, and in addition we insist that the local computations of each processor are performed in polynomial time. The above model is more restrictive than a classical distributed model introduced by Linial in [Li92]. In Linial's model there is no restriction on local computations performed by processors (for example processors can perform computations in exponential time). By default, all processors have different IDs, each proccesor knows $|V|$, the number of vertices in G, and $\Delta(G)$, the maximal degree in G.

In the edge-coloring problem the goal of a distributed algorithm is to properly color the edges of G in a polylogarithmic (in $n = |V|$) number of steps. The main difficulty of designing such an algorithm comes from the fact that in such a "short time" a vertex v can learn only about vertices and edges that are within a "small" distance from v and based on this local information, a proper

[*] This work was supported by KBN GRANT 7 T11C 032 20

F. Meyer auf der Heide (Ed.): ESA 2001, LNCS 2161, pp. 345–355, 2001.
© Springer-Verlag Berlin Heidelberg 2001

coloring of E must be obtained. Let Δ denote the maximum degree of G. By Vizing's theorem there is a proper edge-coloring of G with $\Delta + 1$ colors but the known proofs of this theorem don't lead to a distributed algorithm. It is therefore natural to aim for an algorithm that uses $O(\Delta)$ colors. In [Li92], Linial presented an algorithm which, in $\log^* n$ number of steps, colors vertices of G using $O(\Delta^2)$ colors. Linial's procedure can be used to obtain a $O(\Delta^2)$-edge-coloring of G. Very recently, De Marco and Pelc claimed that an algorithm presented in [MaPe01] colors the vertices of G in $O(\Delta)$ colors. Unfortunately, no complete proof of the main lemma in [MaPe01] has been offered. In addition, in their algorithm the amount of local computations is not polylogarithmically bounded.

In this paper, we present a distributed algorithm which colors edges of graph in $O(\Delta \log n)$ colors. Our approach is based on computing a family of spanners of G. It turns out that this family can be used to color a constant fraction of edges of G using $O(\Delta)$ colors. Iterating this process $O(\log n)$ steps leads to a proper coloring of E. However in each iteration a palette of $O(\Delta)$ new colors is needed. Spanners were previously successfully used by Hańćkowiak, Karoński and Panconesi [HKP99], to design a distributed algorithm for a maximal matching problem.

The rest of this paper is structured as follows. In Section 2, we present a procedure that constructs a family of spanners. Section 3 contains the description of our main algorithm and the proof of its correctness.

2   Family of Spanners

In this section, we present an algorithm that finds a family of spanners that are used to color our graph. The main idea of the algorithm is as follows. Suppose for a moment that all vertices in the graph have degree that are powers of two. In order to color the edges, we find a subgraph such that each vertex has degree (in this subgraph) equal to half of the original degree. We assign one to edges of the subgraph and zero to remaining edges. Next we repeat the procedure in "one-subgraph" and in "zero-subgraph". As a result we obtain a sequence of zeros and ones on each edge. This sequence is a color of an edge. Note that in the distributed model we can not split a graph in such an exact way, but we can do it approximately.

Let us start with some definitions. A bipartite graph $H = (A, B, E)$ is called a D-block if for every vertex $a \in A$,

$$\frac{D}{2} < deg_H(a) \leq D.$$

Definition 1. *An (α, β)-spanner of a D-block $H = (A, B, E)$ is a subgraph $S = (A', B, E')$ of H such that the following conditions are satisfied.*

1. *$|A'| \geq \alpha |A|$.*
2. *For every vertex $a \in A'$, $deg_S(a) = 1$.*
3. *For every vertex $b \in B$, $deg_S(b) < \frac{\beta}{D} deg_H(b) + 1$.*

Procedure FINDSPANNERS finds a family of $O(D)$ edge-disjoined (α, β) -spanners in a D-block for some constants α and β. In each iteration of the main loop of FINDSPANNERS we invoke a procedure SPLITTER that, in parallel, to each edge e assigns the labels $bit(e) \in \{0, 1\}$ and $bad(e) \in \{Yes, No\}$. As a result, on every edge e we obtain a sequence of bits, that will be denoted by $BitsSeq(e)$. In each iteration, we add (concatenate) a new bit to $BitsSeq(e)$ increasing its length by one. During the execution of the algorithm some of the edges will be excluded from further considerations. Such edges will be marked by concatenating letter "N" to the end of $BitsSeq(e)$. When FINDSPANNERS quits then the sequences $BitsSeq$ define a family of spanners. If the sequence $BitsSeq(e)$ does not end with "N" then $BitsSeq(e)$ is the number of a spanner that contains edge e. If the sequence $BitsSeq(e)$ ends with "N" then e does not belong to any spanner of the family. By $S_{<i_1,\ldots,i_k>}$, where $i_j \in \{0, 1\}$, we denote a subgraph of D-block $H = (A, B, E)$ induced by these edges e for which $BitsSeq(e)$ starts with a sequence $(i_1, \ldots, i_k)$. By $S_{<>}$ we denote the whole block H. Let $N_J(v)$ denote the set of neighbors of v in a graph J and let $d_J(v) = |N_J(v)|$.

PROCEDURE FINDSPANNERS

1. For $j := 0, \ldots, \log D - 3$ do:
   In parallel, for every subgraph $J := S_{<i_1,\ldots,i_j>}$,
   where $< i_1, \ldots, i_j >$ is an arbitrary sequence of bits, do:
   - Invoke procedure SPLITTER in J, which determines two functions:
     $bit : E(J) \mapsto \{0, 1\}$
     $bad : E(J) \mapsto \{Yes, No\}$
   - In parallel, for every $v \in A$, do:
     - If the number of edges $\{v, u\}$, $u \in N_J(v)$, such that $bad(v, u) = Yes$ is larger than $d_J(v)/\log n$, then for every $u \in N_J(v)$, do:
       $BitsSeq(v, u) := BitsSeq(v, u) \circ "N"$,
     - else, for every $u \in N_J(v)$, do:
       $BitsSeq(v, u) := BitsSeq(v, u) \circ bit(v, u)$.
2. Let $j := \log D - 4$. In parallel, for every subgraph $J := S_{<i_1,\ldots,i_j>}$,
   where $< i_1, \ldots, i_j >$ is an arbitrary sequence of bits, do:
   For every vertex $v \in V(J) \cap A$ do:
   - If $d_J(v) \geq 2$ then change the jth bit in all but one edges incident to v to "N".
3. The result of this procedure is a set of subgraphs $S_{<i_1,\ldots,i_{\log D-4}>}$,
   where $< i_1, \ldots, i_{\log D-4} >$ is an arbitrary sequence of bits.
   Every such subgraph is a spanner.

Before we describe procedure SPLITTER let us define a *vertex-splitting* operation. For a vertex v let $e_0, \ldots, e_{k-1}$ denote edges incident to v. If k is even then we replace v by vertices $v_0, \ldots, v_{\lfloor k/2 \rfloor - 1}$, where each v_i has degree two. If k is odd then in addition we add one vertex $v_{\lfloor k/2 \rfloor}$ of degree one. Then, for $i \leq \lfloor k/2 \rfloor - 1$, edges e_{2i}, e_{2i+1} are incident to v_i and if k is odd then e_{k-1} is incident to $v_{\lfloor k/2 \rfloor}$.

After splitting the vertices we obtain a graph of maximum degree two, that is a union of paths and cycles. Some of the vertices on a cycle (or path) will be marked as *border* vertices. Paths that connect two consecutive border vertices are called segments. Also the path-segments that connect an endpoint of a path with its closest border vertex are called segments.

In the procedure SPLITTER, given below, we will use procedures from [HKP99] to partition paths and cycles into segments of length at least $\ell := 400 \log^2 n$, in addition, marked by an alternating sequence of bits 0 and 1. Using the algorithms from [HKP99] such a partition can be found distributively in time $O(\log^2 n)$.

PROCEDURE SPLITTER(J)

1. Split every vertex of a graph J into groups of vertices of degree two (and possibly one of degree one) to obtain a union of paths and cycles.
2. Using procedures from [HKP99] (see procedures LONGARROWS and SPLITTER) find in time $O(\log^2 n)$ segments of length which is greater than or equal to $\ell := 400 \log^2 n$. In addition the procedures from [HKP99] assigns bits 0 and 1 to edges of a segment, so that edges that are adjacent in the segment have different bits assigned to them.
3. For every vertex w in the union of paths and cycles that corresponds to some vertex $v \in V(J) \cap B$ in the original graph, do: If w is a border vertex and both edges incident to w have the same bit assigned then flip the value of one of them (i.e., $0 \mapsto 1$ or $1 \mapsto 0$).
4. As a result two functions are obtained:
   $bit : E(J) \mapsto \{1, 0\}$
   $bad : E(J) \mapsto \{Yes, No\}$,
   where $bit(e)$ is equal to the bit assigned to edge e in step 2.
   If e is incident to a border vertex then put $bad(e) = Yes$, else $bad(e) = No$. (In particular, if e is an endpoint of a path then $bed(e) = No$).

An edge e is called *bad* if $bad(e) = Yes$. If edge is not bad then it is called *good*.

A vertex $v \in V(J) \cap A$ is called *nasty* if there are more than $d_J(v)/\log n$ bad edges incident to v. Otherwise a vertex is called *pliable*. Observe that if an edge e is incident to a nasty vertex then FINDSPANNERS assigns "N" to e.

Let $< i_1, \ldots, i_j, i_{j+1} >$ be a sequence of bits (i.e. $i_k \in \{0, 1\}$). Denote by $J := S_{<i_1,\ldots,i_j>}$, $d_j(v) := d_J(v)$, and $d_{j+1}(v) := d_{S_{<i_1,\ldots,i_j,i_{j+1}>}}(v)$. In addition, let $P_j := V(J) \cap A$ and $P_{j+1} := V(S_{<i_1,\ldots,i_j,i_{j+1}>}) \cap A$. Note that P_{j+1} is the set of pliable vertices in graph J obtained by SPLITTER in the jth iteration of FINDSPANNER.

Lemma 1. *For every vertex $v \in P_{j+1}$,*

$$\frac{1}{2}\left(\left(1 - \frac{2}{\log n}\right)d_j(v) - 1\right) \leq d_{j+1}(v) \leq \frac{1}{2}\left(\left(1 + \frac{2}{\log n}\right)d_j(v) + 1\right), \quad (1)$$

and for every $v \in B$,

$$d_{j+1}(v) \leq \frac{1}{2}\left(d_j(v) + 1\right). \quad (2)$$

Proof. Let e_+ and e_- denote the number of good and bad vertices (respectively) incident to v. Of course $d_j(v) = e_+ + e_-$. First we prove (1). Let $v \in P_{j+1}$ and let $v_1, \ldots, v_k$ denote vertices obtained from v after vertex-splitting operation. Then, since v is pliable, $e_- \leq d_j(v)/\log n$. To obtain the upper bound for $d_{j+1}(v)$, observe that the worst case will arise if the following conditions are satisfied: there is no vertex v_i which is incident to two bad edges, vertices from $\{v_1, \ldots, v_k\}$ that are incident to one bad edge have on both edges incident to them bit i_{j+1}, $d_j(v)$ is odd, and the bit on the edge incident to v_k (vertex of degree one) is i_{j+1}. Consequently,

$$d_{j+1}(v) \leq 2e_- + (e_+ - e_- - 1)/2 + 1 \leq \frac{1}{2}\left(\left(1 + \frac{2}{\log n}\right)d_j(v) + 1\right).$$

To obtain a lower bound for $d_{j+1}(v)$, notice that the worst scenario for this case will arise if the following conditions are satisfied: there is no vertex v_i that is incident to two bad edges, vertices from $\{v_1, \ldots, v_k\}$ that are incident to one bad edge have on both edges incident to them bit $(1 - i_{j+1})$, $d_j(v)$ is odd, and the bit on the edge incident to v_k is $(1 - i_{j+1})$. Then,

$$d_{j+1}(v) \geq (e_+ - e_- - 1)/2 \geq \frac{1}{2}\left(\left(1 - \frac{2}{\log n}\right)d_j(v) - 1\right).$$

Inequality (2) follows from step 3 of procedure SPLITTER.

Lemma 2. *In the jth iteration of procedure* FINDSPANNERS *($j = 0, \ldots, \log D - 3$), the following condition is satisfied:*

$$|P_{j+1}| \geq |P_j|\left(1 - \frac{1}{200 \log n}\frac{\Delta_j}{\delta_j}\right),$$

where, $\Delta_j = \max\{d_j(v) : v \in P_j\}$ *and* $\delta_j = \min\{d_j(v) : v \in P_j\}$.

Proof. Proof is the same as the proof of a similar lemma in [HKP99]. Recall that P_{j+1} is the set of pliable vertices obtained after execution of SPLITTER in the jth iteration of procedure FINDSPANNERS. Let N_{j+1} be the set of nasty vertices and let $be[N_{j+1}]$ denote the number of bad edges incident to vertices from N_{j+1}. Observe that $|P_{j+1}| + |N_{j+1}| = |P_j|$. First we establish a lower bound for $be[N_{j+1}]$. Note that if $v \in N_{j+1}$ then there are at least $\delta_j/\log n$ bad edges incident to v. Therefore,

$$be[N_{j+1}] \geq |N_{j+1}|\frac{\delta_j}{\log n}.$$

Observe that $be[N_{j+1}]$ can be bounded from above by the number of all bad edges in graph J. Since $|E(J)| \leq \Delta_j |P_j|$ and every segment has length at least $\ell = 400 \log^2 n$ and contains at most two bad edges,

$$be[N_{j+1}] \leq \frac{2|P_j|\Delta_j}{400 \log^2 n}.$$

Combining last two inequalities yields

$$|N_{j+1}| \leq \frac{1}{200 \log n} \frac{\Delta_j}{\delta_j} |P_j|$$

and the lemma follows.

Theorem 1. *Let $H = (A, B, E)$ be a D-block. Procedure* FINDSPANNERS *finds in $O(\log^3 n)$ rounds a family of $\frac{1}{16}D$, $(\frac{1}{2}, 16)$-spanners of H.*

Proof. There are $O(\log D)$ iterations in procedure FINDSPANNERS. In each iteration, main computations are performed by SPLITTER which runs in $O(\log^2 n)$ rounds. Thus the number of rounds is $O(\log^3 n)$.
Recall that $d_j(v) := d_{S_{<i_1,\dots,i_j>}}(v)$, $P_j := V(S_{<i_1,\dots,i_j>}) \cap A$. Let $k := \log D - 4$. We will show that for every sequence of bits $< i_1, \dots, i_k >$, graph $S := S_{<i_1,\dots,i_k>}$ is a $(\frac{1}{2}, 16)$-spanner. Hence, we have to verify that S satisfies the following three conditions.

- $|P_k| > \frac{1}{2}|P_0|$
- $\forall v \in P_k : d_k(v) = 1$
- $\forall v \in B : d_k(v) < 16 d_0(v) \frac{1}{D} + 1$

First we observe that the third condition is satisfied. By (2) in Lemma 1, we see that for every $v \in B$,

$$d_k(v) \leq \left(\frac{1}{2}\right)^k (d_0(v) - 1) + 1$$

which shows that the third condition is satisfied. Applying Lemma 1 (part (1)) k times shows that for every $v \in P_k$,

$$q_-^k (d_0(v) + 1) - 1 \leq d_k(v) \leq q_+^k (d_0(v) - 1) + 1$$

where

$$q_- = \frac{1}{2}\left(1 - \frac{2}{\log n}\right) , \quad q_+ = \frac{1}{2}\left(1 + \frac{2}{\log n}\right)$$

Since $\forall v \in A : D/2 < d_0(v) \leq D$, we have that for every $v \in P_k$,

$$q_-^k \left(\frac{D}{2} + 1\right) - 1 \leq d_k(v) \leq q_+^k (D - 1) + 1.$$

So, for sufficiently large n,

$$q_-^k \left(\frac{D}{2} + 1\right) - 1 > 8 e^{-2(1 + \frac{2}{\log n})} - 1 > 0$$

and

$$q_+^k (D - 1) + 1 < 16 e^2 + 1 < 120.$$

Thus for every $v \in P_k$, $d_k(v) > 0$. Since in step 2 of FINDSPANNERS we disregard all but one edge incident to v, the second condition is satisfied. Finally we verify that the first condition is satisfied as well. Indeed, apply Lemma 2 k times. Then

$$|P_k| \geq |P_0| \prod_{j=0}^{k-1} \left(1 - \frac{1}{200 \log n} \frac{\Delta_j}{\delta_j}\right).$$

Note that

$$\frac{\Delta_j}{\delta_j} \leq \frac{q_+^j (D-1) + 1}{q_-^j (D/2+1) - 1} \leq 120$$

since the fraction in the middle is an increasing function of j for $j \leq k$. Therefore, for n sufficiently large, we have

$$|P_k| \geq |P_0| \left(1 - \frac{3}{5 \log n}\right)^{\log n} \geq \frac{1}{2}|P_0|.$$

3   Algorithm

In this section, we present and analyze our edge-coloring algorithm. The algorithm runs in $O(\log^4 n)$ rounds and uses $O(\Delta \log n)$ colors. The algorithm can be divided into four smaller procedures: FINDSPANNERS, COLORSPANNER, COLORBIPARTITE, COLORGRAPH. Procedure FINDSPANNERS finds a family of edge-disjoint spanners of a block and was preseneted in a previous section. Let $H = (A, B, E)$ be a D-block, $m = |A| + |B|$, $\alpha = \frac{1}{2}$ and $\beta = 16$. Then, by Theorem 1, FINDSPANNERS finds in $O(\log^3 m)$ rounds a family of D/β edge-disjoint (α, β)-spanners. First, we describe a procedure that colors a family of spanners found by FINDSPANNERS. Note that a single spanner is simply a collection of vertex-disjoint stars, with centers of stars in set B, and so vertices of B (in parallel) can color edges of the spanner provided they have enough colors available. We need some additional notation. Let

$$K := \frac{\beta}{D} \Delta + 1,$$

and for $j := 1, \dots, D/\beta$ let

$$I_j := \{(j-1)K, \dots, jK-1\}.$$

PROCEDURE COLORSPANNER

1. Find a collection of spanners $S_1, \dots, S_{D/\beta}$ using FINDSPANNERS.
2. In parallel for every vertex $b \in B$ do:
   For every $j := 1, \dots, D/\beta$ do:
   – Color edges of S_j incident to b using colors from I_j.

Lemma 3. *Let* $H = (A, B, E)$ *be a* D-*block,* $m = |A| + |B|$. *Procedure* COL-ORSPANNER *colors at least* $\frac{\alpha}{\beta}|E|$ *edges of* H *using* $O(\Delta)$ *colors and runs in* $O(\log^3 m)$ *rounds* .

Proof. Indeed, since every (α, β)-spanner contains at least $\alpha|A|$ vertices of degree one in A, the number of edges in the family of spanners is at least $\alpha|A|\frac{1}{\beta}D \geq \frac{\alpha}{\beta}|E|$. Also, for every $j = 1, \ldots, D/\beta$ and every vertex $b \in B$, $deg_{S_j}(b) \leq K = |I_j|$ and so edges within one spanner are colored properly. Since we use different colors to color different spanners, coloring obtained by COLORSPANNER is proper. Hence $O(\Delta)$ colors is suffices, because at each vertex of B we need $\frac{1}{\beta}DK$ colors and $D \leq \Delta$.

We can now consider bipartite graphs. Let $G = (L, R, E)$ be a bipartite graph with $|L| = |R| = n$. For $e = \{u, v\} \in E$ with $u \in L$, $v \in R$ define a weight of e as

$$\omega(e) := deg_G(u) \tag{3}$$

and let

$$\omega(\bar{E}) := \sum_{e \in \bar{E}} \omega(e) \tag{4}$$

be a total weight of a set $\bar{E} \subset E$. Trivially, we have

$$\omega(E) \leq n^3.$$

Let $D_i := \Delta/2^i$, where $i = 0, \ldots, \log \Delta$, and consider D_i-blocks $H_i = (A_i, B, E_i)$ where

$$A_i := \{u \in L : \frac{D_i}{2} < deg_G(u) \leq D_i\}, \quad B := R,$$

and E_i contains all edges of G that have an endpoint in A_i.

PROCEDURE COLORBIPARTITE

1. In parallel, for $i := 0, \ldots, \log \Delta$, color every H_i using COLORSPANNER.
2. In parallel for every vertex $v \in R$ do:
   For every color c do:
   a) Define $E_{c,v}$ to be the set of edges incident to v that are colored with color c.
   b) Find in $E_{c,v}$ an edge e with maximum weight and uncolor all edges in $E_{c,v}$ except e.

Lemma 4. *Let* $G = (L, R, E)$ *be a bipartite graph with* $n = |L| = |R|$. *Procedure* COLORBIPARTITE *properly colors a set* $\bar{E} \subset E$ *such that* $\omega(\bar{E}) \geq \alpha\omega(E)/(6\beta)$ *using* $O(\Delta)$ *colors and runs in* $O(\log^3 n)$ *rounds.*

Proof. Sets A_i are disjoint and so after step 1 of the algorithm, there will be no edges of the same color that are incident to vertices of L. After step 2, every vertex $v \in R$ has at most one edge of a given color incident to it. Let $\bar{E}_i \subset E_i$ denote the set of edges of the ith block that are colored in step 1. From Lemma

3, $|\bar{E}_i| \geq \alpha|E_i|/\beta$, and since the minimum degree in H_i is larger than $D_i/2$, we have $\omega(\bar{E}_i) \geq \alpha|E_i|D_i/(2\beta)$. Consequently

$$\omega(\bar{E}_0 \cup \ldots \cup \bar{E}_{\log \Delta}) \geq \frac{\alpha}{2\beta} \sum_i |E_i|D_i \geq \frac{\alpha}{2\beta}\omega(E).$$

We need to argue that at least one third of this weight is maintained after "correction" done in the second step of the procedure. Fix a vertex $v \in R$ and color c. Let $M := M(v,c)$ denote the maximum weight of an edge incident to v in color c. Since edges incident to v that have the same color must belong to different blocks, weight of edges incident to v that have color c is less than

$$M + \sum_{i \geq 0} \frac{M}{2^i} < 3M.$$

Therefore, the total weight of edges that remain colored is at least $\frac{\alpha}{6\beta}\omega(E)$.

Before we proceed, we need one additional concept. Let $P = (V, E)$ be a path or cycle on at least two vertices. A partition $V_1, \ldots, V_t$ of V is called *short* if the following conditions are satisfied.

1. Every graph $P_i = P[V_i]$ induced by V_i is a path or cycle.
2. For every $i = 1, \ldots, t$, $2 \leq |V_i| = O(\log|V|)$.
3. For every $i = 1, \ldots, t-1$, $|V_i \cap V_{i+1}| = 1$. If P is a cycle than $|V_0 \cap V_t| = 1$.

Vertices that belong to $V_i \cap V_{i+1}$ for some i are called *border vertices*. In [AGLP89], the following fact is proved.

Lemma 5. *Let $P = (V, E)$ be a path or cycle with $|V| \geq 2$. There is a procedure that finds a short partition of P in $O(\log|V|)$ rounds.*

Let us now consider an arbitrary graph $G = (V, E)$. Our first task is to obtain an auxiliary bipartite graph from G which will be colored using COLORBIPARTITE. Define a bipartite graph $G' := (L, R, E')$ as follows. Every vertex splits itself into two vertices $(v,0)$ and $(v,1)$. Let $L := \{(v,0) : v \in V\}$, $R := \{(v,1) : v \in V\}$ and $\{(u,0), (v,1)\} \in E'$ if and only if $u < v$ and $\{u,v\} \in E$.

PROCEDURE COLORGRAPH

1. Obtain $G' = (L, R, E')$ from $G = (V, E)$ as described above.
2. Apply COLORBIPARTITE to G'.
3. Obtain a coloring of G by merging vertices $(v,0)$, $(v,1)$ that correspond to a single vertex $v \in V$. (This operation can result in some monochromatic paths and cycles.)
4. In parallel for every monochromatic path and cycle P do:
   a) Find a short partition of P into segments $P_1, \ldots, P_t$.
   b) In parallel for every segment P_i do:
      – Find a matching M_1 that saturates all but at most one vertex in P_i and let $M_2 := E(P_i) \setminus M_1$.
      – If $\omega(M_1) \geq \omega(M_2)$ then uncolor edges of M_2. Otherwise uncolor edges of M_1.

c) In parallel for every border vertex v do:
  - If both edges incident to v have the same color then uncolor an edge with a smaller weight.

A weight $\omega(\{u, v\})$ of an edge $\{u, v\} \in E$ with $u < v$ is defined as the number of vertices $w \in V$ such that $u < w$ and $\{u, w\} \in E$. Thus the weight of $\{u, v\}$ is exactly equal to the weight of edge $\{(u, 0), (v, 1)\} \in E'$. In the same way as in (4) we can define $\omega(\bar{E})$ for any set $\bar{E} \subset E$.

Lemma 6. *Let $G = (V, E)$ be a graph with $n = |V|$. Procedure* COLORGRAPH *colors a set $\bar{E} \subset E$ such that $\omega(\bar{E}) \geq \alpha\omega(E)/(24\beta)$ using $O(\Delta)$ colors and runs in $O(\log^3 n)$ rounds.*

Proof. From Lemma 4, we know that after step 2 we colored a set $\bar{E}' \subset E'$ such that $\omega(\bar{E}') \geq \alpha\omega(E')/(6\beta)$. Thus the total weight of edges that remain colored after step 4(b) is at least $\omega(\bar{E}')/2$. Finally, the total weight of edges that remain colored after step 4(c) is at least $\omega(\bar{E}')/4$. Since $\omega(E') = \omega(E)$, procedure COLORGRAPH colors a set $\bar{E}$ such that $\omega(\bar{E}) \geq \alpha\omega(E)/(24\beta)$.

Iterating COLORGRAPH $O(\log n)$ times yields a proper coloring of graph $G = (V, E)$.

PROCEDURE COLOR

1. Run $O(\log n)$ times:
   - Use procedure COLORGRAPH with a pallete of $O(\Delta)$ new colors (different than previously used colors) to properly color set $\bar{E} \subset E$.
   - Let $E := E \setminus \bar{E}$.

Theorem 2. *Let $G = (V, E)$ be a graph with $n = |V|$. Procedure* COLOR *properly colors edges of G using $O(\Delta \log n)$ colors and runs in $O(\log^4 n)$ rounds.*

Proof. By Lemma 6 in each iteration of the procedure $\bar{E}$ is properly colored using $O(\Delta)$ colors. Since COLOR uses different colors in each iteration, obtained coloring is proper.

To complete the proof, we show that E is empty after $O(\log n)$ iterations of the main loop in COLOR. Notice that $E \neq \emptyset$ is equivalent to $\omega(E) \geq 1$. Let $p = (1 - \frac{\alpha}{\beta}\frac{1}{24})^{-1}$ and let ω denote the weight of the edge set of graph G. By Lemma 6, the total weight of edges left after the kth iteration is at most ω/p^k. Since $\omega \leq n^3$, the right hand side of the last inequality is less than one if $k > \frac{3\log n}{\log p}$.

Acknowledgement. This work was supported by KBN GRANT 7 T11C 032 20.

References

[Li92]   N. Linial, *Locality in distributed graph algorithms*, SIAM Journal on Computing, 1992, 21(1), pp. 193-201.

[MaPe01]   G. De Marco, A. Pelc, *Fast distributed graph coloring with $O(\Delta)$ colors*, SODA 2001.

[HKP99]   M. Hańćkowiak, M. Karoński, A. Panconesi, *A faster distributed algorithm for computing maximal matching deterministically*, Proceedings of PODC 99, the Eighteen Annual ACM SIGACT-SIGOPS Symposium on Principles of Distributed Computing, pp. 219-228.

[AGLP89]   B. Awerbuch, A.V. Goldberg, M. Luby, and S. Plotkin, *Network decomposition and locality in distributed computing*, Proceedings of the 30th Symposium on Foundations of Computer Science (FOCS 1989), pp. 364-369.

Modeling Replica Placement in a Distributed File System: Narrowing the Gap between Analysis and Simulation

John R. Douceur and Roger P. Wattenhofer

Microsoft Research, Redmond WA 98052, USA
{johndo,rogerwa}@microsoft.com
http://research.microsoft.com*

Abstract. We examine the replica placement aspect of a distributed peer-to-peer file system that replicates and stores files on ordinary desktop computers. It has been shown that some desktop machines are available for a greater fraction of time than others, and it is crucial not to place all replicas of any file on machines with low availability. In this paper we study the efficacy of three hill-climbing algorithms for file replica placement. Based on large-scale measurements, we assume that the distribution of machine availabilities be uniform. Among other results we show that the MinMax algorithm is competitive, and that for growing replication factor the MinMax and MinRand algorithms have the same asymptotic worst-case efficacy.

1   Introduction

Farsite [10] is a distributed peer-to-peer file system that replicates and stores files on ordinary desktop computers rather than on dedicated storage servers. Multiple replicas are created so that a user can access a file if at least one of the machines holding a replica of that file is accessible. It has been shown [3] that some desktop machines are available for a greater fraction of time than others, and it is crucial not to place all replicas of any file on machines with low availability, or the availability of that file will suffer.

In earlier work [9], we evaluated the efficacy and efficiency of three hill-climbing algorithms for file replica placement, using competitive analysis and simulation. The scenario under consideration was a static problem in which the availability of each machine was fixed, and each replica stably remained on the machine to which the placement algorithm assigned it. Our study found that algorithmic efficiency and efficacy ran counter to each other: The algorithm with the highest rate of improvement yielded a final placement with the poorest quality relative to an optimal placement.

In actual practice, the replica placement problem is not static. The availability of each machine (defined loosely as the fraction of time it is accessible) varies

---

* Due to lack of space we omit most of the proofs in this extended abstract. The complete paper is available as Microsoft Research technical report MSR-TR-2001-62.

F. Meyer auf der Heide (Ed.): ESA 2001, LNCS 2161, pp. 356–367, 2001.

over time as user behavior changes. In addition, file replicas may be evicted from machines by other processes in the system. The replica placement algorithm does not produce a static final placement that thereafter persists; rather, it continuously operates to correct for dynamic changes in the system. Viewed from this dynamic perspective, extensive Monte Carlo simulation shows that the MinMax algorithm consistently out-performs the other two algorithms, even though it was proven [9] to be non-competitive. Hence, our theoretic worst-case competitive analysis opposes use of the algorithm that appears best in practice.

We thus face an apparent dilemma: Either we fail to exploit an algorithm that is demonstrably efficient, or we risk the possibility that our system will encounter a distribution of machine availabilities that renders our algorithm useless. In the present paper, we make stronger assumptions about the algorithm's input, based on large-scale measurement of machine availability [3]. Given these assumptions, which – we stress – are a close approximation of the behavior of actual machines, we show that the MinMax algorithm is competitive for the levels of replication we intend to use in actual deployment. Obtaining these new results requires completely different analysis methods from those used for our earlier general-distribution results, which relied on highly unusual availability distributions utterly dissimilar to those found in real systems.

Furthermore, our earlier studies evaluated competitiveness in terms of the least available file, which is a straightforward quantity to analyze. However, from a systems perspective, a better metric is the effective availability of the overall storage system, which is readily computable in simulation. In the present paper, we show that all worst-case results for minimum file availability are also worst-case results for effective system availability, further legitimizing the relevance of our theoretic analyses.

In our opinion, the significance of this work lies in the fusion of four elements: an important problem from an emerging area of systems research, simulation results that demonstrate the practical performance of a suite of algorithms, large-scale measurements of deployed systems that provide a tractable analytic model, and rigorous theoretic analysis to provide confidence in the algorithm selected for use in the actual system. We consider this an exemplary synergy of systems, simulation, measurement, and theory.

The remainder of the paper is organized as follows. The next section describes the Farsite system and provides some motivation for why file replica placement is an important problem. Section 3 describes the algorithms. In Section 4 we further motivate this paper, followed by a summary of results in Section 5. Section 6 presents a simplified model, which is used in Section 7 to analyze the efficacy of the algorithms. Section 8 compares the two measures of efficacy. In Section 9 we conclude the paper by presenting related work.

2   Farsite

Farsite [10] is a distributed peer-to-peer file system that runs on a networked collection of desktop computers in a large organization, such as a university or corporation. It provides a logically centralized storage repository for the files of all users in the organization. However, rather than storing these files on dedicated server machines, Farsite replicates them and distributes them among all of the client computers sitting on users' desktops. As compared to centralized storage, this architecture yields great savings in hardware capital, physical plant, system administration, and operational maintenance, and it eliminates a single point of failure and single target of attack. The disadvantage of this approach is that user's desktop machines lack the physical security and continuous support enjoyed by managed servers, so the system must be designed to resist the threats to reliability and security that are inherent in a large-scale, distributed, untrusted infrastructure.

For files stored in Farsite, the following properties are maintained: privacy, integrity, persistence, and availability. Data privacy and integrity are ensured by encryption and digital signatures. File persistence is provided by generating R replicas of each file and storing the replicas on different machines. The data will persist as long as one of the replicas resides on a machine that does not suffer a destructive failure, such as a disk head crash. Since it is difficult to estimate the remaining lifetime of a particular disk with any accuracy [16], the degree of data persistence is considered to be determined entirely by the replication factor R and not by any measurable aspect of the particular machines selected for storing the replicas.

In this paper we focus on file availability, meaning the likelihood that the file can be accessed by a user at the time it is requested, which is determined by the likelihood that at least one replica of that file can be accessed at the requested time. The *fractional downtime* of a machine is the mean fraction of time that the machine is unavailable, because it has crashed, has been turned off, has been disconnected from the network, etc. A five-week series of hourly measurements of over 50,000 desktop machines at Microsoft [3] has shown that the times at which different machines are unavailable are not significantly correlated with each other, so the fractional downtime of a file is equal to the product of the fractional downtimes of the machines that store replicas of that file. For simplicity, we express machine and file availability values as the negative logarithm of fractional downtime, so the availability of a file equals the sum of the availabilities of the R machines that store replicas of the file.

The goal of a file placement algorithm is to produce an assignment of file replicas to machines that maximizes an appropriate objective function. We consider two objective functions in this paper: (1) the *minimum file availability* over all files and (2) the *effective system availability* (ESA), defined as the negative logarithm of the expected fractional downtime of a file chosen uniformly at random. When we evaluate the efficacy of a file placement algorithm, we are gauging its ability to maximize one of these objective functions. For our theoretic analyses in both our earlier work [9] and the present paper, we focus on the metric of minimum file availability, because it is more readily tractable. Our simulation results, such as those described in Section 4, relate to ESA because it is more meaningful from a systems perspective. One of our current findings (Section 8) is that all of our theoretic worst-case results for minimum file availability are also theoretic worst-case results for effective system availability.

Measurements of over 10,000 file systems on desktop computers at Microsoft [8] indicate that a replication factor of $R = 3$ is achievable in a real-world setting [3]. Thus, we have a special interest in the case $R = 3$.

3   Algorithms

Files in Farsite are partitioned into disjoint sets, each of which is managed by a small, autonomous group of machines. This imposes the requirement that a file placement algorithm must be capable of operating in a distributed fashion with no central coordination. Farsite is also a highly dynamic system in which files are created and deleted frequently and in which machine availabilities continuously change. This imposes the requirement that a file placement algorithm must be able to incrementally improve an existing placement, rather than require a complete re-allocation of storage resources. These and other considerations [10] have led us to a family of iterative, swap-based algorithms: One group of machines contacts another group (possibly itself), each of which selects a file from the set it manages; the groups then decide whether to exchange the machine locations of one replica from each file. The groups select files according to one of the following algorithms:

- RandRand swaps a replica between two randomly chosen files,
- MinRand swaps a replica between a minimum-availability file and any other file, and
- MinMax swaps a replica between a minimum-availability file and a maximum-availability file.

(We use the particle "Rand" rather than "Any" because this reflects the way files are selected in the system, even though all that matters for our theoretic analysis is the absence of a selection restriction.) The groups swap replicas only if doing so reduces the absolute difference between the availabilities of the two files, which we call a successful swap. If a pair of files has more than one successful swap, the algorithm chooses one with minimum absolute difference between the files' availabilities after the swap (although this does not affect theoretical efficacy). Because the algorithms operate in a distributed fashion, their selection restrictions are weakened, i.e., the MinMax and MinRand algorithms might select files whose availability values are not globally minimum or maximum. For our theoretic analysis, we concentrate on the more restrictive case in which only extremal files are selected.

4   Motivation

If, beginning with a random assignment of replicas to machines, we run each algorithm until it *freezes* (meaning no more swaps can be found), we find that the three algorithms differ substantially in both the efficacy of their final placements and the efficiency with which they achieve those placements. Simulations show that the MinMax algorithm improves the availability of the minimum file more quickly than the other two algorithms. On the other hand, MinMax tends to freeze at a point with lower minimum file availability, since swaps are only considered between the minimum-availability file and the maximum-availability file.

In earlier work [9], we performed a worst-case analysis to determine each algorithm's competitive ratio $\rho = m/m^*$, where m is the availability of a minimum-availability file when the algorithm freezes, and m^* is the availability of a minimum-availability file given an optimal placement, for a worst-case availability distribution. The results were that MinMax (the most efficient algorithm) was not competitive ($\rho = 0$), whereas MinRand and RandRand were 2/3-competitive for $R = 3$.

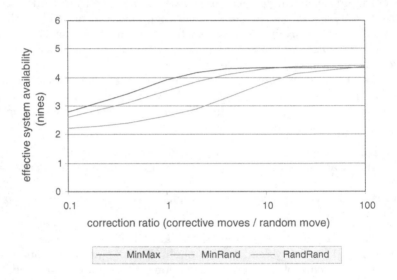

Fig. 1. Steady-state behavior of the algorithms

If we exercise each algorithm in a dynamic scenario that more closely matches the environment in which the Farsite system operates, the results are even more disconcerting. Figure 1 shows the result of a steady-state simulation in which two processes operate concurrently on the placement of replicas. One process (maliciously) moves random replicas to random machines, simulating the dynamic behavior of users and machines. The other process performs one of our three hill-climbing algorithms, trying to repair the damage caused by the random moves. With the exception of unrealistically high correction ratios, MinMax performs significantly better than the other two algorithms.

We are in the unpleasant situation that a theoretical worst-case result ($\rho = 0$ for MinMax) opposes the use of an algorithm that works best for real-world data. In this paper, we begin to address this discrepancy by noting the distribution of availability values found in a large-scale study of desktop machines in a commercial environment [3], reproduced here as Figure 2. This figure shows that, when expressed logarithmically, machine availabilities follow a distribution that is nearly uniform. This finding, coupled with the observation that most of our worst cases [9] need rather unusual distributions of machine availabilities, suggests that we can improve the bounds of the worst-case analysis by making stronger assumptions about the input, namely that we have a uniform distribution of machine availabilities.

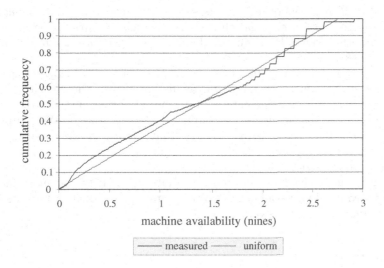

Fig. 2. Distribution of machine availabilities

5 Results

In this paper we will take for granted that the distribution of machine availabilities be uniform. With this assumption we show that the MinMax algorithm is competitive. More surprisingly, when the replication factor R grows the MinRand and MinMax algorithms have the same asymptotic worst-case efficacy. This is counterintuitive when looking at our earlier results [9]. We study the case $R = 3$ with special care, since the real Farsite system is expected to be deployed with $R = 3$. We also give detailed results for $R = 2$ since they are considerably different from $R > 2$. Here is a detailed summary of our results:

Algorithm	general R	$R = 2$	$R = 3$
MinMax	$\rho = 1 - \Theta(1/R)$ (1)	$\rho = 0$ (5)	$\rho = 1/2$ (4)
MinRand	$\rho = 1 - \Theta(1/R)$ (1)	$\rho = 1$ [9] & (2)	$\rho = 22/27$ (2)
RandRand	$\rho = 1 - \Theta(1/R^2)$ (3)	$\rho = 1$ [9] & (2)	$\rho = 8/9$ (3)

6   Model

We are given a set of N unit-size files, each of which has R replicas. We are also given a set of $M = N \cdot R$ machines, each of which has the capacity to store a single file replica. Throughout this paper we assume that machines have (uniformly distributed) *availabilities* $0\gamma, 1\gamma, 2\gamma, \ldots, (M-1)\gamma$, for an arbitrary constant γ.

Let the R replicas of file f be stored on machines with availabilities $a_1, \ldots, a_R$. To avoid notational clutter, we overload a variable to name a file and to give the availability value of the file. Thus, the *availability* of file f is $f = a_1 + \cdots + a_R$.

As in our earlier study [9], we examine the point at which the algorithms freeze. Let m be a file with minimum availability when the algorithm has exhausted all possible improvements. Let m^* be a file with minimum availability given an optimal placement for the same values of N and R. We compute the ratio $\rho = \min m/m^*$ as $N \to \infty$. We say that the algorithm is ρ-competitive. Note that the scale γ of the machine availabilities does not affect ρ; throughout this paper we therefore assume $\gamma = 1$.

If two or more files have minimum availability, or if two or more files have maximum availability, we allow an adversary to choose which of the files will be considered for a potential swap.

7   Analysis

We start this section with a number of undemanding observations which will help us simplify the analysis.

MinMax searches for swap candidates in a subset of MinRand, and similarly MinRand $\subseteq$ RandRand, thus

Lemma 1. $\rho_{\text{MinMax}} \leq \rho_{\text{MinRand}} \leq \rho_{\text{RandRand}}$.

In this paper we study a restricted case (we assume uniform availability) of the general problem that was investigated in [9]. We have immediately:

Lemma 2. *For the same algorithm, we have* $\rho_{general} \leq \rho_{uniform}$.

The next Lemma shows a simple observation: There always is an optimal assignment. This simplifies calculating the competitive ratio ρ.

Lemma 3. *For any $R > 1$ and uniform distribution there is an optimal assignment, where all the files have the same availability $R(M-1)/2$.*

Lemma 4. $\rho_{\text{MinRand}R} \leq 1 - c/R$, *where c is a positive constant. If $R = 3$ then* $c = 5/9$.

Theorem 1. $\rho_{\text{MinRand}R} = \rho_{\text{RandRand}R} = 1 - \Theta(1/R)$.

Theorem 2. $\rho_{\text{MinRand}3} = 22/27$.

Proof. (We include this proof in the extended abstract as a representative for the proof techniques in this paper.) With Lemma 4 we have that $\rho_{\texttt{MinRand3}} \leq 22/27$. For proving the Theorem it is therefore sufficient to show that $\rho_{\texttt{MinRand3}} \geq 22/27$.

The intuition of the proof: We partition the set of machines into five regions. With a detailed case study we show which combinations of regions do not allow successful swaps with the minimum file. Then we classify the valid combinations of regions, and give a combinatorial proof about their quantity which ultimately leads to a lower bound for the availability of the minimum file. For simplicity we omit insignificant constants throughout this proof (i.e. we write M instead of $M - 1$).

Here are the details: Let the minimum file be $m = a_1 + a_2 + a_3$ with $a_1 > a_2 > a_3$. Assume for the sake of contradiction that $m < 11/9 \cdot M$. We define the following sets of machines (see Figure 3): Machines in A have availability less than a_3, machines in B between a_3 and a_2, machines in C between a_2 and $(a_1 + a_2)/2$, machines in D between $(a_1 + a_2)/2$ and a_1, and machines in E more than a_1. With this partitioning the availability of the minimum file m translates into $m = 2|A| + |B| + M - |E|$, and with $m < 11/9 \cdot M$ we get

$$2|A| + |B| - |E| < 2/9 \cdot M = 2/3 \cdot N.$$

Fig. 3. Partition of the machines

Case 1: We consider all the files $f = b_1 + b_2 + b_3$ with $b_1 \in E$, and $b_2 > b_3$. If $b_2 + b_3 > a_2 + a_3$, then we swap the machines b_1 and a_1 and get $m' = b_1 + a_2 + a_3 > a_1 + a_2 + a_3 = m$ and $f' = a_1 + b_2 + b_3 > a_1 + a_2 + a_3 = m$. Thus $b_2 + b_3 \leq a_2 + a_3$, and therefore (with $b_2 > b_3$) $b_3 < (a_2 + a_3)/2$. Since each $b_1 \in E$ needs a b_3, we know that $|E| < (a_2 + a_3)/2$. Since $|A| + |B|/2 = (a_2 + a_3)/2$ and $2|A| + |B| - |E| < 2/9 \cdot M$ we get $(a_2 + a_3)/2 < 2/9 \cdot M$. On the other hand $|E| < (a_2 + a_3)/2$ shows that $(a_1 + a_2)/2 > 1/2 \cdot M$. Since $b_2 + b_3 \leq a_2 + a_3 < 4/9 \cdot M < 1/2 \cdot M < (a_1 + a_2)/2$, we know that $b_2 \in A \cup B \cup C$. If $b_2 > a_2$ then we have $b_3 < a_3$, in other words, if $b_2 \in C$ then $b_3 \in A$. If $b_2 \in A \cup B$ then $b_3 \in A \cup B$. Since we have used all machines in set E in this case, there are no machines in E in the following cases.

Case 2: We identify all remaining files $f = b_1 + b_2 + b_3$ with $b_1 \in C$, and $b_2 > b_3$. If $b_3 \in D$, then we swap machines b_1 and a_2, getting $m' = a_1 + b_1 + a_3 > a_1 + a_2 + a_3 = m$ and $f' = a_2 + b_2 + b_3 > a_2 + b_3 + b_3 \geq a_2 + 2(a_1 + a_2)/2 > m$. Therefore $b_3 \notin D$. Thus for each $b_1 \in C$ we have $b_2 \in A \cup B \cup C \cup D$ and $b_3 \in A \cup B \cup C$. We have used all machines in C; henceforth the sets C and E are taboo.

Case 3: We identify all the remaining files $f = b_1 + b_2 + b_3$ with $b_1 \in B$, and $b_2 > b_3$. If $b_3 \in D$, then we swap machines b_1 and a_3, getting $m' = a_1 + a_2 + b_1 >$

$a_1 + a_2 + a_3 = m$ and $f' = a_3 + b_2 + b_3 > a_3 + b_3 + b_3 \geq a_3 + 2(a_1 + a_2)/2 = m$. Therefore $b_3 \notin D$. Thus for each $b_1 \in B$ we have $b_2 \in A \cup B \cup D$ and $b_3 \in A \cup B$. Henceforth the sets B, C, E are taboo.

Case 4: Finally we identify all the remaining files $f = b_1 + b_2 + b_3$ with $b_1 \in D$, and $b_2 > b_3$. If $b_3 \in D$, then we swap machines b_1 and a_2, getting $m' = a_1 + b_1 + a_3 > a_1 + a_2 + a_3 = m$ and $f' = a_2 + b_2 + b_3 > a_2 + b_3 + b_3 \geq a_2 + 2(a_1 + a_2)/2 > m$. Thus for each $b_1 \in D$ we have $b_2 \in A \cup D$ and $b_3 \in A$.

From above analysis we have seen that a file f can only consist of the these combinations of regions: (Case 1) $E + C + A$ or $E + (A \cup B) + (A \cup B)$ or (Case 2) $C + (A \cup B \cup C \cup D) + (A \cup B \cup C)$ or (Case 3) $B + (A \cup B \cup D) + (A \cup B)$ or (Case 4) $D + A + A$ or $D + D + A$. We define the two functions g_1, g_2:

$$g_1(f) = |C| - |D|$$
$$g_2(f) = 2|A| + |B| - |E|$$

Figure 4 shows all possible files f with respect to the functions $g_1(f)$ and $g_2(f)$.

Note that for all possible files f we have $g_2(f) \geq 0$. We put the files into three classes. Class X are the files f with $g_1(f) < 0$ (the black circles); class Y are the files with $g_1(f) = 0$ (the white circles); class Z are the files with $g_1(f) > 0$ (the grey circles). Note that for files $f \in X$ we have $g_1(f) \geq -2$ and $g_2(f) \geq 2$, and that for files $f \in Y$ we have $g_2(f) \geq 1$.

We have M machines, thus (ignoring the single mimimum file m) $|A| + |B| + |C| + |D| + |E| = M = 3N$. This translates into $|X| + |Y| + |Z| = N$ for the three classes X, Y, Z. The sets C and D were defined such that they exactly split the region of machines between the a_2 and a_1, hence $|C| = |D|$. Using $g_1(f) \geq -2$ for $f \in X$, and $g_1(f) \geq 1$ for $f \in Z$, the constraint $|C| = |D|$ translates into $2|X| \geq |Z|$. Both constraints together, we get $3|X| + |Y| \geq |X| + |Y| + |Z| = N$. We multiply with $2/3$: $2|X| + |Y| \geq 2|X| + 2/3 \cdot |Y| \geq 2/3 \cdot N$. We use this inequality to get:

$$2|A| + |B| - |E| = \sum_{f \in X} g_2(f) + \sum_{f \in Y} g_2(f) + \sum_{f \in Z} g_2(f) \geq 2|X| + |Y| \geq 2/3 \cdot N.$$

(The first equality is the definition of g_2; the middle inequality is because files $f \in X$ have $g_2(f) \geq 2$, files $f \in Y$ have $g_2(f) \geq 1$, and files $f \in Z$ have $g_2(f) \geq 0$.)

This contradicts our assumption that $2|A| + |B| - |E| < 2/3 \cdot N$ and therefore the assumption that $m < 11/9 \cdot M$. Thus $m \geq 11/9 \cdot M$. With Lemma 3 we know that $m^* = 3(M-1)/2$. Thus $\rho = m/m^* \geq 22/27$, as M goes to infinity.

Theorem 3. $\rho_{\text{RandRand}R} = 1 - c/R^2$, *where c is a positive constant. If R is odd then $c = 1$.*

Theorem 4. $\rho_{\text{MinMax3}} = 1/2$.

Theorem 5. $\rho_{\text{MinMax}R} = 1 - 2/R$, *for R even.*

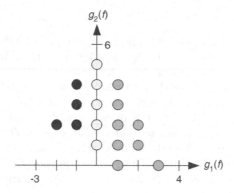

Fig. 4. Possible locations for a file f.

8 Measures of Efficacy

We can show that any worst-case result for minimum file availability is also a worst-case result for effective system availability. We show that the effective system availability can be as low as the minimum file availability, and it cannot be lower.

Theorem 6. *Let b be the base for converting downtime d into availability a, that is $a = -\log_b d$. As $b \to \infty$, the effective system availability (ESA) equals the availability of the minimum file.*

Proof. Let $b = e^c$. Then $a = -\log_b d = -1/c \cdot \ln d$, where $\ln = \log_e$. If $b \to \infty$ then $c \to \infty$. Let m be the availability of the minimum file. Assume that there are $X > 0$ files with availability m and $N - X$ files with availability f_i with $f_i > m$, for $i = 1, \ldots, N - X$. Then, applying the definition of ESA,

$$
\lim_{c \to \infty} \text{ESA} = \lim_{c \to \infty} -\frac{1}{c} \ln \left(\frac{1}{N} \left(X b^{-m} + \sum_{i=1}^{N-X} b^{-f_i} \right) \right)
$$

$$
= \lim_{c \to \infty} -\frac{1}{c} \ln \left(\frac{e^{-cm}}{N} \left(X + \sum_{i=1}^{N-X} e^{c(m-f_i)} \right) \right)
$$

$$
= \lim_{c \to \infty} -\frac{1}{c} \ln \left(\frac{X}{N} e^{-cm} \right) = \lim_{c \to \infty} \left(m - \frac{1}{c} \ln \frac{X}{N} \right) = m.
$$

Similarly,

Theorem 7. *Let b be the positive base for converting uptime into availability. Then, $ESA \geq m$.*

9 Related Work

Other than Farsite, serverless distributed file systems include xFS [2] and Frangipani [17], both of which provide high availability and reliability through distributed RAID semantics, rather than through replication. Archival Intermemory [5] and OceanStore [14] both use erasure codes and widespread data distribution to avoid data loss. The Eternity Service [1] uses full replication to prevent loss even under organized attack, but does not address automated placement of data replicas. A number of peer-to-peer file sharing applications have been released recently: Napster [15] and Gnutella [11] provide services for finding files, but they do not explicitly replicate files nor determine the locations where files will be stored. Freenet [6] performs file migration to generate or relocate replicas near their points of usage.

To the best of our knowledge [9] is the first study of the availability of replicated files, and also the first competitive analysis of the efficacy of a hill-climbing algorithm.

There is a common denominator of our work and the research area of approximation algorithms; especially in the domain of online approximation algorithms [13,4] such as scheduling [12]. In online computing, an algorithm must decide how to act on incoming items without knowledge of the future. This seems to be related our work, in the sense that a distributed hill-climbing algorithm also makes decisions locally, without the knowledge of the whole system. Also, online algorithms research naturally focuses on giving bounds for the efficacy of an algorithm rather than for the efficiency.

Competitive analysis has been criticized as being too crude and unrealistic [4]. In this paper, we have narrowed the gap between theoretical worst-case analysis and real-world simulations, which has emerged because of unusual worst case, by making stronger and more realistic assumptions about the input. This is an approach that is well-known in the area of online algorithms; for an overview, see Chapter 5 in [4] for paging algorithms, and Section 2.3 in [7] for bin packing algorithms.

References

1. Ross Anderson. The eternity service. *Proceedings of Pragocrypt*, 1996.
2. Thomas E. Anderson, Michael Dahlin, Jeanna M. Neefe, David A. Patterson, Drew S. Roselli, and Randolph Wang. Serverless network file systems. *ACM Transactions on Computer Systems*, 14(1):41–79, February 1996.
3. William J. Bolosky, John R. Douceur, David Ely, and Marvin Theimer. Feasibility of a serverless distributed file system deployed on an existing set of desktop PCs. In *Proceedings of the ACM SIGMETRICS International Conference on Measurement and Modeling of Computing Systems*, 2000.
4. Allan Borodin and Ran El-Yaniv. *Online Computation and Competitive Analysis*. Cambridge University Press, 1998.
5. Yuan Chen, Jan Edler, Andrew Goldberg, Allan Gottlieb, Sumeet Sobti, and Peter Yianilos. A prototype implementation of archival intermemory. In *Proceedings of the Fourth ACM International Conference on Digital Libraries*, 1999.

6. Ian Clarke, Oskar Sandberg, Brandon Wiley, and Theodore W. Hong. Freenet: A distributed anonymous information storage and retrieval system, 2000.
7. Edward G. Coffman, M.R. Garey, and David S. Johnson. Appromxiation algorithms for bin packing: A survey. In Dorit S. Hochbaum, editor, *Approximation Algorithms for NP-Hard Problems*. PWS Publishing Company, 1995.
8. John Douceur and William Bolosky. A large-scale study of file-system contents. In *Proceedings of the ACM SIGMETRICS International Conference on Measurement and Modeling of Computing Systems*, pages 59–70, New York, May 1–4 1999.
9. John Douceur and Roger Wattenhofer. Competitive hill-climbing strategies for replica placement in a distributed file system. In *Proceedings of the 15th International Symposium on Distributed Computing*, 2001.
10. John Douceur and Roger Wattenhofer. Optimizing file availability in a serverless distributed file system. In *Proceedings of the 20th Symposium on Reliable Distributed Systems*, 2001. Also see http://research.microsoft.com/sn/farsite/.
11. Gnutella. See http://gnutelladev.wego.com.
12. Leslie A. Hall. Approximation algorithms for scheduling. In Dorit S. Hochbaum, editor, *Approximation Algorithms for NP-Hard Problems*. PWS Publishing Company, 1995.
13. Sandy Irani and Anna R. Karlin. Online computation. In Dorit S. Hochbaum, editor, *Approximation Algorithms for NP-Hard Problems*. PWS Publishing Company, 1995.
14. John Kubiatowicz, David Bindel, Patrick Eaton, Yan Chen, Dennis Geels, Ramakrishna Gummadi, Sean Rhea, Westley Weimer, Chris Wells, Hakim Weatherspoon, and Ben Zhao. OceanStore: An architecture for global-scale persistent storage. *ACM SIGPLAN Notices*, 35(11):190–201, November 2000.
15. Napster. See http://www.napster.com.
16. Roger T. Reich and Doyle Albee. S.M.A.R.T. phase-II. *White paper WP-9803-001, Maxtor Corporation*, February 1998.
17. Chandramohan A. Thekkath, Timothy Mann, and Edward K. Lee. Frangipani: A scalable distributed file system. In *Proceedings of the 16th Symposium on Operating Systems Principles (SOSP-97)*, volume 31,5 of *Operating Systems Review*, pages 224–237, New York, October 5–8 1997.

Computing Cycle Covers without Short Cycles

Markus Bläser and Bodo Siebert*

Institut für Theoretische Informatik, Med. Universität zu Lübeck
Wallstraße 40, 23560 Lübeck, Germany
blaeser/siebert@tcs.mu-luebeck.de

Abstract. A cycle cover of a graph is a spanning subgraph where each node is part of exactly one simple cycle. A k-cycle cover is a cycle cover where each cycle has length at least k. We call the decision problems whether a directed or undirected graph has a k-cycle cover k-DCC and k-UCC. Given a graph with edge weights one and two, Min-k-DCC and Min-k-UCC are the minimization problems of finding a k-cycle cover with minimum weight.

We present factor $4/3$ approximation algorithms for Min-k-DCC with running time $O(n^{5/2})$ (independent of k). Specifically, we obtain a factor $4/3$ approximation algorithm for the asymmetric travelling salesperson problem with distances one and two and a factor $2/3$ approximation algorithm for the directed path packing problem with the same running time. On the other hand, we show that k-DCC is $\mathcal{NP}$-complete for $k \geq 3$ and that Min-k-DCC has no PTAS for $k \geq 4$, unless $\mathcal{P} = \mathcal{NP}$.

Furthermore, we design a polynomial time factor $7/6$ approximation algorithm for Min-k-UCC. As a lower bound, we prove that Min-k-UCC has no PTAS for $k \geq 12$, unless $\mathcal{P} = \mathcal{NP}$.

1 Introduction

A cycle cover of an either directed or undirected graph G is a spanning subgraph C where each node of G is part of exactly one simple cycle of C. Computing cycle covers is an important task in graph theory, see for instance Lovász and Plummer [14], Graham et al. [7], and the vast literature cited there.

A k-restricted cycle cover (or k-cycle cover for short) is a cycle cover in which each cycle has length at least k. To be specific, we call the decision problems whether a graph has a k-cycle cover k-DCC, if the graph is directed, and k-UCC, if the graph is undirected. Since k-DCC and k-UCC are $\mathcal{NP}$-complete for $k \geq 3$ and $k \geq 6$, respectively, we also consider the following relaxation: given a complete loopless graph with edge weights one and two, find a k-cycle cover of minimum weight. Note that a graph $G = (V, E)$ has a k-cycle cover if the corresponding weighted graph has a k-cycle cover of weight $|V|$, where edges get weight one and "nonedges" get weight two in the corresponding complete graph. We call these problems Min-k-DCC and Min-k-UCC. They stand in one-to-one correspondence with simple 2-factors as defined by Hartvigsen [9]. A simple 2-factor is a spanning subgraph that contains only node-disjoint paths and cycles of length at least k. (The paths arise from deleting the weight two edges from the cycles.)

---

* supported by DFG research grant Re 672/3

F. Meyer auf der Heide (Ed.): ESA 2001, LNCS 2161, pp. 368–379, 2001.
© Springer-Verlag Berlin Heidelberg 2001

As our main contribution, we devise approximation algorithms for finding minimum weight k-cycle covers in graphs with weights one and two. Moreover, we provide lower bounds in terms of $\mathcal{NP}$-completeness and nonapproximability, thus determining the computational complexity of these problems for almost all k.

1.1   Previous Results

The problems 2-DCC and Min-2-DCC of finding a (minimum) 2-cycle cover in directed graphs can be solved in polynomial time by reduction to the bipartite matching problem. To our knowledge, nothing is known for values $k \geq 3$.

The problem 3-UCC of finding a 3-cycle cover in undirected graphs can be solved in polynomial time using Tutte's reduction [18] to the classical perfect matching problem in undirected graphs which can be solved in polynomial time (see Edmonds [4]). Also Min-3-UCC can be solved in polynomial time. Hartvigsen [8] has designed a powerful polynomial time algorithm for 4-UCC. This algorithm works for Min-4-UCC, too. He has also presented a polynomial time algorithm that computes a minimum weight 5-cycle cover in graphs where the weight one edges form a bipartite graph [9]. On the other hand, Cornuéjols and Pulleyblank [3] have reported that Papadimitriou showed the $\mathcal{NP}$-completeness of k-UCC for $k \geq 6$.

Let n be the number of nodes of a graph $G = (V, E)$. For $k > n/2$ the problem Min-k-DCC is the asymmetric and Min-k-UCC is the symmetric travelling salesperson problem with distances one and two. These problems are $\mathcal{APX}$-complete [17]. For explicit lower bounds, see Engebretsen and Karpinski [5]. The best upper bound for the symmetric case is due to Papadimitriou and Yannakakis [17]. They give a factor $7/6$ approximation algorithm running in polynomial time. For the asymmetric case, Vishwanathan [19] presents a polynomial time factor $17/12$ approximation algorithm. Exploiting an algorithm by Kosaraju, Park, and Stein [12] for the asymmetric maximum travelling salesperson problem, one obtains an approximation algorithm with performance ratio $88/63 \approx 1.397$ by replacing weights two with weights zero.

Closely related to the travelling salesperson problems with distances one and two is the node-disjoint path packing problem. This problem has various applications, such as mapping parallel programs to parallel architectures and optimization of code, see e.g. Vishwanathan [19] and the pointers provided there. We are given a directed or undirected graph. Our goal is to find a spanning subgraph S consisting of node-disjoint paths such that the number of edges in S is maximized. Utilizing the algorithms of Papadimiriou and Yannakakis [17] and of Kosaraju, Park, and Stein [12], one obtains a polynomial time factor $5/6$ approximation algorithm for the undirected problem and a polynomial time approximation algorithm with performance ratio $38/63 \approx 0.603$ for the directed problem.

1.2   Our Results

We present factor $4/3$ approximation algorithms for Min-k-DCC with running time $O(n^{5/2})$ (independent of k). Specifically, we obtain a factor $4/3$ approximation algorithm for the asymmetric travelling salesperson problem with distances one and two and a factor $2/3$ approximation algorithm for the directed node-disjoint path packing

---

Input: a complete loopless directed graph G with edge weights one and two,
              an integer $k \geq 3$.
Output: a k-cycle cover of G.

1. Compute a minimum weight 2-cycle cover C of G.
2. Form the bipartite graph B and compute the function F.
3. Compute a decomposition of F as in Lemma 2
4. Patch the cycles of C together acccording to the refined patching procedure.

Fig. 1. The algorithm for directed cycle covers.

problem with the same running time, thus improving the results of Vishwanathan and Kosaraju, Park, and Stein. On the other hand, we show that k-DCC is $\mathcal{NP}$-complete for $k \geq 3$ and that Min-k-DCC does not have a PTAS for $k \geq 4$, unless $\mathcal{P} = \mathcal{NP}$. For the undirected case, we design factor $7/6$ approximation algorithms for Min-k-UCC with polynomial running time (independent of k). It includes the algorithm of Papadimitriou and Yannakakis as a special case. As a lower bound, we prove that there is no PTAS for Min-k-UCC for $k \geq 12$, unless $\mathcal{P} = \mathcal{NP}$.

2   Approximation Algorithms for Directed Cycle Covers

In this section, we present approximation algorithms for Min-k-DCC with performance ratio $4/3$ for any $k \geq 3$ running in time $O(n^{5/2})$ (independent of k). Particularly, we obtain a factor $4/3$ approximation algorithm for the asymmetric travelling salesperson problem with distances one and two by choosing $k > n/2$.

Our input consists of a complete loopless directed graph G with node set V of cardinality n, a weight function w that assigns each edge of G weight one or two, and an integer $k \geq 3$. Our aim is to find a k-cycle cover of minimum weight. For the analysis, we assume that a minimum weight k-cycle cover of G has weight $n + \ell$ for some $0 \leq \ell \leq n$. In other words, a minimum weight k-cycle cover consists of $n - \ell$ edges of weight one and ℓ edges of weight two.

Figure 1 gives an overview of our algorithm. A detailed explanation of each of the steps is given in the subsequent paragraphs.

Computing a 2-cycle cover. We first compute an optimal 2-cycle cover C of G. This can be done in polynomial time. Assume that this cover C consists of cycles $c_1, \ldots, c_r$. We denote the set $\{c_1, \ldots, c_r\}$ by $\mathcal{C}$. The lengths of some of these cycles may already be k or larger, but some cycles, say $c_1, \ldots, c_s$ for $s \leq r$, have length strictly less than k. The basic idea of our algorithm is to use *cycle patching* (also called subtour patching when considering travelling salesperson problems, see Lawler et al. [13]). A straight forward way is to discard one edge (if possible of weight two) of each cycle of length strictly less than k and patch the resulting paths arbitrarily together to obtain one long cycle. An easy analysis shows that this yields a factor $3/2$ approximation. We obtain the $4/3$ approximation by refining the patching procedure.

Auxiliary edges. For the refined patching procedure, we form a bipartite graph B as follows: we have the node set $\mathcal{C}^< = \{c_1, \ldots, c_s\}$ on the one side and V on the other. There is an edge (c, v) in B iff v does not belong to the cycle c and there is a node u in c such that (u, v) has weight one in G.

Lemma 1. *B has a matching of cardinality at least $s - \ell$.*

Proof. Consider an optimal k-cycle cover C_{opt} of G. Since the length of the cycles in $\mathcal{C}^<$ are strictly less than k, for each cycle c in $\mathcal{C}^<$ there is an edge (u, v) of C_{opt} such that u belongs to c but v does not. Fix such an edge for each cycle in $\mathcal{C}^<$. At least $s - \ell$ of these edges have weight one, thus appear in B and form a matching.   □

Decomposition of functions. We compute a maximum matching M in B. From M we obtain a directed graph $F = (\mathcal{C}, A)$ with $(c, c') \in A$ whenever (c, u) is an edge of M and u is a node of c'. Each node of F has outdegree at most one, thus F defines a partial function $\mathcal{C} \to \mathcal{C}$ whose domain is a subset of $\mathcal{C}^<$. By abuse of notation, we call this function again F. By the construction of B, we have $F(c) \neq c$ for all $c \in \mathcal{C}$, i.e. F does not contain any loops.

Lemma 2. *Any loopless partial function F has a spanning subgraph S consisting solely of node-disjoint trees of depth one, paths of length two, and isolated nodes such that any node in the domain of F is not an isolated node of S. Such a spanning subgraph S can be found in polynomial time.*

Proof. Every weakly connected component of F is either a cycle possibly with some trees converging into it, a tree (whose root r is not contained in the domain of F), or an isolated node (which is also not contained in the domain of F). It suffices to prove the lemma for each weakly connected component of F.

The case where a component is a cycle with some trees converging into it follows from Papadimitriou and Yannakakis [17, Lem. 2]. In this case, no isolated nodes arise.

In the case of a tree, we take a leaf that has maximum distance from the root r. Let s be the successor of that leaf. If s equals r, then the component considered is a tree of depth one and we are done. Otherwise, we build a tree of height one with root s and all predecessors of s as leaves, remove this tree, and proceed inductively. We end up with a collection of node disjoint trees of height one and possibly one isolated node, the root r of the component. Since r is not contained in the domain of F, this case is completed.

If a node is isolated, then this node is not contained in the domain of F, because F is loopless. Again, we are done.

This decomposition can be computed in polynomial time.   □

A refined patching procedure. We compute a decomposition of the directed graph F according to Lemma 2. Isolated nodes of this decomposition correspond to elements of $\mathcal{C}$ not in the domain of F, i.e. either cycles of length at least k or unmatched cycles from $\mathcal{C}^<$. The former ones, call them $\mathcal{C}^{\geq}_{\text{iso}}$, fulfil the requirements of a k-cycle cover, thus we can ignore them in the subsequent considerations. We denote the latter ones by $\mathcal{C}^<_{\text{iso}}$. The cycles in $\mathcal{C}^<_{\text{iso}}$ have length strictly less than k. We merge those cycles to one long cycle d, breaking an edge of weight two whenever possible.

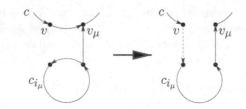

Fig. 2. Trees of height one

Fig. 3. Paths of length two

Next, we consider the trees of height one. Let c be the root of such a tree and $c_{i_1}, \ldots, c_{i_m} \in \mathcal{C}^< \setminus \mathcal{C}^<_{\text{iso}}$ be its leaves. For each cycle c_{i_μ}, there is an edge from c_{i_μ} to a node v_μ of c. By construction, these nodes $v_1, \ldots, v_m$ are pairwise distinct. We merge c_{i_1} and c as depicted in Fig. 2. We call this new cycle again c and incorporate the remaining cycles $c_{i_2}, \ldots, c_{i_m}$ in the same fashion. (The node v in Fig. 2 may be some $v_{\mu'}$ but this does not matter.) After that we merge c with d. In c, we break one of the edges drawn dashed in Fig. 2. In d, we discard an edge that does not belong to a cycle in $\mathcal{C}^<_{\text{iso}}$, i.e. has been added during the merging process. We call this new cycle again d.

Finally, we consider the paths of length two. The three cycles corresponding to such a path are merged as shown in Fig. 3. (The end node of the edge e and the start node of edge f may coincide. Moreover, the two removed edges of the cycle in the middle may coincide. In the latter case, we only incur weight one instead of weight two.) The resulting cycle will be merged with d as described above in the case of a tree.

At the end of this procedure, we are left with the cycle d and the cycles in $\mathcal{C}^{\geq}_{\text{iso}}$. If the cycle d still has length stricly less than k, we break one cycle $b \in \mathcal{C}^{\geq}_{\text{iso}}$ and merge it with d. The resulting cycle has length at least k. If possible, we choose b such that b contains an edge of weight two and break this edge.

Analysis. The algorithm runs in polynomial time. On a unit-cost RAM (with all used numbers bounded by a polynomial in n), the 2–cycle cover and the bipartite matching can be computed in time $O(n^{5/2})$, see e.g. Papadimitriou and Steiglitz [15]. The decomposition of F and the cycle patching can be done in time $O(n^2)$ in a straight forward manner. Thus, the overall running time is $O(n^{5/2})$.

We proceed with estimating the approximation performance.

Lemma 3. *For $n \geq 12$, the k-cycle cover produced by the algorithm has weight no worse than $4/3$ times the weight of an optimal k-cycle cover.*

Proof. Let C_{opt} be an optimal k-cycle cover of G. Since C_{opt} is also a 2-cycle cover of G, we have $w(C) \leq w(C_{\text{opt}}) = n + \ell$. The cost of the k-cycle cover produced by the algorithm is $w(C)$ plus the extra costs due to the mergings.

First, when merging the cycles in $\mathcal{C}_{\text{iso}}^{<}$ to form the cycle d, we incur an extra cost of one for each $c \in \mathcal{C}_{\text{iso}}^{<}$.

Next, we consider the merging as shown in Fig. 2. We charge the costs to v_{μ} and the nodes of c_{μ}. These are at least three nodes. Since the edge of F has weight one, the cost of this merging is at most $1/3$ per node involved. The merging of c with d is free of costs, since we only break edges we have already paid for when forming c and d.

In the case depicted in Fig. 3, we charge the costs of the merging to the nodes of the three cycles. These are at least six nodes. Altogether, the cost of this merging is again at most $1/3$ per node involved. As above, the merging with d is free of costs.

It is clear that each node is only charged once this way. For the moment, assume that the cycle d has length at least k, thus an additional merging is not needed. Let n_2 be the total number of nodes contained in the cycles from $\mathcal{C}_{\text{iso}}^{<}$. The weight $w(C_{\text{apx}})$ of the k-cycle cover C_{apx} produced by the algorithm is at most

$$w(C_{\text{apx}}) \leq n + \ell + \tfrac{1}{3}(n - n_2) + |\mathcal{C}_{\text{iso}}^{<}|. \tag{1}$$

We have $n_2 \geq 2 \cdot |\mathcal{C}_{\text{iso}}^{<}|$ and, by Lemma 1, $|\mathcal{C}_{\text{iso}}^{<}| \leq \ell$. Hence $w(C_{\text{apx}}) \leq \tfrac{4}{3}(n + \ell)$.

If d has length strictly less than k, then one additional merging is needed. This yields an approximation ratio of $4/3 + \epsilon$ for any $\epsilon > 0$. We can get rid of the ϵ by refining the analysis as follows. Either the merging process of $b \in \mathcal{C}_{\text{iso}}^{\geq}$ and d is free of costs, since b contains an edge of weight two, or all cycles in $\mathcal{C}_{\text{iso}}^{\geq}$ consist solely of weight one edges. Since $\mathcal{C}_{\text{iso}}^{\geq}$ is nonempty, these are at least $n/2$ edges. The cycle d contains at least half of the original edges of the merged cycles. Hence d and the cycles in $\mathcal{C}_{\text{iso}}^{\geq}$ contain at least a fraction of $3/4$ of the edges of the 2-cycle cover C. Thus, after the last merging step, we have a cycle cover of weight at most $\tfrac{5}{4}(n + \ell) + 1 \leq \tfrac{4}{3}(n + \ell)$ for $n \geq 12$. $\square$

Theorem 1. *There is a factor $4/3$ approximation algorithm for* Min-k-DCC *running in time $O(n^{5/2})$ for any $k \geq 3$.* $\square$

Corollary 1. *There is a factor $4/3$ approximation algorithm for the asymmetric travelling salesperson problem with distances one and two running in time $O(n^{5/2})$.* $\square$

Corollary 2. *There is a factor $2/3$ approximation algorithm for the node-disjoint path packing problem in directed graphs running in time $O(n^{5/2})$.*

Proof. We transform a given directed graph G into a complete loopless directed graph H with edge weights one and two by assigning edges of G weight one and "nonedges" weight two. The details are spelled out by Vishwanathan [19, Sect. 2]. $\square$

3   Approximation Algorithms for Undirected Cycle Covers

We outline factor $7/6$ approximation algorithms for Min-k-UCC for any $k \geq 5$. In particular, we recover the factor $7/6$ approximation algorithm for the symmetric travelling salesperson problem with distances one and two of Papadimitriou and Yannakakis [17, Thm. 2] by choosing $k > n/2$. The algorithm is quite similar to the directed case, so we confine ourselves to pointing out the differences.

Computing an optimal 4-*cycle cover.* Instead of starting with a minimum weight 2-cycle cover, we exploit Hartvigsen's polynomial time algorithm [8] for computing a minimum weight 4-cycle cover C. This gives us the inequality $n_2 \geq 4 \cdot |C_{iso}^<|$ (instead of $n_2 \geq 2 \cdot |C_{iso}^<|$ in the directed case).

Auxiliary edges. A little more care is necessary when collecting auxiliary edges via the matching in B, since for each weight two edge, we may now only spend an extra amount of $1/6$ instead of $1/3$. We normalize the computed 4-cycle cover C as follows: first we may assume that there is only one cycle t with weight two edges, since we may merge two such cycles without any costs. Second, we may assume that for each weight two edge $\{u, v\}$ of t there is no weight one edge $\{v, x\}$ in G for some node x of a different cycle, because otherwise we may merge this cycle with t at no costs. We now may bound the number of nodes for which we have to charge extra costs of $1/6$ in (1) by $n - n_2 - \ell$ instead of $n - n_2$. This is due to the fact that if t is the root of a tree of height one according to the decomposition of Lemma 2, then at least ℓ nodes of t are unmatched because of the second above mentioned property of t. Altogether, the total weight is
$$w(C_{\mathrm{apx}}) \leq \tfrac{7}{6}n + \tfrac{5}{6}\ell + \tfrac{1}{3}|C_{iso}^<| \leq \tfrac{7}{6}(n + \ell).$$

Decomposition of functions. The decomposition according to Lemma 2 works without any changes in the undirected case.

A refined patching procedure. Here we use the patching procedure as devised by Papadimitriou and Yannakakis [17, Fig. 2], which is only suited for the undirected case. Together with the fact that each involved circle has at least four nodes (instead of two in the directed case) we obtain lower the merging costs of $1/6$ per node.

Applying the above mentioned modifications to our algorithm for the directed case, we get the following theorem.

Theorem 2. *For any* $k \geq 5$, *there is a polynomial time factor* $7/6$ *approximation algorithm for* Min-k-UCC. $\square$

4   Lower Bounds

4.1 $\mathcal{NP}$-Completeness of 3-DCC

To show the $\mathcal{NP}$-completeness of 3-DCC we will reduce 3-Dimensional Matching (3DM) to this problem. Consider a hypergraph $H = (W, X)$ with $W = W_0 \cup W_1 \cup W_2$ and $X \subseteq W_0 \times W_1 \times W_2$. The sets W_0, W_1, W_2 are disjoint and of the same size, $W_k = \{w_1^k, \dots, w_n^k\}$. 3DM is the question whether there exists a subset $X' \subseteq X$ such that each element of W appears in exactly one element of X' (perfect 3-dimensional matching). 3DM is known to be $\mathcal{NP}$-complete (Garey, Johnson [6]).

We construct a graph $G = (V, E)$ such that G has a 3-cycle cover iff H has a perfect 3-dimensional matching. Let $V^\times = (W_0 \times W_1) \cup (W_1 \times W_2) \cup (W_2 \times W_0)$. For $k = 0, 1, 2$ and $j = 1, \dots, n$ let $U_j^k = \{u_j^k[i, q] \mid i = 1, \dots, n-1 \wedge q = 1, 2, 3\}$ be a set of *helper nodes* for w_j^k. The set of nodes V is given by $V = V^\times \cup \left(\bigcup_{k=0}^{2} \bigcup_{j=1}^{n} U_j^k\right)$.

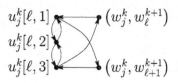

Fig. 4. The subgraph connecting $\left(w_j^k, w_\ell^{k+1}\right)$ and $\left(w_j^k, w_{\ell+1}^{k+1}\right)$ via three helper nodes.

For each edge $\left(w_a^0, w_b^1, w_c^2\right) \in X$ we construct three edges $\left((w_a^0, w_b^1), (w_b^1, w_c^2)\right)$, $\left((w_b^1, w_c^2), (w_c^2, w_a^0)\right)$, $\left((w_c^2, w_a^0), (w_a^0, w_b^1)\right) \in E$ connecting the corresponding elements of $V^\times$. Furthermore, two nodes $\left(w_j^k, w_\ell^{(k+1) \bmod 3}\right)$ and $\left(w_j^k, w_{\ell+1}^{(k+1) \bmod 3}\right)$ are connected via helper nodes as shown in Fig. 4. In the following we write $k+1$ instead of $(k+1) \bmod 3$ for short.

We divide the set $V^\times$ into subsets $\Gamma_j^k = \left\{ (w_j^k, w_\ell^{k+1}) \mid \ell = 1, \ldots, n \right\}$. The subset Γ_j^k contains the nodes that represent w_j^k.

Assume that G has a 3-cycle cover C. We call a helper node $u_j^k[\ell, 2]$ and a node of $V^\times$ *companions* if they are part of the same cycle in C. Due to the construction either $\left(w_j^k, w_\ell^{k+1}\right)$ or $\left(w_j^k, w_{\ell+1}^{k+1}\right)$ is the only companion of $u_j^k[\ell, 2]$. Hence, the following lemma holds.

Lemma 4. *Assume G has a 3-cycle cover G. For any $w_j^k \in V$ exactly $n - 1$ of the n nodes in Γ_j^k have a companion.* $\square$

We say that the only node $\left(w_j^k, w_\ell^{k+1}\right) \in \Gamma_j^k$ that has no companion *participates* for w_j^k. Now we are prepared to prove the $\mathcal{NP}$-completeness of 3-DCC.

Theorem 3. *3-DCC is $\mathcal{NP}$-complete.*

Proof. Given a hypergraph H we construct a graph G as described above.

Assume H has a 3-dimensional matching $X' \subseteq X$. Then G has the following 3-cycle cover. For any $\left(w_a^0, w_b^1, w_c^2\right) \in X'$ let $\left(w_a^0, w_b^1\right)$, $\left(w_b^1, w_c^2\right)$, and $\left(w_c^2, w_a^0\right)$ participate for w_a^0, w_b^1, and w_c^2. These three nodes form a cycle of length 3. Let $\left(w_j^k, w_\ell^{k+1}\right)$ be the node that participates for w_j^k. Then for $\ell' < \ell$ the nodes $u_j^k[\ell', 2]$ and $\left(w_j^k, w_{\ell'}^{k+1}\right)$ and for $\ell' > \ell$ the nodes $u_j^k[\ell' - 1, 2]$ and $\left(w_j^k, w_{\ell'}^{k+1}\right)$ are companions. Thus, for $\ell' < \ell$ the nodes $\left(w_j^k, w_{\ell'}^{k+1}\right)$ and $u_j^k[\ell', q]$ ($q = 1, 2, 3$) and for $\ell' > \ell$ the nodes $\left(w_j^k, w_{\ell'}^{k+1}\right)$ and $u_j^k[\ell' - 1, q]$ ($q = 1, 2, 3$) form cycles each of length 4. Thus all nodes of G are covered by a cycle of length at least 3.

On the other hand assume that G has a 3-cycle cover C. Due to Lemma 4 we only have to take care for the participating nodes in $V^\times$. The participating nodes form cycles whose lengths are multiples of 3. We cut all the edges from $W_1 \times W_2$ to $W_2 \times W_0$. The remaining paths of lengths two yield a 3-dimensional matching for H.

Noting that 3-DCC is in $\mathcal{NP}$ completes the proof. $\square$

By replacing the nodes of $V^\times$ by paths and extending the helper node constructions we obtain the following generalization.

Theorem 4. *The problem k-DCC is $\mathcal{NP}$-complete for any $k \geq 3$.* $\square$

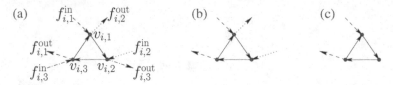

Fig. 5. The clause gadget G_i if c_i consists of (a) three, (b) two, or (c) one literal. The dashed, dotted, and dash-dotted edges correspond to the first, second, and third literal of c_i, respectively.

4.2   Nonapproximability of Min-4-DCC

In this section we show that Min-4-DCC does not have a polynomial time approximation scheme (PTAS, see e.g. Ausiello et al. [2]), unless $\mathcal{NP} = \mathcal{P}$. For this purpose we reduce Max-3SAT(3) to this problem. An instance of Max-3SAT is a set F of disjunctive clauses where each clause consists of at most three literals. Max-3SAT is the problem of finding the maximum number of simultaneously satisfiable clauses. Max-3SAT(3) is the restricted version where each variable occurs at most three times in F. We may assume that each variable occurs at least once positive and at least once negative. Otherwise, we can eliminate this variable by setting it to the appropriate value. In particular, each variable occurs twice or three times. Papadimitriou and Yannakakis [16] have shown that Max-3SAT(3) is MAX $\mathcal{SNP}$-complete. They have presented a reduction from Max-3SAT to Max-3SAT(3) using so called regular expanders (see e.g. Ajtai [1]). A set F of clauses will be called η-satisfiable iff $\eta \cdot |F|$ is the maximum number of satisfiable clauses in F. Håstad [10] has proven that it is $\mathcal{NP}$-hard to distinguish 1- and $(7/8 + \epsilon)$-satisfiable instances of Max-3SAT for any $\epsilon > 0$. The reduction of Papadimitriou and Yannakakis and the result of Håstad yield the following lemma.

Lemma 5. *There exists a constant $\lambda < 1$ such that it is $\mathcal{NP}$-hard to distinguish 1- and λ-satisfiable instances of* Max-3SAT(3). ☐

We reduce Max-3SAT(3) to Min-4-DCC. For this purpose, let $F = \{c_1, \dots, c_t\}$ be a set of disjunctive clauses over variables $U = \{x_1, \dots, x_r\}$. We construct a graph $G = (V, E)$. For each variable x_j we have one node $u_j \in V$. These nodes will be called *variable nodes*. For each clause c_i we have three nodes $v_{i,1}, v_{i,2}, v_{i,3} \in V$. Let $V_i = \{v_{i,1}, v_{i,2}, v_{i,3}\}$ be the set of these nodes.

In the following we describe how the nodes of G are connected via edges with weight one. All other edges have weight two. The nodes in V_i are connected via a cycle as shown in Fig. 5. The subgraph induced by V_i will be called the *clause gadget* G_i. The clause gadgets and the variable nodes are connected as follows. Each variable node u_j has two incoming and two outgoing edges $e_{j,+}^{in}, e_{j,+}^{out}$ representing the literal x_j, and $e_{j,-}^{in}, e_{j,-}^{out}$ representing $\overline{x}_j$ as depicted in Fig. 6a. If c_i is the first clause where the literal x_j appears in, then the edge $e_{j,+}^{out}$ is identical with either $f_{i,1}^{in}$, $f_{i,2}^{in}$, or $f_{i,3}^{in}$ depending on where x_j occurs in c_i. If a literal occurs in more than one clause then one of the outgoing edges of the first gadget and one of the incoming edges of the second gadget are identical according to where the literal appears in these clauses. If c_i is the last clause where the literal x_j appears in, then the edge $e_{j,+}^{in}$ is identical with either $f_{i,1}^{out}$, $f_{i,2}^{out}$, or $f_{i,3}^{out}$

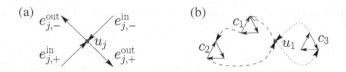

Fig. 6. (a) The edges starting and ending at u_j. (b) Example: x_1 is the first literal in both c_1 and c_2, and $\bar{x}_1$ is the second literal in c_3.

depending on where x_j occurs in c_i. The clauses which contain $\bar{x}_j$ are connected in a similar fashion. An example is shown in Figure 6b.

Let $C = (V, E_C)$ be a cycle cover of G. We introduce a weight $\nu_C(\tilde{V})$ for subsets $\tilde{V} \subseteq V$ of nodes. For a single node $z \in V$ the weight $\nu_C(\{z\})$ of z with respect to the cycle cover C is half of the weight of its incoming edge plus half of the weight of its outgoing edge. For $\tilde{V} \subseteq V$ we have $\nu_C(\tilde{V}) = \sum_{z \in \tilde{V}} \nu_C(\{z\})$. Then $\nu_C(V) = w(C)$.

Let an assignment for the variables U be given that satisfies k of the t clauses of F. We construct a 4-cycle cover C with weight $w(C) = r + 4 \cdot t - k$ as follows. If $x_j = \text{true}$ then $e_{j,+}^{\text{in}}, e_{j,+}^{\text{out}} \in E_C$, otherwise $e_{j,-}^{\text{in}}, e_{j,-}^{\text{out}} \in E_C$. If the j-th literal of c_i is true then $f_{i,j}^{\text{in}}, f_{i,j}^{\text{out}} \in E_C$. For any satisfied clause c_i add some of the edges $(v_{i,1}, v_{i,2})$, $(v_{i,2}, v_{i,3})$, and $(v_{i,3}, v_{i,1})$ if necessary. The clauses that are not satisfied are connected in one big cycle. This yields weight 4 per unsatisfied clause. Every node in C has indegree 1 and outdegree 1. Hence C is a cycle cover. If a clause c_i is satisfied by only one literal, then the cycle passing G_i contains two edges within G_i and both one incoming and one outgoing edge. If a clause is satisfied by more than one literal, then the cycle passes at least two variable nodes. Thus, the obtained cycle cover is a 4-cycle cover. The case in which only one clause is not satisfied by the assignment is a special one since then we cannot form one big loop of the unsatisfied clauses. But this case is negligible for sufficiently large instances of Max-3SAT(3), since we are interested in approximability.

For the other direction consider a 4-cycle cover C of G and a clause gadget G_i with $\nu_C(V_i) = 3$. Such a gadget will be called *satisfying*. Any clause gadget that is not satisfying yields weight at least $7/2$. It holds that G_i is satisfying iff all edges that start or end in G_i have weight one. Since all cycles must have at least length 4 we have the following lemma.

Lemma 6. *Let $C = (V, E_C)$ be an arbitrary 4-cycle cover of G and G_i be a satisfying clause gadget in C. Then the following properties hold:*

1. *At least two of the edges $f_{i,1}^{\text{in}}, f_{i,1}^{\text{out}}, \ldots, f_{i,3}^{\text{out}}$ are in E_C.*
2. *For $j = 1, 2, 3$: $f_{i,j}^{\text{in}} \in E' \Leftrightarrow f_{i,j}^{\text{out}} \in E'$.*   □

A satisfying clause gadget G_i yields a partial assignment for the variables of c_i. If $f_{i,j}^{\text{in}}, f_{i,j}^{\text{out}} \in E'$ then the j-th literal of c_i is set true and the corresponding variable is assigned an appropriate value, otherwise we do not assign a value to this literal. Due to Lemma 6 the obtained partial assignment satisfies c_i.

By considering all satisfying clause gadgets we step by step obtain a partial assignment that satisfies at least those clauses whose gadgets are satisfying. The following lemma assures that the obtained partial assignment is consistent, i.e. we never have to assign both `true` and `false` to one variable.

Lemma 7. *Let C be an arbitrary 4-cycle cover of G. There are no two satisfying clause gadgets G_i and $G_{i'}$ and a variable x_j such that x_j has to be assigned different values according to these clause gadgets.* $\square$

Proof. If x_j has to be assigned different values then x_j occurs positive in c_i and negative in $c_{i'}$ or vice versa. We only consider the first case. By symmetry, we can restrict ourselves to the case where the literal $\overline{x}_j$ occurs exactly once in F and that x_j is the first variable in both clauses. According to $G_{i'}$ the literal $\overline{x}_j$ has to be assigned `true`. Since $c_{i'}$ is the only clause that contains $\overline{x}_j$, the two edges $e_{j,-}^{\text{in}} = f_{i',1}^{\text{out}}$ and $e_{j,-}^{\text{out}} = f_{i',1}^{\text{in}}$ belong to E_C. On the other hand, $f_{i,1}^{\text{out}}$ and $f_{i,1}^{\text{in}}$ belong to E_C. Since x_j occurs at most twice positive in F at least one of the edges $f_{i,1}^{\text{out}}$ and $f_{i,1}^{\text{in}}$ connects V_i to u_j. Thus, u_j has indegree at least 2 or outdegree at least 2, a contradiction. $\square$

Each variable node yields at least weight 1. Each satisfying clause gadget yields weight 3. All other clause gadgets yield at least weight $7/2$.

The following theorem proves that a constant $\xi > 1$ exists such that Min-4-DCC cannot be approximated in polynomial time with performance ratio ξ, unless $\mathcal{NP} = \mathcal{P}$. Thus, Min-4-DCC does not have a PTAS, unless $\mathcal{NP} = \mathcal{P}$.

Theorem 5. *There exists a constant $\xi > 1$ such that it is $\mathcal{NP}$-hard to distinguish instances $G = (V, E)$ of* Min-4-DCC *whose minimum cycle cover has weight $|V|$ and instances whose minimum cycle cover has at least weight $\xi \cdot |V|$.*

Proof. Due to the reduction described above, a 1-satisfiable instance of Max-3SAT(3) yields a graph which has a 4-cycle cover with weight $|V| = r + 3 \cdot t$. On the other hand, every 4-cycle cover of a graph corresponding to a λ-satisfiable instance has at least weight $r + 3 \cdot \lambda \cdot t + (7/2) \cdot (1 - \lambda) \cdot t = K_\lambda$. Since every clause consists of at most three literals and every variable appears at least twice we have $r/t \leq 3/2$. Therefore, the following inequality holds:

$$\frac{K_\lambda}{r + 3 \cdot t} \geq \frac{(3/2 + 3 \cdot \lambda + (7/2) \cdot (1 - \lambda)) \cdot t}{(3/2 + 3) \cdot t} = \frac{10 - \lambda}{9} = \xi > 1.$$

Thus, deciding whether the minimum 4-cycle cover of a graph has weight $|V|$ or at least weight $\xi \cdot |V|$ is at least as hard as distinguishing 1- and λ-satisfiable instances of Max-3SAT(3). This completes the proof due to Lemma 5. $\square$

If we replace the variable nodes by paths of lengths $k - 4$ we obtain the result that Min-k-DCC does not have a PTAS for any $k \geq 4$.

Theorem 6. *For any $k \geq 4$ there exists a constant $\xi_k > 1$ such that* Min-k-DCC *cannot be approximated with performance ratio ξ_k, unless $\mathcal{NP} = \mathcal{P}$.* $\square$

We can transform a directed graph into an undirected graph by replacing each node with three nodes (see e.g. Hopcroft and Ullman [11]). Applying this transformation to the graph constructed in this section we obtain the following theorem.

Theorem 7. Min-k-UCC *does not have a PTAS for any $k \geq 12$, unless $\mathcal{NP} = \mathcal{P}$.* $\square$

5   Conclusions and Open Problems

We have presented factor $4/3$ and $7/6$ approximation algorithms for Min-k-DCC and Min-k-UCC, respectively, with polynomial running time (independent of k). On the other hand, we have shown that k-DCC is $\mathcal{NP}$-complete for $k \geq 3$ and Min-k-DCC does not possess a PTAS for $k \geq 4$, unless $\mathcal{NP} = \mathcal{P}$. The status of Min-3-DCC is open. We strongly conjecture that this problem also has no PTAS, unless $\mathcal{NP} = \mathcal{P}$. In the undirected case, Papadimitriou has shown $\mathcal{NP}$-hardness of k-UCC for $k \geq 6$. The complexity of 5-UCC and the approximability of Min-k-UCC for $5 \leq k \leq 11$ remains open.

References

1. M. Ajtai. Recursive construction for 3-regular expanders. *Combinatorica*, 14(4):379–416, 1994.
2. G. Ausiello, P. Crescenzi, G. Gambosi, V. Kann, A. Marchetti-Spaccamela, and M. Protasi. *Complexity and Approximation*. Springer, 1999.
3. G. Cornuéjols and W. Pulleyblank. A matching problem with side constraints. *Discrete Math.*, 29:135–159, 1980.
4. J. Edmonds. Paths, trees, and flowers. *Canad. J. Math.*, 17:449–467, 1965.
5. L. Engebretsen and M. Karpinski. Approximation hardness of TSP with bounded metrics. Technical Report 00-089, Electronic Colloquium on Comput. Complexity (ECCC), 2000.
6. M. R. Garey and D. S. Johnson. *Computers and Intractability: A Guide to NP-Completeness*. W. H. Freeman and Company, 1979.
7. R. L. Graham, M. Grötschel, and L. Lovász, editors. *Handbook of Combinatorics*, volume 1. Elvsevier, 1995.
8. D. Hartvigsen. *An Extension of Matching Theory*. PhD thesis, Carnegie-Mellon University, 1984.
9. D. Hartvigsen. The square-free 2-factor problem in bipartite graphs. In *7th Int. Conf. on Integer Programming and Combinatorial Optimization (IPCO)*, volume 1620 of *Lecture Notes in Comput. Sci.*, pages 234–241. Springer, 1999.
10. J. Håstad. Some optimal inapproximability results. In *Proc. 29th Ann. Symp. on Theory of Comput. (STOC)*, pages 1–10. ACM, 1997.
11. J. E. Hopcroft and J. D. Ullman. *Introduction to Automata Theory, Languages, and Computation*. Addison-Wesley, 1979.
12. S. R. Kosaraju, J. K. Park, and C. Stein. Long tours and short superstrings. In *Proc. 35th Ann. Symp. on Foundations of Comput. Sci. (FOCS)*, pages 166–177. IEEE, 1994.
13. E. L. Lawler, J. K. Lenstra, A. H. G. Rinnooy Kan, and D. B. Shmoys, editors. *The Traveling Salesman Problem*. Wiley, 1985.
14. L. Lovász and M. D. Plummer. *Matching Theory*. Elsevier, 1986.
15. C. Papadimitriou and K. Steiglitz. *Combinatorial Optimization: Algorithms and Complexity*. Prentice-Hall, 1982.
16. C. H. Papadimitriou and M. Yannakakis. Optimization, approximation, and complexity classes. *J. Comput. System Sci.*, 43(3):425–440, 1991.
17. C. H. Papadimitriou and M. Yannakakis. The traveling salesman problem with distances one and two. *Math. Oper. Res.*, 18:1–11, 1993.
18. W. T. Tutte. A short proof of the factor theorem for finite graphs. *Canad. J. Math.*, 6:347–352, 1954.
19. S. Vishwanathan. An approximation algorithm for the asymmetric travelling salesman problem with distances one and two. *Inform. Process. Lett.*, 44:297–302, 1992.

A Polynomial Time Algorithm for the Cutwidth of Bounded Degree Graphs with Small Treewidth*

Dimitrios M. Thilikos[1], Maria J. Serna[1], and Hans L. Bodlaender[2]

[1] Departament de Llenguatges i Sistemes Informàtics, Universitat Politècnica de Catalunya, Campus Nord – Mòdul C5, c/Jordi Girona Salgado, 1-3. 08034, Barcelona, Spain
{mjserna,sedthilk}@lsi.upc.es

[2] Department of Computer Science, Utrecht University, P.O. Box 80.089, 3508 TB Utrecht, The Netherlands
hansb@cs.uu.nl

Abstract. The *cutwidth* of a graph G is defined as the smallest integer k such that the vertices of G can be arranged in a vertex ordering $[v_1, \ldots, v_n]$ in a way that, for every $i = 1, \ldots, n - 1$, there are at most k edges with the one endpoint in $\{v_1, \ldots, v_i\}$ and the other in $\{v_{i+1}, \ldots, v_n\}$. We examine the problem of computing in polynomial time the cutwidth of a partial w-tree with bounded degree. In particular, we show how to construct an algorithm that, in $n^{O(w^2 d)}$ steps, computes the cutwidth of any partial w-tree with vertices of degree bounded by a fixed constant d. Our algorithm is constructive in the sense that it can be adapted to output the corresponding optimal vertex ordering. Also, it is the main subroutine of an algorithm computing the pathwidth of a bounded degree partial w-tree in $n^{O((wd)^2)}$ steps.

1   Introduction

A wide variety of optimization problems can be formulated as layout or vertex ordering problems. In many cases, such a problem asks for the optimal value of some function defined over all the linear orderings of the vertices or the edges of a graph (for a survey, see [9]). One of the best known problems of this type is the problem to compute the *cutwidth* of a graph. It is also known as the MINIMUM CUT LINEAR ARRANGEMENT problem and has several applications such as VLSI design, network reliability, automatic graph drawing, and information retrieval. Cutwidth has been extensively examined and it appears to be closely related with other graph parameters like pathwidth, linear-width, bandwidth,

---

* The work of all the authors was supported by the IST Program of the EU under contract number IST-99-14186 (ALCOM-FT) and, for the first two authors, by the Spanish CYCIT TIC-2000-1970-CE. The work of the first author was partially supported by the Ministry of Education and Culture of Spain, Grant number MEC-DGES SB98 0K148809.

F. Meyer auf der Heide (Ed.): ESA 2001, LNCS 2161, pp. 380–390, 2001.

and modified bandwidth. Briefly, the *cutwidth* of a graph $G = (V(G), E(G))$ is equal to the minimum k for which there exists a vertex ordering of G such that for any 'gap' (place between two successive vertices) of the ordering, there are at most k edges crossing the gap. Computing cutwidth is an NP-complete problem and it remains NP-complete even if the input is restricted to planar graphs with maximum degree 3. There is a polynomial time approximation algorithm with a ratio of $O(\log |V(G)| \log \log |V(G)|)$ and there is a polynomial time approximation scheme if $E(G) = \Theta(|V(G)|^2)$. Relatively few work has been done on detecting special graph classes where computing cutwidth can be done in polynomial time (for complete references to the aforementioned results, see [9]). In [8], an algorithm was given that computes the cutwidth of any tree with maximum degree bounded by d in $O(n(\log n)^{d-2})$ time. This result was improved in 1983 by Yannakakis [21], who presented an $O(n \log n)$ algorithm computing the cutwidth of any tree. Since then, the only polynomial algorithms reported for the cutwidth of graph classes different from trees, concerned special cases such as hypercubes [12] and b-dimensional c-ary cliques [14]. In this paper, we move one step further presenting an polynomial time algorithm for the cutwidth of bounded degree graphs with small treewidth.

The notions of *treewidth* and *pathwidth* appear to play a central role in many areas of graph theory. Roughly, a graph has small treewidth if it can be constructed by assembling small graphs together in a tree structure, namely a tree decomposition of small width (graphs with treewidth at most w are alternatively called *partial w-trees* – see Subsection 2.1 for the formal definitions). A big variety of graph classes appear to have small treewidth, such as trees, outerplanar graphs, series parallel graphs, and Halin graphs (for a detailed survey of classes with bounded treewidth, see [5]). The pathwidth of a graph is defined similarly to treewidth, but not the tree in its definition is required to be a simple line (path). That way, treewidth can be seen as a "tree"-generalization of pathwidth. Pathwidth and treewidth were introduced by Robertson and Seymour in [15, 16] and served as some of the cornerstones of their lengthy proof of the Wagner conjecture, known now as the Graph Minors Theorem (for a survey see [17]). Treewidth appears to have interesting applications in algorithmic graph theory. In particular, a wide range of otherwise intractable combinatorial problems are polynomially, even linearly, solvable when restricted to graphs with bounded treewidth or pathwidth. In this direction, numerous techniques have been developed in order to construct dynamic programming algorithms making use of the "tree" or "line" structure of the input graph (see e.g. [4]). The results of this paper show how these techniques can be used for constructing a polynomial time algorithm for the cutwidth of partial w-trees with vertices of degrees bounded by fixed constants. Our algorithm is a non trivial extension of the linear time algorithm in [19] concerning the parameterized version of the cutwidth problem.

The parameterized version of the cutwidth problem asks whether the cutwidth of a graph is at most k, where k is a fixed small constant. This problem is known to be solvable in polynomial time. In particular, the first polynomial algorithm for k fixed was given by Makedon and Sudborough in [13] where a

$O(n^{k-1})$ dynamic programming algorithm is described. This time complexity has been considerably improved by Fellows and Langston in [11] where, among others, they prove that, for any fixed k, an $O(n^3)$ algorithm can be constructed checking whether a graph has cutwidth at most k. Furthermore, a technique introduced in [10] (see also [2]) further reduced the bound to $O(n^2)$, while in [1] a general method is given to construct a linear time algorithm that decides whether a given graph has cutwidth at most k, for k constant. Finally, in [19], an explicit constructive linear time algorithm was presented able to output the optimal vertex ordering in case of a positive answer. This algorithm is based on the fact that graphs with small cutwidth have also small pathwidth and develops further the techniques in [6,7,18] in order to use a bounded-width path decomposition for computing the cutwidth of G.

This paper extends the algorithm in [19] in the sense that it uses all of its subroutines and it solves the problem of [19] for graphs with bounded treewidth. Although this extension is not really useful for the parameterized version of cutwidth, it appears that it is useful for solving the *general* cutwidth problem for partial w-trees of bounded degree. This is possible due to the observation that the "hidden constants" of all the subroutines of our algorithm remain polynomial even when we ask whether G has cutwidth at most $O(dk) \cdot \log n$. As this upper bound for cutwidth is indeed satisfied, our algorithm is able to compute in $n^{O(w^2 d)}$ steps the cutwidth of bounded degree partial w-trees.

A main technical contribution of this paper is Algorithm Join-Node in Section 3. This algorithm uses the "small treewidth" property of the input graph. It is used as an important subroutine in the algorithm for the main result. Section 2, contains the main definitions and lemmata supporting the operation of Join-Node. Subsection 2.1 contains the definitions of treewidth, pathwidth and cutwidth. Most of the preliminary results of Subsection 2.2, concern operations on sequences of integers and the definitions of the most elementary of them were introduced in [19] and [6] (see also [7,18]). Also, the main tool for exploiting the small treewidth of the input graph is the notion of the *characteristic of a vertex ordering*, introduced in [19] and defined in Subsection 2.3 of this paper. For the above reasons, we use notation compatible with the one used in [19]. In this extended abstract we omit all the proofs that can be found in [20].

Algorithm Join-Node only helps to compute the cutwidth of a bounded degree partial w-tree G but not to *construct* the corresponding vertex ordering. In the full version of this paper [20], we describe how to transform this algorithm to a constructive one in the sense that we now can output a linear arrangement of G with optimal cutwidth. This uses the analogous constructions of [19] and the procedures Join-Orderings and Construct-Join-Orderings described in Section 3.

An interesting consequence of our result is that the pathwidth of bounded degree partial w-trees can be computed in $n^{O((wd)^2)}$ steps. We mention that the existence of a polynomial time algorithm for this problem, without the degree restriction, has been proved in [6]. However, the time complexity of the involved algorithm appears to be very large and has not been reported. Our technique, described in detail in the full version [20], reduces the computation of pathwidth to the problem of computing the cutwidth on hypergraphs. Then the pathwidth is computed using a generalization of our algorithm for hypergraphs with bounded treewidth. That way, we report more reasonable time bounds, provided that the input graph has bounded degree.

2   Definitions and Preliminary Results

All graphs of this paper are finite, undirected, and without loops or multiple edges (our results can be straightforwardly generalized to the case where the last restriction is altered). We denote the vertex (edge) set of a graph G by $V(G)$ $(E(G))$. A linear ordering of the vertices of G is a bijection, mapping $V(G)$ to the integers in $\{1, \ldots, n\}$. We denote such a vertex ordering by the sequence $[v_1, \ldots, v_n]$.

We proceed with a number of definitions and notations, dealing with finite sequences (i.e., ordered sets) of a given finite set $\mathcal{O}$ (most of the notation in this paper is taken from [19] and [6]). For our purposes, $\mathcal{O}$ can be a set of numbers, sequences of numbers, vertices, or vertex sets. The set of sequences of elements of $\mathcal{O}$ is denoted by $\mathcal{O}^*$. Let ω be a sequence of elements from $\mathcal{O}$. We use the notation $[\omega_1, \ldots, \omega_r]$ to represent ω and we define $\omega[i, j]$ as the subsequence $[\omega_i, \ldots, \omega_j]$ of ω (in case $j < i$, the result is the empty subsequence $[\ ]$). We also denote by $\omega(i)$ the element of ω indexed by i.

Given a set S containing elements of $\mathcal{O}$, and a sequence ω, we denote by $\omega[S]$ the subsequence of ω that contains only those elements of ω that are in S. Given two sequences ω^1, ω^2 from $\mathcal{O}^*$, where $\omega^i = [\omega_1^i, \ldots, \omega_{r_i}^i], i = 1, 2$ we define the *concatenation* of ω_1 and ω_2 as $\omega^1 \oplus \omega^2 = [\omega_1^1, \ldots, \omega_{r_1}^1, \omega_1^2, \ldots, \omega_{r_2}^2]$. Unless mentioned otherwise, we always consider that the first element of a sequence ω is indexed by 1, i.e. $\omega = \omega[1, |\omega|]$.

Let G be a graph and $S \subseteq V(G)$. We call the graph $(S, E(G) \cap \{\{x, y\} \mid x, y \in S\})$ the *subgraph of G induced by S* and we denote it by $G[S]$. We denote by $E_G(S)$ the set of edges of G that have an endpoint in S; we also set $E_G(v) = E_G(\{v\})$ for any vertex v. If $E \subseteq E(G)$ then we denote by $V_G(E)$ the set of all the endpoints of the edges in E i.e. we set $V_G(E) = \cup_{e \in E} e$. The neighborhood of a vertex v in graph G is the set of vertices in G that are adjacent to v in G and we denote it by $N_G(v)$, i.e. $N_G(v) = V_G(E_G(v)) - \{v\}$. If l is a sequence of vertices, we denote the set of its vertices by $V(l)$. If $S \subseteq V(l)$ then we define $l[S]$ as the subsequence of l containing only the vertices of S. If l is a sequence of all the vertices of G without repetitions, then we call it an *vertex ordering of G*.

2.1   Treewidth – Pathwidth – Cutwidth

A *nice tree decomposition* of a graph G is a triple (X, U, r) where U is a tree rooted on r whose vertices we call *nodes* and $X = (\{X_i \mid i \in V(U)\})$ is a collection of subsets of $V(G)$ such that 1. $\bigcup_{i \in V(U)} X_i = V(G)$, 2. for each edge $\{v, w\} \in E(G)$, there is an $i \in V(U)$ such that $v, w \in X_i$, and 3. for each $v \in V(G)$ the set of nodes $\{i \mid v \in X_i\}$ forms a subtree of U, 4. every node of U has at most two children, 5. if a node i has two children j, h then $X_i = X_j = X_h$, 6. if a node i has one child j, then either $|X_i| = |X_j|+1$ and $X_j \subset X_i$ or $|X_i| = |X_j|-1$ and $X_i \subset X_j$. The *width* of a tree decomposition $(\{X_i \mid i \in V(U)\}, U)$ equals $\max_{i \in V(U)}\{|X_i| - 1\}$. The *treewidth* of a graph G is the minimum width over all tree decompositions of G. According to [6,3], it is possible, for any constant $k \geq 1$, to construct a linear time algorithm that for any G outputs – if exists – a nice tree decomposition of G of width $\leq k$ and with at most $O(|V(G)|)$ nodes.

A nice tree decomposition $(\{X_i \mid i \in V(U)\}, U, r)$ contains nodes of the following four possible types. A node $i \in V(U)$ is called "*start*" if i is a leaf of U, "*join*" if i has two children, "*forget*" if i has only one child j and $|X_i| < |X_j|$, "*introduce*" if i has only one child j and $|X_i| > |X_j|$. We may also assume that if i is a *start* node then $|X_i| = 1$: the effect of *start* nodes with $|X_i| > 1$ can be obtained by using a *start* node with a set containing 1 vertex, and then $|X_i| - 1$ *introduce* nodes, which add all the other vertices.

Let $D = (X, U, r)$ be a nice tree decomposition of a graph G. For each node i of U, let U_i be the subtree of U, rooted at node i. For any $i \in V(U)$, we set $V_i = \cup_{v \in V(U_i)} X_v$ and $G_i = G[V_i]$. For any $p \in V(U)$ we refine G_p in a top down fashion as follows. If q is a *join* with children p and p', select one of its two children, say p. Then, for any $i \in U_p$ remove from G_i any edge in the set $E(G_q[X_q])$ (in fact, any partition of $E(G_q[X_q])$ for the edges induced by $G_p[X_p]$ and $G_{p'}[X_{p'}]$ would be suitable for the purposes of this paper). In this construction, we have $V(G_p) = V_p$ for any $p \in V(U)$ and we guarantee that if q is a *join* node with children p and p' then $V(G_p) = V(G_{p'}) = V(G_q)$, $E(G_p[X_q]) \cap E(G'_p[X_q]) = \emptyset$, and $E(G_p) \cup E(G_{p'}) = E(G_q)$. Notice that if r is the root of U, then $G_r = G$. We call G_i the subgraph of G *rooted* at i. We finally set, for any $i \in V(U)$, $D_i = (X^i, U_i, i)$ where $X^i = \{X_v \mid v \in V(U_i)\}$. Observe that for each node $i \in V(U)$, D_i is a tree decomposition of G_i.

A tree decomposition (X, U) is a *path decomposition*, if U is a path. The *pathwidth* of a graph G is defined analogously.

The *cutwidth* of a graph G with n vertices is defined as follows. Let $l = [v_1, \ldots, v_n]$ be a vertex ordering of $V(G)$. For $i = 1, \ldots, n-1$, we define $\theta_{l,G}(i) = E_G(l[1, i]) \cap E_G(l[i+1, n])$ (i.e. $\theta_{l,G}(i)$ is the set of edges of G that have one endpoint in $l[1, i]$ and one in $l[i+1, n]$). The cutwidth of an ordering l of $V(G)$ is $\max_{1 \leq i \leq n-1}\{|\theta_{l,G}(i)|\}$. The cutwidth of a graph is the minimum cutwidth over all the orderings of $V(G)$.

If $l = [v_1, \ldots, v_n]$ is a vertex ordering of a graph G, we set

$$\mathbf{Q}_{G,l} = [[0], [|\theta_{l,G}(1)|], \ldots, [|\theta_{l,G}(n-1)|], [0]].$$

We also assume that the indices of the elements of $\mathbf{Q}_{G,l}$ start from 0 and finish on n, i.e. $\mathbf{Q}_{G,l} = \mathbf{Q}_{G,l}[0, n]$. Clearly, $\mathbf{Q}_{G,l}$ is a sequence of sequences of numbers each containing only one element. It is not hard to prove that any graph G with treewidth at most w and maximum degree at most d, has cutwidth bounded by $(w + 1)d \log |V(G)|$.

2.2   Sequences of Integers

In this section, we give a number of preliminary results on sequences of integers.

We denote the set of all sequences of non-negative integers by $\mathcal{S}$. For any sequence $A = [a_1, \dots, a_{|A|}] \in \mathcal{S}$ and any integer $t \geq 0$ we set $A + t = [a_1 + t, \dots, a_{|A|} + t]$. If $A, B \in \mathcal{S}$ and $A = [a_1, \dots, a_{|A|}]$, we say that $A \sqsubseteq B$ if B is a subsequence of A obtained after applying a number of times (possibly none) the following operations: (i) If for some i, $1 \leq i \leq |A| - 1$, $a_i = a_{i+1}$, then set $A \leftarrow A[1, i] \oplus A[i+2, |A|]$. (ii) If the sequence contains two elements a_i and a_j such that $j - i \geq 2$ and
$$\forall_{i<k<j}\ a_i \leq a_k \leq a_j \text{ or } \forall_{i<k<j}\ a_i \geq a_k \geq a_j, \text{ then set } A \leftarrow A[1, i] \oplus A[j, |A|].$$
We define the *compression* $\tau(A)$ of a sequence $A \in \mathcal{S}$, as the unique minimum length element of $\{B \mid B \sqsubseteq A\}$. Notice that $B = \tau(A)$ is a subsequence $[a_{i_1}, \dots, a_{i_{|B|}}]$ of $A = [a_1, \dots, a_{|A|}]$ such that for any j, $1 \leq j \leq |B| - 1$, either $a_{i_j} \leq a_{i_j+1} \leq \cdots \leq a_{i_{j+1}-1} \leq a_{i_{j+1}}$ or $a_{i_j} \geq a_{i_j+1} \geq \cdots \geq a_{i_{j+1}-1} \geq a_{i_{j+1}}$. We can now define a function $\beta_A : \{1, \dots, |\tau(A)|\} \to \{1, \dots, |A|\}$ where $\beta_A(j) = i_j$ is one of the possible original positions in A of the j-th element in $\tau(A)$. Analogously, we define the function $\beta_A^{-1} : \{1, \dots, |A|\} \to \{1, \dots, \tau(A)\}$ such that $\beta_A^{-1}(j)$ is the unique i such that there exists a function β_A where $\beta_A(i) = j$. For any $A \in \mathcal{S}$, we define $\alpha(A)$ in the same way as $\tau(A)$ with the difference that only operation (i) is considered, i.e., we remove repetitions of a number on successive positions in the sequence. If now A is a typical sequence, we define the *set of extensions* of A as $\mathcal{E}(A) = \{\tilde{A} \in \mathcal{S} \mid \alpha(\tilde{A}) = A\}$. Let $A = [a_1, \dots, a_{r_1}]$ and $B = [b_1, \dots, b_{r_2}]$ be two sequences in $\mathcal{S}$. We say that $A \leq B$ if $r_1 = r_2$ and $\forall_{1 \leq i \leq r_1} a_i \leq b_i$. In general, we say that $A \prec B$ if there exist extensions $\tilde{A} \in \mathcal{E}(A)$, and $\tilde{B} \in \mathcal{E}(B)$ such that $\tilde{A} \leq \tilde{B}$. Suppose now that $\mathbf{A} = [A_1, \dots, A_r]$ and $\mathbf{B} = [B_1, \dots, B_{|r|}]$ are two sequences of typical sequences. We say that $\mathbf{A} \prec \mathbf{B}$ if $\forall_{1 \leq i \leq r} A_i \prec B_i$. For any integer t we set $\mathbf{A} + t = [A_1 + t, \dots, A_{|A|} + t]$ and $\max(\mathbf{A}) = \max_{1 \leq i \leq |A|}\{\max A_i\}$. Finally, for any sequence of typical sequences $\mathbf{A}$ we set $\tau(\mathbf{A}) = \tau(\mathbf{A}(1) \oplus \cdots \oplus \mathbf{A}(|\mathbf{A}|))$.

Let two equal-size sequences A, B of $\mathcal{S}$ where $A = [a_1, \dots, a_r], B = [b_1, \dots, b_r]$. We define $A + B = [a_1 + b_1, \dots, a_r + b_r]$ and we say that $A \sim B$ iff $\forall_{1 \leq i < r} a_i \neq a_{i+1} \Leftrightarrow b_i = b_{i+1}$ (and, therefore, $b_i \neq b_{i+1} \Leftrightarrow a_i = a_{i+1}$).

The *interleaving* $A \otimes B$ of two typical sequences A and B is a set of typical sequences defined as follows

$$A \otimes B = \{\tau(\tilde{A} + \tilde{B}) \mid \tilde{A} \in \mathcal{E}(A), \tilde{B} \in \mathcal{E}(B) \text{ and, } \tilde{A} \sim \tilde{B}\}.$$

The *interleaving* of two sequences of typical sequence $\mathbf{A} = [A_1, \dots, A_w]$ and $\mathbf{B} = [B_1, \dots, B_w]$ where $w = |\mathbf{A}| = |\mathbf{B}|$ is defined as follows:

$$\mathbf{A} \otimes \mathbf{B} = \{[C_1, \dots, C_w] \mid C_i \in A_i \otimes B_i, i = 1, \dots, w\}.$$

Given two sequences B_1 and B_2 where $B_1 \sim B_2$ we define function $\nu_{B_1, B_2} : \{1, \dots, |B_1| - 1\} \to \{1, 2\}$, $\nu_{B_1, B_2}(j) = 1$ if $B_1(j) \neq B_1(j+1)$ and $\nu(j) = 2$ if $B_1(j) = B_1(j+1)$ ($\nu_{B_1, B_2}(j)$ indicates which one of B_1, B_2 changes value between indexes j and $j+1$). When the sequences B_1 and B_2 are obvious, we simply denote ν_{B_1, B_2} by ν.

2.3 Characteristic Pairs

A *characteristic pair* is any pair $(\lambda, \mathbf{A})$ where λ is a sequence over a set $\mathcal{O}$ and $\mathbf{A}$ is a sequence of typical sequences such that $|\mathbf{A}| = |\lambda| + 1$. Notice that for any graph G and any order l of $V(G)$ the pair $(l, \mathbf{Q}_{G,l})$ is a characteristic pair. The *width* of a characteristic pair $(\lambda, \mathbf{A})$, is defined as $\max(\mathbf{A})$.

Procedure Com in Figure 1, defines the *compression* of a characteristic pair relative to a subset of $\mathcal{O}$.

Procedure Com$(l, \mathbf{R}, S)$.
Input: A characteristic pair $(l, \mathbf{R})$ and a set S.
Output: A characteristic pair $(\lambda, \mathbf{A})$.
We assume the notations $l = [v_1, \dots, v_{|l|}]$ and $\lambda = [v_{i_1}, v_{i_2}, \dots, v_{i_\rho}]$.

1: $\lambda \leftarrow l[S]$.
2: $\mathbf{A} \leftarrow [\tau(\mathbf{R}[0, i_1 - 1]), \tau(\mathbf{R}[i_1, i_2 - 1]), \dots, \tau(\mathbf{R}[i_{\rho-1}, i_\rho - 1]), \tau(\mathbf{R}[i_\rho, |l|])]$.
3: Output $(\lambda, \mathbf{A})$.
4: End.

Fig. 1. The procedure Com.

Given a graph G with n vertices, a vertex ordering l of G and $S \subseteq V(G)$, the *S-characteristic* of l is $C_S(G, l) = $ Com$(l, \mathbf{Q}_{G,l}, S)$. Notice that, from the definition of the S-characteristic of a vertex ordering l of a graph G we have that the $V(G)$-characteristic of l is equal to $(l, \mathbf{Q}_{G,l})$, i.e. $C_{V(G)}(G, l) = (l, \mathbf{Q}_{G,l})$ (clearly, Com$(l, \mathbf{Q}_{G,l}, V(G)) = (l, \mathbf{Q}_{G,l})$). We will simply use the term *characteristic* when there is no confusion on the choice of S and l. Given the S-characteristics $(\lambda^i, \mathbf{A}^i), i = 1, 2$, of two different vertex orderings of G we say that $(\lambda^1, \mathbf{A}^1) \prec (\lambda^2, \mathbf{A}^2)$ when $\lambda^1 = \lambda^2$ and $\mathbf{A}^1 \prec \mathbf{A}^2$.

Given a graph G and a vertex subset S, we say that a characteristic pair $(\lambda, \mathbf{A})$ is an *S-characteristic* when $(\lambda, \mathbf{A}) = C_S(l, G)$ for some ordering l of the vertices of G. Notice that for any $S \subseteq V(G)$, l is a vertex ordering of G with width at most k iff the width of $C_S(l, G)$ is at most k.

Assume from now on that we have a graph G and that (X, U) is a nice tree decomposition of G, with width at most w. A set $FS(i)$ of X_i-characteristics of vertex orderings of the graph G_i with cutwidth at most k is called a *full set of characteristics for node i* if for each vertex ordering l of G_i with cutwidth at most k, there is a vertex ordering l' of G_i such that $C_{X_i}(G_i, l') \prec C_{X_i}(G_i, l)$ and $C_{X_i}(G_i, l') \in FS(i)$, i.e. the X_i-characteristic of l' is in $FS(i)$. The following lemma can be derived directly from the definitions. For the proof of Lemma 2, see (see [20].

Lemma 1. *A full set of characteristics for a node i is non-empty if and only if the cutwidth of G_i is at most k. If some full set of characteristics for i is non-empty, then any full set of characteristics for i G_i is non-empty.*

Lemma 2. *Let G be graph with n vertices of degree bounded by d and let (X, U) be a nice tree decomposition of G with width at most w. Then for any node $i \in V(U)$, $|FS(i)| \leq w! \left(\frac{8}{3}\right)^{w+1} n^{2d(w+1)^2}$.*

3   An Algorithm for Cutwidth and Its Consequences

In this section, we give for any pair of integer constants k, w, an algorithm that, given a graph G with maximum degree d and a nice tree decomposition (X, U) of width at most w, decides whether G has cutwidth at most k.

An important consequence of Lemma 1 is that the cutwidth of G is at most k, if and only if any full set of characteristics for the root r is non-empty (recall that $G_r = G$). In [19] there are given algorithms able to construct a full set of characteristics for an *insert* or a *forget* node when a full set of characteristics for the unique child of i is given. In what follows, we show how to compute a full set of characteristics for a *join* node i when two full set of characteristics for its children j_1, j_2 are given.

We now consider the case that node i is a *join* node and $j_h, h = 1, 2$ are the two children of i in U. We observe that $V(G_{j_1}) \cap V(G_{j_2}) = X_i$, $G_{j_1} \cup G_{j_2} = G_i$ and we recall that $E(G_{j_1}[X_i]) \cap E(G_{j_2}[X_i]) = \emptyset$. Given a full set of characteristics $FS(j_1)$ for j_1 and a full set of characteristics F_{j_2} for j_2, the algorithm Join-Node in Figure 2 computes a full set of characteristics $FS(i)$ for i.

Algorithm Join-Node
Input: A full set of characteristics $FS(j_1)$ for j_1 and
         a full set of characteristics $FS(j_2)$ for j_2.
Output: A full set of characteristics $FS(i)$ for i.

1:   Initialize $FS(i) = \emptyset$.
2:   For any pair of X_{j_i}-characteristics $(\lambda, \mathbf{A}_h) \in FS(j_h), h = 1, 2$, **do**
3:   For any $\mathbf{A} \in \mathbf{A}_1 \bigotimes \mathbf{A}_2$, **do**
4:   If $\max(\lambda, \mathbf{A}) \leq k$, set $FS(p) \leftarrow FS(p) \cup \{(\lambda, \mathbf{A})\}$.
5:   Output $FS(p)$.
6:   End.

Fig. 2. The algorithm Join-Node.

Lemma 3. *Let G, G_1 and G_2 be graphs where $G_1 \cup G_2 = G$ and $G_1 \cap G_2 = (S, \emptyset)$. Let also l_1, l_2 be vertex orderings of G_1 and G_2 respectively where $l_1[S] = l_2[S] = \lambda$. If $(C_S(G_i, l_i) = \lambda, \mathbf{A}_i), i = 1, 2$ then, for any $\mathbf{A} \in \mathbf{A}_1 \bigotimes \mathbf{A}_2$, there exists a vertex ordering l of G where $l[V(G_i)] = l_i, i = 1, 2$ and $C_S(G, l) = (\lambda, \mathbf{A})$ and such a vertex ordering can be constructed by procedure Construct-Join-Orderings in Figure 3.*

The proof of the following makes strong use of Lemma 3.

Lemma 4. *If i is a join node with children j_1, j_2, and, for $h = 1, 2$, $FS(j_h)$ is a full set of characteristics for j_h, then the set $FS(i)$ constructed by Algorithm Join-Node in Figure 2 is a full set of characteristics for i.*

Using Lemma 4 it is possible to conclude to the following.

Procedure Construct-Join-Ordering$(G_1, G_2, S, l_1, l_2, \mathbf{A})$.
Input: Two graphs G_1, G_2 and a set S where $G_1 \cap G_2 = (S, \emptyset)$.
            Two vertex orderings l_1 and l_2 of G_1 and G_2, where $l_1[S] = l_2[S] = \lambda$.
            A sequence of typical sequences $\mathbf{A} \in \mathbf{A}_1 \bigotimes \mathbf{A}_2$ where $(\lambda, \mathbf{A}_i) = \text{Com}(G_i, l_i, S), i = 1, 2$.
Output: A vertex ordering l of G where $C_S(G, l) = (\lambda, \mathbf{A})$.
1:     Assume that for $i = 1, 2$, let $l_i = [v_1^i, \ldots, v_{r_i}^i]$
2:     Let $\lambda = [v_{\kappa_1^1}^1, \ldots, v_{\kappa_\rho^1}^1] = [v_{\kappa_1^2}^2, \ldots, v_{\kappa_\rho^2}^2]$ where $\rho = |S|$.
3:     For $i = 1, 2$, set $\kappa_0^i = 0$ and $\kappa_{\rho+1}^i = r_i + 1$.
4:     For $i = 1, 2$, set $Q_i = \mathbf{Q}_{G_i, l_i}(0) \oplus \cdots \oplus \mathbf{Q}_{G_i, l_i}(r_i)$.
5:     For any $h = 0, \ldots, \rho$,
            set $l_1^h = l_1[\kappa_h^1 + 1, \kappa_{i+1}^1 - 1]$ and $l_2^h = l_2[\kappa_h^2 + 1, \kappa_{i+1}^2 - 1]$.
            set $Q_1^h = Q_1[\kappa_h^1, \kappa_{i+1}^1 - 1]$ and $Q_2^h = Q_2[\kappa_h^2, \kappa_{i+1}^2 - 1]$.
            set $w^h = $ Join-Orderings$(l_1^h, l_2^h, Q_1^h, Q_2^h, \mathbf{A}(h))$.
6:     Set $l = w^0 \oplus [\lambda(1)] \oplus w^1 \oplus [\lambda(2)] \oplus w^2 \oplus \cdots \oplus [\lambda(\rho - 1)] \oplus w^{\rho-1} \oplus [\lambda(\rho)] \oplus w^\rho$.
7:     Output l
8:     End.

Procedure Join-Orderings(l_1, l_2, Q_1, Q_2, A).
Input: Two orderings l_1, l_2, two sequences Q_1, Q_2 where $|Q_i| = |l_i| + 1, 1, 2$,
            and a sequence $A \in \tau(Q_1) \otimes \tau(Q_2)$
Output: An ordering l.
1:     Compute B_1, B_2 such that $A = \tau(B_1 + B_2)$, where $B_1 \sim B_2$, and $B_i \in \mathcal{E}(\tau(Q_i)), i = 1, 2$.
2:     Set $w = |B_1| = |B_2|$, and denote $\nu = \nu_{B_1, B_2}$.
3:     For $j = 1, \ldots, w - 1$ set $m_j = l_{\nu(j)}[\beta_{Q_{\nu(j)}}(\beta_{B_{\nu(j)}}^{-1}(j)), \beta_{Q_{\nu(j)}}(\beta_{B_{\nu(j)}}^{-1}(j) + 1)) - 1]$.
4:     Output $m_1 \oplus \cdots \oplus m_{w-1}$.
5:     End.

Fig. 3. Procedures Construct-Join-Ordering and Join-Ordering.

Theorem 1. *An algorithm can be constructed that, given a graph G with n vertices of degree no more than d and a tree decomposition (X, U) of G of $O(n)$ nodes and width at most w, computes the cutwidth of G in $O((w!)^2 (\frac{8}{3})^{2w} n^{4d(2w^2 + 3w + 1) + 1} (wd \log n)^5)$ steps.*

Our algorithm can be turned into a constructive algorithm by making use of Procedures Construct-Join-Ordering and Join-Ordering. (For the details, see [20])

Theorem 2. *An algorithm can be constructed that, given a graph G with n vertices of degree no more than d and a tree decomposition (X, U) of G with $O(n)$ nodes and width at most w, outputs an ordering of $V(G)$ of minimum cutwidth in $O((w!)^2 (\frac{8}{3})^{2w} n^{4d(2w^2 + 3w + 1) + 1} (wd \log n)^5)$ steps.*

According to the main results in [3,6], one can construct an algorithm that, in $O(w^{O(w)} 2^{O(w^3)} n)$ steps, constructs a minimum width nice tree decomposition of any partial w-tree. This algorithm can serve as a preprocessing step to the algorithm of Theorem 2 that with input a partial w-tree G with vertices of degree at most d, outputs a vertex ordering of G of minimum cutwidth.

Our algorithm can be used as the main subroutine of a polynomial time algorithm computing the pathwidth of a bounded degree partial w-tree. We prove the following theorem. (For the details see [20].)

Theorem 3. *An algorithm can be constructed that, given a graph G with n vertices of degree no more than d and a tree decomposition (X, U) of G of $O(n)$ nodes and width at most w, outputs a path decomposition of G with minimum width in $O(((dw)!)^2 (\frac{32}{3})^{2dw} n^{12(dw)^2 + 20dw + 9} (dw \log n)^5)$ steps.*

We mention that the problem of computing the pathwidth of partial w-trees can be solved in polynomial time. The algorithm for the general case was given by Bodlaender and Kloks in [6]. However, the exponent in the complexity of this algorithm is quite large for any practical purpose. The algorithm proposed in Theorem 3 is faster and can serve as a more realistic approach for partial w-trees with bounded degree.

References

1. K. R. Abrahamson and M. R. Fellows. Finite automata, bounded treewidth and well-quasiordering. In N. Robertson and P. Seymour, editors, *Proceedings of the AMS Summer Workshop on Graph Minors, Graph Structure Theory, Contemporary Mathematics vol. 147*, pages 539–564. American Mathematical Society, 1993.
2. H. L. Bodlaender. Improved self-reduction algorithms for graphs with bounded treewidth. *Discrete Appl. Math.*, 54:101–115, 1994.
3. H. L. Bodlaender. A linear-time algorithm for finding tree-decompositions of small treewidth. *SIAM J. Comput.*, 25(6):1305–1317, 1996.
4. H. L. Bodlaender. Treewidth: algorithmic techniques and results. In *Mathematical foundations of computer science 1997 (Bratislava)*, pages 19–36. Springer, Berlin, 1997.
5. H. L. Bodlaender. A partial k-arboretum of graphs with bounded treewidth. *Theoret. Comput. Sci.*, 209(1-2):1–45, 1998.
6. H. L. Bodlaender and T. Kloks. Efficient and constructive algorithms for the pathwidth and treewidth of graphs. *Journal of Algorithms*, 21:358–402, 1996.
7. H. L. Bodlaender and D. M. Thilikos. Computing small search numbers in linear time. Technical Report UU-CS-1998-05, Dept. of Computer Science, Utrecht University, 1998.
8. M. Chung, F. Makedon, I. H. Sudborough, and J. Turner. Polynomial time algorithms for the min cut problem on degree restricted trees. In *Symposium on foundations of computer science*, volume 23, pages 262–271, Chicago, Nov. 1982.
9. J. Díaz, J. Petit, and M. Serna. A survey on graph layout problems. Technical Report LSI-00-61-R, Departament de Llenguatges i Sistemes Informàticss, Universitat Politècnica de Catalunya, Barcelona, Spain, 2000.
10. M. R. Fellows and M. A. Langston. On well-partial-order theory and its application to combinatorial problems of VLSI design. *SIAM J. Disc. Meth.*, 5:117–126, 1992.
11. M. R. Fellows and M. A. Langston. On search, decision and the efficiency of polynomial-time algorithms. *J. Comp. Syst. Sc.*, 49:769–779, 1994.
12. L. H. Harper. Optimal assignments of number to vertices. *SIAM Journal*, 12(1):131–135, 1964.
13. F. S. Makedon and I. H. Sudborough. On minimizing width in linear layouts. *Discrete Appl. Math.*, 23:243–265, 1989.
14. K. Nakano. Linear layouts of generalized hypercubes. In J. van Leewen, editor, *Graph-theoretic concepts in computer science*, volume 790 of *Lecture Notes in Computer Science*, pages 364–375. Springer-Verlag, 1994.
15. N. Robertson and P. D. Seymour. Graph minors. I. Excluding a forest. *J. Comb. Theory Series B*, 35:39–61, 1983.
16. N. Robertson and P. D. Seymour. Graph minors. III. Planar tree-width. *J. Comb. Theory Series B*, 36:49–64, 1984.
17. N. Robertson and P. D. Seymour. Graph minors — a survey. In I. Anderson, editor, *Surveys in Combinatorics*, pages 153–171. Cambridge Univ. Press, 1985.

18. D. M. Thilikos and H. L. Bodlaender. Constructive linear time algorithms for branchwidth. Technical Report UU-CS-2000-38, Dept. of Computer Science, Utrecht University, 2000.
19. D. M. Thilikos, M. J. Serna, and H. L. Bodlaender. Constructinve linear time algorithms for small cutwidth and carving-width. In D. Lee and S.-H. Teng, editors, *Proc. 11th Internatinal Conference ISAAC 2000*, pages 192–203. Springer-Verlag, Lectures Notes in Computer Science, nr. 1969, 2000.
20. D. M. Thilikos, M. J. Serna, and H. L. Bodlaender. A polynomial time algorithm for the cutwidth of bounded degree graphs with small treewidth. Technical Report LSI-01-04-R, Departament de Llenguatges i Sistemes Informàticss, Universitat Politècnica de Catalunya, Barcelona, Spain, 2001.
   http://www.lsi.upc.es/ sedthilk/dlogn.ps.
21. M. Yannakakis. A polynomial algorithm for the min-cut linear arrangement of trees. *J. ACM*, 32:950–988, 1985.

Lower Bounds and Exact Algorithms for the Graph Partitioning Problem Using Multicommodity Flows

Norbert Sensen*

University of Paderborn, Department of Mathematics and Computer Science
Fürstenallee 11, D 33102 Paderborn
sensen@upb.de

Abstract. In this paper new and generalized lower bounds for the graph partitioning problem are presented. These bounds base on the well known lower bound of embedding a clique into the given graph with minimal congestion. This is equivalent to a multicommodity flow problem where each vertex sends a commodity of size one to every other vertex. Our new bounds use arbitrary multicommodity flow instances for the bound calculation, the critical point for the lower bound is the *guaranteed cut flow* of the instances. Furthermore, a branch&bound procedure basing on these bounds is presented and finally it is shown that the new bounds are also useful for lower bounds on classes of graphs, e.g. the Butterfly and Beneš graph.

1 Introduction

Graph Partitioning is the problem of partitioning a set of vertices of a graph into disjoint subsets of a given maximal size such that the number of edges with end points in different subsets is minimized. Graph Partitioning is a very common problem and has a large number of applications. For example circuit layout, compiler design, and load balancing are typical applications in which Graph Partitioning problems appear. Unfortunately, the Graph Partitioning problem is a NP-hard problem. So in the last years a lot of effort has been spent in the development of fast and good heuristics for the problem, a recent survey is given in [3]. These heuristics often can handle rather large graphs with more than a million vertices and deliver good solutions. In contrast to the development of heuristics only a little expense has been done in the development of exact algorithms. From the NP-hardness fact it is clear that generally only relatively small graphs can be solved exactly. Nevertheless, exact solutions are of interest for applications and for the validation of heuristics.

In this paper we present new lower bounds for the Graph Partitioning problem and a branch & bound algorithm for the exact solution of the Graph Partitioning problem using these new lower bounds. The new bounds base on a well known method for proving lower bounds of the graph bisection problem (a special case of the graph partitioning

---

* This work was partially supported by the IST Programme of the EU under contract number IST-1999-14186 (ALCOM-FT), and by the German Science Foundation (DFG) project SFB-376.

problem with two equally sized partitions): If there is an embedding of the Clique graph with n vertices (K_n) into the given graph G with a congestion C, then each bisection of the graph G cuts at least $\frac{n^2}{4C}$ edges. The computation of an embedding with minimal congestion (i.e. the lower bound is maximal) is equivalent to a multicommodity flow problem: Every vertex sends a commodity of size one to every other vertex. Realizing this flow with minimal congestion provides us with the identical lower bound for the graph bisection problem.

The above lower bound is often used for theoretical analysis of the bisection width of a given graph (e.g. see [9]) and it delivers convenient bounds if the graph is quite regular. But the bound is impractical for the construction of a branch & bound algorithm if the graph is more irregular as it is often the fact in practice. In this paper we present generalizations of this bound. The generalizations base on the observation that not every vertex has to send a commodity of identical size to every other vertex. In fact we can compute lower bounds on the graph partitioning problem for every possible combination of commodities. We only have to know the $guaranteed\,CutFlow$, i.e. the amount of flow which crosses the cut of every feasible partitioning of the graph. Furthermore, we have generalized the idea such that it is applicable not only for the graph bisection problem with exactly two equally sized partitions, but also for the general graph partitioning problem with a given number of partitions and a given maximal size for every partition.

In the last years there have been presented a number of different approaches for solving the graph partitioning problem exactly. The most recent approach is presented in [8] by S.E. Karisch, F. Rendl, and J. Clausen. They use a semidefinite relaxation for the computation of a lower bound on the Graph Partitioning problem as the core of a branch & bound algorithm. They address the Graph Partitioning problem with two partitions, edge weights, and a maximal partition size; their approach does not handle vertex weights or more than two partitions. In [5] Ferreira et al. present a branch-and-cut algorithm basing on a variety of separation heuristics. They address the general Graph Partitioning problem with vertex weights, edge weights, an arbitrary number of partitions, and a maximal partition size. In [1] L. Brunetta, M. Conforti, and G. Rinaldi present another branch-and-cut algorithm. They start with a linear program defining the convex hull of all solutions and use several separation procedures to add cuts to the linear program. They address the bisection problem with edge weights. In [7] E. Johnson, A. Mehrotra, and G. Nemhauser present a column generation algorithm for the Graph Partitioning problem. There, the generation of additional columns itself is NP-hard, so they present efficient strategies for the generation. They address the general Graph Partitioning problem with vertex and edge weights, an arbitrary number of partitions and a maximal partition size.

Also related to our paper is the work of F. Shahrokhi and L. Szekely in [15]. They examine bounds basing on the embedding of the clique K_n. They apply this bound to general graphs with a small number of equivalence classes of vertices and use only shortest paths for the flow. Putting this together they get lower bounds on the crossing number, bisection width and edge and vertex expansion of graphs. Their bounds are especially good if there is only one equivalence class of vertices.

In the next section we give basic definitions for the Graph Partitioning problem and Multicommodity Flows. Then in section three we present the new lower bounds for the Graph Partitioning problem basing on more flexible multicommodity flows. In section four the branch & bound algorithm using this bound is described; inside the algorithm we can use the solution of the multicommodity flow instance to force vertices to stay in the same partition. Finally, in section five we give upper bounds on the new lower bounds and apply the new lower bounds to the butterfly graph.

2   Definitions

There have been a couple of slightly different definitions of the Graph Partitioning problem. Inside this paper we are talking about a graph with vertex and edge weights, a given number of partitions and a maximal size for each partition. More formally:

Definition 1. *The* Graph Partitioning problem *has given an undirected graph* $G = (V, E)$, *vertex weights* $g : V \rightarrow I\!N$, *edge weights* $f : E \rightarrow I\!N$, *a number of partitions* $p \in I\!N$, *and a maximal size* $M \in I\!N$. *The problem is to find a partition of the vertices* V *into* p *disjoint sets* $V_1, \ldots, V_p$ *with* $\forall i : \sum_{v \in V_i} g(v) \leq M$ *and minimal* $CutSize :=$ $\min_{V = \bigcup V_i} CutSize(V_1, \ldots, V_p)$, *where*

$$CutSize(V_1, \ldots, V_p) := \sum_{v \in V_i, w \in V_j, i < j} f(\{v, w\}).$$

In the following we use $n = |V|$ and $N = \sum_{v \in V} g(v)$. The above definition is a very general one. If $p = 2$ is used we obtain the well known bisection problem. Often equally sized partitions are requested, i.e. $M = \lceil \frac{N}{p} \rceil$. Notice, that in the case of $p > 2$ this is a less strong requirement than the requirement $\forall i, j : |V_i| - |V_j| \leq 1$ which is also used sometimes.

As mentioned above we want to use Multicommodity Flows in order to compute lower bounds on the Graph Partitioning problem. In the Multicommodity Flow problem there are commodities of specific sizes with a source vertex and a destination vertex. The goal is to fulfill the given set of commodities with minimal congestion. More formally:

Definition 2. *The* Multicommodity Flow problem *has given an undirected graph* $G = (V, E)$, *edge weights* $f : E \rightarrow I\!N$, *and for each pair* $(v, w) \in V^2$ *the size* $d_{v,w} \in I\!R_{\geq 0}$ *with* $\forall v \in V : d_{v,v} = 0$ *of the commodity which has to flow from vertex* v *to vertex* w. *A flow* $h : V \times V \times V \rightarrow I\!R_{\geq 0}$ *is sought with* $h(u, v, w) = 0 \Leftarrow \{v, w\} \notin E$, *and*

$$\forall v, w \in V, v \neq w : \sum_{u \in V} h(v, u, w) - h(v, w, u) = d_{v,w}.$$

The congestion $c : E \rightarrow I\!R$ *of each edge is* $c(\{v, w\}) := \frac{1}{f(\{v,w\})} \sum_{u \in V} h(u, v, w) +$ $h(u, w, v)$. *A flow* h *is searched with minimal total congestion* $C := \max_{e \in E} c(e)$.

The Multicommodity Flow problem is a well known problem. It can be represented as a linear program of polynomial size, so it is solvable in polynomial time. Fast or approximating algorithms for the Multicommodity Flow problem are subject of current research activities, see e.g. [4,11,10].

3   Multicommodity Bounds

General Idea. The main idea for the generalization of the known lower bound on the graph bisection problem into equally sized partitions is the following: The known bound bases on the embedding of a K_n into the given graph with minimal congestion C'. Any bisection of a K_n has a $CutSize$ of at least $\frac{n^2}{4}$. Since each edge of the given graph is used by at most C' edges of the K_n, a $CutSize$ of at least $\frac{n^2}{4C'}$ of the given graph is unavoidable in order to cut the minimal number of edges of the K_n. Computing the minimal congestion of this embedding is equivalent to calculating the minimal congestion C of a Multicommodity Flow problem with $\forall v, w \in V : d_{v,w} = 1$. In this case we can argue equivalently to get a lower bound for the Graph Partitioning problem: Any bisection of the graph has a flow of $\frac{n^2}{2}$ between vertices of different partitions. Since each vertex transports at most C commodities (assuming $\forall e \in E : f(e) = 1$) at least $\frac{n^2}{2C}$ crossing edges are necessary, i.e. $CutSize \geq \frac{n^2}{2C}$. This consideration can be generalized to any Multicommodity Flow instance:

Theorem 1
Given a Graph Partitioning problem with a graph G which has to be divided into p partitions. If a Multicommodity Flow with this graph G, sizes d, and congestion C exists with a CutFlow $\in \mathbb{R}$ such that for any optimal partition $V = \bigcup_{i=1}^{p} V_i$ $\sum_{v \in V_i, w \in V_j, i \neq j} d_{v,w} \geq CutFlow$ holds. Then

$$CutSize \geq \frac{CutFlow}{C}.$$

Proof. Let $V = \bigcup V_i$ be an optimal solution of the Graph Partitioning problem. Then

$$CutFlow \leq \sum_{v \in V_i, w \in V_j, i \neq j} d_{v,w} \leq \sum_{v \in V_i, w \in V_j, i < j} C \cdot f(\{v, w\}) = C \cdot CutSize(V_1, \ldots, V_p)$$

Since we have assumed that $V = \bigcup V_i$ is optimal, $CutSize \geq \frac{CutFlow}{C}$ follows.   □

Different Instantiations. So in principle we could use any Multicommodity Flow instance for the computation of a lower bound on the Graph Partitioning problem. The bound is the better the bigger the guaranteed $CutFlow$ and the smaller the congestion C are. In the following we introduce three different general Multicommodity Flow instances with a different degree of freedom for their choice of the sizes of the commodities. The first one corresponds to the known bound where every vertex sends a commodity of size one to every other vertex. We have adapted it to consider vertex weights, only:

Definition 3. *The 1-1-MC is a Multicommodity Flow instance with a graph $G = (V, E)$, vertex weights $g : V \to \mathbb{N}$, and default sizes of the commodities:*

$$\forall v, w \in V : \quad d_{v,w} = g(v) \cdot g(w)$$

The next Multicommodity Flow instance allows a variable source strength for each vertex. The idea behind this is that more central vertices generate less total load on the edges such that we hope to get a larger $CutFlow$ with a smaller congestion:

Definition 4. *The* VarMC *is a Multicommodity Flow instance with a graph* $G = (V, E)$, *vertex weights* $g : V \rightarrow \mathbb{N}$, *and a free source strength* $s : V \rightarrow \mathbb{R}_{\geq 0}$ *such that*

$$\forall v, w \in V : \quad d_{v,w} = g(w) \cdot s(v).$$

For the computation of a good lower bound on the graph partitioning problem the free source strengths have to be adapted to the given graph such that the bound is maximized. The selecting of the source strengths is not done by hand but can be included into a linear program which also solves the Multicommodity Flow problem. Finally, we introduce a Multicommodity Flow instance in which all sizes of the commodities are free:

Definition 5. *The* MVarMC *is a Multicommodity Flow instance with a graph* $G = (V, E)$, *vertex weights* $g : V \rightarrow \mathbb{N}$, *and a free source strength* $s : V \times V \rightarrow \mathbb{R}_{\geq 0}$ *with* $s(v, v) = 0 \; \forall v \in V$ *such that*

$$\forall v, w \in V : \quad d_{v,w} = g(w) \cdot s(v, w).$$

The critical point for the use of a Multicommodity Flow instance when computing a lower bound on the graph partitioning problem is the guaranteed $CutFlow$. The following Theorem delivers correct values for the MVarMC instance:

Theorem 2
We have given a Graph Partitioning problem with a graph $G = (V, E)$, *vertex weights* g, *and the constants* p *and* M. *A Multicommodity Flow instance MVarMC guarantees a flow over each feasible partition of*

$$CutFlow \geq \sum_{v \in V} \left(\sum_{w \in V} s(v, w) \cdot g(w) - \bar{s}_v (M - g(v)) \right) + \tilde{s} R(M - R)$$

with $\bar{s}_v := max_{w \in V} s(v, w)$, $\tilde{s} := min_{v \in V} \frac{\bar{s}_v}{g(v)}$, *and* $R := N - M \cdot \lfloor \frac{N}{M} \rfloor$.

Proof. We are looking for a guaranteed $CutFlow$ of the commodities with sizes $s(v, w)$. A $CutFlow$ is guaranteed if for any possible partition according to the parameters M and p the actual flow between the partitions is at least as large as $CutFlow$. So let us assume any feasible partition $V = V_1 \cup \ldots \cup V_p$ with $N_i = \sum_{v \in V_i} g(v)$. Then

$$CutFlow = \sum_{i=1}^{p} \sum_{v \in V_i} \sum_{w \in V \setminus V_i} s(v, w) \cdot g(w)$$

$$\geq \sum_{i=1}^{p} \sum_{v \in V_i} \left(\sum_{w \in V} s(v, w) \cdot g(w) - \bar{s}_v \sum_{w \in V_i \setminus \{v\}} g(w) \right)$$

$$= \sum_{v \in V} \left(\sum_{w \in V} s(v, w) g(w) - (M - g(v)) \cdot \bar{s}_v \right) + \sum_{i=1}^{p} (M - N_i) \sum_{v \in V_i} \bar{s}_v$$

$$\geq \sum_{v \in V} \left(\sum_{w \in V} s(v, w) g(w) - (M - g(v)) \cdot \bar{s}_v \right) + \tilde{s} \cdot \sum_{i=1}^{p} (M - N_i) N_i$$

To prove the Theorem it remains to show that $\sum_{i=1}^{p}(M - N_i)N_i \geq R(M - R)$ for every feasible partition, i.e. $\forall i : N_i \leq M \land \sum_{i=1}^{p} N_i = N$:

$$\sum_{i=1}^{p}(M - N_i)N_i = MN - \sum_{i=1}^{p} N_i^2 \geq MN - \lfloor \frac{N}{M} \rfloor M^2 - R^2 = R(M - R)$$

$\square$

Notice that the term $R(M - R)$ corresponds to the problem definition of a maximal size for every partition. If the restriction $\forall i, j : |V_i| - |V_j| \leq 1$ is used, a different (bigger) term is possible.

Basing on the $CutFlow$ of the MVarMC instance it is easy to specify the guaranteed $CutFlow$ of the VarMC and 1-1-MC instance:

Lemma 1. *The VarMC instance guarantees a $CutFlow$ of*

$$CutFlow \geq (N - M) \sum_{v \in V} s(v) + \tilde{s}R(M - R)$$

with $\tilde{s} := \min_{v \in V} \frac{s(v)}{g(v)}$. The 1-1-MC instance guarantees a $CutFlow$ of

$$CutFlow \geq N(N - M) + R(M - R).$$

Proof. Using $\forall v, w \in V : s(v, w) = s(v)$ or $\forall v, w \in V : s(v, w) = g(v)$, respectively, the Lemma follows from the $CutFlow$ of the MVarMC instance. $\square$

So altogether we have introduced three different instances with a different degree of freedom. It is clear that the bound of the MVarMC instance is at least as good as the bound of the VarMC instance which is at least as good as the bound of the 1-1-MC instance. On the other hand the computation of the MVarMC instance should last longer since more variables have to be specified. For practical usage of the bounds we have to compare the quality of the bounds and their running times.

Experimental Results. All three proposed Multicommodity Flow instances can be represented as a linear program. The free source strengths in the VarMC and MVarMC instances are realized by additional variables inside the linear program. A maximal congestion of one is given and the goal is the maximization of the guaranteed $CutFlow$. Thus, solving the linear program contains the selection of the optimal source strengths. The following linear program corresponds to the MVarMC instance:

Maximize $\sum_{v \in V} \left(\sum_{w \in V} s(v, w) \cdot g(w) - \bar{s}_v(M - g(v)) \right) + \tilde{s}R(M - R)$

subject to $\forall \{v, w\} \in E : \quad \sum_{u \in V} h(u, v, w) + h(u, w, v) \leq f(\{v, w\})$

$\forall v, w \in V, v \neq w : \sum_{u \in V} h(v, u, w) - h(v, w, u) \geq s(v, w) \cdot g(w)$

$\forall v, w \in V, v \neq w : \bar{s}_v \geq s(v, w)$

$\forall v \in V : \qquad g(v) \cdot \tilde{s} \leq \bar{s}_v$

Of course the constraints for $\tilde{s}$ can be omitted if $R = 0$.

To solve this linear program we use the barrier algorithm of the CPLEX package ([6]). All experiments have been performed on a Sun Enterprise 450 Model 4400 machine with 1 GB main memory and a Sun UltraSPARC-II 400 MHz processor.

We have done a large amount of tests for different Graph Partitioning problems. Here we show the results of a small representative set of graphs in order to show the typical behavior of the bounds. The "SEd" graph is a shuffle exchange graph of dimension d, the "DBd" graph is a DeBruijn graph of dimension d, "Star 50" is a simple star graph with 50 vertices, "Grid 5x10" is a two-dimensional grid. "ex36a" and "ex36d" are introduces in [8], both having 36 vertices. In the "a" version each edge is in the graph with probability 0.5, in the "d" version each edge has a random weight between 0 and 10. "Rand 0.1" and "Rand 0.05" are random graphs from ourselves, both having 60 vertices; each possible edge is contained in the graph with probability of 0.1 or 0.05, respectively. "RandPlan" is a random maximal planar graph with 100 vertices, constructed from routines of the LEDA library ([12]). "m4", "mc", and "m8" are real world instances from a finite elements method, see [14], also used in [5,1,8]. "cb30" and "cb61" are real world instances from compiler design problems, also used in [5,8], and finally the "wcb$x$" graphs are the "cb$x$" graphs with vertex weights.

The problems in Table 1 with $p = 2$ and $M = \lceil \frac{N}{p} \rceil$ corresponds to the classical bisection problem with equally sized partitions. For each of the three Multicommodity Flow instances the lower bound and the computation time (format hh:mm:ss) is given. The "Opt" column gives the exact result, or in braces the best known upper bound if the exact solution is unknown. The graph with missing 1-1-MC and VarMC results are not connected, so it is not possible for any vertex to send something to all other vertices. Bold faced entries in the 1-1-MC column show instances where the bound is significantly worse than the VarMC bounds. And bold faced entries in the MVarMC column show instances where the bounds are significantly better than the VarMC bounds. Furthermore, results are presented with $p = 2$, $M = \lceil \frac{2}{3}N \rceil$ and $p = 4$, $M = \lceil \frac{1}{4}N \rceil$.

Conclusions from the Experiments

- The 1-1-MC instance is generally faster than the two other instances. The VarMC and MVarMC instances are sometimes equally fast and sometimes the VarMC instance is the faster one.
- The 1-1-MC bounds are generally worse than the bounds from the two other instances. With increasing M the gap of the bounds of the 1-1-MC and the VarMC gets smaller.
- For $p = 2$ and $M = n/2$ VarMC and MVarMC often deliver the same bound while with increasing p or M the gap between the VarMC and MVarMC bounds gets bigger.

4   Branch & Bound Algorithm

Branching Realization. In a branch & bound algorithm a given subproblem, which cannot be bounded, has to be divided into at least two new restricted subproblems. We do this restriction by determining if two vertices $v, w \in V$ stay in the same partition or are separated into different partitions. We call these two possibilities "join" or "split". A join is performed by creating a new graph from the original graph: the two vertices $v, w \in V$, which have to stay in the same partition, are joined into one vertex $\tilde{v}$ with

Table 1. Lower bounds and computation time with the three Multicommodity Flow instances

	Graph	1-1-MC		VarMC		MVarMC		Opt
		Bound	Time	Bound	Time	Bound	Time	
	SE7	**14.26**	21	15.12	49	15.12	1:30	16
	DB7	**27.52**	41	28.98	1:39	28.98	1:45	30
	Star 50	**12.76**	0	25.00	0	25.00	1	25
	Grid 5x10	5.00	1	5.00	2	5.00	3	5
	ex36a	**92.58**	8	104.17	10	104.17	19	117
$p = 2$	ex36d	**1277.49**	41	1385.59	43	1385.59	37	1426
	Rand 0.1	15.25	5	30.00	12	**37.83**	19	42
$M = \lceil \frac{N}{2} \rceil$	Rand 0.05	-	-	-	-	**9.79**	13	10
	RandPlan	**21.37**	15	26.79	22	26.79	50	28
	m4	-	-	-	-	**5.83**	1	6
	mc	**3.69**	3	5.10	8	5.67	9	6
	m8	7.00	28	7.00	24	7.00	45	7
	cb30	**52.23**	0	97.50	0	**213.00**	0	213
	cb61	**170.50**	5	330.00	13	**1906.40**	10	2177
	SE6	7.39	2	7.42	7	7.44	4	8
	DB6	13.98	4	13.99	16	13.99	8	14
	Star 50	**11.10**	0	16.00	0	16.00	1	16
	Grid 5x10	4.35	1	4.35	2	4.35	3	5
	ex36a	**82.29**	8	86.18	10	**88.48**	12	101
$p = 2$	ex36d	**1135.54**	39	1181.32	48	**1212.00**	42	1246
	Rand 0.1	**13.56**	5	20.00	17	**27.30**	17	33
$M = \lceil \frac{2}{3}N \rceil$	Rand 0.05	-	-	-	-	**6.00**	7	6
	RandPlan	**18.90**	14	20.59	31	20.64	26	21
	m4	-	-	-	-	**4.69**	0	6
	mc	3.23	3	3.38	12	3.45	6	4
	m8	6.20	27	6.20	41	6.20	41	7
	cb30	**46.43**	0	65.00	0	**114.00**	0	114
	cb61	**150.33**	6	220.00	14	**717.00**	8	717
	wcb30	**38.85**	0	53.54	0	**126.10**	1	136
	wcb61	**114.31**	8	165.92	16	**826.08**	18	867
	SE6	**12.58**	2	13.37	7	**15.15**	7	18
	DB6	**23.79**	4	25.44	15	**27.17**	10	32
	Star 50	**19.10**	0	37.00	0	37.00	1	37
	Grid 5x10	7.49	1	7.49	2	**13.95**	8	16
	ex36a	**138.86**	7	156.25	9	**159.00**	13	(186)
$p = 4$	ex36d	**1916.23**	38	2078.38	42	**2098.00**	24	(2192)
	Rand 0.1	**22.88**	5	45.00	11	**58.51**	22	(69)
$M = \lceil \frac{N}{p} \rceil$	Rand 0.05	-	-	-	-	**16.56**	13	18
	RandPlan	**32.05**	15	40.20	26	**53.19**	1:09	60
	m4	-	-	-	-	**11.45**	1	12
	mc	**5.53**	3	7.59	9	**13.25**	17	14
	m8	10.50	27	10.50	23	**20.70**	2:36	22
	cb30	**78.00**	0	143.00	0	**430.20**	0	436
	cb61	**255.20**	5	495.00	14	**4487.88**	13	4565

$g(\tilde{v}) = g(v)+g(w)$. The new edge weights are generated accordingly: $\forall u \in V\backslash\{v,w\}$: $f(\{u,\tilde{v}\}) = f(\{u,v\}) + f(\{u,w\})$. It is obvious that this new graph has the same $CutSize$ as the original graph with the restriction that v and w have to be in the same partition.

A split of two vertices $v,w \in V$ is performed by removing the commodity with source v and destination w from the general computation of the $CutFlow$. Instead, this particular flow can be counted into the $CutFlow$ completely. In fact in case of the MVarMC instance the split means only a reinterpretation of $\bar{s}_v$, while in case of the VarMC and 1-1-MC instance an additional variable $s(v,w)$ is included into the linear program. Altogether we manage a set $S \subset V^2$ of split pairs in the branch & bound algorithm

Branching Strategy. A crucial point for a good branch & bound algorithm is the branching strategy: In our application this means the decision, which pair of vertices should be used for the join and split in a given situation. Our branching strategy bases on a simple upper bound of the lower bound of the 1-1-MC instance: By assuming that all flows use only shortest paths and all edges have the same congestion we get a simple upper bound on the lower bound of the 1-1-MC instance. For the branching selection we use a pair of vertices such that the above upper bound is maximized in case of a join of this two vertices. Furthermore, in order to take the split case into consideration, we prefer pairs of vertices with a small distance. This two criteria have to be combined and experiments have shown that a weight of 20 percentage for the upper bound on the lower bound and 80 percentage on the shortest distance gives a quit good branching strategy.

Forcing Moves. Following the subproblems in the search tree, there are some situations where pairs of vertices can be forced to be split or joined. Firstly, if $\exists v \in V : g(v) = M$ this vertex v can be split from all other vertices. Secondly, if $\exists v,w \in V, \{v,w\} \notin S$: $N + g(v) - g(w) - \sum_{u \in V, \{v,u\} \in S} g(u) < N - (p-1)M$ then we can join the vertices v and w, since otherwise vertex v has no possibility to become part of a correct partition. Thirdly, we look at the graph $G_S := (V,S)$: If G_S has two clique subgraphs with p vertices which match in exactly $p - 1$ vertices, then the two remaining vertices can be joined. Finally, the last possibility for forcing a join bases on a given solution of a Multicommodity Flow instance:

Lemma 2. *Given a Graph Partitioning problem with graph G and edge weights f and a Multicommodity Flow problem with graph G, congestions $c : E \to \mathbb{R}$, maximal congestion C and CutFlow CF. An improving solution for the Graph Partitioning problem with CutSize at most L is searched. Then the vertices v,w with $\{v,w\} = e \in E$ can be joined if*

$$\frac{CF + (C - c(e)) \cdot f(e)}{C} > L.$$

This possibility for forcing a join is extremely helpful if the bounds are close to a given feasible solution and the graph is somehow irregular such that there are edges which are not loaded with the maximal congestion.

Experiments. As there are a lot of good heuristics for the Graph Partitioning problem, it is easy to get a good feasible solution at the beginning of the branch & bound procedure. We use the Party library [13] for this purpose. As in most cases the solution from the heuristic is optimal, we do not matter about any best first search but use simple depth first search. Again, CPLEX is used for the computation of the bounds. The barrier algorithm stops if the primary solution is good enough to bound the actual subproblem or if the dual solution shows that we cannot bound the actual subproblem.

In Table 2 a comparison of our branch & bound algorithm with the results from Karisch et al. [8], for our knowledge the best actual code, are presented. The results in [8] are performed on a HP 9000/735 system. The "#B" column gives the number of search nodes in the branch&bound tree, the time is given in hh:mm:ss. For the problems with the DeBruijn graphs we have utilized two specific properties: Firstly, every bisection of the DeBruijn graph has an even CutSize; Karisch has used this fact, also. Secondly, symmetrical parts of the search tree, which result from four automorphisms of the DeBruijn graph, are cut of; this decreases the search tree of the DB8 by a factor of about two. Finally, we want to remark that the bisection problem of the DeBruijn graph of dimension 8 is solved by the presented approach for the first time at all.

Table 2. Results of the branch & bound algorithms with the Graph Bisection problem

Graph	1-1-MC		VarMC		MVarMC		Karisch	
	Time	#B	Time	#B	Time	#B	Time	#B
m4	-	-	-	-	1	1	1	1
mc	2:41	56	7	1	10	1	46	1
m1	-	-	-	-	21	1	18:15	15
m8	27	1	15	1	29	1	5:21	1
cb30	3	23	3	21	0	1	2	1
cb47_99	1:49	238	2:34	237	10	7	10	1
cb47_101	22	53	22	40	7	3	1:52	35
cb61	2:19	78	3:38	72	42	6	20	1
DB7	28:10	44	25	1	1:12	1	8:53:20	195
DB8	>50:00:00	>1000	6:54:35	33	11:23:15	33		
ex36a	3:33:07	10058	22:19	197	41:04	197	3	1
ex36d	39:53:38	67571	45:44	101	40:30	101	11	3

Conclusions from the Experiments

- The MVarMC instance is the best one for the branch & bound algorithm from the three Multicommodity Flow instances.
- Concerning the tested real world applications our approach delivers equally results compared with the results from Karisch. For the DeBruijn graphs our approach is in orders of magnitude better than Karischs approach. This is the other way around for the randomly generated dense "ex" problems.

So we conclude that our approach is quite good for more "sparse" and "regular" graphs while it is less good for "dense" graphs. In most applications of the Graph Partitioning problem, e.g. circuit layout or load balancing, the graphs are relatively "sparse" and "regular", in fact. So our approach is well suited for these applications.

5   Theoretical Analyses

Here we give some theoretical observations on the presented new bounds. Due to the given space constraints we only show some ideas.

First, we look at upper bounds on the presented lower bounds. Assuming a graph $G = (V, E)$ has an infeasible cut $V = V_1 \cup V_2$. Then we can conclude that the 1-1-MC bound has an upper bound of $\frac{N^2}{4N_1 N_2} CutSize(V_1, V_2)$ since a specific number of commodities has to cross the given cut. From the existence of a given infeasible cut we can also conclude an upper bound on the VarMC bound of $\frac{N}{2 \cdot \min\{N_1, N_2\}} CutSize(V_1, V_2)$. In contrast to this we cannot conclude any upper bound for the MVarMC bound since no flow at all is forced to cross the specific cut. So you can see that the three different Multicommodity Flow instances can react differently onto the given restriction and it gets clear that the more flexible the instance is, the potentially better is the bound.

Second, we show that the new bounds can also be used to improve lower bounds on the Graph Partitioning problem with classes of graphs, e.g. the bisection problem with the butterfly graph. The butterfly graph (without wraparound edges) of dimension d has a simple cut with $CutSize = 2^d$. In [2] it has been shown that the bisection width is $2(\sqrt{2} - 1)2^d + o(2^d) \approx 0.83 \cdot 2^d$. The known 1-1-MC bound delivers a lower bound of $(\frac{1}{2} + o(1))2^d$. Now using the possibility of the VarMC instance that not all vertices have to send something, we can improve this lower bound. I.e. only the 2^d vertices of the first level of vertices and the 2^d vertices of the last level send commodities of size one to all other vertices. This schema gives a lower bound of $(\frac{2}{3} + o(1))2^d$.

The butterfly graph is an example of a quite regular graph where the VarMC instance gives a better lower bound on the bisection problem with equally sized partitions than the known 1-1-MC. Furthermore, it is remarkable that the above analysis can be adapted to the Beneš graph, a kind of back-to-back butterfly. The Beneš graph has a similar simple bisection with $CutSize = 2^{d+1}$. To our knowledge the asymptotically exact bisection width is unknown. Using the 1-1-MC we get a lower bound of $\frac{1}{2}2^{d+1}$ while the VarMC instance where only the two most outside levels of vertices and the most inside level of vertices send commodities delivers a lower bound of $\frac{2}{3}2^{d+1}$.

6   Conclusion and Further Work

We have introduces a generalized lower bound on the Graph Partitioning problem. The bound bases on Multicommodity Flow instances with arbitrary sizes of commodities for every pair of source and destination. To get correct lower bounds from a flow instance the guaranteed $CutFlow$ is used. By inserting the sizes of the commodities as variables into the linear program, we get the best selection of these sizes for the given graph automatically. We have compared three different types of Multicommodity Flow instances with a different degree of freedom for the sizes of the commodities. Experiments show the superiority of the instance with the biggest degree of freedom.

Basing on these bounds a branch & bound algorithm has been presented which computes exact solutions for the Graph Partitioning problem. The comparison with other approaches shows that for a lot of graphs the presented approach delivers very good results. For example the DeBruijn graph of dimension eight has been solved exactly for the first time. On the other hand there are graphs, for example the quite dense and random instances introduced by Karisch, where former approaches are better suited for. Finally, it has been shown that the generalized bounds can also be used for theoretical analyses of graphs and can deliver new lower bounds. So altogether the new branch & bound algorithm is of importance for applications, since the algorithm can solve problems, which are unsolved until now. And furthermore, the generalized bounds offer new instruments for theoretical analyses.

For a further speed-up of the branch & bound algorithm three improvements look promising: Firstly, we could use the resulting primal and dual solution of a node of the search tree as starting point for the interior point algorithm of the next nodes in the search tree. Secondly, the branch & bound algorithm could be parallelized. Thirdly, we could use specialized algorithms for the Multicommodity Flow problem instead of using the general tool of an interior point algorithm. But our usage of Multicommodity Flow problems does not correspond to known ones since the sizes of the commodities have to be selected. So existing algorithms must be adapted. Apart of this it is also interesting to use the VarMC and MVarMC instance for theoretical analyses of lower bounds on graphs. For the graph bisection problem the VarMC instance is promising using with quite "regular" graph which are not vertex symmetric. For a partition into more than two partitions the MVarMC approach is promising, even for vertex symmetric graphs.

References

[1] L. Brunetta, M. Conforti, and G. Rinaldi. A branch-and-cut algorithm for the equicut problem. *Mathematical Programming*, 78:243–263, 1997.

[2] C. Bornstein, A. Litman, B. Maggs, R. Sitaraman, and T. Yatzkar. On the Bisection Width and Expansion of Butterfly Networks. In *Proceedings of the 1st Merged International Parallel Processing Symposium and Symposium on Parallel and Distributed Processing (IPPS/SPDP-98)*, pages 144–150. IEEE Computer Society, 1998.

[3] P.-O. Fjällström. Algorithms for graph partitioning: A survey. Linköping Electronic Articles in Computer and Information Science, 1998.

[4] L. K. Fleischer. Approximating Fractional Multicommodity Flow Independent of the Number of Commodities. *SIAM Journal on Discrete Mathematics*, 13(4):505–520, 2000.

[5] C. E. Ferreira, A. Martin, C. C. de Souza, R. Weismantel, and L. A. Wolsey. The node capacitated graph partitioning problem: a computational study. *Mathematical Programming*, 81:229–256, 1998.

[6] ILOG. *CPLEX 7.0 Reference Manual*, 2000.

[7] E. Johnson, A. Mehrotra, and G. Nemhauser. Min-cut clustering. *Mathematical Programming*, 62:133–151, 1993.

[8] S. E. Karisch, F. Rendl, and J. Clausen. Solving graph bisection problems with semidefinite programming. *INFORMS Journal on Computing*, 12(3):177–191, 2000.

[9] F. T. Leighton. *Introduction to Parallel Algorithms and Architectures*. Morgan Kaufman, 1992.

[10] T. Leighton, F. Makedon, S. Plotkin, C. Stein, E. Tardos, and S. Tragoudas. Fast Approximation Algorithms for Multicommodity Flow Problems. *Journal of Computer and System Sciences*, 50(2):228–243, 1995.

[11] R. D. McBride. Progress made in solving the multicommodity flow problem. *SIAM Journal on Optimization*, 8:947–955, 1998.

[12] K. Mehlhorn and S. Näher. LEDA, a library of efficient data types and algorithms. Technical report, University of Saarland, 1989.

[13] R. Preis and R. Dieckmann. The PARTY Partitioning – Library User Guide – Version 1.1. SFB 376 tr-rsfb-96-024, University of Paderborn, 1996.

[14] C. Souza, R. Keunings, L. A. Wolsey, and O. Zone. A new approach to minimising the frontwidth in finite element calculations. *Computer Methods in Applied Mechanics and Engineering*, 111:323–334, 1994.

[15] F. Shahrokhi and L. Szekely. On canonical concurrent flows, crossing number and graph expansion. *Combinatorics, Probability and Computing*, 3:523–543, 1994.

Fast Pricing of European Asian Options with Provable Accuracy: Single-Stock and Basket Options*

Karhan Akcoglu[1]**, Ming-Yang Kao[2]***, and Shuba V. Raghavan[3]

[1] Department of Computer Science, Yale University, New Haven, CT 06520, USA.
[2] Department of EECS, Tufts University, Medford, MA 02155, USA.
[3] Yale Investment Office, Yale University, New Haven, CT 06520, USA.

Abstract. This paper develops three polynomial-time techniques for pricing European Asian options with provably small errors, where the stock prices follow binomial trees or trees of higher-degree. The first technique is the first known Monte Carlo algorithm with analytical error bounds suitable for pricing single-stock options with meaningful confidence and speed. The second technique is a general recursive bucketing-based scheme that enables robust trade-offs between accuracy and run-time. The third technique combines the Fast Fourier Transform with bucketing-based schemes for pricing basket options. This technique is extremely fast, polynomial in the number of days and stocks, and does not add any errors to those already incurred in the companion bucketing scheme.

1 Introduction

A *call* (respectively, *put*) *option* is a contract assigning its *holder* the right, but not the obligation, to buy (respectively, sell) a security at some future time for a specified *strike price* X [10]. If the holder *exercises* her right, the other party in the contract, the *writer*, is obligated to assume the opposite side of the transaction. In exchange for this right, the holder pays the writer an *option price* P. The security in this contract can be any financial asset; for the purpose of this paper, we restrict it to a single stock or a portfolio of stocks. An option in the latter case is commonly called a *basket* option; for clarity, we call an option in the former case a *single-stock* option.

Options are popular financial instruments for a variety of trading strategies. For example, options can be used to hedge risk. As protection from a potential price fall in a stock price, one can purchase a put on the stock, thereby locking in

---

* karhan.akcoglu@yale.edu, kao@eecs.tufts.edu, kao-ming-yang@cs.yale.edu, shuba_raghavan@lycos.com.
** Partially supported by NSF Grant CCR-9896165.
*** Supported in part by NSF Grants CCR-9531028 and CCR-9988376. Part of this work was performed while this author was visiting Department of Computer Science, Yale University.

F. Meyer auf der Heide (Ed.): ESA 2001, LNCS 2161, pp. 404–415, 2001.

a minimum sell price. On the other hand, options can provide additional income for stockholders who write calls on their holdings; of course, this strategy carries the risk of being forced to sell the stock should the calls be exercised.

An option is valid until its *expiry date*. For a *European* option, the holder may exercise it only on the expiry date. For an *American* option, the holder may exercise it on any date up to and including the expiry date. The *payoff* of an option is the amount of money its holder makes on the contract. A European call is worth exercising if and only if $S \geq X$, where S is the stock price on the expiry date. The payoff of the call is $(S - X)^+ = \max(S - X, 0)$. For an American call, S is set to the stock price at the exercise time. An *Asian* option comes in European and American flavors, depending on when it may be exercised. For an European Asian call, if A is the average stock price over the entire life of the contract up to the expiry date, the payoff is $(A - X)^+$. For an American Asian call, A is set to the average stock price up to the exercise date. The payoffs of puts can be symmetrically defined.

The fair price P of an option is the discounted expected value of the payoff with appropriate martingale measure. Because of the popularity of options, pricing techniques for computing P have been extensively researched. Please see the references in [4,3] for details on prior work. Generally, it is more difficult to price a basket option than a single-stock option. To compute P, the price movement of each individual stock needs to be modeled. Typically, it is modeled as Brownian motion with drift. Using a stochastic differential equation, P can then be computed via a closed-form solution to the equation. When a closed-form solution is not known, various approaches are used to find an approximate solution. One class of approaches involves approximating the solution using numerical methods. Other approaches approximate the Brownian motion model with a discrete model, and use this model to approximate P. One such discrete model is the *binomial tree* model, due to Cox, Ross, and Rubinstein [7]; see Section 2 for the definition of the model.

This paper develops three polynomial-time pricing techniques with provably small errors. The remaining discussion makes the following assumptions: (1) The option in question is an European Asian single-stock or basket call. (2) Our task is to price the call at the start of its contract life. (3) The price of each underlying stock follows the binomial tree model. (4) In the case of a basket call, the price of each underlying stock moves independently. Our results generalize easily for puts, for a later time point than the start of the contract life, and for trees with higher degrees than two. The cases of American options and of correlated stocks remain open.

Monte Carlo simulation has been commonly used in the financial community. Despite this popularity, most reported results on error bounds are experimental or heuristic [11,1]. Our first technique is the first known Monte Carlo algorithm that has analytical error bounds suitable for pricing a European Asian call with meaningful confidence and speed. As shown by Theorem 2, the number of simulations required is polynomial in (P1) the logarithm of the inverse of the error probability and (P2) the inverse of the price error relative to the strike price but is exponential in (E1) the square root of the number of underlying stocks and (E2) the volatility of these stocks over the call's life. In particular, the algorithm is reasonably fast and accurate for a single-stock European Asian call with reasonable volatility.

Monte Carlo simulation is a randomized technique, and thus there is always a nonzero probability that the price obtained by polynomial-time Monte Carlo simulation is not accurate enough. The aggregation algorithm of Aingworth, Motwani, and Oldham (AMO) [1] is the first polynomial-time algorithm for pricing single-stock European Asian calls and other path-dependent options with guaranteed worst-case price errors. The AMO algorithm is based on a simple yet powerful idea called *bucketing* [9]. Our second technique is a general recursive bucketing-based scheme that can use the AMO algorithm, Monte-Carlo simulation, and possibly others as the base-case subroutine. This scheme enables robust trade-offs between accuracy and time over subtrees of different sizes. For long-term options or high-frequency price averaging, it can price single-stock European Asian calls with smaller error bounds in less time than the base-case algorithms themselves. In particular, as implied by Theorem 3, given the same runtime, this recursive scheme prices more accurately than the AMO algorithm; similarly, given the same accuracy, the scheme runs faster than the AMO algorithm.

This recursive scheme works for calls written on a single stock. Our third technique combines Fast Fourier Transform (FFT) and bucketing-based schemes to price basket calls and is applicable to European Asian calls as well as others. As shown in Theorem 4, this technique is extremely fast, polynomial in the number of days and the number of stocks, and does not add any errors to those already incurred in the companion bucketing schemes.

The remainder of this paper is organized as follows. Section 2 reviews the binomial tree model and basic definitions. Section 3 describes the new Monte Carlo algorithm. Section 4 details the recursive scheme. Section 5 gives the FFT-based technique for pricing basket calls.

2 The Binomial Tree Model

A *binomial* tree $\mathcal{T}$ is a *recombinant* binary tree. If n is the depth of $\mathcal{T}$, $\mathcal{T}$ has $t + 1$ nodes at depth t, for $0 \leq t \leq n$. For $0 \leq i \leq t$, let $\mathcal{T}[t, i]$ (or simply $[t, i]$ if $\mathcal{T}$ is obvious from context) be the i-th node at level t of $\mathcal{T}$. For $t > 0$, $\mathcal{T}[t, 0]$ and $\mathcal{T}[t, t]$ have one parent each, $\mathcal{T}[t - 1, 0]$ and $\mathcal{T}[t - 1, t - 1]$ respectively. For $0 < i < t$, $\mathcal{T}[t, i]$ has two parents, $\mathcal{T}[t - 1, i - 1]$ and $\mathcal{T}[t - 1, i]$. The number of nodes in $\mathcal{T}$ is $\frac{(n+1)(n+2)}{2}$.

Given a stock in the binomial tree model, the stock price is assumed to follow a geometric random walk through $\mathcal{T}$. Time is divided into n equal periods, with the root $\mathcal{T}[0, 0]$ corresponding to time $t = 0$, when the option is priced, and the leaves $\mathcal{T}[n, \cdot]$ corresponding to time $t = n$, the expiry date of the option. Let $s(\mathcal{T}[t, i])$ (or simply $s(t, i)$) be the stock price at node $\mathcal{T}[t, i]$. At each time step, the stock price $s(t, i)$ rises to $s(t + 1, i + 1) = u \cdot s(t, i)$—an *uptick*—with probability p or falls to $s(t+1,i) = d \cdot s(t,i)$—a *downtick*—with probability $q = 1 - p$. Letting r denote the *risk-free interest rate*, the parameters u and d satisfy $0 < d \leq 1 + r \leq u$ and are typically taken to be $u = \frac{1}{d} = e^{\sigma/\sqrt{n}}$, where σ is the n-period *volatility*, or standard deviation, of the stock price [10]. Although the probability p of an uptick is not known in general, for the purposes of pricing options, we can use the *risk-neutral probability model* [10], which states that $p = \frac{(1+r)-d}{u-d}$, where r is the risk-free interest rate for one period. This makes the expected return on the stock over one period, $pu + (1-p)d$, equal to the risk-free return, $1 + r$.

Let Ω be the sample space of paths $\omega = (\omega_1, \ldots, \omega_n)$ down T, where each $\omega_t \in \{-1,1\}$, with -1 corresponding to a downtick and 1 corresponding to an uptick. Given $\omega \in \Omega$ and $0 \leq t \leq n$, let $T[t,\omega]$ be the unique node at level t that ω passes through. Similar to the notation introduced above, we let $s(T[t,\omega]) = s(t,\omega)$ be the price at node $T[t,\omega]$.

We define the random variables $Y_1, \ldots, Y_n$ on Ω by $Y_t(\omega) = \omega_t$, the t-th component of ω. We define the probability measure Π on Ω to be the unique measure for which the random variables $Y_1, \ldots, Y_n$ are independent, identically distributed (*i.i.d.*) with $P(Y_i = 1) = p$ and $P(Y_i = -1) = q$. We start with an initial fixed stock price $S_0 = s(T[0,0])$. For $1 \leq t \leq n$, the stock price S_t is a random variable defined by $S_t = S_0 u^{\sum_{i=1}^{t} Y_i}$. From the structure of the binomial tree, we have $\Pr(S_t = s(t,i)) = \binom{t}{i} p^i (1-p)^{t-i}$, where $0 \leq i \leq t$. The *running total* of the stock prices is defined by $T_t = \sum_{i=0}^{t} S_i$ and the *running average* is $A_t = T_t/(t+1)$. For $\omega \in \Omega$, we let $S_t(\omega) = S_0 u^{\sum_{i=1}^{t} Y_i(\omega)}$, $T_t(\omega) = \sum_{i=0}^{t} S_i(\omega)$, and $A_t(\omega) = T_t(\omega)/(t+1)$.

Recall that X is the strike price of a European Asian call. Using the above notation, the price of this call is $\mathbb{E}((A_n - X)^+) = \frac{1}{n+1}\mathbb{E}((T_n - (n+1)X)^+) = \frac{1}{n+1}\mathbb{E}(\max(T_n - (n+1)X, 0))$. For large n, it is not known if this quantity can be computed exactly because the stock price can follow exponentially many paths down the tree. Below, we show how to estimate $\mathbb{E}((T_n - (n+1)X)^+)$, from which the price of the option can be easily computed.

3 A New Monte Carlo Algorithm

Monte Carlo simulation methods for asset pricing were introduced to finance by Boyle [2]. They are very popular in pricing complex instruments, particularly path-dependent European-style options. These methods involve randomly sampling paths $\omega \in \Omega$ according to the distribution Π and computing the payoff $(A_n(\omega) - X)^+$ on each sample. Suppose N samples $\omega^1, \ldots, \omega^N$ are taken from Ω. The price estimate of the call is $\mu = \frac{1}{N} \sum_{i=1}^{N} (A_n(\omega^i) - X)^+$. The accuracy of this estimate depends on the number of simulations N and the variance τ^2 of the payoff: the error bound typically guaranteed by Monte Carlo methods is $\mathcal{O}(\tau/\sqrt{N})$. Generally τ^2 is not known, and itself must be estimated to determine the error bound of μ.

A number of techniques are used to reduce the error of μ. For example, [5] uses the control variate technique, which ties the price of the option to the price of another instrument (the control variate) for which an analytically tractable solution is known. The simulation then estimates the difference between the option price and the control variate, which can be determined with greater accuracy than the option price itself. The antithetic variate method [6] follows not only randomly sampled paths ω down the binomial tree, but also the "mirror images" of each ω. None of these techniques have known analytical error bounds for μ.

Below we use concentration of measure results in conjunction with Monte Carlo simulation to estimate the price of a European Asian call and derive analytical error bounds for this estimate. The error bounds are in terms of the strike price X of the option and the maximum volatility σ_{max} of the underlying stocks.

3.1   Analytical Error Bounds for the Single-Stock Case

Let $C = e^{\mathbb{E}(\ln T_n)}$. In this section, we show that if $(n+1)X/C$ is "small", $\mathbb{E}((T_n - (n+1)X)^+)$ is close to $\mathbb{E}(T_n - (n+1)X) = \mathbb{E}(T_n) - (n+1)X$ (Theorem 1(1)). The option is *deep-in-the-money* and will probably be exercised. Since a closed-form formula exists for $\mathbb{E}(T_n)$ [1], $\mathbb{E}(T_n - (n+1)X)$ can be computed exactly, and our algorithm uses it as our estimate for $\mathbb{E}((T_n - (n+1)X)^+)$. On the other hand, if $(n+1)X/C$ is not small, the variance of $(T_n - (n+1)X)^+$ can be bounded from above (Theorem 1(2)) and our algorithm estimates its expectation with bounded error using Monte Carlo simulation.

We first give some theoretical results, then show how these results can be used in our Monte Carlo algorithm, BOUNDEDMC. We begin by finding bounds A and B such that $T_n \in [A, B]$ with high probability.

Lemma 1. Let $C = e^{\mathbb{E}(\ln T_n)}$. For any $\lambda > 0$, we have $\Pr\left(T_n \leq Ce^{-\sigma\lambda} \text{ or } T_n \geq Ce^{\sigma\lambda}\right) \leq 2e^{-\lambda^2/2}$, where σ is the volatility of the stock.

Proof. This follows from Azuma's Inequality [8].

Now, fix $\varepsilon > 0$ and choose $\lambda_0 = \sqrt{2\ln\frac{2}{\varepsilon}}$. Then, by Lemma 1, $\Pr\left(T_n \leq Ce^{-\sigma\lambda} \text{ or } T_n \geq Ce^{\sigma\lambda}\right) \leq \varepsilon$. Theorem 1(1) says that if $(n+1)X < Ce^{-\sigma\lambda_0}$, then $\mathbb{E}((T_n - (n+1)X)^+)$ is close to $\mathbb{E}(T_n - (n+1)X)$. Otherwise, Theorem 1(2) says that $\mathrm{Var}((T_n - (n+1)X)^+)$ is bounded.

Theorem 1. Let $C = e^{\mathbb{E}(\ln T_n)}$. (1) If $(n+1)X < Ce^{-\sigma\lambda_0}$, then $\left|\mathbb{E}((T_n - (n+1)X)^+) - \mathbb{E}(T_n - (n+1)X)\right| \leq \varepsilon(n+1)X$. (2) If $(n+1)X \geq Ce^{-\sigma\lambda_0}$, then $\mathrm{Var}((T_n - (n+1)X)^+) \leq (n+1)^2 X^2 e^{4\sigma\lambda_0}\frac{1+2\sigma\varepsilon}{\lambda_0 - 2\sigma}$.

Proof. Statement 1. Let $\phi(t)$ denote the probability density function of T_n. Note first that $\mathbb{E}((T_n - (n+1)X)^+) = \int_0^\infty (t - (n+1)X)^+\phi(t)dt = \int_{(n+1)X}^\infty (t - (n+1)X)\phi(t)dt = \mathbb{E}(T_n - (n+1)X) - \int_0^{(n+1)X}(t - (n+1)X)\phi(t)dt$. Then $\left|\mathbb{E}((T_n - (n+1)X)^+) - \mathbb{E}(T_n - (n+1)X)\right| = \left|\int_0^{(n+1)X}(t - (n+1)X)\phi(t)dt\right| \leq (n+1)X\int_0^{(n+1)X}\phi(t)dt = (n+1)X\Pr\left(T_n \leq (n+1)X\right) \leq (n+1)X\Pr\left(T_n \leq\right.$

$Ce^{-\sigma\lambda_0}) \leq \varepsilon(n+1)X$, where the second-last inequality follows from the assumption that $(n+1)X < Ce^{-\sigma\lambda_0}$ and the last inequality follows from Lemma 1 and our choice of λ_0.

Statement 2. We arrive at the result by applying several inequalities derived from the theory of "integration-by-parts" to $\mathrm{Var}((T_n - (n+1)X)^+)$ and using the assumption that $(n+1)X \geq Ce^{-\sigma\lambda_0}$. A detailed proof is presented in the full paper.

3.2   The BOUNDEDMC Algorithm

We next use these results in our algorithm. One approach would be to estimate $C = e^{\ln T_n}$ to determine whether we should apply Theorem 1(1) or 1(2). Our algorithm takes a more direct approach. We begin by selecting N samples $\omega^1, \ldots, \omega^N \in \Omega$ and computing $T_n(\omega^1), \ldots, T_n(\omega^N)$. For $1 \leq i \leq N$, define the random variable Z_i as $Z_i = 1$ if $T_n(\omega^i) \leq (n+1)X$, and $Z_i = 0$ otherwise. Let $Z = \sum_{i=1}^N Z_i$.

Theorem 2. Let $0 < \delta < 1$ be given. With $N = \Theta(\log \frac{1}{\delta} + \frac{1}{\varepsilon^2} e^{4\sigma\lambda_0} \frac{1+2\sigma\varepsilon}{\lambda_0-2\sigma})$ trials, the following statements hold.

1. If $\frac{Z}{N} \leq 2\varepsilon$, then, with probability $1 - \delta$, $\mathbb{E}(T_n - (n+1)X)$ estimates $\mathbb{E}((T_n - (n+1)X)^+)$ with error at most $4\varepsilon(n+1)X$. Correspondingly, the price of the call is estimated with error at most $4\varepsilon X$.
2. If $\frac{Z}{N} > 2\varepsilon$, then, with probability $1 - \delta$, $\frac{1}{N}\sum_{i=1}^N (T_n(\omega^i) - (n+1)X)^+$ estimates $\mathbb{E}((T_n - (n+1)X)^+)$ with standard deviation at most $\varepsilon(n+1)X$. Correspondingly, the price of the call is estimated with standard deviation at most εX.

Proof. Statement 1 follows from the proof of Theorem 1(1) and the Chernoff Bound. For Statement 2, we can use the Chernoff Bound and our choice of λ_0 to show that $(n+1)X \geq Ce^{-\sigma\lambda_0}$. We then arrive at our result by applying Theorem 1(2). A detailed proof is presented in the full paper.

Algorithm 1 BOUNDEDMC(δ, ε)
generate $N = \Theta(\log \frac{1}{\delta} + \frac{1}{\varepsilon^2} e^{4\sigma\lambda_0} \frac{1+2\sigma\varepsilon}{\lambda_0-2\sigma})$ *paths,* $\omega^1, \ldots, \omega^N$;
let Z *be the number of paths* ω^i *such that* $T_n(\omega^i) \leq (n+1)X$;
if $Z/N \leq 2\varepsilon$ *return* $\frac{1}{n+1}\mathbb{E}(T_n - (n+1)X) = \frac{1}{n+1}(\mathbb{E}(T_n) - (n+1)X)$;
else return $\frac{1}{N}\sum_{i=1}^N (A_n(\omega^i) - X)^+$.

3.3   Pricing Basket Options Using BOUNDEDMC

The results derived above are applicable for European Asian basket calls as well. If there are m stocks in the basket, each of which has volatility at most $\sigma_{\max}$, we need to take $\Theta(\log \frac{1}{\delta} + \frac{1}{\varepsilon^2} e^{4\sigma_{\max}\lambda_0\sqrt{m}} \frac{1+2\sigma_{\max}\varepsilon\sqrt{m}}{\lambda_0-2\sigma_{\max}\sqrt{m}})$ sample paths. With probability $1 - \delta$, our algorithm will estimate the European Asian basket call with error at most $4\varepsilon X$.

4   A Recursive Bucketing-Based Scheme

The AMO algorithm takes $\mathcal{O}(kn^2)$ time to produce a price estimate in the range $\left[P - \frac{nX}{k}, P\right]$, where P is the exact price of the call and k is any natural number. As in Section 3, this algorithm estimates $\mathbb{E}((T_n - (n+1)X)^+)$, from which the price of the call, $\mathbb{E}((A_n - X)^+)$, can be easily estimated. Our recursive scheme is a generalization of a variant of the AMO algorithm. Below we first describe this variant, called *Bucketed Tree Traversal* (BTT) and then detail our scheme, called *Recursive* Bucketed Tree Traversal (RecBTT).

Given binomial tree $\mathcal{T}$ of depth n, and numbers t, i, m such that $0 \leq t \leq n$, $0 \leq i \leq t$, and $m \leq n - t$, let $\mathcal{T}_m^{[t,i]}$ be the subtree of depth m rooted at node $\mathcal{T}[t,i]$. Given $0 \leq t \leq n$, let $\omega|_t$ be the prefix of ω up to level t of $\mathcal{T}$. Note that $\omega|_n = \omega$. Given $\psi, \omega \in \Omega$, we say that ψ is an *extension* of $\omega|_m$ if, for $0 \leq t \leq m$, we have $\psi_t = \omega_t$. Given another binomial tree $\mathcal{U}$ of depth n we say that $\psi \in \Omega(\mathcal{U})$ is *isomorphic* to $\omega \in \Omega(\mathcal{T})$, if, for all $0 \leq t \leq n$, $\psi_t = \omega_t$.

Like the AMO algorithm, BTT is based on the following simple observation. Suppose that the running total $T_m(\omega) = T_m(\omega|_m)$ of the stock prices on path $\omega \in \Omega$ exceeds the *barrier* $B = (n+1)X$. Then, for any extension ψ of $\omega|_m$, $T_n(\psi)$ also exceeds B and the call will be exercised. If we know the call will be exercised on all extensions of $\omega|_m$, it is easy to compute the payoff of the call on these extensions.

As we travel down a path ω, once the running total $T_m(\omega)$ exceeds B, we can just keep track of the running total on extensions ψ of $\omega|_m$ weighted by $\Pi(\psi)$, from which the value of the option can be computed. Hence, we only need to individually keep track of path prefixes $\omega|_m$ that have running totals $T_m(\omega|_m)$ less than B.

Unfortunately, there may be exponentially many such $\omega|_m$. However, the running totals $T_m(\omega|_m)$ are in the bounded range $[0, B)$. Rather than trying to keep track of each running total individually, we instead group the running totals terminating at each node into *buckets* that subdivide this interval. This introduces some round-off error. Suppose we use k buckets to divide $[0, B)$ into equal-length subintervals and we use the left endpoint of each interval as the representative value of the running totals contained in that bucket. At each step down the tree, when we put a running total into a bucket, an error of at most $\frac{B}{k}$ is introduced. Traveling down n levels of the tree, the total $T_n(\omega)$ of a path ω is underestimated by at most $\frac{nB}{k}$ and the average $A_n(\omega)$ is underestimated by at most $\frac{B}{k}$.

BTT is detailed in Algorithm 2. At each node $v = \mathcal{T}[t,i]$ of $\mathcal{T}$, create $k + 1$ buckets to store partial sums of path prefixes terminating at v. There will be k *core buckets* and one *overflow bucket*. This overflow bucket is the only difference between BTT and the AMO algorithm. For $0 \leq j < k$, core bucket $b_j(v)$ stores the *probability mass* $b_j(v)$.mass of path prefixes that terminate at node v and have running totals in its *range* range$(j) = [j\frac{B}{k}, (j+1)\frac{B}{k})$. The representative value of partial sums in the bucket is denoted by $b_j(v)$.value $= j\frac{B}{k}$. The overflow bucket $b_k(v)$ stores the probability-weighted running total estimates of path prefixes that have estimated running totals exceeding B. This quantity is denoted by $b_k(v)$.value. The probability mass of these path prefixes is denoted by $b_k(v)$.mass. BTT iterates through each of the $k + 1$ buckets of each of the $\frac{(n+1)(n+2)}{2}$ nodes in $\mathcal{T}$, for a total runtime of $\mathcal{O}(kn^2)$.

Algorithm 2 BTT$(\mathcal{T}, k, B)$
  for each node $v \in \mathcal{T}$ and each bucket $b_j(v)$ set $b_j(v)$.mass $\leftarrow 0$;
  take j such that initial price $s(\mathcal{T}[0,0]) \in \text{range}(j)$; $b_j(\mathcal{T}[0,0])$.mass $\leftarrow 1$;
  for $t = 0, \dots, (n-1)$ % iterate through each level
    for $i = 0, \dots, t$ % iterate through each node at level t
      let $v = \mathcal{T}[t,i]$; % shorthand notation for node $\mathcal{T}[t,i]$
      for $w \in \{\mathcal{T}[t+1,i], \mathcal{T}[t+1,i+1]\}$ % for each child of node v
        let $p' \in \{p,q\}$ be the probability of going from node v to w;
        for $b_j(v) \in \{b_0(v), \dots, b_k(v)\}$ % for each bucket at node v
          let $V \leftarrow b_j(v)$.value $+ s(w)$; let $M \leftarrow b_j(v)$.mass $\times p'$;
          add mass M to bucket corresponding to value V;
  return $\sum_{i=0}^{n} b_k(n,i)$.mass $\times (b_k(n,i)$.value $- B)$.
    % return option price estimated from overflow buckets at leaves[1]

We propose RECBTT, a recursive extension of BTT. Consider some level t in our binomial tree $\mathcal{T}$ and assume that the weights of all path prefixes terminating at level t have been put into the appropriate buckets. BTT uses these weights to compute the bucket weights of nodes at level $t + 1$. In contrast, RECBTT recursively solves the problem for subtrees $\mathcal{T}_m^{[t,i]}$, $0 \leq i \leq t$, of some depth $m < n - t$ rooted at node $\mathcal{T}[t,i]$.[2] As each recursive call is complete, RECBTT MERGES the bucket weights at the leaves of $\mathcal{T}_m^{[t,i]}$ into the corresponding nodes at level $t + m$ of $\mathcal{T}$. The advantages of the recursive calls are twofold. (1) They use finer bucket granularity, resulting in improved accuracy. (2) The results of a single recursive call on a particular subtree $\mathcal{T}_m^{[t,i]}$ are used to ESTIMATE the results of other recursive calls to other subtrees $\mathcal{T}_m^{[t,j]}$, where $j > i$, as long as the node prices in $\mathcal{T}_m^{[t,j]}$ are "sufficiently close" to the corresponding node prices in $\mathcal{T}_m^{[t,i]}$. This improves the runtime, since we do not need to make all $t + 1$ of the recursive calls, so there are portions of $\mathcal{T}$ that we do not directly traverse.

4.1   The MERGE Procedure

Consider a recursive call on the subtree $\mathcal{T}_1 = \mathcal{T}_{n_1}^{[t_0,i_0]}$ of depth n_1 rooted at node $v_0 = \mathcal{T}[t_0,i_0]$. A leaf $v_1 = \mathcal{T}_1[n_1,i_1]$ of $\mathcal{T}_1$ $(0 \leq i_1 \leq n_1)$ corresponds to the node $v_2 = \mathcal{T}[t_0+n_1, i_0+i_1]$ of $\mathcal{T}$. MERGE incorporates the bucket weights at v_1 into the bucket weights at v_2. Recall that the recursive call on $\mathcal{T}_1$ is made with finer bucket granularity. Assume we use $k_1 = h_1 k_0$ core buckets instead of just

---

[1] This is the expected payoff at maturity. For the price at the start of the contract, we   must discount the payoff according to the interest rate and length of the contract.
[2] Actually, the trees on which we recursively solve the problem are not exactly the $\mathcal{T}_m^{[t,i]}$. Each is identical to the respective $\mathcal{T}_m^{[t,i]}$, except the price at the root is changed   from $s(\mathcal{T}[t,i])$ to 0. The reason for this is explained in Remark 1.

$k_0 = k$. We first combine each group of h_1 buckets at v_1 into a single bucket, so that we are left with k_0 buckets to merge into v_2. When we refer to a bucket $b_{j_1}(v_1)$ below, $0 \le j_1 < k_0$, we mean one of these k_0 combined buckets.

The MERGE procedure is described in Algorithm 3. Consider first the core buckets. Let $0 \le j_0, j_1, j_2 < k_0$ denote core bucket indices. Bucket $b_{j_0}(v_0)$ contains the mass of path prefixes in $\mathcal{T}$ terminating at v_0 whose running total estimates fall into the interval range$(j_0) = [j_0\frac{B}{k_0}, (j_0+1)\frac{B}{k_0})$. Bucket $b_{j_1}(v_1)$ contains the mass of full paths in $\mathcal{T}_1$ terminating at node v_1 whose running total estimates fall into the interval range(j_1). This is equal to the mass of partial paths in $\mathcal{T}$ starting at node v_0 and terminating at node v_2 whose running total estimates fall into the interval range(j_1). Merging $\mathcal{T}_1$ into $\mathcal{T}$ involves merging each leaf v_1 of $\mathcal{T}_1$ into the corresponding node v_2 of $\mathcal{T}$. Once the merging procedure is done, $b_{j_2}(v_2)$.mass is updated to contain the weight of path prefixes in $\mathcal{T}$ passing through node v_0 and terminating at node v_2 that have running total estimates in the interval range(j_2). The overflow buckets are handled similarly.

Algorithm 3 MERGE$(\mathcal{T}, \mathcal{T}_1 = \mathcal{T}_{n_1}^{[t_0, i_0]})$
  let $v_0 \leftarrow \mathcal{T}[t_0, i_0]$;
  for $i_1 = 0, \ldots, n_1$ % for each leaf of $\mathcal{T}_1$
    let $v_1 \leftarrow \mathcal{T}_1[n_1, i_1]$, $v_2 \leftarrow \mathcal{T}[t_0 + n_1, i_0 + i_1]$;
    for $j_0 = 0, \ldots, k_0$ % buckets in v_0
      for $j_1 = 0, \ldots, k_0$ % buckets in v_1
        let $V \leftarrow b_{j_0}(v_0)$.value $+ b_{j_1}(v_1)$.value; let $M \leftarrow b_{j_0}(v_0)$.mass $\times b_{j_1}(v_1)$.mass;
        add mass M to bucket corresponding to value V.

Remark 1. Notice that the price at v_0 is counted twice: once in the path prefix from the root $\mathcal{T}[0,0]$ to v_0 and once in the partial path between v_0 and v_2. To address this issue, when we recursively solve the problem on the subtree $\mathcal{T}_1$, we set the price $s(\mathcal{T}_1[0,0])$ at the root to be 0, ensuring that this price is counted once. This modification does not change our algorithms.

Lemma 2. *For an arbitrary node v, let $E(v)$ be the maximum amount by which running totals terminating at node v are underestimated by the bucket values. Using the MERGE algorithm, we have $E(v_2) \le E(v_0) + E(v_1) + \frac{B}{k}$.*

Proof. When the paths of merged trees are concatinated, the errors are summed together. An extra error of one bucket size, $\frac{B}{k}$, may be introduced since merged buckets cover the range of two buckets, but are put into a single bucket.

Lemma 3. MERGE *can be made to run in $\mathcal{O}(n_1 k \log k)$ time.*

Proof. In the implementation of Algorithm 3, MERGE runs in $\mathcal{O}(n_1 k^2)$ time. However, the core buckets can be merged with a faster technique. Core bucket masses at v_2 are the colvolution of core bucket masses at v_0 and v_1 with an appropriate offset. This product can be computed in $\mathcal{O}(k \log k)$ time with the Fast Fourier Transform (FFT).

4.2   The ESTIMATE Procedure

Let $\mathcal{T}_1 = T_m^{[t_0,i_1]}$ and $\mathcal{T}_2 = T_m^{[t_0,i_2]}$ be two subtrees of $\mathcal{T}$, where $i_2 > i_1$ We now describe the ESTIMATE$(\mathcal{T}_2, \mathcal{T}_1)$ procedure, which estimates the weights in the leaf buckets of $\mathcal{T}_2$ from the weights in the leaf buckets of $\mathcal{T}_1$. This saves us the work of recursively solving the problem on $\mathcal{T}_2$.

ESTIMATE is described in Algorithm 4. It uses the following fact. Given any node $v_1 = \mathcal{T}_1[t, i]$ in $\mathcal{T}_1$, let $v_2 = \mathcal{T}_2[t, i]$ be the corresponding node in $\mathcal{T}_2$. Notice that there is a constant $\alpha > 1$ such that for all (v_1, v_2) pairs, $s(v_2) = \alpha s(v_1)$. Hence, for any path $\psi \in \Omega(\mathcal{T}_2)$, we have $T_m(\psi) = \alpha T_m(\omega)$, where $\omega \in \Omega(\mathcal{T}_1)$ is isomorphic to ψ.

Algorithm 4 ESTIMATE$(\mathcal{T}_2 = T_m^{[t_0,i_2]}, \mathcal{T}_1 = T_m^{[t_0,i_1]})$
  for $i = 0, \dots, m$ % **go through the leaf buckets of** $\mathcal{T}_1$ **and** $\mathcal{T}_2$
    let $v_1 \leftarrow \mathcal{T}_1[m, i]$, $v_2 \leftarrow \mathcal{T}_2[m, i]$;
    for $j_1 = 0, \dots, k$ % **go through each bucket at** v_1
      let $V \leftarrow \alpha b_{j_1}(v_1)$.value; *let* $M \leftarrow b_{j_1}(v_1)$.mass;
      add mass M *to bucket corresponding to value* V.

Lemma 4. *Suppose that* $\alpha \leq 2$ *and assume that the total path sums in* $\mathcal{T}_1$ *are underestimated by our bucketing scheme by at most* E. *ESTIMATE underestimates the total path sums in* $\mathcal{T}_2$ *by at most* $2E + 2\frac{B}{k}$.

Proof. The original error is amplified by $\alpha < 2$ in the estimation, accounting for the first error term. The second error term comes from the fact that the range of each bucket in $\mathcal{T}_1$ expands to cover at most 3 buckets in $\mathcal{T}_2$, all of whose masses are put into a single bucket.

Lemma 5. *Suppose that we would like to determine the leaf bucket weights of the subtrees* $T_m^{[t,i]}$, *where* $0 \leq i \leq t$. *We need only call* RECBTT *at most once for every* $\Theta(\frac{\sqrt{n}}{\sigma})$ *subtrees, and use the* ESTIMATE *procedure to estimate the leaf bucket weights of the other subtrees with bounded error.*

Proof. From the structure of the binomial tree, we can ESTIMATE $T_m^{[t,i]}$ from $T_m^{[t,0]}$ for i as large as $\frac{\sqrt{n}\ln 2}{2\sigma} = \Theta(\frac{\sqrt{n}}{\sigma})$ while keeping the error bounded.

4.3   Error and Runtime Analysis

Using the results of Sections 4.1 and 4.2 we can derive recursive expressions for the error and runtime of RECBTT. Suppose there are n_0 trading periods and we use a total of k_0 buckets per node. For $i > 0$, let n_i be the number of trading periods and k_i the number of buckets per node we use in the i-th subproblem (at the i-th level into the recursion). Theorem 3 shows us how to choose n_i and k_i for an error/runtime tradeoff that is stronger than that of the AMO algorithm.

Theorem 3. *Given integer* $R > 2$, *let* $\gamma = \frac{1}{R}$, *and for* $i > 0$, *let* $n_i = (\frac{n_0}{\sigma^2})^{1/2-i\gamma}$ *and* $k_i = 4^i k_0 (\frac{n_0}{\sigma^2})^{i\gamma}$, *where* σ *is the volatility of the stock.* RECBTT *under-estimates* $\mathbb{E}((T_n - (n+1)X)^+)$ *by at most* $\mathcal{O}(\frac{Bn_0^{1/2+\gamma}\sigma^{1-2\gamma}}{k_0})$ *and takes time* $\mathcal{O}(2^{1/\gamma}n_0^2 k_0(\frac{1}{\gamma} + \log\frac{k_0 n_0}{\sigma^2}))$.

4.4   Customization of the RECBTT Scheme

A key strength of the RECBTT scheme is that the number of recursive calls, the size of each recursive call, and the pricing method by which we solve the base case of the recursion (the final recursive call) can be custom tailored for the application. The base case of the recursion can be solved with any option pricing scheme, including BOUNDEDMC, other variants of the Monte Carlo algorithm, the AMO algorithm, or exhaustive path traversal.

In practice, larger values of n appear in several applications. Long-term options contracts, LEAPS [10], are negotiated for exercise dates several years in advance. Companies also offer stock options to employees over periods of up to five years or more. Finally, in *Istanbul* options contracts, rather than computing the average based on daily prices, the average is computed over prices taken at higher frequencies. In each of these cases, n is sufficiently large that several recursive calls are required to reduce the problem to a manageable size.

5   Pricing Basket Options

The bucketed tree structure created by BTT and RECBTT can be used to price various kinds of European basket options as well. Here, we describe how to price European Asian basket options. A basket option [10] is composed of m stocks, $z_1, \ldots, z_m$. For each stock z_i, we construct a binomial tree (according to the respective stock's volatility), as described in Section 2. For $1 \leq t \leq n$ and $1 \leq i \leq m$, let S_t^i be the random variable denoting the price of stock z_i on day t and define $S_t = \sum_{i=1}^m S_t^i$ to be the random variable denoting the total of the stock prices on day t. Recall that the payoff of a European basket call with strike price X is $\mathbb{E}((S_n - X)^+)$. Letting $A_n = \frac{1}{n+1} \sum_{t=0}^n S_t$ be the average total stock price, the payoff of a European Asian basket call with strike price X is $\mathbb{E}((A_n - X)^+)$. There is additional complexity with pricing basket options that does not appear in their single-stock counterparts: the number of paths that the total basket price can follow is exponential, not only in the number of trading periods n but also in that of stocks m. Basket options are usually priced using traditional Monte Carlo methods. The scheme we describe here is the first polynomial time, in both the number of stocks and trading periods, pricing scheme for any kind of basket option with provably small error bounds.

We will call our European Asian basket call pricing algorithm BASKETBTT. Let $B = (n + 1)X$, where X is the strike price of the basket option. For each stock z_i, $1 \leq i \leq m$, use RECBTT to construct the bucketed binomial tree structure T^i described in Section 4, this time using B as the barrier; should the running total of any z_i exceed B, the basket option will always be exercised, regardless of what the other stocks do. For each stock z_i, we construct $k + 1$ *superbuckets* β_j^i, $0 \leq j \leq k$, where β_j^i is the combination of buckets $b_j(v)$ for all leaves $v \in T^i$. For the core buckets β_j^i, $0 \leq j < k$, let β_j^i.value $= j\frac{B}{k}$ and β_j^i.mass $= \sum_{\ell=0}^n b_j(T^i[n, \ell])$.mass, where this summation ranges over all leaves $T^i[n, \ell]$ of T^i. For the overflow bucket β_k^i, let β_k^i.mass $= \sum_{\ell=0}^n b_k(T^i[n, \ell])$.mass and β_k^i.value $= \frac{1}{\beta_k^i.\text{mass}} \sum_{\ell=0}^n b_k(T^i[n, \ell])$.value $\times b_j(T^i[n, \ell])$.mass.

Handling overflow superbuckets: If the running total of a stock z_i reaches the overflow superbucket β_k^i, the option will be exercised regardless of what the other stocks do. Given this, we can determine the value of the option exactly, since $\mathbb{E}\big((T_n - (n+1)X)^+\big) = \mathbb{E}\big(T_n - (n+1)X\big) = \mathbb{E}(T_n) - (n+1)X = \beta_k^i.\text{value} + \sum_{i' \neq i} \mathbb{E}(T_n^{i'}) - (n+1)X$, where $T_n^{i'}$ is the random variable denoting the running total of stock $z_{i'}$ up to day n.

Handling core superbuckets: Consider now the core superbuckets β_j^i, $0 \leq j < k$. Let $f_i(x) = \sum_{j=0}^{k-1} \beta_j^i.\text{mass} \cdot x^j$ be the polynomial representation of the core bucket masses of stock z_i and let $f(x) = \prod_{i=1}^{m} f_i(x)$. This product can be computed efficiently, using the Fast Fourier Transform. Notice that $f(x)$ has the form $f(x) = b_0 x^0 + b_1 x^1 + \cdots + b_{m(k-1)} x^{m(k-1)}$. From the definition of $f(x)$, observe that b_j is just the probability that the sum (over all stocks z_i) of running totals T_n^i from the core buckets falls in the range $[j\frac{B}{k}, (j+1)\frac{B}{k})$. Hence, the contribution to the option price from the core buckets can be estimated by $\sum_{j=k}^{m(k-1)} b_j(j\frac{B}{k} - (n+1)X)$.

Theorem 4. *Given n, m, k, $R > 2$, $\gamma = \frac{1}{R}$, $\sigma_{\min}$ and $\sigma_{\max}$, if we apply* RecBTT *as described in Theorem 3 to construct the bucketed binomial tree for each stock,* BasketBTT *has an error of $\mathcal{O}(m\frac{Bn^{1/2+\gamma}\sigma_{\max}^{1-2\gamma}}{k})$ and runs in time*

$$\mathcal{O}\big(2^{1/\gamma}n^2 mk(\tfrac{1}{\gamma} + \log \tfrac{kn}{\sigma_{\min}^2}) + mk\log m \log k + mk\log^2 m\big).$$

References

1. D. Aingworth, R. Motwani, and J. D. Oldham. Accurate approximations for asian options. In *Proceedings of the 11th Annual ACM-SIAM Symposium on Discrete Algorithms*, pages 891–900, 2000.
2. P. P. Boyle. Options: A Monte Carlo approach. *Journal of Financial Economics*, 4:323–338, 1977.
3. P. Chalasani, S. Jha, and I. Saias. Approximate option pricing. *Algorithmica*, 25(1):2–21, 1999.
4. P. Chalasani, S. Jha, and A. Varikooty. Accurate approximations for European-Asian options. *Journal of Computational Finance*, 1(4):11–29, 1999.
5. L. Clewlow and A. Carverhill. On the simulation of contingent claim. *The Journal of Derivatives*, pages 66–74, Winter 1994.
6. C. Costantini. Variance reduction by antithetic random numbers of Monte Carlo methods for unrestricted and reflecting diffusions. *Mathematics and Computers in Simulation*, 51(1-2):1–17, 1999.
7. J. C. Cox, S. A. Ross, and M. Rubinstein. Option pricing: A simplified approach. *Journal of Financial Economics*, 7(3):229–263, September 1979.
8. A. Frieze. On the length of the longest monotone subsequence in a random permutation. *The Annals of Applied Probability*, 1(2):301–305, May 1991.
9. J. Hull and A. White. Efficient procedures for valuing European and American path-dependent options. *Journal of Derivatives*, 1(1):21–31, 1993.
10. J. C. Hull. *Options, Futures, and Other Derivatives*. Prentice-Hall, Upper Saddle River, NJ, fourth edition, 2000.
11. C. P. Robert and G. Casella. *Monte Carlo statistical methods*. Springer-Verlag, New York, 1999.

Competitive Auctions for Multiple Digital Goods

Andrew V. Goldberg[1] and Jason D. Hartline[2]

[1] STAR Laboratory, InterTrust Technologies Corp.
4750 Patrick Henry Dr., Santa Clara, CA 95054, USA
goldberg@intertrust.com
[2] Computer Science Department, University of Washington.
hartline@cs.washington.edu

Abstract. Competitive auctions encourage consumers to bid their utility values while achieving revenue close to that of fixed pricing with perfect market analysis. These auctions were introduced in [6] in the context of selling an unlimited number of copies of a single item (e.g., rights to watch a movie broadcast). In this paper we study the case of multiple items (e.g., concurrent broadcast of several movies). We show auctions that are competitive for this case. The underlying auction mechanisms are more sophisticated than in the single item case, and require solving an interesting optimization problem. Our results are based on a sampling problem that may have other applications.

1 Introduction

Consider an airplane flight where passengers have individual movie screens and can choose to view one out of a dozen movies that are broadcast simultaneously. The flight is only long enough for one movie to be seen. The airline wants to price movies to maximize its revenue. Currently, airlines charge a flat fee for movies. Even if the fee is based on a careful marketing study, passenger demographics may vary from one flight to another, and individual utilities can vary with flight route, time of the year, etc. Therefore a non-adaptive pricing is unlikely to be optimal for the seller. We investigate adaptive pricing via auctions.

We consider the problem of selling several items, with each item available in unlimited supply. By unlimited supply we mean that either the seller has at least as many items as there are consumers, or that the seller can reproduce items on demand at negligible marginal cost. Of particular interest are digital and broadcast items. With unlimited supply, consumer *utilities*, the maximum price a consumer is willing to pay for an item, are the sole factor determining sale prices and number of items sold. We assume that each consumer has potentially different utilities for different items, and needs one item only. The seller's goal is to set prices to maximize total revenue.

In the scarce supply case, multiple item auctions have been studied by Shapley and Shubik [15]. (See [13] for a survey of the area.) Results for the scarce case, however, do not directly apply to the unlimited supply case. Consider the

F. Meyer auf der Heide (Ed.): ESA 2001, LNCS 2161, pp. 416–427, 2001.

case where each item for sale is unique – for example the real estate market considered in [15]. In this case consumers will bid heavily for highly desirable items, which will sell for a high price. In contrast, in the unlimited supply case the seller can in principle give every consumer a copy of the item the consumer desires most. However, in such an auction, the consumer has no incentive to bid high. Thus a good auction mechanism must in some cases limit the number of copies of each item.

A consumer's utility value for an item is the most they are willing to pay for that item. We would like to develop auctions in which rational consumers bid their utilities. In game theory, such auctions are called *truthful* and are special cases of strategyproof mechanisms, which have been studied for a long time. For example, the Vickrey–Clarke–Groves mechanism [3,8,17] maximizes the general welfare of a system. The Shapley Value [14] mechanism shares costs among the participants. Recent work in the Computer Science community combines economic or game-theoretic questions with computational questions or techniques; see e.g., [5,10,9,12].

Our previous work [6,7], addressed a special case of the unlimited supply auction problem for a single item. In particular, we introduced *competitive auctions* which are truthful and at the same time attain revenues close to that of fixed pricing with perfect market analysis. As the term suggests, competitive analysis of auctions is similar in spirit to the analysis of on-line algorithms; see, e.g., [1,16]. We introduced several randomized auctions which are competitive under certain assumptions and showed some impossibility results, including the nonexistence of deterministic competitive auctions.

In this paper we extend some of these results to multiple item auctions. In particular, we develop competitive auctions based on random sampling. These auction mechanisms are intuitive but more sophisticated than in the single item case. We introduce a multiple item variant of the random sampling auction and of the dual price auction and show that these auctions are competitive under certain assumptions. We also discuss a deterministic auction. Although this auction is not competitive in the worst-case, its single item variant worked well in most cases in the experimental study [6,7]) and show that these auctions are competitive under certain assumptions. We also discuss a deterministic auction. Although this auction is not competitive in the worst-case, its single item variant worked well in most cases in the experimental study [6,7].

Our work uses the relationship between multiple item auctions and mathematical programming pointed out by Shapley and Shubik. For our random sampling auction we need to solve the following subproblem, which is interesting on its own: given the consumer's utilities, find item prices that maximize seller's revenue. We state this problem as a nonlinear mathematical program.

One of our main results is on a sampling problem that may be of independent interest. A variant of the sampling problem is as follows. Suppose we have n applicants and m tests. Each applicant takes each test and gets a real-valued score. We have to select k applicants based on the results of these scores. Furthermore suppose that we choose a random subset of the applicants, call the applicants in the subset red, and call the remaining applicants blue. After the results of the tests are known and the subset is selected, an adversary selects the k winning applicants while obeying the following restriction: If an applicant x is accepted and for every test, applicant y get a score that is at least as good as the score of x, then y must be accepted as well. Adversary's goal is to bias the admission in favor of red applicants. Although we study a slightly different problem, our techniques can be used to show that if $k = o(m^2 \log n)$, then with high probability the ratio of the number of red applicants to the number of blue applicants is bounded by a constant.

This problem seems natural.One can view candidates as points in m-dimensional space, and view the adversary as selecting a shift of the positive quadrant so that the shifted quadrant contains k points total and as many red points as possible.

In on-line markets, with rapid changes and the availability of computer trading tools and agents, pricing using auctions is sometimes attractive. Competitive auctions for multiple unlimited supply items may be useful in some of these scenarios.

2   Background

The input to an auction is a number of bidders, n, a number of items, m and a set of bids $\{a_{ij}\}$. We assume that all bids are nonnegative and that there is no collusion among the bidders. We study the case when each bidder wants only a single item.

Given a set of bids, the outcome of an auction is an assignment of a subset of (winning) bidders to items. Each bidder i in the subset is assigned a single item j and a sales price of at most a_{ij}. An item can be assigned to any number of bidders. A *deterministic auction mechanism* maps auction inputs to auction outcomes. A *randomized auction mechanism* maps inputs to probability distributions on auction outcomes. We use $\mathcal{R}$ to denote the auction *revenue* for a particular auction mechanism and set of bids. $\mathcal{R}$ is the sum of all sale prices. For randomized auctions, $\mathcal{R}$ is a random variable. We will assume that the m-th item is a dummy item of no value and that all bidders have utility of zero for this item ($a_{im} = 0$ for all i). Losing is then equivalent to being assigned the dummy item at cost zero.

We say that an auction is *single-price* if the sale prices for copies of the same item are the same, and *multiple-price* otherwise.

Next we define *truthful auctions*, first introduced by Vickrey [17]. Let u_{ij} be bidder i's utility value for item j. Define a *bidder's profit* to be the difference between the bidder's utility value for the item won and the price the bidder pays if they win the auction, or zero if they lose. An auction is truthful if bidding u_{ij} is a dominant strategy for bidder i. In other words, the bidder's profit (or expected profit, for randomized auctions), as a function of the bidder's bids $(a_{i1}, \ldots, a_{im})$, is maximized at the bidder's utility values $(u_{i1}, \ldots, u_{im})$, for any fixed values of the other bidders' bids. Truthfulness is a strong condition for auctions: bidding utility maximizes the profit of the bidder no matter what the other bidders' strategies are. When considering truthful auctions, we assume that $a_{ij} = u_{ij}$, unless mentioned otherwise.

To enable analysis of auction revenue we define several parameters of an input set of bids. The revenue for optimal fixed pricing is $\mathcal{F}$. Note that $\mathcal{F}$ can also be interpreted as the revenue due to the optimal nontruthful single-price auction. Other parameters that we use in analysis are ℓ, the lowest bid value, and h, the highest bid value. Because bids can be arbitrarily scaled, we assume, without loss of generality, that $\ell = 1$, in which case h is really the ratio of the highest bid to the lowest bid.

Analogous to on-line algorithm theory, we express auction performance relative to that for the optimal nontruthful auction, as ratios $\mathcal{R}/\mathcal{F}$. However, we solve a maximization problem, while on-line algorithms solve minimization problems. Thus, positive results, which are lower bounds on $\mathcal{R}/\mathcal{F}$, are expressed using "Ω".

Note that h and $\mathcal{F}$ are used only for analysis. Our auctions work without knowing their values in advance.

As shown in [6], if we do not impose any restrictions on h, we get the upper bound of $\mathcal{R}/\mathcal{F} = O(1/h)$. To prevent this upper bound on auction revenue we can make the assumption that the optimal revenue $\mathcal{F}$ is significantly larger than h, the highest bid. With this assumption, optimal fixed pricing sells many items.

We say that an auction is *competitive* under certain assumptions if when the assumptions hold, the revenue is $\Omega(\mathcal{F})$.

For convenience, we assume that the input bids are *non-degenerate*, i.e., all input bids values a_{ij} are distinct or zero. This assumption can be made without loss of generality because we can always apply a random perturbation or use lexicographic tie-breaking to achieve it.

As shown for the single-commodity case [6], no deterministic auction is competitive in the worst case. Our competitive auctions are randomized. We use the following lemma, which is a variation of the Chernoff bound (see e.g. [2,11]), as the main tool in our analysis.

Lemma 1. *Consider a set A and its subset $B \subset A$. Suppose we pick an integer k such that $0 < k < |A|$ and a random subset (sample) $S \subset A$ of size k. Then for $0 < \delta \le 1$ we have*

$$\mathbf{Pr}[|S \cap B| < (1 - \delta)|B| \cdot k/|A|] < \exp(-|B| \cdot k\delta^2/(2|A|)).$$

Proof. We refer to elements of A as points. Note that $|S \cap B|$ is the number of sample points in B, and its expected value is $|B| \cdot k/|A|$. Let $p = k/|A|$. If instead of selecting a sample of size exactly k we choose each point to be in the sample independently with probability p then the Chernoff bound would yield the lemma.

Let $A = \{a_1, \ldots, a_n\}$ and without loss of generality assume that $B = \{a_1, \ldots, a_k\}$. We can view the process of selecting S as follows. Consider the elements of A in the order induced by the indices. For each element a_i considered, select the element with probability p_i, where p_i depends on the selections made up to this point.

At the point when a_{i+1} is considered, let t be the number currently selected points. Then $i - t$ is the number of points considered but not selected. Suppose that $t/i < p$. Then $p_{i+1} > p$.

We conclude that when we select the sample as a random subset of size k, the probability that the number of sample points in B is less than the expected value is smaller than in the case we select each point to be in the sample with probability p. ∎

3   Fixed Price Auction and Optimal Prices

Consider the following *fixed price auction*. The bidders supply the bids and the seller supplies the *sale prices*, r_j, $1 \leq j \leq m$. Define $c_{ij} = a_{ij} - r_j$. The auction assigns each bidder i to the item j with the maximum c_{ij}, if the maximum is nonnegative, and to no item otherwise. In case of a tie, we chose the item with the maximum j. If a bidder i is assigned item j, the corresponding sale price is r_j.

Lemma 2. *Suppose the sale prices are set independently of the input bids. Then the fixed price auction is truthful.*

Proof. If bidder i gets object j, the bidder's price is at least r_j and the bidder's profit is at most $a_{ij} - r_j$. The best possible profit for i is $\max_j(u_{ij} - r_j)$. If the bidder bids $a_{ij} = u_{ij}$, this is exactly the profit of the bidder. ∎

Remark Although we assume that the bidders do not see sale prices before making their bids, the lemma holds even if the bidders do see the prices.

Now consider the following *optimal pricing* problem: Given a set of bids, find the set of prices such that the fixed price auction brings the highest revenue. Suppose an auction solves this problem and uses the resulting prices. We call this auction the *optimal nontruthful single-price auction* and denote its revenue by $\mathcal{F}$. We can interpret $\mathcal{F}$ as the revenue of fixed pricing using perfect market analysis or as the revenue of the optimal nontruthful single-price auction. The prices depend on the input bids, and one can easily show this auction is nontruthful.

We use $\mathcal{F}$ to measure performance of our truthful auctions. Although one might think that being a single-price auction is a serious restriction, in the single-item auction case this is not so. In this case, the revenue of the optimal single-price auction is at least as big as the expected revenue of any reasonable[1] (possible multiple-price) truthful auction; see [6].

Next we state the optimal pricing problem as a mathematical programming problem. We start by stating the problem of finding a bidder-optimal object assignment given the bids and the sale prices as an integer programming problem. This problem is a special case of the b-matching problem [4] (bipartite, weighted, and capacitated, with unit node capacities on one side and infinite capacities on

---

[1] See [6] for the precise definition of reasonable. The intuition is that we preclude auctions that are taylored to specific inputs. Such an auction would perform well these specific inputs, but poorly on all others.

the other). For the limited supply case, when only one copy of an item is available, the classical paper [15] takes a similar approach. For our case, this problem is easy to solve by taking the maximum as in the previous section. However, we then treat sale prices as variables to get a mathematical programming formulation of the optimal pricing problem.

One can show that the optimal price problem is equivalent to the following mathematical programming problem; we omit details.

$$
\begin{aligned}
\max \quad & \sum_j \sum_i x_{ij} r_j && \text{subject to} && (1) \\
& r_m = 0 \\
& \sum_j x_{ij} \le 1 && 1 \le i \le n \\
& x_{ij} \ge 0 && 1 \le i \le n,\ m \le j \le m \\
& p_i + r_j \ge a_{ij} && 1 \le i \le n,\ m \le j \le m \\
& \sum_i p_i = \sum_j \sum_i x_{ij} \cdot (a_{ij} - r_j)
\end{aligned}
$$

This problem has quadratic objective function; some constraints are linear while other constraints are quadratic. Here x_{ij} is one exactly when bidder i gets item j and p_i's are profits of the corresponding bidders.

Since $\sum_j \sum_i x_{ij} r_j = \sum_j \sum_i x_{ij} a_{ij} - \sum_i p_i \le \sum_j \sum_i a_{ij}$, the objective function is bounded. Since the feasibility region is closed, it follows that (1) always has an optimal solution.

We omit proofs of the next two results.

Lemma 3. *For any solution of (1) with fractional x_{ij}'s there is a solution with $x_{ij} \in \{0,1\}$ and an objective function value that is at least as good.*

Theorem 1. *Consider sale prices defined by an optimal solution of (1). The revenue of the fixed price auction that uses these prices and has bids a_{ij} in the input is equal to the objective function value of the optimal solution.*

Recall that we use the problem (1) to find a set of prices that maximizes the fixed price auction revenue. In the rest of the paper we assume that we can compute such prices and leave open the question of how to do this efficiently. Note that we could also use an approximate solution.

4   The Random Sampling Auction

We use random sampling to make the optimal single-price auction truthful.

The *random sampling auction* works as follows.

1. Pick a random sample S of the set of bidders. Let N be the set of bidders not in the sample.
2. Compute the optimal sale prices for S as outlined in the previous section.
3. The result of the random sampling auction is then just the result of running the fixed-price auction on N using the sale prices computed in the previous step. All bidders in S lose the auction.

The sample size is a tradeoff between how well the sample represents the input and how much potential revenue is wasted because all bidders in the sample lose. Unless mentioned otherwise, we assume that the sample size is $n/2$ or, if n is odd, the floor or the ceiling of $n/2$ with probability $1/2$.

The facts that the bidders who determine the prices lose the auction and that the fixed price auction is truthful imply the following result.

Lemma 4. *The random sampling auction is truthful.*

Remark Another natural way of sampling is to sample bids instead of bidders. However, this does not seem to lead to a truthful auction, because bidder's bids selected in the sample may influence the price used to satisfy the bidder's remaining bids.

Next we show that, under certain assumptions, the auction's revenue $\mathcal{R}$ is within a constant factor of $\mathcal{F}$. Without loss of generality, for every $1 \le i \le n, 1 \le j \le m$, if a_{ij} is undefined (not in the input) we define a_{ij} to be zero. For every bidder i, we view $(a_{i1}, \ldots, a_{im})$ as a point in the m-dimensional space and denote this point by v_i. Thus v_i is in the quadrant Q of the m-dimensional space where all coordinates are nonnegative. We denote the set of all input points by B.

For a fixed m and a set of sale prices $r_1, \ldots, r_m$, let R_j be a region in the m-dimensional space such that if $v_i \in R_j$, then i prefers j to any other item, i.e., for any $1 \le k \le m$, $c_{ij} \ge c_{ik}$ (recall that $c_{ij} = a_{ij} - r_j$). We would like $\{R_j : 1 \le j \le m\}$ to be a partitioning of Q. We achieve this by assigning every boundary point to the highest-index region containing the point. (This is consistent with our tie-breaking rule for the fixed price auction.) R_j is a convex (and therefore connected) region in Q. In fact, the region R_j is as follows:

$$R_j = \{x : x_j \ge r_j \ \& \ x_j - r_j \ge x_k - r_k \ \forall k \ne j\}. \tag{2}$$

Figure 1 shows a two item auction with prices r_1 and r_2 for items 1 and 2 respectively. These prices induce the regions $R_1 = R_1' \cup R_1''$ and $R_2 = R_2' \cup R_2''$. Arrows point to selling prices for the bidders in each region.

Thus sampling and computing r_j's partitions Q into the regions, and each bidder i in N gets the item corresponding to the region that i is in. Intuitively, our analysis says that if a region has many sample points, it must have a comparable number of nonsample points – even though the regions are defined based on the sample. The latter fact makes the analysis difficult by introducing conditioning. Intuitively, we deal with the conditioning by considering regions defined by the input independently of the sample.

For a given input, let $q_1, \ldots, q_m$ be a set of optimal prices for the input bids that yield revenue $\mathcal{F}$. These prices induce the regions discussed above. Bidders in region R_j pay q_j for the item j. If we sample half of the points, the expected number of sample points in a region R_j is half of the total number of points in the region, and for the prices $q_1, \ldots, q_m$, the expected revenue is $\mathcal{F}/2$. The optimal fixed pricing on the sample does at least as well. Thus the expected revenue of optimal fixed pricing of the sample, $\mathbf{E}[\mathcal{F}_s]$, is at least $\mathcal{F}/2$. However, we need a high-probability result. Our goal is to show that with high probability

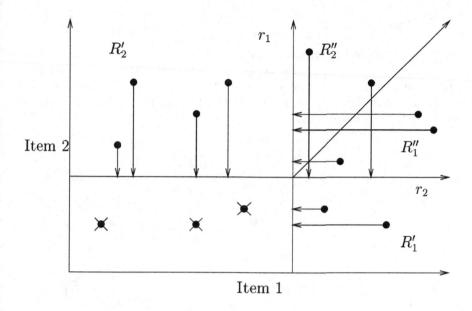

Fig. 1. Two item auction with regions R_1 and R_2

$\mathbf{E}[\mathcal{F}_s]$ is close to $\mathcal{F}/2$ and that $\mathbf{E}[\mathcal{R}]$ is close to $\mathbf{E}[\mathcal{F}_s]$, where $\mathcal{R}$ is the revenue of the random sampling auction.

We say that a set $A \subseteq B$ is t-*feasible* if A is nonempty and for some set of sale prices, A is exactly the set of points in R_t. For each feasible set A, we define its signature $S_A = (s_1, \ldots, s_m)$ such that s_i's are (not necessarily distinct) elements of A and, for a fixed t, different t-feasible sets have different signatures. In the following discussion, s_{ij} denotes the j-th coordinate of s_i.

We construct signatures as follows. Let R_t be a region defining A. R_t is determined by a set of prices $(r_1, \ldots, r_m)$. We first increase all r_j's by the same amount (moving R_t diagonally) until some point in A is on the boundary of R_t. Note that since we change all prices by the same amount, the limiting constraint from (2) is $x_t \geq r_t$. Thus the stopping is defined by $x_t = r_t$, and the point on the boundary has the smallest t-th coordinate among the points in A. We set s_t to this point.

Then for $j \neq t$, we move the the region starting at its current position down the j-th coordinate direction by reducing r_j until the first point hits the boundary. The boundary we hit is defined by $x_t - r_t = x_j - r_j$, and the point that hits it first has the minimum $x_j - x_t + s_{tt}$ among the points in A. Observe that the point s_t remains on the boundary $x_t = r_t$, and therefore we stop before r_j becomes negative. When we stop, we take a point that hits the boundary and assign s_j to it.

Consider the set of points in the signature, $S_A = \{s_1, \ldots, s_m\}$. Define R to be the region we got at the end of the procedure that computed S_A. R is defined

by
$$R = \{x : x_t \geq s_{tt} \ \& \ x_t - s_{tt} \geq x_j - s_{jj} \ \forall j \neq t\}.$$

It follows that R can be constructed directly from S_A.

Figure 2 shows the signatures we get from the prices r_1 and r_2. The points on the boundary of the shaded region are the signature of that region. Note, for example, that there are no points in R_1 that are not inside the boundary induced by the signature for R_1.

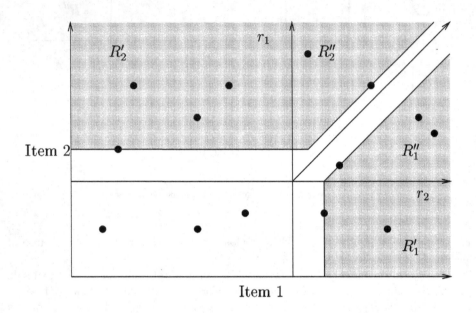

Fig. 2. Signatures in a two item auction

Suppose two feasible sets have the same signature S and let R be the region defined by the signature. Then the two sets are exactly the set of points in R, and are thus identical.

The next two lemmas are simple, so we omit the proofs.

Lemma 5. *For each t, $1 \leq t \leq m$, there are at most n^m t-feasible sets.*

Lemma 6. *For every t-feasible subset C of the sample S there is a t-feasible subset A of the input such that $C = A \cap S$.*

For $k \geq 1$ and $0 < \delta < 1$, we say that a sample S is (k, δ)-*balanced* if for every $1 \leq t \leq m$ and for every t-feasible subset of the input, A, such that $|A| \geq k$, we have
$$(1 - \delta) \leq (|A \cap S|)/(|A \cap N|) \leq 1/(1 - \delta).$$

Lemma 7. *The probability that a sample containing half of the input points is (k, δ)-balanced is at least $1 - 2mn^m \exp(-k\delta^2/8)$.*

Proof. Lemma 1 implies that the probability that for a set A with $|A| \geq k$,

$$\mathbf{Pr}[|A \cap S| < (1 - \delta)|A \cap N|] < \exp(-k\delta^2/8)$$

and

$$\mathbf{Pr}[|A \cap N| < (1 - \delta)|A \cap S|] < \exp(-k\delta^2/8).$$

Note that the fact that the number of sample points in one subset is close to its expectation makes it no less likely that the number of sample points in another subset is close to expectation. Thus the conditioning we get is favorable. By Lemma 6, there are at most n^m t-feasible subsets for every t, so the total number of feasible subsets is mn^m. These observations imply the lemma. ■

Theorem 2. *Assume* $\alpha h m^2 \ln n \leq \mathcal{F}$ *and* $m \geq 2$. *Then* $\mathcal{R} \geq \mathcal{F}/24$ *with probability of at least* $1 - \exp(-\alpha/1728)$ *(for some constant* $\alpha > 1$*).*

Proof. Consider Lemma 7 with $\delta = 1/2$ and $k = \alpha m \log n/12$. The probability that the sample is (k, δ)-balanced is

$$1 - 2mn^m \exp(-k\delta^2/8) = 1 - 2mn^m \exp(-\alpha m \log n/864) \geq 1 - \exp(-\alpha/1728)$$

for $m \geq 2$. For the rest of the proof we assume that the sample is (k, δ)-balanced; we call this the balanced sample assumption.

Next we show that the revenue of the auction on the sample, $\mathcal{F}_s$, satisfies $\mathcal{F}_s \geq \mathcal{F}/6$. Let Q_i be the set of bidders who get item i when computing $\mathcal{F}$ on the entire bid set. Consider sets Q_i containing less than $(\alpha m \log n)/2$ bidders. The total contribution of such sets to $\mathcal{F}$ is less then $\mathcal{F}/2$. This is because there are at most m such sets and each bid is at most h giving a maximum possible revenue of $\alpha h m^2 \log n/2 = \mathcal{F}/2$. Thus the contribution of the sets with at least $(\alpha m \log n)/2$ bidders is more than $\mathcal{F}/2$, and we restrict our attention to such sets. By the balanced sample assumption, each such set contains at least $1/3$ sample points, and thus $\mathcal{F}_s \geq (1/3)\mathcal{F}/2 = \mathcal{F}/6$.

Finally we show that $\mathcal{R} \geq \mathcal{F}/24$ using a similar argument. Let R_i be the regions defined by the prices computed by the auction on the sample. Consider the regions containing less than $(\alpha m \log n)/12$ sample points. The total contribution of such sets to the revenue is less then $\mathcal{F}/12$. The remaining regions contribute at least $\mathcal{F}/12$ (out of $\mathcal{F}/6$). Each remaining region contains at least $(\alpha m \log n)/12$ sample points. By the balanced sample assumption, each such region contains at least one nonsample point for every two sample point, and thus $\mathcal{R} \geq \mathcal{F}/24$. ■

Lemma 4 and Theorem 2 imply that if the assumptions of the theorem hold, the random sampling auction is competitive.

4.1   The Dual Price Auction

The random sampling auction is wasteful in the sense that all bidders in the sample lose the auction. The *dual price auction* eliminates the waste by treating

S and N symmetrically: S is used to compute sale prices for N and vice versa. Note that for each item, the two sale prices used are, in general, different; this motivates the name of the auction.

By symmetry, the expected revenue of the dual price auction is twice the expected revenue of the single price auction with $n/2$ sample size. Thus, under conditions of Theorem 2 the dual price auction is competitive.

5   A Deterministic Auction

The following auction is a generalization of the *deterministic optimal threshold auction* introduced in [6] to the multi-item case. Although not competitive in general, the single-item variant of this auction works well when the input is non-pathological, e.g., when bidder utilities are selected independently from the same distribution.

The deterministic auction determines what item, if any, the bidder i gets as follows. It deletes i from B, computes optimal prices for the remaining bidders, and then chooses the most profitable item for i under these prices. This is done independently for each bidder. This auction is truthful but, as we have mentioned, not competitive in some cases.

6   Concluding Remarks

Our analysis of the random sampling auction is somewhat brute-force, and a more careful analysis may lead to better results, both in terms of constants and in terms of asymptotic bounds. In particular, the assumption $\alpha h m^2 \ln n \leq \mathcal{F}$ in Theorem 2 may be stronger than necessary. One can prove that $\mathcal{F}_s = \Omega(\mathcal{F})$ assuming $\alpha h m \leq \mathcal{F}$. We wonder if the theorem holds under this weaker assumption.

Although our theoretical bounds require m to be small compared to n and the optimal fixed price solution to contain a large number of items, it is likely that in practice our auctions will work well for moderately large m and moderately small optimal fixed price solutions. This is because our analysis is for the worst-case. In many real-life applications, bidder utilities for the same item are closely correlated and our auctions perform better.

The optimal fixed pricing problem has a very special form that may allow one to solve this problem efficiently. Note that if one uses an approximation algorithm to solve the problem (say within 2% of the optimal) and our auctions remain truthful. (This is in contrast to combinatorial auctions [9].) It is possible that in practice this problem can be solved approximately, in reasonable time, using general nonlinear optimization techniques. We leave an existence of such an algorithm as an open problem.

Another open problem is a generalization of our results. One possible generalization is to the case when some items are in fixed supply. Another generalization is to the case when consumer i wants up to k_i items.

References

1. A. Borodin and R. El-Yaniv. *Online Computation and Competitive Analysis.* Cambridge University Press, 1998.
2. H. Chernoff. A Measure of Asymptotic Efficiency for Test of a Hypothesis Based on the Sum of Observations. *Anals of Math. Stat.*, 23:493–509, 1952.
3. E. H. Clarke. Multipart Pricing of Public Goods. *Public Choice*, 11:17–33, 1971.
4. J. Edmonds and E. L. Johnson. Matching, a Well-Solved Class of Integer Linear Programs. In R. Guy, H. Haneni, and J. Schönhein, editors, *Combinatorial Structures and Their Applications*, pages 89–92. Gordon and Breach, NY, 1970.
5. J. Feigenbaum, C. Papadimitriou, and S. Shenker. Sharing the Cost of Multicast Transmissions. In *Proc. of 32nd Symposium Theory of Computing*, pages 218–226. ACM Press, New York, 2000.
6. A. V. Goldberg, J. D. Hartline, and A. Wright. Competitive auctions and digital goods. Technical Report STAR-TR-99.09.01, STAR Laboratory, InterTrust Tech. Corp., Santa Clara, CA, 1999. Available at URL http://www.star-lab.com/tr/tr-99-01.html.
7. A. V. Goldberg, J. D. Hartline, and A. Wright. Competitive Auctions and Digital Goods. In *Proc. 12th Symp. on Discrete Alg.*, pages 735–744. ACM/SIAM, 2001.
8. T. Groves. Incentives in Teams. *Econometrica*, 41:617–631, 1973.
9. D. Lehmann, L. I. O'Callaghan, and Y. Shoham. Truth Revelation in Approximately Efficient Combinatorial Auctions. In *Proc. of 1st ACM Conf. on E-Commerce*, pages 96–102. ACM Press, New York, 1999.
10. N. Linial. Game Theoretic Aspects of Computing. In *Handbook of Game Theory*, volume 2, pages 1339–1395. Elseveir Science Publishers B.V., 1994.
11. R. Motwani and P. Raghavan. *Randomized Algorithms.* Cambridge University Press, 1995.
12. N. Nisan and A. Ronen. Algorithmic Mechanism Design. In *Proc. of 31st Symposium on Theory of Computing*, pages 129–140. ACM Press, New York, 1999.
13. A. E. Roth and M. A. Oliveira Sotomayor. *Two-Sided Matching: A Study in Game-Theoretic Modeling and Analysis.* Cambridge University Press, 1990.
14. L. S. Shapley. Core of Convex Games. *Int. J. of Game Theory*, 1:11–26, 1971.
15. L. S. Shapley and M. Shubik. The Assignment Game I: The Core. *Int. J. of Game Theory*, 1:111–130, 1972.
16. D. D. Sleator and R. E. Tarjan. Amortized Efficiency of List Update and Paging Rules. *Communications of the ACM*, 28:202–208, 1985.
17. W. Vickrey. Counterspeculation, Auctions, and Competitive Sealed Tenders. *J. of Finance*, 16:8–37, 1961.

Algorithms for Efficient Filtering in Content-Based Multicast

Stefan Langerman, Sachin Lodha, and Rahul Shah

Department of Computer Science,
Rutgers University, New Brunswick, NJ, USA
{lfalse,lodha,sharahul}@cs.rutgers.edu

Abstract. Content-Based Multicast is a type of multicast where the source sends a set of different classes of information and not all the subscribers in the multicast group need all the information. Use of filtering publish-subscribe agents on the intermediate nodes was suggested [5] to filter out the unnecessary information on the multicast tree. However, filters have their own drawbacks like processing delays and infrastructure cost. Hence, it is desired to place these filters most efficiently. An $O(n^2)$ dynamic programming algorithm was proposed to calculate the best locations for filters that would minimize overall delays in the network [6]. We propose an improvement of this algorithm which exploits the geometry of *piecewise linear functions* and fast merging of sorted lists, represented by height balanced search trees, to achieve $O(n \log n)$ time complexity. Also, we show an improvement of this algorithm which runs in $O(n \log h)$ time, where h is the height of the multicast tree. This problem is closely related to p-median and uncapacitated facility location over trees. Theoretically, this is an uncapacitated analogue of the p-inmedian problem on trees as defined in [9].

1   Introduction

There has been a surge of interest in the delivery of personalized information to users as the amount of information readily available from sources like the WWW increases. When the number of information recipients is large and there is sufficient commonality in their interests, it is worthwhile to use multicast rather than unicast to deliver information. But, if the interests of recipients are not sufficiently common, there could be huge redundancy in traditional IP multicast. As the solution to this *Content-Based Multicast* (CBM) was proposed [5,6] where extra content filtering is performed at the interior nodes of the multicast tree so as to reduce network bandwidth usage and delivery delay. This kind of filtering is performed either at the IP level or, more likely, at the software level e.g. in applications such as publish-subscribe [2] and event-notification systems [3].

Essentially, CBM reduces network bandwidth and recipient computation at the cost of increased computation in the network. CBM at the application level is increasingly important, as the quantity and diversity of information being disseminated in information systems and networks like the Internet increases,

F. Meyer auf der Heide (Ed.): ESA 2001, LNCS 2161, pp. 428–439, 2001.

and users suffering from information overload desire personalized information. A form of CBM is also useful at the middleware level [3,1] and network signaling level [4]. Previous work applies CBM to address issues in diverse areas [3,1,4].

[1] addresses the problem of providing an efficient matching algorithm suitable for a content based subscription system. [2] addresses the problem of matching the information being multicast with that being desired by leaves. [5] proposes mobile filtering agents to perform filtering in CBM framework. They consider four main components of these systems: subscription processing, matching, filtering and efficiently moving the filtering agents within the multicast tree.

[6] evaluates the situations in which CBM is worthwhile. It assumes that the multicast tree has been set up using appropriate methods, and concentrates on efficiently placing the filters within that multicast tree. It also gives the mathematical modeling of optimization framework. The problem considered is that of placing the filters under two criteria :

- Minimize the total bandwidth utilization in the multicast tree, with the restriction that at most p filters are allowed to be placed in the tree. This is similar to p-median problem on trees. An optimum $O(pn^2)$ dynamic programming algorithm was described.
- Minimize total delivery delay over the network, with no restriction on number of filters, assuming that the filters introduce their own delays F and the delay on the link of multicast tree is proportional to the amount of traffic on that particular link. That means although filters have their own delays, they could effectively reduce traffic and hence delays. This problem is similar to uncapacitated facility location on trees. An optimum $O(n^2)$ dynamic programming algorithm was described.

In this paper, we consider the second formulation, (minimizing delay) and show that the complexity of the dynamic programming algorithm can be improved. We do not concern ourselves with the construction of the multicast tree, or with processing of the subscriptions. Also, we assume that minimum required amount of traffic at each node of multicast tree is known. We consider this to be a part of subscription processing. This could be done by taking of unions of subscription list bottom up on the multicast tree, or by probabilistic estimation. Given these we focus on the question of where to place filters to minimize the total delay. We also assume that the filters do as much possible filtering as they could and do not allow any extra information traffic on subsequent links.

In section 2 , we show the formulation of the problem. We discuss the preliminaries and assumptions. We go over the dynamic programming algorithm described in [6] and give the intuition which motivates faster algorithms. In sections 3 and 4, we describe two algorithms, the first being an improvement of the dynamic programming and the second being an improvement of the first. We also include the analysis of their running time . In section 5, we describe the piecewise linear functions data structure and the required operations on it along with the complexity bounds. In section 6, we consider further extensions and related problems.

2   Preliminaries

2.1   Notations

A filter placement on a rooted multicast tree $M = (V, E)$ with vertex set V and edge set $E \subset V \times V$ is a set $S \subset V$ where filters are placed at all vertices in S and on no vertex in $V - S$. Let $|V| = N$, and so $|E| = N - 1$. We denote the root of M by r. $Tree(v)$ denotes the subtree rooted at vertex $v \in V$. For example, $Tree(r) = M$. Let us denote the height of tree M by H.

For simplicity of writing, we will use some functional notations. We denote size of $Tree(v)$, that is the number of vertices in $Tree(v)$, by $n(v)$. Thus, $|Tree(r)| = n(r) = N$. $c(v)$ denotes the number of children of vertex $v \in V$, while $s(v)$ denotes the number of leaves in $Tree(v)$. For example, $c(v) = 0$ and $s(v) = 1$ if vertex v is a leaf in M.

$f(v)$ is the total size of information requested in $Tree(v)$. For a leaf $v \in V$, $f(v)$ denotes the size of the information requested from that user. In other words, $f(v)$ is also the amount information that node v gets from its parent, if the parent has a filter. We assume that $f(v)$ for each v is known.

2.2   Assumptions

We make the following assumptions in our model.

- The delay on a link is proportional to the length of the message transmitted across the link, ignoring propagation delay. Thus if m is the length of the message going across a link (or an edge), then the delay on that link is mL units, where the link delay per unit of data is L, a constant.
- The delay introduced by an active filter is a constant. We denote it by F. It is a (typically) a big constant.[1]
- Each internal vertex of M waits and collects all incoming information before forwarding it to its children. But this time is much smaller than the delay rate over the link.

2.3   Recurrence and Dynamic Programming

Our objective is to minimize the average delay from the instant the source multicasts the information to the instant that a leaf receives it. Since number of leaves is a constant for a given multicast tree M, we can think of minimizing the total delay, where total is made over all leaves in M.

Let $A(v)$ stand for the lowest ancestor of v whose parent has a filter. For example, $A(v) = v$ if parent of v has a filter.

Now consider a CBM with a required flow f known for each vertex in the tree. For a vertex v, let $D(v, p)$ denote the minimum total delay in $Tree(v)$,

---

[1] It is not necessary to assume that L and F are same constants for each link and each filter locations, they could be different constants for different links and location as in general formulation of p-median problem[7]. This will not change the algorithm.

assuming $A(v) = p$. Let $v_1, v_2, \ldots, v_{c(v)}$ be children of vertex v. Let $C_v = s(v)F + L\sum_{i=1}^{c(v)} f(v_i)s(v_i)$ and $E_v = Ls(v)$. C_v and E_v are constants and can be computed for each vertex v in $O(N)$ time by bottoms up calculation.

Then the minimum total delay can be expressed by the following recurrence relation

> **if** v is a leaf **then**
> $\quad D(v, p) = 0$ for all p
> **else**
> $\quad D(v, p) = \min\{$
> $\qquad C_v + \sum_{i=1}^{c(v)} D(v_i, v_i)$, if v has a filter
> $\qquad f(p).E_v + \sum_{i=1}^{c(v)} D(v_i, p)$, otherwise
> $\quad \}$
> **end if**

The optimal placement can be found using dynamic programming as noted in [6]. But a naive implementation of it would take time $O(NH)$.

We will "undiscretize" the above recurrence relation and write it as a function of a real number p, which is now the incoming information flow into v. To make it clear that this function is specific for a vertex v, we denote it as D_v. Now our recurrence relations takes a new look

> **if** v is a leaf **then**
> $\quad D_v(p) = 0$ for all p
> **else**
> $\quad D_v(p) = \min\{$
> $\qquad C_v + \sum_{i=1}^{c(v)} D_{v_i}(f(v_i))$, if v has a filter
> $\qquad p.E_v + \sum_{i=1}^{c(v)} D_{v_i}(p)$, otherwise
> $\quad \}$
> **end if**

Notice that we can still compute $D(v, p)$ by plugging the discrete value $f(p)$ in D_v. Intuitively, D_v is a function of real value p which is incoming flow to the $Tree(v)$. It is a piecewise linear non-decreasing function. Each break point in the function indicates a change in the arrangement of filters in the subtree $Tree(v)$. This change occurs in order to reduce the rate of increase of D_v (slope) for higher values of p. The slope of each segment is lesser than the previous, and the slope of final segment (infinite ray) is zero because this would correspond to filter at v. Once, a filter is placed at v, the value of variable p no longer matters. Therefore, D_v is a piecewise linear non-decreasing concave function. We will use $|D_v|$ notation to denote the number of break-points (or number of linear pieces) in D_v.

The big advantage of above formulation is that it allows us to store D_v as a height balanced binary search tree which in turn allows efficient probing and merging, so that we can implement above recurrence and find optimal filter placements in quicker time.

Before we proceed with actual description of algorithms and data-structures, we present two simple lemmas which prove useful properties of D_v as claimed above.

Lemma 1. D_v is a piecewise linear non-decreasing function and $\exists p_v, t_v$ such that $D_v(p) = t_v$ for all $p \geq p_v$.

Proof : By induction on height[2] of v. Claim is trivially true for a leaf v. $D_v = 0$. So $p_v = 0$ and $t_v = 0$. Let's assume the claim is true for all $c(v)$ children of v. Also let $L_v(p) = p.E_v$. L_v is a line passing through origin. Hence, it is a piecewise linear non-decreasing function.

Let $W_v(p) = C_v + \sum_{i=1}^{c(v)} D_{v_i}(f(v_i))$. W_v is a constant and hence a piecewise linear non-decreasing function. Let $F_v(p) = L_v(p) + \sum_{i=1}^{c(v)} D_{v_i}(p)$. Hence, F_v is a piecewise linear non-decreasing function. $D_v(p) = min\{W_v(p), F_v(p)\}$. Therefore, D_v is a piecewise linear non-decreasing function because minimum preserves piecewise linear non-decreasing property. $t_v = W_v(p)$ and p_v is the value of p where W_v and F_v intersect. □

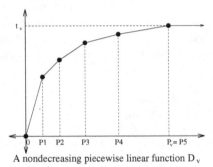

A nondecreasing piecewise linear function D_v

Lemma 2. $|D_v| \leq n(v)$.

Proof : By induction on height of v. Claim is trivially true for v if its a leaf, that is its height is 0. If claim were true for each of the $c(v)$ children of v, then each of D_{v_i} is a piecewise linear function made up of at most $n(v_i)$ different linear pieces. D_v is a minimum of sum total of D_{v_i} and a constant. It can have at most one more extra piece added to it. So the number of linear pieces in it cannot be more than $1 + \sum_{i=1}^{c(v)} n(v_i)$. But that is precisely $n(v)$. □

It is apparent from the above proof that each break-point in D_v is introduced by some node in $Tree(v)$ as a result of "min" operation.

3   Algorithm-1

We are now ready to present our first algorithm, *Algorithm-1*. Let I be the total amount of the incoming information at r. The function $A(r)$ returns the piecewise linear function D_r at root r. D_v is stored as a balanced binary search tree whose size is equal to the number of break-points in D_v.

---

[2] Height of $v = 1+$ max{Height of $u \mid u$ is a child of v}. Height of a leaf is 0.

3.1 Algorithm

Algorithm-1 {
  A(r);
  M-DFS(r, I);
}
M-DFS (v, p) {
  if $c(v) == 0$ **then**
    return;
  end if
  if $p > p_v$ **then**
    place filter at v;
    for $i = 1$ to $c(v)$ **do**
      M-DFS($v_i, f(v_i)$);
    end for
  else
    for $i = 1$ to $c(v)$ **do**
      M-DFS(v_i, p);
    end for
  end if
  return;
}

A(v) {
  if $c(v) == 0$ **then**
    $p_v = +\infty$;
    return create(0,0);
  else
    for $i = 1$ to $c(v)$ **do**
      $q_i = A(v_i)$;
    end for
    $t_v = C_v$;
    for $i = 1$ to $c(v)$ **do**
      $t_v = t_v + $probe($q_i, f(v_i)$);
    end for
    $z = $create($E_v, 0$);
    for $i = 1$ to $c(v)$ **do**
      $z = $add_merge($z, q_i$);
    end for
    $p_v = $truncate($z, t_v$);
    return z;
  end if
}

3.2 Data Structure Operations

The data structure supports the following operations:

$create(a, b)$: Returns a new function with equation $y = ax + b$ in time $O(1)$.

$probe(q, t)$: returns $q(t)$ in $O(\log|q|)$ time.

$add\_merge(q_1, q_2)$: Returns a piecewise linear function which is the sum of q_1 and q_2. Assuming without loss of generality $|q_1| \geq |q_2| \geq 2$, the running time is $O(|q_2|\log(\frac{|q_1|+|q_2|}{|q_2|}))$. q_1 and q_2 are destroyed during this operation and the new function has size $|q_1| + |q_2|$.

$truncate(q, t)$: This assumes that some z s.t. $q(z) = t$ exists. Modifies q to a function q' which is equal to $q(x)$ for $x \leq z$, and t for $x > z$. This destroys q. It returns z. q' has at most one more breakpoint than q. All the breakpoints in q after z are deleted (except at $+\infty$). The running time is $O(\log|q|)$ for search plus time $O(\log|q|)$ per each deletion.

3.3 Analysis of Algorithm-1

Algorithm-1 first recursively builds up the piecewise linear function D_r , bottom up, by calling A(r). It uses p_v values stored for each v in the tree and runs simple linear time Depth-First-Search algorithm to decide filter placement at each vertex of the tree.

We will now show that the total running time of algorithm A, and therefore Algorithm-1, is $O(N \log N)$. There are three main operations which constitute the running time: *probe, truncate and add_merge*. Over the entire algorithm, we do N probes each costing time $\leq \log N$ because each probe is nothing but a search in a binary search tree. *truncate* involves N search operations and $\leq N$ deletions (because each break-point is deleted only once and there is only one break-point each node in the multicast tree can introduce) each costing $\leq \log N$ time. Therefore, *truncate* and *probe* cost $O(N \log N)$ time over the entire algorithm. We still need to show that total time taken by all the merge operations is $O(N \log N)$. The following lemma proves this.

Lemma 3. *Total cost of merge in calculation of D_v is at most $n(v) \log n(v)$.*

Proof : We proceed by induction on height of v. The claim is trivially true for all leaf nodes since there are no merge operations to be done. At any internal node v, to obtain D_v we merge $D_{v_1}, D_{v_2}, \ldots, D_{v_{c(v)}}$ sequentially. Let $s_i = \sum_{j=1}^{i} n(v_i)$. Now, again by induction (new induction on the number of children of v) assume that the time to obtain the merged function of first i D_{v_j}'s is $s_i \log s_i$. The base case when $i = 1$ is true by induction (previous induction). Then, assuming without loss of generality $s_i \geq n(v_{i+1})$, the total time to obtain merged function of first $i + 1$ D_{v_j}'s is at most $s_i \log s_i + n(v_{i+1}) \log n(v_{i+1}) + n(v_{i+1}) \log ((s_i + n(v_{i+1}))/n(v_{i+1}))$ which is at most $s_{i+1} \log s_{i+1}$. Therefore time taken to merge all the children at v and obtain D_v is at most $n(v) \log n(v)$.   □

4   Algorithm-2

4.1   Motivation

We observe that lemma 2 suggests a bound of $n(v)$ on the number of different linear pieces in D_v. On the other hand, we need to probe and evaluate D_v at at most H different values (that is the number of ancestors v can have !). This suggests that we can gain more if we "convert" our functions which grow "bigger" and have more than H breakpoints and reduce them to at most H breakpoints.

For example, consider the case of multicast tree M being a balanced binary tree. Let Y be the set of nodes at depth $\log \log N$. For each $v \in Y$ the subtree size $n(v)$ is roughly $\log N$. $|Y|$ is roughly $N/\log N$ and computing D_v at each such v takes $\log N \log \log N$ time. This makes it $N \log \log N$ over all $v \in Y$. Now, we convert D_v into array form as in dynamic programming and resume the previous dynamic programming algorithm in [6]. This dynamic programming calculations occur at roughly $N/\log N$ nodes each taking $\log N$ ($H = \log N$) time. Hence we achieve an enhancement in total running time, taking it down to $N \log \log N$ which is essentially $N \log H$. However, to achieve this in general case, we still have to stick to the binary search tree representation in the second phase. The advantage is that the size of the binary search tree never grows more than H.

4.2   Data Structure Operations

Before we proceed with this new algorithm, we consider "converted" functions, since some of the functions would be evaluated only for a small number of values. A "converted" function $q_X(x)$ for $q(x)$ with respect to (sorted) set $X = x_1, x_2, ..., x_k$ is a piecewise linear function such that $q(x_i) = q_X(x_i)$ for $x_i \in X$, and piecewise linear in between those values. We define the following operations:

$convert(q, X)$: Returns the "converted" function q_X in $O(k \log(|q|/k))$. Assumes
   X is sorted.
$add\_dissolve(q_X, g)$: Adds function g to the converted function q_X, and returns
   the resulting function dissolved w.r.t. set X. Running time : $O(|g| \log |q_X|)$
$add\_collide(q_{X_1}, g_{X_2})$: Adds two converted functions q_{X_1} and g_{X_2}. Creates new
   converted function only on $X_1 \cap X_2$.
$truncate\_converted(f_X, t)$: Almost the same as $truncate$. Does not cause any
   deletions. It uses some "mark" to keep track of invalid values in data struc-
   ture. Running time: $O(\log |f_X|)$.

4.3   Description of Algorithm

Using this, we modify the implementation of Algorithm-1. We continue building D_v's bottom up as in algorithm A(v). Suppose we are at some v whose all $c(v)$ children, namely $v_1, v_2, \ldots, v_{c(v)}$, have less than H breakpoints in their respective data-structures. We now start building D_v by merging $D_{v_1}, D_{v_2}, \ldots$ one by one. Suppose after merging D_{v_1} through D_{v_i} for first i children, we find that the number of breakpoints for function q constructed so far exceeds H (and it's trivially less than $2H$), then we call function $convert(f, X)$. Here X is the sorted list of values $f(p)$ of v's ancestors. Note that X is very easy to maintain due to recursive top-down calls in A(v) and monotonicity of $f(p)$ values along a path. Then, for the remaining children of v, we use $add\_dissolve(q, D_{v_j})$, where $j \in \{i+1, \ldots, c(v)\}$. Once the function is "converted", it always remains "converted". For add operation involving one "converted" function q_1 and one "unconverted" function q_2 we use $add\_dissolve(q_1, q_2)$ which runs in $O(q_2 \log H)$ and for two "converted" functions q_1, q_2 we use $add\_collide(q_1, q_2)$ which runs in $O(H)$ since the size of the bigger data structure is now restricted by H.

Thus, we don't store more than required information about D_v, while main-taining enough of it to calculate the required parts of D_u, where u is v's parent.

4.4   Analysis of Algorithm-2

Let Y be the set of all nodes $v \in V$ such that we needed to use the $convert$ for finding D_v. Use of $convert$ function implies $n(v) = |Tree(v)| \geq H$ in the light of lemma 2.$|Y| \leq \frac{N}{H}$ because for any two $u, v \in Y$, $Tree(u) \cap Tree(v) = \phi$.

Let W be union of Y and all ancestors of nodes in Y. Subgraph of M on set W forms an upper subtree of M. Let U be the set of children of nodes in Y.

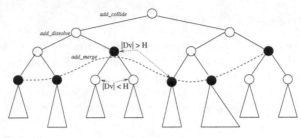

● indicates nodes where convert function is used

For all nodes $v \in U$, we will run the normal *add_merge* procedure. Since final D_v's constructed for each $v \in U$ have sizes less than H, the total cost of building them would be $|U| \log H \leq N \log H$.

For each node $v \in Y$, we do a few *add_merge* operations, followed by a *convert* operation, followed by few *add_dissolve* and *add_collide* operations. Let X be the sorted set of $f(p)$ values for the ancestors of v. $1 \leq |X| \leq H$. Since the overall effect of *add_merge* operations leads to a data-structure of size at most $2H$, total cost is $O(H \log H)$. *convert* will cost at most $|X| \log \frac{2H}{|X|} = O(H)$.

If we sum over all *add_dissolves* performed during the run of the algorithm, it is easy to see that at most N breakpoints will be *dissolved* in data-structure of size H. So the total cost of *add_dissolves* is at most $N \log H$.

Further, there are at most N/H "converted" functions, each *add_collide* takes $O(H)$ time and causes one less "converted" function. Hence, the total cost of *add_collide* is $O(N)$.

Thus, the overall cost is at most $N \log H + \frac{N}{H} \cdot (H \log H + H) + N \log H + N$, that is $O(N \log H)$.

Theorem 1. *Given the multicast tree M having N nodes and height H with source at root r disseminating I amount of information, along with values $f(v)$ which is the minimum amount of information required at node v of the multicast tree, the placement of filters in the multicast tree to minimize total delay can be computed in $O(N \log H)$ time.*

5   Data-Structures

In the previous sections, we have assumed the existence of a data structure to maintain non-decreasing piecewise-linear functions. Here, we describe the data structure along with the implementation of the operations.

The data structure will maintain the breakpoints (or each value in X for a converted function) sorted by x coordinate in an AVL tree [11,12]. An AVL tree is a balanced binary search tree in which for any node, the height difference between its left and right subtrees is at most one. Along with the x coordinate of the breakpoint, each node will also contain two real numbers a and b such that the linear segment to the left of the breakpoint is of equation $y = Ax + B$ where A (resp. B) is the sum of all the a (resp. b) values on the path from the node to the root of the tree. A dummy breakpoint at $x = +\infty$ will be included in the tree to encode the rightmost linear piece of the function.

Each node will also contain a mark for handling truncated parts of a function. A node is invalid (i.e. its a and b values are not correct) if itself or a node on the path to the root is marked. The linear function at the x value of an invalid node will be the same as the function of the first valid node that appears after it in the tree inorder. The node at $x = +\infty$ will always be valid. Every time we visit a marked node during any of the operations, we unmark it, correct its a and b values and mark its two children. This ensures that the only invalid nodes are the ones the algorithm doesn't see. This marking scheme will be necessary to implement *truncate_converted* which is "truncate" on "converted" functions, since we cannot delete the nodes in that case.

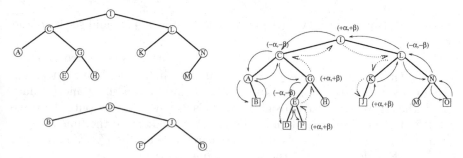

Sorted lists represented as height-balanced trees   Merging by sequential insertions (square nodes have been inserted)

The data structure will use the AVL tree merging algorithm of Brown and Tarjan [10] to implement add_merge, convert and add_dissolve. Given two AVL trees T_1 with n elements and T_2 with m elements, $m \le n$, we will search/insert the elements of T_2 into T_1 in sorted order, but instead of performing each search from the root of the tree, we start from the last searched node, climb up to the first parent(LCA) having the next element to search in its subtree, and continue searching down the tree from there. Brown and Tarjan show that the total number of operations performed during this procedure is $O(m \log((n + m)/m))$. This method can also be used to search or visit m sorted values in T_1 within $O(m \log (n/m))$ time.

In order to add two functions, while merging the corresponding AVL trees using Brown and Tarjan's method, and we will need to update the a and b values of the nodes. First, when we insert a new node in the tree, we find its "inherited" A and B values and adjust its a and b values accordingly. Then we consider the effect of adding the linear piece $y = \alpha x + \beta$ to its right in the previous data structure where it came from. This can be done along the same walk in the tree. While walking in the tree from an element u to the next element to be inserted v, we will need to add the piecewise linear function joining them, say $\alpha x + \beta$ to all the nodes between u and v. To do that, add α and β to the a and b values of the least common ancestor (LCA) of u and v. Now, the function values for all the nodes between u and v have been increased correctly, but some nodes outside of that range might have been increased as well. To correct that, we walk down from the LCA to u. This is a series of right child and left child choices, the first being left. In this series, whenever we choose right child after some (non-empty) sequence of left child choices, we subtract tuple (α, β) at that node. Similarly, whenever we choose left child after (non-empty) sequence of right child choices, we add tuple (α, β) to the node where choice is made. Also, similarly (vice-versa) a and b values can be adjusted along the path LCA to v. Thus, updates are only required along the Brown and Tarjan's search path. To complete the argument, it can be verified that the validity of the a and b values of the nodes can be preserved during a rotations and double rotations in the tree for the AVL insertion. The figure illustrates the insert path along with updates due to linear segment $\alpha x + \beta$ between inserted points F and J.

We now outline the workings of the different operations:

create(a, b): Create a new AVL tree with 1 node at $x = +\infty$, and set its a and b values.

add_merge(f_1, f_2): Use the method of Brown and Tarjan as described above.

truncate(f, t): Find z such that $f(z) = t$ by performing a search in the tree. As we go down the tree, we maintain A and B, the sum of the a and b values of all the ancestors of the current node. This way, we can compute the value of $f(x)$ for the x value of each of the nodes visited. Since the function f is non-decreasing, the tree is also a binary search tree for the $f(x)$ values. Once the search is done, we find the linear segment for which $Az + B = t$, and thus find z. We then insert a new breakpoint in the tree at $x = z$, and delete all the break-points in f which come after z (except the one at $+\infty$) one-by-one, using usual AVL-tree deletion. Add the line segment $(0, t)$ between z and $+\infty$.

probe(f, t): Search in the tree the successor for t in the x values. Compute the A and B sums on the path, returns $f(t) = At + B$.

convert(f, X): Use the method of Brown and Tarjan to find the successors of all $x_i \in X$ in $O(k \log(n/k))$. Evaluate $f(x_i)$ at each of those values, and construct a new AVL tree for a piecewise linear function with x_i values as breakpoints, and joining each adjacent breakpoints with an ad-hoc linear function.

add_dissolve(f_X, g): Just like in *add_merge*, but we do not insert the break-points, we just update the a and b values of the existing breakpoints.

add_collide(f_{X_1}, g_{X_2}): Find the values of f and g on $X_1 \cap X_2$ and construct a new AVL tree as in *convert*.

truncate_converted(f_X, t): As in *truncate*, we find z. But in this case we do not insert the new break-point. Also we do not delete the break-points in f_X after z. We invalidate the (a, b) values of the remaining points by marking the right child of the nodes traversed from the left and also we adjust (a, b) value at these nodes so that the linear function reflects the line $y = t$, while walking up from the position of z to the root. Once at the root, we walk down to $+\infty$, validating the a and b values on that path. We then set the a and b values at $+\infty$ such that $A = 0$ and $B = t$. It returns z.

6 Related Problems and Future Work

Future work in this area is to design an algorithm based on similar methods for the first model of [6]. Here, the dynamic programming algorithm runs in $O(pn^2)$ time, since the objective function at each node involves two parameters. This model, where only p filters are allowed, is a variant of p-median on tree where all the links are directed away from the root. The problem is called p-inmedians in [9].When $p \geq n$ it reduces to our problem. The dynamic programming optimal algorithm known for the uncapacitated facility location (which is p-median with $p \geq n$) over trees takes $O(n^2)$ time. Interesting future work is to find faster algorithms for uncapacitated facility location [9] and p-median [7]. That could also give us faster algorithms for p-forest [8] and tree partitioning problems [9].

Acknowledgments. We would like to thank John Iacono and Michael Fredman for useful suggestions regarding choice of data structures and Farooq Anjum and Ravi Jain for providing useful background on the problem.

References

1. Aguilera et al, "Matching Events in a Content-based Subscription System", *http://www.research.ibm.com/gryphon*
2. Banavar et al, "An efficient multicast protocol for content-based publish-subscribe systems", *Technical report*, IBM 1998.
3. Carzaniga et al, "Design of Scalable Event Notification Service: Interface and Architecture", *Tech Report CU-CS-863-98, University of Colorado, Dept. of Computer Science*, 1998.
4. Kasera et al, "Scalable Fair Reliable Multicast Using Active Services",*IEEE Network Magazine*, Jan/Feb 2000.
5. F. Anjum and R. Jain, "Generalized Multicast Using Mobile Filtering Agents", *Internal Report, Telcordia Tech, Morristown*, Mar 00.
6. F. Anjum, R. Jain, S. Rajagopalan and R. Shah, "Mobile Filters for Efficient Dissemination of Personalized Information Using Content-Based Multicast", *submitted*, 2001.
7. A. Tamir, "An $O(pn^2)$ algorithm for the p-median and related problems on tree graphs", *Operations Research Letters, 19:59-94*, 1996.
8. A. Tamir and T. Lowe, "The generalized p-forest problem on a tree network", *Networks* 22, 217-230, 1992.
9. G. Cornuejols, G.L. Nemhauser and L.A. Wosley, "The uncapacitated facility location problem", in P.B. Mirchandani and R.L. Francis(eds), *Discrete Location Theory*, Wiley, New York, 1990, pp. 119-171.
10. M. Brown and R. Tarjan, "A Fast Merging Algorithm",*Journal of ACM, 26(2), pp 211-225*, Apr 79.
11. G. Adel'son-Vel'skii and Y. Landis, "An algorithm for the organization of information", *Dokl. Akad. Nauk SSSR 146*, 263-266, (in Russian) English translation in *Soviet Math. Dokl.*, 3-1962, pp1259-1262.
12. C. Crane, "Linear lists and priority queues as balanced binary trees", *PhD Thesis, Stanford University*, 1972.

Approximation Algorithms for Minimum-Time Broadcast under the Vertex-Disjoint Paths Mode

Pierre Fraigniaud

CNRS-LRI, Université Paris-Sud, `pierre@lri.fr`.

Abstract. We give a polynomial-time $O(\frac{\log n}{\log OPT})$-approximation algorithm for minimum-time broadcast and minimum-time multicast in n-node networks under the single-port vertex-disjoint paths mode. This improves a previous approximation algorithm by Kortsarz and Peleg. In contrast, we give an $\Omega(\log n)$ lower bound for the approximation ratio of the minimum-time multicast problem in directed networks. This lower bound holds unless $NP \subset DTIME(n^{\log \log n})$. An important consequence of this latter result is that the Steiner version of the Minimum Degree Spanning Tree (MDST) problem in digraphs cannot be approximated within a constant ratio, as opposed to the undirected version. Finally, we give a polynomial-time $O(1)$-approximation algorithm for minimum-time gossip (i.e., all-to-all broadcast).

Keywords: Approximation Algorithms, Graph and Network Algorithms, Broadcasting, Multicasting, Gossiping, Minimum Degree Spanning Tree.

1   Introduction

Given a node s of a network, and a set of nodes D, *multicasting* from s to D consists to transmit a piece of information from s to all nodes in D using the communication facilities of the network. *Broadcasting* is the particular case in which the destination-set is composed of all nodes. Given a set of nodes D, *gossiping* in D consists to perform multicast from s to D, for every node $s \in D$. These communication patterns are basic operations upon which network applications are frequently based, and they hence gave rise to a vast literature, covering both applied and fundamental aspects of the problem (cf. [10,25] and [18], respectively).

A standard communication model assumes the network to be a connected graph $G = (V, E)$. Transmissions proceed by synchronous *calls* between the nodes of the network. It is generally assumed that (1) a call involves exactly two nodes, (2) a node can participate to at most one call at a time (*single-port* constraint), and (3) the duration of a call is 1 (assuming that the two nodes involved in the call exchange a constant amount of information). Two main variants of this model have then been investigated: the *local* model and the *line* model. The former states that calls can be placed between neighboring nodes only, whereas the latter allows calls to be placed between non-neighboring nodes

F. Meyer auf der Heide (Ed.): ESA 2001, LNCS 2161, pp. 440–451, 2001.

(i.e., a call is a path in the graph G, whose two extremities are the "caller" and the "callee"). The local model aims to model switching technologies such as store-and-forward, whereas the line model aims to model "distance-insensitive" switching technologies such as circuit-switching, wormhole, single-hop WDM in optical networks, and virtual paths in ATM networks. For both variants, the notion of time is captured by counting communication *rounds*, where round t is defined as the set of all calls performed between time $t-1$ and time t, $t = 1, 2, \ldots$ A multicast (resp., broadcast, gossip) *protocol* is simply described by the list of calls performed in the graph to complete multicast (resp., broadcast, gossip).

Notation. We denote by $m_s(G, D)$ the minimum number of rounds required to perform multicast from s to $D \subseteq V$ in $G = (V, E)$. Similarly, we denote by $b_s(G)$ the minimum number of rounds required to perform broadcast from s in G, that is $b_s(G) = m_s(G, V)$. Finally, we denote by $g(G, D)$ the minimum number of rounds required to perform gossip among the nodes of D. If $D = V$, $g(G, V)$ is simplified in $g(G)$.

Of course, these numbers depend on whether we consider the local or the line model. This will be specified later.

Definition 1. *The* multicast problem *is defined as follows. Given any graph $G = (V, E)$, any source-node $s \in V$, and any destination-set $D \subseteq V$, compute a multicast protocol from s to D performing in $m_s(G, D)$ rounds. The* broadcast *and* gossip problems *are defined similarly.*

These problems are inherently difficult, and only approximated solutions can be expected in polynomial time.

Definition 2. *An algorithm for the multicast problem is a ρ-approximation algorithm if, for any instance (G, s, D) of the problem, it returns a multicast protocol from s to D in G which completes in at most $\rho \cdot m_s(G, D)$ rounds. Approximation algorithms for the broadcast and gossip problems are defined similarly.*

A large literature has been devoted to the description of broadcast and gossip protocols performing in a small number of rounds. Under the local model, many authors have considered specific topologies (see, e.g., [4,15,18,19]), but lots of efforts have also been made to derive approximation algorithms for arbitrary topologies [1,2,24,27,28]. The line model has also been very much investigated, in its two variants: the *edge-disjoint paths mode* [3,5,7,8,9,11,13,22,26], and the *vertex-disjoint paths mode* [5,6,20,21,23,24]. The former mode specifies that the paths joining the participants of simultaneous calls must be pairwise edge-disjoint, whereas the latter specifies that they must be pairwise vertex-disjoint.

The vertex-disjoint paths mode is motivated by several studies which pointed out that avoiding node-congestion is a critical issue, especially in the context of several multicasts occurring simultaneously. The remaining of this paper is entirely dedicated to this latter communication mode. Let us hence recall the model so that no ambiguity should result from it.

Definition 3. *(Vertex-disjoint paths mode.) Communications proceed by sequence of synchronous calls of duration 1. A call involves exactly two nodes,* the caller and the callee, and a unit piece of information can be transmitted from the caller to the callee during the call. The two participants of a call can be at distance greater than one in the network, i.e., a call is a path (non necessarily minimal) between the caller and the callee. Any two simultaneous calls (i.e., paths) must be vertex-disjoint. A round, i.e., a set of calls performed simultaneously, is therefore a set of pairs of nodes that are matched by pairwise vertex-disjoint paths.

Although the vertex-disjoint paths mode allows fast broadcast in some networks (e.g., $\lceil \log_2 n \rceil$ rounds in the ring C_n), there are networks which require $\Omega(n)$ rounds to perform broadcast (e.g., the star S_n of n nodes requires $n - 1$ rounds). The vertex-disjoint paths mode can hence be very slow or very fast depending on the network, and this is true even if we restrict the study to specific families of graphs such as trees or interval graphs. The multicast, broadcast and gossip problems are actually NP-complete in arbitrary graphs, and hence lots of efforts have been devoted to specific topologies (cf. [20,21]) and to trees (cf. [5,6]). For arbitrary graphs, lower bounds for the gossip problem if calls can transmit an unbounded number of pieces of information can be found in [23]. As far as the broadcast problem is concerned, Kortsarz and Peleg [24] have shown that there exists a polynomial-time $O(\frac{\log n}{\log \log n})$-approximation algorithm.

The main contributions of this paper are the following:

1. We derive a polynomial-time $O(\frac{\log |D|}{\log OPT})$-approximation algorithm for the multicast problem. This improves the best previously known approximation algorithm for minimum-time broadcast [24], and yields a constant approximation ratio for graphs of broadcast time $OPT = \Omega(n^\epsilon)$, $\epsilon > 0$. Our algorithm is based on an algorithm by Fürer and Raghavachari [17] for the Minimum Degree Spanning Tree (MDST) problem.
2. We show that the nature of the multicast problem considerably changes from undirected to directed networks. In the latter case, we prove that, unless $NP \subset DTIME(n^{\log \log n})$, optimal solutions of the multicast problem cannot be approximated in polynomial time within less than an $\Omega(\log n)$ multiplicative factor. This result is obtained from an approximation threshold for the Minimum Set Cover problem due to Feige [12]. Nevertheless, we extend our approximation algorithm to directed graphs, though for the broadcast problem only. We note that this extension does not hold with the protocol in [24].
3. Beside the study of information dissemination problems, a direct consequence of our lower bound on the approximation ratio for the multicast problem is that the optimal solution of the Steiner version of the MDST problem in digraphs cannot be approximated in polynomial time within less than an $\Omega(\log n)$ multiplicative factor, nor within less than an $\Omega(\log n \log \log n)$ additive factor, again unless problems in NP can be solved in slightly more than polynomial time.
4. Finally, we show that the minimum gossip time can be approximated within a constant multiplicative factor in undirected networks.

Section 2 presents our $O(\frac{\log n}{\log OPT})$-approximation algorithm for the multicast problem. Section 3 revisits the multicast problem in directed networks, an includes the lower bounds on the approximation ratio of the MDST problem. Section 4 presents our $O(1)$-approximation algorithm for the gossip problem. Finally, Section 5 contains some concluding remarks.

2   Approximation Algorithm

Let us start by computing a lower bound on the number of rounds required for multicasting.

Notation. Given any graph $G = (V, E)$, and any set $D \subseteq V$, let $\Delta_{\min}(G, D)$ be the smallest integer k such that there exists a tree of maximum degree k, spanning D in G. If $D = V$, then $\Delta_{\min}(G, D)$ is simplified in $\Delta_{\min}(G)$. $\Delta_{\min}(G, D)$ is abbreviated in $\Delta_{\min}$ if no confusion can arise from this simplification.

In the remaining of the paper, we assume, w.l.o.g., that $s \in D$.

Lemma 1. *Let $G = (V, E)$, $s \in V$, and $D \subseteq V$. We have*
$$m_s(G, D) \geq \max\{\lceil \log_2 |D| \rceil, \lceil \Delta_{\min}/2 \rceil\}.$$

Proof. The single-port constraint implies that at least $\lceil \log_2 |D| \rceil$ rounds are required because the maximum number of informed nodes can at most double at each round. Let $\mathcal{M}$ be an optimal multicast protocol from s to D in G, that is a protocol performing in $m_s(G, D)$ rounds. Let H be the graph induced by the union of all the calls, i.e., all the paths, of $\mathcal{M}$. H is a connected subgraph of G which spans D. Let Δ be the maximum degree of the nodes in H. The vertex-disjoint constraint implies that $\mathcal{M}$ completes in at least $\lceil \Delta/2 \rceil$ rounds. Therefore, since $\Delta_{\min} \leq \Delta$, $\mathcal{M}$ completes in at least $\lceil \Delta_{\min}/2 \rceil$ rounds.   □

It is trivial observation that multicasting in Hamiltonian graphs can be done in $\lceil \log_2 |D| \rceil$ rounds by just using an Hamiltonian cycle. Actually, multicasting can also be achieved in $\lceil \log_2 |D| \rceil$ rounds if there is an Hamiltonian partial subgraph of G spanning D. Our approximation algorithm for the multicast problem is based on approximated solutions for the Minimum Degree Spanning Tree problem (MDST for short) defined as follows. We are given a graph $G = (V, E)$, and we are looking for a spanning tree whose maximum degree is $\Delta_{\min}(G)$. In the Steiner version of the problem, we are given a graph $G = (V, E)$, and a vertex-set $D \subseteq V$, and we are looking for a tree spanning D whose maximum degree is $\Delta_{\min}(G, D)$. A (ρ, r)-approximation algorithm for the MDST problem is an algorithm which, given G and D, returns a tree spanning D and of maximum degree at most $\rho \cdot \Delta_{\min}(G, D) + r$.

Lemma 2. (Fürer and Raghavachari [17]) *There exists a polynomial-time $(1, 1)$-approximation algorithm for the Steiner version of the MDST problem in graphs.*

Lemma 3. *Let $t(n)$ be the complexity of an $(1, 1)$-approximation algorithm for the Steiner version of the MDST problem in n-node graphs. For any m, $2 \leq m < n$, there exists a $t(n)$-time algorithm which, for any n-node graph $G = (V, E)$, any set $D \subseteq V$, and any node $s \in V$, returns a multicast protocol from s to D which performs in $O\left((m + \log |D| + \Delta_{\min}) \frac{\log |D|}{\log m}\right)$ rounds.*

Proof. Let G be an n-node graph, and $2 \leq m < n$. Let T be a tree of degree $\Delta_{\min} + 1$ spanning D in G. Consider T as rooted at s. The proof simply consists to show how to construct a multicast protocol from s to D in T which performs in $O\left((m + \log |D| + \Delta_{\min}) \frac{\log |D|}{\log m}\right)$ rounds. We look for subtrees of T containing at most $|D|/m$ destination nodes. For that purpose, we perform a depth-first search (DFS) traversal of T, starting from s, with the following restriction: we visit a child of the current node x if and only if the subtree T_x of T, rooted at x, contains at least $|D|/m$ destination nodes. The visited nodes form a subtree T' of T. The leaves of T' are roots of subtrees of T containing at least $|D|/m$ destination nodes. Let T'' be the subtree of T composed of T' plus all the children of the nodes in T'. We describe a 3-phase multicast protocol from s to the leaves of T'', performing in at most $O(m + \log |D| + \Delta_{\min})$ rounds.

Phase 1 is a multicast from s to all the leaves of T'. Node s proceeds sequentially by calling successively every leaf. The number of leaves of T' is at most m because every leaf is the root of a subtree of T of at least $|D|/m$ destination nodes, and there are $|D|$ destination nodes in total. Therefore, Phase 1 takes at most m rounds.

Phase 2 consists to inform specific nodes of T', from its leaves. For that purpose, observe that, given any tree, one can decompose the tree in a set $\mathcal{P}$ of pairwise vertex-disjoint paths so that every internal node is linked to exactly one leaf by exactly one path in $\mathcal{P}$. This operation is called a *path-decomposition* of the tree. Given a path-decomposition $\mathcal{P}$ of T', Phase 2 is performed as follows. For every leaf x, let P_x be the path of $\mathcal{P}$ containing x. Each leaf x performs a multicast in P_x to all nodes of P_x that either belong to D, or have at least one child in T''. Since there are at most $|D|$ such nodes along each path P_x, Phase 2 completes in at most $\lceil \log_2 |D| \rceil$ rounds.

Phase 3 consists to inform leaves of T'' from nodes of T'. For that purpose, every node of T' which is aware of the message after Phases 1 and 2 informs all its children in T''. Since every node in T has a degree at most $\Delta_{\min} + 1$, Phase 3 takes at most $\Delta_{\min} + 1$ rounds.

Once this 3-phase multicast protocol is completed, we are let with multicast problems in subtrees of T containing at most $|D|/m$ destination nodes. Multicasts in these trees can be performed in parallel. The whole multicast protocol results of at most $\frac{\log |D|}{\log m}$ repetitions of this strategy.

The construction of the protocol requires $t(n)$ time to construct the tree T. We use $O(n)$ time to count the number of destination nodes in every subtree T_x of T, $x \in V(T)$. The path-decomposition $\mathcal{P}$ can be constructed in $O(|V(T')|)$ time, and each multicast protocol along $P_x \in \mathcal{P}$ can be computed in $O(|P_x|)$ time. It results that, once T is set up, the whole multicast protocol can be computed in $O(n)$ time. So the total complexity of the construction is dominated by the extraction of the MDST of D in G. □

For $D = V$, by choosing $m = \log n$, lemmas 1, 2 and 3 together show that there exists a polynomial-time $O(\frac{\log n}{\log \log n})$-approximation algorithm for the broadcast problem, as already shown by Kortsarz and Peleg [24], using other techniques. For the multicast problem, $m = \log |D|$ yields a polynomial-time $O(\frac{\log |D|}{\log \log |D|})$-

approximation algorithm. More interestingly, Lemma 3 allows m to be tuned to get a smaller approximation ratio.

Theorem 1. *Let $t(n)$ be the complexity of an $(1,1)$-approximation algorithm for Steiner version of the MDST problem in n-node graphs. There exists a $t(n)$-time $O(\frac{\log |D|}{\log OPT})$-approximation algorithm for the multicast problem.*

Proof. Let $G = (V, E)$, $s \in V$, and $D \subseteq V$, and let $OPT = m_s(G, D)$. Applying Lemma 3 with $m = \log |D|$ results in a multicast protocol $\mathcal{M}$ from s to D in G which completes in time $t = O(\frac{\log |D|}{\log \log |D|} \cdot OPT)$. Thus

$$\frac{1}{c} t \, \frac{\log \log |D|}{\log |D|} \le OPT \le t$$

where c is a positive constant. There are two cases:

– **Case 1:** $t \le c \, \frac{\log^3 |D|}{\log \log |D|}$. In this case, $\log OPT = \Theta(\log \log |D|)$, and thus $\mathcal{M}$ completes in at most $O(\frac{\log |D|}{\log OPT} \cdot OPT)$ rounds.

– **Case 2:** $t > c \, \frac{\log^3 |D|}{\log \log |D|}$. In this case, $OPT > \log^2 |D|$. Then we re-apply Lemma 3 with $m = \frac{1}{c} t \, \frac{\log \log |D|}{\log |D|} \le OPT$. The resulting multicast protocol $\mathcal{M}'$ completes in

$$O\left(\frac{OPT \, \log |D|}{\log t - \log \log |D| + \log \log \log |D| - \log c}\right)$$

rounds. We have $\log t \ge \log OPT$, and thus $\mathcal{M}'$ completes in at most $O(\frac{\log |D|}{R} \cdot OPT)$ rounds where

$$R = \log OPT - \log \log |D| + \log \log \log |D| - \log c.$$

Since $OPT > \log^2 |D|$, $\frac{1}{2} \log OPT > \log \log |D|$, and thus $R > \frac{1}{2} \log OPT + \log \log \log |D| - \log c$, that is $R = \Omega(\log OPT)$. Therefore, $\mathcal{M}'$ completes in at most $O(\frac{\log |D|}{\log OPT} \cdot OPT)$ rounds. $\square$

Remark. From Lemma 2, $t(n)$ can be chosen to be polynomial.

We note that trees deserve a specific attention in the context of broadcast and multicast. Indeed, most of the multicast protocols for Internet are based on a tree connecting the destination nodes [10,25]. Unfortunately, we were not able to prove that the multicast problem in trees is NP-complete, nor to prove that it is in P. In particular, the merging method presented in [7,9] does not seem to apply in this context. However, one can prove the following:

Theorem 2. *There exists an $O(n \log n)$-time 3-approximation algorithm for the multicast problem in trees. More precisely, the approximation algorithm returns a protocol that is optimal up to an additive factor of $2\lceil \log_2 |D| \rceil$.*

Due to lack of space, the proof is omitted, but can be found in [14].

3   Broadcast and Multicast in Directed Networks

A natural question that arises in this context, is how far can we extend Theorem 1 to directed networks. Surprisingly, we show that such an extension is not doable unless all problems in NP can be solved in time slightly larger than polynomial. The proof is based on an inapproximability threshold for the Minimum Set Cover problem (MSC for short). Let us recall that, in the MSC problem, we are given a collection $\mathbf{C}$ of subsets of a set S, and we are looking for the smallest set $\mathbf{C}_{\min} \subseteq \mathbf{C}$ such that every element of S belongs to at least one set of $\mathbf{C}_{\min}$. Feige [12] has shown the following:

Lemma 4. (Feige [12]) *Unless* $\mathrm{NP} \subset \mathrm{DTIME}(n^{\log \log n})$, *the optimal solution of the MSC problem is not approximable in polynomial time within* $(1 - \epsilon) \ln |S|$ *for any* $\epsilon > 0$.

Theorem 3. *Unless* $\mathrm{NP} \subset \mathrm{DTIME}(n^{\log \log n})$, *the optimal solution of the minimum-time multicast problem in digraphs is not approximable in polynomial time within* $(1 - \epsilon) \ln |D|$ *for any* $\epsilon > 0$.

Proof. Let $(S, \mathbf{C})$ be an instance of the Minimum Set Cover problem. Let $G = (V, E)$ be the directed graph of $O(k \cdot |\mathbf{C}| \cdot |S|)$ vertices obtained from $(S, \mathbf{C})$ as follows (k will be specified later). G consists of k "branches" attached to a center node v. Each branch B is a copy of a graph of $O(|S| \cdot |\mathbf{C}|)$ vertices constructed as follows. Let $B' = (V', E')$ where

$$V' = \{v\} \cup \mathbf{C} \cup S,$$

and

$$E' = \{(v, C),\ C \in \mathbf{C}\} \cup \{(C, s),\ C \in \mathbf{C},\ s \in C\}.$$

B is obtained from B' by replacing, for every $C \in \mathbf{C}$, the star $\{(C, s), s \in C\}$ by a binomial tree of at most $2|C|$ nodes, whose root is C, whose leaves are the nodes $s \in C$, and whose edges are directed from the root toward the leaves. (If $|C|$ is not a power of 2, then the binomial tree is pruned so that it contains $|C|$ leaves.) Since $|C| \leq |S|$, B has at most $O(|S| \cdot |\mathbf{C}|)$ vertices, and the whole digraph G has at most $O(k \cdot |S| \cdot |\mathbf{C}|)$ vertices. The construction of the multicast instance is completed by setting the source-node to v and the destination-set D to the $k|S|$ nodes of the k copies of S (one copy in each branch). We have

$$m_v(G, D) \leq k|\mathbf{C}_{\min}| + \lceil \log_2 |S| \rceil. \tag{1}$$

Indeed, given $\mathbf{C}_{\min}$, the source v sequentially informs the nodes of $\mathbf{C}_{\min}$ in each of the k branches. This takes $k|\mathbf{C}_{\min}|$ rounds. Once informed, every node $C \in \mathbf{C}_{\min}$ multicasts to all nodes $s \in C$ in $\lceil \log_2 |C| \rceil$ rounds using the binomial tree rooted at C.

Let $\mathcal{A}$ be a ρ-approximation algorithm for the multicast problem. Applied on (G, v, D), $\mathcal{A}$ returns a multicast protocol $\mathcal{M}$ from v to D which performs in at most $\rho \cdot m_v(G, D)$ rounds. Moreover, $\mathcal{M}$ determines a collection $\mathbf{C}_1^*, \ldots, \mathbf{C}_k^*$ of subsets of $\mathbf{C}$, one for each of the k branches: $\mathbf{C}_i^*$ is defined as the subset of nodes in the ith copy of $\mathbf{C}$ which received a call, or which were traversed by a call, during the execution of $\mathcal{M}$. By construction, each $\mathbf{C}_i^*$ covers S. Let $\mathbf{C}^*$ be such that $|\mathbf{C}^*| = \min_{1 \leq i \leq k} |\mathbf{C}_i^*|$. The single-port constraint implies that $\mathcal{M}$ completes in at least $k|\mathbf{C}^*|$ rounds by definition of $\mathbf{C}^*$. Therefore, from Inequality 1, $\mathcal{A}$ computes $\mathbf{C}^*$ with

$$k|\mathbf{C}^*| \leq \rho \cdot (k|\mathbf{C}_{\min}| + \lceil \log_2 |S| \rceil).$$

Therefore,

$$\frac{|\mathbf{C}^*|}{|\mathbf{C}_{\min}|} \leq \rho \Big(1 + \frac{\lceil \log_2 |S| \rceil}{k} \Big).$$

In other words, for any $\epsilon > 0$, by choosing $k \simeq \log |S|/\epsilon$, $\mathcal{A}$ allows to approximate $|\mathbf{C}_{\min}|$ within $\rho(1 + \epsilon)$. Lemma 4 then yields $\rho \geq (1 - \epsilon) \ln |S|$ for any $\epsilon > 0$. We have $|D| = k|S| + 1$, and thus $\ln |D| = \ln |S| + o(\log |S|)$, which completes the proof.   □

Remark. One can check in [12] that Feige's Lemma still holds for instances $(S, \mathbf{C})$ such that $|\mathbf{C}| \leq |S|$. The proof of Theorem 3 hence yields an $\Omega(\log n)$ lower bound for the approximation ratio of the multicast problem.

Despite the negative result of Theorem 3, we extend Theorem 1 to directed graphs, though for the broadcast problem only. The MDST problem in directed graphs is defined as follows: we are given a digraph $G = (V, E)$ and a node $s \in V$, and we are looking for a directed spanning tree rooted at s (i.e., whose edges are directed from the root towards the leaves) whose maximum degree is the smallest among all directed spanning trees rooted at s. The following lemma is a variant of Lemma 2 dedicated to directed graphs.

Lemma 5. (Fürer and Raghavachari [16]) *There exists a polynomial-time $(1, \log n)$-approximation algorithm for the MDST problem in digraphs.*

Note that this lemma does not apply to the Steiner version of the MDST problem. Nevertheless, it allows to prove the following:

Theorem 4. *Let $t(n)$ be the complexity of an $(1, \log n)$-approximation algorithm for the MDST problem in n-node directed graphs. There exists an $t(n)$-time $O(\frac{\log n}{\log OPT})$-approximation algorithm for the broadcast problem in directed graphs.*

Proof. First note that, using the same technique as in the proof of Lemma 1, we get that the broadcast time from any source s is at least $\lceil \log_2 n \rceil$, and at least the out-degree of a MDST rooted at s. The proof then follows the same guidelines as the proofs of Theorem 1 and Lemma 3, by replacing the use of Lemma 2 by Lemma 5. The proof of Lemma 3 must also be slightly modified because Phase 2 involves calls from the leaves toward the root, that is calls proceeding upward the tree. We use the following result due to Kortsarz and Peleg:

Lemma 6. (Kortsarz and Peleg [24]) *Given any directed tree T rooted at s and whose edges are directed from s toward the leaves, broadcasting from s to T requires at most $2L + \lceil \log_2 n \rceil$ rounds, where L is the number of leaves of T. Moreover, a $(2L + \lceil \log_2 n \rceil)$-round protocol can be computed in $O(n)$ time.*

Using that lemma, the proof of Theorem 4 works the same as the proofs of Lemma 3 and Theorem 1.   □

Fürer and Raghavachari let as an open problem in [17] the possible extension to directed graphs of their $(1,1)$-approximation algorithm for the Steiner version of the MDST problem in undirected graphs. Theorems 3 and 4 show that the answer is negative. Indeed, we have:

Corollary 1. *Unless* $\text{NP} \subset \text{DTIME}(n^{\log\log n})$, *any polynomial-time* $(1,r)$-*approximation algorithm for the Steiner version of the MDST problem in directed graphs satisfies* $r = \Omega(\log|D|\log\log|D|)$.

Proof. Given a $(1, \alpha \cdot \log|D|)$-approximation algorithm for the Steiner version of the MDST problem in directed graphs, the construction of Theorem 4 with $m = \log|D|$ yields a ρ-approximation algorithm for the multicast problem with $\rho = O\left(\alpha \cdot \frac{\log|D|}{\log\log|D|}\right)$. From Theorem 3, $\alpha \geq \Omega(\log\log|D|)$. $\square$

The proof of Corollary 1 also shows that any polynomial-time ρ-approximation algorithm for the Steiner version of the MDST problem in directed graphs satisfies $\rho \geq \Omega(\log\log|D|)$. The next result shows how to obtain a better lower bound with an *ad hoc* construction.

Theorem 5. *Unless* $\text{NP} \subset \text{DTIME}(n^{\log\log n})$, *the optimal solution of the Steiner version of the MDST problem in directed graphs is not approximable in polynomial time within* $(1-\epsilon)\ln|D|$ *for any* $\epsilon > 0$.

Proof. By simple reduction from MSC (see [14] for more details). $\square$

Remark. Again, since Feige's Lemma still holds for instances $(S, \mathbf{C})$ such that $|\mathbf{C}| \leq |S|$, Theorem 5 yields an $\Omega(\log n)$ lower bound for the approximation ratio of the Steiner version of the MDST problem in digraphs.

Up to our knowledge, there is no polynomial-time $O(\log n)$-approximation algorithm for the Steiner version of the MDST problem in digraphs.

4   Gossip Problem

Trivially, we have $g(G, D) \geq |D| - 1$ since every node must receive $|D| - 1$ information. The next lemma improves this lower bound.

Lemma 7. *Given* $G = (V, E)$ *and* $D \subseteq V$, $g(G, D) \geq \frac{1}{2}|D|(\Delta_{\min} - 1)$.

Proof. Let $\mathcal{P}$ be an optimal gossip protocol in D. For every source $s \in D$, let G_s be the subgraph of G induced by the union of all the calls of $\mathcal{P}$ carrying the message of s. G_s spans D, and $\mathcal{P}$ completes in at least $\frac{1}{2}\max_{v \in V}\sum_{s \in D}\deg_{G_s}(v)$ rounds. Let us show that this sum is in $\Omega(|D|\Delta_{\min})$. For that purpose, for any $S \subset V$, define $c(G \backslash S)$ as the number of connected components of $G \backslash S$ containing at least one node in D. Then let $S^* \subset V$ be such that

$$c(G \backslash S^*)/|S^*| = \min_{S \neq \emptyset} c(G \backslash S)/|S|.$$

It is shown in [16] that $c(G \backslash S^*)/|S^*| \leq \Delta_{\min} \leq 1 + c(G \backslash S^*)/|S^*|$. Let $\mathcal{F}$ be any family of $|D|$ subgraphs of G, each of them spanning D. We have

$$\sum_{H \in \mathcal{F}}\sum_{v \in S^*}\deg_H(v) \geq |D|\,c(G \backslash S^*).$$

Therefore there exists $v^* \in S^*$ such that

$$\sum_{H \in \mathcal{F}} \deg_H(v^*) \geq |D| \, c(G \setminus S^*)/|S^*| \geq |D|(\Delta_{\min} - 1).$$

Therefore $\max_{v \in V} \sum_{s \in D} \deg_{G_s}(v) \geq |D|(\Delta_{\min} - 1)$ which completes the proof.
□

Theorem 6. *Let $t(n)$ be the complexity of an $(1,1)$-approximation algorithm for the MDST problem in n-node graphs. There exists a $t(n)$-time $O(1)$-approximation algorithm for the gossip problem.*

Again, by Lemma 2, $t(n)$ can be chosen to be polynomial.

Proof. Let $G = (V, E)$ be any graph, and let $D \subseteq V$. Let T be a tree of degree Δ that spans D in G. Let s be any node of D. Gossiping in D can be performed in two phases: Phase 1: Accumulation in s of all the messages of D; Phase 2: Broadcasting from s to D of the $|D|$ messages accumulated during Phase 1. Phase 1 requires $|D| - 1$ rounds. By pipelining along the edges of a tree T of maximum degree $\Delta \leq \Delta_{\min}+1$, Phase 2 can be performed in $O(|D|\Delta_{\min})$ rounds. From Lemma 7, this protocol is optimal within a constant multiplicative factor. (For more details, see [14]).
□

5 Conclusion and Further Research

Many problems remain open: (1) Can we approximate the optimal solution of the multicast problem in graphs within a constant factor? (2) Can we approximate the optimal solution of the multicast problem in digraphs within an $O(\log|D|)$ factor? (3) can we describe a polynomial approximation scheme for the gossip problem in graphs? In addition to these problems, we want to point out that, although approximation algorithms are usually preferred to heuristics, there are known heuristics for broadcasting and gossiping in the local model that perform very well in general. We therefore believe that the following heuristic for broadcasting under the vertex-disjoint paths mode should be worth to experiment:

While $|D| > 1$ do
    (1) Find the maximum number of vertex-disjoint paths connecting
        pairs of nodes in D;
    (2) For each of these paths do select one extremity of the path and
        remove the other from D;
End.

Note that Instruction (1) can be performed in polynomial time by using the algorithm in [29]. A broadcast protocol is obtained from this iteration by reversing the process, i.e., if D_i is the set of nodes selected at the ith iteration, and if $D_k = \{v\}$ is the set of nodes remaining after the last iteration, then broadcasting from u consists of (1) u calls v, and (2) for $i = k$ down to 1, every node in D_i calls its "matched node" in D_{i-1}.

Acknowledgements. The author is thankful to Lali Barrière for her helpful comments on preliminary versions of this paper. The author is also thankful to Peter Widmayer for early discussions on the topic of this paper. This work was partially done while the author was visiting UPC in Barcelona, Carleton U. in Ottawa, and Concordia U. in Montréal. Additional supports from Carleton U., Concordia U., NATO, and RNRT project ROM.

References

1. A. Bar-Noy, S. Guha, J. Naor, and B. Schieber. Multicasting in heterogeneous networks. In *30th ACM Symposium on Theory of Computing (STOC '98)*, pages 448–453, 1998.
2. D. Barth and P. Fraigniaud. Approximation algorithms for structured communication problems. In *9th ACM Symposium on Parallel Algorithms and Architectures (SPAA '97)*, pages 180–188, 1997. (Tech. Rep. LRI-1239, Univ. Paris-Sud. http://www.lri.fr/~pierre).
3. J.-C. Bermond, A. Bonnecaze, T. Kodate, S. Pérennes, and P. Sole. Broadcasting in hypercubes under the circuit-switched model. In *IEEE Int. Parallel and Distributed Processing Symp. (IPDPS)*, 2000.
4. J.-C. Bermond, L. Gargano, A. Rescigno, and U. Vaccaro. Fast gossiping by short messages. In *22nd International Colloquium on Automata, Languages and Programming (ICALP '95)*, volume 944 of *Lecture Notes in Computer Science*, pages 135–146, 1995.
5. B. Birchler, A.-H. Esfahanian, and E. Torng. Toward a general theory of unicast-based multicast communication. In *21st Workshop on Graph-Theoretic Concepts in Computer Science (WG '95)*, volume 1017 of *LNCS*, pages 237–251, 1995.
6. B. Birchler, A.-H. Esfahanian, and E. Torng. Information dissemination in restricted routing networks. In *International Symposium on Combinatorics and Applications*, pages 33–44, 1996.
7. J. Cohen. Broadcasting, multicasting and gossiping in trees under the all-port line model. In *10th ACM Symposium on Parallel Algorithms and Architectures (SPAA '98)*, pages 164–171, 1998.
8. J. Cohen, P. Fraigniaud, J.-C. Konig, and A. Raspaud. Broadcasting and multicasting in cut-through routed networks. In *11th IEEE Int. Parallel Processing Symposium (IPPS '97)*, pages 734–738, 1997.
9. J. Cohen, P. Fraigniaud, and M. Mitjana. Scheduling calls for multicasting in tree-networks. In *10th ACM-SIAM Symp. on Discrete Algorithms (SODA '99)*, pages 881–882, 1999.
10. C. Diot, W. Dabbous, and J. Crowcroft. Multipoint communication: a survey of protocols, functions, and mechanisms. *IEEE Journal on Selected Areas in Communications*, 15(3):277–290, 1997.
11. A. Farley. Minimum-time line broadcast networks. *Networks*, 10:59–70, 1980.
12. U. Feige. A threshold of $\ln n$ for approximating set cover. *Journal of the ACM*, 45:634–652, 1998.
13. P. Fraigniaud. Minimum-time broadcast under the edge-disjoint paths mode. In *2nd Int. Conference on Fun with Algorithms* (FUN '01), Carleton Scientific, 2001.
14. P. Fraigniaud. Approximation algorithms for collective communications with limited link and node-contention. Tech. Rep. LRI-1264, Université Paris-Sud, France, 2000.

15. P. Fraigniaud and E. Lazard. Methods and Problems of Communication in Usual Networks. *Discrete Applied Mathematics*, 53:79–133, 1994.
16. M. Fürer and B. Raghavachari. Approximating the minimum degree spanning tree within one from the optimal degree. In *3rd Annual ACM-SIAM Sumposium on Discrete Algorithms (SODA '92)*, pages 317–324, 1992.
17. M. Fürer and B. Raghavachari. Approximating the minimum-degree steiner tree to within one of optimal. *Journal of Algorithms*, 17:409–423, 1994.
18. S. Hedetniemi, S. Hedetniemi, and A. Liestman. A survey of gossiping and broadcasting in communication networks. *Networks*, 18:319–349, 1986.
19. J. Hromković, R. Klasing, B. Monien, and R. Peine. Dissemination of information in interconnection networks (broadcasting and gossiping). In Ding-Zhu Du and D. Frank Hsu, editors, *Combinatorial Network Theory*, pages 125–212. Kluwer Academic, 1995.
20. J. Hromković, R. Klasing, and E. Stohr. Dissemination of information in vertex-disjoint paths mode. *Computer and Artificial Intelligence*, 15(4):295–318, 1996.
21. J. Hromković, R. Klasing, E. Stohr, and Wagener H. Gossiping in vertex-disjoint paths mode in d-dimensional grids and planar graphs. *Information and Computation*, 123(1):17–28, 1995.
22. J. Hromković, R. Klasing, W. Unger, and H. Wagener. Optimal algorithms for broadcast and gossip in the edge-disjoint paths mode. *Information and Computation*, 133(1):1–33, 1997.
23. R. Klasing. The relationship between the gossip complexity in vertex-disjoint paths mode and the vertex-bisection width. *Discrete Applied Maths*, 83(1-3):229–246, 1998.
24. G. Kortsarz and D. Peleg. Approximation algorithms for minimum time broadcast. *SIAM Journal on Discrete Mathematics*, 8(3):401–427, 1995.
25. W. Mostafa and M. Singhal. A taxonomy of multicast protocols for internet applications. *Computer Communications*, 20:1448–1457, 1998.
26. J. Peters and M. Syska. Circuit-switched broadcasting in torus networks. *IEEE Transactions on Parallel and Distributed Systems*, 7(3):246–255, 1996.
27. R. Ravi. Rapid rumor ramification: approximating the minimum broadcast time. In *35-th IEEE Symposium on Foundations of Computer Science*, pages 202–213, 1994.
28. C. Schindelhauer. On the inapproximability of broadcasting time. In *3rd Int. Workshop on Approximation Algorithms for Combinatorial Optimization Problems (APPROX)*, LNCS 1913. Springer-Verlag, 2000.
29. H. Yinnone. Maximum number of disjoint paths connecting specified terminals in a graph. *Disc. Appl. Maths.* 55:183–195, 1994.

Round Robin Is Optimal for Fault-Tolerant Broadcasting on Wireless Networks*
(Extended Abstract)

Andrea E.F. Clementi[1], Angelo Monti[2], and Riccardo Silvestri[2]

[1] Dipartimento di Matematica, Università di Roma "Tor Vergata",
clementi@mat.uniroma2.it
[2] Dipartimento di Scienze dell'Informazione, Università di Roma "La Sapienza",
{monti, silvestri}dsi.uniroma1.it

Abstract. We study the completion time of broadcast operations on *Static Ad-Hoc Wireless Networks* in presence of *unpredictable* and *dynamical* faults. As for *oblivious* fault-tolerant distributed protocols, we provide an $\Omega(Dn)$ lower bound where n is the number of nodes of the network and D is the source eccentricity in the fault-free part of the network. Rather surprisingly, this lower bound implies that the simple *Round-Robin* protocol, working in $O(Dn)$ time, is an *optimal* fault-tolerant oblivious protocol. Then, we demonstrate that networks of $o(n/\log n)$ maximum in-degree admit faster oblivious protocols. Indeed, we derive an oblivious protocol having $O(D \min\{n, \Delta \log n\})$ completion time on any network of maximum in-degree Δ. Finally, we address the question whether *adaptive* protocols can be faster than oblivious ones. We show that the answer is negative at least in the general setting: we indeed prove an $\Omega(Dn)$ lower bound when $D = \Theta(\sqrt{n})$. This clearly implies that no (*adaptive*) protocol can achieve, in general, $o(Dn)$ completion time.

1 Introduction

Static ad-hoc wireless networks (in short, wireless networks) have been the subject of several works in recent years due to their potential applications in scenarios such as battlefields, emergency disaster relief, and in any situation in which it is very difficult (or impossible) to provide the necessary infrastructure [19, 21]. As in other network models, a challenging task is to enable fast and *reliable* communication.

A wireless network can be modeled as a directed graph G where an edge (u, v) exists if and only if u can communicate with v in one hop. Communication between two stations that are not adjacent can be achieved by *multi-hop* transmissions. A useful (and sometimes unavoidable) paradigm of wireless communication is the structuring of communication into synchronous *time-slots*. This

---

* Research partially supported by the Italian MURST Project "REACTION" and the EC RTN Project "ARACNE".

F. Meyer auf der Heide (Ed.): ESA 2001, LNCS 2161, pp. 452–463, 2001.

paradigm is commonly adopted in the practical design of protocols and hence its use in theoretical analysis is well motivated [2,4,11,20]. In every time-slot, each active node may perform local computations and either transmit a message along all of its outgoing edges or try to recover messages from all its incoming edges (the last two operations are carried out by means of an omnidirectional antenna). This feature is extremely attractive in its *broadcast* nature: a single transmission by a node could be received by all its neighbors within one time-slot. However, since a single radio frequence is typically used, when two or more neighbors of a node are transmitting at the same time-slot, a *collision* occurs and the message is lost. So, a node can recover a message from one of its incoming edges if and only if this edge is the only one bringing in a message.

One of the fundamental tasks in wireless network communication is the *broadcast* operation. It consists in transmitting a message from one source node to all the nodes. Most of the proposed broadcast protocols in wireless networks concern the case in which the network is *fault-free*. However, wireless networks are typically adopted in scenarios where unpredictable node and link faults happen very frequently.

Node failures happen when some hardware or software component of a station does not work, while link failures are due to the presence of a new (artificial or natural) hurdle that does not allow the communication along that link. Typically, while it is reasonable to assume that nodes know the initial topology of the network, they know nothing about the duration and the location (in the network) of the faults. Such faults may clearly happen at any instant, *even during the execution of a protocol*. In the sequel, such kind of faults will be called *dynamical faults* or, simply, faults.

The (worst-case) completion-time of a fault-tolerant broadcasting protocol on a graph G is defined as the maximum number (over all possible fault patterns) of time-slots required to *inform* all nodes in the fault-free part of the network which are reachable from the source (a more formal definition will be given in Sect. 1.3). The aim of this paper is thus to investigate the completion time of such broadcast protocols in presence of *dynamical-faults*.

1.1   Previous Results

(Fault-free) broadcasting. We will mention only the best results which are presently known for the fault-free model. An $O(D + \log^5 n)$ upper bound on the completion time for n-node networks of *source eccentricity D* is proved in [12]. The source eccentricity is the maximum oriented distance (i.e. number of hops) from the source s to a reachable node of the network. Notice that D is a trivial lower bound for the broadcast operation.

In [7], the authors give a protocol that completes broadcasting within time $O(D \log \Delta \log(n/D))$, where Δ denotes the maximal in-degree of the network. This bound cannot be improved in general: [1] provides an $\Omega(\log^2 n)$ lower bound that holds for graphs of maximal eccentricity $D = 2$.

In [3], the authors show that scheduling an optimal broadcast is NP-hard. The APX-hardness of the problem is proved in [7].

Permanent-fault tolerant broadcasting. A node has a *permanent* fault if it never sends or receives messages since the beginning of the execution of the protocol. In [14], the authors consider the broadcasting operation in presence of permanent unknown node faults for two restricted classes of networks: linear and square (or hexagonal) meshes. They consider both *oblivious* and *adaptive* protocols. In the former case, all transmissions are scheduled in advance; in particular, the action of a node in a given time-slot is independent of the messages received so far. In the latter case, nodes can decide their action also depending on the messages received so far. For both cases, the authors assume the existence of a bound t on the number of faults and, then, they derive a $\Theta(D + t)$ bound for oblivious protocols and a $\Theta(D + \log \min\{\Delta, t\})$ bound for adaptive protocols, where, in this case (and in the sequel), D denotes the source eccentricity in the fault-free part of the network, i.e., the *residual* eccentricity.

More recently, the issue of permanent-fault tolerant broadcasting on general networks has been studied in [4,5,6,8]. Indeed, in these papers, several lower and upper bounds on the completion time of broadcasting are obtained on the *unknown* fault-free network model. A wireless network is said to be *unknown* when every node knows nothing about the network but its own label. Even though it has never been observed, it is easy to show that a broadcasting protocol for unknown fault-free networks is also a permanent-fault tolerant protocol for general (known) networks and viceversa. So, the results obtained in the unknown model immediately apply to the permanent-fault tolerance issue. In particular, one of the results in [8] can be interpreted as showing the existence of an infinite family of networks for which any permanent-fault tolerant protocol is forced to perform $\Omega(n \log D)$ time-slots to complete broadcast. The best general upper bound for permanent-fault tolerant protocols is $O(n \log^2 n)$ [6]. This protocol is thus almost optimal when $D = \Omega(n^\alpha)$ for any constant $\alpha > 0$. In [8], the authors provide a permanent-fault tolerant protocol having $O(D\Delta \log^2 n)$ completion time on any network of maximum in-degree Δ.

Other models. A different kind of fault-tolerant broadcasting is studied in [17]: they in fact introduce fault-tolerant protocols that work under the assumption that all faults are *eventually* repaired. The protocols are not analyzed from the point of view of worst-case completion time. Finally, in [15], the case in which broadcasting messages may be corrupted with some probability distribution is studied.

Dynamical-fault tolerance broadcasting. We observe that permanent faults are special cases of dynamical faults and, moreover, we emphasize that *all* the above protocols *do not work* in presence of dynamical faults. This is mainly due to the collisions yielded by any unpredictable wake-up of a faulty node/link during the protocol execution.

To the best of our knowledge, broadcasting on arbitrary wireless networks in presence of dynamical faults has never been studied before.

1.2   Our Results

Oblivious protocols. A simple oblivious *(dynamical-)Fault-tolerant Distributed Broadcasting* (FDB) protocol relies on the *Round Robin* scheduling: given an n-

node network G and a source node s, the protocol runs a sequence of consecutive identical *phases*; each phase consists of n time-slots and, during the i-th time slot ($i = 1, \ldots, n$), node i, if *informed*[1], acts as transmitter while all the other nodes work as receivers. It is not hard to show that, for any fault pattern yielding residual source eccentricity D, the Round Robin protocol completes broadcasting on G, after D phases (so, after Dn time-slots). One may think that this simple oblivious FDB protocol is not efficient (or, at least, not optimal) since it does *never* exploit simultaneous transmissions.

Rather surprisingly, we show that, for any n and for any $D < n$, it is possible to define an n-node network G, such that *any* oblivious FDB protocol requires $\Omega(Dn)$ time-slots to complete broadcast on G. It thus follows that the Round Robin protocol is *optimal* on general networks. The proof departs significantly from the techniques used in all previous related works (such as those used for the case of permanent-faults and based on *selective families* [4,8]): it in fact relies on a tight lower bound on the length of D-*sequences* (see Def. 1), a combinatorial tool that might have further applications in scheduling theory.

We then show that a broad class of wireless networks admits an oblivious FDB protocol which is faster than the Round Robin protocol. Indeed, we exploit small *ad-hoc strongly-selective* families, a variant of strongly-selective families (also known as *superimposed codes* [10,8,13]), in order to develop an oblivious FDB protocol that completes broadcasting within $O(D \min\{n, \Delta \log n\})$ time-slots, where Δ is the maximum in-degree of the input network. This protocol is thus faster than the Round Robin protocol for all networks such that $\Delta = o(n/\log n)$ and it is almost optimal for constant Δ.

Adaptive protocols. In adaptive FDB protocols, nodes have the ability to decide their own scheduling as a function of the messages received so far. A natural and interesting question is whether adaptive FDB protocols are faster than oblivious ones. We give a partial negative answer to this question. We strengthen the connection between strong-selectivity and the task of fault-tolerant broadcasting: we indeed exploit the tight lower bound on the size of strongly-selective families, given in [8], to derive an $\Omega(Dn)$ lower bound for adaptive FDB protocols, when $D = \Theta(\sqrt{n})$. This implies that *no* (adaptive) FDB protocol can achieve $o(Dn)$ completion time on arbitrary networks.

1.3   Preliminaries

The aim of this subsection is to formalize the concept of FDB protocol and its completion time.

According to the fault-tolerance model adopted in the literature [14,18], an FDB protocol for a graph G is a broadcasting protocol that, for any source s, and for any (node/link) *fault pattern* F, guarantees that every node, which is reachable from s in the *residual subgraph* G^F, will receive the source message. A *fault pattern* F is a function that maps every time-slot t to the subset $F(t)$ of nodes and links that are faulty at time slot t. The residual subgraph G^F is

---

[1] A node is *informed* during a time-slot t if it has received the source message in some time slot $t' < t$.

the graph obtained from G by removing all those nodes and links that belong to $F(t)$, for some time-slot t during the execution of the protocol. The completion time of the protocol on a graph G and source s is the maximal (over all possible fault patterns) number of time-slots to perform the above task.

This definition implies that nodes that are not reachable from the source in the residual subgraph are not considered in the analysis of the completion time of FDB protocols. We emphasize that any attempt to consider a larger residual subgraph makes the worst-case completion time of *any* FDB protocol unbounded.

2   Oblivious Fault-Tolerant Protocols

2.1   Lower Bound

An oblivious protocol for an n-node network can be represented as a sequence $\mathbf{S} = (S_1, S_2, \ldots, S_l)$ of transmissions, where $S_i \subseteq [n]$ is the set of nodes that transmit during the i-th time-slot and l denotes the worst-case (w.r.t. all possible fault patterns) completion time. Wlog, we also assume that, if a node belongs to S_t (so it should transmit at time-slot t) but it has not received the source message during the first $t-1$ time-slots, then it will send no message at time-slot t.

In order to prove the lower bound, we consider the complete directed graphs K_n, $n \geq 1$. We first show that any oblivious FDB protocol $\mathbf{S}$ on K_n must satisfy the following property

Definition 1. *A sequence* $\mathbf{S} = (S_1, S_2, \ldots, S_l)$ *of subsets of* $[n]$ *is called a D-sequence for* $[n]$ *if, for each subset H of $[n]$ with $|H| \leq D$ and each permutation $\pi = (\pi_1, \pi_2 \ldots, \pi_{|H|})$ of H, there exists a subsequence $(S_{i_1}, S_{i_2}, \ldots, S_{i_{|H|-1}})$ of $\mathbf{S}$ such that*

$$\pi_j \in S_{i_j} \quad and \quad S_{i_j} \subseteq H, \quad for \ \ 1 \leq j < |H|.$$

Lemma 1. *For every n and every $D < n$, if $\mathbf{S}$ is an oblivious protocol, which completes broadcast on the graph K_n for every fault pattern yielding a residual source eccentricity at most D, then $\mathbf{S}$ is a D-sequence for $[n]$.*

Proof. Let $\mathbf{S} = (S_1, S_2, \ldots, S_l)$ be an oblivious FDB protocol which completes broadcast on K_n for every fault pattern yielding a residual subgraph of source eccentricity at most D. Let us consider a subset H of $[n]$ with $|H| \leq D$ and a permutation $\pi = (\pi_1, \pi_2 \ldots, \pi_{|H|})$ of H. We will define a fault pattern F of K_n and a source node π_0 such that

1. π_0 has residual eccentricity $|H|$;
2. the fact that $\mathbf{S}$ completes broadcast for the pattern F implies the existence of a subsequence $(S_{i_1}, \ldots, S_{i_{|H|-1}})$ of $\mathbf{S}$ such that $\pi_j \in S_{i_j}$ and $S_{i_j} \subseteq H$, for $1 \leq j < |H|$.

Let us choose the source π_0 as a node in $[n] \setminus H$ and consider the set of (directed) edges

$$A = \bigcup_{0 \leq i < |H|} (\pi_i, \pi_{i+1}).$$

The pattern F is defined as follows:

for any $i \geq 1$ and for any $u \in [n]$ with $u \neq \pi_{i-1}$, the edge (u, π_i) is faulty at time slot t if and only if $\pi_{i-1} \notin S_t$ or π_{i-1} is not yet informed at that time slot. In other words, the edge (u, π_i) is faulty whenever π_{i-1} cannot inform π_i.

Observe that the edges in A are never faulty. Moreover, it is easy to verify that the residual source eccentricity is $|H|$. Since, by definition, the protocol **S** completes the broadcast on any residual subgraph of source eccentricity at most D, then the protocol **S** completes the broadcast on the graph K_n^F. So, for any $i > 0$, there is a time-slot in which π_{i+1} gets informed. By definition of F, the only node that can inform π_{i+1} is π_i. Since all nodes in $[n] \setminus H$ are informed during the first time-slot in which the source π_0 transmits, then π_i can inform π_{i+1} at a time slot t only if $S_t \subseteq \{\pi_1, \pi_2 \ldots, \pi_{|H|}\}$. Thus, there must exist a subsequence $(S_{i_1}, S_{i_2}, \ldots, S_{i_{|H|-1}})$ of **S** such that $\pi_j \in S_{i_j}$ and $S_{i_j} \subseteq H$, for $1 \leq j < |H|$.

$\square$

We now prove a tight lower bound on the length of a D-sequence. To this aim, we need the following technical result.

Lemma 2. *If* **S** *is a D-sequence for $[n]$, then*

$$\sum_{S \in \mathbf{S}} |S| = \Omega(Dn).$$

Proof. Let $\mathbf{S} = (S_1, S_2, \ldots, S_l)$ be a D-sequence for $[n]$ and consider the sequence $(k_1, k_2, \ldots, k_{D-1})$ defined (by induction) as follows:

$$k_1 = \min\{h \mid [n] = \cup_{1 \leq j \leq h} S_j\}.$$

By definition of D-sequence, k_1 must exist and so there exists (at least one) element in the set

$$S_{k_1} \setminus \cup_{1 \leq j < k_1} S_j.$$

Then, let π_1 be any of such elements. We now assume that the indices $k_1, k_2, \ldots, k_i$ and the elements $\pi_1, \pi_2, \ldots, \pi_i$ are already defined, then

$$k_{i+1} = \min\{h \mid ([n] - \{\pi_1, \ldots, \pi_i\}) \subseteq \bigcup_{k_i < j \leq h} S_j\}$$

(again, we notice that k_{i+1} must exist since **S** is a D-sequence and $i + 1 < D$) and let π_{i+1} be any element in

$$([n] - \{\pi_1, \ldots, \pi_i\}) \bigcap (S_{k_{i+1}} \setminus \bigcup_{k_i < j < k_{i+1}} S_j).$$

By definition of the above sequence, it holds that

$$\sum_{j=1}^{k_1} |S_j| = n \text{ and } \sum_{k_i < j \leq k_{i+1}} |S_j| \geq n - i, \text{ for any } i = 1, \ldots, D - 1.$$

It thus follows that

$$\sum_{S \in \mathbf{S}} |S| \geq \sum_{j=1}^{k_1} |S_j| + \sum_{i=1}^{D-2} \sum_{k_i < j \leq k_{i+1}} |S_j| \geq n + \sum_{i=1}^{D-2} (n - i) = \Omega(Dn).$$

$\square$

Lemma 3. If $\mathbf{S}$ is a D-sequence for $[n]$ then $|\mathbf{S}| = \Omega(Dn)$.

Proof. We count in two different ways the following number

$$N = |\{(H, (S, x)) \mid H \subseteq [n], \ S \in \mathbf{S}, \ S \subseteq H \text{ and } x \in S\}|.$$

Let us consider H and the subsequence $\mathbf{S_H}$ of $\mathbf{S}$ obtained by deleting from $\mathbf{S}$ all the sets S such that $S \not\subseteq H$. By definition of D-sequence $\mathbf{S_H}$ must form a D-sequence for H. Hence, from Lemma 2, we have at least $cD|H|$ ways of choosing (S, x), for a constant $c > 0$. Thus

$$N \geq \sum_{H \subseteq [n]} cD|H| \ = \ cD \sum_{i=1}^{n} \binom{n}{i} i \ > \ cD \sum_{i=\lceil \frac{n}{2} \rceil}^{n} \binom{n}{i} i \ > \ \frac{cDn}{4} 2^n. \tag{1}$$

Now let (S, x) be fixed. There are $2^{n-|S|}$ subsets H of $[n]$ such that $S \subseteq H$. Thus

$$N \ = \ \sum_{S \in \mathbf{S}} |S| \cdot 2^{n-|S|} \ = \ 2^n \sum_{S \in \mathbf{S}} \frac{|S|}{2^{|S|}} \ \leq \ 2^n \sum_{S \in \mathbf{S}} \frac{1}{2} \ = \ \frac{2^n}{2} |\mathbf{S}|. \tag{2}$$

Finally, by comparing Eq.s 1 and 2, we derive

$$\frac{2^n}{2} |\mathbf{S}| \ > \ \frac{cDn}{4} 2^n, \quad \text{so} \quad |\mathbf{S}| > \frac{cDn}{2}.$$

$\square$

We are now able to show the following

Theorem 1. Let $n > 0$ and $1 \leq D \leq n-1$. For every oblivious FDB protocol on the graph K_n, there exist a source s and a fault pattern F that force the protocol to perform $\Omega(Dn)$ time-slots.

Proof. Let $\mathbf{S}$ be any oblivious FDB protocol for K_n and let T be the maximum completion-time of $\mathbf{S}$ over all possible residual subgraphs of source eccentricity D. Then, from Lemma 1, the first T transmission sets of $\mathbf{S}$ must be a D-sequence for $[n]$. From Lemma 3, it must hold that $T = \Omega(Dn)$.

$\square$

Corollary 1. For any $n > 0$, any oblivious FDB protocol completes broadcasting on the graph K_n in $\Omega(n^2)$ time-slots.

2.2 Efficient Protocols for Networks of "Small" In-Degree

In this subsection, we show that networks of maximum in-degree $\Delta = o(n/\log n)$ admit oblivious FDB protocols which are faster than the Round-Robin one. To this aim, we need the following combinatorial tool.

Definition 2. *Let $\mathcal{S}$ and $\mathcal{N}$ be families of sets. The family $\mathcal{S}$ is strongly-selective for $\mathcal{N}$ if for every set $N \in \mathcal{N}$ and for every element $x \in N$ there exists a set $S \in \mathcal{S}$ such that $N \cap S = \{x\}$.*

Strongly-selective families for the family of all subsets of $[n]$ having size at most Δ have been recently used to develop multi-broadcast protocols on the unknown model [8,9]. The following protocol instead uses strong selectivity for the family $\mathcal{N}$ consisting of the sets of in-neighbors of the nodes of the input network G. In fact, for each node v of G, let $N(v) \subseteq [n]$ be the set of its in-neighbors and let

$$\mathcal{N} = \{N(v) \mid v \in V\}.$$

Let $\mathcal{S} = \{S_1, S_2, \dots, S_m\}$ be any (arbitrarily ordered) strongly-selective family for $\mathcal{N}$.

Description of Protocol BROAD. The protocol consists of a sequence of phases.

- In the first phase the source sends its message.
- All successive phases are identical and each of them consists of m time-slots. At time-slot j of every phase, any informed node v sends the source message if and only if it belongs to S_j; All the remaining nodes act as receivers.

Lemma 4. *For any (dynamical) fault pattern F, at the end of phase i, every node at distance i, from the source s in the residual subgraph G^F, is informed. So, BROAD completes broadcasting within Dm time-slots, where the D is the residual source eccentricity.*

Proof. The proof is by induction on the distance i. For $i = 1$ it is obvious. We thus assume that all nodes at distance i have received the source message during the first i phases. Consider a node v at distance $i + 1$ in the residual subgraph and a node $u \in N(v)$ at distance i in the residual subgraph. Notice that, since v is at "residual" distance $i + 1$ from s, such an u must exist. Moreover, $N(v)$ belongs to $\mathcal{N}$ and $\mathcal{S}$ is strongly-selective for $\mathcal{N}$, so there will be a time-slot in phase $i + 1$ in which only u (among the nodes in $N(v)$) transmits the source message and v will successfully receive it.

It is now clear that the total number of time-slots required by the protocol to complete the broadcast is Dm.

$\square$

The above lemma motivates our interest in finding strongly-selective families of small size since the latter is a factor of the completion time of our protocol.

The probabilistic construction of strongly-selective families in the proof of the next lemma can be efficiently (i.e. in polynomial time) de-randomized by means of a suitable application of the method of *conditional probabilities* [16].

This technique has been recently applied [7] to a weaker version of selectivity. Furthermore, as for strong selectivity, we can also use the deterministic efficient construction of *superimposed codes* given in [10] that yield strongly-selective families of size equivalent to that in our lemma.

Lemma 5. *For any family $\mathcal{N}$ of sets, each of size at most Δ, there exists a strongly selective family $\mathcal{S}$ for $\mathcal{N}$ such that $|\mathcal{S}| = O(\Delta \max\{\log |\mathcal{N}|, \log \Delta\})$.*

Proof. We assume, without loss of generality, that the sets in $\mathcal{N}$ are subsets of the ground set $[n]$ and that $\Delta \geq 2$ (for $\Delta = 1$ the family $\mathcal{S} = \{[n]\}$ trivially proves the lemma).

We use a probabilistic argument: construct a set S by picking every element of $[n]$ with probability $\frac{1}{\Delta}$. For fixed $N \in \mathcal{N}$ and $x \in N$ it holds that:

$$\Pr[N \cap S = \{x\}] = \frac{1}{\Delta}\left(1 - \frac{1}{\Delta}\right)^{|N|-1} \geq \frac{1}{\Delta}\left(1 - \frac{1}{\Delta}\right)^{\Delta} \geq \frac{1}{4\Delta} \qquad (3)$$

(where the last inequality holds since $\Delta \geq 2$). Consider now a family $\mathcal{S} = \{S_1, S_2, \ldots, S_m\}$ where each set S_i is constructed, independently, as above. From Ineq (3), it follows that the probability that $\mathcal{S}$ is not strongly-selective for fixed $N \in \mathcal{N}$ and $x \in N$ is at most

$$\left(1 - \frac{1}{4\Delta}\right)^m \leq e^{-\frac{m}{4\Delta}}$$

(the above bound follows from the well-known inequality $1 - t \leq e^{-t}$ that holds for any real t). It thus follows that

$$\Pr[\, \mathcal{S} \text{ is not strongly-selective for } \mathcal{N}] \leq \sum_{N \in \mathcal{N}} \sum_{x \in N} e^{-\frac{m}{4\Delta}} \leq |\mathcal{N}| \Delta e^{-\frac{m}{4\Delta}}$$

and the last value is less than 1 for $m > 8\Delta \max(\log |\mathcal{N}|, \log \Delta)$.

$\square$

Theorem 2. *The oblivious FDB protocol* BROAD *completes broadcast within $O(D \min\{\Delta \log n, n\})$ time-slots on any n-node graph G with maximum in-degree Δ.*

Proof. Since the graph has maximum in-degree Δ, the size of any subset in $\mathcal{N}$ is at most Δ. Hence, from Lemma 5, there exists a strongly-selective family $\mathcal{S}$ for $\mathcal{N}$ of size $|\mathcal{S}| \leq \min\{c\Delta \log n, n\}$, for some constant $c > 0$ (the bound n is due to the fact that a family of n singletons is always strongly-selective). The theorem is thus an easy consequence of the above bound and Lemma 4.

$\square$

3   Adaptive Fault-Tolerant Protocols

In this section, a lower bound on the completion-time of adaptive FDB protocols is given. To this aim, we consider strongly-selective family for the family of all subsets of $[n]$ having size at most Δ (in short, (n, Δ)-*strongly-selective families*). As for the size of such families, the following lower bound is known.

Theorem 3. [8]. *If $\mathcal{S}$ is an (n, Δ)-strongly-selective family then it holds that*

$$|\mathcal{S}| = \Omega \left(\min \left\{ \frac{\Delta^2}{\log \Delta} \log n, n \right\} \right)$$

We will adopt the general definition of (adaptive) distributed broadcast protocols introduced in [2]. In such protocols, the action of a node in a specific time-slot is a function of its own label, the number of the current time-slot t, the input graph G, and the messages received during the previous time-slots.

Theorem 4. *For any $n > 0$, any FDB protocol, that completes broadcasting on the graph K_n, requires $\Omega(n\sqrt{n})$ time-slots.*

Sketch of the proof. Given any protocol P, we define two sets of faults: a set of edges of K_n that suffer a permanent fault (i.e., they will be forever down since the beginning of the protocol) and a set of dynamical node faults. The permanent faults are chosen in order to yield a layered graph G^P which consists of $D + 1$ levels $L_0, L_1, \ldots, L_D$ where $D = \lfloor \sqrt{n}/2 \rfloor$. Level L_0 contains only the source s, level L_j, $j < D$, consists of at most Δ nodes with $\Delta = \sqrt{n}$ and L_D consists of all the remaining nodes. All nodes of L_{j-1} in G^P have (only) outgoing edges to all nodes in L_j.

Both permanent and dynamical faults will be determined depending on the actions of P. As for permanent faults, instead of describing the set of faulty edges, we provide (in the proof of the next claim), for any $j = 1, \ldots, D$, the set of nodes that belongs to L_j. This permanent fault pattern will be combined with the dynamical fault pattern (which is described below) in such a way that the protocol is forced to execute $\Omega(\frac{\Delta^2}{\log \Delta} \log n) = \Omega(n)$ time-slots in order to successfully transmit the initial message between two consecutive levels.

From Thm. 3, there exists a constant $c > 0$ such that, any $(\lceil n/2 \rceil, \Delta)$-strongly-selective family must have size at least T, where $T \geq cn$. The theorem is thus an easy consequence of the following

> CLAIM   For any $j \geq 0$, there exists a node assignment to L_j and a pattern of dynamical node faults in the first j levels such that P does not broadcast the source message to level L_{j+1} before the time-slot jT.
>
> *Proof.* The proof is by induction on j. For $j = 0$, the claim is trivial. We thus assume the thesis be true for $j - 1$. Let us define
>
> $$R = \{\text{node not already assigned to levels } L_0, \ldots, L_{j-1}\}.$$
>
> Notice that $|R| \geq \lceil n/2 \rceil$. Let L be an arbitrary subset of R. Consider the following two cases: *i)* L_j is chosen as L; *ii)* L_j is chosen as R (i.e., all the remaining nodes are assigned to the $j + 1$-th level). In both cases, the predecessor[2] subgraph G_u^P of any node $u \in L$ is that induced by $L_0 \cup L_1 \cup \ldots L_{j-1} \cup \{u\}$ in G^P. It follows that the behavior

---

[2] Given a graph G, the predecessor subgraph G_u of a node u is the subgraph of G induced by all nodes v for which there exists a directed path from v to u.

of node u, according to protocol P, is the same in both cases. We can thus consider the behavior of P when $L_j = R$. Then, we define

$$S_t = \{u \in R \mid u \text{ acts as transmitter at time-slot } (j-1)T + t\}.$$

and the family $\mathcal{S} = \{S_1, \ldots, S_{T-1}\}$ of subsets from R. Since $|\mathcal{S}| < T$, $\mathcal{S}$ is not $(\lceil n/2 \rceil, \Delta)$-strongly-selective; so, a subset $L \subset R$ exists such that $|L| \leq \Delta$ and L is not strongly selected by $\mathcal{S}$ (and thus by P) in any time-slot t such that $(j-1)T + 1 \leq t \leq jT - 1$. Let thus u be a node in L which is not selected. Then, the proof is completed by considering a suitable pattern of dynamical faults in such a way that: i) all the outgoing edges of u are always fault-free, and ii) no node (in particular, the nodes different from u) in L will successfully transmit during those $T-1$ time slots.

$\square$

Since the residual graph yielded by the above proof has residual eccentricity $\Theta(\sqrt{n})$, it also follows that

Corollary 2. *No FDB protocol can achieve an $o(Dn)$ completion time on general n-node networks.*

Acknowledgement. A significant credit goes to Paolo Penna for helpful discussions. More importantly, Paolo suggested us to investigate the issue of fault-tolerance in wireless networks.

References

1. N. Alon,, A. Bar-Noy, N. Linial, and D. Peleg (1991), A lower bound for radio broadcast, *JCCS*, 43, 290-298 (An extended abstract appeared also in *ACM - STOC* 1989).
2. R. Bar-Yehuda, O. Goldreich, and A. Itai (1992), On the time-complexity of broadcast in multi-hop radio networks: An exponential gap between determinism and randomization, *JCSS*, 45, 104-126 (preliminary version in *6th ACM PODC*, 1987).
3. I. Chlamtac and S. Kutten (1985), On Broadcasting in Radio Networks - Problem Analysis and Protocol Design, *IEEE Transactions on Communications* 33, 1240-1246.
4. B.S. Chlebus, L. Gąsieniec, A.M. Gibbons, A. Pelc, and W. Rytter (2000), Deterministic broadcasting in unknown radio networks, *Proc. of 11th ACM-SIAM SODA*, 861-870.
5. B. S. Chlebus, L. Gąsieniec, A. Ostlin, and J. M. Robson (2000), Deterministic radio broadcasting, *Proc. of 27th ICALP*, LNCS 1853, 717-728.
6. M. Chrobak, L. Gąsieniec, and W. Rytter (2000), Fast Broadcasting and Gossiping in Radio Networks, *Proc. of 41st IEEE FOCS*.
7. A.E.F. Clementi, P. Crescenzi, A. Monti, P. Penna, and R. Silvestri (2001), On Computing Ad-Hoc Selective Families, to appear on *Proc. of 5th RANDOM*, LNCS.
8. A.E.F. Clementi, A. Monti, and R. Silvestri (2001), Selective Families, Superimposed Codes, and Broadcasting in Unknown Radio Networks, *Proc. of 12th ACM-SIAM SODA*, 709-718.

9. A.E.F. Clementi, A. Monti, and R. Silvestri (2001), Distributed Multi-Broadcast in Unknown Radio Networks, to appear in *Proc. of 20th ACM-PODC*.

10. R. Cole, R. Hariharan, and P. Indyk (1999), Tree Pattern matching and subset matching in deterministic $O(n \log^3 n)$-time, *Proc. of 10th ACM-SIAM SODA*.

11. R. Gallager (1985), A Perspective on Multiaccess Channels, *IEEE Trans. Inform. Theory*, 31, 124-142.

12. I. Gaber and Y. Mansour (1995), Broadcast in Radio Networks, *Proc. of 6th ACM-SIAM SODA*, 577-585.

13. P. Indyk (1997), Deterministic Superimposed Coding with Application to Pattern Matching, *Proc. of IEEE 38th FOCS*, 127-136.

14. E. Kranakis, D. Krizanc, and A. Pelc (1998), Fault-Tolerant Broadcasting in Radio Networks, *Proc. of 6th ESA*, LNCS 1461, 283-294.

15. E. Kushilevitz and Y. Mansour (1998), Computation in Noisy Radio Networks, *Proc. of 9th ACM-SIAM SODA*, 236-243.

16. R. Motwani and P. Raghavan (1995), *Randomized Algorithms*, Cambridge University Press.

17. E. Pagani and G. Rossi (1997), Reliable Broadcast in Mobile Multihop Packet Networks, *Proc. of 3rd ACM-IEEE MOBICOM*, 34-42.

18. A. Pelc (2000), Broadcasting in Radio Networks, unpublished manuscript.

19. T. S. Rappaport (1996), *Wireless Communications: Principles and Practice*, Prentice Hall.

20. L.G. Roberts (1972), Aloha Packet System with and without Slots and Capture, ASS Notes 8, Advanced Research Projects Agency, Network Information Center, Stanford Research Institute.

21. J. E. Wieselthier, G. D. Ngyuyen, and A. Ephremides (2000), "On the Construction of energy-Efficient Broadcast and Multicast Trees in Wireless Networks", *Proc. of 19th IEEE INFOCOM*.

Online and Offline Distance Constrained Labeling of Disk Graphs

(Extended Abstract)

Jiří Fiala[*,1,2], Aleksei V. Fishkin [**,1], and Fedor V. Fomin[* * *,3]

[1] CAU, Institute of Computer Science and Applied Mathematics, Olshausenstr. 40, 24098, Kiel, Germany. {jfi,avf}@informatik.uni-kiel.de
[2] Charles University, Faculty of Mathematics and Physics, Institute for Theoretical Computer Science (ITI), Malostranské nám. 2/25, 118 00, Prague, Czech Republic.
[3] Faculty of Mathematics and Mechanics, St. Petersburg State University, Bibliotechnaya sq. 2, St. Petersburg, 198904, Russia. fomin@gamma.math.spbu.ru

Abstract. A disk graph is the intersection graph of a set of disks in the plane. We consider the problem of assigning labels to vertices of a disk graph satisfying a sequence of distance constrains. Our objective is to minimize the distance between the smallest and the largest labels. We propose an on-line labeling algorithm on disk graphs, if the maximum and minimum diameters are bounded. We give the upper and lower bounds on its competitive ratio, and show that the algorithm is asymptotically optimal. In more detail we explore the case of distance constraints $(2, 1)$, and present two off-line approximation algorithms. The last one we call robust, i.e. it does not require the disks representation and either outputs a feasible labeling, or answers the input is not a unit disk graph.

1 Introduction

The *frequency assignment problem* arises in *radio* or *mobile telephony* networks when different radio *transmitters* interfere to each other while operating on the same or closely related frequency channels. The most common model for an instance of frequency assignment is the *interference graph*. Each vertex of the interference graph represents a transmitter. If simultaneous broadcasting of two transmitters may cause an interference, then they are connected in the inter- ference graph by an edge. One can associate the coverage area of a transmitter with a disk of a particular diameter, and afterwards, to model the interference

* Partially supported by EU ARACNE project HPRN-CT-1999-00112, by GAČR 201/99/0242 and by the Ministry of Education of the Czech Republic as project LN00A056.
** Partially supported by the DFG-Graduiertenkolleg "Effiziente Algorithmen und Mehrskalenmethoden".
* * * The work of this author was done while he was a visiting postdoc at DIMATIA-ITI partially supported by GAČR 201/99/0242 and by the Ministry of Education of the Czech Republic as project LN00A056. Also supported by Netherlands Organization for Scientific Research (NWO grant 047.008.006.)

F. Meyer auf der Heide (Ed.): ESA 2001, LNCS 2161, pp. 464–475, 2001.

graph as a graph whose edges connect transmitters at "close" distance, e.g. when their scopes intersect. Then the underlying interference graph is a *disk graph*, i.e. the intersection graph of disks in the plane, or (when all disks are of the same diameter) a *unit disk graph* [10].

With the assumption that a pair of "close" transmitters should be assigned different frequencies, the frequency assignment is equivalent to the problem of coloring the interference graph. However, in [10] it was observed that the signal propagation may affect the interference even in distant regions (but with decreasing intensity). Hence not only "close" transmitters should get different frequencies, but also frequencies used at some distance should be appropriately separated. In this case, the frequency assignment can be modeled as the problem of *distance constrained labeling*, or so called $L_{(p_1,...,p_k)}$-*labeling* of the interference graph [12]:

Definition 1. *Let $p_1, ..., p_k$ be a sequence of positive integers called* distance constraints. *The $L_{(p_1,...,p_k)}$-labeling of a graph G is a mapping $c: V(G) \rightarrow \{1, ..., \lambda\}$ such that the following holds:*

$$\forall i : 1 \leq i \leq k, \forall u, v \in V(G) : dist_G(u,v) \leq i \Rightarrow |c(u) - c(v)| \geq p_i.$$

The minimum number for which an $L_{(p_1,...,p_k)}$-labeling of G exists, is denoted by $\chi_{(p_1,...,p_k)}(G)$. Notice that $\chi_{(1)}(G) = \chi(G)$, where $\chi(G)$ is the chromatic number of G. Also for $p_1 = p_2 = \cdots = p_k = 1$, $\chi_{(p_1,...,p_k)}(G) = \chi(G^k)$, where G^k is the k-th power of G, i.e. a graph which arise from G by adding edges connecting vertices at distance at most k.

Related works on graph labeling problems. A special case $k = 2$ and the distance constraints $(p_1, p_2) = (2, 1)$ was intensively studied for different graph classes in [1,3,7,8,16]. The exact value of $\chi_{(2,1)}$ can be derived for cycles and paths, and there are polynomial algorithms which compute the value $\chi_{(2,1)}$ for trees and co-graphs [3]. The problem of recognizing graphs such that $\chi_{(2,1)} \leq \lambda$ is *NP*-complete for all fixed $\lambda \geq 4$ [6]. For planar graphs, the problem of deciding $\chi_{(2,1)} \leq 9$ was shown to be *NP*-complete in [1].

Related works on disk graphs. The disk graph and unit disk graph recognition problem is *NP*-hard [2,11]. Hence algorithms that require the corresponding disk graph representation are substantially weaker than those which work only with graphs.

In [4] it is shown that the 3-coloring of unit disk graph is *NP*-complete (even if the input is a set of unit disks). However for unit disk graph coloring problem there is a 5-competitive on-line algorithm [13,14], and in the case of given disks representation, there is a 3-approximation off-line algorithm [2,14]. Furthermore, there is a 5-approximate off-line algorithm for the disk graph coloring problem, which uses the first-fit technique on disks that are ordered from those with the biggest diameter to the smallest one [13]. From another side, one cannot expect the existence of an on-line coloring algorithm of a constant competitive ratio in the case of general disk graphs, since there is no such one even for planar graphs [9], and every planar graph is also a disk graph.

Our results. For fixed distance constraints $(p_1, ..., p_k)$ and fixed diameter ratio σ we present a constant-competitive on-line labeling algorithm which require the disk representation. This is the first constant-competitive on-line labeling algorithm known to our knowledge. We prove that when the representation is not given or when the diameter ratio is not bounded then no on-line algorithm with fixed competitive ratio exists.

For our algorithm we present upper and lower bounds on its competitive ratio, and show that in the case of unit disk graphs the algorithm is asymptotically optimal.

Finally, we explore the case of distance constraints $(2, 1)$, and present two off-line approximation algorithms for unit disk graphs. We present robust labeling algorithm, i.e. an algorithm that does not require the disk representation and either outputs a feasible labeling, or answers the input is not a unit disk graph.

The following table summarizes the known and new upper bounds on the competitive or performance ratio on coloring and labeling problems on unit disk graphs (UDG), on disk graphs with diameter ratio bounded by σ (DG$_\sigma$) and on general disk graphs (DG).

Execution	Off-line		On-line	
Repres.	+	−	+	−
Coloring:				
UDG	3 [14]	3[1]	5 [13,14]	5[1]
DG$_\sigma$	5 (as DG)	5 (as DG)	Theorem 2 [*]	YES [5]
DG	5 [13]	5[1]	NO [5]	NO [9]
$L_{(2,1)}$-labeling:				
UDG [2]	$12 \rightarrow 9$ [*]	$10.6 \rightarrow 10$ [*]	$25 \rightarrow 12.5$ [*]	NO
$L_{(p_1,...p_k)}$-labeling:				
UDG	Th. 2[4], C. 3[3] [*]	Corollary 4[3] [*]	Theorem 2 [*]	NO [*]
DG$_\sigma$	Theorem 2[4] [*]	?	Theorem 2 [*]	NO [*]
DG	?	?	NO [*]	NO [*]

Positive results are marked either by "YES" or by the appropriate statement. "NO" means that no algorithm with fixed performance/competitive ratio exists. The sign "?" marks an open problem. The results presented in this paper are highlighted by "[*]". The list of further explanations follows:

[1] The algorithm working without the disk representation can be derived from those which uses the representation as shown in Section 4.2.

[2] This rows shows the values of the upper bound in the worst case. Since our results are better for graphs with large cliques, we give also the limit of the upper bound as the clique size grows to infinity.

[3] Here we prove only the case $k = 2, (p_1, p_2) = (2, 1)$.

[4] Every on-line algorithm can be executed off-line.

Throughout the extended abstract various proofs are omitted or postponed into appendix due to space restriction.

2   Preliminaries

For a set of geometric objects, the corresponding *intersection graph* is the undirected graph whose vertices are objects and an edge connects two vertices if the corresponding objects intersect. Let $\mathcal{E}$ be a 2-dimensional Euclidean plane with the coordinates x, y. Let $\mathcal{D} = \{D_1, ..., D_n\}$ be a set of n disks in $\mathcal{E}$, where each D_i is uniquely determined by its center in (x_i, y_i) and by the diameter $d_i \in \mathbb{R}_+$. The intersection graph $G_\mathcal{D}$ of the set $\mathcal{D}$ is called a *disk graph*, more formally $V(G_\mathcal{D}) = \mathcal{D}, E(G_\mathcal{D}) = \{(D_i, D_j) \in \binom{\mathcal{D}}{2} : D_i \cap D_j \neq \emptyset\}$. When all disks of $\mathcal{D}$ have unit diameter, i.e. $d_i = 1$ for all $D_i \in \mathcal{D}$, then $G_\mathcal{D}$ is called a *unit disk graph*. In both cases, $\mathcal{D}$ is called the *disk representation* of $G_\mathcal{D}$. The value $\sigma(\mathcal{D}) = \frac{\max d_i}{\min d_i}$ is called the *diameter ratio* of $\mathcal{D}$.

We say that an algorithm $\mathcal{A}$ is a ρ-*approximation off-line* $L_{(p_1,...,p_k)}$-*labeling algorithm* if for a given graph $G_\mathcal{D}$ it runs in polynomial time and outputs an $L_{(p_1,...,p_k)}$-labeling of G such that the maximum label that $\mathcal{A}$ uses is at most $\rho \chi_{(p_1,...,p_k)}(G_\mathcal{D})$. The value ρ is called the *approximation ratio* of $\mathcal{A}$.

We say that an algorithm $\mathcal{B}$ is an *on-line* $L_{(p_1,...,p_k)}$-*labeling algorithm* if it labels the vertices of a graph $G_\mathcal{D}$ in an externally determined sequence $D_1 \prec ... \prec D_n$. At the time t the algorithm $\mathcal{B}$ has to irrevocably assign a label to D_t, while it can only see the edges connecting vertices $D_1, ..., D_t$. Such an algorithm $\mathcal{B}$ is a ρ-*competitive on-line* $L_{(p_1,...,p_k)}$-*labeling algorithm* if for every graph $G_\mathcal{D}$ and any ordering $\prec$ on $V(G_\mathcal{D})$ it always outputs an $L_{(p_1,...,p_k)}$-labeling with the maximum label at most $\rho \chi_{(p_1,...,p_k)}(G_\mathcal{D})$. We call the constant ρ the *competitive ratio* of $\mathcal{B}$.

Our on-line algorithm is based on the following partition of the plane $\mathcal{E}$. We call a *hexagonal tiling* on $\mathcal{E}$ the partition of the plane into the set $\mathcal{C}$ of hexagonal cells of diameter one. Each cell $C_{i,j} \in \mathcal{C}$ is characterized by two integer coordinates i, j, and is a simplex delimited by the following lines:

$$2i - j - 1 < \tfrac{4}{3}\sqrt{3}x \leq 2i - j + 1$$
$$i + j - 1 < \tfrac{2}{3}(\sqrt{3}x + 3y) \leq i + j + 1$$
$$-i + 2j - 1 < \tfrac{2}{3}(-\sqrt{3}x + 3y) \leq -i + 2j + 1.$$

Each point of the plane belongs to exactly one cell $C_{i,j}$. Observe also, that each cell contains exactly two adjacent corners of the bounding hexagon, and the distance between every two points inside the same cell is at most one.

We denote the *plane distance* of two points $p, p' \in \mathcal{E}$ by $dist_\mathcal{E}(p, p')$. Similarly the plane distance of cells $C_{i,j}$ and $C_{k,l}$ is defined as

$$dist_\mathcal{E}(C_{i,j}, C_{k,l}) = \inf\{dist_\mathcal{E}(p, p') : p \in C_{i,j}, p' \in C_{k,l}\}.$$

3   Online Distance Constrained Labeling

In this section we explore the on-line distance constrained labeling problem for disk graphs.

We start with the following observation. Let $(p_1, ..., p_k)$ be a sequence of distance constraints ($k \geq 2$ and $p_2 \geq 1$) and let $\mathcal{B}$ be an arbitrary on-line

$L_{(p_1,...,p_k)}$-labeling algorithm for disk graphs. Let $\mathcal{D}$ be a set of mutually disjoint disks, i.e. $G_\mathcal{D}$ has no edges.

Consider the two following cases. First assume that the disk representation is not a part of the input. Secondly, $\mathcal{B}$ the disk representation is given as a part of the input, but the diameter ratio $\sigma(\mathcal{D})$ is not known during the execution of $\mathcal{B}$.

In both cases, the algorithm $\mathcal{B}$ labels all vertices of $G_\mathcal{D}$ by distinct labels. Otherwise any new vertex may create a path of length two between an arbitrary pair of already labeled vertices (in the second case by extending $\mathcal{D}$ by a disk of large diameter). The maximal label used by $\mathcal{B}$ on $G_\mathcal{D}$ is at least $|\mathcal{D}|$, although $\chi_{(p_1,...,p_k)}(G_\mathcal{D}) = 1$.

Due to the above arguments, we consider the case when the disk representation $\mathcal{D}$ of $G_\mathcal{D}$ is a part of the input and when the corresponding diameter ratio $\sigma(\mathcal{D})$ is known, i.e. it is bounded by a constant $\sigma \geq 1$.

First we introduce a special circular labeling of cells and show how our on-line labeling algorithm uses such labeling. Later we derive the upper bound on its competitive ratio and show that the algorithm is asymptotically optimal for the class of unit disk graphs with at least one edge. Finally, we illustrate the algorithm performance for distance constraints $(2, 1)$.

3.1   Circular Labeling

Let $\mathcal{D}$ be a set of disks with the diameter ratio $\sigma(\mathcal{D})$ and $G_\mathcal{D}$ be the corresponding disk graph. We suppose that the coordinates of plane $\mathcal{E}$ are scaled such that the smallest disk has the unit diameter. Moreover, through this section we assume that the diameter ratio σ and distance constraints $(p_1, ..., p_k)$ are fixed, i.e. for every selection of these parameters we design a specific algorithm.

Let $\mathcal{C}$ be the set of hexagonal cells. We say that a mapping $\varphi : \mathcal{C} \rightarrow \{1, 2, ..., l\}$ is a *circular l-labeling of $\mathcal{C}$* (with respect to $(p_1, ..., p_k)$ and $\sigma \geq 1$) if the following condition

$$dist_\mathcal{E}(C, C') \leq i \cdot \sigma \Rightarrow \min\{|\varphi(C) - \varphi(C')|, l - |\varphi(C) - \varphi(C')|\} \geq p_i,$$

is satisfied for all cells $C, C' \in \mathcal{C}$ and all $i \in \{1, ..., k\}$.

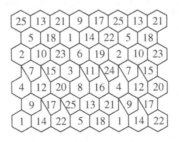

Fig. 1. An example of a 25-circular labeling $((p_1, p_2) = (2, 1), \sigma = \frac{\sqrt{7}}{2})$

An example of a circular 25-labeling (with respect to $(p_1, p_2) = (2, 1), \sigma = \frac{\sqrt{7}}{2}$) is depicted in Figure 1. In fact, the cells with equal labels are at plane distance at least $2\sqrt{3}$ and those pairs labeled by consecutive labels have plane distance at least $\frac{\sqrt{7}}{2} \doteq 1.32$.

For general case, the existence of such labelings is guaranteed by the following theorem.

Theorem 1. *For every k-tuple of distance constraints $(p_1, ..., p_k)$, every $\sigma \geq 1$ and*

$$l^* = 1 + 6\left(2p_1 - 1 + \sum_{i=2}^{\lfloor \frac{4k\sigma+4}{3} \rfloor} i \cdot (2p_{\lceil \frac{3i-4}{4\sigma} \rceil} - 1)\right),$$

there exists a circular l^*-labeling of $\mathcal{C}$ and can be found in $O(l^* \sigma^4 k^4)$ time.

3.2   The Algorithm

Assume now that the parameters of algorithm, i.e. distance constraints $(p_1, ..., p_k)$ and diameter ratio $\sigma \geq 1$ are fixed.

Algorithm On-line disks labeling (ODL)
Input: Set of disks $\mathcal{D}$, in an arbitrarily order $D_1 \prec ... \prec D_n$.
Output: An $L_{(p_1, ..., p_k)}$-labeling $c : \mathcal{D} \to \mathbb{Z}^+$ of $G_{\mathcal{D}}$.

    1. Find a circular l-labeling $\varphi : \mathcal{C} \to \{1, ..., l\}$
    (with respect to $(p_1, ..., p_k)$ and σ).
    2. For each cell $C_{i,j}$ put $\mathcal{D}_{i,j} = \emptyset$.
    ($\mathcal{D}_{i,j}$ is the set of already labeled disks of $\mathcal{D}$ with centers in a cell $C_{i,j} \in \mathcal{C}$.)
    3. For each k from 1 to n:
        3a. Decide into which cell $C_{i,j}$ the disk D_k belongs.
        3b. Define the label $c(D_k) = \varphi(C_{i,j}) + l \cdot |\mathcal{D}_{i,j}|$.
        3c. Set $\mathcal{D}_{i,j} = \mathcal{D}_{i,j} \cup \{D_k\}$.

The next lemma follows directly from the properties of circular l-labeling.

Lemma 1. *Suppose that at the first step the algorithm ODL finds a circular l-labeling (with respect to $(p_1, ..., p_k)$ and σ). Then for any set of disks $\mathcal{D}$ of diameter ratio $\sigma(\mathcal{D}) \leq \sigma$, the algorithm ODL produces a feasible $L_{(p_1,...,p_k)}$-labeling of $G_{\mathcal{D}}$ and the maximal label used by this algorithm is at most $l\omega(G_{\mathcal{D}})$.*

The analysis of our algorithms is based on the following fact: If a set of vertices X induces a complete subgraph of a graph G, then the maximal label used for vertices of X in any labeling is at least $p_1(|X| - 1) + 1$.

Lemma 2. *For any k-tuple $(p_1, ..., p_k)$ of distance constraints and any set of disks $\mathcal{D}$,*

$$\chi_{(p_1,...,p_k)}(G_{\mathcal{D}}) \geq p_1(\omega(G_{\mathcal{D}}) - 1) + 1 \geq p_1(\max_{i,j}\{|V(G_{\mathcal{D}_{i,j}})|\} - 1) + 1,$$

where $\omega(G_{\mathcal{D}})$ is the size of the maximum clique in $G_{\mathcal{D}}$.

Proof. Observe that the subgraph $G_{\mathcal{D}_{i,j}}$ of $G_{\mathcal{D}}$ is isomorphic to a complete graph and the number of disks in $|\mathcal{D}_{i,j}|$ is at most $\omega(G_{\mathcal{D}})$. $\qquad\Box$

Combining Theorem 1 with Lemmas 1 and 2, we get the main result of this section.

Theorem 2. *For every $(p_1, ..., p_k)$ and every $\sigma \geq 1$ the competitive ratio of algorithm ODL is bounded by*

$$\rho \leq \max_{\mathcal{D}} \frac{\omega(G_{\mathcal{D}}) \cdot l^*}{(\omega(G_{\mathcal{D}}) - 1) \cdot p_1 + 1},$$

where the maximum is taken over all sets of disks of diameter ratio at most σ.

Therefore, ODL is an l^*-competitive on-line $L_{(p_1,...,p_k)}$-labeling algorithm for the class of disks graphs $G_{\mathcal{D}}$ of the diameter ratio $\sigma(\mathcal{D}) \leq \sigma$. Moreover, if disk graphs have at least one edge then the competitive ratio $\rho \leq \frac{2l^*}{p_1+1}$, and if $\omega(G_{\mathcal{D}}) \to \infty$ then $\rho \to \frac{l^*}{p_1}$.

If we apply the above theorem on the labeling depicted in Fig. 1, we get:

Corollary 1. *For the distance constraints $(2, 1)$, the algorithm ODL is $\frac{50}{3} \doteq 16.67$-competitive on disk graphs with at least one edge and $\sigma \leq 1.32$. Moreover, the competitive ratio tends to 12.5 when $\omega(G_{\mathcal{D}}) \to \infty$.*

Now we present a lower bound on the competitive ratio of any on-line algorithm solving the distance constrained labeling problem for disk graphs, which have a representation of bounded diameter ratio.

Theorem 3. *For any k-tuple $(p_1, ..., p_k)$ of distance constraints $(k \geq 2)$, any $\sigma \geq 1$ and any $\varepsilon > 0$, there is no $(\bar{\rho} - \varepsilon)$-competitive on-line $L_{(p_1,...,p_k)}$-labeling algorithm for the class of disk graphs with representation of diameter ratio at most σ, where*

$$\bar{\rho} = 1 + \frac{\sigma^2}{9} \max_{i=2,...,k} \{i^2 p_i\}.$$

Theorems 2 and 3 imply that:

Corollary 2. *For any $(p_1, ..., p_k)$, $k \geq 2$ and any $\sigma \geq 1$, the competitive ratio of algorithm ODL on disk graph (with given representation, diameter ratio at most σ and with at least one edge) is at most $O(\log k)$ times larger than the competitive ratio of any on-line $L_{(p_1,...,p_k)}$-labeling algorithm. Therefore, under these restricted conditions and for fixed number of distance constraints, the algorithm ODL is asymptotically optimal.*

As the last remark we propose a slightly different strategy with a smaller channel separation used for disks in the same location, though it does not lead to a better competitive ratio of the final algorithm. The idea is as follows. First scale the cells of $\mathcal{C}$ such that every cell has diameter σ. Then, find a feasible circular l-labeling $\varphi : \bar{\mathcal{C}} \to \{1, ..., l\}$. After that, run the algorithm ODL on the set $\mathcal{D}$ of non-unit disks using the scaled cells. As the result, we find a feasible

$L_{(p_1,\ldots,p_k)}$-labeling of the disk graph $G_{\mathcal{D}}$, but the disks belonging to the same cell do not induce a complete graph. On the other hand, the area of any cell of diameter σ can be covered by at most $\sigma^2 + t\sigma$ unit cells, where t is a suitable constant that does not depend on σ. Then the number of vertices in any $G_{\mathcal{D}_{i,j}}$ is at most $\sigma^2 + t\sigma$ times more than the maximum clique size $\omega(G_{\mathcal{D}})$. Thus, it follows that

$$\chi_{(p_1,\ldots,p_k)}(G_{\mathcal{D}_{i,j}}) \geq p_1(\omega(G_{\mathcal{D}_{i,j}}) - 1) \geq p_1\left(\left\lceil \frac{V(G_{\mathcal{D}_{i,j}})}{\sigma^2 + t\sigma} \right\rceil - 1\right)$$

and we use the same arguments as above.

4   Offline Labeling with Constraints (2,1)

In this section we explore the off-line distance labeling problem only for one particular selection of distance constrains $(p_1, p_2) = (2, 1)$. First we discuss the case when the graph and its disc representation are known. Then we present a robust algorithm which do not require the disc representation.

4.1   The Approximation Algorithm

We first discuss the case that the entire graph $G_{\mathcal{D}}$ is given together with the representation $\mathcal{D}$ as the part of the input. The main idea of the off-line approximation algorithm is rather simple: We "cut" the plane into strips of small width, label unit disk graphs induced by strips and combine the labelings of strip graphs.

An unit disk graph $G_{\mathcal{D}}$ is a $\frac{1}{\sqrt{2}}$-*strip graph* if there is a disk representation $\mathcal{D}$ of $G_{\mathcal{D}}$ such that the centers of disks in $\mathcal{D}$ are in a strip of width $\frac{1}{\sqrt{2}}$. In other words, there is a mapping $f\colon V(G_{\mathcal{D}}) \to \mathbb{R} \times [0, \frac{1}{\sqrt{2}}]$ such that $(u, v) \in E(G_{\mathcal{D}})$ if and only if $dist_{\mathcal{E}}(f(u), f(v)) \leq 1$.

Lemma 3. *Let G be a $\frac{1}{\sqrt{2}}$-strip graph and let v be a vertex such that the unit disk corresponding to v (in some representation $\mathcal{D}$) has the smallest x-coordinate. Then the cardinality of the vertex set*

$$N_{G^2}(v) = \{u \in V(G) - \{v\}\colon dist_G(u, v) \leq 2\}$$

is at most $3\omega(G) - 1$.

A vertex ordering $v_1 \prec \ldots \prec v_n$ of a $\frac{1}{\sqrt{2}}$-strip graph G is *increasing* if there is a disk representation $\mathcal{D}$ such that $i < j$ if and only if x-coordinate of a disk in $\mathcal{D}$ corresponding to vertex v_i is at most x-coordinate of a disk corresponding to v_j.

By Lemma 3, the first-fit coloring algorithm that process vertices of the second power G^2 of a $\frac{1}{\sqrt{2}}$-strip graph G in an increasing order, uses at most $3\omega(G)$ distinct colors. This coloring of G^2 is equivalent to an $L_{(1,1)}$-labeling of

G. By multiplying all labels of $L_{(1,1)}$-labeling by 2, we obtain an $L_{(2,2)}$-labeling of G. Finally,

$$\chi_{(2,1)}(G) \leq \chi_{(2,2)}(G) \leq 2\chi_{(1,1)}(G) = 2\chi(G^2) \leq 6\omega(G).$$

Using the first-fit approach, the $L_{(2,1)}$-labeling of G can be obtained in $O(nm)$ time ($n = |V(G)|, m = |E(G)|$). Notice, that all labels used for this labeling are *even*.

Now we are ready to describe the approximate labeling algorithm for unit disk graphs and constraints $(2,1)$. Without loss of generality we may assume that G is connected and has at least two vertices.

Algorithm Distance Labeling (DL)
Input: Unit disk representation of a unit disk graph G.
Output: Labeling of c of $V(G)$.

1. Partition a plane into $k = O(n)$ strips $S_1, ..., S_k$ of width $\frac{1}{\sqrt{2}}$, numbered from top to bottom, such that S_1 contains a disk with maximal y-coordinate and S_k the one with minimal. This partition induces partition of G into $\frac{1}{\sqrt{2}}$-strip graphs $G_1, ..., G_k$. (In a case of disks with centers in two strips ties are broken arbitrarily.)
2. For each $i \in \{1, ..., k\}$ find an $L_{(2,1)}$-labeling of G_i using even labels and with the maximum label bounded by $6\omega(G_i) \leq 6\omega(G)$.
3. Change the labels of graph G_i by increasing them by the number $\sharp_{(i \bmod 6)}$, where

$$(\sharp_0, ..., \sharp_5) = (0, \ 6\omega(G), \ 12\omega(G), \ -1, \ 6\omega(G) - 1, \ 12\omega(G) - 1)$$

Theorem 4. *For any unit disk graph G, algorithm DL produces an $L_{(2,1)}$-labeling and the maximal label used by this algorithm is at most $18\omega(G)$.*

Corollary 3. *The approximation ratio of the algorithm DL is bounded by 12 and tends to 9 as $\omega(G)$ grows to infinity.*

Observe that $\frac{1}{\sqrt{2}}$-strips were used in the description of the algorithm to simplify the explanation. To avoid irrationality, $\frac{1}{\sqrt{2}}$-strips in the algorithm can be replaced by c-strips, where c is any rational number between $\frac{2}{3}$ and $\frac{1}{\sqrt{2}}$. Also the algorithm can be generalized easily to an algorithm producing an $L_{(p,1)}$-labeling ($p \geq 1$) with the maximal label used by the algorithm at most $9p\omega(G)$.

4.2   Robust Approximation Algorithm

The main purpose of this section is to present the approximation labeling algorithm which doesn't need a geometric representation of a unit disk graph as part of input. (Let us remind that it is NP-hard to recognize unit disk graphs.)

In [15] the following notion of *robust algorithms* is discussed: an algorithm which solves an optimization problem on class is called robust if it satisfies the following conditions.

1. Whenever the input is in class C, the algorithm finds the correct solution.
2. If the input is not in the class C, then the algorithm either finds the correct solution, or answers that the input is not in the class.

Based on the ideas of [4], a robust algorithm computing the maximal clique of a unit disk graph is given in [15].

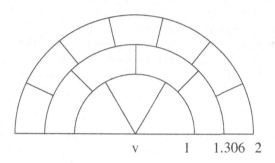

$$v \qquad 1 \qquad 1.306 \quad 2$$

Fig. 2. The plane partition around v

Lemma 4. *Every unit disk graph G has a vertex v such that the set*

$$N_G(v) = \{u \neq v \colon \{u, v\} \in E(G)\}$$

has cardinality at most $3\omega(G) - 3$ and the set

$$N_{G^2}(v) - N_G(v)$$

contains at most $11\omega(G)$ vertices.

Proof. The proof of the lemma is based on the plane partition around vertex v shown in Fig. 2. □

We say that a vertex ordering $v_1 \prec \ldots \prec v_n$ of G is *good* if for every $2 \leq i \leq n$

(i) $|N_G(v_i) \cap \{v_1, ..., v_{i-1}\}| \leq 3\omega(G) - 3$;
(ii) $|(N_{G^2}(v_i) - N_G(v_i)) \cap \{v_1, ..., v_{i-1}\}| \leq 11\omega(G)$.

Notice, that by Lemma 4 every unit disk graph has a good vertex ordering. Also for a graph G with n vertices one can in $O(n^3)$ time either find a good vertex ordering, or conclude that there is no good ordering for G.

Now we are ready to describe algorithm RDL which can be regarded as robust distance labeling approximation algorithm. It doesn't use the geometric representation of a graph G and either concludes that G is not unit disk graph, or (we prove it below) it finds an $L_{(2,1)}$-labeling of G with the maximum label at most $\leq 20\omega(G) - 8$.

Algorithm Robust Distance Labeling (RDL)
Input: Graph G in adjacency list.
Output: Either $L_{(2,1)}$-labeling c of $V(G)$, or conclusion that G is not unit disk graph.

1. Run the robust algorithm to compute $\omega(G)$. This algorithm either computes $\omega(G)$, or concludes that G is not unit disk graph.
2. Find a good vertex ordering $v_1 \prec \ldots \prec v_n$. If there is no such an ordering, then conclude that G is not unit disk graph.
3. Label vertices sequentially in order $\prec$ as follows:
   3a. Assume that vertices $v_1, \ldots, v_{i-1}$ are already labeled.
   3b. Let $\lambda \geq 1$ be the smallest integer which is not used as a label of vertices in $N_{G^2}(v_i) \cap \{v_1, \ldots, v_{i-1}\}$ nor is a member of the set
   $$\bigcup_{j \in \{1, \ldots, i-1\} : v_j \in N_G(v_i)} \{c(v_j) - 1, c(v_j), c(v_j) + 1\}.$$
   3c. Label v_i by $c(v_i) = \lambda$.

Theorem 5. *For any graph G, Algorithm RDL either produces an $L_{2,1}$-labeling with maximum label $\leq 20\omega(G) - 8$, or concludes that G is not unit disk graph.*

Proof. Suppose that algorithm output that G is not unit disk graph. If it was done after the first step, then G has no edge ordering $\prec_e$ and therefore is not unit disk graph. If the algorithm halts at the second step, then its conclusion is verified by Lemma 4.

Suppose that RDL outputs a labeling. Let us first show that the maximum label used by the algorithm is $\leq 20\omega(G) - 8$. We proceed by induction. The vertex v_1 is labeled by 1, since both sets declared in 3b are empty. Suppose that we have labeled vertices $v_1, \ldots, v_{i-1}$. When we assign a label to v_i, then by (i) it has at most $3\omega(G) - 3$ labeled neighbors and at most $9\omega(G) - 9$ labels are unavailable because of these neighbors. (If one neighbor of v_i has label x then labels $x - 1, x$ and $x + 1$ are forbidden for v_i.) By (ii), at most $11\omega(G)$ labeled vertices are at distance two from v_i. Therefore, the number of unavailable labels is at most $20\omega(G) - 9$. Since we have $20\omega(G) - 8$ labels, $c(v_i) \leq 20\omega(G) - 8$. $\square$

Corollary 4. *The approximation ratio of the algorithm RDL is bounded by $\frac{32}{3} \doteq 10, 67$ and tends to 10 as $\omega(G)$ grows to infinity.*

Finally, we observe that for the construction of on-line algorithms with constant competitive ratio on unit disk graphs the knowledge of geometrical representation is crucial. For the coloring problem similar robust algorithm on unit disk graphs can be turned into on-line coloring algorithm (with worse competitive ratio). However this is not the case for the labeling problem. The main reason why RDL cannot be turned into "first-fit" algorithm is that at the moment when we have to select a suitable label for vertex v_i we need information about all vertices from the set $\{v_1, \ldots, v_{i-1}\}$ that are at distance two from v_i in G. Unfortunately this information cannot be fully derived from G restricted onto $\{v_1, \ldots, v_i\}$, moreover any robust on-line algorithm should use at least $|V(G)| - 1$ labels, even if G is an independent set: Every pair of vertices, except maybe the last one, should get different labels because there is always a possibility that the last vertex will be adjacent to both of them and they will be at distance two.

Acknowledgments. We thank the referee for helpful comments and suggestions.

References

1. H. L. BODLAENDER, T. KLOKS, R. B. TAN, AND J. VAN LEEUWEN, *λ-coloring of graphs*, in Proceedings of the 17th Annual Symp. on Theoretical Aspects of Computer Science (STACS 2000), H. Reichel and S. Tison, eds., Springer Verlag, Lecture Notes in Computer Science, vol. 1770, 2000, pp. 395–406.
2. H. BREU AND D. G. KIRKPATRICK, *Unit disc graph recognition is NP-hard*, Computational Geometry: Theory and Applications, 9 (1998), pp. 3–24.
3. G. J. CHANG AND D. KUO, *The L(2, 1)-labeling problem on graphs*, SIAM Journal of Discrete Mathematics, 9 (1996), pp. 309–316.
4. B. N. CLARK, C. J. COLBOURN, AND D. S. JOHNSON, *Unit disk graphs.*, Discrete Math., 86 (1990), pp. 165–177.
5. personal communication to Thomas Erlebach.
6. J. FIALA, J. KRATOCHVÍL, AND T. KLOKS, *Fixed-parameter tractability of λ-colorings*, in Graph-Theoretical Concepts in Computer Science, 25th WG '99, Ascona, no. 1665 in Lecture Notes in Computer Science, Springer Verlag, 1999, pp. 350–363.
7. J. P. GEORGES AND D. W. MAURO, *On the size of graphs labeled with a condition at distance two*, Journal of Graph Theory, 22 (1996), pp. 47–57.
8. J. R. GRIGGS AND R. K. YEH, *Labelling graphs with a condition at distance 2*, SIAM Journal of Discrete Mathematics, 5 (1992), pp. 586–595.
9. A. GYÁRFÁS AND J. LEHEL, *On-line and first fit colourings of graphs*, Jornal of Graph Theory, 12 (1988), pp. 217–227.
10. W. K. HALE, *Frequency assignment: Theory and applications*, Proc. of the IEEE, 68 (1980), pp. 1497–1514.
11. P. HLINĚNÝ AND J. KRATOCHVÍL, *Representing graphs by disks and balls.* to appear in Discrete Math.
12. R. A. LEESE, *Radio spectrum: a raw material for the telecommunications industry.* 10th Conference of the European Consortium for Mathematics in Industry, Goteborg, 1998.
13. E. MALESIŃSKA, *Graph theoretical models for frequency assignment problems*, PhD thesis, Technical University of Berlin, 1997.
14. R. PEETERS, *On coloring j-unit sphere graphs*, tech. rep., Dept. of Economics, Tilburg University, 1991.
15. V. RAGHAVAN AND J. SPINRAD, *Robust algorithms for restricted domains* manuscript submitted to Special Issue of Journal of Algorithms.
16. J. VAN DEN HEUVEL, R. A. LEESE, AND M. A. SHEPHERD, *Graph labeling and radio channel assignment*, Journal of Graph Theory, 29 (1998), pp. 263–283.

Approximate Distance Labeling Schemes
(Extended Abstract)

Cyril Gavoille[1], Michal Katz[2], Nir A. Katz[2],
Christophe Paul[1], and David Peleg[3]

[1] LaBRI, Université Bordeaux I, 351, cours de la Libération, 33405 Talence Cedex,
France. {gavoille,paul}@labri.fr.
[2] Dept. of Applied Mathematics, Bar Ilan University, Ramat Gan, 52900, Israel.
[3] Dept. of Computer Science and Applied Mathematics, The Weizmann Institute of
Science, Rehovot, 76100 Israel. peleg@wisdom.weizmann.ac.il.

Abstract. We consider the problem of labeling the nodes of an n-node
graph G with short labels in such a way that the distance between
any two nodes u, v of G can be approximated efficiently (in constant
time) by merely inspecting the labels of u and v, without using any
other information. We develop such constant approximate distance labeling schemes for the classes of trees, bounded treewidth graphs, planar
graphs, k-chordal graphs, and graphs with a dominating pair (including
for instance interval, permutation, and AT-free graphs). We also establish lower bounds, and prove that most of our schemes are optimal in
terms of the length of the labels generated and the quality of the approximation.

Keywords: Local representations, approximate distance, labeling
schemes, distributed data structures.

1 Introduction

1.1 Motivation

Common network representations are usually global in nature; in order to derive
a useful piece of information, one must have access to a global data structure
representing the entire network, even if the sought piece of information is local,
pertaining to only few nodes.

In contrast, the notion of *labeling schemes* [2,1,12,17,11] [10] involves using
a more *localized* representation scheme for the network. The idea is to label the
nodes in a way that will allow one to infer information concerning any two nodes
directly from their labels, without using *any* additional information sources.

Clearly, for such a labeling scheme to be useful, it should use relatively *short*
labels (say, of length polylogarithmic in n), and yet allow efficient (say, polylogarithmic time) information deduction. Recently, this natural idea was studied
with respect to capturing *distance* information. This has led to the notion of
distance labeling schemes, which are schemes possessing the ability to determine
the distance between two nodes efficiently (i.e., in polylogarithmic time) given

F. Meyer auf der Heide (Ed.): ESA 2001, LNCS 2161, pp. 476–487, 2001.

their labels [17]. The applicability of distance labeling schemes in the context of communication networks has been illustrated in [17], and various aspects of such schemes were further studied in [10].

Observe that efficient *exact* distance labeling schemes may not exist for *every* graph family. In particular, for a family of $\Omega(\exp(n^{1+\epsilon}))$ non-isomorphic n-vertex graphs, any distance labeling scheme must use labels whose total combined length is $\Omega(n^{1+\epsilon})$, hence at least one label must be of $\Omega(n^\epsilon)$ bits. Specifically, for the class of all unweighted graphs, any distance labeling scheme must label some n-vertex graphs with labels of size $\Omega(n)$ [10].

This raises the natural question of whether more efficient labeling schemes be constructed if we abandon the ambitious goal of capturing *exact* information, and settle for obtaining *approximate* estimates. This leads to the notion of *approximate distance* labeling schemes, which are the topic of the current paper.

1.2   Labeling Schemes for Approximate Distance

Let us define the notion of approximate distance labeling schemes more precisely. Given a connected undirected graph G and two nodes u and v, let $d_G(u, v)$ denote the distance between u and v in G. A *node-labeling* for the graph G is a non-negative integer function L that assigns a label $L(u, G)$ (in the form of a binary string) to each node u of G.

A *distance decoder* is an integer function f responsible for distance computation; given two labels λ_1, λ_2 (not knowing which graph they are taken from), it returns $f(\lambda_1, \lambda_2)$. We say that the pair $\langle L, f \rangle$ is a *distance labeling* for G if $f(L(u, G), L(v, G)) = d_G(u, v)$ for any pair of nodes $u, v \in V(G)$. We say that $\langle L, f \rangle$ is an (s, r)-*approximate* distance labeling for G if

$$d_G(u, v) \;\leqslant\; f(L(u, G), L(v, G)) \;\leqslant\; s \cdot d_G(u, v) + r \qquad (1)$$

for any pair of nodes $u, v \in V(G)$. More generally, $\langle L, f \rangle$ is an (s, r)-approximate *distance labeling scheme* for the graph family $\mathcal{G}$ if it is an (s, r)-approximate distance labeling for every graph $G \in \mathcal{G}$. This paper concerns the existence of approximate distance labeling schemes which use short labels. Let $|L(u, G)|$ denote the length of the binary label $L(u, G)$ associated with u in $G \in \mathcal{G}$, and denote

$$\ell_{\langle L, f \rangle}(\mathcal{G}) \;=\; \max_{G \in \mathcal{G}} \max_{u \in V(G)} |L(u, G)| \quad \text{and} \quad \ell_{(s, r)}(\mathcal{G}) \;=\; \min_{\langle L, f \rangle} \left\{ \ell_{\langle L, f \rangle}(\mathcal{G}) \right\}$$

where the minimum is taken over all (s, r)-approximate distance labeling schemes $\langle L, f \rangle$ for $\mathcal{G}$.

We focus on two important special cases of (s, r)-approximate distance labeling schemes. An $(s, 0)$-approximate distance labeling scheme is referred to as a s-*multiplicative* distance labeling scheme. Analogously, a $(1, r)$-approximate distance labeling scheme is referred to as an r-*additive* distance labeling scheme. A 0-additive (or 1-multiplicative) distance labeling is termed an *exact* distance labeling scheme.

One may consider two variations on the distance labeling definition: the *distinct* label model and the *non-distinct* label model. The latter allows two different vertices to have the same labels, and thus lets us build schemes with sublogarithmic label size. In the extreme case, graphs of constant diameter D enjoy a D-additive distance labeling scheme with no labels at all in the non-distinct label model: it suffices to return the estimate D to satisfy Eq. (1). In contrast, $\log n$ is a lower bound on the size of labels in the distinct label model, even in the case of an n-node clique. We remark that any distance labeling scheme in the non-distinct label model can be transformed into a scheme in the distinct label model (with the same estimate quality) by adding a unique $\log n$ bit identifier to each of the nodes. The notations $\ell_{\langle L,f \rangle}(\mathcal{G})$ and $\ell_{(s,r)}(\mathcal{G})$ are defined for the non-distinct label model, and thus differ from the distinct label model by an additive $\log n$ factor.

We are also interested in the *query time*, i.e., the worst-case time complexity of the distance decoder. We assume a ω-bit word RAM model of computation. Each memory word can contain an integer in the range $[0, 2^\omega)$. The instruction set available consists of standard arithmetic, bitwise logical and comparison operations on words, all assumed to require constant time. For the use of bucketing and perfect hashing functions, we assume that the arithmetic operation set contains integer multiplication and integer division. Denoting the size of the input graph by n, we assume that $\omega \geqslant \log n$.

1.3   Related Work

An exact distance labeling scheme for weighted n-node trees with integral weights from the range $[0, W)$ using $O(\log^2 n + \log n \log W)$ bit labels has been given in [17], and $O(\log^2 n)$ bit labeling schemes for interval graphs and permutation graphs were presented in [11], all with $O(\log n)$ query time. The *Squashed Cube Conjecture* of Graham and Pollak [8], proved by Winkler [19], implies an exact distance labeling scheme for arbitrary n-node graphs with $n \log_2(3) \approx 1.58n$ bit labels for general graphs, although with a prohibitive $\Theta(n)$ query time to decode the distance. With a different approach, [10] presented a distance labeling scheme with label size $11n + o(n)$ and with $O(\log \log n)$ query time. An $8k$-multiplicative distance labeling scheme, for each integer $k \geqslant 1$, is built in [17], using $O(k\, n^{1/k} \log n\, \log W)$ bit labels with query time linear in the label size, where W stands for the integral weighted diameter. For unweighted n-node graphs, this yields a $O(\log n)$-multiplicative distance labeling scheme using $O(\log^3 n)$ bit labels with polylog query time.

Some bounds on the size of the labels are established in [10] for several families of n-node graphs. In particular, it is shown that $\ell_{(s,0)}(\mathcal{G}_n) = \Omega(n)$ for every $s < 3$, where $\mathcal{G}_n$ denotes the family of all connected graphs. For the family $\mathcal{G}_{r(n),n}$ of graphs of treewidth bounded by $r(n)$, it is shown that $\ell_{(s,0)}(\mathcal{G}_{r(n),n}) = \Omega(r(n))$, for all $s < 3$ and $r(n) < n/2$. On the other hand, an exact distance labeling scheme is proposed for $\mathcal{G}_{r(n),n}$, with $O(R(n) \log n)$ bit labels and $O(R(n))$ query time, where $R(n) = \sum_{i=1}^{\log n} r(n/2^i)$. (We have $R(n) \leqslant r(n) \log n$, and for monotone $r(n) \geqslant n^\epsilon$ with constant $\epsilon > 0$, $R(n) = O(r(n))$.) It follows, for instance, that planar graphs support an exact distance labeling scheme with $O(\sqrt{n}\log n)$ bit labels, and that trees and bounded treewidth graphs support exact distance labeling schemes with $O(\log^2 n)$ bit labels.

It is also proved in [10] that $\ell_{(1,0)}(\mathcal{B}_n) = \Omega(\sqrt{n})$ for the family $\mathcal{B}_n$ of bounded degree graphs, and that $\ell_{(1,0)}(\mathcal{P}_n) = \Omega(n^{1/3})$ for the family $\mathcal{P}_n$ of bounded degree planar graphs. (For the family $\mathcal{P}_{W,n}$ of weighted bounded degree planar graphs with weights in the range $[0, \sqrt{n}]$, it is proved that $\ell_{(1,0)}(\mathcal{P}_{W,n}) = \Omega(\sqrt{n})$.) Finally, for the family $\mathcal{T}_n$ of binary trees, it is shown that $\ell_{(1,0)}(\mathcal{T}_n) \geqslant \frac{1}{8}\log^2 n - O(\log n)$. (For the family $\mathcal{T}_{W,n}$ of weighted binary trees with weights in the range $[0, W)$, it is proved that $\ell_{(1,0)}(\mathcal{T}_{W,n}) \geqslant \frac{1}{2}\log n \log W - O(\log W)$.)

1.4   Our Contribution

Section 2 deals with bounds on s-multiplicative distance labeling schemes. It is first shown that planar graphs have a 3-multiplicative distance labeling scheme with $O(n^{1/3}\log n)$ bit labels. This should be contrasted with the upper and lower bounds of [10] mentioned above for *exact* distance labeling schemes for planar graphs.

Then, we propose a $(1 + 1/\log W)$-multiplicative distance labeling scheme using $O(\log n \cdot \log\log W)$ bit labels for the family of weighted n-node trees with weighted diameter bounded by W. More generally, we show that the family of n-node graphs of treewidth at most $r(n)$ and weighted diameter bounded by W has a $(1 + 1/\log W)$-multiplicative distance labeling scheme using $O(R(n)\log\log W)$ bit labels and with $O(r(n))$ query time, assuming that $W = n^{O(1)}$. Hence, unweighted trees (and bounded treewidth graphs) enjoy a $(1 + 1/\log n)$-multiplicative distance labeling scheme using $O(\log n \cdot \log\log n)$ labels with constant query time.

We then turn to lower bounds on s-multiplicative distance labeling schemes. We establish, for every $s \geqslant 1$, a lower bound on such schemes on general n-node graphs. Specifically, it is shown that for general graphs, $\Omega(n^{1/(3s/4-O(1))})$ bit labels are required in the worst-case. The current upper bound is $O(s\, n^{1/\lceil s/8 \rceil}\log^2 n)$, derived from the result of [17] mentioned previously. Also, for trees, we show a lower bound of $\Omega(\log n \cdot \log\log n)$ bit labels for every $(1 + 1/\log n)$-multiplicative scheme, thus proving that the scheme establishing our upper bound is optimal. For the class of weighted trees, whose weights are in the range $[0, Z)$, we show for sufficiently large Z a lower bound of $\Omega(\log n \cdot \log\log Z)$ for $(1 + 1/\log Z)$-multiplicative schemes (which is again an optimal bound). For lack of space, the lower bounds for trees are not presented in this abstract. Full proofs can be founded in [9].

In Section 3.1, we turn to r-additive distance labeling schemes. We consider the family of k-chordal n-node graphs of diameter D, and show that they enjoy a $\lfloor k/2 \rfloor$-additive distance labeling scheme with labels of size $O(\log n \log D)$. In particular, the family of n-node chordal graphs has an 1-additive distance labeling scheme with $O(\log^2 n)$ bit labels. Moreover, these schemes are polynomial-time constructible, and have $O(1)$ query time. We then consider the class of graphs of diameter D which have a dominating pair. This class includes, for instance, AT-free, permutation and interval graphs. We show that this class enjoys a 2-additive scheme, and even a 1-additive scheme for interval graphs, with $\log D + O(1)$ bit labels in the non-distinct label model, and $O(\log n)$ bit labels otherwise.

We then turn to lower bounds for r-additive distance labeling schemes. We show, for every $r \geqslant 0$, a lower bound of $\Omega(\sqrt{n/(r+1)})$ on the required label size for r-additive distance labeling schemes over general n-node graphs. The bound is $\Theta(n)$ if $r < 2$. We also show a $\Omega(\log^2(n/(r+1)))$ lower bound on the label size for r-additive schemes on k-chordal graphs, proving that our scheme for k-chordal graphs is optimal in the label size and in the quality of the approximation. We also notice that exact distance labeling schemes for AT-free graphs or k-chordal graphs require $\Omega(n)$ bit labels.

2   Multiplicative Approximate Schemes

2.1   A Scheme for Planar Graphs

This section presents a 3-multiplicative distance labeling scheme with $O(n^{1/3} \log n)$ bit labels for the family of planar graphs. Let us start with some background concerning partitions of planar graphs. A *region* is a connected subgraph of a planar graph. One can distinguish two kinds of nodes: internal (belonging to only one region) and boundary nodes (that belong to two or more regions). The following decomposition lemma has been established in [5] using the $O(\sqrt{n})$-separator theorem.

Lemma 1. [5] *For every n-node planar graph G and integer $k > 0$, it is possible (in polynomial time) to partition the nodes of G into k regions, each of $O(n/k)$ nodes and with $O(\sqrt{n/k})$ boundary nodes, such that any path connecting an internal node of one region to an internal node of another must go through at least one boundary node of each region.*

Setting $k = \lceil n^{1/3} \rceil$, we get $O(n^{1/3})$ regions $R_1, \ldots, R_k$, each with $O(n^{2/3})$ nodes and $O(n^{1/3})$ boundary nodes. Each region R_i is partitioned again into $O(n^{1/3})$ subregions S_j of $O(n^{1/3})$ (internal and boundary) nodes. Each node is given a unique integer identifier $I(v) \in \{1, \ldots, n\}$, as well as a pair $r(v) = (i, j)$, indicating its region number i and its subregion number j. (For boundary nodes, choose a valid pair (i, j) arbitrarily.)

Consider a node u that belongs to a region R_i and to the subregion S_j within R_i. For every other region $R_{i'}$ of G, $i' \neq i$, let $p_u(R_{i'})$ denote the closest node to u in $R_{i'}$. Note that $p_u(R_{i'})$ is necessarily a boundary node in $R_{i'}$. Similarly, for every other subregion $S_{j'}$ of R_i, $j' \neq j$, let $p_u(S_{j'})$ denote the closest node to u in $S_{j'}$.

The label $L(u, G)$ assigned to the node u consists of the following fields:

[a] its identifier $I(u)$ and pair $r(u) = (i, j)$;
[b] the distance from u to all the nodes in S_j;

[c] the identifier of $p_u(R_{i'})$ and the distance $d_G(u, p_u(R_{i'}))$, for every region $R_{i'}$ of G, $i' \neq i$;

[d] the identifier of $p_u(S_{j'})$ and the distance $d_G(u, p_u(S_{j'}))$, for every subregion $S_{j'}$ of R_i, $j' \neq j$;

[e] the distance from u to all the boundary nodes in its region R_i.

The number of bits in the resulting label is bounded by $O(n^{1/3} \log n)$. The distance between u and v is computed from the labels $L(u, G)$ and $L(v, G)$ as follows:

1. Extract the region and subregion numbers of u and v from field [a] of the two labels.

2. If u and v belong to the same subregion S_j in the same region R_i, then do:
   [a] extract $d_G(u, v)$ from field [b] of $L(v, G)$ using $I(v)$;
   [b] set $\tilde{d}(u, v) = d_G(u, v)$.

3. If u and v are in the same region R_i but in different subregions, say $u \in S_j$ and $v \in S_{j'}$, then do:
   [a] extract $z = p_u(S_{j'})$ and $d_G(u, z)$ from field [d] of $L(u, G)$;
   [b] extract $d_G(z, v)$ from field [b] of $L(v, G)$;
   [c] set $\tilde{d}(u, v) = d_G(u, z) + d_G(z, v)$.

4. If u and v belong to different regions, say $u \in R_i$ and $v \in R_{i'}$, then do:
   [a] extract $z = p_u(R_{i'})$ and $d_G(u, z)$ from field [c] of $L(u, G)$;
   [b] extract $d_G(z, v)$ from field [e] of $L(v, G)$;
   [c] set $\tilde{d}(u, v) = d_G(u, z) + d_G(z, v)$.

5. Return $\tilde{d}(u, v)$.

Theorem 1. *There exists a 3-multiplicative distance labeling scheme using labels of size $O(n^{1/3} \log n)$ for the family of n-node planar graphs. Moreover, the labels are polynomial time constructible and the distance decoder is $O(1)$-time complexity.*

(All proofs are omitted and can be founded in [9].)

2.2   A Scheme for Trees and Bounded Treewidth Graphs

A pair of integer functions $\langle \lambda, \phi \rangle$ is an (s, r)-*estimator* of $\{1, \ldots, W\}$ if $\lambda : \{1, \ldots, W\} \to \{1, \ldots, 2^\alpha\}$ (where typically $2^\alpha \ll W$), $\phi : \{1, \ldots, 2^\alpha\} \to \mathbb{N}$, and for every $x \in \{1, \ldots, W\}$, $x \leqslant \phi(\lambda(x)) \leqslant s \cdot x + r$. Intuitively, we think of λ as a function "compacting" x, and of ϕ as a function attempting to reconstruct x from $\lambda(x)$. The *size* of the estimator is α, and its *time complexity* is the worst-case time complexity of the function ϕ.

Lemma 2. *For every $k \leqslant O(\omega)$, and for every $m \in \{0, \ldots, k\}$, there exists a constant time $(1 + 2^{-m}, 0)$-estimator of $\{1, \ldots, 2^k\}$ of size $\alpha = m + \lceil \log(k - m + 1) \rceil$.*

Given a binary string S of length k, let $\text{rank}_S(i)$ denote the function that returns the number of 1's up to and including position i, for $i \in \{1, \ldots, k\}$, and let $\text{lsb}_k(S)$ denote the position p of the least significant bit in S set to 1, $p \in \{0, \ldots, k\}$

($p = 0$ if S is null). The following two results require constant time integer multiplications and divisions on $O(\log n)$-bit words, which are allowed in our computational model.

Lemma 3.   *1. [16,15] For every integer $k \leqslant n$, and for every binary string S of length k, the operation $\text{rank}_S(\cdot)$ can be performed in $O(1)$ worst-case time with $o(k)$ pre-computed auxiliary bits, and $k + o(k)$ bits in total.*
2. [6] For every integer $k \leqslant O(\log n)$, the operation $\text{lsb}_k(\cdot)$ can be performed in $O(1)$ worst-case time with $O(k)$ pre-computed auxiliary bits.

We state the main result for weighted trees. Given an integer $W > 0$, a *W-tree* is a weighted tree whose weighted diameter is bounded by W. Note that an unweighted tree T of diameter D is a particular case of a W-tree, for $W \geqslant D$.

It is well-known that every n-node tree has a node, hereafter called a *separator*, which splits tree into connected components each of at most $n/2$ nodes.

Let T be any n-node W-tree, and $u \in V(T)$, and let $\langle \lambda, \phi \rangle$ be an (s,r)-estimator of $\{1, \ldots, W\}$. Apply the above separator property to T, and let s_1 be the obtained separator. With each connected component F of $T \setminus \{s_1\}$, we associate a unique label $c(F) \in \{1, \ldots, \deg(s_1)\}$ such that for every two components A, B of $T \setminus \{s_1\}$, if $|V(A)| \geqslant |V(B)|$ then $c(A) \leqslant c(B)$. Let T_1 be the connected component of $T \setminus \{s_1\}$ containing u, and let $c_1 = c(T_1)$.

We recursively apply this decomposition scheme to T_i, $i \geqslant 1$, in order to obtain s_{i+1}, T_{i+1} and c_{i+1} such that $u \in V(T_{i+1})$ and $c_{i+1} = c(T_{i+1})$, until we have $u = s_{i+1}$. Let h be the index such that $u = s_{h+1}$. Note that $h + 1 \leqslant \log n$. We set the label of u in T to be

$$L(u,T) = \langle (c_1, \ldots, c_h), (\lambda(d_T(u, s_1)), \ldots, \lambda(d_T(u, s_h))) \rangle .$$

Given two nodes u, u' of labels $\langle (c_1, \ldots, c_h), (\lambda_1, \ldots, \lambda_h) \rangle$ and $\langle (c'_1, \ldots, c'_{h'}), (\lambda'_1, \ldots, \lambda'_{h'}) \rangle$ respectively, we compute the distance as follows:

1. compute the lowest index i_0 such that $c_{i_0} \neq c'_{i_0}$;
2. return $\phi(\lambda_{i_0}) + \phi(\lambda'_{i_0})$.

Theorem 2. *Let α be the size of an (s,r)-estimator of $\{1, \ldots, W\}$, and let t denote its time complexity. Then the above scheme is an $(s, 2r)$-approximate distance labeling scheme using labels of size $\alpha \log n + O(\log n)$ for the family of n-node W-trees. Moreover, the distance decoder is $O(t)$-time complexity.*

In particular, choosing $k = \lceil \log W \rceil$ and $m = \lceil \log \log W \rceil$ in Lemma 2, we have the following (noting that if $W = n^{O(1)}$, then $k = O(\omega)$ and thus all the distances can be estimated in constant time).

Corollary 1. *1. There exists a $(1 + 1/\log W)$-multiplicative distance labeling scheme using labels of size $O(\log n \cdot \log \log W)$ for the family of n-node W-trees. Moreover, the distance decoder is $O(1)$-time complexity for $W = n^{O(1)}$.*
2. There exists a $(1 + 1/\log n)$-multiplicative distance labeling scheme using labels of size $O(\log n \cdot \log \log n)$ for the family of n-node unweighted trees. Moreover, the distance decoder is $O(1)$-time complexity.

The scheme for trees can be applied together with the tree-decomposition scheme for bounded treewidth graphs [18] (see also Section 3.1). In [10] it is shown that graphs of treewidth bounded by $r(n)$ have exact distance labeling scheme using $O(R(n)\log n)$ bit labels, for every monotone function $r(n)$, and with $O(R(n))$ query time, where $R(n) = \sum_{i=0}^{\log n} r(n/2^i)$. Using the above scheme for trees, this result can be extended as follows.

Corollary 2. *There exists a $(1+1/\log W)$-multiplicative (resp., exact) distance labeling scheme using labels of size $O(R(n)\log\log W)$ (resp., $O(R(n)\log W)$) for the family of weighted n-node graphs of treewidth at most $r(n)$ and of weighted diameter W. Moreover, the distance decoder is $O(r(n))$-time complexity for $W = n^{O(1)}$.*

It follows that n-node trees and bounded treewidth graphs enjoy exact distance labeling with $O(\log^2 n)$ bit labels with constant query time (improving the query time of the scheme of [17]).

2.3   A Lower Bound for Multiplicative Schemes

To prove a lower bound on s-multiplicative distance labeling schemes on general graphs, we need the following concept introduced in [10]. For every graph family $\mathcal{F}$ under consideration, we assume that each n-node graph of $\mathcal{F}$ is a labeled graph on the set of nodes $V_n = \{1, \ldots, n\}$. Let $A \subseteq V_n$, and let $k > 1$ be a real number (k can be a function of n). $\mathcal{F}$ is an (A, k)-*family* if for every two distinct graphs $G, H \in \mathcal{F}$ there exist $x, y \in A$ such that $d_G(x, y) \geqslant k \cdot d_H(x, y)$ or $d_H(x, y) \geqslant k \cdot d_G(x, y)$. The following lemma is shown in [10] and is useful to prove Lemma 5.

Lemma 4. [10] *Let $\mathcal{F}$ be an (A, k)-family for $k > 1$. Then for any $s < k$, $\ell_{(s,0)}(\mathcal{F}) \geqslant (\log |\mathcal{F}|)/|A|$.*

Lemma 5. *Let G be any connected graph with n nodes, m edges and girth g. Then the family $\mathcal{S}_G$ composed of all the n-node connected subgraphs of G satisfies, for every real number $1 \leqslant s < g - 1$, $\ell_{(s,0)}(\mathcal{S}_G) \geqslant m/n - O(1)$.*

In particular, Lemma 5 implies that, for every $s < g-1$, $\ell_{(s,0)}(\mathcal{G}_n) \geqslant m(g,n)/n - O(1)$, where $m(g, n)$ denotes the maximum number of edges in an n-node graphs of girth g. In [4] it is conjectured that $m(2k + 2, n) = \Omega(n^{1+1/k})$, proving later for $k = 1, 2, 3$ and 5. In [14], it is shown that $m(4k + 2, n) = \Omega(n^{1+1/(3k-1)})$ and that $m(4k, n) = \Omega(n^{1+1/(3(k-1))})$. Therefore,

Theorem 3. *For every $s \geqslant 1$, $\ell_{(s,0)}(\mathcal{G}_n) = \Omega\left(n^{1/(3s/4 - O(1))}\right)$.*

3   Additive Approximate Schemes

3.1   A Scheme for k-Chordal Graphs

A graph is k-*chordal* if it does not contain any chordless cycles of length larger than k (where a *chord* is an edge joining two non-neighbors of the cycle). *Chordal* graphs are exactly 3-chordal graphs.

We use the notion of tree-decomposition used by Roberston and Seymour in their work on graphs minors [18]. A *tree-decomposition* of a graph G is a tree T whose nodes are subsets of $V(G)$, such that

1. $\bigcup_{X \in V(T)} X = V(G)$;
2. for every $\{u, v\} \in E(G)$, there exists $X \in V(T)$ such that $u, v \in X$; and
3. for all $X, Y, Z \in V(T)$, if Y is on the path from X to Z in T, then $X \cap Z \subseteq Y$.

For every $S \subseteq V(G)$, denote $\text{diam}_G(S) = \max_{u,v \in S} d_G(u, v)$. We denote by $G \setminus S$ the subgraph of G induced by the set of nodes $V(G) \setminus S$.

S is a *separator* of G if $G \setminus S$ is composed of two or more connected components. Moreover, S is said to be *minimal* if every proper subset of S if not a separator of G. Given $x, y \in V(G)$, S is an x, y-*separator* if x and y belongs to two distinct connected components in $G \setminus S$. We have the following.

Lemma 6. *For every k-chordal graph G, there exists a tree-decomposition T such that $|V(T)| \leq |V(G)|$, and such that, for every $X \in V(T)$, $\text{diam}_G(X) \leq k/2$. Moreover, T is polynomial-time constructible.*

Using the scheme for trees, we have the following result:

Theorem 4. *There exists a $\lfloor k/2 \rfloor$-additive distance labeling scheme with labels of size $O(\log n \log D)$ for the family of k-chordal n-node graphs with (weighted) diameter D. Moreover, the scheme is polynomial-time constructible, and the distance decoder is $O(1)$-time complexity.*

Corollary 3. *There exists a 1-additive distance labeling scheme with labels of size $O(\log^2 n)$ for the family of n-node chordal graphs. Moreover, the scheme is polynomial-time constructible, and the distance decoder is $O(1)$-time complexity.*

The previous bound is optimal with respect to both the approximation ratio and the label size. Indeed, as seen later in Section 3.3, every 1-additive distance labeling scheme on the family of trees (that are chordal) requires some labels of size $\Omega(\log^2 n)$. Moreover, every exact distance labeling scheme on chordal graphs of diameter 3 requires label of size $\Omega(n)$ (cf. Theorem 6).

3.2   Dominating Pair

Our next goal is to show that in certain cases (including for instance interval, permutation, and AT-free graphs), Theorem 4 can be improved upon. A pair of nodes $\{x, y\}$ in a graph G is called a *dominating pair* if for every path P in G between x and y, and for every $u \in V(G)$, there exists a node of P at distance at most 1 of u (we say that P dominates u).

Consider any graph G having a dominating pair $\{x, y\}$, and let $\delta = d_G(x, y)$. The distance labeling scheme for graphs having a dominating pair is based on the decomposition by distance layers from node x. Let us define $L_t = \{u \in V(G) \mid d_G(x, u) = t\}$, for every $t \geq 0$. Note that for every $0 < t < \delta$, L_t is a x, y-separator. Let S_t denote a minimal x, y-separator in L_t, for every $0 < t < \delta$. We set $S_0 = \{x\}$, $S_\delta = \{y\}$ and $S_{\delta+1} = \varnothing$. An x, y-*path* is a loop-free path between x and y. We establish a number of claims useful for our analysis.

Claim. Let C_x and C_y be the connected components of $G \setminus S_t$, for $0 < t < \delta$, containing respectively x and y. For any node $z \in S_t$, there exists an x, y-path contained in $C_x \cup C_y \cup \{z\}$, composed of a shortest x, z-path and of an induced z, y-path (i.e., a chordless z, y-path) called hereafter an x, z, y-*path*.

Claim. (1) Let t, t' such that $0 < t \leqslant t' < \delta$. Let P be a x, z, y-path with $z \in S_t$. Then, every $w' \in L_{t'} \setminus S_{t'}$ has no neighbors in $P \cap L_{t'+1}$.
(2) For every $t \geqslant 1$, and every $w \in L_t$, w has a neighbor in S_{t-1}.

Let us now consider two distinct nodes u and v, and let $t_u = d_G(x, u)$ and let where $t_v = d_G(x, v)$. W.l.o.g., assume that $t_u \leqslant t_v$. The next claim gives approximations of $d_G(u, v)$ depending on the respective positions of u and v in the distance layer decomposition. It allows us to prove Theorem 5.

Claim. (1) If $t_u = 0$, then $d_G(u, v) = t_v$.
(2) If $0 < t_u = t_v \leqslant \delta + 1$, then $1 \leqslant d_G(u, v) \leqslant 2$.
(3) If $0 < t_u < t_v \leqslant \delta + 1$, then $t_v - t_u \leqslant d_G(u, v) \leqslant t_v - t_u + 2$.

Theorem 5. *Let G be a graph of diameter D with a dominating pair. Then G has a 2-additive distance labeling with labels of size $\log D + O(1)$. Moreover, the distance decoder is $O(1)$-time complexity.*

An *asteroidal triple* of a graph G is an independent triple of nodes, each two of which are joined by a path of G that avoids the neighborhood of the third node [13]. Graphs without asteroidal triple are termed *AT-free graphs*. This class includes, in particular, the classes of interval, permutation, bounded tolerance and co-comparability graphs (see [7]). It is well-known that AT-free graphs have a dominating pair [3] that can be founded in linear time. We thus have the following.

Corollary 4. *There exists a 2-additive distance labeling scheme with labels of size $\log D + O(1)$ for the family of AT-free graphs of diameter D. Moreover, the scheme is polynomial-time constructible, and the distance decoder is $O(1)$-time complexity.*

To get a 1-additive scheme for a subclass of AT-free, we examine in more detail the situation described by the last claim in case (3), namely, $t_u < t_v$, i.e., $d_G(x, u) < d_G(x, v)$. Indeed, thanks to the case (1) of the same claim, we already have a 1-additive scheme for $t_u = t_v$. We can show (details can be founded in [9]):

Lemma 7. *Let G be a graph that has a dominating pair $\{x, y\}$. For any two nodes u and v of G, with $d_G(x, u) \leqslant d_G(x, v)$, if u does not belong two a chordless cycle C_k, $k \geqslant 4$, then $d_G(u, v) \leqslant \tilde{d}(u, v) \leqslant d_G(u, v) + 1$.*

Since interval graphs are 3-chordal AT-free graphs, the following corollary holds:

Corollary 5. *There exists a 1-additive distance labeling scheme with labels of size $\log D + O(1)$ for interval graphs of diameter D. Moreover, the scheme is polynomial-time constructible, and the distance decoder is $O(1)$-time complexity.*

3.3   Lower Bounds for Additive Schemes

In this section we establish some lower bounds on r-additive distance labeling schemes. For general graphs we show that labels of size $\Omega(\sqrt{n/r})$ are required in the worst-case (the lower bound can be improved to $\Theta(n)$ for $r < 2$), and that every r-additive distance labeling on k-chordal graphs must use labels of length $\Omega(\log^2 n)$, for every constant $r \geqslant 0$. Moreover, there is no exact distance labeling scheme with labels shorter than $\Omega(n)$ bits, proving that the scheme presented in Section 3.1 is optimal (cf. Theorem 4 and Corollary 3).

Let us now show that the labeling scheme of Section 3.1 for k-chordal graphs with bounded k is optimal in terms of the length of its labels. First, let us show that there is no exact distance labeling scheme for chordal graphs using "short" labels. Consider the family $\mathcal{S}_n$ of connected *split* graphs, namely, all the n-node graphs composed of a clique C, and of an independent set I of $n - |C|$ nodes, such that each node of I is connected to at least one node of C. Let $\mathcal{C}_{n,k}$ be the class of connected n-node k-chordal graphs. Clearly, $\mathcal{S}_n$ is a subclass of $\mathcal{C}_{n,3}$ (because every cycle of 4 or more nodes has a chord belonging to C) and $\mathcal{S}_n$ is connected (in fact, it contains only graphs of diameter at most three).

Recall that $\ell_{(s,r)}(\cdot)$ is related to the non-distinct label model.

Theorem 6. *For every $k \geqslant 3$ and $s < 2$, $\ell_{(s,0)}(\mathcal{C}_{n,k}) \geqslant \ell_{(s,0)}(\mathcal{S}_n) \geqslant n/4 - O(1)$. On the other hand, the family $\mathcal{S}_n$ supports a 1-additive distance labeling scheme with 1 bit labels (or $\lceil \log n \rceil + 1$ bit label, if we insist on distinct labels).*

Similarly, we have the following for AT-free graphs.

Theorem 7. *Let $\mathcal{A}_n$ be the class of connected n-node AT-free graphs. For every $s < 2$, $\ell_{(s,0)}(\mathcal{A}_n) \geqslant n/4 - O(1)$.*

We now give a bound of $\Omega(\sqrt{n/r})$ on the label size for general graphs. We start with a simple observation: $\ell_{(s,r)}(\mathcal{F}) \geqslant \ell_{(s+r,0)}(\mathcal{F})$, for all $s \geqslant 1$, $r \geqslant 0$, and graph family $\mathcal{F}$. Thus, for general graphs, since $\ell_{(s,0)}(\mathcal{G}_n) = \Theta(n)$ for every $s < 3$ (cf. [10]), we have $\ell_{(1,r)}(\mathcal{G}_n) = \Theta(n)$ for every $r < 2$.

We complete Theorem 3, by showing that even for $r > \lfloor k/2 \rfloor$, every r-additive scheme for k-chordal graphs requires labels of length $\Omega(\log^2 n)$. Let $\mathcal{T}$ be the family of trees, and let $\mathcal{T}_n \subset \mathcal{T}$ denote n-node trees. Since $\mathcal{T}_n \subseteq \mathcal{C}_{n,k}$, it suffices to show that $\ell_{(1,r)}(\mathcal{T}_n) = \Omega(\log^2 n)$ for every constant r. Using subdivision of edges, one can easily show:

Corollary 6. *For every $r \geqslant 0$, $\ell_{(1,r)}(\mathcal{T}_n) = \Omega(\log^2(n/(r+1))$, and thus, for every $k \geqslant 3$, $\ell_{(1,r)}(\mathcal{C}_{n,k}) = \Omega(\log^2(n/(r+1)))$.*

We also improve the lower bound on r-additive schemes on general graphs.

Corollary 7. *$\ell_{(1,r)}(\mathcal{G}_n) = \Theta(n)$ if $r < 2$, and $\ell_{(1,r)}(\mathcal{G}_n) = \Omega(\sqrt{n/r})$, if $r \geqslant 2$.*

Acknowledgment. We would like to thank Stéphane Pérennes for his help in developing the upper bound for trees, and Douglas West for his bibliographic support.

References

1. M. A. Breuer and J. Folkman. An unexpected result on coding the vertices of a graph. *J. of Mathematical Analysis and Applications*, 20:583–600, 1967.
2. M. A. Breuer. Coding the vertexes of a graph. *IEEE Trans. on Information Theory*, IT-12:148–153, 1966.
3. D. G. Corneil, S. Olariu, and L. Stewart. Asteroidal triple-free graphs. *SIAM Journal on Discrete Mathematics*, 10(3):399–430, Aug. 1997.
4. P. Erdös, A. Rényi, and V. T. Sós. On a problem of graph theory. In *Studia Sci. Math. Hungar.*, vol. 1, pp. 215–235, 1966.
5. G. N. Frederickson and R. Janardan. Efficient message routing in planar networks. *SIAM Journal on Computing*, 18(4):843–857, Aug. 1989.
6. M. L. Fredman and D. E. Willard. Surpassing the information theoric bound with fusion trees. *J. of Computer and System Sciences*, 47:424–436, 1993.
7. M. C. Golumbic. *Algorithmic Graph Theory and Perfect Graphs*. Academic Press, Harcourt Brace Jovanovich, Academic Press edition, 1980.
8. R. L. Graham and H. O. Pollak. On embedding graphs in squashed cubes. *Lecture Notes in Mathematics*, 303:99–110, 1972.
9. C. Gavoille, M. Katz, N. A. Katz, C. Paul, and D. Peleg. Approximate distance labeling schemes. TR RR-1250-00, LaBRI, University of Bordeaux, 351, cours de la Libération, 33405 Talence Cedex, France, Dec. 2000.
10. C. Gavoille, D. Peleg, S. Pérennes, and R. Raz. Distance labeling in graphs. In *12th Symposium on Discrete Algorithms (SODA)*, pp. 210–219. ACM-SIAM, Jan. 2001.
11. M. Katz, N. A. Katz, and D. Peleg. Distance labeling schemes for well-separated graph classes. In *17th Annual Symposium on Theoretical Aspects of Computer Science (STACS)*, vol. 1770 of LNCS, pp. 516–528. Springer, Feb. 2000.
12. S. Kannan, M. Naor, and S. Rudich. Implicit representation of graphs. In *20th Annual ACM Symposium on Theory of Computing (STOC)*, pp. 334–343, Chicago, IL, May 1988.
13. C. G. Lekkerkerker and J. Ch. Boland. Representation of a finite graph by a set of intervals on the real line. *Fund. Math.*, 51:45–64, 1962.
14. F. Lazebnik, V. A. Ustimenko, and A. J. Woldar. A new series of dense graphs of high girth. *Bulletin of American Mathematical Society (New Series)*, 32(1):73–79, 1995.
15. J. I. Munro and V. Raman. Succinct representation of balanced parentheses, static trees and planar graphs. In *38th Symposium on Foundations of Computer Science (FOCS)*, pp. 118–126. IEEE Comp. Society Press, 1997.
16. J. I. Munro. Tables. In *16th FST&TCS*, vol. 1180 of LNCS, pp. 37–42. Springer-Verlag, 1996.
17. D. Peleg. Proximity-preserving labeling schemes and their applications. In *25th International Workshop, Graph - Theoretic Concepts in Computer Science (WG)*, vol. 1665 of LNCS, pp. 30–41. Springer, June 1999.
18. N. Robertson and P. D. Seymour. Graph minors. II. Algorithmic aspects of tree-width. *Journal of Algorithms*, 7:309–322, 1986.
19. P. Winkler. Proof of the squashed cube conjecture. *Combinatorica*, 3(1):135–139, 1983.

On the Parameterized Complexity of Layered Graph Drawing*

V. Dujmović[1], M. Fellows[2], M. Hallett[1], M. Kitching[1],
G. Liotta[3], C. McCartin[4], N. Nishimura[5], P. Ragde[5],
F. Rosamond[2], M. Suderman[1], S. Whitesides[1], and D.R. Wood[6]

[1] McGill University, Canada.
[2] University of Victoria, Canada.
[3] Università di Perugia, Italy.
[4] Victoria University of Wellington, New Zealand.
[5] University of Waterloo, Canada.
[6] The University of Sydney, Australia.

Abstract. We consider graph drawings in which vertices are assigned to layers and edges are drawn as straight line-segments between vertices on adjacent layers. We prove that graphs admitting crossing-free h-layer drawings (for fixed h) have bounded pathwidth. We then use a path decomposition as the basis for a linear-time algorithm to decide if a graph has a crossing-free h-layer drawing (for fixed h). This algorithm is extended to solve a large number of related problems, including allowing at most k crossings, or removing at most r edges to leave a crossing-free drawing (for fixed k or r). If the number of crossings or deleted edges is a non-fixed parameter then these problems are NP-complete. For each setting, we can also permit downward drawings of directed graphs and drawings in which edges may span multiple layers, in which case the total span or the maximum span of edges can be minimized. In contrast to the so-called Sugiyama method for layered graph drawing, our algorithms do not assume a preassignment of the vertices to layers.

1 Introduction

Layered graph drawing [28,5,26] is a popular paradigm for drawing graphs, and has applications in visualization [6], in DNA mapping [29], and in VLSI layout [21]. In a layered drawing of a graph, vertices are arranged in horizontal layers, and edges are routed as polygonal lines between distinct layers. For acyclic digraphs, it may be required that edges point downward.

---

* Research initiated at the International Workshop on Fixed Parameter Tractability in Graph Drawing, Bellairs Research Institute of McGill University, Holetown, Barbados, Feb. 9-16, 2001, organized by S. Whitesides. Contact author: P. Ragde, Dept. of Computer Science, University of Waterloo, Waterloo, Ontario, Canada N2L 2P9, e-mail `plragde@uwaterloo.ca`. Research of Canada-based authors is supported by NSERC. Research of D. R. Wood supported by ARC and completed while visiting McGill University. Research of G. Liotta supported by CNR and MURST.

F. Meyer auf der Heide (Ed.): ESA 2001, LNCS 2161, pp. 488–499, 2001.
© Springer-Verlag Berlin Heidelberg 2001

The quality of layered drawings is assessed in terms of criteria to be minimized, such as the number of edge crossings; the number of edges whose removal eliminates all crossings; the number of layers; the maximum *span* of an edge, i.e., the number of layers it crosses; the total span of the edges; and the maximum number of vertices in one layer. Unfortunately, the question of whether a graph G can be drawn in two layers with at most k crossings, where k is part of the input, is NP-complete [11,12], as is the question of whether r or fewer edges can be removed from G so that the remaining graph has a crossing-free drawing on two layers [27,10]. Both problems remain NP-complete when the permutation of vertices in one of the layers is given [11,10].

When, say, the maximum number of allowed crossings is small, an algorithm whose running time is exponential in this parameter but polynomial in the size of the graph may be useful. The theory of parameterized complexity (surveyed in [7]) addresses complexity issues of this nature, in which a problem is specified in terms of one or more parameters. A parameterized problem with input size n and parameter size k is *fixed parameter tractable*, or in the class *FPT*, if there is an algorithm to solve the problem in $f(k) \cdot n^\alpha$ time, for some function f and constant α.

In this paper we present fixed parameter tractability results for a variety of layered graph drawing problems. To our knowledge, these problems have not been previously studied from this point of view. In particular, we give a linear time algorithm to decide if a graph has a drawing in h layers (for fixed h) with no crossings, and if so, to produce such a drawing. We then modify this basic algorithm to handle many variations, including the k-*crossings* problem (for fixed k, can G be drawn with at most k crossings?), and the r-*planarization* problem (for fixed r, can G be drawn so that the deletion of at most r edges removes all crossings?). The exact solution of the r-planarization problem for $h \geq 3$ layers is stated as an open problem in a recent survey [23], even with vertices preassigned to layers. Our algorithm can be modified to handle acyclic directed graphs whose edges must be drawn pointing downward. We also consider drawings whose edges are allowed to span multiple layers. In this case, our algorithm can minimize the total span of the edges, or alternatively, minimize the maximum span of an edge.

We do not assume a preassignment of vertices to layers. In this regard, our approach is markedly different from the traditional method for producing layered drawings, commonly called the *Sugiyama* algorithm, which operates in three phases. In the first phase, the graph is *layered*; that is, the vertices are assigned to layers to meet some objective, such as to minimize the number of layers or the number of vertices within a layer [6, Chapter 9.1]. In the second phase the vertices within each layer are permuted to reduce crossings among edges, typically using a layer-by-layer sweep algorithm [26]. In this method, for successive pairs of neighbouring layers, the permutation of one layer is fixed, and a good permutation of the other layer is determined. The third phase of the method assigns coordinates to the vertices [2].

A disadvantage of the Sugiyama approach is that after the vertices have been assigned to layers in the first phase, these layer assignments are not changed during the crossing minimization process in the second phase. In contrast, our algorithms do not assume a preassignment of vertices to layers. For example, in linear time we can determine whether a graph can be drawn in h layers with at most k edge crossings (for fixed h and k), taking into account all possible assignments of vertices to the layers.

The second phase of the Sugiyama algorithm has received much attention in the graph drawing literature. Notable are polynomial time algorithms to test if a layered graph admits a crossing-free drawing [15,17], and if so, to produce such a drawing with straight line-segments [16], even for edges which span multiple layers [9]. Integer linear programming formulations have been developed for crossing minimization in layered graphs [15,18,19], and for 2-layer planarization [22,24]. The special case of two layers is important for the layer-by-layer sweep approach. Junger and Mutzel [19] summarize the many heuristics for 2-layer crossing minimization. Our companion paper [8] addresses the 2-layer case.

The remainder of this paper is organized as follows. Section 2 gives definitions and discusses pathwidth, a key concept for our algorithms. The overall framework for our algorithms is presented in Section 3, where we consider the problem of producing layered drawings with no crossings. The r-planarization problem, the k-crossings problem, and further variants are considered in Section 4. Section 5 concludes with some open problems. Many proofs are omitted or sketched due to space limitations.

2   Preliminaries

We denote the vertex and edge sets of a graph G by $V(G)$ and $E(G)$, respectively; we use n to denote $|V(G)|$. Unless stated otherwise, the graphs considered are simple and without self-loops. For a subset $S \subseteq V(G)$, we use $G[S]$ to denote the subgraph of G induced by the vertices in S. In order to structure our dynamic programming algorithms, we make use of the well-known graph-theoretic concepts of *path decomposition* and *pathwidth*.

A *path decomposition* P of a graph G is a sequence $P_1, \ldots, P_p$ of subsets of $V(G)$ that satisfies the following three properties: (1) for every $u \in V(G)$, there is an i such that $u \in P_i$; (2) for every edge $uv \in E(G)$, there is an i such that both $u, v \in P_i$; and (3) for all $1 \leq i < j < k \leq p$, $P_i \cap P_k \subseteq P_j$. The *width* of a path decomposition is defined to be $\max\{|P_i| - 1 : 1 \leq i \leq p\}$. The *pathwidth* of a graph G is the minimum width w of a path decomposition of G. Each P_i is called a *bag* of P. It is easily seen that the set of vertices in a bag is a separator of the graph G. For fixed w, path decompositions of graphs of pathwidth w can be found in linear time [4].

A path decomposition $P = P_1, \ldots, P_p$, of a graph G of pathwidth w is a *normalized path decomposition* if (1) $|P_i| = w + 1$ for i odd; (2) $|P_i| = w$ for i even; and (3) $P_{i-1} \cap P_{i+1} = P_i$ for even i. Given a path decomposition, a normalized path decomposition of the same width (and $\Theta(n)$ bags) can be found in linear time [13].

A *proper h-layer drawing* of a (directed or undirected) graph G consists of a partition of the vertices $V(G)$ into h *layers* $L_1, L_2, \ldots L_h$ such that for each edge $uv \in E(G)$, $u \in L_i$ and $v \in L_j$ implies $|i - j| = 1$; vertices in layer L_i, $1 \le i \le h$, are positioned at distinct points in the plane with a Y-coordinate of i, and edges are represented by straight line-segments.

Edge crossings in layered drawings do not depend on the actual assignment of X-coordinates to the vertices, and we shall not be concerned with the determination of such assignments. For our purposes, a layered drawing can be represented by the partition of the vertices into layers and linear orderings of the vertices within each layer. In a layered drawing, we say a vertex u is to the *left* of a vertex v and v is to the *right* of u, if u and v are in the same layer and $u < v$ in the corresponding linear ordering.

An (a, b)-*stretched h-layer drawing* of a graph G is a proper h-layer drawing of a graph G' obtained from G by replacing each edge of G by a path of length at most $a + 1$ such that the total number of "dummy" vertices is at most b, and all edges have monotonically increasing or decreasing Y-coordinates. Of course, proper h-layer drawings are $(0, 0)$-stretched. A graph is said to be an (a, b)-*stretchable h-layer graph* if it admits an (a, b)-*stretched h'-layer drawing* for some $h' \le h$.

A layered drawing with at most k crossings is said to be k-*crossing*, where a crossing is counted every time a pair of edges cross. A 0-crossing h-layer drawing is called an h-*layer plane drawing*. A graph is $((a, b)$-*stretchable*) h-*layer planar* if it admits an $((a, b)$-stretched) plane h'-layer drawing for some $h' \le h$. A layered drawing in which r edges can be deleted to remove all crossings is said to be r-*planarizable*, and a graph which admits an $((a, b)$-stretched) r-planarizable h-layer drawing is said to be an $((a, b)$-*stretchable*) r-*planarizable h-layer graph*.

For an acyclic digraph G, an $((a, b)$-stretched) h-layer drawing if G is called *downward* if for each edge $(u, v) \in E(G)$, $u \in L_i$ and $v \in L_j$ implies $i > j$.

3   Proper *h*-Layer Plane Drawings

In this section we present an algorithm for recognizing proper h-layer planar graphs. Our algorithm, which performs dynamic programming on a path decomposition, relies on the fact that an h-layer planar graph has bounded pathwidth.

Lemma 1. *If G is an h-layer planar graph, then G has pathwidth at most h.*

Proof Sketch. First consider a proper h-layer planar graph G. Given an h'-layer plane drawing of G for some $h' \le h$, we form a normalized path decomposition of G in which each bag of size h' contains exactly one vertex from each layer. Let S be the set of leftmost vertices on each layer, where $s_i \in S$ is the vertex on layer i. Initialize the path decomposition to consist of the single bag S, and repeat the following step until S consists of the rightmost vertices on each layer: find a vertex $v \in S$ such that all neighbours of v are either in S or to the left of vertices in S. Let u be the vertex immediately to the right of v. Append the bag $S \cup \{u\}$, followed by $S \cup \{u\} \setminus \{v\}$ to the path decomposition, and set the current bag $S \leftarrow S \cup \{u\} \setminus \{v\}$. It is not hard to show that this yields the required decomposition. To handle stretchable h-layer planar graphs, insert dummy nodes, apply the above algorithm, replace dummy vertices on an edge uv by u, and remove all but one of a sequence of duplicate bags.   □

As stated in Section 2, we can obtain a normalized width-h path decomposition of any graph for which one exists in time $O(n)$ (for fixed h) [4,13]. Applying this algorithm to an h-layer planar graph will in general not result in a "nice" decomposition like that in Lemma 1 (where bags of size h contain exactly one vertex from each layer), but we can use the fact that each bag is a separator in order to obtain a dynamic programming algorithm. In the remainder of this section we prove the following result.

Theorem 1. *There is an $f(h) \cdot n$ time algorithm that decides whether a given graph G on n vertices is proper h-layer planar, and if so, produces a drawing.*

By applying the algorithm of Bodlaender [4], we can test if G has a path decomposition of width at most h. If G does not have such a path decomposition, by Lemma 1, G is not h-layer planar. Otherwise, let $P = P_1, \ldots, P_p$ be the normalized path decomposition of G given by the algorithm of Gupta *et al.* [13]. Let $w \leq h$ be the width of this path decomposition. (Our algorithm, in fact, works on any path decomposition of fixed width w, and in subsequent sections we present modifications of this procedure where the path decomposition has width $w > h$.)

Our dynamic programming is structured on the path decomposition, where for each bag P_i in turn, we determine all possible assignments of the vertices of P_i to layers and, for each assignment, all possible orderings of the vertices on a particular layer such that $G[\cup_{j \leq i} P_j]$ is proper h-layer planar.

The key to the complexity of the algorithm is the fact that the number of proper h-layer plane drawings of $G[\cup_{j \leq i} P_j]$ can be bounded as a function only of the parameters h and w. In order to ensure this bound, we need a way of representing not only edges between vertices in P_i but also edges with one or more endpoints in $\cup_{j < i} P_j \setminus P_i$. Since representing each edge will require a prohibitively large amount of information, we instead label horizontal "intervals" between vertices in P_i on the same layer.

More formally, for a subset S of vertices in a proper h-layer drawing of a graph, we add a set S' of $2h$ boundary vertices to the extreme left and extreme right of each layer, and use the term *interval*, denoted by $\mathrm{int}(u, v)$, to refer to the horizontal line-segment between any pair of consecutive vertices u and v in $S \cup S'$ on the same layer. The set of all intervals for $S \cup S'$ is denoted by $\mathrm{int}(S)$. It is not difficult to see that the number of intervals is linear in $h + |S|$.

A crucial observation is that, for the purposes of deciding where vertices and edges can be inserted into a partial drawing of G, the edges with endpoints in $\mathrm{int}(u, v)$ can be represented by the sets of vertices in the layers above and below that can "see" a particular "subinterval" of $\mathrm{int}(u, v)$.

In a proper h-layer plane drawing, let u and v be a pair of vertices appearing at layer ℓ, $1 < \ell < h$, (where other vertices may appear between u and v on layer ℓ), T be any subsequence of the vertices at layer $\ell+1$, and B be any subsequence of the vertices at layer $\ell - 1$, where the vertices in a particular layer are ordered from left to right. A point x in $\text{int}(u, v)$ is *visible* to a subsequence T' of T and a subsequence B' of B if it is possible to draw line-segments between x and every point in T' and B' without creating any crossings. A *subinterval* I of $\text{int}(u, v)$ is a sequence of points in $\text{int}(u, v)$ visible to the same subsequences T' of T and B' of B. We say I is *visible* to these subsequences, and we label I by the pair (T', B'), its *visibility label with respect to T and B*, or more succinctly, its *visibility label* (see Fig. 1). By extension, a sequence of visibility labels for subintervals of $\text{int}(u, v)$, with each consecutive sequence of identical labels replaced by a single copy, forms a *visibility label of* $\text{int}(u, v)$, and for any subset S of vertices in a layered graph, the visibility labels of all subintervals of intervals in $\text{int}(S)$ forms a *visibility label of* $\text{int}(S)$.

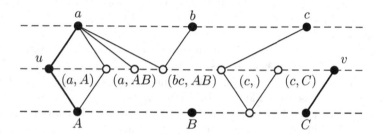

Fig. 1. An example of visibility labelling. Solid vertices are in the set S, and edges with both ends in S are bold. The interval $\text{int}(u, v)$ is divided into five subintervals of differing visibility, each with visibility labels shown. The label of each subinterval indicates how a vertex added to that subinterval could be connected to vertices in S.

This definition can be extended to label intervals on layers 1 (where B is empty) and h (where T is empty). The following lemmas are direct consequences of the definition of visibility.

Lemma 2. *Given vertices x and y on consecutive layers,*

1. *if there is an edge xy then no vertex to the left (resp. right) of x is visible to any vertex to the right (resp. left) of y, and*
2. *if x is not visible to y, then there must exist an edge $x'y'$ such that either x' is to the right of x and and y' is to the left of y, or x' is to the left of x and y' is to the right of y.* □

Lemma 3. *For each visibility label (T', B'), T' is a consecutive subsequence of T and B' is a consecutive subsequence of B.* □

In a proper h-layer plane drawing, the visibility labels satisfy the following lexicographic ordering. For a sequence $X = (x_1, x_2, \ldots, x_{|X|})$ and two consecutive subsequences y_1 and y_2 of members of X, where y_1 contains x_a through x_b and y_2 contains $x_{a'}$ through $x_{b'}$, $y_1 \prec y_2$ if one of the following three cases holds: $a < a'$, $a = a'$ and $b < b'$, or exactly one of y_1 and y_2 is empty. Similarly, $y_1 \preceq y_2$ if either $y_1 \prec y_2$ or $a = a'$ and $b = b'$.

Based on the above definition and Lemma 2, it is straightforward to prove the following lemma.

Lemma 4. *For any vertices u and v in a proper h-layer plane drawing, the number j of visibility labels of subintervals of $\mathrm{int}(u,v)$ with respect to T and B is $O(|T||B|)$. Moreover, the labels $(T_1, B_1), (T_2, B_2), \ldots, (T_j, B_j)$ satisfy the following conditions:*

1. *for any $1 \leq i < j$, $T_i \preceq T_{i+1}$, $B_i \preceq B_{i+1}$;*
2. *for any $1 \leq i < j$, either $T_i \prec T_{i+1}$, $B_i \prec B_{i+1}$, or both hold; and*
3. *for any w to the right of v, if (T_j, B_j) is the last label for $\mathrm{int}(u,v)$ and (T_1', B_1') is the first label for $\mathrm{int}(v,w)$, then $T_j \preceq T_1'$ and $B_j \preceq B_1'$.*   □

We define an *ordered layer assignment* A of P_i to be an assignment of the vertices in P_i to layers such that there is an ordering imposed on the vertices of P_i in each layer. The drawing of $G[P_i]$ obtained by placing vertices according to A and including all edges in $E(G[P_i])$ is called the *drawing of $G[P_i]$ induced by A*. As a consequence of Lemma 3, the number of ordered layer assignments and visibility labels for these assignments are functions of h and w.

In our dynamic programming table, the entry $\mathsf{TABLE}\langle i, A, L \rangle$ indicates whether or not it is possible to obtain a proper h-layer plane drawing of $G[\cup_{j \leq i} P_j]$ with ordered layer assignment A of P_i and visibility label L of $\mathrm{int}(P_i)$. The notation $\mathsf{TABLE}\langle i, A, * \rangle$ is used to denote entries for i, A, and any L. The order of evaluation is by increasing i. For a bag P_i, ordered layer assignment A of P_i, and visibility label L, a necessary condition for $\mathsf{TABLE}\langle i, A, * \rangle$ to be YES is that no edges in $G[P_i]$ violate the conditions of a proper h-layer plane drawing or of part 1 of Lemma 2. In this case, we call P_i, A, and L *consistent*.

For $i = 1$, the base case of our dynamic programming recurrence, we consider all possible ordered layer assignments of the vertices in P_1. If P_1, A, and L are inconsistent, we set $\mathsf{TABLE}\langle 1, A, L \rangle$ to NO. In general, a labeling L represents visibility due to edges not only in $G[P_i]$ but also with endpoints in $G[\cup_{j<i} P_j]$; since for $i = 1$ the latter graph is empty, we impose the additional constraint that L corresponds to exactly the edges in $G[P_i]$. That is, for a given P_1 and A, there is exactly one labeling L such that P_1, A, and L are consistent and such that part 2 of Lemma 2 is satisfied for edges in $G[P_1]$; for this L, we set $\mathsf{TABLE}\langle 1, A, L \rangle$ to YES, and set $\mathsf{TABLE}\langle 1, A, L' \rangle$ to NO for each $L' \neq L$.

To define the general recurrence, we need to consider two different cases, namely for $|P_i| = w + 1$ (i odd) and $|P_i| = w$ (i even). In the first case, P_{i-1} is missing exactly one vertex that is in P_i, and the algorithm checks the single table entry defined by removing that vertex. In the second case, P_{i-1} has exactly one vertex x that is not in P_i, and we consider all possible placements of x. Both of these tasks can be done in constant time (for fixed h and w). In each case, we first check if P_i, A, and L are consistent, and if not, enter NO. Details of the remaining steps are given below.

To compute $\mathsf{TABLE}\langle i, A, L \rangle$ for odd i, we note that $|P_i| = w + 1$ and $|P_{i-1}| = w$. Let x be the single vertex in $P_i \setminus P_{i-1}$. The algorithm computes A' and

L', the ordered layer assignment and visibility label of $\mathsf{int}(P_{i-1})$ that would result by removing x from a drawing corresponding to A and L, and then sets $\mathsf{TABLE}\langle i, A, L\rangle \leftarrow \mathsf{TABLE}\langle i-1, A', L'\rangle$.

To compute A', we simply remove x from A. The computation of L' is only slightly more complicated. Observe that for ℓ, the layer containing x, the only visibility labels that will be altered are those in layers $\ell+1$, ℓ, and $\ell-1$. Part 3 of Lemma 4 shows that for t and t', the vertices appearing immediately before and after x in layer ℓ in A, there is an order imposed on the last label of $\mathsf{int}(t, x)$ and the first label of $\mathsf{int}(x, t')$. Since parts 1 and 2 establish orderings within the intervals, the sequence formed by concatenating these labels and removing x forms a labeling of $\mathsf{int}(t, t')$ that satisfies Lemma 4, parts 1 and 2.

Next, for each visibility label at layer $\ell+1$, the algorithm removes x from all B_j's, and then removes any adjacent duplicate pairs of subinterval labels that result. Explicit adjustments will only occur when x appears as an endpoint of a subinterval. The T_j's at layer $\ell-1$ are adjusted in the same fashion.

To compute $\mathsf{TABLE}\langle i, A, L\rangle$ for even i, we note that $|P_i| = w$ and $|P_{i-1}| = w+1$. Let x be the single vertex in $P_{i-1} \setminus P_i$. The algorithm computes the set $\mathcal{S}$ of pairs (A', L') representing the ordered layer assignment and visibility label of $\mathsf{int}(P_{i-1})$ that would result from adding x to a drawing corresponding to A and L in all possible subintervals I, and then sets $\mathsf{TABLE}\langle i, A, L\rangle$ to YES if and only if $\mathsf{TABLE}\langle i-1, A', L'\rangle$ is YES for some $(A', L') \in \mathcal{S}$.

To add the new vertex x to a particular subinterval I, we first add x to A in the appropriate position, along with all its edges. If the neighbours of x in P_i do not form a subset of the visibility label of I, $\mathsf{TABLE}\langle i, A, L\rangle$ is set to NO.

If instead the neighbours form a subset of the visibility label of I, to form L', define $m_t = \min\{i'|t_{i'}x$ is an edge$\}$, $M_t = \max\{i'|t_{i'}x$ is an edge$\}$, $m_b = \min\{i'|x, b_{i'}$ is an edge$\}$, and $M_b = \max\{i'|xb_{i'}$ is an edge$\}$, where $t_1, \ldots t_{|T|}$ is the sequence of vertices of P_{i-1} at layer $\ell+1$ and $b_1, \ldots, b_{|B|}$ is the sequence of vertices of P_{i-1} at layer $\ell-1$.

Split I into two identical subintervals, one on each side of x. Then, for all visibility labels of subintervals to the left of x, remove t_i for all $i > m_t$, and remove b_j for all $j > m_b$. Similarly, for all visibility labels to the right of x, remove t_i for all $i < M_t$ and remove b_j for all $j < M_b$.

A straightforward induction on i proves the following statement and therefore the correctness of the algorithm: the entry $\mathsf{TABLE}\langle i, A, L\rangle$ is YES if and only if it is possible to obtain a proper h-layer plane drawing of $G[\cup_{j \leq i} P_j]$ with ordered layer assignment A of P_i and visibility label L of $\mathsf{int}(P_i)$.

Obviously, G has a proper h-layer plane drawing if and only if some entry $\mathsf{TABLE}\langle p, *, *\rangle$ is YES. To determine the complexity of the algorithm, we recall that the number of bags is in $O(n)$ and thus the number of table entries is $g(h, w) \cdot n$ for some function g. Each table entry can be computed in time $e(h, w)$ for some function e, and $w = h$. Thus the total running time is $f(h) \cdot n$. (The precise nature of the functions of the parameters is discussed in Section 5.) An actual drawing can be obtained by tracing back through the table in standard dynamic programming fashion. This concludes the proof of Theorem 1.

4   Edge Removals, Crossings, and Other Variants

Consider a proper r-planarizable h-layer graph. The h-layer planar subgraph obtained by removing the appropriate r edges has a path decomposition of width at most h by Lemma 1. By placing both endpoints of each of the r removed edges in each bag of the path decomposition, we form a path decomposition of the original graph of width at most $h+2r$. In a proper k-crossing h-layer drawing we can delete at most k edges to remove all crossings. By the same argument as above we obtain the following lemma.

Lemma 5. *If G is a k-crossing h-layer graph (r-planarizable h-layer graph), then G has pathwidth at most $h + 2k$ ($h + 2r$).*   □

Theorem 2. *There is a $f(h,r) \cdot n$ time algorithm to determine whether a given graph on n vertices is a proper r-planarizable h-layer graph, and if so, produces a drawing.*

Proof Sketch. The main change to the dynamic programming algorithm described in Section 3 is an additional dimension to the table representing an edge removal budget of size at most r. The entry TABLE$\langle i, A, L, c \rangle$ indicates whether or not it is possible to obtain a proper h-layer plane drawing of $G[\cup_{j \leq i} P_j]$ with ordered layer assignment A of P_i and visibility label L of $\text{int}(P_i)$ by removing at most $r - c$ edges.   □

Theorem 3. *There is a $f(h,k) \cdot n$ time algorithm to determine whether a given graph on n vertices is a proper k-crossing h-layer graph, and if so, produces a drawing.*

Proof Sketch. We proceed in a fashion similar to Theorem 2. We modify the graph representation of the previous section to include two different types of edges, *black* edges not involved in crossings, and up to $2k$ *red* edges which may be involved in crossings. Since we can obtain more than k crossings using $2k$ edges, we also need to keep a budget of crossings. In our algorithm, the entry TABLE$\langle i, A, L, R, c \rangle$ indicates whether or not it is possible to obtain a proper h-layer drawing of $G[\cup_{j \leq i} P_j]$ with ordered layer assignment A of P_i, visibility label L of $\text{int}(P_i)$ such that a subset of edges map to red edges in the set R, the only crossings in the graph involve red edges, and the total number of crossings is $k - c$.   □

We now describe how the algorithms above can be modified to take into account the directions of edges and stretch. For a downward drawing of a digraph, the directions of edges can easily be verified in the consistency check. Suppose a graph G is stretchable h-layer planar. By Lemma 1, the graph G' (defined in Section 2) has pathwidth at most h. Since graphs of bounded pathwidth are closed under edge contraction, G also has bounded pathwidth. To handle stretch in the dynamic programming, we consider placements not only of new vertices but also of dummy vertices. The total number of possibilities to consider at any step of the dynamic programming is still a function only of h and w (the bag size). The bound on the total number of dummy vertices, if used, need not be a parameter, though the running time is multiplied by this bound.

Theorem 4. *For each of the following classes of graphs, there are FPT algorithms that decide whether or not an input graph belongs to the class, and if so, produces an appropriate drawing, with the parameters as listed:*

1. *h-layer planar, k-crossing or r-planarizable graphs, (h, ∞)-stretched, (a, ∞)-stretched, or (a, b)-stretched, with parameters h, k, and r;*
2. *radial graphs (drawn on h concentric circles), with k crossings or r edges removed, (h, ∞)-stretched, (a, ∞)-stretched, or (a, b)-stretched, with parameters h, k, and r;*
3. *digraph versions of the above classes such that the drawings are downward;*
4. *multigraph versions of the above classes, where edges can be drawn as curves; and*
5. *versions of any of the above classes of graphs where some vertices have been preassigned to layers and some vertices must respect a given partial order.*

Proof Sketch. These classes have bounded pathwidth, and the basic dynamic programming scheme can be modified to deal with them.   □

5   Conclusions and Open Problems

Mutzel [23] writes "The ultimate goal is to solve these [layered graph drawing] problems not levelwise but in one step". In this paper we employ bounded pathwidth techniques to solve many layered graph drawing problems in one step *and* without the preassignment of vertices to layers.

A straightforward estimation of the constants involved in our linear-time algorithms shows that if $s = h + 2k + 2r$ is the sum of the parameters, then the dynamic programming can be completed in time $s^{cs}n$ for some small c. However, the cost of finding the path decomposition on which to perform the dynamic programming dominates this; it is $2^{32s^3}n$. Hence our algorithms should be considered a theoretical start to finding more practical FPT results. In a companion paper [8] we use other FPT techniques to shed light on the case of 2-layer drawings, obtaining better constants.

Improving the general result might involve finding more efficient ways to compute the path decomposition (perhaps with a modest increase in the width) for the classes of graphs under consideration.

Another approach to laying out proper h-layer planar graphs is to use the observation that such graphs are k-outerplanar, where $k = \lceil \frac{1}{2}(h + 1) \rceil$. One could use an algorithm to find such an embedding in $O(k^3 n^2)$ time [3], and then apply Baker's approach of dynamic programming on k-outerplanar graphs [1]. However, this approach depends heavily on planarity, and so does not appear to be amenable to allowing crossings or edge deletions.

If we relax the requirement of using h layers, recent work gives an $f(k) \cdot n^2$ algorithm for recognizing graphs that can be embedded in the plane with at most k crossings [14]. A very similar approach would work for deleting r edges to leave a graph planar. Unfortunately, the approach relies on deep structure theorems from the Robertson-Seymour graph minors project, and so is even more impractical. Nevertheless, since the maximum planar subgraph problem is of considerable interest to the graph drawing community, this should provide additional incentive to consider FPT approaches.

If we relax the requirement of planarity, we ask only if r edges can be deleted from a DAG to leave its height at most h. This is easily solved in time $O(((h + 1)(r + 1))^r + n)$; find a longest directed path (which cannot have length more than $(h + 1)(r + 1)$), and recursively search on each of the graphs formed by deleting one edge from this path to see if it requires only $r - 1$ deletions.

There is a linear-time test for h-layer planar graphs when the assignment of vertices to layers is specified [17]. Is recognizing h-layer planar graphs without such a specification NP-complete, if h is not fixed?

Acknowledgements. We would like to thank the Bellairs Research Institute of McGill University for hosting the International Workshop on Fixed Parameter Tractability in Graph Drawing, which made this collaboration possible.

References

1. B. S. Baker. Approximation algorithms for NP-complete problems on planar graphs. *J. ACM*, 41(1):153–180, 1994.
2. C. Buchheim, M. Jünger, and S. Leipert. A fast layout algorithm for k-level graphs. In J. Marks, editor, *Proc. Graph Drawing: 8th International Symposium (GD'00)*, volume 1984 of *Lecture Notes in Comput. Sci.*, pages 229–240. Springer, 2001.
3. D. Bienstock and C. L. Monma. On the complexity of embedding planar graphs to minimize certain distance measures. *Algorithmics*, 5:93–109, 1990.
4. H. L. Bodlaender. A linear-time algorithm for finding tree-decompositions of small treewidth. *SIAM Journal on Computing*, 25(6):1305–1317, 1996.
5. M. J. Carpano. Automatic display of hierarchized graphs for computer aided decision analysis. *IEEE Trans. Syst. Man Cybern.*, SMC-10(11):705– 715, 1980.
6. G. Di Battista, P. Eades, R. Tamassia, and I. G. Tollis. *Graph Drawing: Algorithms for the Visualization of Graphs.* Prentice-Hall, 1999.
7. R. G. Downey and M. R. Fellows. *Parameterized complexity.* Springer, 1999.
8. V. Dujmović, M. Fellows, M. Hallett, M. Kitching, G. Liotta, C. McCartin, N. Nishimura, P. Ragde, F. Rosemand, M. Suderman, S. Whitesides, and D. R. Wood. A fixed-parameter approach to two-layer planarization. Submitted.
9. P. Eades, Q. W. Feng, and X. Lin. Straight-line drawing algorithms for hierarchical graphs and clustered graphs. In North [25], pages 113–128.
10. P. Eades and S. Whitesides. Drawing graphs in two layers. *Theoret. Comput. Sci.*, 131(2):361–374, 1994.
11. P. Eades and N. C. Wormald. Edge crossings in drawings of bipartite graphs. *Algorithmica*, 11(4):379–403, 1994.
12. M. R. Garey and D. S. Johnson. Crossing number is NP-complete. *SIAM J. Algebraic Discrete Methods*, 4(3):312–316, 1983.
13. A. Gupta, N. Nishimura, A. Proskurowski, and P. Ragde. Embeddings of k-connected graphs of pathwidth k. In M. M. Halldorsson, editor, *Proc. 7th Scandinavian Workshop on Algorithm Theory (SWAT'00)*, volume 1851 of *Lecture Notes in Comput. Sci.*, pages 111–124. Springer, 2000.

14. M. Grohe. Computing crossing numbers in quadratic time. In *Proc. 32nd Annual ACM Symposium on Theory of Computing (STOC'01)*, 2001. To appear.
15. P. Healy and A. Kuusik. The vertex-exchange graph and its use in multi-level graph layout. In Kratochvil [20], pages 205–216.
16. M. Jünger and S. Leipert. Level planar embedding in linear time. In Kratochvil [20], pages 72–81.
17. M. Jünger, S. Leipert, and P. Mutzel. Level planarity testing in linear time. In S. Whitesides, editor, *Proc. Graph Drawing: 6th International Symposium (GD'98)*, volume 1547 of *Lecture Notes in Comput. Sci.*, pages 224–237. Springer, 1998.
18. M. Jünger, E. Lee, P. Mutzel, and T. Odenthal. A polyhedral approach to the multi-layer crossing minimization problem. In G. Di Battista, editor, *Proc. Graph Drawing: 5th International Symposium (GD'97)*, volume 1353 of *Lecture Notes in Comput. Sci.*, pages 13–24. Springer, 1998.
19. M. Jünger and P. Mutzel. 2-layer straightline crossing minimization: performance of exact and heuristic algorithms. *J. Graph Algorithms Appl.*, 1(1):1–25, 1997.
20. J. Kratochvil, editor. *Proc. Graph Drawing: 7th International Symposium (GD'99)*, volume 1731 of *Lecture Notes in Comput. Sci.* Springer, 1999.
21. T. Lengauer. *Combinatorial Algorithms for Integrated Circuit Layout*. John Wiley, 1990.
22. P. Mutzel. An alternative method to crossing minimization on hierarchical graphs. In North [25], pages 318–333. To appear in *SIAM Journal on Optimization*.
23. P. Mutzel. Optimization in leveled graphs. In P. M. Pardalos and C. A. Floudas, editors, *Encyclopedia of Optimization*. Kluwer, 2001. To appear.
24. P. Mutzel and R. Weiskircher. Two-layer planarization in graph drawing. In K. Y. Chwa and O. H. Ibarra, editors, *Proc. 9th International Symposium on Algorithms and Computation (ISAAC'98)*, volume 1533 of *Lecture Notes in Comput. Sci.*, pages 69–78. Springer, 1998.
25. S. North, editor. *Proc. Graph Drawing: Symposium on Graph Drawing (GD'96)*, volume 1190 of *Lecture Notes in Comput. Sci.* Springer, 1997.
26. K. Sugiyama, S. Tagawa, and M. Toda. Methods for visual understanding of hierarchical system structures. *IEEE Trans. Systems Man Cybernet.*, 11(2):109–125, 1981.
27. N. Tomii, Y. Kambayashi, and S. Yajima. On planarization algorithms of 2-level graphs. *Papers of tech. group on elect. comp., IECEJ*, EC77-38:1–12, 1977.
28. J. N. Warfield. Crossing theory and hierarchy mapping. *IEEE Trans. Systems Man Cybernet.*, SMC-7(7):505–523, 1977.
29. M. S. Waterman and J. R. Griggs. Interval graphs and maps of DNA. *Bull. Math. Biol.*, 48(2):189–195, 1986.

A General Model of Undirected Web Graphs

Colin Cooper[1] and Alan M. Frieze[2]

[1] School of Mathematical and Computing Sciences, Goldsmiths College, University of London, London SE14 6NW, UK.
[2] Department of Mathematical Sciences, Carnegie Mellon University, Pittsburgh PA15213, U.S.A.

Abstract. We describe a general model of a random graph process whose degree sequence obeys a power law. Such laws have recently been observed in graphs associated with the world wide web.

1 Introduction

We describe the evolution of an undirected random (multi-)graph $G(t)$ which is an example of the type of model referred to as a web graph. Such graphs (or digraphs) evolve by the addition of new vertices and/or edges at each step t. See eg: [1], [2], [3], [4], [5], [7], [10], [11] for further details.

Initially, at step $t = 0$, there is a single vertex v_0. At step $t = 1, 2, ..., T, ...$ either a procedure NEW is followed with probability $1 - \alpha$, or a procedure OLD is followed with probability α. In procedure NEW, a new vertex v is added to $G(t-1)$ with one or more edges added between v and $G(t-1)$. In procedure OLD, an existing vertex v is selected and extra edges are added at v.

The recipe for adding edges typically permits the choice of initial vertex v (in the case of OLD) and the terminal vertices (in both cases) to be made either u.a.r or according to vertex degree, or a mixture of these two based on further sampling. The number of edges added to vertex v at step t by the procedures (NEW, OLD) is given by distributions specific to the procedure. The details of these choices are given below.

A question arises in our model: Should we regard the edges as directed or undirected in relation to the sampling procedures NEW, OLD? We note that the edges have an intrinsic direction arising from the way they are inserted, which we can ignore or not as we please. In the undirected model, the variable we estimate is $\widehat{d_k}$, the expected proportion of vertices of degree k at step t, and we use sampling procedures based on vertex degree. Due to limitations of space we will only consider the case of the undirected model here. For equivalent directed versions of this model see [6].

For the undirected model, we prove that $\widehat{d_k}$ is asymptotically independent of t (provided $t \to \infty$), and provided some degree based sampling occurs, $\widehat{d_k}$ is asymptotic (in k) to Ck^{-x} where $x > 1$ is a function of the parameters of the model. The explicit functional form of x is given by (1). If all the vertex sampling is u.a.r then $\widehat{d_k}$ is a geometric series in powers of k. This model is well understood and we do not consider it further here.

F. Meyer auf der Heide (Ed.): ESA 2001, LNCS 2161, pp. 500–511, 2001.

Our undirected model $G(t)$ has sampling parameters $\alpha, \beta, \gamma, \delta, \boldsymbol{p}, \boldsymbol{q}$ whose meaning is given below:

Choice of procedure at step t.

α: Probability that an OLD node generates edges.

$1 - \alpha$: Probability that a NEW node is created.

Procedure NEW

$\boldsymbol{p} = (p_i : i \geq 1)$: Probability that new node generates i new edges.

β: Probability that choices of terminal vertices are made uniformly.

$1 - \beta$: Probability for choice of terminal vertices according to degree.

Procedure OLD

$\boldsymbol{q} = (q_i : i \geq 1)$: Probability that old node generates i new edges.

δ: Probability that the initial node is selected uniformly.

$1 - \delta$: Probability that the initial node is selected according to degree.

γ: Probability that choices of terminal vertices are made uniformly.

$1 - \gamma$: Probability for choice of terminal vertices according to degree.

The models we study here require $\alpha < 1$, always. We also assume a *finiteness* condition for the distributions $\{p_j\}$, $\{q_j\}$. This means that there exists j_0, j_1 such that $p_j = 0$, $j \geq j_0$ and $q_j = 0$, $j > j_1$.

The model creates edges in the following way: An initial vertex v is selected. If the terminal vertex w is chosen u.a.r, we regard the newly inserted edge vw as directed (v, w) and say v is *assigned* to w. If the terminal vertex w is chosen according to its vertex degree, we say v is *copied* to w.

The papers [10], [11] introduce an alternative copying model in which a new vertex v chooses an old vertex w and selects (copies) a randomly chosen set of out-neighbours of w to be its own out-neighbours. This construction leads to a larger number of small complete bipartite subgraphs than would be obtained by purely random selection of endpoints. This suggests an explanation the occurrence of the number of small complete bipartite subgraphs found in trawls of the web.

Since our focus is on the degree sequence and since the construction above does not lead to a fundamentally different recurrence (see [6] for details) we will not explicitly use this method of generating edges. We will continue to use the term copy to mean that vertices are chosen with probability proportional to degree.

Let $\mu_p = \sum_{j=0}^{j_0} j p_j$, $\mu_q = \sum_{j=0}^{j_1} i q_j$ and let $\theta = 2((1 - \alpha)\mu_p + \alpha\mu_q)$. To simplify subsequent notation, we transform the parameters as follows:

$$a = 1 + \beta\mu_p + \frac{\alpha\gamma\mu_q}{1 - \alpha} + \frac{\alpha\delta}{1 - \alpha},$$

$$b = \frac{(1 - \alpha)(1 - \beta)\mu_p}{\theta} + \frac{\alpha(1 - \gamma)\mu_q}{\theta} + \frac{\alpha(1 - \delta)}{\theta},$$

$$c = \beta\mu_p + \frac{\alpha\gamma\mu_q}{1 - \alpha},$$

$$d = \frac{(1 - \alpha)(1 - \beta)\mu_p}{\theta} + \frac{\alpha(1 - \gamma)\mu_q}{\theta},$$

$$e = \frac{\alpha\delta}{1 - \alpha}, \qquad f = \frac{\alpha(1 - \delta)}{\theta}.$$

Note that $c + e = a - 1$ and $b = d + f$.

The main quantity we study is the random variable $D_k(t)$, the number of vertices of degree k at step t. We let $\overline{D}_k(t) = \mathbf{E}(D_k(t))$ and then $\widehat{d}_k = t^{-1}\mathbf{E}(D_k(t))$.

Theorem 1. *Assume that* $t \to \infty$. *Then provided* $k = o(\log t/\log\log t)$, *we have that with probability 1-o(1),*

(i) $Ak^{-\gamma} \leq \widehat{d}_k \leq Bk^{-\gamma/j_1}$ *where* $\gamma = (1 + d + f\mu_q)/(d + f)$.
(ii) *If* $j_1 = 1$ *then* $\widehat{d}_k \sim Ck^{-(1+1/(d+f))}$.
(iii) *If* $f = 0$ *then* $\widehat{d}_k \sim Ck^{-(1+1/d)}$.
(iv) *If the* SOLUTION CONDITIONS *hold then*

$$\widehat{d}_k = C\left(1 + O\left(\frac{1}{k}\right)\right)k^{-x},$$

where C *is constant and*

$$x = 1 + \frac{1}{d + f\mu_q}. \tag{1}$$

We say that $\{q_j : j = 1, ..., j_1\}$ is *periodic* if there exists $m > 1$ such that $q_j = 0$ unless $j \in \{m, 2m, 3m, ...\}$.
Let

$$\phi_1(y) = y^{j_1} - \left(\frac{d + q_1 f}{b}y^{j_1 - 1} + \frac{q_2 f}{b}y^{j_1 - 2} + \cdots + \frac{q_{j_1} f}{b}\right).$$

Our SOLUTION CONDITIONS are:

S(i) $f > 0$ and either (a) $d + q_1 f > 0$ or (b) $\{q_j\}$ is not periodic.
S(ii) The polynomial $\phi_1(y)$ has no repeated roots.

We can also prove the following concentration result (see [6] for details).
Theorem 2. *For any* $u > 0$,

$$\mathbf{Pr}(|D_k(t) - \overline{D}_k(t)| \geq u) \leq \exp\left\{-\frac{2u^2}{tj_{\max}^2}\right\}$$

where $j_{\max} = \max\{j_0, j_1\}$.

2   Evolution of the Degree Sequence the Graph $G(t)$

Let $\nu(t) = |V(t)|$ be the number of vertices and let $\eta(t) = |2E(t)|$ be the total degree of the graph at the end of step t. $\mathbf{E}\nu(t) = (1 - \alpha)t$ and $\mathbf{E}\eta(t) = \theta t$. The random variables $\nu(t)$, $\eta(t)$ are sharply concentrated provided $t \to \infty$. Indeed $\nu(t)$ has binomial distribution $B(t, 1 - \alpha)$ and so by the Chernoff bounds,

$$\mathbf{Pr}(|\nu(t) - (1 - \alpha)t| \geq t^{1/2}\log t) = O(t^{-K}) \tag{2}$$

for any constant $K > 0$.

Similarly, $\eta(t)$ has expectation θt and is the sum of t independent random variables, each bounded by $\max\{j_0, j_1\}$. Hence, by Hoeffding's theorem [8],

$$\mathbf{Pr}(|\eta(t) - \theta t| \geq t^{1/2} \log t) = O(t^{-K}) \tag{3}$$

for any constant $K > 0$.

$D_k(t)$ is produced by t random choices and changing any one of them will change $D_k(t)$ by at most $\max\{j_0, j_1\} + 1$ and so applying the Azuma-Hoeffding martingale tail inequality, see for example Janson, Łuczak and Ruciński [9] we see that for $k \geq 0$,

$$\mathbf{Pr}(|D_k(t) - \overline{D}_k(t)| \geq t^{1/2} \log t) = O(t^{-K}) \tag{4}$$

for any constant $K > 0$.

Our main focus here is on giving asymptotic expressions for $\overline{D}_k(t)$, where $k \leq \epsilon \log t / \log \log t$ for ϵ sufficiently small.

Here $\overline{D}_0(t) = 0$ for all t, $\overline{D}_1(0) = 1$, $\overline{D}_k(0) = 0$, $k \geq 2$. Write $\widehat{D}_l = \overline{D}_l(t-1)$. Then, after using (2), (3) and (4) we see that

$$\overline{D}_k(t) = \overline{D}_k(t-1) + (1-\alpha)p_k + O(t^{-1/2}\log t) \tag{5}$$

$$+ (1-\alpha)\sum_{j=1}^{j_0} p_j \left(\frac{\beta j \widehat{D}_{k-1}}{(1-\alpha)t} - \frac{\beta j \widehat{D}_k}{(1-\alpha)t} + (1-\beta)\left(\frac{j(k-1)\widehat{D}_{k-1}}{\theta t} - \frac{jk\widehat{D}_k}{\theta t} \right) \right) \tag{6}$$

$$- \alpha\left(\frac{\delta \widehat{D}_k}{(1-\alpha)t} + \frac{(1-\delta)k\widehat{D}_k}{\theta t} \right) + \alpha \sum_{j=1}^{k-1} q_j \left(\frac{\delta \widehat{D}_{k-j}}{(1-\alpha)t} + \frac{(1-\delta)(k-j)\widehat{D}_{k-j}}{\theta t} \right) \tag{7}$$

$$+ \alpha \sum_{j=1}^{j_1} q_j \left(\frac{\gamma j \widehat{D}_{k-1}}{(1-\alpha)t} - \frac{\gamma j \widehat{D}_k}{(1-\alpha)t} + (1-\gamma)\left(\frac{j(k-1)\widehat{D}_{k-1}}{\theta t} - \frac{jk\widehat{D}_k}{\theta t} \right) \right). \tag{8}$$

Here (6), (7), (8) are (respectively) the main terms of the change in the expected number of vertices of degree k due to the effect on: terminal vertices in NEW, the initial vertex in OLD and the terminal vertices in OLD. Rearranging the right hand side, we find:

$$\overline{D}_k(t) = \widehat{D}_k + (1-\alpha)p_k + O(t^{-1/2}\log t)$$

$$- \frac{\widehat{D}_k}{t}\left(\beta\mu_p + \frac{\alpha\gamma\mu_q}{1-\alpha} + \frac{\alpha\delta}{1-\alpha} + \frac{(1-\alpha)(1-\beta)\mu_p k}{\theta} + \frac{\alpha(1-\gamma)\mu_q k}{\theta} + \frac{\alpha(1-\delta)k}{\theta} \right)$$

$$+ \frac{\widehat{D}_{k-1}}{t}\left(\beta\mu_p + \frac{\alpha\gamma\mu_q}{1-\alpha} + \frac{(1-\alpha)(1-\beta)\mu_p(k-1)}{\theta} + \frac{\alpha(1-\gamma)\mu_q(k-1)}{\theta} \right)$$

$$+ \sum_{j=1}^{j_1-1} q_j \frac{\widehat{D}_{k-j}}{t}\left(\frac{\alpha\delta}{1-\alpha} + \frac{\alpha(1-\delta)(k-j)}{\theta} \right).$$

Thus

$$\overline{D}_k(t) = \hat{D}_k + (1 - \alpha)p_k + O(t^{-1/2}\log t)$$

$$+ \frac{1}{t}\left((1 - (a + bk))\hat{D}_k + (c + d(k-1))\hat{D}_{k-1} + \sum_{j=1}^{j_1-1} q_j(e + f(k-j))\hat{D}_{k-j}\right). \tag{9}$$

Now define the sequence $(d_0, d_1, ..., d_k, ...)$ by $d_0 = 0$ and for $k \geq 1$

$$d_k(a + bk) = (1 - \alpha)p_k + (c + d(k-1))d_{k-1} + \sum_{j=1}^{k-1}(e + f(k-j))q_j d_{k-j}. \tag{10}$$

Since $a > 0$, this system of equations clearly has a unique solution.
We now prove that for small k, $\overline{D}_k(t) \approx d_k t$ as $t \to \infty$.

Lemma 1. *There exists an absolute constant $M > 0$ such that for $k = 1, 2, \ldots$,*

$$|\overline{D}_k(t) - td_k| \leq M^k k! t^{1/2}\log t.$$

Proof. Let $\Delta_k(t) = \overline{D}_k(t) - td_k$. It follows from (9) and (10) that

$$\Delta_k(t) = \Delta_k(t-1)\left(1 - \frac{a + bk - 1}{t}\right) + O(t^{-1/2}\log t) +$$

$$t^{-1}\left((c + d(k-1))\Delta_{k-1}(t-1) + \sum_{j=1}^{k-1}(e + f(k-j))q_j\Delta_{k-j}(t-1)\right). \tag{11}$$

Let L denote the hidden constant in $O(t^{-1/2}\log t)$ and assume inductively that $\Delta_\kappa(\tau) \leq M^\kappa \kappa! \tau^{1/2}\log\tau$ for $\tau < t$ and $\kappa < \omega(\tau)$. Also assume that t, k are sufficiently large (we can adjust M to deal with small values of t, k). Then (11) implies that for M large,

$$|\Delta_k(t)| \leq M^k k!(t-1)^{1/2}\log t + Lt^{-1/2}\log t$$

$$+ \frac{\log t}{t^{1/2}}\left((c + d(k-1))M^{k-1}(k-1)! + \sum_{j=1}^{k-1}(e + f(k-j))q_j M^{k-j}(k-j)!\right)$$

$$\leq M^k k! t^{1/2}\log t\left(1 - \frac{1}{t}\right) + Lt^{-1/2}\log t + 2M^{k-1}k!t^{-1/2}\log t$$

$$\leq M^k k! t^{1/2}\log t$$

This completes the induction. $\qquad\square$

Analysis of (10)

Re-writing (10) we see that for $k \geq j_0$, d_k satisfies

$$d_k = d_{k-1}\frac{c + d(k-1)}{a + bk} + \sum_{j=1}^{j_1} d_{k-j}q_j\frac{e + f(k-j)}{a + bk}, \tag{12}$$

which is a linear difference equation with rational coefficients [12]. The general solution for d_k is a power law, i.e. there are constants x_L, x_U such that $Ak^{-x_L} \leq d_k \leq Bk^{-x_U}$. This is established in Lemma 2.

In the cases where $j_1 = 1$ (a new vertex generates a single edge) or $f = 0$ (old initial vertices are chosen u.a.r) a direct solution to (12) can easily be found. In general however, when $d > 0$ or $d = 0$ and $\{q_j\}$ is non-periodic, we use classical results on the solution of Laplace's difference equation, (of which (10) is an example) given in [12].

A General Power Law Bound for d_k

Lemma 2. Let $p_j = 0, j \geq j_0$ and $q_j = 0, j > j_1$.

(i) For all $k \geq j_0$, $d_k > 0$.
(ii) then, for $k \geq j_0 + j_1$, there exist constants $C, D > 0$ such that

$$Ck^{-(1+d+f\mu_q)/b} \leq d_k \leq Dk^{-(1+d+f\mu_q)/bj_1}.$$

Proof. Let I be the first index such that $p_I > 0$, so that, from (10), $d_I > 0$. As it is impossible for both c and d to be zero, the coefficient of d_{k-1} in (10) is non-zero and thus $d_k > 0$ for $k \geq I$.
For $k \geq j_0$ the recurrence (10) satisfies (12), that is

$$d_k = d_{k-1}\frac{c+d(k-1)}{a+bk} + \sum_{j=1}^{j_1} d_{k-j}q_j\frac{e+f(k-j)}{a+bk}.$$

Let $y = 1 + d + f\mu_q$, then

$$\frac{c+d(k-1)}{a+bk} + \sum_{j=1}^{j_1} q_j\frac{e+f(k-j)}{a+bk} = 1 - y/(a+bk) \geq 0$$

and thus

$$\left(1 - \frac{y}{a+bk}\right)\min\{d_{k-1}, ..., d_{k-j_1}\} \leq d_k \leq \left(1 - \frac{y}{a+bk}\right)\max\{d_{k-1}, ..., d_{k-j_1}\}.$$

$$(13)$$

It follows that

$$d_{j_0}\prod_{j=j_0}^{k}\left(1 - \frac{y}{a+bj}\right) \leq d_k \leq \prod_{s=0}^{\lfloor(k-(j_0+j_1))/j_1\rfloor}\left(1 - \frac{y}{a+b(k-sj_1)}\right). \quad (14)$$

The LHS is clear. For the RHS note that $d_k \leq 1$ (as can be seen by using induction and the upper bound in (13)). When iterating d_j backwards on the RHS, we must make at least $\lfloor(k-(j_0+j_1))/l\rfloor$ iterations, and at least one value of j falls in each interval $[k-(s+1)l, k-sj_1)$. For that value of j we upper bound $(1 - y/(a+bj))$ by $(1 - y/(a+b(k-sj_1)))$.

Now consider the product in the LHS of (14).

$$\log\left(\prod_{j=j_0}^{k}\left(1-\frac{y}{a+bj}\right)\right)=\sum_{j=j_0}^{k}\left(-\frac{y}{a+bj}-\frac{1}{2}\left(\frac{y}{a+bj}\right)^{2}-\cdots\right)$$

$$=O(1)-\sum_{j=j_0}^{k}\frac{y}{a+bj}.$$

This justifies the lower bound of the lemma and the upper bound follows similarly for the upper bound of (14).   □

The Case $j_1 = 1$

When $q_1 = 1$, $p_j = 0, j \geq j_0 = \Theta(1)$ the general value of d_k, $k \geq j_0$ can be found directly, by iterating the recurrence (10). Thus

$$d_k = \frac{1}{a+bk}\left(d_{k-1}\left((a-1)+b(k-1)\right)\right)$$

$$= d_{k-1}\left(1-\frac{1+b}{a+bk}\right)$$

$$= d_{j_0}\prod_{j=j_0}^{k}\left(1-\frac{1+b}{a+jb}\right).$$

Thus, for some constant C,

$$d_k \sim C(a+bk)^{-x}$$

where

$$x = 1 + \frac{1}{b} = 1 + \frac{2}{a(1-\delta)+(1-\alpha)(1-\beta)+\alpha(1-\gamma)}.$$

The Case $f = 0$

The case $(f = 0)$ arises in two ways. Firstly if $\alpha = 0$ so that a new vertex is added at each step. Secondly, if $\alpha \neq 0$ but $\delta = 1$ so that the initial vertex of an OLD choice is sampled u.a.r.

We first prove that for a sufficiently large absolute constant $A > 0$ and for all sufficiently large k, that

$$\frac{d_k}{d_{k-1}} = 1 - \frac{1+d}{a+dk} + \frac{\xi(k)}{k^2} \tag{15}$$

where $|\xi(k)| \leq A$.

We use induction and re-write (10) as

$$\frac{d_k}{d_{k-1}} = \frac{c+d(k-1)}{a+dk} + \sum_{j=1}^{j_1}\frac{e}{a+dk}\prod_{t=k-j}^{k-2}\frac{d_t}{d_{t-1}}. \tag{16}$$

Now use induction to write

$$\prod_{t=k-j}^{k-2} \frac{d_t}{d_{t-1}} = 1 - \frac{j-1}{a+dk} + \frac{\xi'(j,k)}{k^2} \tag{17}$$

where $|\xi'(j,k)| \le Aj$.
Substituting (17) into (16) gives

$$\frac{d_k}{d_{k-1}} = \frac{c+d(k-1)}{a+dk} + \frac{e}{a+dk} - \frac{e\mu_q(d+1)}{(a+dk)^2} + \frac{\xi''(k)}{(a+dk)k^2}$$

where $|\xi''(k)| \le 2A\mu_q$.
Equation (15) follows immediately and on iterating this we see that

$$d_k \sim Ck^{-\left(1+\frac{1}{d}\right)}.$$

3   Analysis of the General Undirected Model

Linear Difference Equations with Rational Coefficients: The Method of Laplace

This section summarizes Chapter XV (pages 478-503) of *The Calculus of Finite Differences* by I. M. Milne-Thomson [12].
The equation (12) is an example of a linear difference equation with rational coefficients. It can equivalently be written as,

$$d_k(a+bk) - d_{k-1}(c+d(k-1)) - \sum_{j=1}^{k-1} d_{k-j}q_j(e+f(k-j)) = 0. \tag{18}$$

Laplace's difference equation is the name given to the equation whose coefficients are linear functions of a real variable w and an integer l. The general form of the homogeneous equation is

$$\sum_{j=0}^{l} [A_{l-j}(w+l-j) + B_{l-j}]u(w+l-j) = 0. \tag{19}$$

Thus (18) is a special case of (19) with $l = j_1$, $w = k - j_1$.
A method of solving difference equations with rational coefficients in general, and equation (19) in particular is to use the substitution

$$u(w) = \oint_C t^{w-1}v(t)dt.$$

The function $v(t)$ is obtained as the solution of the differential equation (22), given below, and C is a suitable contour of integration.

Let

$$\phi_1(t) = A_l t^l + A_{l-1} t^{l-1} + \cdots + A_1 t + A_0 \tag{20}$$
$$\phi_0(t) = B_l t^l + B_{l-1} t^{l-1} + \cdots + B_1 t + B_0, \tag{21}$$

where $\phi_1(t)$ is the *characteristic equation*. The differential equation referred to, is

$$t\phi_1(t) \frac{dv(t)}{dt} - \phi_0(t) v(t) = 0. \tag{22}$$

The general method of solution requires (19) to be of the *Normal type*, namely:

N(i) Both A_l and A_0 are non-zero.
N(ii) The differential equation (22) satisfied by $v(t)$ is of the Fuchsian type.

Let the roots of the characteristic equation be $a_1, ..., a_l$ (with repetition). The condition that $v(t)$ is of the Fuchsian type, requires that $\phi_0(t)/\phi_1(t)$ can be expressed as a convergent power series of t for some $t > 0$. Thus either the roots $a_1, ..., a_l$ of the characteristic equation must be distinct, or if a is repeated ν times, then a is a root of $\phi_0(t)$ at least $\nu - 1$ times.
Assuming the roots are distinct,

$$\frac{v'(t)}{v(t)} = \frac{\phi_0(t)}{t\phi_1(t)}$$
$$= \frac{-\alpha_0}{t} + \frac{\beta_1}{t - a_1} + \cdots + \frac{\beta_l}{t - a_l}, \tag{23}$$

and $\phi_0(t)/\phi_1(t)$ has the required series expansion. The general solution is

$$v^*(t) = t^{-\alpha_0} (t - a_1)^{\beta_1} ... (t - a_l)^{\beta_l}.$$

As long as there are no repeated roots, a system of fundamental solutions $(u_j(w), j = 1, ..., l)$ is given by

$$u_j(w) = \frac{1}{2\pi i} \oint_{C_j} t^{w-1-\alpha_0} (t - a_1)^{\beta_1} \cdots (t - a_l)^{\beta_l} dt,$$

where C_j is a contour containing 0 and a_j but excluding the other roots. If β_j is integer the contour integral is replaced by the integral from 0 to a_j.
A specific solution for $u_j(w)$, valid for $\Re(w) > \alpha_0$, can be obtained as

$$u_j(w) = (a_j)^w \sum_{m=0}^{\infty} C_m B\left(\frac{w - \alpha_0 + \theta - 1}{\theta}, \beta_j + m + 1\right)$$

where $B(p, q) = \Gamma(p)\Gamma(q)/\Gamma(p + q)$.
The variable $\theta > 1$ measures the angular separation, about the origin, of the root a_j from the other roots in the transformation $a_j z^{1/\theta} = t$ used to expand the transformed integral about $z = 1$ and obtain the above solution.
Now using the fact that $\Gamma(x) \sim \sqrt{2\pi} e^{-x} x^{x-1/2}$, as $w \to \infty$,

$$u_j(w) \sim C_j a_j^w w^{-(1+\beta_j)} (1 + O(1/w)). \tag{24}$$

Application of the Technique

For convenience we let $l = j_1$ for the rest of this section.
Considering the equation (18) we see that

$$\phi_1(y) = y^l - \left(\frac{d + q_1 f}{b} y^{l-1} + \frac{q_2 f}{b} y^{l-2} + \cdots + \frac{q_l f}{b}\right)$$

$$\phi_0(y) = \frac{a}{b} y^l - \left(\frac{c + q_1 e}{b} y^{l-1} + \frac{q_2 e}{b} y^{l-2} + \cdots + \frac{q_l e}{b}\right).$$

We assume that $f > 0$ so that N(i) is satisfied. Let the roots of the characteristic equation be ordered in decreasing size so that $|a_1| \geq |a_2| \geq \cdots \geq |a_l|$. Because of the SOLUTION CONDITIONS we see from Lemma 3, given below, that

$$a_1 = 1$$

and all other roots are either negative or complex and satisfy $|a| < 1$. Considering the partial fraction expansion (23) we see that

$$\phi_0(0) = -\alpha_0 \phi_1(0),$$

so that $\alpha_0 = -e/f$. Also

$$\phi_0(1) = \beta_1 \psi(1),$$

where $\psi(y) = \phi_1(y)/(y-1)$ is given by

$$\psi(z) = z^{l-1} + (1 - \alpha_1)z^{l-2} + (1 - \alpha_1 - \alpha_2)z^{l-3} + \cdots +$$
$$(1 - \alpha_1 - \cdots - \alpha_{l-2})z + (1 - \alpha_1 - \cdots - \alpha_{l-1}), \quad (25)$$

and where

$$\alpha_1 = \frac{d + q_1 f}{b}, \qquad \alpha_2 = \frac{q_2 f}{b}, \ldots, \alpha_l = \frac{q_l f}{b}.$$

Now $\phi_0(1) = 1/b$ as $c + d = a - 1$ and $\psi(1) = (d + f\mu_q)/b$. Thus $\beta_1 = 1/(d + f\mu_q)$. The other β_j require detailed knowledge of the roots of $\phi_1(t)$ and are not relevant to the asymptotic solution.
The solutions $u_j(w)$ are valid for $\Re(w) > \alpha_0 = -e/f$ which includes all $k \geq 0$. Thus considering the root $a_1 = 1$ we see that

$$u_1(k) = Ck^{-(1+\beta_1)}\left(1 + O\left(\frac{1}{k}\right)\right)$$

where $\beta_1 = \phi_0(1)/\psi(1) = \frac{1}{d + f\mu_q}$, and $1 + \beta_1$ is the parameter x of our degree sequence.
For $j \geq 2$, we use (24), giving

$$u_j(k) \to (a_j)^k k^{-(1+\beta_j)} \to 0,$$

faster than $o(1/k)$, if $|a_j| < 1$.

The specific solution for the sequence $(d_1, d_2, ..., d_k, ...)$ is

$$d_k = b_1 u_1(k) + \cdots + b_l u_l(k),$$

where $u_1(w), ..., u_l(w)$ are the fundamental solutions corresponding to the roots $a_1, ..., a_l$. We note that $b_1 \neq 0$. Indeed from Lemma 2, we know d_k obeys a power law, whereas if $b_1 = 0$, then d_k would decay exponentially as $|a_2|^k$.

Thus the error in the approximation of d_k is $O(1/k)$ from the non-asymptotic expansion of $u_1(w)$, and we conclude

$$d_k = Ck^{-\left(1+\frac{1}{d+f\mu_q}\right)} \left(1 + O\left(\tfrac{1}{k}\right)\right).$$

In the case where $\phi_1(t)$ has other solutions $|a_j| = 1$, $j = 2, ..., j', j' \leq l$, then the asymptotic solution d_k will be a linear combination of k-th powers of these roots.

Roots of the Characteristic Equation

Lemma 3. *Let $\alpha_1 = (d + q_1 f)/(d + f)$ and for $2 \leq j \leq l$, let $\alpha_j = q_j f/(d + f)$, and let*

$$\phi_1(z) = z^l - \alpha_1 z^{l-1} - \alpha_2 z^{l-2} - \cdots - \alpha_l.$$

Provided $\alpha_1 > 0$ or $\{q_j\}$ is not periodic, then the solutions of $\phi_1(z) = 0$ are

i) *An un-repeated root at $z = 1$,*

ii) *$l - 1$ other (possibly repeated) roots λ satisfying $|\lambda| < 1$.*

Proof. We note the following (see Pólya & Szegő [13] p106 16,17). A polynomial $f(z)$ of the form

$$f(z) = z^n - p_1 z^{n-1} - p_2 z^{n-2} - \cdots - p_{n-1} z - p_n,$$

where $p_i \geq 0$, $i = 1, ..., n$ and $p_1 + \cdots + p_n > 0$ has just one positive zero ζ. All other zeroes z_0 of $f(z)$ satisfy $|z_0| \leq \zeta$.

Now $\alpha_i \geq 0$ and $\sum \alpha_i = 1$, and so $\phi_1(1) = 0$ and all other zeros, z_0, of $\phi_1(z)$ satisfy $|z_0| \leq 1$.

Let $\psi(z) = \phi_1(z)/(z - 1)$ be as in (25). Now $\psi(1)$ is given by

$$1 + (1 - \alpha_1) + (1 - \alpha_1 - \alpha_2) + \cdots + (1 - \alpha_1 - \cdots - \alpha_{l-1}) = \frac{d + f\mu_q}{d + f}, \quad (26)$$

and thus $\psi(1) \neq 0$, so that $z = 1$ is not a repeated root of ϕ_1.

Let z satisfy $\phi_1(z) = 0$, $|z| = 1$, $z \neq 1$, and let $w = 1/z$; then $\phi_1(z) = 0$ is equivalent to $h(w) = 1$, where

$$h(w) = \alpha_1 w + \alpha_2 w^2 + \cdots + \alpha_l w^l.$$

Suppose there exists $w \neq 1$, on the unit circle satisfying $h(w) = 1$. Let $T = \{w, w^2, ..., w^l\}$ then all elements of T are points on the unit circle. As $w \neq 1$, $\Re(w) < 1$ and $\Re(w^j) \leq 1$, $j = 2, ..., l$.

Now, by S(i), either $\alpha_1 > 0$ or $\alpha_1 = 0$ but $\{q_j\}$ is not periodic. If $\alpha_1 > 0$, then

$$\sum \alpha_j \Re(w^j) \leq \alpha_1 \Re(w) + \alpha_2 + \cdots \alpha_l < 1,$$

and the conclusion, that $h(w) \neq 1$ follows.

Suppose $\alpha_1 = 0$. If $1 \notin T$, then the real part of w^j satisfies $\Re(w^j) < 1$, contradicting $h(w) = 1$. If $1 \in T$ then $w = e^{2\pi i/m}$ for some integer $m > 1$. However, as $\{q_j\}$ is not periodic, the conclusion that $h(w) < 1$ follows as before. $\square$

References

1. M. Adler and M. Mitzenmacher, *Toward Compressing Web Graphs*, To appear in the 2001 Data Compression Conference.
2. R. Albert, A. Barabasi and H. Jeong. *Diameter of the world wide web.* Nature 401:103-131 (1999) see also http://xxx.lanl.gov/abs/cond-mat/9907038
3. B. Bollobás, O. Riordan and J. Spencer, *The degree sequence of a scale free random graph process*, to appear.
4. B. Bollobás and O. Riordan, *The diameter of a scale free random graph*, to appear.
5. A. Broder, R. Kumar, F.Maghoul, P. Raghavan, S. Rajagopalan, R. Stata, A. Tomkins and J. Wiener. *Graph structure in the web.*
   http://gatekeeper.dec.com/pub/DEC/SRC/publications/stata/www9.htm
6. C. Cooper, A. M. Frieze. *General models of web graphs.* Submitted to Random Structures and Algorithms (2001).
7. E. Drinea, M. Enachescu and M. Mitzenmacher, *Variations on random graph models for the web.*
8. W. Hoeffding, *Probability inequalities for sums of bounded random variables*, Jornal of the American Statistical Association.
9. S.Janson, T.Łuczak and A.Ruciński, *Random Graphs*, John Wiley and Sons, 2000.
10. R. Kumar, P. Raghavan, S. Rajagopalan, D. Sivakumar, A. Tomkins and E. Upfal. *The web as a graph.* www.almaden.ibm.com
11. R. Kumar, P. Raghavan, S. Rajagopalan, D. Sivakumar, A. Tomkins and E. Upfal. *Stochastic models for the web graph.* www.almaden.ibm.com
12. I. M. Milne-Thomson. *The Calculus of Finite Differences.* Macmillian, London (1951).
13. G. Pólya and G. Szegő. *Problems and Theorems in Analysis I.* Springer, Berlin (1970).

Packing Cycles and Cuts in Undirected Graphs

Alberto Caprara[1], Alessandro Panconesi[2], and Romeo Rizzi[3]

[1] DEIS, University of Bologna
Viale Risorgimento 2, I-40136 Bologna, Italy
acaprara@deis.unibo.it
[2] DSI, *La Sapienza*
via Salaria 113, 00198 Roma, Italy
ale@dsi.uniroma1.it
[3] Dipartimento di Matematica, Universitá di Trento
via Sommarive 14, I-38050 Povo, Trento, Italy
rrizzi@science.unitn.it

Abstract. We study the complexity and approximability of CUT PACK-ING and CYCLE PACKING. For CYCLE PACKING, we show that the problem is $\mathcal{APX}$-hard but can be approximated within a factor of $O(\log n)$ by a simple greedy approach. Essentially the same approach achieves constant approximation for "dense" graphs. We show that both problems are $\mathcal{NP}$-hard for planar graphs. For CUT PACKING we show that, given a graph G the maximum cut packing is always between $\alpha(G)$ and $2\alpha(G)$. We then derive new or improved polynomial-time algorithms for CUT PACKING for special classes of graphs.

Introduction

Several combinatorial optimization problems in computational molecular biology are related to the following natural packing problem, dubbed here CYCLE PACKING: given an undirected graph G with n nodes and m edges, find a largest collection of *edge-disjoint* cycles in G. In particular, SORTING BY REVERSALS, a basic problem arising in the reconstruction of evolutionary trees [20,3], is closely related to the following variant of CYCLE PACKING. The input is a graph whose edge set is *bicolored*, i.e. partitioned into, say, *grey* and *black* edges. One is to find the largest collection of edge-disjoint *alternating* cycles. A cycle is alternating if it has even length and its edges are alternately black and grey [3]. The connection of this problem with CYCLE PACKING is discussed in [9].

Despite its connections and basic character, very little seems to be known about the complexity and approximability of CYCLE PACKING. To the best of our knowledge, the only known result is the $\mathcal{NP}$-hardness of the problem, implied by an old result of Holyer concerning the packing of triangles [18].

CYCLE PACKING seems to be related to another basic packing problem, CUT PACKING. Here, given an undirected graph G, we are to find a largest collection of *edge-disjoint* cuts in G. The two problems are the same for planar graphs, since CYCLE PACKING on a planar graph G is equivalent to CUT PACKING on the *dual*,

F. Meyer auf der Heide (Ed.): ESA 2001, LNCS 2161, pp. 512–523, 2001.
© Springer-Verlag Berlin Heidelberg 2001

with respect to any planar embedding, of G. Colbourn [10] has shown that CUT PACKING is $\mathcal{NP}$-hard by establishing a very close connection between the problem and INDEPENDENT SET, which implies that the approximation thresholds of the two problems coincide. More precisely, any lower-bound for INDEPENDENT SET is also a lower-bound for CUT PACKING. Note that we are considering the case of undirected graphs, for the problem of packing directed cuts in *digraphs* can be solved in polynomial time (see [16] and the references therein).

Motivated by the above mentioned equivalence for the case of planar graphs, in this paper we study the approximability of CYCLE PACKING and CUT PACKING.

As far as CYCLE PACKING is concerned our results are as follows. We show that the problem is $\mathcal{APX}$-hard, but that it can be approximated within a factor of $O(\log n)$ by a simple greedy approach. Whether this bound is tight we do not know and leave it as an interesting open problem. What we are able to show is that the performance of greedy is not constant. More precisely, it must be $\Omega\left(\sqrt{\frac{\log n}{\log \log n}}\right)$. We also show that CYCLE PACKING, and therefore CUT PACKING, is $\mathcal{NP}$-hard for planar graphs. We finally show that the problem can be approximated in polynomial time for "dense" graphs where by dense we mean graphs defined as follows. Given two positive constants α and β, $\mathcal{G}(\alpha, \beta)$ consists of those graphs such that $m \geq \alpha n^{1+\beta}$. For each $\alpha, \beta > 0$, we can achieve an approximation arbitrarily close to $\frac{2(1+\beta)}{3\beta}$ (i.e. independent of α) in polynomial time. In particular, for any $\varepsilon > 0$, this yields a $\frac{4}{3} + \varepsilon$ approximation for graphs such that $\beta = 1$. The running time is $O(c_{\alpha\beta\varepsilon} + p(n))$, where $c_{\alpha\beta\varepsilon}$ is a (huge) constant depending on $\alpha, \beta, \varepsilon$, and $p(n)$ is a polynomial independent of $\alpha, \beta, \varepsilon$.

On the other hand, our study of CUT PACKING leads to the following results. First, we illustrate a further connection between CUT PACKING and INDEPENDENT SET. We show that *for the same graph G*, the size of the largest cycle packing of G is at least the independence number of G and at most twice that number. Besides giving an alternative proof that CUT PACKING is at least as hard to approximate as INDEPENDENT SET, this implies that any approximation guarantee for INDEPENDENT SET extends to CUT PACKING by loosing a factor 2. We then switch to the study of the complexity of CUT PACKING for special classes of graphs. First, we give a simple variation of the algorithm given in [2] to solve CUT PACKING on Seymour graphs, a class containing bipartite and series-parallel graphs. The resulting running time is a somewhat significant improvement over that of [2], from $O(\min\{mn^3 \log n, n^5\})$ to $O(\min\{mn^2, n^{3.5} \log^{1.5} n \sqrt{\alpha(n^2, n)}\})$, where α is the inverse Ackermann function. We also point out how a specialization of an algorithm of Frank [14] can be used to solve CUT PACKING on bipartite graphs in $O(mn)$ time. Second, we give a combinatorial dual for CUT PACKING, similar in flavor to CLIQUE COVER, the combinatorial dual of INDEPENDENT SET. This strong duality leads to a linear-time algorithm for triangulated (or chordal) graphs. Finally, as mentioned above, we also show that CUT PACKING is $\mathcal{NP}$-hard for planar graphs (see below).

The main open question remains, namely whether there exists a constant factor approximation algorithm for CYCLE PACKING. Another intriguing question concerning CYCLE PACKING is to characterize the exact performance of greedy.

Basic Definitions and Notation

Consider a maximization problem P. Given a parameter $\rho \geq 1$, a *ρ-approximation algorithm* for P is a polynomial-time algorithm which returns a solution whose value is at least $\frac{1}{\rho} \cdot opt$, where opt denotes the optimal solution value. We will also say that the *approximation guarantee* of the algorithm is ρ. Similarly, given a parameter k and a function $f(\cdot)$, an $O(f(k))$-approximation algorithm is a polynomial-time algorithm that returns a solution whose value is $\Omega(\frac{1}{f(k)} \cdot opt)$ (with approximation guarantee $O(f(k))$). A problem P is *$\mathcal{APX}$-hard* if there exists some constant $\sigma > 1$ such that there is no σ-approximation algorithm for P unless $\mathcal{P} = \mathcal{NP}$. We remark that this definition is non-standard but it is adopted here because it simplifies the exposition, while at the same time it allows us to reach the conclusions we are interested in.

Although our results apply to undirected multigraphs, we shall focus on *connected, simple* graphs. Connectedness can be assumed without loss of generality, for both problems we study can be solved separately on each connected component and the solutions be pasted together. As for multiple edges they can be replaced by a single edge in the case of CUT PACKING and removed in pairs for CYCLE PACKING, counting one cycle per removed pair.

Given an undirected graph $G = (V, E)$, we will let $n_G := |V|$ and $m_G := |E|$. The subscript G will be omitted when no ambiguity arises.

Definition 1 *A* cycle *of* G *is a sequence of edges* $v_1 v_2, v_2 v_3, \ldots, v_{k-1} v_k, v_k v_1$, $k \geq 2$, *such that* $v_i \neq v_j$ *for* $i \neq j$.

Sometimes, for convenience, the cycle will be denoted simply by the sequence of nodes it visits, namely $v_1, v_2, \ldots, v_{k-1}, v_k$. A *packing of cycles* is a collection of edge-disjoint cycles, and CYCLE PACKING is the problem of finding a packing of cycles in G of maximum cardinality.

Definition 2 ψ_G *denotes the maximum size of a packing of cycles in* G, d_G *denotes the minimum degree of* G, *and* g_G *the* girth *of* G, *i.e. the length of the shortest cycle of* G.

Note that $g_G \geq 3$, as we are dealing with simple graphs only.

Let S be a *nontrivial* subset of V, i.e. $S \neq \emptyset, V$.

Definition 3 *The cut* $\delta_G(S)$ *is the set of the edges in* E *with one endpoint in* S *and the other in* $V \setminus S$. *Given* $v \in V$, *the cut* $\delta(\{v\})$ *is called the* star of v *and simply denoted by* $\delta(v)$.

Again the subscript will be omitted when no ambiguity arises. Note that $\delta(S) = \delta(V \setminus S)$. The problem of packing the maximum number of stars is the well studied INDEPENDENT SET.

A *packing of cuts* is a collection of edge-disjoint cuts. CUT PACKING is the problem of finding a packing of cuts in G of maximum cardinality. The following simple facts are recorded for later use.

Fact 1 *Given any packing of cuts* $\delta(S_1), \ldots, \delta(S_k)$ *and any node* $z \in V$, *we can always assume that* $z \notin S_1, \ldots, S_k$.

The fact follows from the observation that, for every i, it is always possible to replace S_i with $V \setminus S_i$.

Fact 2 *Let* $\{v_1, \ldots, v_k\}$ *be an independent set of* G. *Then* $\delta_{G'}(v_1), \ldots, \delta_{G'}(v_k)$ *is a packing of cuts in* G'.

Definition 4 γ_G *denotes the maximum size of a packing of cuts in* G, *and, as customary,* α_G *denotes the size of a largest independent set of* G.

The following contraction operation will be used repeatedly in the paper.

Definition 5 *Given a set of edges* $F \subseteq E$ *which induced a connected subgraph of* G, *say* $G(F)$, *the graph obtained by* contracting F, *denoted by* G/F, *is the graph obtained by identifying all nodes in* $G(F)$ *into a single node.*

Finally, the notation log without subscripts stands for the logarithm base 2.

1 On the Approximability of CYCLE PACKING

In this section we show that CYCLE PACKING is $\mathcal{APX}$-hard in general and $\mathcal{NP}$-hard for planar graphs. The reductions are from variants of SATISFIABILITY. Then we present an $O(\log n)$-approximation algorithm for general graphs and a constant factor approximation algorithm for dense graphs.

1.1 The Hardness of CYCLE PACKING

For the $\mathcal{APX}$-hardness proof, the reduction is from MAX-2-SAT-3. The input to this problem is a boolean formula φ in conjunctive normal form in which each clause is the OR of at most 2 literals. Each literal is a variable or the negation of a variable taken from a ground set of boolean variables $X := \{x_1, \ldots, x_n\}$, with the additional restriction that each variable appears in at most 3 of the clauses, counting together both positive and negative occurrences. The optimization problem calls for a truth assignment that satisfies as many clauses as possible. It is known that MAX-2-SAT-3 is $\mathcal{APX}$-hard [6].

Theorem 1 CYCLE PACKING *is* $\mathcal{APX}$*-hard, even for graphs with maximum degree 3.*

We conclude this section with the $\mathcal{N P}$-hardness proof for the planar case, reducing the following PLANAR 3-SAT problem to CYCLE PACKING. As customary $X = \{x_1, \ldots, x_n\}$ and $C = \{c_1, \ldots, c_m\}$ denote, respectively, the set of variables and clauses in a boolean formula φ in conjunctive normal form, where each clause has *exactly* 3 literals. Consider the bipartite graph $G_\varphi = (X \cup C, E_\varphi)$, with color classes X and C and edge set $E_\varphi = \{xc$: variable x occurs in clause $c\}$. The boolean formula φ is called *planar* when G_φ is planar. PLANAR 3-SAT is the problem of finding, if any, a truth assignment that satisfies all clauses in a planar boolean formula, where each clause has exactly three literals. It is known that PLANAR 3-SAT is $\mathcal{N P}$-complete [21].

Theorem 2 CYCLE PACKING *is $\mathcal{N P}$-complete for planar graphs with maximum degree 3.*

Corollary 1 CUT PACKING *is $\mathcal{N P}$-hard for planar graphs.*

1.2   An $O(\log n)$ Approximation Algorithm

Given the results of the previous section, we investigate the approximability of the problem. Recalling that the shortest cycle in a graph can be found efficiently, consider the following intuitive idea: repeatedly pick a cycle of smallest length in the solution and remove the associated edges. The resulting algorithm will be called *basic greedy*. As such, this algorithm does not work well. The following simple example demonstrates that the approximation guarantee can be as bad as $\Omega(\sqrt{n})$.

A *sunflower* is a graph S consisting of a *core* cycle $C = v_1, \ldots, v_p$ and of p *petals*, namely p cycles $P_i = v_i, u_1^i, \ldots, u_{p-2}^i, v_{i+1}$, where indices are modulo p. Note that $n_S = \Theta(p^2)$. The optimum packing consists of the p petals, therefore $\psi_S = p$. Basic greedy on the other hand, may select the core first. Thereafter there is only the *external* cycle remaining, $E := v_1, u_1^1, \ldots, u_{p-2}^1, v_2, u_1^2, \ldots, u_{p-2}^2, v_3, \ldots, v_1$. The approximation guarantee is therefore $\Omega(\sqrt{n})$. However, we next show that a small modification of basic greedy, called *modified greedy* works "reasonably well". We use the following well-known result (see for example [7]):

Fact 3 *For a graph G with $d_G \geq 3$, $g_G \leq 2\lceil \log n \rceil$.*

Modified greedy iteratively performs the following three steps: (1) while G contains a node w of degree 1, w is removed from G along with the incident edge; (2) while G contains a node w of degree 2, with neighbors u and v, w is removed from G along with the incident edges, and edge uv is added to G (note that edge uv may already be present, in which case a cycle of length 2 is formed, which is removed and added to the solution); (3) an arbitrarily chosen shortest cycle C of G is added to the solution and the corresponding edges are removed from G. Steps (1), (2) and (3) are repeated until there are no edges left in G.

Theorem 3 *Modified greedy is a $O(\log n)$-approximation algorithm for* CYCLE PACKING.

Proof: Let $\mathcal{C}$ be the set of cycles found by modified greedy and let $\mathcal{S}$ be any maximum packing of cycles. We show that each $C \in \mathcal{C}$ intersects at most $O(\log n)$ cycles of $\mathcal{S}$. Let G_C be the remaining graph when greedy selects C. When this happens, $d_{G_C} \geq 3$ because Steps (1) and (2) systematically eliminate nodes of degree less than 3. By Theorem 3 and the fact that $n_{G_C} \leq n_G = n$, C has $O(\log n)$ edges. Hence, removing the edges in C decreases the optimal CYCLE PACKING value by at most $O(\log n)$. The proof of the approximation guarantee is concluded by observing that Steps (1) and (2) do not change the optimal CYCLE PACKING value. □

If one wants to get rid of the $O(\cdot)$ notation in the approximation ratio, it is easy to see that the approximation guarantee is $(2 + o(1)) \log n$.

We do not know of any examples for which the approximation ratio achieved by modified greedy is $\Omega(\log n)$. We remark that if g_G is $\Omega(\log n)$ it is fairly easy to show that the approximation guarantee of modified greedy is constant. The same holds if g_G is bounded by a constant during the whole execution of modified greedy. What we are able to show is the following. Regrettably, for lack of space, the proof is omitted from this extended abstract.

Theorem 4 *The approximation guarantee of modified greedy is* $\Omega\left(\sqrt{\frac{\log n}{\log\log n}}\right)$.

1.3   Constant Approximation Guarantee for "Nonsparse" Graphs

In the full paper we analize the performance of a slight modification of the greedy algorithm dubbed *dense greedy*, and show that for "dense graphs" it achieves constant approximation. More precisely we show the following.

Theorem 5 *Given constants $\alpha, \beta \in (0, 1)$, let $\mathcal{G}(\alpha, \beta)$ be the family of graphs with n nodes and m edges such that $\alpha n^{1+\beta} \leq m \leq n^{1+\beta}$. For any fixed $\varepsilon > 0$ and $\alpha \in (0, 1)$,* DENSE GREEDY *is a ρ-approximation algorithm for* CYCLE PACKING *for graphs in $\mathcal{G}(\alpha, \beta)$, where $\rho := (1 + \varepsilon)(2(1 + \beta)/3\beta)$. The running time is $O(c_{\alpha\beta\varepsilon} + p(n))$, where $c_{\alpha\beta\varepsilon}$ is a constant depending on $\alpha, \beta, \varepsilon$ and $p(n)$ is a polynomial independent of $\alpha, \beta, \varepsilon$.*

Therefore for dense graphs, i.e. graphs for which $\beta = 1$, the approximation achieved is arbitrarily close to $\frac{4}{3}$. We remark that $c_{\alpha\beta\varepsilon}$ is huge.

2   On the Complexity of CUT PACKING

In this section, we show that there is a tight connection between γ_G and α_G and show how γ_G can be computed efficiently for the class of Seymour and triangulated (chordal) graphs.

2.1   The Relationship between CUT PACKING and INDEPENDENT SET

The following simple reduction from INDEPENDENT SET, given in [10], establishes the $\mathcal{NP}$-hardness of CUT PACKING. Given a graph $G = (V, E)$, let G' be
the graph obtained from G by adding a node $z \notin V$ with z adjacent to all nodes
in V. Namely,

$$G' := (V \cup \{z\}, E \cup \{zv : v \in V\}).$$

It is not hard to prove that wlog all packings of G' can be considered to be
packings of stars, and that there is a cost-preserving bijection between independent sets of G and packings of stars of G'. The reduction also implies that CUT
PACKING is at least as hard to approximate as INDEPENDENT SET. In this section we show that the relationship between the two problems is in in some sense
tighter. We prove that, for every graph G, $\alpha_G \leq \gamma_G \leq 2\,\alpha_G$ i.e., *for the same
graph G, the size of a largest cut packing is between the independence number
and twice that number.* We begin with a simple technical lemma.

Lemma 1 *Let F be a forest on n nodes. Then $\alpha_F \geq \lceil \frac{n}{2} \rceil$.*

Proof: F is bipartite.   □

To prove the new result of this section, we employ the above lemma together
with some well known facts about uncrossing and nested families.
    A family of subsets $S_1, \ldots, S_k$ of a certain set V is *nested* if any two sets in
the family are either disjoint or one is contained in the other. A nested family
can be represented as a *directed forest* on nodes $S_1, \ldots, S_k$, where $S_i S_j$ is an arc
of the forest if and only if S_i is the smallest subset in the family containing S_j.
In this case, we say that S_i is the *father* of S_j and S_j is a *child* of S_i.
    Our next ingredient is the simple fact that there always exists a maximum
packing of cuts which is *laminar*, i.e. it can be assumed without loss of generality
that the sets realizing the maximum cut packing form a nested family in which
the minimal sets are singletons.
    The proof of the next lemma is based on a technique called *uncrossing* (for
further details we refer the interested reader to [22,16]).

Lemma 2 *Given any packing of cuts $\mathcal{F}$, there exists a laminar packing of cuts
$\mathcal{F}'$ with $|\mathcal{F}'| = |\mathcal{F}|$ in which the minimal sets of the nested family are singletons.*

Theorem 6 *For any graph G, $\alpha_G \leq \gamma_G \leq 2\alpha_G$.*

Proof: Let $\{v_1, \ldots, v_k\}$ be an independent set of G. Then $\delta(v_1), \ldots, \delta(v_k)$ is a
packing of cuts in G and $\alpha_G \leq \gamma_G$ follows.
    Let $\delta(S_1), \ldots, \delta(S_{\gamma_G})$ be a maximum packing of cuts in G. By Lemma 2,
we can assume that $\mathcal{S} := \{S_1, \ldots, S_{\gamma_G}\}$ is a nested family. Let F be the forest
representing the nested family $\mathcal{S}$. By Lemma 1, F has an independent set
$\mathcal{T} := \{T_1, \ldots, T_k\}$ with $k \geq \lceil \frac{\gamma_G}{2} \rceil$. For $i = 1, \ldots, k$, pick a node v_i which is
contained in T_i but not contained in any child of T_i. We claim that the set

$I := \{v_i : 1 \leq i \leq k\}$ is an independent set. Consider two nodes $v_i, v_j \in I$ and look at the sets they come from. We now show that it is not possible that $v_i v_j \in E$. If $T_i \cap T_j = \emptyset$ and $v_i v_j \in E$ then $\delta(T_i) \cap \delta(T_j) \neq \emptyset$, contradicting the fact that T_i and T_j belong to a packing of cuts. Suppose on the other hand that $T_i \cap T_j \neq \emptyset$ and $v_i v_j \in E$ and assume without loss of generality that $T_i \subset T_j$, where $\subset$ denotes strict containment. Since T is itself an independent set, there exist a set $X \in S$ such that $T_i \subset X \subset T_j$. Now, if $v_i v_j \in E$ then $\delta(T_i) \cap \delta(X) \neq \emptyset$, again contradicting the fact that S is a packing of cuts. The claim follows.   □

Besides showing that CUT PACKING cannot be approximated better than INDEPENDENT SET, Theorem 6 also shows that any approximation guarantee for INDEPENDENT SET immediately extends to CUT PACKING by loosing a factor 2.

2.2   CUT PACKING on Seymour Graphs

Recently, Ageev [2] showed that CUT PACKING can be solved in polynomial time for *Seymour graphs*. This class, containing bipartite graphs among others, is formally defined below. In this section, we outline the approach in [2] and make a few additional observations that allow us to reduce the asymptotic running time from the $O(\min\{mn^3 \log n, n^5\})$ of [2] to $O(\min\{mn^2, n^{3.5} \log^{1.5} n \sqrt{\alpha(n^2, n)}\})$, where α is the inverse of Ackermann's function. Moreover, we point out how an algorithm of Frank [14] can be specialized to solve CUT PACKING on bipartite graphs in $O(mn)$ time.

Consider a node subset $T \subseteq V$ of even cardinality. A *T-cut* is a cut $\delta(S)$ such that $|S \cap T|$ is odd. A *T-join* is an edge subset $J \subseteq E$ such that every node $v \in V$ is incident with an odd number of edges in J if and only if $v \in T$. It is easy to verify that, if J is a T-join and C is a T-cut, then $|J \cap C| \geq 1$. Therefore, whenever J is a T-join, $|J|$ is an upper bound on the size of a packing of T-cuts. A graph G is called a *Seymour graph* if for every $T \subseteq V$ with $|T|$ even, the minimum cardinality of a T-join equals the maximum size of a packing of T-cuts. Seymour graphs include bipartite graphs and series parallel graphs and have been partially characterized by Ageev, Kostochka and Szigeti [1].

The following simple lemma relates optimal cut packings to optimal T-cut packings.

Lemma 3 *Let $G = (V, E)$ be any graph and $\mathcal{F}$ be a laminar packing of cuts of G. Then there exists a node subset $T \subseteq V$ with $|T|$ even and such that all cuts in $\mathcal{F}$ are T-cuts.*

Note that the above proof does not work if the cuts in the family are not edge disjoint, as there is no guarantee that $S \setminus \bigcup_{i=1}^{k} S_i \neq \emptyset$ in this case.

In [14], Frank gave a polynomial time algorithm to find, for any given graph G, a subset T of V (with $|T|$ even) such that the minimum size of a T-join is as large as possible, along with an associated T-join J. His algorithm runs in $O(nm)$ time. Hence, by the above lemma, given a Seymour graph G, Frank's algorithm computes γ_G in $O(mn)$ time. The following result, which is a main contribution of [2], shows how to turn this into a poyonomial time algorithm for CUT PACKING. Recall the contraction operation defined in the introduction.

Lemma 4 *Given a Seymour graph $G = (V, E)$ and a node $v \in V$, $G/\delta(v)$ is a Seymour graph.*

We know by Lemma 2 that there exists an optimal cut packing in which one of the cuts is a star $\delta(v)$. For a given $v \in V$, let $G' := G/\delta(v)$ be the graph obtained by contracting $\delta(v)$. Then, $\delta(v)$ is in an optimal cut packing if and only if $\gamma_{G'} = \gamma_G - 1$. Moreover, since by the lemma above G' is a Seymour graph, we can compute $\gamma_{G'}$ in $O(mn)$ time (this simple observation is missing in [2]).

This suggests the following algorithm. Consider a vertex $v \in V$ and try to contract $\delta(v)$, obtaining G'. If $\gamma_{G'} = \gamma_G - 1$, then add $\delta(v)$ to the optimal cut packing and iterate the procedure on G' (stopping when G' is a single node). Otherwise, $\delta(v)$ is not in an optimal cut packing, and hence another vertex is considered, *avoiding* to try to contract $\delta(v)$ in the following iterations (this was also not noted in [2]). The running time of this algorithm is bounded by the time to execute the test $\gamma_{G'} = \gamma_G - 1$ for all the star contraction attempts. Each test can be carried out in $O(mn)$ time by Frank's algorithm, whereas the number of attempts cannot exceed $2n$ because every attempt tries to contract a different cut of a laminar family of the original graph G. Hence, this simple algorithm runs in $O(mn^2)$ time. In the following, we describe an alternative implementation along the same lines of [2].

Let T be the set found by Frank's algorithm applied to G and J be the associated T-join. Moreover, for a given $v \in V$, let $G' := G/\delta(v)$, $J' := J \setminus \delta(v)$ and T' be the set of vertices in G' incident with an odd number of edges in J'. The following lemma (whose proof follows from the discussion above) is stated in [2].

Lemma 5 *There exists an optimal cut packing containing $\delta(v)$ if and only if J' is a minumum cardinality T'-join of G'.*

Then, in the algorithm above, testing if $\gamma_{G'} = \gamma_G - 1$ can be replaced by testing if J' is a minumum cardinality T'-join. This can be done by finding a minimum cost perfect matching in the complete graph with node set T' in which the cost of each edge uv is given by the minimum number of edges in a path between u and v in G'. The costs can easily be computed in $O(n^2)$ time and are all between 0 and n. Therefore an optimal perfect matching can be found in $O(n^{2.5} \log^{1.5} n \sqrt{\alpha(n^2, n)})$ time by the algorithm of Gabow and Tarjan [15]. It follows that the running time is $O(n^{3.5} \log^{1.5} n \sqrt{\alpha(n^2, n)})$. Therefore the following result obtains.

Theorem 7 CUT PACKING *on Seymour graphs can be solved in $O(\min\{mn^2, n^{3.5} \log^{1.5} n \sqrt{\alpha(n^2, n)}\})$ time.*

We conclude this section by outlining how the above mentioned algorithm by Frank [14] can be specialized to solve CUT PACKING on bipartite graphs.

Consider a connected biparite graph $G = (U \cup V, E)$. Let M be a maximum matching of G and N be a corresponding minimum node cover (each node in N is matched), letting $U' := N \cap U$ and $V' := N \cap V$. By possibly exchanging U and V we can assume without loss of generality $U \setminus U' \neq \emptyset$ and $V' \neq \emptyset$ (excluding the trivial case in which G has at most one node). For every node v' in V' let $u(v')$ be the node such that $v'u(v') \in M$. Consider the directed graph D on node set V' containing arc (s, t) if and only if $u(s)t$ is an edge of G. By contracting each maximal strongly connected component in D into a single node, one obtains an acyclic directed graph D'. Consider a sink node in D' (i.e. a node with no outgoing arc) and the associated node set in D, say $\tilde{V} \subseteq V'$. Moreover, let $\tilde{U} \subset U$ be the set of nodes in G which are adjacent only to nodes in $\tilde{V}$ (i.e. $\delta(\tilde{U}) \subseteq \delta(\tilde{V})$). Finally, let G' be the bipartite graph obtained by removing the nodes in $\tilde{U}$ and identifying all the nodes in $\tilde{V}$ into a single node. Although not stated explicitly, the following result is implicit in [14].

Lemma 6 *An optimal cut packing for G can be obtained from an optimal cut packing for G' by adding all cuts in $\{\delta(\tilde{u}) : \tilde{u} \in \tilde{U}\}$.*

By simple implementation considerations (also given in [14]) the following result follows.

Theorem 8 CUT PACKING *on bipartite graphs can be solved in $O(mn)$ time.*

Note that for bipartite graphs, any maximal collection of edge-disjoint cuts induces a *partition* of the edge set. It is easy to see that the same happens with a maximal collection of edge-disjoint cycles in a *Eulerian* graph. While CUT PACKING is easy on bipartite graphs, CYCLE PACKING is NP-hard for Eulerian graphs, even if the maximum degree is 4 [9]. However, since the dual of a bipartite planar graph is Eulerian, we have the following.

Corollary 2 CYCLE PACKING *on planar Eulerian graphs can be solved in $O(mn)$ time.*

2.3   A Combinatorial Dual

In this section, we introduce a combinatorial problem in minimization form whose optimal solution is an upper bound for the optimal CUT PACKING solution. This will lead to a case of CUT PACKING that can be solved efficiently, as the two optima coincide. A *clique* K of G is a set of pairwise adjacent nodes in G. Let $E(K)$ denote the set of edges joining two nodes in the clique.

Lemma 7 *For every packing of cuts $\mathcal{F}$ and every clique K of G, $\delta(S) \cap E(K) \neq \emptyset$ for at most one cut $\delta(S) \in \mathcal{F}$.*

Proof: Given a cut $\delta(S) \in \mathcal{F}$, consider the bipartite graph $(K \cap S, K \cap (V \setminus S); E(K) \cap \delta(S))$. Note that this graph is complete, i.e. every two nodes belonging to different sides of the bipartition are adjacent. Therefore, in case $E(K) \cap \delta(S) \neq \emptyset$, any cut $\delta(S')$ such that $\delta(S) \cap \delta(S') = \emptyset$

cannot contain an edge uv for $u \in K \cap S$, $v \in K \cap (V \setminus S)$. But this means that either $K \cap S' = \emptyset$ or $K \cap S' = K$, i.e. $E(K) \cap \delta(S') = \emptyset$.   □

A *clique sequence* of G is a non-empty sequence of node subsets $K_1, \ldots, K_p$ associated with a sequence of graphs $G_0 := G, G_1, \ldots, G_p$ such that K_i is a clique of G_{i-1} (for $i = 1, \ldots, p$), and $G_i = G_{i-1}/E(K_i)$ is the graph obtained from G_{i-1} by contracting the edges in K_i (for $i = 1, \ldots, p - 1$). The *length* of the clique sequence is p. We say that the clique sequence is *complete* if G_p is the graph consisting of a single node. Denote by ρ_G the length a *shortest* complete clique sequence of G.

Theorem 9 *For any graph G, $\gamma_G \leq \rho_G$.*

Proof: Let $\mathcal{F}$ be any packing of cuts. By Fact 7, $E(K_1) \cap C \neq \emptyset$ for at most one cut $C \in \mathcal{F}$. Hence, each cut in $\mathcal{F}$, except at most one, is a cut of $G/E(K_1)$. The proof follows by induction.   □

This combinatorial upper bound is analogous to $\bar{\chi}_G$, the minimum number of cliques needed to cover the nodes of G, which is a well-known upper bound on α_G. Recall that a graph G is called *perfect* if $\alpha_{G'} = \bar{\chi}_{G'}$ for all node induced subgraphs G' of G. Unfortunately, the class of graphs G such that $\gamma_G = \rho_G$ seems to be much poorer than the class of perfect graphs. In particular, this class does not contain the bipartite graphs, as for the even cycle C_{2k} we have $\gamma_{C_{2k}} = k$ and $\rho_{C_{2k}} = 2k - 2$. Moreover, C_6 shows that this class is not closed under taking the complement. However, there is a wide family of graphs contained in the class.

A graph is *triangulated* (or *chordal*) if it contains no chordless cycle of length 4 or more.

Theorem 10 *For a triangulated graph G, $\gamma_G = \rho_G$. Moreover,* CUT PACKING *on triangulated graphs can be solved in $O(m)$ time.*

Proof: Dirac [11] observed that every triangulated graph G has a *simplicial node*, i.e. a node v such that the neighbors $N(v)$ of v are a clique in G. Take the cut $\delta(v)$ and obtain G' from G by contracting the clique $N(v) \cup \{v\}$. Note that G' is triangulated. Apply induction. The running time follows from [24]. □

Acknowledgments. We gratefully acknowledge the hospitality of BRICS where most of this work was done. The first author was also partially supported by CNR and MURST, Italy. We would like to thank Zoltan Szigeti and Maxim Sviridenko for helpful discussions and email exchanges.

References

1. A.A. Ageev, A.V. Kostochka, Z. Szigeti, A Characterization of Seymour Graphs. *J. Graph Theory* 24 (1997) 357–364.
2. A.A. Ageev, On Finding the Maximum Number of Disjoint Cuts in Seymour Graphs. *Proceedings of the 7th European Symposium on Algorithms (ESA '99)*, Lecture Notes in Comput. Sci., 1643, Springer, Berlin (1999) 490–497.

3. V. Bafna and P.A. Pevzner, Genome Rearrangements and Sorting by Reversals. *SIAM J. on Computing* 25 (1996) 272–289.

4. B.S. Baker, Approximation Algorithms for $\mathcal{NP}$-Complete Problems on Planar Graphs. *J. ACM* 41 (1994) 153–180.

5. P. Berman, T. Fujito, On Approximation Properties of the Independent Set Problem for Low Degree Graphs. *Theory of Computing Systems* 32 (1999) 115–132.

6. P. Berman and M. Karpinski, On Some Tighter Inapproximability Results. ECCC Report No. 29, University of Trier (1998).

7. B. Bollobás, *Extremal Graph Theory*, Academic Press, New-York (1978).

8. R. Boppana, M.M. Halldórsson, Approximating Maximum Independent Sets by Excluding Subgraphs. *Bit* 32 (1992) 180–196.

9. A. Caprara, Sorting Permutations by Reversals and Eulerian Cycle Decompositions. *SIAM J. on Discrete Mathematics* 12 (1999) 91–110.

10. C.J. Colbourn, *The Combinatorics of Network Reliability*. Oxford University Press (1986).

11. G.A. Dirac, On Rigid Circuit Graphs. *Abh. Math. Sem. Univ. Hamburg* 25 (1961) 71–76.

12. P. Erdös and L. Pósa. On the Maximal Number of Disjoint Circuits of a Graph. *Publ. Math. Debrecen* 9 (1962) 3–12.

13. P. Erdös and H. Sachs. Regulare Graphen Gegebener Taillenweite mit Minimaler Knotenzahl. *Wittenberg Math. – Natur. Reine* 12 (1963) 251–257.

14. A. Frank, Conservative Weightings and Ear-Decompositions of Graphs. *Combinatorica* 13 (1993) 65–81.

15. H.N. Gabow and R.E. Tarjan, Faster Scaling Algorithms for General Matching Problems. *J. ACM* 38 (1991) 815–853.

16. M. Grötschel, L. Lovász, A. Schrijver, *Geometric algorithms and combinatorial optimization*, Second edition: Algorithms and Combinatorics, 2. Springer-Verlag, Berlin, (1993). ISBN: 3-540-56740-2.

17. J. Håstad, Clique is Hard to Approximate within $n^{1-\varepsilon}$. *Acta Mathematica* 182 (2000) 105–142.

18. I. Holyer, The $\mathcal{NP}$-Completeness of Some Edge-Partition Problems. *SIAM J. on Computing* 10 (1981) 713–717.

19. H.B. Hunt III, M.V. Marathe, V. Radhakrishnan, S.S. Ravi, D.J. Rosenkrantz and R.E. Stearns, A Unified Approach to Approximation Schemes for $\mathcal{NP}$- and $\mathcal{PSPACE}$-Hard Problems for Geometric Graphs. *Proceedings of the 2nd Euoropean Symposium on Algorithms (ESA '94)*, Lecture Notes in Comput. Sci., 855, Springer, Berlin (1994) 424-435.

20. J. Kececioglu and D. Sankoff, Exact and Approximation Algorithms for Sorting by Reversals, with Application to Genome Rearrangement. *Algorithmica* 13 (1995) 180–210.

21. D. Lichtenstein, Planar formulae and their uses. *SIAM J. on Computing* 11 (1982) 329–343.

22. L. Lovász, M.D. Plummer, *Matching Theory*, Akadémiai Kiadó (1986)

23. C.H. Papadimitriou and M. Yannakakis (1991), Optimization, Approximation, and Complexity Classes *J. Comput. System Sci.* 43 (1991) 425–440.

24. D.J. Rose, R.E. Tarjan and G.S. Lueker, Algorithmic Aspects of Vertex Elimination on Graphs. *SIAM J. on Computing* 5 (1976) 266–283.

Greedy Algorithms for Minimisation Problems in Random Regular Graphs

Michele Zito

Department of Computer Science, University of Liverpool, Liverpool, L69 7ZF, UK.

Abstract. In this paper we introduce a general strategy for approximating the solution to minimisation problems in random regular graphs. We describe how the approach can be applied to the minimum vertex cover (MVC), minimum independent dominating set (MIDS) and minimum edge dominating set (MEDS) problems. In almost all cases we are able to improve the best known results for these problems. Results for the MVC problem translate immediately to results for the maximum independent set problem. We also derive lower bounds on the size of an optimal MIDS.

1   Introduction

This paper is concerned with graphs generated uniformly at random according to the $\mathcal{G}(n, r\text{-reg})$ model [19, Chap. 9]. Let n urns be given, each containing r balls (with rn even): a set of $rn/2$ pairs of balls is chosen at random. To get a random graph $G = (V, E)$, identify the n urns with the graph's n vertices and let $\{i, j\} \in E$ if and only if there is a pair with one ball belonging to urn i and the other one belonging to urn j. The maximum degree of a vertex in G is at most r. Moreover, for every integer $r > 0$, there is a positive fixed probability that the random pairing contains neither pairs with two balls from the same urn nor couples of pairs with balls coming from just two urns. In this case the graph is r-regular (all vertices have the same degree r). Notation $G \in \mathcal{G}(n, r\text{-reg})$ will signify that G is selected according to the model $\mathcal{G}(n, r\text{-reg})$. An event $\mathcal{E}_n$, describing a property of a random graph depending on a parameter n, holds *asymptotically almost surely* (a.a.s.), if the probability that $\mathcal{E}_n$ holds tends to one as n tends to infinity.

Heuristics for approximating the solution to minimisation problems in random graphs are often based on algorithms for maximisation problems. A small vertex cover or a good vertex colouring can be found by running an independent set algorithm (see for example [17]). The minimum k-center problem [18] can be solved optimally a.a.s. by using a greedy heuristic for a generalisation of the maximum independent set problem [23]. In this paper we introduce a general strategy for approximating the solution to minimisation problems in random regular graphs. The approach has a number of advantages over previously known heuristics. First of all it is more natural to these problems: it does not use as a subroutine any algorithm for an associated maximisation problem. Secondly, it is simple to implement. It is based on a greedy algorithm that repeatedly

F. Meyer auf der Heide (Ed.): ESA 2001, LNCS 2161, pp. 524–536, 2001.

picks vertices of given degree in the current graph, updates the partial solution obtained so far and then removes few vertices from the graph. The proposed algorithm is also quite simple to analyse. At the price of performing a larger case analysis, the only mathematical tool needed to estimate the size of the final solution in each case is a theorem of Wormald [26]. The analyses given in other papers [3,23,25] use other results in the theory of random processes (e.g. random walks or birth-death processes). No such a tool is needed here. A further advantage of the approach described in this paper is that it uniformly extends to random r-regular graphs for each fixed $r \geq 3$. Finally, using this approach, we are able to give best known results in almost all cases considered.

The rest of the paper is organised as follows: in Section 2 we define the problems of interest and present a summary of the results proved in this paper, in Section 3 we describe our algorithmic approach, present the analysis method along with the statement of the main result of this paper; in Section 4 we complete the proof of the main theorem for each of the problems considered; some conclusions are drawn in Section 5.

2 Problems and Results

We first define the problems of interest and give a short survey of the relevant literature. The reader may find elsewhere [15,4,9,12] more detailed bibliographic notes. The tables at the end of the section give, for each problem and for the first few values of r, a lower bound $\lambda = \lambda(r)$ on the optimal size of the structure of interest divided by n, the best known upper bound $\mu = \mu(r)$ on the same quantity, and the bound $\sigma = \sigma(r)$ obtained in this paper. In all cases the bounds hold a.a.s. for $G \in \mathcal{G}(n, r\text{-reg})$.

Minimum Vertex Cover. (MVC) A *vertex cover* in a graph G is a set of vertices that intersects every edge of G. Let $\tau(G)$ be the size of the smallest vertex covers in G. The MVC problem asks for a vertex cover of size $\tau(G)$.

The problem is NP-hard to approximate within some fixed constant larger than one for planar cubic graphs [16,2]. It is approximable within $\frac{7}{6}$ for graphs of maximum degree three [5]. The author knows of no result explicitly stated in terms of vertex covers on any random graph model, but results on small vertex covers can be derived from results on large independent sets. The values of λ and μ in the table MVC below are implicitly obtained in [22,25].

Maximum Independent Set. (MIS) An *independent set* in a graph G is a set of vertices containing no edge of G. Let $\alpha(G)$ be the size of the largest independent sets in G. The MIS problem asks for an independent set of size $\alpha(G)$.

The problem is NP-hard to solve optimally for planar cubic graphs [16], and NP-hard to approximate within some fixed constant for bounded degree graphs [6]. On the same class of graphs the problem is approximable within a constant factor [5]. Many results are known on the most likely value of $\alpha(G)$ in random graphs [7,13]. If $G \in \mathcal{G}(n, r\text{-reg})$, for any $\epsilon > 0$ there is an r_ϵ such that

$$\left| \alpha(G) - \tfrac{2n}{r}(\log r - \log\log r + 1 - \log 2) \right| \leq \tfrac{\epsilon n}{r}$$

a.a.s. for any $r \geq r_\epsilon$ [14]. For smaller values of r the best bounds are proved in [22] and [25].

Minimum Independent Dominating Set. (MIDS) An *independent dominating set* in a graph G is an independent set U such that all vertices in $V \setminus U$ are adjacent to at least one element of U. Let $\gamma(G)$ be the size of the smallest independent dominating sets in G. The MIDS problem asks for an independent dominating set of size $\gamma(G)$.

The problem is NP-hard [15] to solve optimally for graphs of maximum degree three. Moreover, for each r, there are constants $1 < b_1 < b_2$ such that the problem is approximable within b_2 [1] and NP-hard to approximate within b_1 [20] in graphs of max degree r. The values of $\lambda(3)$ and $\mu(3)$ in the table for MIDS were obtained in [11] whereas all other values of μ follow from the fact that almost all regular graphs are hamiltonian [24]. All values of $\lambda(r)$ for $r \geq 4$ are proved for the first time in Theorem 2.

Minimum Edge Dominating Set. (MEDS) An *edge dominating set* in a graph G is a set of edges F such that every $e \in E \setminus F$ is incident to at least an element of F. Let $\beta(G)$ be the size of the smallest edge dominating sets in G. The MEDS problem asks for an edge dominating set of size $\beta(G)$.

The problem was first shown to be NP-hard to solve optimally by Yannakakis and Gavril [27]. It is NP-hard to solve within some constant factor of the optimal for cubic graphs [28]. Any maximal matching has at most $2\beta(G)$ edges [21]. Recently Duckworth and Wormald [10] gave the first non-trivial approximation result for (worst-case) cubic graphs (this gives $\mu(3)$). The values of λ in the table below were proved in [29].

MVC				MIDS				MEDS		
r	λ	μ	σ	r	λ	μ	σ	r	λ	μ	σ
3	0.5446	0.5672	0.5708	3	0.2641	0.2794	0.281	3	0.3158	$0.45 + o(1)$	0.3528
4	0.5837	0.6099	0.6122	4	0.2236	$0.\overline{3}$	0.2453	4	0.315		0.3683
5	0.6156	0.6434	0.6459	5	0.1959	$0.\overline{3}$	0.2197	5	0.3176		0.3797
6	0.642	0.6704	0.6721	6	0.1755	$0.\overline{3}$	0.2001	6	0.3212		0.3881
7	0.6643	0.6929	0.6966	7	0.1596	$0.\overline{3}$	0.1848	7	0.3321		0.3978

3   Algorithm and Analysis Method

Next we describe the algorithm (template) that will be used to solve the problems above. The procedure is parameterised to the problem Π, the degree of the input graph r, a constant $\epsilon \in (0,1)$, and a sequence of numbers $c_1, \ldots, c_r$ where c_r can be safely set to be one, and details on how to choose all other c_j's will be given towards the end of this section. Let $V_i = \{v : \deg_G v = i\}$. We call a *step* one iteration of the while loop and a *phase* one complete execution of such loop. The algorithm runs for r phases: Phase 0, Phase 1, etc. In the following discussion G always denotes the subgraph of the input graph still to be dealt with after a certain number of steps. In each step a further portion of G is dealt with.

Algorithm DescendingDegree$_{\Pi,r,\epsilon,c_1,\ldots,c_r}(G)$:
Input: an r-regular graph $G = (V, E)$ on n vertices.
   $\mathcal{S} \leftarrow \emptyset$;
   for $j = 0$ **to** $r - 1$
      while $|\mathcal{S}| \leq \lceil c_{j+1}n \rceil$
         pick at random a vertex u of degree $r - j$;
         update$_{\Pi,r}(\mathcal{S})$;
         shrink$_{\Pi,r}(G)$;
         if $(j < r - 1) \wedge (\exists i : V_i = \emptyset)$ **return** FAILURE;
         else if $(j = r - 1) \wedge \sum_{i=1}^{r} i|V_i| \leq \lceil n^\epsilon \rceil$
            clean-up$_{\Pi,r}(\mathcal{S}, G)$;
            output$(\mathcal{S})$;
         else if $(\exists i : V_i = \emptyset)$ **return** FAILURE;
   if $|E| > 0$ **return** FAILURE.

In all cases the output structure $\mathcal{S}$ is updated as follows. Let $N(u) = \{v \in G : \{u, v\} \in E\}$. For each Π, function *update* probes the degree of each $v \in N(u)$ and decides what to add to $\mathcal{S}$ based on the sequence $\boldsymbol{d} \equiv (d_1, d_2, \ldots, d_{\deg u})$ where $r - 1 \geq d_1 \geq d_2 \geq \ldots \geq d_{k_u} > d_{k_u+1} = \ldots = d_{\deg u} = 0$ are the degrees of u's neighbours in $G \setminus u$, the graph obtained from G by removing u and all its incident edges. In all cases $|\mathcal{S}|$ increases by one if $k_u < \deg u$ (i.e. u has at least one neighbour of degree one) and by at least one if $k_u = \deg u$ (i.e. all neighbours of u have degree at least two). We call *Type 1* (resp. *Type 2*) *configurations* the sequences $\boldsymbol{d}$ of the first (second) kind. Function *shrink* will then remove u and all its incident edges from G. The choice of further vertices to be removed depends on Π and details are given in Section 4. Each step may end in one of two legal conditions. If $V_i \neq \emptyset$ for all i, the algorithm proceeds to the next step (possibly starting a new phase). If $j = r - 1$ and the $\sum_{i=1}^{r} i|V_i|$ is less than $\lceil n^\epsilon \rceil$ the algorithm is (almost) ready to output the structure $\mathcal{S}$. The function *clean-up*, in each case, completes $\mathcal{S}$ according to some fixed (but otherwise arbitrary) greedy heuristic (e.g. *clean-up*$_{\mathrm{MEDS},r}(\mathcal{S}, G)$ might return any maximal matching in G). Any other condition is illegal and the algorithm terminates with a "FAILURE" message.

The major outcome of this paper is that although in principle, for a given choice of the c_j's, the evolution of the $|V_i|$'s could be quite arbitrary, for the problems considered it is possible to "fine-tune" the c_j's to control the $|V_i|$'s (in particular these numbers remain positive throughout the algorithm execution, at least a.a.s.) and hence the size of the structure of interest. For $j \geq 0$, let $Y_i = \check{Y}_i^j(t)$ be the size of V_{r-i} after step t of Phase j, with $Y_0^0(0) = n$ and $Y_i^0(0) = 0$ for all $i > 0$, and $Y_i^{j+1}(0) = Y_i^j(\lceil c_j n \rceil)$ for all $i \geq 0$ and $j \in \{1, \ldots, r - 1\}$. Notice in particular that $|Y_i^j(t+1) - Y_i^j(t)| \leq r^{O(1)}$ for all t, i, and j. Finally, let $X^j(t) = \sum_{i=0}^{r-1}(r - i)Y_i^j(t)$. From now on the dependency on t and j will be omitted unless ambiguity arises. The key ingredient in the analysis of the algorithm above is the use (in each Phase) of Theorem 5.1 in [26] which provides tight asymptotics for the most likely values of Y_i (for each $i \in \{0, \ldots, r\}$), as the algorithm progresses through successive Phases.

Note that it would be fairly simple to modify the graph generation process described in Section 1 to incorporate the decisions made by the algorithm DescendingDegree: the input random graph and the output structure S would be generated at the same time. In this setting one would keep track of the degree sequence of the so called *evolving graph* H_t (see for instance [25]). H_0 would be empty, and a number of edges would be added to H_t to get H_{t+1} according to the behaviour of algorithm DescendingDegree at step $t + 1$. The random variables Y_i also denote the number of vertices of degree i in the evolving graph H_t. We prefer to give our description in terms of the original regular graph.

Let $E(\Delta Y_i)$ denote the expected change of Y_i during step $t + 1$ in some given Phase, conditioned to the history of the algorithm execution from the start of the Phase until step t. This is asymptotically

$$\sum_d E(\Delta Y_i \mid d) \Pr[d] \tag{1}$$

where $E(\Delta Y_i \mid d)$ is the expected change of Y_i conditional to the degrees of u's neighbours in $G \setminus u$ being described by d, $\Pr[d]$ denotes the probability that d occurs in u's neighbourhood conditioned to the algorithm history so far and the sum is over all possible configurations in the given Phase. Also, in each case, the expected change in the size of the structure output by the algorithm, $E(\Delta|S|)$, is asymptotically

$$\sum_{d \in \text{Type1}} \Pr[d] + c_\Pi \sum_{d \in \text{Type2}} \Pr[d] \tag{2}$$

where c_Π is a problem specific constant that will be defined in Section 4. Setting $x = t/n$, $y_i^j(x) = Y_i^j/n$ (again dependency on j will be usually omitted), and $\sigma(x) = |S|/n$, the following system of differential equations is associated with each Phase,

$$\frac{dy_i}{dx} = \tilde{E}(\Delta Y_i)\big|_{t=xn, Y_i=y_i n} \qquad \frac{d\sigma}{dx} = \tilde{E}(\Delta|S|)\big|_{t=xn, Y_i=y_i n} \tag{3}$$

where $\tilde{E}(\Delta Y_i)$ and $\tilde{E}(\Delta|S|)$ denote the asymptotic expressions for the corresponding expectations obtained from (1) and (2) using the estimates on $\Pr[d]$ and $E(\Delta Y_i \mid d)$ given later on. Since x does not occur in $\tilde{E}(\Delta Y_i)$ and $\tilde{E}(\Delta|S|)$, we actually solve, one after the other, the systems

$$\frac{dy_i}{d\sigma} = \frac{\tilde{E}(\Delta Y_i)}{\tilde{E}(\Delta|S|)}\big|_{Y_i=y_i n} \tag{4}$$

where differentiation is w.r.t. σ, setting $y_0^0(0) = 1$ and $y_i^0(0) = 0$ for all $i \geq 1$ and using $y_i^j(c_j)$, the final conditions of Phase j, as initial conditions of Phase $j + 1$ for each $j > 0$. Theorem 5.1 in [26] can be applied (as long as X is not too small) to conclude that a.a.s. the solutions to the systems (3) multiplied by n approximate the values of the variables Y_i and $|S|$. Furthermore, in each case, for any given $\epsilon \in (0, 1)$, there is a value $\sigma = \sigma_\epsilon$ such that X becomes less than $\lceil n^\epsilon \rceil$ during Phase $r - 1$ for $|S| > \sigma n$. After this point G has less than n^ϵ vertices of positive degree and the contribution to $|S|$ coming from running the algorithm DescendingDegrees to completion is only $o(n)$. The following statement summarises our results:

Theorem 1. *Let $\Pi \in \{\text{MVC}, \text{MIDS}, \text{MEDS}\}$. For each integer $r \geq 2$ and $\epsilon \in (0, 1)$ there is a sequence of values $c_1, \ldots, c_r$, and a constant σ such that the algorithm DescendingDegree$_{\Pi, r, \epsilon, c_1, \ldots, c_r}$ returns a structure S with $|S| \leq \sigma n + o(n)$ a.a.s.*

In the remainder of this section a formula is proved for the number of configurations in each Phase (its proof gives a method to list them which is needed to compute $\tilde{E}(\Delta Y_i)$ and $\tilde{E}(\Delta|\mathcal{S}|)$), details are given on how to compute asymptotic expressions for $\Pr[\boldsymbol{d}]$ and $E(\Delta Y_i \mid \boldsymbol{d})$ for each $\boldsymbol{d}$, and some comments are made on how to solve the systems in (4) and how to choose the constants c_j. Section 4 provides a definition for all problem specific quantities needed to complete the proof of Theorem 1.

Listing configurations. Let $A_=(r,k)$ (resp. $A_\le(r,k)$) be the number of configurations having exactly (resp. at most) k vertices and whose largest positive degree is r. Then $r + \sum_{j=2}^{r} \sum_{i=1}^{r-1} A_\le(i,j)$ is the total number of configurations of Type 1 and $\sum_{j=1}^{r} \sum_{i=1}^{r-1} A_=(i,j)$ is the total number of configurations of Type 2. More precisely, in Phase $r-j$ there are $1 + \sum_{i=1}^{r-1} A_\le(i,j-1)$ configurations of Type 1 (this reduces to 1 for $j=1$, assuming $A_\le(i,0)=0$) and $\sum_{i=1}^{r-1} A_=(i,j)$ configurations of Type 2.

Lemma 1. $A_\le(r,k) = A_=(r+1,k) = \binom{k+r-1}{r}$, *for every positive integer* r,k *and* $A_=(1,k) = 1$ *for every positive integer* k.

Proof. The equality for $A_\le(r,k)$ follows from the one for $A_=(r,k)$ by induction on k since a sequence of at most k numbers starting with a r can either be a sequence of length at most $k-1$ starting with a r (there are $A_\le(r,k-1)$ of them) or it can be a sequence of exactly k numbers starting with a r. $A_=(r,k) = \binom{k+r-2}{r-1}$ can be proved by induction on $r+k$. A sequence of length exactly k beginning with an r can continue with an $i \in \{1,\ldots,r\}$ (in $A_=(i,k-1)$ possible ways). Therefore $A_=(r,k) = A_=(r,k-1) + A_=(r-1,k)$ with initial condition $A_=(r,1) = 1$ for every $r \ge 1$. Assuming $A_=(i,j) = \binom{j+i-2}{i-1}$ for every i and j with $i+j < r+k$ the equality for $A_=(r,k)$ follows from $\binom{a}{b} + \binom{a}{b-1} = \binom{a+1}{b}$. $\square$

The proof of Lemma 1 implicitly gives a method for listing all configurations of given type.

Probability of a configuration. The formula for $\Pr[\boldsymbol{d}]$ is better understood if we think of the algorithm DescendingDegree as embedded in the graph generation process. A configuration $\boldsymbol{d} \equiv (d_1, d_2, \ldots, d_{\deg u})$ occurs at the neighbourhood of a given u if the $\deg u$ balls still available at the given step in the urn U associated with u are paired up with balls from random urns containing respectively $d_1 + 1, d_2 + 1, \ldots, d_{\deg u} + 1$ free balls. The probability of pairing up one ball from U with a ball from an urn with $r-i$ free balls is $P_i = \frac{(r-i)Y_i}{X}$ for $i \in \{0,\ldots,r\}$ at the beginning of a step, and this only changes by a $o(1)$ factor during each step due to the multiple edge selections which are part of a step. Hence $\Pr[\boldsymbol{d}]$ is asymptotically equal to $\binom{\deg u}{m_1,\ldots,m_{\deg u}} P_{r-(d_1+1)} \cdot P_{r-(d_2+1)} \cdot \cdots \cdot P_{r-(d_{\deg u}+1)}$ where m_l are the multiplicities of the possibly $\deg u$ distinct values occurring in $\boldsymbol{d}$.

Conditional expectations. Note that for each problem Π, the sequence $\boldsymbol{d}$ contains all the information needed to compute asymptotic expressions for all the conditional expected changes of the variables Y_i. There will be a function $\mathrm{rm}_{\Pi,r}(i, \boldsymbol{d})$, giving, for each d in a given Phase, the number of vertices of degree $r-i$ removed from the subgraph of G induced by $u \cup N(u)$. Then

$$E(\Delta Y_i \mid d) = -rm_{\Pi,r}(i, d) - \text{Spec}_{\Pi,r}(i, d)$$

where Spec is a problem specific function depending on a wider neighbourhood around u. In each case Spec may involve $R_i = P_i - P_{i-1}$ for $i \in \{0, \ldots, r\}$ (with $P_{-1} = 0$) where $-R_i$ describes asymptotically the contribution to $E(\Delta Y_i \mid d)$ given by the removal from G of one edge incident to some $v \in N(u)$ (different from $\{u, v\}$), or

$$Q_i = \sum_{k=0}^{r-1} P_k[\delta_{ik} + (r - k - 1)R_i], \text{ for } i \in \{0, 1, \ldots, r\},$$

where $-Q_i$ is (asymptotically) the contribution to $E(\Delta Y_i \mid d)$ given by the removal from G of a vertex adjacent to some $v \in N(u)$ (different from u).

Computational aspects. The performances of algorithm DescendingDegree depend heavily on a good choice of the points c_j. Unfortunately, although in any given Phase j the systems in (4) are sufficiently well-behaved to guarantee the existence of a unique solution in a domain $\mathcal{D}_j \subseteq \{(\sigma, z^{(1)}, \ldots, z^{(a)}) \in [0, 1]^{a+1}, \sum_{i=1}^{r-1} (r - i) z^{(i)} \geq \delta\}$ (for some $\delta > 0$), where $a = r$ (resp. $r - 1$) for MIDS (for MVC and MEDS), it is not easy to find a closed expression (depending on the c_j's) for the functions y_i^j and consequently optimise the choice of the c_j's. However an alternative approach based on solving the relevant systems numerically using a Runge-Kutta method [8], gives remarkably good results.

Given a positive integer M and an upper bound c_r on σ, let $t_0 = 0$ and $t_h = t_{h-1} + \frac{c_r}{M}$ for all $h \in \{1, \ldots, M\}$. Any $(r-1)$-set $\{s_1, \ldots, s_{r-1}\}$ over $\{1, \ldots, M-1\}$ defines a subsequence of t_h. The system for Phase j is then solved numerically using a variable step Runge-Kutta method in the interval $[0, t_{s_{j+1}} - t_{s_j}]$ (with $s_0 = 0$). The initial values for Phase 0 are $y_0^0(0) = 1$ and $y_i^0(0) = 0$ and the final values for the functions in Phase j are the initial values for the functions in Phase $j + 1$, for $j \in \{1, \ldots, r - 1\}$. The *cost* of a given $(r - 1)$-set $\{s_1, \ldots, s_{r-1}\}$ is defined to be the first index x such that $y_i(t_{x-1}) \neq 0$ and $y_i(t_x) = 0$ for all $i \in \{0, \ldots, r - 2\}$ (with the cost being ∞ if the transition from non-zero to zero does not occur for all i at the same time or it does not occur during Phase $r - 1$). The parameters c_j are defined as $c_j = t_{\hat{s}_j}$ for $j \in \{1, \ldots, r - 1\}$ where $\{\hat{s}_1, \ldots, \hat{s}_{r-1}\}$ is a minimum cost $(r - 1)$-set.

4   Specific Problems

In this Section we complete the description of algorithm DescendingDegree by giving full definition of all problem specific quantities. In each case a picture is drawn to represent the neighbourhood of the chosen vertex u when a given configuration $(d_1, d_2, \ldots)$ occurs. The circled vertices (resp. the thick black edges) are added to the output structure in each case. Dark vertices are removed from G at the end of the given step. White vertices have their degree decreased by one. Question-marked square vertices represent vertices whose degree is not probed at the given step (but which nonetheless are affected by the incurring graph transformation). Note that, since graphs in $\mathcal{G}(n, r\text{-reg})$ do not contain small cycles a.a.s. [19, Theorem 9.5], the question marked vertices are a.a.s. all distinct. All edges in the pictures are removed at the end of the given step (the arrows being just a shorthand for a number of edges). A table is included in each of the following sections giving, for the first few values of r, the sequence of values $c_1, \ldots, c_{r-1}$ that allow the algorithm to attain the bounds given in Section 2.

4.1   Vertex Cover

Each of the r systems is formed by r equations, since there is no need to keep track of Y_r. The Figure below shows the two types of configurations around vertex u along with the choices of the vertices added to the cover C.

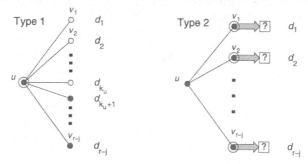

In a given Phase j, for configurations of Type 1, $\mathrm{rm}_{\mathrm{MVC},r}(i, \boldsymbol{d})$ will have a contribution of one if $i = j$, for the degree of u and a contribution of one for each of the $\deg u - k_u$ vertices that are only connected to u, if $i = r - 1$. For configurations of Type 2, all neighbours of u are removed from G. Therefore $\mathrm{rm}_{\mathrm{MVC},r}(i, \boldsymbol{d})$ will have the term $\delta_{i,j}$ and a possible contribution from each neighbour of u in V_{r-i}. In symbols this is

$$\mathrm{rm}_{\mathrm{MVC},r}(i, \boldsymbol{d}) = \begin{cases} \delta_{i,j} + (r - j - k_u)\delta_{i,r-1} & \boldsymbol{d} \in \text{ Type 1} \\ \delta_{i,j} + \sum_{h=1}^{r-j} \delta_{i,r-(d_h+1)} & \boldsymbol{d} \in \text{ Type 2} \end{cases}$$

where $\delta_{xy} = 1(0)$ if $x = y$ (resp. $x \neq y$). Furthermore, for each $i \in \{0, r-1\}$ and $\boldsymbol{d}$ in the given Phase,

$$\mathrm{Spec}_{\mathrm{MVC},r}(i, \boldsymbol{d}) = \begin{cases} \sum_{h=1}^{k_u}(\delta_{i,r-(d_h+1)} - \delta_{i,r-d_h}) & \boldsymbol{d} \in \text{ Type 1} \\ R_i \sum_{h=1}^{r-j} d_h & \boldsymbol{d} \in \text{ Type 2} \end{cases}$$

For configurations of Type 1 a vertex of degree $r - i$ is created (resp. removed) if the h-th neighbour (of degree at least two) of u has indeed degree $r - i + 1$ (resp. $r - i$) before the current step takes place. This event is accounted for by the term $(\delta_{i,r-(d_h+1)} - \delta_{i,r-d_h})$. For configurations of Type 2, $R_i \sum_{h=1}^{r-j} d_h$ accounts for all the vertices of degree $r - i$ that are either created or removed when one of the edges incident to some $v \in N(u)$ is removed. Finally, the expected change in the size of the cover is obtained from (2) setting $c_{\mathrm{MVC}} = r - j$, since one vertex (resp. $r - j$ vertices) is (are) added to C if a configuration of Type 1 (resp. Type 2) is met.

Note that the algorithm DescendingDegree$_{\mathrm{MVC},r}$ can be easily modified to output the independent set $V \setminus C$. Therefore, as announced previously, all results for the MVC problem translate immediately to results for the MIS problem.

r	c_1	c_2	c_3	c_4	c_5	c_6
3	0.001624	0.336168				
4	0.002294	0.045756	0.479508			
5	0.002145	0.01443	0.175305	0.5694		
6	0.00816	0.01972	0.08636	0.29036	0.60452	
7	0.00686	0.02107	0.06517	0.17647	0.37751	0.68754

4.2 Independent Dominating Set

The Figure below shows the two types of configurations around vertex u along with the choices of the vertices to add to the dominating set D. The function $\text{rm}_{\text{MIDS},r}(i, \boldsymbol{d})$ is defined as follows

$$\text{rm}_{\text{MIDS},r}(i, \boldsymbol{d}) = \begin{cases} \delta_{i,j} + \sum_{h=1}^{r-j} \delta_{i,r-(d_h+1)} & \boldsymbol{d} \in \text{ Type 1} \\ \delta_{i,j} + \delta_{i,r-(d_1+1)} & \boldsymbol{d} \in \text{ Type 2} \end{cases}$$

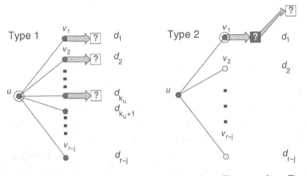

When applied to the MIDS problem, the algorithm DescendingDegree may generate a number of isolated vertices that are not covered by a vertex in D. This may happen at any of the white question marked vertices in the Figure above. The number of these vertices must be accounted for in assessing the quality of a solution. Therefore one needs to keep track of the conditional expected changes in the variables Y_i, for $i \in \{0, \ldots, r\}$, where Y_r counts the number of uncovered isolated vertices. Let $\text{rm}_{\text{MIDS},r}(r, \boldsymbol{d}) = 0$. For each $i \in \{0, \ldots, r\}$ and each $\boldsymbol{d}$ in the given Phase, function $\text{Spec}_{\text{MIDS},r}$ takes the following expressions

$$\text{Spec}_{\text{MIDS},r} = \begin{cases} R_i \sum_{h=1}^{r-j} d_h & \boldsymbol{d} \in \text{ Type 1} \\ \sum_{h=2}^{r-j} (\delta_{i,r-(d_h+1)} - \delta_{i,r-d_h}) + d_1 Q_i & \boldsymbol{d} \in \text{ Type 2} \end{cases}$$

Finally the expected change in the size of the dominating set is given by (2) with $c_{\text{MIDS}} = 1$.

r	c_1	c_2	c_3	c_4	c_5	c_6
3	0.00015	0.0558817				
4	0.0005	0.0062501	0.079565			
5	0.00184	0.00552	0.0220828	0.0858274		
6	0.0014	0.0056	0.0098	0.0378094	0.0862792	
7	0.002856	0.0057122	0.0085692	0.0229683	0.0474604	0.0838873

Theorem 2. *For each $r \geq 3$ there exists a constant $\lambda > 0$ such that $\gamma(G) > \lambda n$ for $G \in \mathcal{G}(n, r\text{-reg})$ a.a.s.*

Proof. The expected number of independent dominating sets of size s in a random pairing on n urns each containing r balls is

$$\binom{n}{s}(sr)^r - 1)^{n-s}\frac{N([r(n-2s)]/2)}{N(nr/2)}$$

where $N(x) = \frac{(2x)!}{x!\,2^x}$ and $(sr)^r - 1)^{n-s}$ counts the ways of pairing the sr balls in s chosen urns in such a way that each of the remaining $n - s$ urns is connected to at least one of the urns in the chosen set (if $p(x)$ is a polynomial then $[x^k]p(x)$ is the standard notation of the coefficient of the monomial of degree k). This number is at most $(sr)!\frac{((1+x)^r-1)^{n-s}}{x^{sr}}$. Using Stirling's approximation to the factorial and setting $s = \lambda n$ implies that the expectation above is at most

$$n^{O(1)}\left\{\frac{(1-2\lambda)^{\frac{r(1-2\lambda)}{2}}}{\lambda^{\lambda}(1-\lambda)^{1-\lambda}r^{r\lambda}}(\lambda r)^{\lambda r}\frac{((1+x)^r-1)^{1-\lambda}}{x^{\lambda r}}\right\}^n$$

where x takes the value that minimises $\frac{((1+x)^r-1)^{1-\lambda}}{x^{\lambda r}}$. For every r there exists a positive $\lambda = \lambda(r)$ such that the expected number of independent dominating sets of size $s < \lambda n$ in a random pairing tends to zero and therefore, by the Markov inequality, with high probability a random r-regular graph will not contain an independent dominating set of this size. The values of λ are reported in the second column of the table for MIDS in Section 2.   □

4.3   Edge Dominating Set

The Figure below shows the two types of configurations around vertex u along with the choices of the edges to add to the edge dominating set M. The function $\mathrm{rm}_{\mathrm{MEDS},r}(i,\boldsymbol{d})$ is defined as follows

$$\mathrm{rm}_{\mathrm{MEDS},r}(i,\boldsymbol{d}) = \begin{cases} \delta_{i,j} + \delta_{i,r-(d_1+1)} + (r-j-k)\delta_{i,r-1} & \boldsymbol{d} \in \text{Type1} \\ \delta_{i,j} + \sum_{h=1}^{r-j}\delta_{i,r-(d_h+1)} & \boldsymbol{d} \in \text{Type2} \end{cases}$$

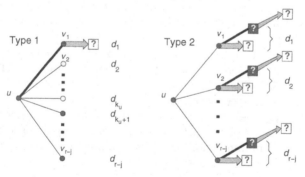

Also, for each $i \in \{0,\ldots,r-1\}$ and $\boldsymbol{d}$ in the given Phase, let

$$\mathrm{Spec}_{\mathrm{MEDS},r}(i,\boldsymbol{d}) = \begin{cases} \sum_{h=2}^{k_u}(\delta_{i,r-(d_h+1)} - \delta_{i,r-d_h}) & \boldsymbol{d} \in \text{Type 1} \\ R_i\sum_{h=1}^{r-j}(d_h-1) + Q_i(r-j) & \boldsymbol{d} \in \text{Type 2} \end{cases}$$

(with the convention that the sum in the first line is empty if $k_u = 1$).
Finally, the expected change in the size of the edge dominating set is obtained from formula (2) setting $c_{\mathrm{MEDS}} = r - j$, since one vertex (resp. $r - j$ vertices) is (are) added to M if a configuration of Type 1 (resp. Type 2) is met. Again we solve r systems of r equations in the variables $y_0,\ldots,y_{r-1}$.

r	c_1	c_2	c_3	c_4	c_5	c_6
3	0.00054	0.123696				
4	0.00038	0.0152	0.22078			
5	0.008848	0.015408	0.06864	0.26692		
6	0.003485	0.010045	0.033005	0.12505	0.28946	
7	0.0066	0.0154	0.0286	0.0814	0.1694	0.33022

5   Conclusions

In this paper we presented a new approach for the approximate solution to minimisation problems in random regular graphs. We defined a greedy algorithm that, in a number of successive Phases, picks vertices of given degree in the current graph, updates the partial solution produced so far and removes few vertices from the graph. Although in principle it is not clear that all Phases (as defined in Section 3) can be successfully completed, one after the other, we prove that, for the problems at hand under a suitable choice of the algorithm's parameters this is indeed the case. This in turns leads to improved approximation performances in almost all cases considered.

Many questions are left open by this work. First of all, the results in the paper rely on a numerical integration procedure based on a second order Runge-Kutta method. A tedious analysis of the first few partial derivatives of the functions defining the systems in each case would give a quantitative measure of their stability and a-priori bounds on the quality of the numerical approximation. Esperimentally, successive runs of the integration procedure with ever decreasing step sizes confirmed that our results are accurate to the first four decimal digits. Secondly, no serious attempt was made to optimise the values of the constants c_j. The value of M was set to 10000 for $r = 3$ but to much smaller values for all other r. Slight improvements on the values of σ for all problems may be possible by using a more careful search through the various multi-dimensional parameter spaces.

The algorithm performances could also be improved by expanding the neighbourhood of the vertex considered at each step or analysing more carefully the set of configurations in each Phase. The price for this will be an increased complexity in finding the problem specific parameters $c_1, c_2, \ldots, c_{r-1}$.

Finally, in principle, the same approach can be exploited in other random graph models, such as $\mathcal{G}(n, p)$ or $\mathcal{G}(n, m)$ or even in the design of greedy algorithms for solving the same minimisation problems making no assumption on the input distribution.

References

1. P. Alimonti, T. Calamoneri. Improved approximations of independent dominating set in bounded degree graphs. In *Proc. 22nd WG*, pp 2-16, LNCS 1197, Springer-Verlag, 1997.
2. P. Alimonti, V. Kann. Hardness of approximating problems on cubic graphs. In *Proc. 3rd CIAC*, pp 288-298. LNCS 1203, Springer-Verlag, 1997.

3. J. Aronson, A. Frieze, B. G. Pittel. Maximum matchings in sparse random graphs: Karp-Sipser revisited. *RSA*, 12:111-178, 1998.
4. G. Ausiello, P. Crescenzi, G. Gambosi, V. Kann, A. Marchetti-Spaccamela, M. Protasi. *Complexity and Approximation*. Springer-Verlag, 1999.
5. P. Berman, T. Fujito. Approximating independent sets in degree 3 graphs. In *Proc. WADS'95*, pp 449-460. LNCS 955, Springer-Verlag, 1995.
6. P. Berman, M. Karpinski. On some tighter inapproximability results. Technical Report TR98-29, ECCC, 1998.
7. B. Bollobás, P. Erdős. Cliques in random graphs. *Math. Proc. Camb. Phil. Soc.*, 80:419-427, 1976.
8. R. L. Burden, J. D. Faires, A. C. Reynolds. *Numerical Analysis*. Wadsworth Int., 1981.
9. P. Crescenzi, V. Kann. A compendium of NP optimization problems. Available at http://www.nada.kth.se/~viggo/wwwcompendium/, 2000.
10. W. Duckworth, N. C. Wormald. Linear programming and the worst case analysis of greedy algorithms for cubic graphs. To be submitted. Pre-print available from the authors.
11. W. Duckworth, N. C. Wormald. Minimum independent dominating sets of random cubic graphs. Submitted to RSA, 2000.
12. A. Frieze, C. McDiarmid. Algorithmic theory of random graphs. *RSA*, 10:5-42, 1997.
13. A. M. Frieze. On the independence number of random graphs. *Disc. Math.*, 81:171-175, 1990.
14. A. M. Frieze, T. Łuczak. On the independence and chromatic number of random regular graphs. *J. Comb. Theory*, B 54:123-132, 1992.
15. M. R. Garey, D. S. Johnson. *Computer and Intractability, a Guide to the Theory of NP-Completeness*. Freeman & Company, 1979.
16. M. R. Garey, D. S. Johnson, L. Stockmeyer. Some simplified *NP*-complete graph problems. *TCS*, 1:237-267, 1976.
17. G. R. Grimmett, C. J. H. McDiarmid. On colouring random graphs. *Math. Proc. Camb. Phil. Soc.*, 77:313-324, 1975.
18. D. S. Hochbaum. Easy solutions for the k-center problem or the dominating set problem on random graph. In *Annals of Disc. Math.*, 25:189-210, 1985.
19. S. Janson, T. Łuczak, A. Ruciński. *Random Graphs*. John Wiley & Sons, 2000.
20. V. Kann. *On the Approximability of NP-complete Optimization Problems*. PhD thesis, Royal Institute of Technology, Stockholm, 1992.
21. B. Korte, D. Hausmann. An analysis of the greedy heuristic for independence systems. *Annals of Disc. Math.*, 2:65-74, 1978.
22. B. D. McKay. Independent sets in regular graphs of high girth. *Ars Combinatoria*, 23A:179-185, 1987.
23. T. Nierhoff. *The k-Center Problem and r-Independent Sets*. PhD thesis, Humboldt-Universität zu Berlin, Berlin, 1999.
24. R. W. Robinson, N. C. Wormald. Almost all regular graphs are hamiltonian. *RSA*, 5:363-374, 1994.
25. N. C. Wormald. Differential equations for random processes and random graphs. *Annals of Applied Prob.*, 5:1217-1235, 1995.
26. N. C. Wormald. The differential equation method for random graph processes and greedy algorithms. In M. Karoński, H. J. Prömel, editors, *Lectures on Approximation and Randomized Algorithms*, pages 73–155. PWN, Warsaw, 1999.
27. M. Yannakakis, F. Gavril. Edge dominating sets in graphs. *SIAM J. Applied Math.*, 38(3):364-372, June 1980.

28. M. Zito. *Randomised Techniques in Combinatorial Algorithmics*. PhD the-sis,University of Warwick, 1999.
29. M. Zito. Small maximal matchings in random graphs. In *Proc. LATIN 2000*, pp 18-27. LNCS 1776, Springer-Verlag, 2000.

Author Index

Lecture Notes in Computer Science

For information about Vols. 1–2048
please contact your bookseller or Springer-Verlag